D0301311

INTRODUCTION TO THE MECHANICS OF A CONTINUOUS MEDIUM

PRENTICE-HALL SERIES IN
ENGINEERING OF THE PHYSICAL SCIENCES

James B. Reswick and Warren M. Rohsenow, editors

HANSEN *Similarity Analyses of Boundary Value Problems in Engineering*
HURTY AND RUBINSTEIN *Dynamics of Structures*
JACKSON *Equilibrium Statistical Mechanics*
LANGILL *Control Systems Engineering*
LANGILL *Advanced Control Systems Engineering*
LONG *Mechanics of Solids and Fluids*
MALVERN *Introduction to the Mechanics of a Continuous Medium*
MARK *Thermodynamics*
PLAPP *Engineering Fluid Mechanics*
RESWICK AND TAFT *Introduction to Dynamics Systems*
ROHSENOW AND CHOI *Heat, Mass, and Momentum Transfer*
RUBINSTEIN *Matrix Computer Analysis of Structures*
SAAD *Thermodynamics for Engineers*

PRENTICE-HALL INTERNATIONAL, INC., *London*
PRENTICE-HALL OF AUSTRALIA PTY. LTD., *Sydney*
PRENTICE-HALL OF CANADA, LTD., *Toronto*
PRENTICE-HALL OF INDIA PRIVATE LTD., *New Delhi*
PRENTICE-HALL OF JAPAN, INC., *Tokyo*

INTRODUCTION TO THE MECHANICS OF A CONTINUOUS MEDIUM

Lawrence E. Malvern

Professor of Mechanics
College of Engineering
Michigan State University

Prentice-Hall, Inc.

Englewood Cliffs, New Jersey

Current printing (last digit):

10 9 8 7

13-487603-2
Library of Congress Catalog Card Number 69-13712
Printed in the United States of America

Preface

This book offers a unified presentation of the concepts and general principles common to all branches of solid and fluid mechanics, designed to appeal to the intuition and understanding of advanced undergraduate or first-year postgraduate students in engineering or engineering science.

The book arose from the need to provide a general preparation in continuum mechanics for students who will pursue further work in specialized fields such as viscous fluids, elasticity, viscoelasticity, and plasticity. Originally the book was introduced for reasons of pedagogical economy—to present the common foundations of these specialized subjects in a unified manner and also to provide some introduction to each subject for students who will not take courses in all of these areas. This approach develops the foundations more carefully than the traditional separate courses where there is a tendency to hurry on to the applications, and moreover provides a background for later advanced study in modern nonlinear continuum mechanics.

The first five chapters devoted to general concepts and principles applicable to all continuous media are followed by a chapter on constitutive equations, the equations defining particular media. The chapter on constitutive theory begins with sections on the specific constitutive equations of linear viscosity, linearized elasticity, linear viscoelasticity, and plasticity, and concludes with two sections on modern constitutive theory. There are also a chapter on fluid mechanics and one on linearized elasticity to serve as examples of how the general principles of the first five chapters are combined with a constitutive equation to formulate a complete theory. Two appendices on curvilinear tensor components follow, which may be omitted altogether or postponed until after the main exposition is completed.

Although the book grew out of lecture notes for a one-quarter course for first-year graduate students taught by the author and several colleagues during the past 12 years, it contains enough material for a two-semester course and is written at a level suitable for advanced undergraduate students. The only prerequisites are the basic mathematics and mechanics equivalent to that usually taught in the first two or three years of an undergraduate engineering program. Chapter 2 reviews vectors and matrices and introduces what tensor methods are needed. Part of this material may be postponed until needed, but it is collected in Chap. 2 for reference.

The last 15 to 20 years have seen a great expansion of research and publication in modern continuum mechanics. The most notable developments have been in the theory of constitutive equations, especially in the formulation of very general principles restricting the possible forms that constitutive equations can take. These new theoretical developments are especially addressed to the formulation of nonlinear constitutive equations, which are only briefly touched upon in this book. But the new developments have also pointed up the limitations of some of the widely used linear theories. This does not mean that any of the older linear theories must be discarded, but the new developments provide some guidance to the conditions under which the older theories can be used and the conditions where they are subject to significant error. The last two sections of Chap. 6 survey modern constitutive theory and provide references to original papers and to more extended treatments of the modern theory than that given in this introductory text.

The book is a carefully graduated approach to the subject in both content and style. The earlier part of the book is written with a great deal of illustrative detail in the development of the basic concepts of stress and deformation and the mathematical formulation used to represent the concepts. Symbolic forms of the equations, using dyadic notation, are supplemented by expanded Cartesian component forms, matrix forms, and indicial forms of the same equations to give the student abundant opportunity to master the notations. There are also many simple exercises involving interpretation of the general ideas in concrete examples. In Chaps. 4 and 5 there is a gradual transition to more reliance on compact notations and a gradual increase in the demands on the reader's ability to comprehend general statements.

Until the end of Sec. 4.2, each topic considered is treated fairly completely and (except for the brief section on stress resultants in plate theory) only concepts that will be used repeatedly in the following sections are introduced. Then there begin to appear concepts and formulations whose full implementation is beyond the scope of the book. These include, for example, the relative description of motion, mentioned in Sec. 4.3 and also in some later sections, and the finite rotation and stretch tensors of Sec. 4.6, which are important in some of the modern developments referred to in the last two sections of Chap. 6. The aim in presenting this material is to heighten the reader's awareness that the subject of continuum mechanics is in a state of rapid development, and to encourage his reading of the current literature. The chapters on fluids and on elasticity also refer to published methods and results in addition to those actually presented.

The sections on the constitutive equations of viscoelasticity and plasticity are introduced by accounts of the observed responses of real materials in order to motivate and also to point up the limitations of the idealized representations that follow. The second section on plasticity includes work-hardening, a part of the theory not in a satisfactory state, but so important in engineering applications that it was believed essential to mention and point out some of the shortcomings of the available formulations.

A one-quarter course might well include most of the first five chapters, only part of Chap. 6, and either Chap. 7 on fluids or Chap. 8 on elasticity. Section 3.6 on stress resultants in plates and those parts of Secs. 5.3 and 5.4 treating couple stress can be omitted without destroying the continuity, as also can Secs. 6.5 and 6.6 on plasticity. Section 4.6 can be given only minor emphasis, or omitted altogether if the last two sections of Chap. 6 are not to be covered. The second appendix, presenting only physical components in orthogonal curvilinear coordinates might be included if time permits; although not needed in the text, it is useful for applications.

A two-term course could include the first appendix on general curvilinear tensor components, useful as a preparation for reading some of the modern literature. There is sufficient textual material in the book for a full year course, but it should probably be supplemented with some challenging applications problems. Most of the exercises in the text are teaching devices to illuminate the theory, rather than applications.

The book is a textbook, designed for classroom teaching or self-study, not a treatise reporting new scientific results. Obviously the author is indebted to hundreds of investigators over a period of more than two centuries as well as to earlier books in the field or in its specialized branches. Some of these investigators and authors are named in the text, but the bibliography at the end of the book includes only the twentieth-century writings cited. Extensive bibliographies may be found in the two *Encyclopedia of Physics* treatises; "The Classical Field Theories," by C. Truesdell and R. A. Toupin, Vol. III/1, pp. 226–793 (1960), and "The Non-Linear Field Theories of Mechanics," by C. Truesdell and W. Noll, Vol. III/3 (1965), published by Springer-Verlag, Berlin. These two valuable comprehensive treatises are among the references for collateral reading cited at the end of the introduction. Many of the historical allusions in the text are based on these two sources.

The author is indebted to several colleagues at Michigan State University who have used preliminary versions of the book in their classes. These include Dr. C. A. Tatro (now at the Lawrence Radiation Laboratory, Livermore, California) and Professors M. A. Medick, R. W. Little, and K. N. Subramanian. Professors John Foss and Merle Potter read the first version of the material on fluid mechanics. Encouragement and helpful criticism have been provided by these colleagues and also by the dozens of students who have taken the course.

The author is also indebted to Michigan State University for sabbatical leave during 1966–67 to work on the book and to Prentice-Hall, Inc., for their cooperation and assistance in preparing the final text and illustrations.

Finally, thanks are due to the author's wife for inspiration, encouragement and forbearance.

LAWRENCE E. MALVERN

East Lansing, Michigan

Contents

INTRODUCTION TO THE MECHANICS OF A CONTINUOUS MEDIUM

CHAPTER 1

Introduction

1.1 The continuous medium

The mechanics of a continuous medium is that branch of mechanics concerned with the stresses in solids, liquids, and gases and the deformation or flow of these materials. The adjective ***continuous*** refers to the simplifying concept underlying the analysis: we disregard the molecular structure of matter and picture it as being without gaps or empty spaces. We further suppose that all the mathematical functions entering the theory are continuous functions, except possibly at a finite number of interior surfaces separating regions of continuity. This statement implies that the derivatives of the functions are continuous too, if they enter the theory, since all functions entering the theory are assumed continuous. This hypothetical continuous material we call a ***continuous medium or continuum.***

The concept of a continuous medium permits us to define stress ***at a point***, a geometric point in space conceived as occupying no volume, by a mathematical limit like the definition of the derivative in differential calculus. This approach immediately makes the powerful methods of calculus available for the study of nonuniform distributions of stress, and at the same time provides an easily visualized physical model which agrees with the testimony of everyday observation of matter in the large. Thus this approach allows the mathematical analysis to be guided by intuition. The theories of elasticity, plasticity, and fluid mechanics based on the concept of continuous material lead moreover to quantitative predictions which agree closely with experience over a wide range of conditions. These theories and the further simplified engineering theories of beams and plates and shells, which are also based on the continuum concept of matter, are adequate for analysis of stress and deformation in most engineering problems.

There are exceptions of course. It would be too much to expect that a theory which disregards the molecular nature of matter could account for all the observed properties of matter, even in the large. There is, for example, nothing in the theories that accounts for the formation of a fatigue crack in a body after many cycles of reversed loading. Problems in super aerodynamics at altitudes where the air is extremely rarefied may require molecular theories. Even in marginal situations like these, where continuum mechanics alone is inadequate, it is sometimes possible to use the continuum theory in combination with empirical information or with information derived from a physical theory based on the molecular nature of the material.

The use of the continuum concept to construct a working theory unifying a large body of observational knowledge and permitting the deduction of useful conclusions obviously has nothing to do with any assumption as to the real nature of matter. The existence of areas in which the theory is not applicable does not destroy its usefulness in other areas.

In most of the analyses in continuum mechanics, two further assumptions about the nature of the material are made, namely, that it is homogeneous and isotropic. It should be clearly understood that the three assumptions are completely independent.

Continuity. A material is continuous if it completely fills the space that it occupies, leaving no pores or empty spaces, and if furthermore its properties are describable by continuous functions.

Homogeneity. A homogeneous material has identical properties at all points.

Isotropy. A material is isotropic with respect to certain properties if these properties are the same in all directions.

It is quite possible, for example, to conceive of the atmosphere as a continuous fluid but still suppose that its density decreases with altitude. It is also possible to conceive of a rolled steel plate that is both continuous and homogeneous but has different tensile strengths or different elastic moduli in tension specimens cut from the plate, one with its axis in the direction of rolling and one with its axis at right angles to the direction of rolling. This kind of anisotropy can be introduced into continuum mechanics, and has led to useful results. The developments of the concepts of stress and strain in Chaps. 3 and 4 and the general principles of Chap. 5 use only the basic assumption of continuity. The other two simplifying assumptions are not needed until constitutive equations (stress-strain relations) or strength properties of a material are considered.

It is sometimes said that solutions to engineering problems obtained through the simplifying assumptions of the engineering theory of beams are only approximations to the "exact" solutions of the theory of elasticity. The foregoing paragraphs should have made it clear that the adjective "exact" is not justified. There may be exact mathematical solutions to the equations formulated in elasticity or in other branches of continuum mechanics, but the equations themselves are not exact descriptions of nature. In this respect the difference between the elementary theory and the advanced theory is one of degree rather than of kind. When the elementary theory is formulated consistently and logically, it is just as respectable as the advanced theory from a mathematical or logical point of view. And from a practical point of view it is just as good in those areas where its predictions agree closely enough with experience. The bounds of applicability of these elementary theories are determined by experience, either from experimental verification, or from comparisons with predictions of the more advanced continuum theories.

The last fifteen years have seen a spectacular growth in general continuum theory as well as in its specialized branches. The theory divides logically into three parts: (1) ***general principles***, assumptions and consequences applicable to all continuous media, (2) ***constitutive equations*** defining the particular idealized material, and (3) ***specialized theories*** of each individual idealized material built on the foundations of the general principles and the constitutive equations of that material to the point where boundary-value problems are formulated and solved for application to specific problems. The third part would require a book at least as long as this one for each specialized theory. Moreover, despite some superficial resemblance between the application of Laplace's equation in fluid mechanics and its application in some parts of linearized elasticity, there is very little common ground in the way the complete specialized theories are developed and applied. Hence specialized treatises must be consulted for the individual materials. Two small samples are included in Chap. 7 on fluid mechanics and Chap. 8 on linearized elasticity, to give some introduction to the way these two theories are put together and applied. References to specialized treatises on fluids and elasticity are given in those chapters. Some works on viscoelasticity and plasticity are also cited in the sections dealing with the constitutive equations of viscoelasticity and plasticity in Chap. 6. Until recently continuum mechanics was presented and studied almost exclusively in its separate specialized branches, but during the last ten years a definite trend has developed toward a unified treatment of the first two parts of the theory, the general principles and the constitutive theory.

The general principles are best studied within the framework of a general continuum theory. This approach emphasizes their general applicability, which is easily lost from sight when the general principles are interspersed with results applicable only to a specific theory. Some pedagogical economy is also achieved by presenting them only once instead of in two or three courses that might be taken by the same student. But an even more important advantage is that the general principles can be presented more carefully, avoiding the tendency to hurry through them to get to the applications.

Most of the material in the first five chapters of this book, covering the general principles, is not new. Important concepts date back to Euler (1770) and much of the formal development was carried out during the early part of the nineteenth century by such pioneers as Cauchy, Navier, and Green, to mention only a few. The foundations of the theory have been subjected to extensive recent critical review. C. Truesdell (1952) published an important study, "The Mechanical Foundations of Elasticity and Fluid Mechanics," reprinted in book form in *Continuum Mechanics*, Vol. I (1966). Two monumental treatises, *Classical Field Theories* by Truesdell and R. Toupin (1960), and *Nonlinear Field Theories of Mechanics* by Truesdell and W. Noll (1965), have included comprehensive accounts of the foundations of the theory as well as of the newer developments in constitutive theory, with the most complete historical and bibliographical information on the subject up to that time. Because of the availability of these reference works no attempt will be made in this book to give a complete historical account, especially of the early work. For those early scientists mentioned in this text in connection with developments attributed to them, complete bibliographical information can be found in the three works cited above. These invaluable reference works and a few recent books at various levels on continuum mechanics are included in the list of references at the end of this chapter.

The most noteworthy recent developments in continuum mechanics have been in the theories of constitutive equations. The classical constitutive equations presented in the first six sections of Chap. 6 were developed independently before the advent of the modern theories of constitutive equations, although a few modifications to them have been made as a result of the modern theories. The modern theories are briefly referred to in Secs. 6.7 and 6.8, but their full development and application, especially to nonlinear constitutive equations, is beyond the scope of this introductory book. The modern theories are in essence theories of theories, setting some limitations on the forms the individual theories can take, once the proposed independent and dependent variables have been chosen. This reduces the amount of experimentation and intuition required to furnish the information needed for completing the definition of the individual idealized material. Historically the linear theories and the theory of plasticity developed by a different route, beginning with assumptions based on intuition and experience and only later being subjected to critical scrutiny to determine whether the form of the theory satisfies the requirements of a theory of theories. Some new constitutive equations will no doubt continue to spring up in this fashion, growing out of the needs of engineers and scientists. But the modern constitutive theories promise valuable aid in formulating new theories, especially of nonlinear material behavior.

Some of the pioneering contributions to the modern theories, especially the work of R. S. Rivlin and his associates and of W. Noll, dating from about 1956, are cited in Secs. 6.7 and 6.8. Many of the early papers and subsequent papers continuing the development of the theory have been reprinted in *Continuum Mechanics*, Vols. I, II, III, IV (1965–66), a four-volume series reprinting 49 papers published by 18 authors between 1945 and 1961, edited by C. Truesdell, referenced at the end of this chapter. The most complete account of the development growing out of the work of Noll is in the *Nonlinear Field Theories of Mechanics* already cited. For more introductory recent accounts see Jaunzemis (1967), Truesdell (1965), Leigh (1968), and Eringen (1967).

The general principles of Chap. 5 applicable to all continuous media, and constitutive equations defining the particular material, for example one of those given in Chap. 6, form the field equations of a continuum theory. The mechanical variables appearing in these equations are the stress and kinematic variables such as strain and rate of deformation. In elementary mechanics of materials, tensile strain is defined as elongation per unit length of the test specimen and is a dimensionless measure of the deformation. Stress is the internal force brought into play by the deformation, per unit cross-sectional area: it is equal to the externally applied force, per unit cross-sectional area. The stress tensor of a three-dimensional combined-stress field is a generalization of the elementary concept of stress, which will be developed in Chap. 3. The small-strain tensor of Secs. 4.1 and 4.2, a generalization of the elementary concept of strain, is the only kinematic variable appearing in the constitutive equation of linearized elasticity. But additional kinds of kinematic variables are needed for more general continua. Some of these additional kinematic concepts, including rate of deformation and finite strain and rotation tensors, will be developed in Chap. 4. The stress tensor will be treated first, since it is simpler and easier to visualize for most engineers, and it affords an opportunity to develop some skill in the use of tensors before the more complicated kinematic tensor development is encountered.

As a preparation for the stress tensor development of Chap. 3 and the kinematic tensor development of Chap. 4, a summary of vectors and tensors is given in Chap. 2, which can serve either as a quick review for the reader already acquainted with tensors or as a self-contained introduction to tensors. Additional material on curvilinear components of tensors is given in Appendices I and II, but these are not needed for the main body of the text. Some additional vector and tensor concepts are also presented where they are first needed in Chaps. 3, 5, and 7, and the reader to whom the tensor methods are new will find that the repeated use of them required throughout the book and the many cross references to Chap. 2 will greatly strengthen his understanding beyond that acquired in a first reading of Chap. 2.

Selected references for collateral reading

COLLECTIONS OF ORIGINAL PAPERS AND TREATISES

Truesdell, C., ed., *Continuum Mechanics* (reprints). New York: Gordon and Breach, v. I (1966), v. II, III, IV (1965).

Truesdell, C. and R. A. Toupin, "The Classical Field Theories," in *Encyclopedia of Physics*, v. III/1, ed. S. Flügge. Berlin: Springer-Verlag, 1960.

Truesdell, C. and W. Noll, "The Non-Linear Field Theories of Mechanics," in *Encyclopedia of Physics*, v. III/3, ed. S. Flügge. Berlin: Springer-Verlag, 1965.

INTRODUCTORY TEXTBOOKS NOT INCLUDING MODERN CONSTITUTIVE THEORIES

Aris, R., *Vectors, Tensors, and the Basic Equations of Fluid Mechanics*. Englewood Cliffs, N. J.: Prentice-Hall, Inc., 1962.

Frederick, D. and T. S. Chang, *Continuum Mechanics*. Boston: Allyn and Bacon, Inc., 1965.

Fung, Y. C., *Foundations of Solid Mechanics*. Englewood Cliffs, N. J.: Prentice-Hall, Inc., 1965.

Long, R. R., *Mechanics of Solids and Fluids*. Englewood Cliffs, N. J.: Prentice-Hall, Inc., 1961.

Prager, W., *Introduction to Mechanics of Continua*. New York: Ginn and Company, 1961.

Scipio, L. A., *Principles of Continua with Applications*. New York: John Wiley & Sons, Inc., 1967.

TEXTBOOKS INCLUDING MODERN CONSTITUTIVE THEORIES

Eringen, A. C., *Mechanics of Continua*. New York: John Wiley & Sons, Inc., 1967.

Eringen, A. C., *Nonlinear Theory of Continuous Media*. New York: McGraw-Hill Book Company, 1962.

Jaunzemis, W., *Continuum Mechanics*. New York: The Macmillan Company, 1967.

Leigh, D. C., *Nonlinear Continuum Mechanics*. New York: McGraw-Hill Book Company, 1968.

Truesdell, C., *The Elements of Continuum Mechanics*. New York: Springer-Verlag, 1965.

CHAPTER 2

Vectors and Tensors

2.1 Introduction

Physical laws, if they really describe the physical world, should be independent of the position and orientation of the observer. That is, if two scientists using different coordinate systems observe the same physical event, it should be possible to state a physical law governing the event in such a way that if the law is true for one observer, it is also true for the other. For this reason, the equations of physical laws are ***vector equations*** or ***tensor equations***, since vectors and tensors transform from one coordinate system to another in such a way that if a vector or tensor equation holds in one coordinate system, it holds in any other coordinate system not moving relative to the first one, i.e., in any other coordinate system in the same reference frame. Invariance of the form of a physical law referred to two frames of reference in accelerated motion relative to each other is more difficult and requires the apparatus of general relativity theory, tensors in four-dimensional space-time. For simplicity, we limit ourselves in this book to tensors in three-dimensional Euclidean space and for most of the discussion use only rectangular Cartesian coordinates for the components of tensors and vectors. The way in which the physical laws defining material response transform from one frame of reference to another frame moving relative to the first will be explained in Sec. 6.7 by means of the postulated ***principle of material frame indifference***.

Most readers will have used vectors before; however, we review in Secs. 2.2 to 2.5 and Sec. 5.1 some of the ideas of vector algebra and calculus. The concept of a tensor as a linear vector operator will be presented in this introduction and further developed in Sec. 2.4, where some properties of tensors are summarized and Cartesian tensor components are defined. The Cartesian component formulation is sufficient for the development of the theory of continuum mechanics and will be used throughout the main body of the text. For the solution of specific problems, orthogonal curvilinear coordinates suitable to the geometry of the problem may lead to simplification of the analysis. Appendix II presents the physical components of vectors and tensors in orthogonal curvilinear coordinates and summarizes the equations of continuum mechanics in this form. It is recommended that on a first reading the material of this appendix be omitted or postponed until the continuum theory has been completed. Appendix I presents the more complicated formulation of general tensor components referred to an arbitrary curvilinear coordinate system; this general formulation is not actually necessary for the development of continuum mechanics, and is included only because many modern writers on the subject have chosen to use this language. The advanced student can profit from some acquaintance with the language of general tensors in order to read the literature, but it is recommended that App. I be omitted in an introductory course.

Tensors as linear vector operators. In three-dimensional space, a ***vector*** may be visualized as an arrow that has length equal to the magnitude of the vector and that is pointing in the direction of the vector; it thus is seen to have an existence quite independent of any coordinate system in which it might be observed. Although a tensor cannot be visualized simply as an arrow, it also has an existence independent of the coordinate system. Tensors frequently arise as physical entities whose components are the coefficients of a linear relationship between vectors. For example, in Chap. 3 we will find that for every direction at a point, specified by a vector $\hat{\mathbf{n}}$, there may be associated a second vector $\mathbf{t}$, the traction vector (force per unit area) acting at the point across the surface whose unit normal is $\hat{\mathbf{n}}$. At a given point, $\mathbf{t}$ is thus a vector function of $\hat{\mathbf{n}}$; symbolically $\mathbf{t} = \mathbf{t}(\hat{\mathbf{n}})$. The functional dependence in this case is an example of a ***linear vector function***. (A vector function $\mathbf{F}$ is said to be a linear vector function if $\mathbf{F}(a\mathbf{u} + b\mathbf{v}) = a\mathbf{F}(\mathbf{u}) + b\mathbf{F}(\mathbf{v})$ for every choice of the vectors $\mathbf{u}$, $\mathbf{v}$, and real numbers a, b.) ***Since the argument $\hat{\mathbf{n}}$ of the function and the functional value $\mathbf{t}$ associated with it are both vectors, and therefore have an existence independent of any coordinate system, the functional relationship must then be independent of any coordinate system.*** Actually, in practice, we observe the two vectors in a particular coordinate system and express the functional relationship by means of coefficients referred to that coordinate system. These coefficients will be the components of the tensor that relates one vector to the other. And we will derive tensor transformation formulas giving the tensor components in any other coordinate system, such that when these components are used as coefficients of the vector function, the function yields the same vector as before, but the vector is represented by its components in the new coordinate system.

In any one rectangular coordinate system, the linear vector functional relationship may be represented as a matrix equation involving the square matrix of tensor components either premultiplied by the argument vector expressed as a row matrix or postmultiplied by the argument vector expressed as a column matrix. (See Part 4 of Sec. 2.4, for a review of elementary matrix concepts.) This is a very important kind of relationship in all branches of physics. Some examples are:

1. T is stress tensor (square matrix).
   n is unit vector normal of surface (row matrix).
   t is traction vector on surface (row matrix):

$$t = nT.$$

2. I is inertia tensor of a rigid body (square matrix).
   ω is angular velocity vector of body (column matrix).
   h is angular momentum vector of body (column matrix):

$$h = I\omega.$$

3. E is small strain tensor (square matrix).
   dX is infinitesimal initial relative position vector of adjacent material particles (column matrix).
   du is relative displacement vector produced by straining (column matrix):

$$du = E\,dX.$$

4. K is thermal conductivity of anisotropic medium (square matrix).
   q is heat flux vector (column matrix).
   $\nabla\theta$ is gradient of the temperature θ (column matrix with components $\frac{\partial\theta}{\partial x}, \frac{\partial\theta}{\partial y}, \frac{\partial\theta}{\partial z}$):

$$q = -K\,\nabla\theta.$$

5. Σ is electrical conductivity of anisotropic medium (square matrix).
   e is electric field intensity vector (column matrix).
   i is current density vector (column matrix):

$$i = \Sigma\, e.$$

These are just a few of the many physical laws in which a tensor is used as the function associating one vector with another. Remember that the relationship between vectors must be independent of the coordinate system, although we usually observe it in some one system; and remember that it is precisely this independence of the coordinate system that motivates us to study tensors at all.

Before beginning our study of tensors we first review (in Secs. 2.2 and 2.3) some of the notions of elementary vector algebra.

2.2 Vectors;† vector addition; vector and scalar components; indicial notation; summation convention; finite rotations are not vectors

In a three-dimensional Euclidean space, ***vector quantities*** (or briefly, ***vectors***) are entities possessing both magnitude and direction and obeying certain laws; a vector is conventionally represented by an arrow that is pointing in the direction associated with the vector and that has a length proportional to the magnitude of the vector. The simplest example is a directed line joining two points. Other familiar examples are force, velocity, and acceleration. A ***scalar*** quantity, on the other hand, does not have any direction associated with it, although some scalar quantities may have positive and negative values, e.g., Fahrenheit temperature. In printed text, a vector is usually denoted by a boldface letter. The magnitude of the vector is then usually denoted by the same letter in lightface italic type. In typewritten or handwritten text, the wavy underscore used by copyreaders to designate boldface type is often used for both vectors and tensors, or the vector may be identified by an arrow above the letter, or by underlining. The magnitude of vector $\mathbf{a}$ is denoted by a (the same letter in italic instead of boldface and not underlined), or by $|\mathbf{a}|$, and is always a positive number or zero. In script $a = |\underset{\sim}{a}|$.

Two ***vectors are equal*** if they have the same direction and the same magnitude. Thus a vector is unchanged if it is moved parallel to itself. In some applications, however, the point or the line of action of a vector quantity is important. For example, a force is a vector quantity with a definite point of action; it is a ***fixed*** or ***bound vector***. In rigid-body dynamics, only the line of action of the force is important, not the point; force is then a ***sliding vector***. When we say that two forces are equal, however, we mean only that they have the same magnitude and direction; this does not imply that they are mechanically equivalent.

Vector addition. Any two vector quantities of the same kind (e.g., two forces or two velocities) may be represented as two vectors $\mathbf{a}$ and $\mathbf{b}$ so placed that the initial point of $\mathbf{b}$ coincides with the terminal point of $\mathbf{a}$, as in Fig. 2.1. The ***sum*** of $\mathbf{a}$ and $\mathbf{b}$ is then defined as the vector $\mathbf{c}$ extending from the initial point of $\mathbf{a}$ to the terminal point of $\mathbf{b}$:

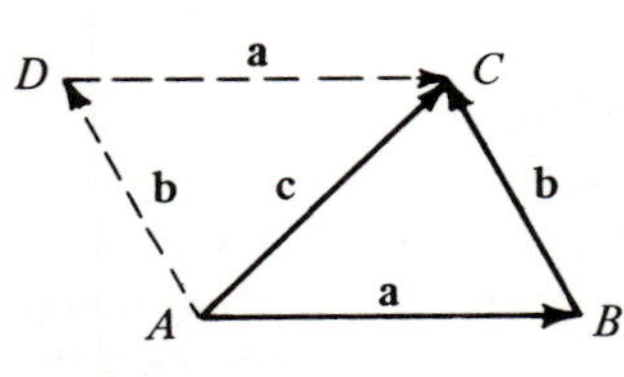

Fig. 2.1 Addition of vectors.

$$\mathbf{c} = \mathbf{a} + \mathbf{b}. \tag{2.2.1}$$

†Only those definitions and formulas used in the text are summarized here; they are so important in mechanics that they should be mastered by every serious student of the subject. For additional details, or for an axiomatic development of the theory, see books on vector analysis and vector spaces.

The actual addition of two vectors is most conveniently performed by using their rectangular components in some coordinate system as in Eq. (8), but the definition and properties of vector addition do not depend on the introduction of a coordinate system.

From the definition, it follows that

$$\mathbf{b} + \mathbf{a} = \mathbf{a} + \mathbf{b},$$

as is indicated by the dashed arrows in Fig. 2.1. Thus vector addition obeys the ***commutative law.*** It also follows from the definition that the ***associative law***

$$(\mathbf{a} + \mathbf{b}) + \mathbf{c} = \mathbf{a} + (\mathbf{b} + \mathbf{c})$$

is satisfied. The common value of the two expressions is written as the sum $\mathbf{a} + \mathbf{b} + \mathbf{c}$ of the three vectors.

If a vector is reversed in direction with no change in magnitude, the resulting vector is called the ***negative*** of the original vector. For example, $\overrightarrow{BA} = -\overrightarrow{AB}$ in Fig. 2.1. To subtract $\mathbf{b}$ from $\mathbf{a}$, add the negative of $\mathbf{b}$ to $\mathbf{a}$:

$$\mathbf{a} - \mathbf{b} = \mathbf{a} + (-\mathbf{b}). \tag{2.2.2}$$

A vector $\mathbf{a}$ may be ***multiplied by a scalar*** c to yield a new vector, $c\mathbf{a}$ or $\mathbf{a}c$, of magnitude $|ca|$. If c is positive, $c\mathbf{a}$ has the same direction as $\mathbf{a}$; if c is negative, $c\mathbf{a}$ has the direction of $-\mathbf{a}$. If c is zero, the product is the ***zero vector*** with zero magnitude and undefined direction. Thus $(0)\mathbf{a} = \mathbf{0}$. Sometimes $\mathbf{0}$ is written as 0, not in boldface and not underlined in script.

A vector of unit length is called a ***unit vector***. Any vector may be written as the product of the scalar magnitude of the given vector by a dimensionless unit vector in the given direction. Thus $\mathbf{a} = a\hat{\mathbf{u}}$, where $\hat{\mathbf{u}}$ is a dimensionless unit vector in the direction of $\mathbf{a}$. A unit vector may be signified by the caret (^) placed over its letter symbol. The caret is often omitted, when the context makes it clear that the vector is a unit vector.

Vector components and scalar components. Two or more vectors whose sum is a given vector are called ***vector components*** of the given vector. It is often convenient to resolve a given vector into three components parallel to three noncoplanar ***base vectors***; the resolution is unique for a given vector and given base vectors. The set of base vectors is called a ***basis***. If the basis consists of mutually orthogonal unit vectors, it is called an ***orthonormal basis***. In this book, the base vectors are usually specified as unit vectors in the positive directions of a set of rectangular Cartesian coordinate axes. The components are then called the ***rectangular vector components*** of the given vector in that coordinate system. The name rectangular components is also used for the numerical coefficients of the base vectors, that is, for a_x, a_y and a_z in Eq. (4) below.

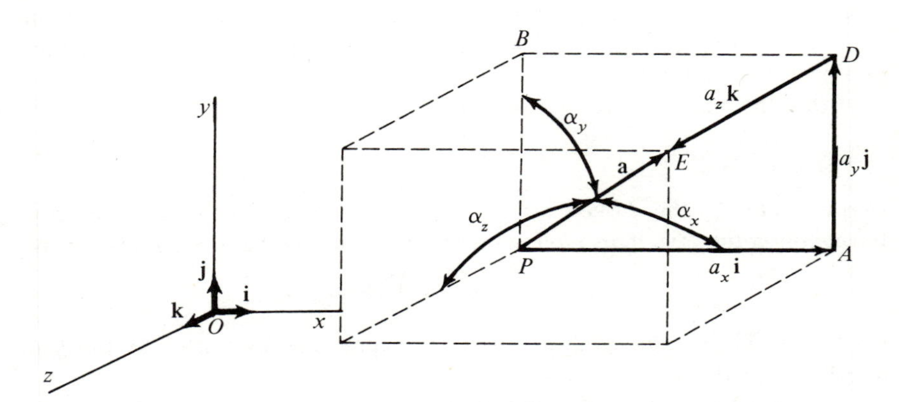

Fig. 2.2 Rectangular components and unit vectors.

In Fig. 2.2, for example, $\overrightarrow{PA}$, $\overrightarrow{AD}$, and $\overrightarrow{DE}$ are the rectangular vector components of the vector $\overrightarrow{PE}$, in the directions of the unit base vectors **i**, **j**, **k**. We denote the vector by **a**, its magnitude by a, and the angles it makes with the positive coordinate directions by $\alpha_x, \alpha_y, \alpha_z$. Then the three ***rectangular components*** of the vector, that is, the numerical coefficients of the base vectors, are

$$a_x = a \cos \alpha_x, \qquad a_y = a \cos \alpha_y, \qquad a_z = a \cos \alpha_z, \tag{2.2.3}$$

and the rectangular vector components are $a_x\mathbf{i}$, $a_y\mathbf{j}$, $a_z\mathbf{k}$ so that

$$\mathbf{a} = a_x\mathbf{i} + a_y\mathbf{j} + a_z\mathbf{k}. \tag{2.2.4}$$

Note that a_x is a number (positive, negative, or zero), while $a_x\mathbf{i}$ is a vector, i.e., a directed physical quantity parallel to **i**. The rectangular components are the orthogonal projections of the given vector onto the three coordinate axes.

The word ***component*** is also often used as a synonym for "projection on an arbitrary direction." For example, a unit vector $\hat{\mathbf{n}}$ may be given (not necessarily parallel to a coordinate axis); "the component a_n of **a** in the direction of $\hat{\mathbf{n}}$" then simply means the orthogonal projection of **a** onto a line having the direction of $\hat{\mathbf{n}}$, namely

$$a_n = a \cos \theta, \tag{2.2.5}$$

where θ is the angle between the directions of **a** and $\hat{\mathbf{n}}$. This equation applies whether or not **a** and $\hat{\mathbf{n}}$ are drawn from the same point, but the angle θ is more easily visualized if one of the vectors is moved parallel to itself to the same initial point as the other.

The three cosines, $\cos \alpha_x$, $\cos \alpha_y$, $\cos \alpha_z$, are the ***direction cosines*** of the vector. By repeated application of the theorem of Pythagoras we see that the magnitude of the vector is given in terms of the scalar components by

$$a^2 = a_x^2 + a_y^2 + a_z^2. \tag{2.2.6}$$

Substitution of Eqs. (3) into this relationship yields

$$\cos^2 \alpha_x + \cos^2 \alpha_y + \cos^2 \alpha_z = 1, \tag{2.2.7}$$

which expresses the fact that the sum of the squares of the direction cosines of any direction is unity.

The addition of two or more vectors is most simply performed in terms of their rectangular components. For example, if

$$\mathbf{c} = \mathbf{a} + \mathbf{b} \quad \text{and} \quad \mathbf{d} = \mathbf{a} - \mathbf{b}$$

then the components of **c** and **d** are given by

$$\begin{aligned} c_x &= a_x + b_x & d_x &= a_x - b_x \\ c_y &= a_y + b_y & d_y &= a_y - b_y \\ c_z &= a_z + b_z & d_z &= a_z - b_z. \end{aligned} \tag{2.2.8}$$

Equations (8) are special cases of the relationship

$$c_n = a_n + b_n, \tag{2.2.9}$$

which is valid for the components in an arbitrary direction $\hat{\mathbf{n}}$ (projections), as is evident from Fig. 2.3, where $\mathbf{c} = \mathbf{a} + \mathbf{b}$ and where $a_n = OA$, $b_n = AB$ and $c_n = OB$.

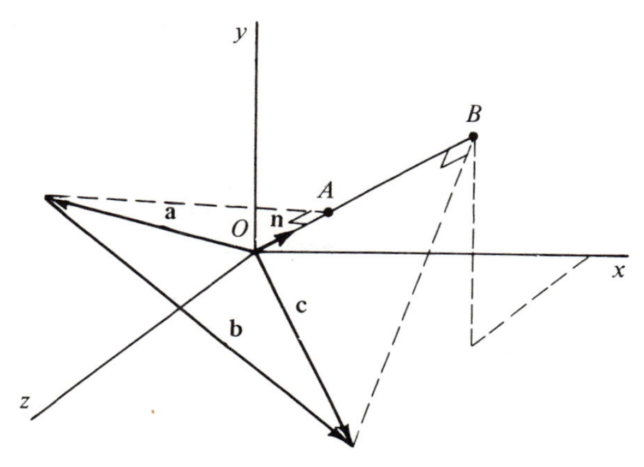

Fig. 2.3 Projection OB of sum **c** equals sum of projections.

Indicial notation. Instead of denoting the rectangular Cartesian coordinates of a point by (x, y, z), we may call them (x_1, x_2, x_3) as is indicated by the labeling of the coordinate axes in Fig. 2.4.

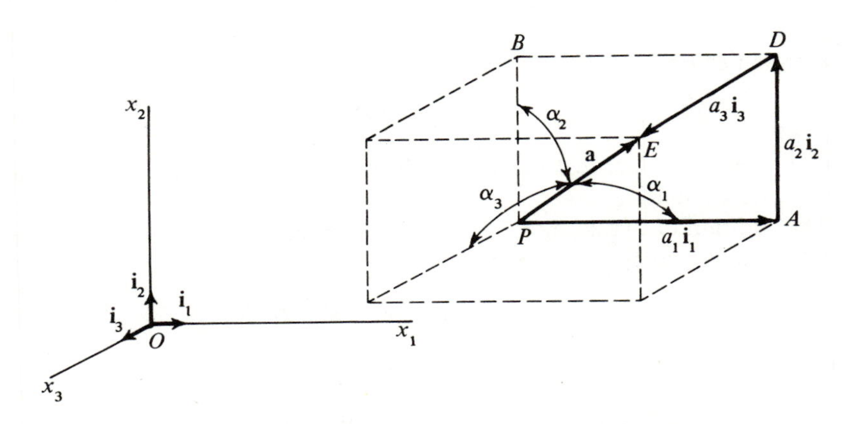

Fig. 2.4 Indicial notation: x_1-, x_2-, x_3-axes. Vector **a**.

The components of a vector **a** are then denoted by a_i, $i = 1, 2, 3$. The three components are (a_1, a_2, a_3). The symbol a_i is used for a typical one of the components or for the set of three components. It is also sometimes used instead of **a** to represent the vector itself; the context prevents serious misunderstanding of this usage. Then if $\hat{\mathbf{n}}$ is a unit vector in the direction of the vector **a**, its components are

$$n_i = \cos \alpha_i,$$

where the α_i are the angles between **a** and the x_i-axes. Hence the components of the vector **a** are given by

$$a_i = a n_i. \tag{2.2.10}$$

Equation (10) represents a set of three equations, obtained by giving i, successively, the values 1, 2, 3. In fact Eq. (10) is a compact form of Eq. (3). Similarly the three equations (8) for the components of **c** can be written compactly as

$$c_i = a_i + b_i.$$

Summation convention. The indicial notation permits us to write many of the formulas of continuum mechanics in a much shorter form than would otherwise be possible. This brevity makes the formulas easier to remember and also easier to understand, once the notation is learned. Many of our formulas require a summation over all values of the index. For example, the magnitude a of a vector a_i is given by Eq. (6) as

$$a^2 = a_1^2 + a_2^2 + a_3^2 = \sum_{i=1}^{3} a_i^2. \tag{2.2.11}$$

We adopt the ***summation convention*** which states that in Cartesian coordinates whenever the same letter subscript occurs twice in a term, that subscript is to be given all possible values and the results added together. The symbol $\sum$ may then be omitted because the repeated subscript indicates the summation. For example, in three dimensions,

$$\begin{aligned} a_i a_i &= a_1^2 + a_2^2 + a_3^2, \\ a_{kk} &= a_{11} + a_{22} + a_{33}, \\ a_{j2} b_{3j} &= a_{12} b_{31} + a_{22} b_{32} + a_{32} b_{33}. \end{aligned}$$

When general tensor components in curvilinear coordinates are used, summation is only implied by repeated indices if one of the repeated indices appears as a superscript and one as a subscript; e.g.,

$$a^m b_m = a^1 b_1 + a^2 b_2 + a^3 b_3.$$

Equation (11) can now be written as

$$a^2 = a_i a_i. \tag{2.2.12}$$

This is a single equation, while Eq. (10) is a set of three equations, because Eq. (10) has a free subscript, i.e., a letter subscript not repeated twice in the same term. Note that $c_i = a_i + b_i$ is also a set of three equations because the same subscript does not occur twice in one term. The notation $c_{ij} = a_{ij} + b_{ij}$ represents nine equations, one for each of the nine possible combinations of numerical values 1, 2, 3 that the free subscripts i and j can have.

Finite rotations are not vectors. Not all entities representable as directed lines are vectors; to qualify as vectors they must satisfy the laws of vectors. In particular, they must combine according to the laws of vector addition. For example, it is one of the basic postulates of Newtonian mechanics that two forces applied at the same point may be combined according to these laws, i.e., the vector sum applied at the point produces the same effect as the two forces would produce. And this postulate is crucial in establishing that forces are vector quantities. Finite rigid-body rotations are not vector quantities, since they do not combine vectorially, as the following counterexample shows, despite the fact that any one rigid-body rotation about a fixed axis may be represented as an arrow laid off to scale along the axis of rotation.

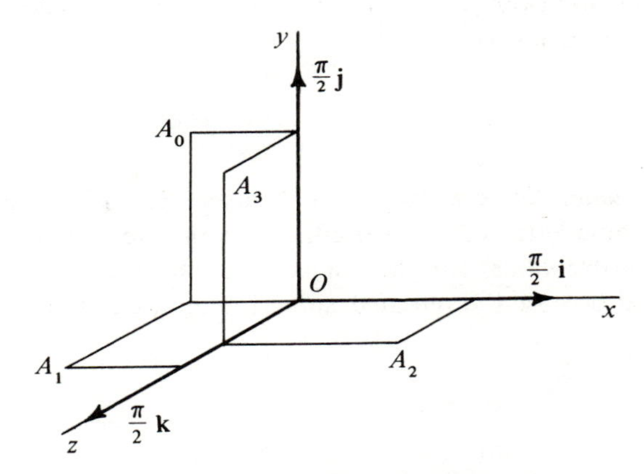

Fig. 2.5 Example showing that finite rotations are not vectors.

In Fig. 2.5, the rigid body is taken to be the rectangle initially lying in the xy-plane with diagonal OA_0. It is given the following sequence of 90-degree rotations, each about a fixed axis:

1. About x-axis to position OA_1, represented as an arrow $(\pi/2)\mathbf{i}$ in the x-direction,
2. About y-axis to position OA_2, represented by $(\pi/2)\mathbf{j}$,
3. About z-axis to position OA_3, represented by $(\pi/2)\mathbf{k}$.

If finite rotations were vector quantities, then the single rotation about a fixed axis, defined by the vector sum $(\pi/2)(\mathbf{i} + \mathbf{j} + \mathbf{k})$ of the three individual rotations, would take the rectangle from the initial position OA_0 to the final position OA_3. But this is not the case, since evidently the single rotation which takes OA_0 to OA_3 is a 90-degree rotation about the y-axis, represented as $(\pi/2)\mathbf{j}$, which is not the vector sum of the three individual rotations. This shows that finite rotations are not vectors.

If a vector quantity is conceived of as a physical entity (e.g., force or velocity), it is clear that it has an existence independent of the coordinate system. If another rectangular Cartesian system is used, the components of the vector in the second system are different from those in the first, ***but it is still the same vector***. It must therefore be possible to express the vector's components in the new coordinate system in terms of those in the old one; the formulas for such a change of axes will be given in Sec. 2.4. Thus, if the components of a vector are given in one coordinate system, the vector is completely specified. Before considering change of axes in Sec. 2.4, we will review the notions of scalar product and vector product: two different ways of multiplying two vectors.

2.3 Scalar product and vector product

In Sec. 2.2, multiplication of a vector by a scalar was defined to yield a vector. We now define two different kinds of multiplication of one vector by another. In one case, the product is not a vector but a scalar; it is called the ***scalar product*** or ***dot product***. The second case yields a vector; it is called the ***vector product*** or ***cross product***. A third kind of multiplication yielding a tensor is defined in Eq. (2.4.21).

The ***scalar product*** is defined as the product of the two magnitudes times the cosine of the angle between the vectors:

$$\mathbf{a}\cdot\mathbf{b} = ab\cos\theta. \tag{2.3.1}$$

It follows immediately from the definition that if m and n are scalars (numbers), then

$$(m\mathbf{a})\cdot(n\mathbf{b}) = mn(\mathbf{a}\cdot\mathbf{b}). \tag{2.3.2}$$

It is also evident from the definition that the scalar product is commutative,

$$\mathbf{a}\cdot\mathbf{b} = \mathbf{b}\cdot\mathbf{a}. \tag{2.3.3}$$

We may interpret $ab\cos\theta$ either as b times the projection of $\mathbf{a}$ onto the direction of $\mathbf{b}$ or as a times the projection of $\mathbf{b}$ onto $\mathbf{a}$.

The scalar product is also distributive, i.e.,

$$\mathbf{a}\cdot(\mathbf{b}+\mathbf{c}) = (\mathbf{a}\cdot\mathbf{b}) + (\mathbf{a}\cdot\mathbf{c}), \tag{2.3.4}$$

as the following argument shows. Since the projection of $\mathbf{b}+\mathbf{c}$ onto $\mathbf{a}$ equals the sum of the projections of $\mathbf{b}$ and $\mathbf{c}$ onto $\mathbf{a}$, according to Eq. (2.2.9); and since $\mathbf{a}\cdot(\mathbf{b}+\mathbf{c})$ may, by the scalar product definition, be interpreted as the product of a by the projection of $\mathbf{b}+\mathbf{c}$ onto $\mathbf{a}$, Eq. (4) takes the form

$$a(b_a + c_a) = ab_a + ac_a,$$

which is known to be true by the distributive property of the arithmetic product of numbers. Here b_a and c_a are the components (projections) of $\mathbf{b}$ and $\mathbf{c}$ in the direction of $\mathbf{a}$. The distributive property of the scalar product permits us to derive a formula for the scalar product in terms of the rectangular components by using the unit base vectors $\mathbf{i}, \mathbf{j}, \mathbf{k}$.

From the defining equation (1), it follows that the scalar product of any two different unit base vectors is zero, since $\cos 90° = 0$, while the scalar product of any unit vector by itself equals unity.

$$\begin{aligned} \mathbf{i}\cdot\mathbf{i} = \mathbf{j}\cdot\mathbf{j} = \mathbf{k}\cdot\mathbf{k} = 1 \\ \mathbf{i}\cdot\mathbf{j} = \mathbf{j}\cdot\mathbf{k} = \mathbf{k}\cdot\mathbf{i} = 0. \end{aligned} \tag{2.3.5}$$

Then the scalar product

$$\mathbf{a}\cdot\mathbf{b} = (a_x\mathbf{i} + a_y\mathbf{j} + a_z\mathbf{k})\cdot(b_x\mathbf{i} + b_y\mathbf{j} + b_z\mathbf{k})$$

yields

$$\boxed{\begin{aligned} &\mathbf{a}\cdot\mathbf{b} = a_x b_x + a_y b_y + a_z b_z, \\ &\text{or} \\ &\mathbf{a}\cdot\mathbf{b} = a_i b_i \qquad \text{in indicial notation.} \end{aligned}} \tag{2.3.6}$$

Remember this very important formula. It furnishes the method by which we actually evaluate scalar products. But do not forget Eq. (1), which gives physical significance to what we accomplish by using Eq. (6).

One of the most common applications of the scalar product is the projection of a given vector onto a line having a given direction. If **a** is the given vector and $\hat{\mathbf{n}}$ is a unit vector in the given direction, then the projection a_n is given by

$$a_n = \mathbf{a}\cdot\hat{\mathbf{n}} = a_x n_x + a_y n_y + a_z n_z \quad \text{or} \quad a_n = a_i n_i \tag{2.3.7}$$

The three components (n_x, n_y, n_z) of the unit vector $\hat{\mathbf{n}}$ are the direction cosines of the given direction.

Another important geometrical application is the determination of the angle between two vectors given in terms of their components. From Eq. (1), we obtain

$$\cos\theta = \frac{\mathbf{a}\cdot\mathbf{b}}{ab}. \tag{2.3.8}$$

The numerator of this formula is given by Eq. (6), and the two magnitudes in the denominator may be calculated in terms of the components by Eq. (2.2.6). It follows from Eq. (8), that a necessary and sufficient condition for two nonzero vectors to be perpendicular is

$$\mathbf{a}\cdot\mathbf{b} = 0. \tag{2.3.9}$$

Another important application of the scalar product is the calculation of the work done by a force. If force **f** acts on a point of application which moves through the infinitesimal displacement $d\mathbf{r}$, then the work dW done is

$$dW = \mathbf{f}\cdot d\mathbf{r} = f\cos\theta\, ds, \tag{2.3.10}$$

where θ is the angle between **f** and $d\mathbf{r}$, and where f and ds are the magnitudes of **f** and $d\mathbf{r}$, respectively. Since, in general, the force may vary both in magnitude and direction during any finite motion, it will be necessary to express the force as a function of position and evaluate the total work as a line integral along the path of the motion. When **f** is constant and the motion is along a straight line, this results in a total work W given by

$$W = \mathbf{f} \cdot \Delta\mathbf{r} \tag{2.3.11}$$

where $\Delta\mathbf{r}$ is the finite displacement. We note again that the work W is a scalar quantity, obtained by multiplying two vectors, using the scalar product. We now consider another way of multiplying two vectors; this time the product will itself be a vector.

The second kind of product of one vector multiplied by another is the ***vector product*** or ***cross product***:

$$\mathbf{c} = \mathbf{a} \times \mathbf{b}, \tag{2.3.12}$$

defined as a vector **c** perpendicular to both **a** and **b** in the sense that makes **a, b, c** a right-handed system. Magnitude c is given by

$$c = ab \sin\theta. \tag{2.3.13}$$

(This always yields a nonnegative value for c because the angle is between zero and 180°.) Because of the sense prescription, vector multiplication is not commutative. We have instead

$$\mathbf{b} \times \mathbf{a} = -(\mathbf{a} \times \mathbf{b}), \tag{2.3.14}$$

as illustrated in Fig. 2.6. The magnitudes are the same; in either case the magnitude c is numerically equal to the area $ab \sin\theta$ of the parallelogram formed with the two given vectors as sides; but the senses are opposite.

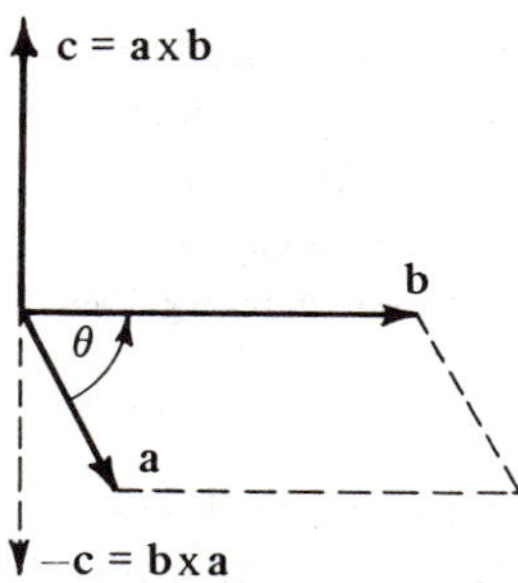

Fig. 2.6 Vector products.

For example, the unit vectors **i, j, k** of a right-handed system form the following products:

$$\begin{aligned} &\mathbf{i} \times \mathbf{i} = \mathbf{j} \times \mathbf{j} = \mathbf{k} \times \mathbf{k} = \mathbf{0} \quad &&\text{(the zero vector)}, \\ &\mathbf{i} \times \mathbf{j} = \mathbf{k}, \quad \mathbf{j} \times \mathbf{k} = \mathbf{i}, \quad &&\mathbf{k} \times \mathbf{i} = \mathbf{j}, \\ &\mathbf{j} \times \mathbf{i} = -\mathbf{k}, \quad \mathbf{k} \times \mathbf{j} = -\mathbf{i}, \quad &&\mathbf{i} \times \mathbf{k} = -\mathbf{j}. \end{aligned} \tag{2.3.15}$$

The algebraic signs for the products (15) can be remembered by the scheme shown in the figure on page 20. If the two factors occur in the same sequence

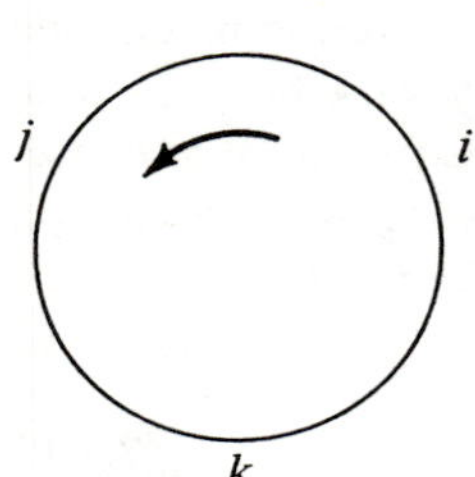

as i, j, k on the circle, the product is positive. If the order is reversed, the product is negative. (All the signs are reversed for a left-handed system.)

The vector product is also distributive, but we omit the proof.† Thus

$$\mathbf{a} \times (\mathbf{b} + \mathbf{c}) = (\mathbf{a} \times \mathbf{b}) + (\mathbf{a} \times \mathbf{c}). \tag{2.3.16}$$

The distributive character of the product permits us to evaluate it in terms of the components of the given vectors. This is done by writing each vector in the form given in Eq. (2.2.4), multiplying the two vectors together, and simplifying by means of Eqs. (15). The result is, for a right-handed system,

$$\mathbf{a} \times \mathbf{b} = (a_y b_z - a_z b_y)\mathbf{i} + (a_z b_x - a_x b_z)\mathbf{j} + (a_x b_y - a_y b_x)\mathbf{k}, \tag{2.3.17}$$

which may be remembered conveniently as the formal expansion of the determinant

$$\mathbf{a} \times \mathbf{b} = \begin{vmatrix} \mathbf{i} & \mathbf{j} & \mathbf{k} \\ a_x & a_y & a_z \\ b_x & b_y & b_z \end{vmatrix} \tag{2.3.18}$$

in terms of the minors of the elements in the first row. For a left-handed coordinate system a minus sign is prefixed to the formulas. The cross product would be an example of an ***axial vector*** if we did not employ this convention of prefixing a minus sign to Eqs. (17) and (18) in a left-handed system. With this convention, the cross product behaves as a ***polar vector*** or true vector under transformation of coordinates (see Sec. 2.4). When the vectors have several zero components, it is often more convenient to apply Eq. (15) directly instead of using Eqs. (17) or (18). For example,

$$2\mathbf{i} \times (3\mathbf{j} - 2\mathbf{k}) = 6\mathbf{i} \times \mathbf{j} - 4\mathbf{i} \times \mathbf{k} = 6\mathbf{k} + 4\mathbf{j}.$$

Here and in the derivation of Eq. (17), we have tacitly assumed in addition to the distributive property that

$$(m\mathbf{a}) \times (n\mathbf{b}) = mn(\mathbf{a} \times \mathbf{b}) \tag{2.3.19}$$

for arbitrary scalars m and n, a consequence immediately evident from the definition of the vector product. However, the vector product is not associative

$$\mathbf{a} \times (\mathbf{b} \times \mathbf{c}) \neq (\mathbf{a} \times \mathbf{b}) \times \mathbf{c}.$$

†See, for example, Nathaniel Coburn, *Vector Analysis* (New York: The Macmillan Company, 1955), p. 14.

Hence parentheses must be used to indicate the order of multiplication in the vector triple product.

With the ***scalar triple product*** $\mathbf{a} \times \mathbf{b} \cdot \mathbf{c}$, parentheses are not needed, since this product has meaning only if the cross product is evaluated first. However, the dot and cross may be interchanged without changing the (scalar) value of the product. You can easily verify this by writing each vector in the standard component form of Eq. (2.2.4) and carrying out the multiplications for a right-handed coordinate system:

$$\mathbf{a} \times \mathbf{b} \cdot \mathbf{c} = \mathbf{a} \cdot \mathbf{b} \times \mathbf{c} = \begin{vmatrix} a_x & a_y & a_z \\ b_x & b_y & b_z \\ c_x & c_y & c_z \end{vmatrix}. \tag{2.3.20}$$

In fact if **a, b, c** are oriented so that they form a right-handed system (so that this product is positive), the scalar triple product of Eq. (20) gives the volume of the parallelepiped formed with **a, b** and **c** as three edges.

Applications of the vector product include the calculation of the moment of a force, the determination of the displacements produced by a small rotation, or the particle velocities induced by rotation. One important geometric application is the determination of a direction perpendicular to two given directions. A unit vector $\hat{\mathbf{n}}$ perpendicular to both **a** and **b** is obtained as

$$\hat{\mathbf{n}} = \frac{\mathbf{a} \times \mathbf{b}}{|\mathbf{a} \times \mathbf{b}|}. \tag{2.3.21}$$

The rectangular components thus obtained will be the direction cosines of the common perpendicular. This is especially useful in relating new coordinate axes to the old ones. An example will be given in Chap. 3; when we are finding the principal axes of stress, we will obtain the third axis as the common perpendicular to the first two.

In order to write the vector product in indicial notation, we introduce the ***permutation symbol*** e_{mnr}, defined to have the value 0, +1, or −1 as follows:

$$e_{mnr} = \begin{cases} 0 & \text{when any two indices are equal;} \\ +1 & \text{when } m, n, r \text{ are } 1, 2, 3 \text{ or an even permutation of } 1, 2, 3; \\ -1 & \text{when } m, n, r \text{ are an odd permutation of } 1, 2, 3. \end{cases} \tag{2.3.22}$$

Thus, for example,

$$e_{123} = e_{231} = e_{312} = 1$$
$$e_{132} = e_{213} = e_{321} = -1$$
$$e_{112} = e_{122} = e_{222} = 0, \quad \text{etc.}$$

The sign convention can be remembered by marking the numbers 1, 2, 3 on a circle. Then any even permutation will be in ***cyclic order***, i.e., it will go around the circle in the same sense as 1, 2, 3, while an odd permutation will go in the reverse direction or ***acyclic order***. Then the three components c_p are given in a right-handed coordinate system by

$$\boxed{\mathbf{c} = \mathbf{a} \times \mathbf{b} \quad c_p = e_{pqr} a_q b_r = e_{qrp} a_q b_r,} \tag{2.3.23}$$

as you can verify by writing out the sums represented in Eq. (23) and comparing the results with Eq. (17). Note that because there is one free subscript p, there are three Eqs. (23), and each equation contains nine terms: one for each possible choice of the two summation indices q and r. But when these nine terms are written out, all but two will have a repeated index on the permutation symbol, which means that there are only two nonzero terms in each nine-term sum—one requiring a plus sign and one a minus sign. The vector equation (17) may be written for a right-handed system as

$$\mathbf{a} \times \mathbf{b} = e_{pqr} a_q b_r \mathbf{i}_p \qquad \text{(vector sum on } p\text{)}. \tag{2.3.24}$$

For actual numerical evaluations, the determinant scheme of Eq. (18) or direct use of the unit vector products, as in the example after Eq. (18), is more convenient than the indicial Eqs. (23) or (24), but the indicial expressions are useful in theoretical derivations because they can be substituted conveniently into other equations.

The Kronecker delta δ_{pq} is another symbol useful to simplify and shorten equations in index notation. It is defined by

$$\text{Kronecker delta} \quad \delta_{pq} = \begin{cases} 1 \text{ if } p = q \\ 0 \text{ if } p \neq q \end{cases} \tag{2.3.25}$$

where p and q can each have any positive integer value appearing in the discussion, that is 1, 2, or 3 for the usual case of three dimensions. As an example of the use of the Kronecker delta, we note that the statement that $\mathbf{i}_1$, $\mathbf{i}_2$, $\mathbf{i}_3$ form an orthonormal set, i.e. a set of three mutually orthogonal vectors each of unit magnitude, can be expressed by the equation

$$\mathbf{i}_p \cdot \mathbf{i}_q = \delta_{pq}. \tag{2.3.26}$$

By writing out the summations implied by the repeated indices, and using the definition of Eq. (25) we can easily establish the following useful identities.

$$\text{(a)} \quad \delta_{mm} = 3 \qquad \text{(b)} \quad \delta_{mn}\,\delta_{mn} = 3$$
$$\text{(c)} \quad u_m\,\delta_{mn} = u_n \qquad \text{(d)} \quad T_{mn}\,\delta_{mn} = T_{kk} \tag{2.3.27}$$

The e-δ identity of Eq. (28) below can also be verified by writing out the expression, although this requires a more lengthy enumeration of the independent choices of the free indices.

$$\boxed{e\text{-}\delta \text{ identity} \quad e_{ijk}\,e_{irs} = \delta_{jr}\,\delta_{ks} - \delta_{js}\,\delta_{kr}} \tag{2.3.28}$$

This identity relating the three-index symbols e_{mnr} of Eq. (22) to the Kronecker delta of Eq. (25) is useful for proving many vector identities. For example, the following identity for triple vector product

$$\mathbf{a} \times (\mathbf{b} \times \mathbf{c}) = (\mathbf{a}\cdot\mathbf{c})\mathbf{b} - (\mathbf{a}\cdot\mathbf{b})\mathbf{c} \tag{2.3.29}$$

may be proved as follows. Let v_k denote the rectangular Cartesian components of the triple product on the left. Then

$$\begin{aligned} v_k &= e_{kji}\,a_j(e_{irs}\,b_r\,c_s) \quad \text{by Eq. (23)} \\ &= -e_{ijk}e_{irs}a_j b_r c_s = (\delta_{js}\delta_{kr} - \delta_{jr}\delta_{ks})a_j b_r c_s \quad \text{by Eq. (28).} \end{aligned}$$

Hence

$$\begin{aligned} v_k &= (\delta_{js}\,a_j\,c_s)\delta_{kr}\,b_r - (\delta_{jr}\,a_j\,b_r)\delta_{ks}\,c_s \\ &= (a_p\,c_p)b_k - (a_p\,b_p)c_k \quad \text{by Eqs. (27c) and (27d).} \end{aligned}$$

But this is the component form of the right side of Eq. (29), which establishes the required identity. A similar derivation shows that

$$(\mathbf{a} \times \mathbf{b}) \times \mathbf{c} = (\mathbf{a}\cdot\mathbf{c})\mathbf{b} - (\mathbf{b}\cdot\mathbf{c})\mathbf{a}, \tag{2.3.30}$$

which is a completely different vector from $\mathbf{a} \times (\mathbf{b} \times \mathbf{c})$. The vector product therefore is not associative, so that the parentheses must not be omitted on the left sides of Eqs. (29) and (30).

We have emphasized the physical or geometric character of a vector, which exists independently of any coordinate system. However, most of our actual computations are made with rectangular components, which require the introduction of a coordinate system. We now turn our attention to the relationship between the components of a vector or a tensor in one rectangular coordinate system and the components in any other rectangular coordinate system.

EXERCISES

1. Express in terms of **i**, **j**, **k** a unit vector parallel to $\mathbf{a} = 4\mathbf{i} - 2\mathbf{j} + 4\mathbf{k}$.
2. A vector of magnitude 100 is directed along the line from point $A(10, -5, 0)$ to point $B(9, 0, 24)$. Express the vector in terms of **i**, **j**, **k**.
3. Find: $\mathbf{u}\cdot\mathbf{v}$, $\mathbf{u} \times \mathbf{v}$, and $\mathbf{v} \times \mathbf{u}$ for

   (a) $\mathbf{u} = 6\mathbf{i} - 4\mathbf{j} - 6\mathbf{k}, \quad \mathbf{v} = 4\mathbf{i} - 2\mathbf{j} - 8\mathbf{k}$

   (b) $\mathbf{u} = 3\mathbf{i} + \mathbf{j} - 2\mathbf{k}, \quad \mathbf{v} = -\mathbf{i} + 2\mathbf{j} - 3\mathbf{k}.$
4. Find the projection of $\mathbf{c} = 18\mathbf{i} - 27\mathbf{j} + 81\mathbf{k}$ onto **d**, if

   (a) $\mathbf{d} = -\mathbf{i} - 2\mathbf{j} + 2\mathbf{k}$, (b) $\mathbf{d} = \mathbf{i} + 2\mathbf{j} - 2\mathbf{k}$.
5. Show that the vector $A\mathbf{i} + B\mathbf{j} + C\mathbf{k}$ is normal to the plane whose equation is $Ax + By + Cz = D$. *Hint*: Show that it is perpendicular to the vector joining any two points (x_1, y_1, z_1) and (x_2, y_2, z_2) lying in the plane.
6. Find the component of the force $\mathbf{f} = 600\mathbf{i} + 1200\mathbf{j} - 900\mathbf{k}$ pounds in the direction of the normal to the plane $2x - 2y + z = 3$, choosing the positive normal to be on the opposite side of the plane from the origin.
7. Find the equation of the locus of all points (x, y, z) such that a vector from point $(2, -1, 4)$ to (x, y, z) is perpendicular to the vector from $(2, -1, 4)$ to $(3, 3, 2)$.
8. Find the angle between the two vectors from the origin to the two points $A(4, 3, 2)$ and $B(-2, 4, 3)$.
9. Find the area of the parallelogram formed on the two vectors of Ex. 8.
10. Find the unit vector perpendicular to both vectors of Ex. 8.
11. Find a unit vector parallel to the line $x - z = y = 2z$. *Hint*: The line is the intersection of the two planes $x - y - z = 0$ and $y - 2z = 0$.
12. New $\bar{x}, \bar{y}, \bar{z}$ axes are chosen so that the new $\bar{x}$-axis makes equal angles with the positive x- and y-axes and the ***negative*** z-axis, while the new $\bar{y}$-axis lies in the xz-plane and bisects the angle between the old x- and z-axes. Show that these new $\bar{x}$- and $\bar{y}$-axes are perpendicular, and find the direction cosines of the new $\bar{z}$-axis for a right-handed system.
13. Write out the following expressions:

   (a) $t_i = T_{ji}\, n_j$ (d) $T'_{ij} = 2G\, \epsilon'_{ij}$

   (b) $e = \epsilon_{kk}$ (e) $\dfrac{\partial T_{sr}}{\partial x_s} + \rho b_r = \rho a_r.$

   (c) $2W = T_{ij}\, \epsilon_{ij}$
14. Verify the identities of Eqs. (2.3.27).
15. Verify Eq. (2.3.20).
16. Verify that Eq. (2.3.23) gives the same results as Eq. (2.3.18), when Eq. (2.3.18) is written with $\mathbf{i}_1, \mathbf{i}_2, \mathbf{i}_3$ for **i**, **j**, **k** and a_1 for a_x, etc.

17. Prove the identity of Eq. (2.3.30).
18. Prove the e-δ identity $e_{ijk}e_{irs} = \delta_{jr}\delta_{ks} - \delta_{js}\delta_{kr}$ of Eq. (2.3.28). *Hint*: Two of the four free indices $jkrs$ must be equal. Show first that if $j = k$ or $r = s$, then both sides vanish identically. [For example, $e_{i(k)(k)} = 0$, and $\delta_{(k)r}\delta_{(k)s} - \delta_{(k)s}\delta_{(k)r} = 0$, where the parentheses around k are used to indicate that there is no sum on k.] Then show that even with $j \neq k$ and $r \neq s$, both sides vanish when $j = r$ unless also $k = s$, and then prove that

$$e_{i(r)(k)}e_{i(r)(k)} = \delta_{(r)(r)}\delta_{(k)(k)} - \delta_{(r)(k)}\delta_{(r)(k)} \quad \text{for } r \neq k.$$

What cases remain to consider?
19. If $\mathbf{v}$ is any vector and $\hat{\mathbf{n}}$ is any unit vector, show that

$$\mathbf{v} = (\mathbf{v}\cdot\hat{\mathbf{n}})\hat{\mathbf{n}} + \hat{\mathbf{n}} \times (\mathbf{v} \times \hat{\mathbf{n}}).$$

This gives a resolution of $\mathbf{v}$ into vector components parallel and perpendicular to $\hat{\mathbf{n}}$.
20. Show that three concurrent vectors $\mathbf{a}, \mathbf{b}, \mathbf{c}$ are coplanar if $e_{pqr}a_pb_qc_r = 0$.

2.4 Change of orthonormal basis (rotation of axes); tensors as linear vector functions; rectangular Cartesian tensor components; dyadics; tensor properties; review of elementary matrix concepts

Part 1. Change of Orthonormal Basis

A change of rectangular coordinate systems can be made by a translation of the axes without rotation followed by a rigid-body rotation of the axes keeping the origin fixed (and possibly followed by a reflection in a coordinate plane if we allow transformations between right-handed and left-handed systems). The Cartesian components of a vector are not changed by translation of axes, since the projections of a vector onto parallel axes are the same in both magnitude and sense. Hence we need to consider only transformations with fixed origin.

Rotation of axes. If new right-handed rectangular Cartesian $\bar{x}_1, \bar{x}_2, \bar{x}_3$-axes are chosen with the same origin as the original right-handed x_1, x_2, x_3-axes, as in Fig. 2.7, the direction of each new axis can be specified by giving its direction cosines referred to the old axes, denoted† $a_k^r = \cos(\bar{x}_k, x_r)$. The subscript refers to a new (barred) axis $\bar{x}_k$, while the superscript refers to the original (unbarred) axis x_r. The direction cosines defining the change of axes are conveniently tabulated as in Table 2-1. The components of the new unit vector

†Some authors use a_{rk}, and some use a_{kr}, for the direction cosine. We denote it by a_k^r to emphasize that it is not a Cartesian tensor component.

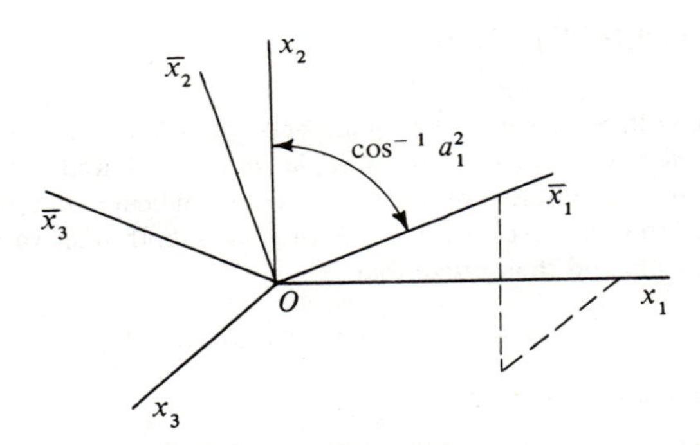

Fig. 2.7 Rotation of axes.

$\bar{\mathbf{i}}_2$, for example, referred to the old axes, are found in the column below the symbol $\bar{\mathbf{i}}_2$.

TABLE 2-1

DIRECTION COSINES FOR ROTATION OF AXES

	$\bar{\mathbf{i}}_1$	$\bar{\mathbf{i}}_2$	$\bar{\mathbf{i}}_3$
$\mathbf{i}_1$	a_1^1	a_2^1	a_3^1
$\mathbf{i}_2$	a_1^2	a_2^2	a_3^2
$\mathbf{i}_3$	a_1^3	a_2^3	a_3^3

Each new unit vector $\bar{\mathbf{i}}_r$ is thus given as a linear combination of the old unit vectors $\mathbf{i}_s$ with coefficients listed in the column. Inversely, each old unit vector is a linear combination of the new unit vectors with coefficients given in a row. Thus†

$$\bar{\mathbf{i}}_r = a_r^s \mathbf{i}_s \quad \text{and} \quad \mathbf{i}_s = a_r^s \bar{\mathbf{i}}_r \tag{2.4.1}$$

(vector sum on s in the first case and on r in the second). Until you learn that the subscript refers to the barred system, you can write the cosines as $a_{\bar{r}}^s$.

Because of the orthogonality of the unit vectors we must have

$$\bar{\mathbf{i}}_p \cdot \bar{\mathbf{i}}_q = 0 \quad \text{for} \quad p \neq q, \quad \text{while} \quad \bar{\mathbf{i}}_p \cdot \bar{\mathbf{i}}_p = 1.$$

This is conveniently expressed by introducing the ***Kronecker delta*** δ_{pq}, defined

†In the curvilinear-component notation of App. I, the second Eq. (2.4.1) would be written as $\mathbf{i}^s = a_r^s \bar{\mathbf{i}}^r$, so that s would appear as a superscript on both sides of the equation. But in Cartesian components the free indices need not appear on the same level, and we replace $\mathbf{i}^s$ by $\mathbf{i}_s$.

to be zero when p and q differ and to be unity when $p = q$. See Eq. (2.3.25).

$$\delta_{pq} = \begin{cases} 1 & \text{if} \quad p = q \\ 0 & \text{if} \quad p \neq q. \end{cases} \tag{2.4.2}$$

Then the fact that each ***basis*** (set of base vectors) is ***orthonormal,*** i.e., composed of mutually orthogonal unit vectors can be expressed by the ***orthogonality conditions***

$$\bar{\mathbf{i}}_p \cdot \bar{\mathbf{i}}_q = \delta_{pq} \quad \text{and} \quad \mathbf{i}_m \cdot \mathbf{i}_n = \delta_{mn} \tag{2.4.3a}$$

or

$$a_p^s a_q^s = \delta_{pq} \quad \text{and} \quad a_r^m a_r^n = \delta_{mn}, \tag{2.4.3b}$$

where Eqs. (3b) follow from Eqs. (3a) by use of Eqs. (1). [Note that the left side of each of the Eqs. (3b) is a sum of three terms because of the repeated index s or r.] An additional identity involving the direction cosines follows from the fact that the determinant det $|a_n^m|$ with elements a_n^m in the mth row and nth column is equal to the scalar triple product $\mathbf{i}_1 \times \mathbf{i}_2 \cdot \mathbf{i}_3$ of the old unit vectors in terms of their components referred to the new axes, according to Eq. (2.3.20). But this triple product equals unity, the volume of the unit cube whose edges are the three unit vectors $\mathbf{i}_1$, $\mathbf{i}_2$, $\mathbf{i}_3$, provided the old axes are right-handed. The triple product equals -1 if the old axes are left-handed. Hence

$$\text{for rotation:} \qquad \det |a_n^m| = 1; \tag{2.4.4a}$$

while

$$\text{for rotation and reflection:} \qquad \det |a_n^m| = -1. \tag{2.4.4b}$$

A linear change of basis of the form of Eqs. (1) is called an ***orthogonal transformation*** if the coefficients a_i^j satisfy Eqs. (3b). Every orthogonal transformation satisfies one of the two Eqs. (4). If it satisfies Eq. (4a) [positive determinant], it is called a ***proper orthogonal transformation;*** if the determinant is negative, the transformation is improper. Hence rotation of axes is represented by a proper orthogonal transformation, while reflection is represented by an improper orthogonal transformation. For most of our purposes we need only consider proper orthogonal transformations, but occasionally an improper transformation is useful. [See, for example, Eqs. (6.2.22).]

Polar vectors and axial vectors, transformation of Cartesian components. When we consider how the rectangular Cartesian components of a vector transform under the change of basis given above, and admit the possibility of improper orthogonal transformations, it is necessary to distinguish two kinds of vectors called ***polar vectors*** and ***axial vectors.*** These vectors are distinguished by the way their components change under an improper orthogonal transformation. The axial vector transformation formula differs in having an extra minus sign prefixed if the transformation is improper. No distinction is needed if only proper orthogonal transformations are used. An example of an axial vector would be furnished by the cross product, if we used the definition of Eqs. (2.3.17) and (2.3.23) without prefixing a minus sign in a left-handed system. But with the convention we adopted in Sec. 2.3, the components of the cross product will transform as a polar vector.

The ***polar vector*** is the typical vector or ***true vector.*** Its component transformation formula follows from the requirement that the representations of the vector in the two systems have exactly the same form whether or not the transformation is proper. Thus the vector equality

$$\bar{v}_r \bar{\mathbf{i}}_r = \mathbf{v} = v_s \mathbf{i}_s \tag{2.4.5}$$

must hold between the two representations. Substituting in Eq. (5) the second Eq. (1) and collecting terms, we obtain

$$(\bar{v}_r - v_s a_r^s)\bar{\mathbf{i}}_r = \mathbf{0}.$$

Since all the components of the zero vector $\mathbf{0}$ must be zero, the expression in parentheses must be zero for each r, whence we obtain the three equations

Polar Vector $\mathbf{v}$

$$\underset{\text{sum on } s}{\bar{v}_r = a_r^s v_s} \quad \text{or inversely} \quad \underset{\text{sum on } r}{v_s = a_r^s \bar{v}_r}. \tag{2.4.6a}$$

The inverse equations follow in a similar fashion from substituting the first Eq. (1) on the left side of Eq. (5). We see that the transformation equations for the vector components (6a) have exactly the same form as the transformation equations for the base vectors (1). Despite the similarity in form, the two transformation equations [(1) and (6a)] are quite different, since Eqs. (6a) give the components of the ***same vector*** $\mathbf{v}$ in two different systems, while the vector equations (1) pertain to two different sets of vectors, one set for each system. Equations (6a) may be written in the following matrix form (see Part 4 of this section for a review of matrix notation):

$$\bar{v} = A^T v \qquad v = A\bar{v}, \tag{2.4.6b}$$

where A is the matrix shown in the array of Table 2-1, with element a_r^s in row s and column r, A^T is the transpose of A, and $\bar{v}$ and v are column matrices with elements $\bar{v}_r$ and v_s. ***Note that in the matrix formulation the matrix A is the matrix used for the inverse transformation from $\bar{v}$ to v.*** For the orthogonal transformation, the inverse A^{-1} of the matrix A is equal to its transpose A^T. In the indicial form, the second Eq. (6a) represents multiplication by the matrix A, since the summation is on the column index r, while in the first Eq. (6a) the summation is on the row index s.

An ***axial vector*** differs from a polar vector in that for an improper orthogonal transformation of an axial vector a minus sign must be introduced in the defining Eq. (5), which becomes $\bar{v}_r \bar{\mathbf{i}}_r = -v_s \mathbf{i}_s$. Since the minus sign is not needed for a proper orthogonal transformation, both cases may be included by prefixing $(\det |a_n^m|)$ in place of the minus sign. Thus

$$\text{axial vector } \mathbf{v}: \qquad \bar{v}_r \bar{\mathbf{i}}_r = (\det |a_n^m|) v_s \mathbf{i}_s,$$

whence as in the derivation of Eqs. (6) we obtain the component transformations

Axial Vector $\mathbf{v}$

$$\bar{v}_r = (\det |a_n^m|) a_r^s v_s \qquad v_s = (\det |a_n^m|) a_r^s \bar{v}_r; \tag{2.4.7a}$$

or in matrix form

$$\bar{v} = (\det |a_n^m|) A^T v \qquad v = (\det |a_n^m|) A \bar{v}. \tag{2.4.7b}$$

We have already noted that one example of an axial vector would be the vector product of two polar vectors if we did not prefix a minus sign to Eqs. (2.3.17) and (2.3.23) in a left-handed system. Another example would be the *curl* of a vector field, to be discussed in Sec. 2.5, if we did not employ the same convention about the minus sign as in the cross-product formulas. With our conventions, all the vectors we encounter will transform as polar vectors. But the reader should be warned that some writers do not use the minus sign in the cross-product formulas; then the cross product is an axial vector.

Coordinate transformation formulas for the change of axes under rotation and reflection are merely special cases of Eqs. (6) in which the vector $\mathbf{v}$ is the position vector $\mathbf{r}$ with components $\bar{x}_r$ or x_s in the two systems. If a translation is added, the most general transformation of rectangular Cartesian coordinates has the form

$$\bar{x}_r = a_r^s (x_s - b_s) \quad \text{or inversely} \quad x_s = b_s + a_r^s \bar{x}_r, \tag{2.4.8}$$

where the a_r^s satisfy Eqs. (3b) and either (4a) or (4b).

The transformation property of rectangular components under rotation of axes is sometimes made a part of the formal definition of a vector quantity as follows. A physical entity, defined by three components in every system of rectangular Cartesian coordinates, is a vector quantity if

1. the three components in any one system are related to the three components in any other (meaning every other) system by Eqs. (6), and if
2. two such physical entities of the same type, say **a** and **b** with components a_i and b_i in one coordinate system, together produce the same effect as would be produced by the single one, **c** with components $c_i = a_i + b_i$ in the same system.

In short, to be a vector quantity, the physical entity must transform like a vector and compound like a vector. We remark that it is possible to invent physical entities with three components not satisfying these two conditions. We might take the triplet (p, v, t) representing the pressure, specific volume, and temperature at a point in a gas. These values do not transform as the components of a vector under rotation of axes; each one maintains its value, i.e., each is a scalar point function under rotation of axes. We could always construct a vector artificially by specifying its components in one system of axes and prescribing that its components in any other system should be given by the vector transformation formulas. When we are dealing with physical entities, however, we do not have this much freedom of definition, as is shown by the absurd example (p, v, t).

Part 2. Second-Order Tensors as Linear Vector Functions (Transformations)

A vector function is a correspondence which associates with each argument vector **v** in a certain set of vectors (the domain of definition of the function) another vector, which is the value of the function. The function is also called a ***transformation***, since it transforms one set of vectors (the domain) into another set of vectors. A vector function **F** is a ***linear vector function*** if it satisfies the two requirements that for any vectors **u** and **v** in the domain and any real number c,

$$\mathbf{F}(\mathbf{u} + \mathbf{v}) = \mathbf{F}(\mathbf{u}) + \mathbf{F}(\mathbf{v}) \quad \text{and} \quad \mathbf{F}(c\mathbf{u}) = c\mathbf{F}(\mathbf{u}). \tag{2.4.9}$$

For the purposes of this book we define a ***second-order tensor*** (also called a ***second-rank tensor***) to be a linear vector function. A vector is a first-order tensor, while a scalar is a tensor of order zero. Higher-order tensors will also be defined later, but the second-order tensors are the ones of primary interest in this book. For brevity, we will often use the word ***tensor*** to mean ***second-order tensor***. Since the two sets of vectors have an existence quite independent of any coordinate system in which they might happen to be observed, the tensor correspondence or pairing of each ***antecedent*** element in the domain with its ***image*** element in the second set must also have an existence independent of any coordinate system. In order to make explicit the way in which any particular tensor operates on the argument vector, however, we will introduce a coordinate system. For simplicity, we limit ourselves for the time being to rectangular Cartesian coordinate systems, but the tensor itself is not limited to any coordinate system. We introduce a notation for the second-order tensor itself (as distinct from any possible component representation). A boldface capital letter symbol will be used in printed text for a second-order tensor, while a boldface lower-case letter will identify a vector. Some exceptions to this distinction between capitals and lower-case symbols may be made with commonly used tensors or vectors whose physical nature will help us to remember whether they are vectors or second-order tensors. In typed or handwritten text, the second-order tensor will be identified by wavy underscores, e.g., $\utilde{T}$ or $\utilde{E}$. In the following, the word "tensor" will mean "second-order tensor" unless otherwise stated.

Linear transformations of rectangular Cartesian vector components, tensor components. The second Eq. (9) requires that the zero vector be transformed into the zero vector (put $c = 0$ to see this). In any one rectangular Cartesian system, the most general linear transformation of the components v_j of the argument vector **v** into the components u_i of the image vector **u**, which always transforms the zero vector into the zero vector, has the form

$$u_i = T_{ij}v_j \quad \text{with matrix† notation} \quad u = Tv \tag{2.4.10a}$$

or the form, for $S = T^T$ and $T = S^T$,

$$u_i = S_{ji}v_j \quad \text{with matrix notation} \quad u = vS. \tag{2.4.10b}$$

If Eqs. (10b) always give the same output vector components as are given by (10a) for an arbitrary input vector **v**, then the two forms are not fundamentally different, since they differ only in how the same nine coefficients are labeled: S_{ji} in (b) being equal to T_{ij} in (a), i.e., in the matrix notation the matrix S in the second equation is the transpose of the one which should be used in the first equation. Each of the two tensors **T** and **S** with rectangular Cartesian component matrices T and $S = T^T$ is said to be the ***tensor transpose*** of the other, denoted

$$\mathbf{S} = \mathbf{T}^T \quad \text{and} \quad \mathbf{T} = \mathbf{S}^T. \tag{2.4.10c}$$

When we do not plan to write both forms for the same tensor, we let T denote the matrix of components in the form which we actually use. The nine coefficients are the rectangular Cartesian components of the tensor **T**. They are denoted by the same letter printed in ordinary type instead of boldface (without

†A brief review of some elementary matrix properties is given in Part 4 of this section.

the wavy underscore in handwritten text) and with two subscripts attached, e.g., T_{11}, T_{12}, etc. Some exceptions will be made to the rule that components are denoted with the same letter as the tensor. For example, the strain tensor is often denoted by **E** with components ϵ_{ij}. The linear transformation of Eq. (10) is the representation of the tensor **T** (linear vector function) in one system of Cartesian components. This transformation of the components of one vector into the components of another vector referred to the same system should not be confused with the change of components of one vector under rotation of axes, given by Eq. (6a).

Symmetric tensors and skew or antisymmetric tensors. A tensor is said to be symmetric, if its matrix of rectangular Cartesian components is symmetric, i.e., if $T_{ji} = T_{ij}$. If the matrix of a tensor is symmetric in one Cartesian coordinate system, it is symmetric in all such systems; thus symmetry is really a tensor property. A skew or antisymmetric tensor has $T_{ji} = -T_{ij}$ in any Cartesian system, which implies that the diagonal elements, T_{11}, T_{22}, T_{33} are all zero. We will now derive the rule for change of components of a tensor under rotation of axes.

Rotation of axes, change of tensor components. The rule for changing second-order tensor components under rotation of axes follows from the rule for changing vector components when the transformation (10) is required to apply in each of the two systems. For example, Eq. (6a) yields

$$\begin{aligned} \bar{u}_i &= a_i^j u_j \\ &= a_i^j T_{jq} v_q && \text{by Eq. (10a)} \\ &= a_i^j T_{jq} a_p^q \bar{v}_p && \text{by Eq. (6a).} \end{aligned}$$

Also $\bar{u}_i = \bar{T}_{ip}\bar{v}_p$ by Eq. (10a) in the barred system. By equating these two different expressions for $\bar{u}_i$ and by collecting terms on one side of the equation, we conclude that

$$(\bar{T}_{ip} - a_i^j a_p^q T_{jq})\bar{v}_p = 0$$

for every possible choice of the argument vector **v**. Hence the quantity in parentheses on the left must vanish, whence

$$\bar{T}_{ip} = a_i^j a_p^q T_{jq} \quad \text{and inversely} \quad T_{jq} = a_i^j a_p^q \bar{T}_{ip}, \tag{2.4.11}$$

or in matrix notation

$$\bar{T} = A^T T A \qquad\qquad T = A\bar{T}A^T, \tag{2.4.12}$$

where for the proper orthogonal transformation $A^{-1} = A^T$. The inverse relation can be derived in a way similar to the derivation of the first Eq. (11). Comparing these formulas for the change of components of a second-order tensor with Eqs. (6a) for a polar vector (first-order tensor), we see that they follow a similar pattern. There is on the right side one free index on each direction cosine, corresponding to the free indices on the left, and each direction cosine also has a summation index corresponding to one on a tensor component on the right. Each of the nine first equations (11) has a sum of nine terms on the right, corresponding to the nine possible values of the pair of indices jq. For example,

$$\begin{aligned}\bar{T}_{23} &= a_2^1 a_3^1 T_{11} + a_2^1 a_3^2 T_{12} + a_2^1 a_3^3 T_{13} \\ &+ a_2^2 a_3^1 T_{21} + a_2^2 a_3^2 T_{22} + a_2^2 a_3^3 T_{23} \\ &+ a_2^3 a_3^1 T_{31} + a_2^3 a_3^2 T_{32} + a_2^3 a_3^3 T_{33}.\end{aligned} \tag{2.4.13}$$

We remark that exactly the same change-of-component formulas apply whether the components are taken in the form of Eq. (10a) or of Eq. (10b), provided the same choice is made in both coordinate systems.

These change-of-component formulas are sometimes made the basis for the definition of a "Cartesian tensor." Briefly, a second-order Cartesian tensor would be characterized as an entity with nine components in any Cartesian coordinate system, such that the components in any one system are related to the components in any other system by Eqs. (11) and (12). ***Higher-order tensors*** can be similarly defined, for example, with three subscripts T_{ijk}, and with change-of-axis formulas similar to Eq. (11) but with three direction cosines in each term. We will be concerned primarily with vectors and second-order tensors. For our purposes it is more convenient to define the second-order tensor as a linear vector function and to deduce that Eqs. (11) and (12) relate its components in any two rectangular Cartesian systems. We postpone until the appendix any consideration of how the tensor would be represented by components in a curvilinear coordinate system, but our basic definition will still apply, since it is not dependent on any coordinate system.

Dyadic (single-dot) product notation for a second-order tensor. Many of the general equations containing tensors have a simpler appearance when written in terms of the symbolic representation **T** instead of in terms of the components. We use the Gibbs dyadic notation (see Part 3 of this section) to give a symbolic representation independent of any choice of coordinate system for the linear vector operation represented in rectangular Cartesians by Eqs. (10).

Dyadic Notation for Tensors as Linear Vector Operators

$$\mathbf{u} = \mathbf{T} \cdot \mathbf{v} \tag{2.4.14a}$$

$$\mathbf{u} = \mathbf{v} \cdot \mathbf{S} \qquad \text{where } \mathbf{S} = \mathbf{T}^T \tag{2.4.14b}$$

The symbolic representation $\mathbf{T}\cdot\mathbf{v}$ is not merely another notation for the matrix product Tv of Eq. (10a). $\mathbf{T}\cdot\mathbf{v}$ symbolizes an ***operation*** pairing one physical or geometric vector with another one, while Tv is the mathematical product of appropriate matrices of components of $\mathbf{T}$ and of $\mathbf{v}$ in one coordinate system. In general curvilinear tensor components some care is necessary in evaluating $\mathbf{T}\cdot\mathbf{v}$ as a matrix product [see App. I, Eq. (I.2.14)]. For example, the matrix product of two matrices of covariant components is not in general equal to the matrix of covariant components of $\mathbf{u}$. The covariant component matrix of $\mathbf{u}$ can be obtained as the matrix product of the covariant components of $\mathbf{T}$ by the contravariant components of $\mathbf{v}$ instead of by the covariant components of $\mathbf{v}$. In rectangular coordinates there is no difference between the different kinds of components, and the matrix forms of Eqs. (10) can always be used to represent in any one coordinate system the operation symbolized by Eqs. (14).

Some authors† use a symbolic operational product notation without the dot. For example the relationship we have expressed by Eq. (14a) would be written

$$\mathbf{u} = \mathbf{T}\mathbf{v}, \tag{2.4.15}$$

an operation completely independent of any component representation.

Some properties of second-order tensors. The ***sum*** $\mathbf{S} = \mathbf{T} + \mathbf{U}$ of two tensors is the single tensor which operates on any vector $\mathbf{v}$ to yield the sum of the two vectors obtained by operating with $\mathbf{T}$ and $\mathbf{U}$ separately; in rectangular Cartesian components, $S_{ij} = T_{ij} + U_{ij}$. The ***zero tensor*** $\mathbf{0}$ assigns the zero vector as image to every argument vector $\mathbf{v}$; its components are all zero in every coordinate system. ***Multiplication by a scalar*** c is defined, for example, by the requirement that $(c\mathbf{T})\cdot\mathbf{v} = c(\mathbf{T}\cdot\mathbf{v})$ for every vector $\mathbf{v}$. The rectangular Cartesian components of $c\mathbf{T} = \mathbf{T}c$ are $cT_{ij} = T_{ij}c$. Tensor addition and multiplication by a scalar obey the following four addition axioms and four scalar-multiple axioms, characteristic of a generalized vector space; hence the set of all tensors is a vector space (see Sec. I.1).

ADDITION AXIOMS

(a) $\mathbf{T} + \mathbf{U} = \mathbf{U} + \mathbf{T}$ commutative

(b) $\mathbf{T} + (\mathbf{U} + \mathbf{V}) = (\mathbf{T} + \mathbf{U}) + \mathbf{V}$ associative

(c) $\mathbf{T} + \mathbf{0} = \mathbf{T}$ (2.4.16)

(d) For each $\mathbf{T}$, there exists another, called $-\mathbf{T}$, such that $\mathbf{T} + (-\mathbf{T}) = \mathbf{0}$.

†See, for example, Truesdell and Noll (1965) and Jaunzemis (1967).

SCALAR-MULTIPLE AXIOMS

(Note $a\mathbf{T} \equiv \mathbf{T}a$)

$$
\begin{aligned}
&\text{(a)}\quad a(b\mathbf{T}) = (ab)\mathbf{T}\\
&\text{(b)}\quad 1\mathbf{T} = \mathbf{T}\\
&\text{(c)}\quad (a+b)\mathbf{T} = a\mathbf{T} + b\mathbf{T}\\
&\text{(d)}\quad a(\mathbf{T}+\mathbf{U}) = a\mathbf{T} + a\mathbf{U}
\end{aligned}
\tag{2.4.17}
$$

The scalar product of two tensors is a scalar, denoted $\mathbf{T} : \mathbf{U}$, which can be calculated in terms of components of the two tensors in any one rectangular Cartesian system, by

$$\mathbf{T} : \mathbf{U} = T_{ij} U_{ij}, \tag{2.4.18a}$$

i.e., by the sum of the nine products of components. This formula applies only to rectangular Cartesian components, but the result, a scalar, is independent of any coordinate system. Some authors define the scalar product as $T_{ij} U_{ji}$ instead of $T_{ij} U_{ij}$ and some use the notation $\mathbf{T} \cdot\cdot \mathbf{U}$ instead of $\mathbf{T} : \mathbf{U}$. In this book, $\mathbf{T} : \mathbf{U}$ will always mean $T_{ij} U_{ij}$ in rectangular Cartesians, while

$$\mathbf{T} \cdot\cdot \mathbf{U} = T_{ij} U_{ji} \tag{2.4.18b}$$

will represent another possible scalar product, different from $\mathbf{T} : \mathbf{U}$ unless one of the two tensors $\mathbf{T}$ or $\mathbf{U}$ is symmetric. The scalar products satisfy the following four axioms, which with (16) and (17) characterize an ***inner-product space***. See also Eq. (25b) below, where the scalar products are expressed as traces of the product of two tensors.

INNER-PRODUCT AXIOMS

$$
\begin{aligned}
&\text{(a)}\quad \mathbf{T} : \mathbf{U} = \mathbf{U} : \mathbf{T}\\
&\text{(b)}\quad \mathbf{T} : (\mathbf{U} + \mathbf{V}) = \mathbf{T} : \mathbf{U} + \mathbf{T} : \mathbf{V}\\
&\text{(c)}\quad a(\mathbf{T} : \mathbf{U}) = (a\mathbf{T}) : \mathbf{U} = \mathbf{T} : (a\mathbf{U})\\
&\text{(d)}\quad \mathbf{T} : \mathbf{T} > 0 \quad \text{unless} \quad \mathbf{T} = \mathbf{0}
\end{aligned}
\tag{2.4.19}
$$

The identity tensor, also called the ***unit tensor*** and denoted by $\mathbf{1}$, is defined by the requirement that

$$\mathbf{1} \cdot \mathbf{v} = \mathbf{v} \cdot \mathbf{1} = \mathbf{v} \tag{2.4.20}$$

for all vectors $\mathbf{v}$. The rectangular Cartesian components of $\mathbf{1}$ are δ_{ij}; its matrix of rectangular Cartesian components is the identity matrix with zero in each off-diagonal position and one in each diagonal position.

The tensor product or open product of two vectors, denoted **ab**, is a tensor called a ***dyad***, defined by the requirement that

$$(\mathbf{ab})\cdot\mathbf{v} = \mathbf{a}(\mathbf{b}\cdot\mathbf{v}) \qquad (2.4.21a)$$

for all vectors **v**. That is, if $\mathbf{T} = \mathbf{ab}$, then $\mathbf{T}\cdot\mathbf{v} = \mathbf{a}(\mathbf{b}\cdot\mathbf{v})$ for all vectors **v**; in rectangular Cartesian components

$$T_{ij} = a_i b_j. \qquad (2.4.21b)$$

Some writers use the notation $\mathbf{a} \otimes \mathbf{b}$ instead of **ab**.

The product of two second-order tensors, denoted $\mathbf{T}\cdot\mathbf{U}$ in this book, means the composition of the two operations **T** and **U**, with **U** performed first, defined by the requirement that

$$(\mathbf{T}\cdot\mathbf{U})\cdot\mathbf{v} = \mathbf{T}\cdot(\mathbf{U}\cdot\mathbf{v}) \qquad (2.4.22a)$$

for all vectors **v**. If

$$\boxed{\mathbf{P} = \mathbf{T}\cdot\mathbf{U}, \quad \text{then} \quad P_{ij} = T_{ik}U_{kj} \quad \text{or} \quad P = TU} \qquad (2.4.22b)$$

in matrix notation, where P, T, and U are the matrices of components in any one rectangular Cartesian system. Some authors omit the dot and write the tensor product as

$$\mathbf{P} = \mathbf{TU}. \qquad (2.4.22c)$$

The product satisfies the linearity rules and is therefore also a second-order tensor. It also satisfies the following five axioms, which together with (16) and (17) characterize the set of all second-order tensors as an algebra. Similar definitions and axioms apply to $\mathbf{v}\cdot(\mathbf{U}\cdot\mathbf{T}) = (\mathbf{v}\cdot\mathbf{U})\cdot\mathbf{T}$.

TENSOR ALGEBRA AXIOMS

$$\begin{aligned}
&(a) \quad (\mathbf{T}\cdot\mathbf{U})\cdot\mathbf{R} = \mathbf{T}\cdot(\mathbf{U}\cdot\mathbf{R}) \\
&(b) \quad \mathbf{T}\cdot(\mathbf{R} + \mathbf{U}) = \mathbf{T}\cdot\mathbf{R} + \mathbf{T}\cdot\mathbf{U} \\
&(c) \quad (\mathbf{R} + \mathbf{U})\cdot\mathbf{T} = \mathbf{R}\cdot\mathbf{T} + \mathbf{U}\cdot\mathbf{T} \\
&(d) \quad a(\mathbf{T}\cdot\mathbf{U}) = (a\mathbf{T})\cdot\mathbf{U} = \mathbf{T}\cdot(a\mathbf{U}) \\
&(e) \quad \mathbf{1}\cdot\mathbf{T} = \mathbf{T}\cdot\mathbf{1} = \mathbf{T}
\end{aligned} \qquad (2.4.23)$$

The product is ***not commutative***; in general $\mathbf{U}\cdot\mathbf{T} \neq \mathbf{T}\cdot\mathbf{U}$. Instead, Eq. (49) below, when applied to the matrices of rectangular Cartesian components, shows that

$$\mathbf{B} \cdot \mathbf{A} = [\mathbf{A}^T \cdot \mathbf{B}^T]^T \tag{2.4.24}$$

where the superscript T denotes the transpose as in Eqs. (10). In exceptional cases, when $\mathbf{T} \cdot \mathbf{U} = \mathbf{U} \cdot \mathbf{T}$, we say that $\mathbf{T}$ and $\mathbf{U}$ commute. For example, the unit tensor $\mathbf{1}$ commutes with any other tensor.

The trace, denoted tr$\mathbf{T}$, of a second-order tensor $\mathbf{T}$ is a scalar invariant function of the tensor, having the same numerical value in all coordinates systems. In rectangular Cartesians (see Sec. 3.3)

$$\operatorname{tr} \mathbf{T} = T_{kk}. \tag{2.4.25a}$$

Thus tr $\mathbf{T}$ is equal to the trace of the matrix of rectangular Cartesian components, the sum of the elements on the main diagonal.

The scalar products of Eqs. (18) can be represented as traces of tensor products. It is left as an exercise for the student to show that

$$\begin{aligned} \mathbf{A} \cdot \cdot \mathbf{B} &= \operatorname{tr} (\mathbf{A} \cdot \mathbf{B}) \\ \mathbf{A} : \mathbf{B} &= \operatorname{tr} (\mathbf{A} \cdot \mathbf{B}^T) = \operatorname{tr} (\mathbf{A}^T \cdot \mathbf{B}) \\ \operatorname{tr} (\mathbf{A} \cdot \mathbf{B}) &= \operatorname{tr} (\mathbf{B} \cdot \mathbf{A}) \\ \operatorname{tr} (\mathbf{A} \cdot \mathbf{B} \cdot \mathbf{C}) &= \operatorname{tr} (\mathbf{B} \cdot \mathbf{C} \cdot \mathbf{A}) = \operatorname{tr} (\mathbf{C} \cdot \mathbf{A} \cdot \mathbf{B}) \end{aligned} \tag{2.4.25b}$$

The last line demonstrates the ***cyclic property*** of the trace of the continued tensor product of any number of factors; the trace is unaltered in value by any cyclic rearrangement of the factors.

A test for tensor character. An entity $\mathbf{T}$ defined by nine components in any coordinate system is a tensor, if and only if $\mathbf{T} : (\mathbf{uv})$ is a scalar invariant for arbitrary vectors $\mathbf{u}$ and $\mathbf{v}$, where the symbol ":" denotes the scalar product of Eq. (18a), while $\mathbf{uv}$ denotes the tensor product of Eq. (21). This means that in rectangular Cartesian components

$$\bar{T}_{ij}\bar{u}_i\bar{v}_j = T_{pq}u_p v_q \tag{2.4.26}$$

in every pair of Cartesian systems, for every choice of the vectors $\mathbf{u}$ and $\mathbf{v}$. If the components being tested satisfy the conditions for symmetry of a tensor in all coordinate systems, it is sufficient to show that

$$\mathbf{T} : (\mathbf{vv}) \text{ is a scalar invariant for arbitrary } \mathbf{v} \text{ and symmetric } \mathbf{T}, \tag{2.4.27}$$

which means that

$$\bar{T}_{ij}\bar{v}_i\bar{v}_j = T_{pq}v_p v_q \quad \text{and} \quad \bar{T}_{ij} = \bar{T}_{ji} \quad T_{pq} = T_{qp} \tag{2.4.28}$$

in every pair of Cartesian systems for arbitrary choice of $\mathbf{v}$.

Part 3. Algebra of Dyads and Polyads; Dyadics

In Eq. (21), an operational definition was given for a ***dyad*** or open product of two vectors. Higher-order ***polyads*** can be formed by open products of more than two vectors, e.g., triads **abc** or tetrads **abcd**. Without assigning any physical significance to these polyads for the moment, we postulate that open multiplication obeys all the rules of ordinary algebra except that ***open multiplication is not commutative.***

$$\mathbf{ab} \neq \mathbf{ba}. \tag{2.4.29a}$$

We define the ***conjugate dyad*** $(\mathbf{ab})_c$ to **ab** by

$$(\mathbf{ab})_c = \mathbf{ba}. \tag{2.4.29b}$$

A linear combination of dyads with real coefficients is called a ***dyadic***, e.g., $m\mathbf{ab} + n\mathbf{cd}$.

Contraction: single-dot product, double-dot (scalar) product of dyads. If in a polyad we replace one of the open products by a scalar product of the two vectors, we obtain a polyad of order two less than that of the original polyad. This process is called ***contraction***. For example, with the tetrad **abcd**, contraction can be done in three ways with adjacent vectors to yield ***three different dyadics***:

$$\begin{aligned} \mathbf{a}\cdot\mathbf{bcd} &= (\mathbf{a}\cdot\mathbf{b})\mathbf{cd} \\ \mathbf{ab}\cdot\mathbf{cd} &= (\mathbf{b}\cdot\mathbf{c})\mathbf{ad} \\ \mathbf{abc}\cdot\mathbf{d} &= (\mathbf{c}\cdot\mathbf{d})\mathbf{ab}, \end{aligned} \tag{2.4.30}$$

where the dot product in parentheses is a scalar factor multiplying the dyad. The second of these contractions defines the **single-dot** product of the dyad **ab** by the dyad **cd**. Note that

$$\mathbf{ab}\cdot\mathbf{cd} \neq \mathbf{cd}\cdot\mathbf{ab} \tag{2.4.31}$$

so that (unlike the dot product of two vectors) ***the single-dot product of two dyads is not commutative.***

The ***scalar product*** of two dyads (or ***double-dot product***) denoted $\mathbf{ab} : \mathbf{cd}$ is defined as the scalar obtained by multiplying together the two scalar products $\mathbf{a}\cdot\mathbf{c}$ and $\mathbf{b}\cdot\mathbf{d}$. Note that the first vector of the first dyad multiplies the first vector of the second dyad, and the second vector of the first dyad multiplies the second vector of the second dyad. ***The scalar product is commutative.***

$$\mathbf{ab} : \mathbf{cd} = (\mathbf{a}\cdot\mathbf{c})(\mathbf{b}\cdot\mathbf{d}) = (\mathbf{c}\cdot\mathbf{a})(\mathbf{d}\cdot\mathbf{b}) = \mathbf{cd} : \mathbf{ab}. \tag{2.4.32}$$

Some authors denote the scalar product of Eq. (32) by $\mathbf{ab} \cdot\cdot\, \mathbf{cd}$ instead of $\mathbf{ab} : \mathbf{cd}$. Also, some authors define the scalar product differently, multiplying the two outside vectors together and the two inside vectors together. In this book, the double-dot notation with the two dots on the same line will denote the product obtained by multiplying the two outside vectors together and the two inside vectors together.

$$\mathbf{ab} \cdot\cdot\, \mathbf{cd} = (\mathbf{a}\cdot\mathbf{d})(\mathbf{b}\cdot\mathbf{c}). \tag{2.4.33}$$

Dyadics as second-order tensors. Since the single-dot product of a dyad $\mathbf{ab}$ by a vector $\mathbf{v}$ yields another vector, as in Eq. (21), and since this operation is a linear vector function, a dyad operates as a second-order tensor. Is the converse true? Can every second-order tensor be represented as a dyad? The answer is no, but every second-order tensor can be represented as a ***dyadic***, a linear combination of the dyads formed with the base vectors. Thus, in rectangular Cartesians,

$$\left.\begin{aligned} \mathbf{T} = {} & T_{11}\mathbf{i}_1\mathbf{i}_1 + T_{12}\mathbf{i}_1\mathbf{i}_2 + T_{13}\mathbf{i}_1\mathbf{i}_3 \\ & + T_{21}\mathbf{i}_2\mathbf{i}_1 + T_{22}\mathbf{i}_2\mathbf{i}_2 + T_{23}\mathbf{i}_2\mathbf{i}_3 \\ & + T_{31}\mathbf{i}_3\mathbf{i}_1 + T_{32}\mathbf{i}_3\mathbf{i}_2 + T_{33}\mathbf{i}_3\mathbf{i}_3 \quad \text{or} \\ \mathbf{T} = {} & T_{rs}\mathbf{i}_r\mathbf{i}_s. \end{aligned}\right\} \tag{2.4.34}$$

The algebra of dyads and polyads, as outlined in Eqs. (29) through (33) can then be shown to imply the results for tensors given earlier, when the tensors are written in the form of Eq. (34). Moreover, if the change of basis Eqs. (1) are substituted into Eq. (34), and the tensor is required to have the same form in the new coordinate system, i.e.,

$$\bar{T}_{pq}\bar{\mathbf{i}}_p\bar{\mathbf{i}}_q = \mathbf{T} = T_{rs}\mathbf{i}_r\mathbf{i}_s, \tag{2.4.35}$$

it can be shown that this implies that the tensor component transformation formulas of Eq. (11) must hold between the coefficients $\bar{T}_{pq}$ and T_{rs}. Thus the coefficients of the dyadic representation of Eq. (34) are simply the rectangular Cartesian components of the tensor.

Section I.3 considers dyadic representation with an arbitrary basis $\mathbf{b}_r$ instead of the orthonormal basis $\mathbf{i}_r$. The dyadic representation is especially convenient for introducing derivatives of tensor fields. In Cartesian coordinates, since $\mathbf{i}_r$ is independent of position, $\partial\mathbf{T}/\partial x_p = (\partial T_{rs}/\partial x_p)\mathbf{i}_r\mathbf{i}_s$, but in curvilinear coordinates the unit base vector orientations vary with position. Thus in curvilinear coordinates with unit base vectors $\hat{\mathbf{e}}_r$ it is necessary to differentiate each term of the dyadic representation

$$\mathbf{T} = T^{\langle rs\rangle}\hat{\mathbf{e}}_r\hat{\mathbf{e}}_s \tag{2.4.36}$$

as a product of three functions. For orthogonal curvilinear coordinates this is discussed in App. II, which avoids the general tensor complications of App. I. (It is not necessary to read either appendix before proceeding with the text.)

Part 4. Review of Elementary Matrix Concepts

Definition of a matrix. A matrix is a rectangular array of elements. We will consider here only matrices whose elements are real numbers.

$$A = \begin{bmatrix} A_{11} & A_{12} & A_{13} \\ A_{21} & A_{22} & A_{23} \end{bmatrix} \tag{2.4.37}$$

A_{ij} is the element in row i and column j. (We will sometimes denote the element A^i_{j}.) Two matrices are equal only if their elements are equal term by term. Unlike a determinant, a matrix does not have a numerical value. Also it may not have the same number of rows and columns. An $m \times n$ matrix has m rows and n columns. The example above is a 2×3 matrix. Other common notations are || || or ( ) instead of [ ].

Addition of matrices. Matrices of the same order add term by term.

$$A + B = \begin{bmatrix} A_{11} + B_{11} & A_{12} + B_{12} & A_{13} + B_{13} \\ A_{21} + B_{21} & A_{22} + B_{22} & A_{23} + B_{23} \end{bmatrix} \tag{2.4.38}$$

Multiplication by a numerical factor. Multiply *every term*.

$$kA = \begin{bmatrix} kA_{11} & kA_{12} & kA_{13} \\ kA_{21} & kA_{22} & kA_{23} \end{bmatrix} \tag{2.4.39}$$

Multiplication of two matrices. Matrices must be ***conformable*** or else they cannot be multiplied. Matrices A and B are conformable and can be multiplied in the order AB if the first matrix has the same number of columns as the second matrix has rows. For example,

$$A = \begin{bmatrix} 2 & -1 & 3 \\ 4 & 0 & 2 \end{bmatrix}, \quad \text{and} \quad B = \begin{bmatrix} 6 \\ 4 \\ 8 \end{bmatrix}$$

are conformable in the order AB, but not in the order BA. The product of the 2×3 matrix A by the 3×1 matrix B is a 2×1 matrix C.

$$AB = \begin{bmatrix} 2 & -1 & 3 \\ 4 & 0 & 2 \end{bmatrix} \begin{bmatrix} 6 \\ 4 \\ 8 \end{bmatrix} = \begin{bmatrix} (2)(6) + (-1)(4) + (3)(8) \\ (4)(6) + \quad (0)(4) + (2)(8) \end{bmatrix} = \begin{bmatrix} 32 \\ 40 \end{bmatrix} = C$$

The element in the first row and first column of C is obtained as the inner product (like a vector scalar product) of the first row of A by the first column of B, as indicated above. The element in the second row and first column of the product matrix C is similarly obtained from the inner product of the second row by the first column.

Formulas for the elements of the product matrix. Consider the product AB, schematically shown as

$$AB = [m \times n][n \times s] = [m \times s] = C.$$

The two matrices are conformable, since the first matrix has the same number of columns as the second matrix has rows, namely n. The element C_{ij} in the ith row and jth column of the product is

$$C_{ij} = A_{ik}B_{kj} \qquad \text{(summed).} \tag{2.4.40}$$

If $s = m$, it is possible to multiply in the opposite order, and

$$BA = [B_{ik}A_{kj}] \qquad \text{(summed).} \tag{2.4.41}$$

In general $BA \neq AB$.

Square matrices and linear transformations. A square matrix has the same number of rows as columns. The system of equations

$$\left.\begin{aligned} M_{11}x_1 + M_{12}x_2 + M_{13}x_3 &= y_1 \\ M_{21}x_1 + M_{22}x_2 + M_{23}x_3 &= y_2 \\ M_{31}x_1 + M_{32}x_2 + M_{33}x_3 &= y_3 \end{aligned}\right\} \tag{2.4.42}$$

may be abbreviated as

$$Mx = y, \tag{2.4.43}$$

where M is the square matrix of coefficients and x and y represent the two ***column matrices***

$$x = \begin{bmatrix} x_1 \\ x_2 \\ x_3 \end{bmatrix} \qquad y = \begin{bmatrix} y_1 \\ y_2 \\ y_3 \end{bmatrix}. \tag{2.4.44}$$

Equations (42) and (43) are said to transform the column matrix x into the column matrix y. This is an example of a ***linear transformation***. Solving (43) consists of finding the column matrix x transformed into a given column matrix y by the given square matrix M. The example illustrated is for a 3×3 system, but the ideas apply for any $n \times n$ system.

A diagonal matrix is a square matrix with nonzero elements only on the main diagonal $M_{11}, M_{22}, M_{33}, \ldots$. If M is diagonal, the system (42) may be solved immediately, one equation at a time. When M is not diagonal, we may try to diagonalize it by performing on it a succession of operations equivalent to the kind of operations performed in solving (42) by elimination.

An ***identity matrix***, denoted by I, is a diagonal matrix with each diagonal element equal to unity. The elements of the identity matrix are δ_{ij}. An identity matrix transforms a column vector into itself (the identity transformation).

$$Ix = x, \tag{2.4.45}$$

and similarly for any ***row matrix*** x, i.e., any matrix with only one row,

$$xI = x. \tag{2.4.46}$$

The ***inverse of a square matrix*** M (if it has an inverse) is denoted by M^{-1} and defined by

$$MM^{-1} = M^{-1}M = I. \tag{2.4.47}$$

Multiplying (43) by M^{-1} (from the left), we get

$$M^{-1}Mx = M^{-1}y$$

or

$$Ix = M^{-1}y$$

whence

$$x = M^{-1}y. \tag{2.4.48}$$

Thus, finding the inverse of M is equivalent to solving the system of equations, which can be done provided the determinant is nonzero. A square matrix with a nonzero determinant is called ***nonsingular***. Thus, every nonsingular square matrix has an inverse. A formula for the inverse is given in Eq. (51).

The transpose M^T of a square matrix M is the matrix obtained by interchanging rows and columns. For example,

$$M^T = \begin{bmatrix} M_{11} & M_{21} & M_{31} \\ M_{12} & M_{22} & M_{32} \\ M_{13} & M_{23} & M_{33} \end{bmatrix}.$$

A symmetric matrix is a square matrix with equal elements in the positions symmetrically placed with respect to the main diagonal, i.e., with $M_{ij} = M_{ji}$ so that $M^T = M$. A ***skew*** or ***antisymmetric*** matrix has $M_{ij} = -M_{ji}$, so that $M^T = -M$, which implies that the diagonal elements are all zero. For square matrices A and B

$$BA = (A^T B^T)^T. \tag{2.4.49}$$

The proof of Eq. (49) is left as an exercise.

Formula for the inverse. For a nonsingular matrix M, if

$$D = \det M$$

is the determinant of the matrix, the solution of Eq. (42) by Cramer's rule gives, for example,

$$x_1 = \frac{1}{D}\begin{vmatrix} y_1 & M_{12} & M_{13} \\ y_2 & M_{22} & M_{23} \\ y_3 & M_{32} & M_{33} \end{vmatrix} = \frac{y_1}{D}\begin{vmatrix} M_{22} & M_{23} \\ M_{32} & M_{33} \end{vmatrix} - \frac{y_2}{D}\begin{vmatrix} M_{12} & M_{13} \\ M_{32} & M_{33} \end{vmatrix} + \frac{y_3}{D}\begin{vmatrix} M_{12} & M_{13} \\ M_{22} & M_{23} \end{vmatrix}$$

or

$$x_1 = \frac{D^{11}y_1 + D^{21}y_2 + D^{31}y_3}{D} = \frac{D^{n1}y_n}{D}$$

where D^{mn} ***is the cofactor*** (signed minor subdeterminant) of the element M_{mn} in the matrix M. Thus, the solution of Eqs. (42) is

$$x_m = \frac{D^{nm}}{D} y_n \tag{2.4.50a}$$

or, in matrix form,

$$[x_m] = \left[\frac{D^{mn}}{D}\right]^T [y_n]. \tag{2.4.50b}$$

Hence the inverse matrix M^{-1} of Eq. (48) is given by

$$M^{-1} = \left[\frac{D^{mn}}{D}\right]^T \qquad \text{with element} \quad \frac{D^{nm}}{D} \tag{2.4.51}$$

in row n and column m, where D^{mn} is the cofactor of M_{mn}. This method of finding the inverse is prohibitively complicated for a large matrix. For numerical methods of inverting a large matrix, see a text on numerical analysis.

Similar matrices. If M and N are two $n \times n$ matrices, then N ***is said to be similar to*** M if there exists a nonsingular matrix A such that

$$N = A^{-1}MA \tag{2.4.52}$$

Then M is also similar to N, since

$$M = ANA^{-1} = (A^{-1})^{-1}N(A^{-1}).$$

Orthogonal matrix. A matrix Q is said to be orthogonal if

$$Q^T Q = QQ^T = I \quad \text{whence} \quad Q^T = Q^{-1}. \tag{2.4.53}$$

Equations (3b) show that the matrix A of direction cosines for change of rectangular Cartesian axes is orthogonal. Hence the matrices $\bar{T}$ and T of rectangular Cartesian components of a tensor $\mathbf{T}$ are similar matrices under the class of orthogonal transformation matrices, according to Eqs. (12) and (52).

Eigenvectors and eigenvalues of a matrix. The transformation $y = Mx$ of Eqs. (42) and (43) may be interpreted as associating to each point $P(x_1, x_2, x_3)$ another point $Q(y_1, y_2, y_3)$. Since the transformation is homogeneous, it associates to any other point (rx_1, rx_2, rx_3) on the line OP another point (ry_1, ry_2, ry_3) on the line OQ, and we may consider the transformation to be a ***transformation of the line*** OP ***into the line*** OQ. We may ask whether there are any lines not changed by this transformation, i.e., any lines transformed into themselves so that (x_1, x_2, x_3) transforms into $(\lambda x_1, \lambda x_2, \lambda x_3)$ for some scalar λ. Any line transformed into itself is called an ***eigenvector*** of the matrix. (Other names for this are ***proper vector, characteristic vector,*** or ***latent vector.***) If there is such an eigenvector x, then $y = \lambda x$ and

$$Mx = \lambda x = \lambda I x$$

or

$$(M - \lambda I)x = 0, \tag{2.4.54}$$

and the eigenvector is given by the solution of the set of homogeneous algebraic equations (54). A nontrivial solution will exist if and only if the determinant vanishes:

$$|M - \lambda I| = 0. \tag{2.4.55}$$

When the 3×3 determinant is expanded, Eq. (55) is a polynomial equation, a cubic in three dimensions, whose roots $\lambda_1, \lambda_2, \lambda_3$ are called the ***eigenvalues*** (***proper numbers, characteristic values,*** or ***latent roots***) of the matrix. When λ

is put equal to one of these eigenvalues, Eq. (54) can be solved for the x_r, which are determined only up to a scalar multiple. They are determined up to a factor of ± 1, if we require in addition that the x_r be the components of a unit vector. The whole discussion can be generalized to n dimensions.

A real symmetric matrix has only real eigenvalues. For, if there were a complex root λ of the polynomial equation (54) (with real coefficients), then $Mx = \lambda x$, whence $M\bar{x} = \bar{\lambda}\bar{x}$ (where overbars denote complex conjugates). Multiplying the first of these equations by $\bar{x}^T$ and the second by x^T, we have

$$\bar{x}^T Mx = \lambda \bar{x}^T x \quad \text{and} \quad x^T M\bar{x} = \bar{\lambda} x^T \bar{x}. \tag{2.4.56}$$

But $\bar{x}^T x = x^T \bar{x} = x_1\bar{x}_1 + x_2\bar{x}_2 + \cdots + x_n\bar{x}_n$; and, since M is symmetric $x^T M\bar{x} = \bar{x}^T Mx$. Hence, subtracting the two Eqs. (56) we find

$$(\lambda - \bar{\lambda})\bar{x}^T x = 0.$$

Since x is nontrivial $\bar{x}^T x \neq 0$, and it follows that

$$\lambda = \bar{\lambda} \tag{2.4.57}$$

so that λ must be real. It is also known that when no two of the eigenvalues are equal, the eigenvectors are uniquely determined up to a scalar multiple of each vector, so that the three lines in space (one associated with each eigenvalue) are uniquely determined. Furthermore, if any two eigenvalues are distinct, say $\lambda_1 \neq \lambda_2$, then the two associated eigenvectors must be orthogonal. This will be demonstrated in Chap. 3 for the particular example where the eigenvectors are the principal axes of the stress tensor. If all three eigenvalues are equal, then every direction is an eigenvector; in this case the eigenvectors are not unique, but it is still possible to choose a set of three mutually orthogonal eigenvectors. If two are equal, say $\lambda_2 = \lambda_3 \neq \lambda_1$, then the eigenvector associated with λ_1 is unique (up to a scalar multiple) and any direction orthogonal to it is an eigenvector, so that the other two are not uniquely determined. In any case, it is possible to choose a set of mutually orthogonal eigenvectors. If the three orthogonal eigenvectors are normalized to be unit vectors, then the square matrix L whose columns are the column matrices of the orthonormal eigenvectors will be an orthogonal matrix, such that

$$D = L^T ML \tag{2.4.58}$$

is a diagonal matrix whose elements are the eigenvalues. Thus every real symmetric matrix M is similar to a diagonal matrix whose elements are the eigenvalues, and the transformation matrix L required for the similarity relation is an orthogonal matrix whose columns are a set of mutually orthonormal eigenvectors of M.

This completes our introduction to tensor algebra. Some additional tensor properties having to do with principal axes of tensors and the algebraic invariants of a tensor will be presented in Sec. 3.3 in connection with the development of the stress tensor. Although developed specifically for the stress tensor, these properties are also relevant to other tensors. Also presented in Sec. 3.3 will be the separation of a tensor into the sum of a spherical and a deviatoric tensor. And in Sec. 4.6 the concept of the inverse of a tensor and of positive, negative, and fractional powers of a tensor will be presented. The differential calculus of vectors and tensors, which will be needed in our study of strain and deformation, is considered in Sec. 2.5.

EXERCISES

1. New right-handed coordinate axes are chosen with new origin at (4, −1, −2) and with $\bar{\mathbf{i}}_1 = (2\mathbf{i}_1 + 2\mathbf{i}_2 + \mathbf{i}_3)/3$ and $\bar{\mathbf{i}}_2 = (\mathbf{i}_1 - \mathbf{i}_2)/\sqrt{2}$.
   (a) Express $\bar{\mathbf{i}}_3$ in terms of the $\mathbf{i}_k$.
   (b) If $\mathbf{t} = 10\mathbf{i}_1 + 10\mathbf{i}_2 - 20\mathbf{i}_3$, express $\mathbf{t}$ in terms of the new basis.
   (c) Express the old coordinates x_1, x_2, x_3 in terms of $\bar{x}_1, \bar{x}_2, \bar{x}_3$.

2. Complete the following table of direction cosines for rotation of axes. If $f_x = f_y = f_z = 1000$ lb, find the components of $\mathbf{f}$ in the barred system.

	$\bar{x}$	$\bar{y}$	$\bar{z}$
x	0		$\frac{3}{5}$
y	0		$-\frac{4}{5}$
z	1		0

3. Verify that Eq. (2.4.6b) gives the same results as Eq. (6a).
4. Carry out the details of deriving the inverse transformation of the second Eq. (2.4.6a).
5. Derive the first Eq. (2.4.7a).
6. If $(\mathbf{ab})\cdot\mathbf{v}$ means $\mathbf{a}(\mathbf{b}\cdot\mathbf{v})$, show that its rectangular Cartesian components are the same as those of $\mathbf{T}\cdot\mathbf{v}$, where $\mathbf{T}$ is the tensor with components $a_i b_j$.
7. Write out the nine-term sum of the first Eq. (2.4.11) for $\bar{T}_{31}$, for $\bar{T}_{22}$.
8. Rotated $\bar{x}_m$-axes are chosen making angles with the x_r-axes as shown in the following table.

	$\bar{x}$	$\bar{y}$	$\bar{z}$
x	90°	45°	135°
y	45°	60°	60°
z	45°	120°	120°

(a) Verify that the new axes are a right-handed orthogonal system, and display the matrix A of direction cosines.

(b) If tensor **T** has the components shown in the matrix below, referred to the x_r-axes, compute the matrix of $\bar{x}_m$-components of **T**.

$$\begin{bmatrix} 2 & -2 & 0 \\ -2 & 6 & 0 \\ 0 & 0 & 4 \end{bmatrix}$$

9. Derive the second Eq. (2.4.11).

10. Show that the first Eq. (2.4.12) is equivalent to the first Eq. (11).

11. Show that the two scalar products $\mathbf{T} : \mathbf{U}$ and $\mathbf{T} \cdot\cdot\, \mathbf{U}$, defined in Eqs. (2.4.18), are equal if **U** is symmetric.

12. Show that $\mathbf{T} : \mathbf{W} = 0$, if **T** is symmetric and **W** is skew.

13. Show by use of Eqs. (2.4.6a) that the test of Eq. (26) implies the first transformation of Eq. (11).

14. Show by use of Eqs. (2.4.6a) that the test of Eq. (28) implies the first Eq. (11).

15. Show that if Eq. (2.4.35) holds for arbitrary rotations of axes, then the $\bar{T}_{pq}$ are related to the T_{rs} by Eq. (11).

16. Find the coefficients a_r^s for the rotation of axes, if the new $\bar{x}_r$-axes are obtained by a counterclockwise rotation through angle θ about the old x_3-axis.

17. Find the rotation matrix A for the following coordinate changes: (a) rotate 180° about x_3-axis, (b) rotate 45° about x_1-axis, (c) rotate 90° about x_1-axis and then rotate 90° about the position assumed by the x_2-axis after the first rotation.

18. Determine the product matrices AB and BA, if A and B are the matrices shown.

$$A = \begin{bmatrix} 1 & 1 & 0 \\ 0 & 0 & 0 \\ 0 & 1 & 0 \end{bmatrix} \qquad B = \begin{bmatrix} 0 & 0 & 0 \\ 0 & 0 & 0 \\ 1 & 0 & 0 \end{bmatrix}$$

19. Multiply AB, if

$$A = \begin{bmatrix} 1 & 3 & 1 \\ 2 & 1 & -1 \end{bmatrix} \qquad B = \begin{bmatrix} 1 & 2 \\ 3 & 0 \\ 2 & -1 \end{bmatrix}.$$

Can you multiply as BA?

20. Multiply AB and BC.

$$A = [\tfrac{2}{3} \quad -\tfrac{2}{3} \quad \tfrac{1}{3}] \qquad B = \begin{bmatrix} 21 & -12 & 6 \\ -12 & 9 & 0 \\ 6 & 0 & 6 \end{bmatrix} \qquad C = \begin{bmatrix} \tfrac{2}{3} \\ -\tfrac{2}{3} \\ \tfrac{1}{3} \end{bmatrix}.$$

21. Verify that, if

$$A = \begin{bmatrix} 3 & 2 & 1 \\ 4 & 6 & 5 \\ 3 & 8 & 4 \end{bmatrix} \qquad \text{then} \qquad A^{-1} = \begin{bmatrix} \frac{4}{9} & 0 & -\frac{1}{9} \\ \frac{1}{36} & -\frac{1}{4} & \frac{11}{36} \\ -\frac{7}{18} & \frac{1}{2} & -\frac{5}{18} \end{bmatrix}.$$

22. Show that if u, v, w are orthonormal eigenvectors of matrix M, corresponding to eigenvalues $\lambda_1, \lambda_2, \lambda_3$, respectively, and L is the square matrix whose columns are u, v, w, then $L^T(ML)$ is a diagonal matrix, as asserted in Eq. (2.4.58).

23. and 24. Determine the eigenvalues and eigenvectors of the matrices shown below, and verify that the orthogonal matrix L of Eq. (2. 4. 58) diagonalizes the given matrix in each case.

$$23.\quad \begin{bmatrix} 11 & 4 & 0 \\ 4 & 5 & 0 \\ 0 & 0 & 7 \end{bmatrix} \qquad 24.\quad \begin{bmatrix} 7 & 0 & -2 \\ 0 & 5 & 0 \\ -2 & 0 & 4 \end{bmatrix}$$

25. Verify Eq. (2.4.49).

26. Verify the first two Eqs. (2.4.25b) for $\mathbf{A} \cdot\cdot \mathbf{B}$ and $\mathbf{A} : \mathbf{B}$.

27. Verify the last two lines of Eqs. (2. 4. 25b).

2.5 Vector and tensor calculus; differentiation; gradient, divergence, and curl†

The preceding sections have dealt with the algebra of vectors at a point. The physical entities defined as vectors, in general, vary from point to point and with time. The vector quantity is then a ***vector field*** or ***vector function***. Study of the rate of change of such a vector function leads naturally to the study of derivatives with respect to space coordinates. A brief summary of vector differential calculus is given in this section; for more details and proofs see a textbook on vector analysis. Some integral theorems will be given in Sec. 5.1, since these are useful in deriving the general principles of continuum mechanics introduced there. We introduce the total time derivative of a vector function of t by an example, the differentiation of the position vector of a moving particle.

If $\mathbf{p}(t)$ is the position vector at time t of a particle (material point), then its velocity $\mathbf{v}$ is given by $d\mathbf{p}/dt$, which is defined in the same way as the derivative of a scalar function, namely

†This section may be postponed until after Chap. 3, since it will not be needed until the development of the concepts of strain and deformation in Chap. 4.

$$\frac{d\mathbf{p}}{dt} = \lim_{\Delta t \to 0} \frac{\Delta \mathbf{p}}{\Delta t} \tag{2.5.1}$$

If the particle moves on a smooth path, as in Fig. 2.8, then

$$\left|\frac{\Delta \mathbf{p}}{\Delta t}\right| = \frac{|\Delta \mathbf{p}|}{\Delta s}\frac{\Delta s}{\Delta t}, \tag{2.5.2}$$

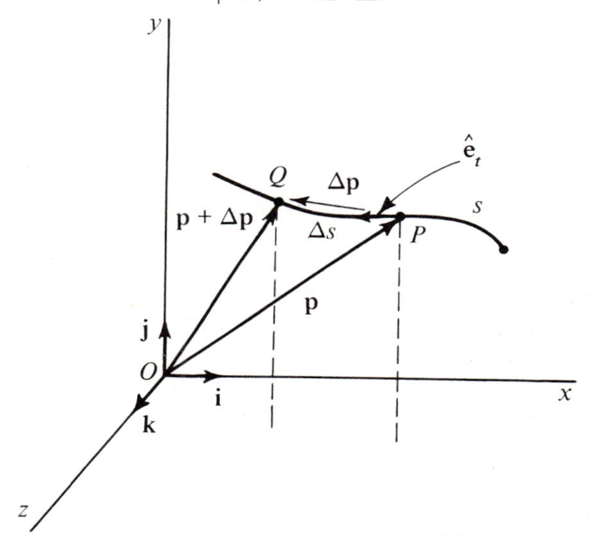

Fig. 2.8 Differentiation of position vector **p**.

and in the limit, assuming that $|\Delta \mathbf{p}|/\Delta s$ approaches unity and that the direction of the secant *PQ* approaches that of the tangent to the path at *P*

$$\mathbf{v} = \frac{d\mathbf{p}}{dt} = \frac{ds}{dt}\,\hat{\mathbf{e}}_t, \tag{2.5.3}$$

where $\hat{\mathbf{e}}_t$ is the unit vector tangent to the path. For the evaluation of the derivative, if we express the Cartesian coordinates of the particle in a "fixed" system of reference as functions of t,

$$\mathbf{p} = x(t)\mathbf{i} + y(t)\mathbf{j} + z(t)\mathbf{k}, \tag{2.5.4}$$

we obtain

$$\frac{d\mathbf{p}}{dt} = \frac{dx}{dt}\mathbf{i} + \frac{dy}{dt}\mathbf{j} + \frac{dz}{dt}\mathbf{k} = \dot{x}_m \mathbf{i}_m, \tag{2.5.5a}$$

since **i**, **j**, **k** do not change with time. Inversely the time integral is

$$\mathbf{p} - \mathbf{p}_0 = \int_{t_0}^{t} \mathbf{v}\,dt = \mathbf{i}\int_{t_0}^{t} v_x\,dt + \mathbf{j}\int_{t_0}^{t} v_y\,dt + \mathbf{k}\int_{t_0}^{t} v_z\,dt. \qquad (2.5.5b)$$

The time derivative of any other vector **a** is evaluated in the same way,

$$\frac{d\mathbf{a}}{dt} = \frac{da_x}{dt}\mathbf{i} + \frac{da_y}{dt}\mathbf{j} + \frac{da_z}{dt}\mathbf{k} = \dot{a}_m \mathbf{i}_m, \qquad (2.5.6)$$

and a similar interpretation to that of Fig. 2.8 can be made if we plot successive values of **a** in a space with coordinates (a_x, a_y, a_z).

It is easy to show that the derivatives of a product of a scalar by a vector, of the dot product of two vectors, of the cross product of two vectors, and of the tensor products all obey the usual calculus rule for the derivative of a product. Proofs will not be given here; the essential ingredients for the proofs are the distributive character of the products considered and certain limit theorems—such as that the limit of a product is the product of the limits when both limits exist. Under the assumption that the vector functions considered have continuous derivatives, the whole algebra of derivatives of scalar functions carries over to the derivatives of vector functions. In applying the product rule to the cross product of two vectors, however, it it necessary to keep the two factors always in the same order, since the cross product is not commutative. Thus

$$\frac{d}{dt}(\mathbf{u} \times \mathbf{v}) = \mathbf{u} \times \frac{d\mathbf{v}}{dt} + \frac{d\mathbf{u}}{dt} \times \mathbf{v}, \qquad (2.5.7)$$

and it would be incorrect to write the last term as $\mathbf{v} \times (d\mathbf{u}/dt)$ in analogy with the way the scalar formula is often written.

If A and B in Fig. 2.9 are two points on a rigid body with instantaneous angular velocity $\boldsymbol{\omega}$, the relative velocity $\mathbf{v}_{B/A}$ of B relative to A is given entirely by the rotation. Thus, the magnitude

$$|\mathbf{v}_{B/A}| = \frac{d\phi}{dt}|CB| = \frac{d\phi}{dt}|\mathbf{r}_{B/A}|\sin\theta, \qquad (2.5.8)$$

and the direction of $\mathbf{v}_{B/A}$ is tangent to the circle of radius CB in a plane perpendicular to the axis of instantaneous relative rotation drawn through A. Hence

$$\mathbf{v}_{B/A} = \boldsymbol{\omega} \times \mathbf{r}_{B/A}. \qquad (2.5.9)$$

If any other vector of constant magnitude, say **a**, is fixed to a rigid body, then it follows also that

$$\frac{d\mathbf{a}}{dt} = \boldsymbol{\omega} \times \mathbf{a}, \qquad (2.5.10)$$

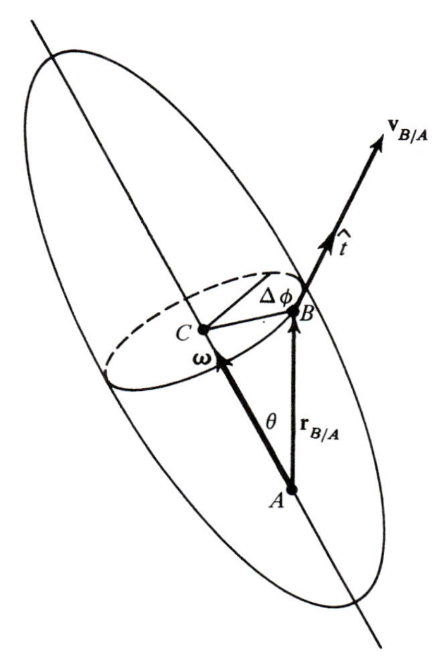

Fig. 2.9 Relative velocity $\mathbf{v}_{B/A}$ for two points on a rigid body.

a result which can be used to express vector derivatives in curvilinear coordinates, or with respect to moving axes.

In cylindrical coordinates (r, θ, z) (Fig. 2.10), where

$$x = r\cos\theta, \qquad y = r\sin\theta,$$

the position vector $\mathbf{p}$ is given by

$$\mathbf{p} = r\hat{\mathbf{e}}_r + z\hat{\mathbf{e}}_z \tag{2.5.11}$$

where $\hat{\mathbf{e}}_r$ and $\hat{\mathbf{e}}_z$ are unit vectors in the direction of increasing r and z, respectively. Note that $\hat{\mathbf{e}}_z = \mathbf{k}$ is constant, but that in a general motion $\hat{\mathbf{e}}_r$ changes with time as does the third unit vector, $\hat{\mathbf{e}}_\theta$, in the transverse direction of increasing θ.

If we differentiate Eq. (11), we obtain

$$\mathbf{v} = \frac{d\mathbf{p}}{dt} = \frac{dr}{dt}\hat{\mathbf{e}}_r + r\frac{d\hat{\mathbf{e}}_r}{dt} + \frac{dz}{dt}\hat{\mathbf{e}}_z. \tag{2.5.12}$$

The triad of unit vectors $\hat{\mathbf{e}}_r$, $\hat{\mathbf{e}}_\theta$, $\hat{\mathbf{e}}_z$ may be considered as a rigid body moving with the particle; since $\hat{\mathbf{e}}_r$ always intersects the z-axis at a right angle, the

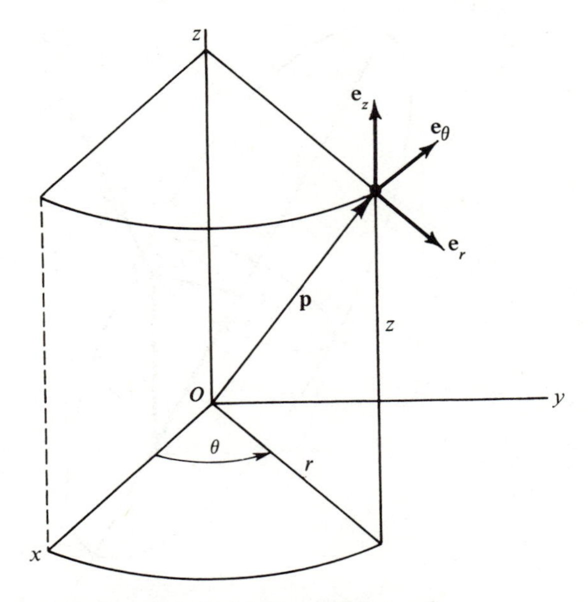

Fig. 2.10 Vector **p** in cylindrical coordinates r, θ, z.

angular velocity $\boldsymbol{\omega}$ of this little rigid body is $(d\theta/dt)\hat{\mathbf{e}}_z$, and the rates of change of the unit vectors are given by Eq. (10) as

$$\left.\begin{aligned} \frac{d\hat{\mathbf{e}}_r}{dt} &= \boldsymbol{\omega} \times \hat{\mathbf{e}}_r = \frac{d\theta}{dt}\,\hat{\mathbf{e}}_z \times \hat{\mathbf{e}}_r = \frac{d\theta}{dt}\,\hat{\mathbf{e}}_\theta \\ \frac{d\hat{\mathbf{e}}_\theta}{dt} &= \boldsymbol{\omega} \times \hat{\mathbf{e}}_\theta = \frac{d\theta}{dt}\,\hat{\mathbf{e}}_z \times \hat{\mathbf{e}}_\theta = -\frac{d\theta}{dt}\,\hat{\mathbf{e}}_r \\ \frac{d\hat{\mathbf{e}}_z}{dt} &= 0 \end{aligned}\right\} \tag{2.5.13}$$

since

$$\hat{\mathbf{e}}_\theta = \hat{\mathbf{e}}_z \times \hat{\mathbf{e}}_r, \quad \text{while} \quad \hat{\mathbf{e}}_z \times \hat{\mathbf{e}}_\theta = -\hat{\mathbf{e}}_r.$$

Hence

$$\mathbf{v} = \frac{dr}{dt}\,\hat{\mathbf{e}}_r + r\,\frac{d\theta}{dt}\,\hat{\mathbf{e}}_\theta + \frac{dz}{dt}\,\hat{\mathbf{e}}_z. \tag{2.5.14}$$

This last equation may be differentiated again to yield the acceleration, using once more Eqs. (13) for the derivatives of the unit vectors. For further examples of this kind see a textbook on vector dynamics. We turn our attention now to differentiation with respect to space coordinates.

Scalar point function. A scalar point function has a single component in any coordinate system, which has the same numerical value in all coordinate systems. Thus

$$\bar{F}(\bar{x}, \bar{y}, \bar{z}) = F(x, y, z), \tag{2.5.15}$$

where, by definition,

$$\bar{F}(\bar{x}, \bar{y}, \bar{z}) \equiv F[x(\bar{x}, \bar{y}, \bar{z}), y(\bar{x}, \bar{y}, \bar{z}), z(\bar{x}, \bar{y}, \bar{z})]. \tag{2.5.16}$$

It is common practice to write Eq. (15) as

$$F(\bar{x}, \bar{y}, \bar{z}) = F(x, y, z),$$

but it must be remembered that F does not represent the same function of the barred coordinates as it does of the unbarred coordinates. For example, if $F(x, y, z) = 25(x^2 - y^2) + z^2$, and if the transformation is $x = (4\bar{x} - 3\bar{y})/5$, $y = (3\bar{x} + 4\bar{y})/5$, and $z = \bar{z}$, then $\bar{F}(\bar{x}, \bar{y}, \bar{z}) = 7(\bar{x}^2 - \bar{y}^2) - 48\bar{x}\bar{y} + \bar{z}^2$, an apparently quite different function from $F(x, y, z)$ although it has the same value at every point.

The gradient of a scalar F is defined to be the vector, denoted ∇F or grad F, such that for any unit vector $\hat{\mathbf{u}}$, the directional derivative dF/ds in the direction of $\hat{\mathbf{u}}$ is given by the scalar product

$$\boxed{\frac{dF}{ds} = \nabla F \cdot \hat{\mathbf{u}},} \tag{2.5.17}$$

where

$$\hat{\mathbf{u}} = \frac{d\mathbf{p}}{ds}.$$

Note that this definition makes no reference to any coordinate system. The gradient is thus a ***vector invariant*** independent of the coordinate system. To find its components in any rectangular Cartesian coordinate system we note that

$$\hat{\mathbf{u}} = \frac{d\mathbf{p}}{ds} = \frac{dx_k}{ds}\mathbf{i}_k, \tag{2.5.18}$$

while, by the chain rule of calculus,

$$\frac{dF}{ds} = \frac{\partial F}{\partial x_k}\frac{dx_k}{ds}. \tag{2.5.19}$$

If $(\nabla F)_k$ denotes the x_k-component of ∇F, then Eq. (17) yields

$$\frac{dF}{ds} = (\nabla F)_k \frac{dx_k}{ds}.$$

Comparing this with Eq. (19), we see that

$$\left[(\nabla F)_k - \frac{\partial F}{\partial x_k}\right]\frac{dx_k}{ds} = 0$$

for arbitrary choice of the dx_k/ds (arbitrary direction $\hat{\mathbf{u}}$). Hence

$$\boxed{\begin{aligned} &(\nabla F)_k = \frac{\partial F}{\partial x_k} \quad \text{or} \\ &\text{grad } F = \nabla F = \frac{\partial F}{\partial x_k}\mathbf{i}_k. \end{aligned}} \tag{2.5.20a}$$

In indicial notation partial derivatives are frequently denoted by a subscript preceded by a comma: $\partial F/\partial x_k = F_{,k}$. Another notation places the operator ∂_k ahead of the function. Thus

$$\partial_k F \equiv F_{,k} \equiv \frac{\partial F}{\partial x_k} \qquad \text{and} \qquad \nabla F = \mathbf{i}_k \partial_k F = \mathbf{i}_k F_{,k}. \tag{2.5.20b}$$

A geometric interpretation of the gradient follows from Eq. (17) if we take $\hat{\mathbf{u}} = \hat{\mathbf{t}}$ as a tangent vector to the level surface $F(x, y, z) =$ constant.

Then $dF/ds = 0$, and Eq. (17) implies that

$$(\nabla F)\cdot\hat{\mathbf{t}} = 0. \tag{2.5.21}$$

Since according to Eq. (21), the gradient vector is perpendicular to every tangent vector at the point, the gradient vector must be normal to the level surface through the point. Moreover the rate of change of the function F with respect to distance in the direction $\hat{\mathbf{n}}$ of increasing F normal to the surface $F =$ constant, namely, $(\nabla F)\cdot\hat{\mathbf{n}}$, is equal to $|\nabla F|$, since ∇F is parallel to the normal and since $(\nabla F)\cdot\hat{\mathbf{n}}$ is positive because $\hat{\mathbf{n}}$ was taken as the direction of increasing F. The unit normal vector $\hat{\mathbf{n}}$ to the level surface, taken in the direction of increasing F, is therefore given by

$$\hat{\mathbf{n}} = \frac{\text{grad } F}{|\text{grad } F|}. \tag{2.5.22}$$

Other important applications of the gradient in mechanics relate the force vector to the gradient of the potential of a conservative force field and the velocity vector to the gradient of a velocity potential in irrotational fluid flow.

The symbolic vector operator del, denoted by ∇,

$$\nabla = \mathbf{i}\frac{\partial}{\partial x} + \mathbf{j}\frac{\partial}{\partial y} + \mathbf{k}\frac{\partial}{\partial z} = \mathbf{i}_m\frac{\partial}{\partial x_m} = \mathbf{i}_m\,\partial_m \tag{2.5.23}$$

may be used formally to obtain Eq. (20)

$$\text{grad}\, F = \nabla F = \left(\mathbf{i}\frac{\partial}{\partial x} + \mathbf{j}\frac{\partial}{\partial y} + \mathbf{k}\frac{\partial}{\partial z}\right)F = \mathbf{i}\frac{\partial F}{\partial x} + \mathbf{j}\frac{\partial F}{\partial y} + \mathbf{k}\frac{\partial F}{\partial z}. \tag{2.5.24}$$

The operator operates on the quantity written to the right of it. When the quantity to the right is a scalar function F, the result is grad F. When the quantity to the right is a vector function, it is necessary to specify how the vector operator operates on it. Two common applications are $\nabla\cdot\mathbf{v}$ and $\nabla\times\mathbf{v}$, the first yielding a scalar, the second a vector. These will be introduced here briefly and formally; applications and physical interpretations are given later. $\nabla\mathbf{v}$ without dot or cross, is a tensor; see Eq. (31).

Divergence $\nabla\cdot\mathbf{v}$. When the vector operator ∇ operates in a manner analogous to scalar multiplication, the result is a scalar point function, div $\mathbf{v}$, called the ***divergence of the vector field*** $\mathbf{v}$. In rectangular Cartesians,

$$\text{div}\,\mathbf{v} = \nabla\cdot\mathbf{v} = \left(\mathbf{i}\frac{\partial}{\partial x} + \mathbf{j}\frac{\partial}{\partial y} + \mathbf{k}\frac{\partial}{\partial z}\right)\cdot(\mathbf{i}v_x + \mathbf{j}v_y + \mathbf{k}v_z) \quad \text{or}$$

$$\text{div}\,\mathbf{v} = \frac{\partial v_x}{\partial x} + \frac{\partial v_y}{\partial y} + \frac{\partial v_z}{\partial z} = \frac{\partial v_i}{\partial x_i} \equiv \partial_i v_i \equiv v_{i,i}. \tag{2.5.25a}$$

A physical interpretation of div $\mathbf{v}$ is given in Sec. 5.2, where it is shown that if $\mathbf{v}$ is the velocity field in a flow of a continuous medium, then div $\mathbf{v} = -(1/\rho)(d\rho/dt)$, where ρ is density. Thus the divergence of the velocity vector measures the rate of flow of material away from the particle and is equal to the unit rate of decrease of density ρ in the neighborhood of the particle.

The Laplacian Operator $\nabla^2 \equiv \nabla\cdot\nabla$ gives a scalar point function when it operates on a twice-differentiable scalar field

$$\nabla^2 F \equiv \nabla\cdot\nabla F = F_{,kk} \equiv \frac{\partial^2 F}{\partial x^2} + \frac{\partial^2 F}{\partial y^2} + \frac{\partial^2 F}{\partial z^2} \tag{2.5.25b}$$

in rectangular Cartesian coordinates. For orthogonal curvilinear coordinates, see Eq. (II.2.14). For a general-tensor component form, see Eq. (I.6.28). For the Laplacian $\nabla^2\mathbf{v}$ of a vector $\mathbf{v}$, see Eq. (2.5.35).

Curl $\nabla\times\mathbf{v}$. When the vector operator ∇ operates in a manner analogous to vector multiplication, the result is a vector, curl $\mathbf{v}$, called the curl of the vector field $\mathbf{v}$ (sometimes called the rotation and denoted Rot $\mathbf{v}$).

$$\left.\begin{aligned}\text{curl } \mathbf{v} = \nabla \times \mathbf{v} &= \begin{vmatrix} \mathbf{i} & \mathbf{j} & \mathbf{k} \\ \dfrac{\partial}{\partial x} & \dfrac{\partial}{\partial y} & \dfrac{\partial}{\partial z} \\ v_x & v_y & v_z \end{vmatrix} \\ &= \mathbf{i}\left(\frac{\partial v_z}{\partial y} - \frac{\partial v_y}{\partial z}\right) + \mathbf{j}\left(\frac{\partial v_x}{\partial z} - \frac{\partial v_z}{\partial x}\right) + \mathbf{k}\left(\frac{\partial v_y}{\partial x} - \frac{\partial v_x}{\partial y}\right)\end{aligned}\right\} \qquad (2.5.26a)$$

or in indicial notation,

$$\text{if} \quad \mathbf{q} = \nabla \times \mathbf{v}, \quad \text{then} \quad q_i = e_{ijk}\frac{\partial}{\partial x_j}(v_k) \equiv e_{ijk}\partial_j v_k, \qquad (2.5.26b)$$

using Eq. (2.3.24). The notation $\partial_j v_k$ for $\partial v_k/\partial x_j$ is preferable to $v_{k,j}$, which has the disadvantage of reversing the order of the indices and causing confusion in relation to Eq. (2.3.24). Thus

$$\nabla \times \mathbf{v} = e_{ijk} v_{k,j} \mathbf{i}_i. \qquad (2.5.26c)$$

When $\mathbf{v}$ is the velocity field curl $\mathbf{v}$ is twice the angular-velocity vector $\mathbf{w}$ giving the instantaneous angular velocity of the volume element, defined in Sec. 4.4. A vector field satisfying the condition curl $\mathbf{v} = 0$ everywhere is called an ***irrotational vector field***. The reader can easily verify that if the vector field $\mathbf{v}$ is the gradient of a scalar function, say F, then it is irrotational, i.e., curl (grad F) = 0.

A number of useful vector identities involving the ∇ operator are summarized in Eq. (2.5.27). Proofs are omitted here. Most of them can be readily verified by writing out the components; indeed the indicial representation in rectangular Cartesians makes several of the identities almost self-evident, assuming that the functions involved are sufficiently differentiable.

VECTOR IDENTITIES INVOLVING ∇

$$\left.\begin{aligned}
\nabla\cdot(\nabla F) &= \nabla^2 F && \text{(a)}\\
\nabla(FG) &= F\nabla G + G\,\nabla F && \text{(b)}\\
\nabla^2(FG) &= F\nabla^2 G + 2(\nabla F)\cdot(\nabla G) + G\,\nabla^2 F && \text{(c)}\\
\nabla\cdot(F\mathbf{v}) &= (\nabla F)\cdot\mathbf{v} + F\,\nabla\cdot\mathbf{v} && \text{(d)}\\
\nabla\cdot(F\,\nabla G) &= F\nabla^2 G + \nabla F\cdot\nabla G && \text{(e)}\\
\nabla\times(\nabla F) &= 0 && \text{(f)}\\
\nabla\cdot(\nabla\times\mathbf{v}) &= 0 && \text{(g)}\\
\nabla\cdot(\mathbf{a}\times\mathbf{b}) &= (\nabla\times\mathbf{a})\cdot\mathbf{b} - \mathbf{a}\cdot(\nabla\times\mathbf{b}) && \text{(h)}\\
\nabla\times(F\mathbf{v}) &= \nabla F\times\mathbf{v} + F\,\nabla\times\mathbf{v} && \text{(i)}\\
\nabla\times(\nabla\times\mathbf{v}) &= \nabla(\nabla\cdot\mathbf{v}) - \nabla^2\mathbf{v} && \text{(j)}
\end{aligned}\right\} \qquad (2.5.27)$$

where $\nabla^2\mathbf{v} \equiv \vec{\nabla}\cdot\vec{\nabla}\mathbf{v}$ [see Eq. (2.5.35)].

The scalar potential function. We saw in the discussion following Eq. (26c) that if a vector field $\mathbf{v}$ is the gradient of a scalar field, then curl $\mathbf{v} = 0$. [This was repeated as the identity of Eq. (27f).] Conversely, it can be proved that if curl $\mathbf{v} = 0$ throughout a simply connected region (one such that every closed path in it can be continuously deformed to a point without passing out of the region), then $\mathbf{v}$ is the gradient of a potential function ϕ defined by

$$\phi = \int_C \mathbf{v} \cdot d\mathbf{r} = \int_C v_x\, dx + v_y\, dy + v_z\, dz, \tag{2.5.28}$$

a line integral along any curve C from an arbitrary point P_0 in the region to a variable point P of the region. Since there are an infinite number of possible paths from P_0 to P it might be thought that different values would be obtained for different paths. But the component equations of the vector equation curl $\mathbf{v} = 0$ are precisely the necessary and sufficient conditions that the line integral of Eq. (28) be independent of the path in a simply connected region in which $\mathbf{v}$ and curl $\mathbf{v}$ are continuous. Another way of saying this is that the conditions curl $\mathbf{v} = 0$ are the necessary and sufficient conditions that the integrand be a perfect differential of some function, which we call ϕ. Thus

$$d\phi = v_x\, dx + v_y\, dy + v_z\, dz \tag{2.5.29a}$$

and

$$\mathbf{v} = \operatorname{grad} \phi. \tag{2.5.29b}$$

In a simply connected region where curl $\mathbf{v} = 0$, the function ϕ is uniquely determined by the vector field $\mathbf{v}$, according to Eq. (28), except for an arbitrary additive constant; the arbitrary constant appears in Eq. (28) in that the initial point P_0 is arbitrary. This result, that every irrotational vector field is the gradient of a scalar field, called a ***potential function***, has many important applications in mechanics. The result follows from Stokes theorem as shown in Sec. 5.1.

Gradient of a vector field is a tensor. The gradient of a vector field $\mathbf{v}$ is defined to be the second-order tensor (linear vector function) $\mathbf{T}$, which at any point associates with each argument unit vector $\hat{\mathbf{u}}$ another vector $d\mathbf{v}/ds$—the rate of change of $\mathbf{v}$ with respect to distance in the direction of $\hat{\mathbf{u}}$. In dyadic notation, two different symbols are used to denote the gradient of a vector according to the way it is to operate on $\hat{\mathbf{u}}$. If the operation is to be performed in the order $\hat{\mathbf{u}} \cdot \mathbf{T} = d\mathbf{v}/ds$, the notation $\overrightarrow{\nabla}\mathbf{v}$ is used for $\mathbf{T}$. If it is to be performed in the order $\mathbf{T} \cdot \hat{\mathbf{u}} = d\mathbf{v}/ds$, then the notation used is $\mathbf{v}\overleftarrow{\nabla}$. Thus $\mathbf{v}\overleftarrow{\nabla}$ is the transpose of the tensor $\overrightarrow{\nabla}\mathbf{v}$.

In rectangular Cartesians, $\mathbf{v}\overleftarrow{\nabla}$ can be obtained formally as the open product of the vector $\mathbf{v} = v_k \mathbf{i}_k$ by the symbolic operator $\overleftarrow{\nabla}$.

$$\overleftarrow{\nabla} = \overleftarrow{\partial}_m \mathbf{i}_m$$
$$\mathbf{v}\overleftarrow{\nabla} = (v_k \mathbf{i}_k)(\overleftarrow{\partial}_m \mathbf{i}_m) = v_{k,m} \mathbf{i}_k \mathbf{i}_m \tag{2.5.30a}$$

The arrows on $\overleftarrow{\nabla}$ and $\overleftarrow{\partial}_m$ indicate that the differential operator acts on the preceding quantity. The resulting product $v_{k,m}\mathbf{i}_k\mathbf{i}_m$ is a dyadic or second-order tensor with rectangular Cartesian components $v_{k,m}$. Similarly,

$$\overrightarrow{\nabla}\mathbf{v} = (\mathbf{i}_k \overrightarrow{\partial}_k)(v_m \mathbf{i}_m) = \overrightarrow{\partial}_k v_m \mathbf{i}_k \mathbf{i}_m \tag{2.5.30b}$$

is a second-order tensor with components $\partial_k v_m$. (The arrows are often omitted when the operator acts to the right.) The matrices of rectangular Cartesian components are

$$[\mathbf{v}\overleftarrow{\nabla}] = \begin{bmatrix} \dfrac{\partial v_x}{\partial x} & \dfrac{\partial v_x}{\partial y} & \dfrac{\partial v_x}{\partial z} \\ \dfrac{\partial v_y}{\partial x} & \dfrac{\partial v_y}{\partial y} & \dfrac{\partial v_y}{\partial z} \\ \dfrac{\partial v_z}{\partial x} & \dfrac{\partial v_z}{\partial y} & \dfrac{\partial v_z}{\partial z} \end{bmatrix} \tag{2.5.31a}$$

and

$$[\overrightarrow{\nabla}\mathbf{v}] = \begin{bmatrix} \dfrac{\partial v_x}{\partial x} & \dfrac{\partial v_y}{\partial x} & \dfrac{\partial v_z}{\partial x} \\ \dfrac{\partial v_x}{\partial y} & \dfrac{\partial v_y}{\partial y} & \dfrac{\partial v_z}{\partial y} \\ \dfrac{\partial v_x}{\partial z} & \dfrac{\partial v_y}{\partial z} & \dfrac{\partial v_z}{\partial z} \end{bmatrix}. \tag{2.5.31b}$$

Thus

$$(\mathbf{v}\overleftarrow{\nabla})\cdot\hat{\mathbf{u}} = \frac{d\mathbf{v}}{ds} \quad \text{or} \quad [v_{i,j}]\left[\frac{dx_j}{ds}\right] = \left[\frac{dv_i}{ds}\right] \tag{2.5.32a}$$

gives the column matrix of components dv_i/ds, while

$$\hat{\mathbf{u}}\cdot(\overrightarrow{\nabla}\mathbf{v}) = \frac{d\mathbf{v}}{ds} \quad \text{or} \quad \left[\frac{dx_j}{ds}\right][\partial_j v_i] = \left[\frac{dv_i}{ds}\right] \tag{2.5.32b}$$

gives the row matrix of components dv_i/ds. (Here $\partial_j v_i$ is the element in the jth row and ith column of the square matrix of $\overrightarrow{\nabla}\mathbf{v}$, while $v_{i,j}$ is the element in the ith row and jth column of the matrix of $\mathbf{v}\overleftarrow{\nabla}$.)

Examples of the application of the gradient of a vector field are the tensors of displacement gradient and velocity gradient discussed in Chap. 4.

Divergence of a second-order tensor field. The divergence of a second-order tensor field $\mathbf{T}$ is a vector field, usually defined to be given in a rectangular Cartesian system by Eq. (33a). Another divergence is given in Eq. (33b). (note: $\partial_k T_{pq}\mathbf{i}_k\cdot\mathbf{i}_p = \partial_k T_{pq}\delta_{kp} = \partial_p T_{pq}$)

$$\vec{\nabla}\cdot\mathbf{T} = (\partial_k\mathbf{i}_k)\cdot(T_{pq}\mathbf{i}_p\mathbf{i}_q) = \partial_p T_{pq}\mathbf{i}_q = \frac{\partial T_{pq}}{\partial x_p}\mathbf{i}_q. \tag{2.5.33a}$$

$$\mathbf{T}\cdot\overleftarrow{\nabla} = (T_{pq}\mathbf{i}_p\mathbf{i}_q)\cdot(\overleftarrow{\partial}_k\mathbf{i}_k) = \frac{\partial T_{pq}}{\partial x_q}\mathbf{i}_p. \tag{2.5.33b}$$

The two divergences obtained in (33a) and (33b) are, in general, different, but they will be the same if the tensor $\mathbf{T}$ is symmetric.

Gradients of a second-order tensor $\mathbf{T}$ are the quantities $\vec{\nabla}\mathbf{T}$ and $\mathbf{T}\overleftarrow{\nabla}$ such that

$$\frac{d\mathbf{T}}{ds} = (\mathbf{T}\overleftarrow{\nabla})\cdot\hat{\mathbf{u}} = \hat{\mathbf{u}}\cdot(\vec{\nabla}\mathbf{T}), \tag{2.5.34a}$$

where $d\mathbf{T}/ds$ is the directional derivative in the direction of the unit vector $\hat{\mathbf{u}}$. Since $d\mathbf{T}/ds = (dT_{km}/ds)\mathbf{i}_k\mathbf{i}_m$ in rectangular coordinates, where the base vectors are constants, it may be shown that $\mathbf{T}\overleftarrow{\nabla}$ and $\vec{\nabla}\mathbf{T}$ are triadics or third-order tensors with rectangular Cartesian components

$$(\mathbf{T}\overleftarrow{\nabla})_{kmq} = T_{km,q} \quad \text{and} \quad (\vec{\nabla}\mathbf{T})_{kmq} = \partial_k T_{mq}. \tag{2.5.34b}$$

The Laplacian of a vector, denoted $\nabla^2\mathbf{v}$, is a vector. In rectangular Cartesian components it is easy to verify that (assuming sufficient differentiability) the vector $\vec{\nabla}\cdot(\vec{\nabla}\mathbf{v})$ has the same components as the vector $(\mathbf{v}\overleftarrow{\nabla})\cdot\overleftarrow{\nabla}$. The resulting vector is called the Laplacian of $\mathbf{v}$.

$$\left.\begin{aligned} \nabla^2\mathbf{v} &= \vec{\nabla}\cdot\vec{\nabla}\mathbf{v} = \mathbf{v}\overleftarrow{\nabla}\cdot\overleftarrow{\nabla} = v_{r,ss}\mathbf{i}_r \\ \text{or} \qquad \nabla^2\mathbf{v} &= \nabla^2 v_x\mathbf{i} + \nabla^2 v_y\mathbf{j} + \nabla^2 v_z\mathbf{k} \end{aligned}\right\} \tag{2.5.35}$$

in rectangular Cartesians, where, for example,

$$\nabla^2 v_x = \frac{\partial^2 v_x}{\partial x^2} + \frac{\partial^2 v_x}{\partial y^2} + \frac{\partial^2 v_x}{\partial z^2}.$$

Curvilinear-component evaluation of $\vec{\nabla}\cdot\vec{\nabla}\mathbf{v}$ or $\mathbf{v}\overleftarrow{\nabla}\cdot\overleftarrow{\nabla}$ is quite complicated. [See, for example, Eq. (II.2.16) for the divergence $\vec{\nabla}\cdot\mathbf{T}$ of a tensor $\mathbf{T}$ in orthogonal curvilinear coordinates.] It is usually more convenient to express $\nabla^2\mathbf{v}$ in terms of $\nabla\times\mathbf{v}$ and $\nabla\cdot\mathbf{v}$ by using the identity of Eq. (27j), as will be seen in Eq. (8.1.10).

Curls of a second-order tensor. $\vec{\nabla} \times \mathbf{T}$ and $\mathbf{T} \times \overleftarrow{\nabla}$ are second-order tensors obtained formally by operating with the symbolic vector operators $\vec{\nabla}$ and $\overleftarrow{\nabla}$ on the dyadic $\mathbf{T}$ given in rectangular Cartesians by $\mathbf{T} = T_{km}\mathbf{i}_k\mathbf{i}_m$. Thus

$$\vec{\nabla} \times \mathbf{T} = (\mathbf{i}_p \partial_p) \times (T_{km}\mathbf{i}_k\mathbf{i}_m) = \partial_p T_{km}(\mathbf{i}_p \times \mathbf{i}_k)\mathbf{i}_m$$

or, since $\mathbf{i}_p \times \mathbf{i}_k = e_{pkq}\mathbf{i}_q$

$$\vec{\nabla} \times \mathbf{T} = e_{pkq}\partial_p T_{km}\mathbf{i}_q\mathbf{i}_m \tag{2.5.36}$$

Examples of rectangular components of $\vec{\nabla} \times \mathbf{T}$ are†

$$\left.\begin{aligned} (\vec{\nabla} \times \mathbf{T})_{11} &= \frac{\partial T_{31}}{\partial x_2} - \frac{\partial T_{21}}{\partial x_3} \\ (\vec{\nabla} \times \mathbf{T})_{21} &= \frac{\partial T_{11}}{\partial x_3} - \frac{\partial T_{31}}{\partial x_1} \\ (\vec{\nabla} \times \mathbf{T})_{12} &= \frac{\partial T_{32}}{\partial x_2} - \frac{\partial T_{22}}{\partial x_3}. \end{aligned}\right\} \tag{2.5.37}$$

The curl operator from the right $\mathbf{T} \times \overleftarrow{\nabla}$ similarly yields

$$\mathbf{T} \times \overleftarrow{\nabla} = (T_{rs}\mathbf{i}_r\mathbf{i}_s) \times (\overleftarrow{\partial}_p\mathbf{i}_p) = T_{rs,p}\mathbf{i}_r(\mathbf{i}_s \times \mathbf{i}_p)$$

whence

$$\mathbf{T} \times \overleftarrow{\nabla} = T_{rs,p}e_{spq}\mathbf{i}_r\mathbf{i}_q \tag{2.5.38}$$

with typical components†

$$\left.\begin{aligned} (\mathbf{T} \times \overleftarrow{\nabla})_{11} &= \frac{\partial T_{12}}{\partial x_3} - \frac{\partial T_{13}}{\partial x_2} \\ (\mathbf{T} \times \overleftarrow{\nabla})_{12} &= \frac{\partial T_{13}}{\partial x_1} - \frac{\partial T_{11}}{\partial x_3} \\ (\mathbf{T} \times \overleftarrow{\nabla})_{21} &= \frac{\partial T_{22}}{\partial x_3} - \frac{\partial T_{23}}{\partial x_2}. \end{aligned}\right\} \tag{2.5.39}$$

Comparison with Eq. (37) shows that

$$\mathbf{T} \times \overleftarrow{\nabla} = -[\vec{\nabla} \times (\mathbf{T}^T)]^T. \tag{2.5.40}$$

†The other diagonal components [22 and 33] may be written down by ***cyclic permutation*** of the indices in the first Eq. (37), replacing 1 by 2, 2 by 3, and 3 by 1. Similarly the 23 and 31 components may be written down by cyclic permutation of the indices in the 12 component. However, the 13 and 32 components are obtained by ***acyclic permutation*** of the indices in the 21 component, replacing 1 by 3, 2 by 1, and 3 by 2.

This differs from the curl of a vector $\mathbf{v}$ where we always have

$$\mathbf{v} \times \overleftarrow{\nabla} = -\overrightarrow{\nabla} \times \mathbf{v}. \tag{2.5.41}$$

The divergences and curls have been introduced formally by operations with Cartesian forms of the symbolic operators $\overrightarrow{\nabla}$ and $\overleftarrow{\nabla}$, but the results obtained, for example, the scalar div $\mathbf{v}$ or the vector curl $\mathbf{v}$, have an existence independent of the coordinate system. The component forms in orthogonal curvilinear coordinates are given in App. II. They are considerably more complicated than the Cartesian forms because the base vectors are not constant, so that when the differential operator acts on a vector $\mathbf{v} = v_r\hat{\mathbf{e}}_r + v_\theta\hat{\mathbf{e}}_\theta + v_z\hat{\mathbf{e}}_z$ in cylindrical coordinates, for example, it is necessary to treat each term as a product and differentiate the base unit vectors $\hat{\mathbf{e}}_r$, $\hat{\mathbf{e}}_\theta$, $\hat{\mathbf{e}}_z$ as well as their coefficients v_r, v_θ, v_z.

This completes our introduction to vector and tensor differential calculus. Discussion of integral theorems will be postponed to Sec. 5.1, since they are not needed for the development of the concepts of stress, strain, and rate of deformation in Chaps. 3 and 4.

EXERCISES

1. Verify by using the chain rule of calculus that the three partial derivatives of the scalar point function $\bar{F}(\bar{x}, \bar{y}, \bar{z})$ of Eq. (2.5.16) are related to the partial derivatives of $F(x, y, z)$ by the vector component transformation formulas.

2. A force of magnitude F acts in a direction radially away from the origin at point $(a/3, 2b/3, 2c/3)$ on the surface of the ellipsoid $(x^2/a^2) + (y^2/b^2) + (z^2/c^2) = 1$. Determine the component of the force in the direction of the normal to the surface.

3. Show that if the A_{mn} are constants, $\nabla(A_{mn}x_m x_n) = (A_{pn} + A_{np})x_n \mathbf{i}_p$. *Hint:* Note that $\partial x_s/\partial x_p = \delta_{sp}$.

4. (a) through (j). Verify the identities of Eq. (2.5.27). *Hint for* (27j): Use the e-δ relation of Eq. (2.3.28) with components of $\nabla \times \mathbf{u}$ given by $e_{irs}\partial_r u_s$.

5. If $\mathbf{v} = (-y\mathbf{i} + x\mathbf{j})/(x^2 + y^2)$, show that curl $\mathbf{v} = 0$, but that $\oint \mathbf{v} \cdot d\mathbf{r} \neq 0$ around the circle $x^2 + y^2 = 1$. Why does this not violate the statement in connection with Eq. (2.5.28) that curl $\mathbf{v} = 0$ implies the line integral to be independent of the path between any two points and therefore to be zero around any closed path?

6. The velocity of point B in Fig. 2.9 is $\mathbf{v} = \mathbf{v}_A + \mathbf{v}_{B/A}$, where $\mathbf{v}_A$, the velocity of the reference point A, is not a function of the variable point B. Show that $\nabla \times \mathbf{v} = 2\boldsymbol{\omega}$.

7. Show that the rectangular components of the vector $\mathbf{v}\cdot\vec{\nabla}\mathbf{u}$, where the tensor $\vec{\nabla}\mathbf{u}$ is the gradient of $\mathbf{u}$ [see Eq. (2.5.32)], are the same as result from applying to $\mathbf{u}$ the operator obtained formally as the product $\mathbf{v}\cdot\nabla$, with ∇ defined by Eq. (23).
8. Write out, without summation conventions, the three component equations of

$$\vec{\nabla}\cdot\mathbf{T} + \rho\mathbf{b} = \rho\mathbf{a}.$$

9. Show that the triadics with components given by Eqs. (2.5.34b) yield by contraction Eq. (33).
10. Without using Eq. (2.5.40), verify that for rectangular Cartesians $(\mathbf{S} \times \overleftarrow{\nabla})_{31} = -(\vec{\nabla} \times \mathbf{S})_{13}$, if $\mathbf{S}$ is symmetric.
11. If $\mathbf{M} = \vec{\nabla} \times \mathbf{E}$, write out the rectangular Cartesian components M_{11}, M_{22}, M_{13}, and M_{31}.
12. Use the results of Ex. 11 to show that, if $\mathbf{S} = (\vec{\nabla} \times \mathbf{E}) \times \overleftarrow{\nabla}$, then $S_{12} = S_{21}$, if $\mathbf{E}$ is symmetric.
13. If $\mathbf{M}^* = \mathbf{E} \times \overleftarrow{\nabla}$, write out the rectangular Cartesian components M^*_{11}, M^*_{22}, M^*_{31}, M^*_{32}.
14. (a) Use results of Ex. 13 to show that if $\mathbf{S}^* = \vec{\nabla} \times (\mathbf{E} \times \overleftarrow{\nabla})$, then $S^*_{12} = S^*_{21}$ if $\mathbf{E}$ is symmetric.
    (b) Show also that $S^*_{12} = S_{12}$ when $\mathbf{E}$ is symmetric, where $\mathbf{S}$ is as defined in Ex. 12.

    Remark: In fact, for symmetric $\mathbf{E}$, the tensors $\mathbf{S}$ and $\mathbf{S}^*$ are equal, so that the parentheses in their definitions may be omitted, as in Eq. (4.7.5b).

Selected references on vectors and tensors

Aris, R., *Vectors, Tensors and the Basic Equations of Fluid Mechanics*. Englewood Cliffs, N. J.: Prentice-Hall, Inc., 1962.

Block, H. D., *Introduction to Tensor Analysis*. Columbus, Ohio: Charles E. Merrill Books, Inc., 1962.

Brillouin, L., *Tensors in Mechanics and Elasticity*. New York: Academic Press, Inc., 1964.

Coburn, N., *Vector and Tensor Analysis*. New York: The Macmillan Company, 1955.

Drew, T. B., *Handbook of Vector and Polyadic Analysis*. New York: Reinhold Publishing Corp., 1961.

Ericksen, J. L., "Tensor Fields," Appendix pp. 794-858, in *Encyclopedia of Physics*, v. III/1, ed. S. Flügge. Berlin: Springer-Verlag, 1960.

Greub, W., *Linear Algebra* (2nd. ed.). Berlin: Springer-Verlag, 1963.

Halmos, P. R., *Finite Dimensional Vector Spaces* (2nd ed.). Princeton, N. J.: Princeton University Press, 1958.

Hawkins, G. A., *Multilinear Analysis for Students in Engineering and Science.* New York: John Wiley & Sons, Inc., 1963.

Hay, G. E., *Vector and Tensor Analysis.* New York: Dover Publications, Inc., 1953.

Jaunzemis, W., *Continuum Mechanics,* Chapter 1. New York: The Macmillan Company, 1967.

Lass, H., *Vector and Tensor Analysis.* New York: McGraw-Hill Book Company, 1950.

Levi-Civita, T., *The Absolute Differential Calculus.* London and Glasgow: Blackie and Son, Ltd., 1927.

Lichnerowicz, A., *Elements of Tensor Analysis.* London: Methuen & Co., Ltd.; New York: John Wiley & Sons, Inc., 1962.

Michal, A. D., *Matrix and Tensor Calculus.* New York: John Wiley & Sons, Inc., 1947.

Moon, P., and D. E. Spencer, *Vectors.* Princeton, N. J.: D. Van Nostrand Co., Inc., 1965.

Morse, P. M., and H. Feshbach, *Methods of Mathematical Physics.* New York: McGraw-Hill Book Company, 1953.

Spiegel, M. R., *Theory and Problems of Vector Analysis and an Introduction to Tensor Analysis.* New York: Schaum Publishing Co., 1959.

Synge, J. L., and A. Schild, *Tensor Calculus.* Toronto: University of Toronto Press, 1959.

Weatherburn, C. E., *Advanced Vector Analysis with Applications to Mathematical Physics.* London: G. Bell & Sons, Ltd., 1928.

Weatherburn, C. E., *Elementary Vector Analysis with Applications to Geometry and Physics.* London: G. Bell & Sons, Ltd., 1931.

Wilson, E. B., *Vector Analysis.* New Haven: Yale University Press, 1901.

CHAPTER 3

Stress

3.1 Body forces and surface forces

Forces may be classified as external forces acting on a body or internal forces acting between two parts of the body. Such a classification is used, for example, to distinguish between the loads applied to a tension member and the internal forces which resist the tendency for one part of the member to be pulled away from another part. However, by a suitable choice of a free body imagined to be cut out of the member, any internal force in the original member may become an external force on the isolated free body.

The term ***free body*** is not limited to a solid body but may be used to denote a portion of fluid instantaneously bounded by an arbitrary closed surface. This closed surface may consist in part of an actual free surface of a liquid or an actual container wall, but it may be wholly or in part an imaginary surface within the fluid, as in the case of the bounding surface of a free body imagined to be cut out of a solid member.

The external forces acting at any instant on a chosen free body are classified in continuum mechanics in two kinds: body forces and surface forces.

Body forces act on the elements of volume or mass inside the body, e.g., gravity. These are "action-at-a-distance" forces. In the equations to be developed, these forces will usually be reckoned per unit mass or sometimes per unit volume.

Surface forces are contact forces acting on the free body at its bounding surface; these will be reckoned per unit area of the surface across which they act.

In mechanics, real forces are always exerted by one body on another body (possibly by one part of a body acting on another part), regardless of whether they are body forces or surface forces. Two bodies are always involved, and by Newton's Third Law the force exerted by one body on a second body is equal in magnitude and opposite in direction to the force exerted by the second body on the first. The so-called ***inertia forces*** used to create a fictitious state of equilibrium in dynamics are not real forces, since they are not exerted by bodies; Newton's Third Law does not apply to these fictitious forces. When the inertia-force method is used in continuum mechanics the fictitious inertia forces are included as body forces.

The body force per unit mass acting on an infinitesimal volume element dV of the body is denoted by the vector $\mathbf{b}$. The body force on the volume dV is $\rho\mathbf{b}\,dV$ with rectangular components $\rho b_k\,dV$, where ρ is the mass per unit volume.† Since the volume element dV is infinitesimal, the point of application of the force $\rho\mathbf{b}\,dV$ may be taken to be any point in the element of volume. The vector sum of all the body forces acting on a finite volume V is then given by the space integral over the volume

$$\int_V \rho\mathbf{b}\,dV = \mathbf{i}_k \int_V \rho b_k\,dV = \mathbf{i}\int_V \rho b_x\,dV + \mathbf{j}\int_V \rho b_y\,dV + \mathbf{k}\int_V \rho b_z\,dV. \quad (3.1.1)$$

In general, the vector $\mathbf{b}$ varies from point to point in the body at any given time and may also vary with time at any given point. The space integral gives the vector sum at the instant of time. The three component integrals may be evaluated when the components are known functions of x, y, z by placing suitable limits on the three integrals and writing $dV = dx\,dy\,dz$. Numerical integration may be required if these functions are not simple or if the shape of the finite volume is complicated. Fortunately, in many applications the body forces are either negligible or may be considered uniform, e.g., gravity. When the inertia-force method is used in dynamics problems, however, the fictitious inertia forces are usually neither negligible nor uniformly distributed.

The surface force per unit area, or ***traction***, will be denoted by the vector $\mathbf{t}$; the force exerted across an element dS of surface is then $\mathbf{t}\,dS$ and the vector sum of the forces across a finite portion S of the bounding surface will be given by the vector surface integral

$$\int_S \mathbf{t}\,dS = \mathbf{i}_k \int_S t_k\,dS = \mathbf{i}\int_S t_x\,dS + \mathbf{j}\int_S t_y\,dS + \mathbf{k}\int_S t_z\,dS. \quad (3.1.2)$$

When t_x, t_y, t_z are known as functions of suitable variables on the bounding surface, the integration over the surface can be evaluated as an iterated double integral. This is, however, not easy unless the geometry of the surface is simple. When the bounding surface is made up of parts of coordinate surfaces of a system of curvilinear (or rectangular) coordinates, the evaluation is usually feasible. Such surfaces are loci in space where one of the coordinates is constant. An example is one of the concentric spheres of a system of spherical polar coordinates, where the radial coordinate r is constant. The quantities t_x, t_y, t_z on the

†Another notation often used is $\mathbf{f}$ for body force per unit mass and $\mathbf{F}=\rho\mathbf{f}$ for body force per unit volume, but in this book we will denote the body force per unit volume by $\mathbf{f}=\rho\mathbf{b}$.

sphere may be expressed in terms of the azimuth angle ϕ and the colatitude θ, and the element of surface area is $dS = r^2 \sin\theta \, d\theta \, d\phi$. It may be more convenient to express the vector in terms of curvilinear components (see App. II). An even simpler example is the case where part of the surface is a portion of a plane $x =$ constant. On such a plane t_x, t_y, t_z are functions of y and z, and $dS = dy\,dz$.

Moments of body forces and surface forces. The moment **m** of a force **f** about a point A is defined by the vector product

$$\mathbf{m} = \mathbf{r} \times \mathbf{f} \qquad \mathbf{r} = \mathbf{p} - \mathbf{p}_A, \tag{3.1.3}$$

where **r** is the vector from the moment-center A to any point on the line of action of the force, and **p** denotes the position vector. If the moment-center is chosen as the origin, the moment is given by $\mathbf{p} \times \mathbf{f}$. The turning moment m_n about any axis through the moment-center, defined by a unit vector $\hat{\mathbf{n}}$ along the axis, is given by

$$m_n = \mathbf{m} \cdot \hat{\mathbf{n}}. \tag{3.1.4}$$

The moment of a force about any axis is equal to the algebraic sum of the moments of its components about that axis. For example, in Fig. 3.1 the mo-

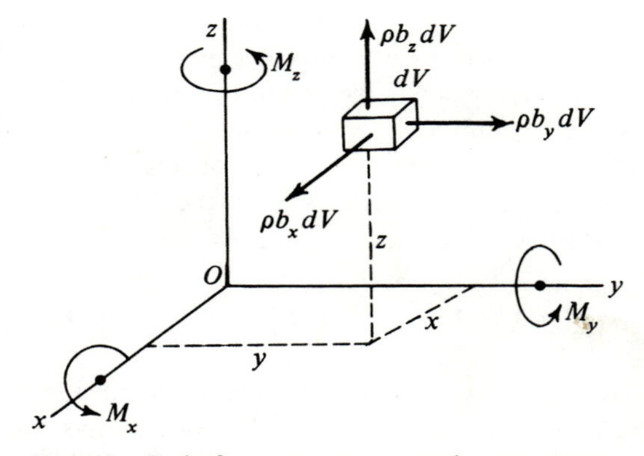

Fig. 3.1 Body-force components and moment-arms for volume element dV.

ment of the force $\rho \mathbf{b} dV$ about the z-axis is the algebraic sum of the moments of the components $\rho b_x dV$ and $\rho b_y dV$, since the component $\rho b_z dV$ is parallel to the z-axis and therefore has no moment with respect to the z-axis. If the moment is reckoned positive when it is counterclockwise as seen from the positive end of a coordinate axis, then the moments about the three axes are given by

$$(yb_z - zb_y)\rho\, dV, \quad (zb_x - xb_z)\rho\, dV, \quad (xb_y - yb_x)\rho\, dV,$$

which are equal to the rectangular components of the vector moment

$$(\mathbf{p} \times \mathbf{b})\rho\, dV$$

about the origin O. The total moment of all the body forces acting on a finite volume V is given by

$$\int_V (\mathbf{p} \times \mathbf{b})\rho\, dV = \mathbf{i}\int_V (yb_z - zb_y)\rho\, dV + \mathbf{j}\int_V (zb_x - xb_z)\rho\, dV + \mathbf{k}\int_V (xb_y - yb_x)\rho\, dV \tag{3.1.5}$$

or

$$\int_V (\mathbf{p} \times \mathbf{b})\rho\, dV = \mathbf{i}_r \int_V e_{rmn} x_m b_n \rho\, dV. \tag{3.1.6}$$

The moment of the surface forces on an element of area dS can be expressed in a similar fashion, and the total moment of the distributed force on a finite surface S can be obtained by means of a surface integral. The result for the vector moment about O is

$$\int_S (\mathbf{p} \times \mathbf{t})\, dS = \mathbf{i}_r \int_S e_{rmn} x_m t_n\, dS. \tag{3.1.7}$$

The total moment of all the exernal forces on a given body V bounded by a closed surface S is the sum of the two moment expressions for the body and surface forces, respectively.

Remarks on other notations and on the assumptions implied by our definitions. The notations (X, Y, Z) or (F_x, F_y, F_z) are often used for the components of body force per unit volume, and $(\bar{X}, \bar{Y}, \bar{Z})$ for surface force per unit area (traction) in place of our (t_x, t_y, t_z). If (X, Y, Z) are the components of body force ***per unit mass***, then $(\rho X, \rho Y, \rho Z)$ are the components per unit volume, where ρ is the density.

When applied to a given solid body, the volume and surface integrals should be taken over the space occupied instantaneously by the deformed configuration of the body. In many applications of elasticity and plasticity theories, however, the deformations and displacements are so small that with negligible error the integration can be carried out over the initial or undeformed configuration, and the density factor ρ may then also be taken as the initial density for calculating body forces per unit mass. Alternatively, the integrals may always be applied to a given fixed volume of space, which at different times does not usually contain the same material; this is the usual procedure in fluid mechanics, where the (possibly imaginary) closed surface is called a ***control surface***.

In this chapter, we consider only distributed body force per unit mass and surface forces per unit area. We exclude the possibility of ***distributed body couples*** in the interior and ***distributed surface couples*** per unit area. These will be considered briefly in Sec. 5.3; they do not occur in the usual classical formulations of continuum mechanics, where the only moments are the moments of forces. However, distributed couples as well as "couple-stress" have been important in some recent developments.

When the actual molecular structure of the material is considered on a submicroscopic scale, the forces we have called surface forces are seen also to be "action-at-a-distance" forces involving the short-range intermolecular forces of molecular particles near the imaginary boundary surface; the body forces involve longer-range forces acting on the particles in the interior of the volume. As long as the volume element is not too small and the number of particles in the element is large, our resultant body and surface forces may be interpreted as sums of these molecular forces. When we pass, however, to the infinitesimal limit dV or dS, we are dealing with a hypothetical concept, a ***continuum*** or ***continuous medium***, whose justification depends not on any study of actual materials in the small, but rather on the efficacy and utility of the concept in enabling us to describe and predict the behavior of actual materials in the large, i.e., the ***macroscopic*** behavior. Our definition of body force per unit volume ***at a point*** involves the tacit ***postulate*** that if a sequence of volumes is considered each containing the point and such that the volume ΔV of the nth member of the sequence approaches zero as n increases without bound (in such a way that the maximum dimension also tends to zero), then, if $\Delta \mathbf{f}$ is the total body force on ΔV, the quotient $\Delta \mathbf{f}/\rho\Delta V$ approaches a definite limiting value $\mathbf{b}$.

$$\lim_{n\to\infty}\left[\frac{\Delta \mathbf{f}}{\rho\Delta V}\right]_{(n)} = \mathbf{b}. \tag{3.1.8}$$

The postulate assumes that this limit exists at each point of the continuous medium and that the limit is independent of the particular sequence of volumes considered, i.e., we may use a sequence of concentric spheres or concentric cubes or any other sequence of volumes whose maximum dimension tends to zero, so that the volume shrinks to the point and not to a curve or a surface.

A similar postulate is implied by our definition of surface traction; in fact, an even stronger assumption is made here, as will be pointed out in the next section, where it will also be shown that the traction on each of the infinitely many interior surfaces passing through a point can be found if we know the tractions on only three mutually perpendicular surfaces through the point.

3.2 Traction or stress vector; stress components

The vector **t** of surface force per unit area acting on an element dS as discussed in Sec. 3.1, is called the ***traction***† or ***stress vector*** acting on the element. If the surface element is an imaginary one within the continuous medium, it will separate two portions of the medium, each exerting a traction on the other. We arbitrarily isolate a free body on one side of the surface element and fix our attention on the traction exerted on this body across the infinitesimal element dS. We select as the positive direction of the unit normal $\hat{\mathbf{n}}$ to dS the outward direction from the free body. (This does not mean that the force is necessarily outward; the vector $\hat{\mathbf{n}}$ is merely the geometric normal.)

The concept of the traction vector acting on a surface element dS with normal $\hat{\mathbf{n}}$ ***at a point*** Q in a continuous medium may be clarified by the following discussion. Consider the surface force $\Delta\mathbf{f}$ acting on a portion ΔS of the surface S of the volume V shown in Fig. 3.2. The vector $\Delta\mathbf{f}$ is the vector sum

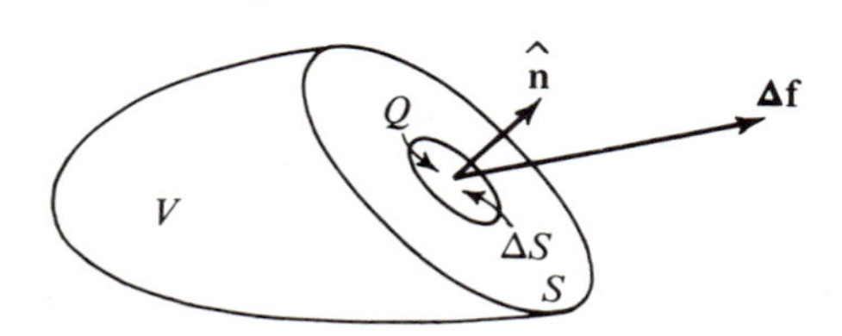

Fig. 3.2 Surface force $\Delta\mathbf{f}$ on area ΔS.

of the distributed force acting on ΔS. The ***average traction*** on ΔS is the vector $\Delta\mathbf{f}/\Delta S$, a vector having the same direction as $\Delta\mathbf{f}$ but with magnitude equal to $|\Delta\mathbf{f}|/\Delta S$. A sequence of portions of the surface S is selected so that the areas $\Delta S_1, \Delta S_2, \Delta S_3, \ldots$ approach zero with each area containing the point Q and with the greatest dimension also tending to zero, so that the area shrinks down to the point Q and not to a line or a curve. To each element ΔS of the sequence of areas corresponds a different force vector $\Delta\mathbf{f}$ acting on the area. The successive force vectors differ from each other, in general, not only in magnitude but also in direction; we assume, however, that the average traction vectors approach a definite limit which is defined to be the traction vector **t** (force per unit area) on the surface S ***at the point*** Q.

†Some authors limit the term ***traction*** to an actual bounding surface of a body and use only the term ***stress vector*** for an imaginary interior surface, but we prefer the term ***traction*** also for interior surfaces, since we wish to emphasize that the state of stress at a point is a tensor instead of a vector.

$$\mathbf{t} = \lim_{k\to\infty} \left[\frac{\Delta \mathbf{f}}{\Delta S}\right]_{(k)} \tag{3.2.1}$$

The numerator and denominator of the fraction each approach zero, but the fraction, in general, approaches a finite limit. It is a basic postulate of continuum mechanics that such a limit exists and is independent of the particular sequence of areas used. This postulate was tacitly assumed in the discussion of surface forces in Sec. 3.1, and a similar assumption was made in the discussion of body force per unit volume given there.

An even stronger hypothesis is made about the limit approached at Q by the surface force per unit area. Consider several different surfaces passing through Q all having the same normal $\hat{\mathbf{n}}$ at Q.

This is illustrated in Fig. 3.3, which shows a section through Q with each surface represented by its trace in the section. We can select the sequence of areas $\Delta S_1, \Delta S_2, \ldots$ on the surface S or on the surface S' or on S''. The assumption is that so long as the surfaces are all smooth with the same outward drawn normal at Q, the same limit $\mathbf{t}$ is approached. The limit is thus independent not only of the sequence of areas chosen on any one of the surfaces, but is also

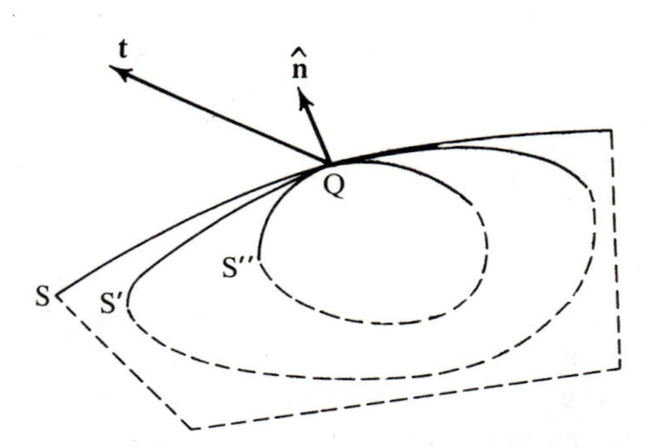

Fig. 3.3 Traction vector at Q is the same on surfaces S, S', S'' which have the same normal at Q.

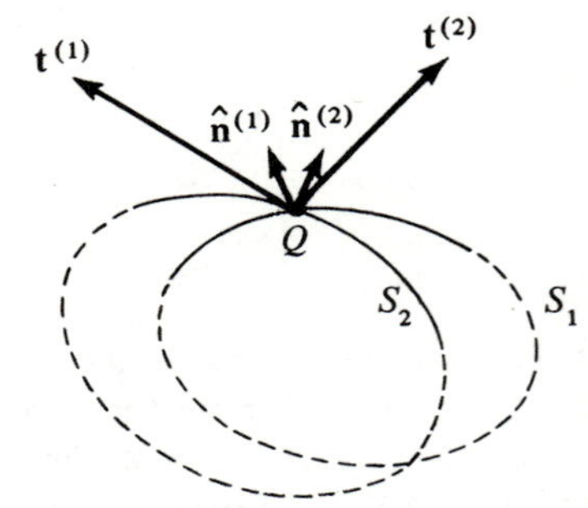

Fig. 3.4 Different traction vectors at Q on two surfaces having different normals.

independent of the surface chosen so long as they all have the same normal. The traction vector at the point Q does depend on the direction of the normal to the surfaces. If a new surface is chosen through Q with a different normal direction, the traction vector on the second surface at Q will be different from that on the first. This is illustrated in Fig. 3.4, which shows two surfaces S_1 and S_2 passing through Q with different normals, $\hat{\mathbf{n}}^{(1)}$ and $\hat{\mathbf{n}}^{(2)}$, and different traction vectors, $\mathbf{t}^{(1)}$ and $\mathbf{t}^{(2)}$, acting at Q. $\mathbf{t}^{(1)}$ is the force per unit area exerted at Q by the material on the positive side of S_1 acting on the material on the negative side (inside) of S_1. $\mathbf{t}^{(2)}$ is the force per unit area exerted at Q by the material on the positive side of S_2 acting on the material on the negative side of S_2.

We have considered only the force per unit area transmitted across a surface at a point and not a possible couple. When a continuous distribution of force acts across a finite area ΔS, the resultant of the distribution is, in general, a force and a couple. If this result is divided by ΔS and the limit taken as ΔS tends to zero, it is found that the couple per unit area produced at the point by the continuous distribution of force is zero. This does not, of course, preclude the possibility that there might also be a continuous distribution of couple, whose limit (per unit area) would be different from zero, a couple stress. Such couple stresses have in fact been proposed in continuum mechanics; a brief introduction to couple stresses will be given in Sec. 5.3. For the present, we will assume that there are no couple stresses, as in classical continuum mechanics, and that the action of one body on another across an infinitesimal surface area is adequately represented by the stress vector.

The traction vectors on planes perpendicular to the coordinate axes are especially useful because, when the vectors acting at a point on three such mutually perpendicular planes are given, the stress vector at that point on any other plane inclined arbitrarily to the coordinate axes can be expressed in terms of these three given special vectors. We denote by $\mathbf{t}^{(1)}$ the vector acting on a plane with normal pointing in the positive x_1-direction, while $\mathbf{t}^{(2)}$ and $\mathbf{t}^{(3)}$ denote stress vectors on planes with positive normals in the x_2- and x_3-directions. Figure 3.5 shows three such vectors acting at three different points on three faces of a block.

It should be remembered that the vector $\mathbf{t}^{(1)}$ acts on the positive x_1 side of the element, i.e., it is exerted by material lying on the positive x_1 side of the element. The stress vector acting on the negative side of the element, say on the left side of the cube in Fig. 3.5, will be denoted by $-\mathbf{t}^{(1)}$, since when the block has shrunk to a point Q, the vector on the left side at Q will be equal in magnitude and opposite in direction to the vector $\mathbf{t}^{(1)}$ at Q, by Newton's Third Law.

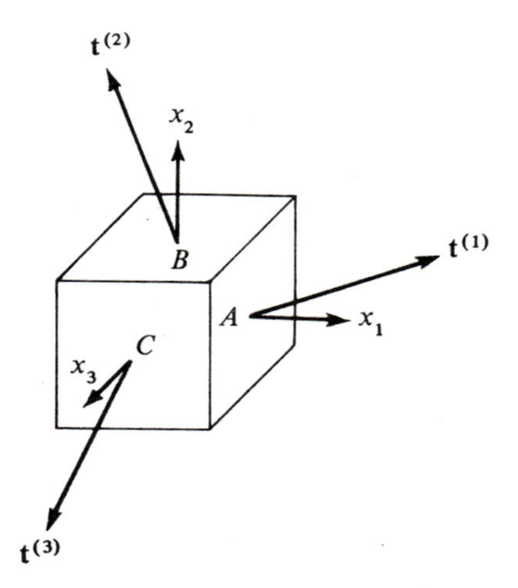

Fig. 3.5 Traction vectors on three planes perpendicular to coordinate axes.

Stress components. We shall see below that the stress vector on an arbitrary plane with normal $\hat{\mathbf{n}}$ at a point can be expressed in terms of the three stress vectors $\mathbf{t}^{(k)}$. These three vectors are thus a representation of the stress tensor $\mathbf{T}$ at the point, the linear vector function which associates with each argument unit vector $\hat{\mathbf{n}}$ the traction vector $\mathbf{t}^{(\mathbf{n})} = \hat{\mathbf{n}} \cdot \mathbf{T}$ acting across the surface whose normal is $\hat{\mathbf{n}}$. In fact, the nine rectangular components T_{ij} of $\mathbf{T}$ turn out to be the three sets of three vector components. Let (T_{11}, T_{12}, T_{13}) denote the components of $\mathbf{t}^{(1)}$, while (T_{21}, T_{22}, T_{23}) are the components of $\mathbf{t}^{(2)}$, and (T_{31}, T_{32}, T_{33}) are the components of $\mathbf{t}^{(3)}$, as illustrated in Fig. 3.6.

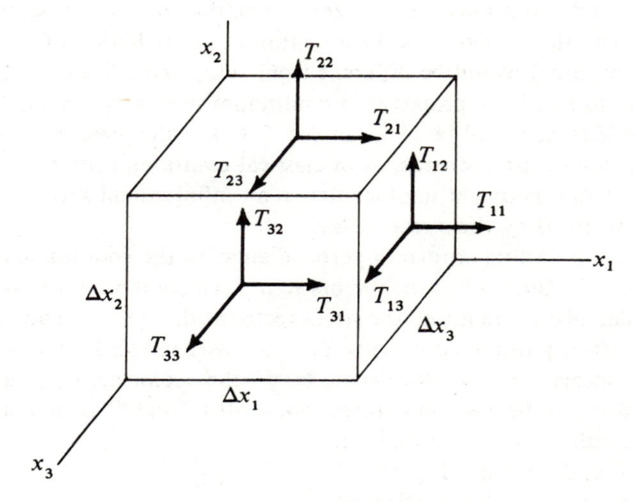

Fig. 3.6 Stress components.

Sign convention. When (T_{11}, T_{12}, T_{13}) are positive numbers, then the vector components of the vector $\mathbf{t}^{(1)}$ have the positive senses illustrated in Fig. 3.6. If any one of the components T_{11}, T_{12}, T_{13} is a negative number, the corresponding vector component points in the sense opposite to that illustrated. Similar sign conventions apply for the components of $\mathbf{t}^{(2)}$ and $\mathbf{t}^{(3)}$. Remember that the vector components on the negative sides of the block will have senses opposite to those on the positive side in any case. Thus when T_{11} is positive, the normal force pulls to the right on the right side of the block and pulls to the left on the left side; it is, therefore, a ***tensile*** stress. When T_{11} is negative, the normal stress is ***compressive.*** The algebraic sign of a tangential component (shear component) does not have an intrinsic physical significance like that of the normal stress; tension and compression are fundamentally different kinds of loading, while positive and negative shear stresses represent the same kind of loading in different directions. Positive T_{12}, for example, represents an upward-acting vector component on the right side and a downward-acting component on the left side of the block in Fig. 3.6. Negative T_{12}, on the other hand, would act down on the right side and up on the left.†

†In soil mechanics, it is common to reckon compressive stresses as positive and tension as negative. Then in order to get a consistent sign convention for all of the components the shear component sign convention is usually also reversed.

Stress state at a point. In Figs. 3.5 and 3.6, the three vectors are shown acting at three different points. They may be taken to represent the average traction vectors on the respective block faces, or alternatively they may represent the local stress at the point shown. In general, the traction vector at a different point on the same face will differ both in magnitude and direction from the one at the center of the face. Furthermore, if we imagine that the block shrinks down toward its center point Q (not shown in the figures), on successive positions of the block faces the stress vectors will be different even at the centers of the faces. The limit approached by the vector on each face will be the traction vector ***at the point*** Q on a plane perpendicular to one of the coordinate axes. The components of the three vectors at Q are usually displayed in one array or matrix T as follows:

$$T = \begin{bmatrix} T_{11} & T_{12} & T_{13} \\ T_{21} & T_{22} & T_{23} \\ T_{31} & T_{32} & T_{33} \end{bmatrix}. \tag{3.2.2}$$

Here the first row of the matrix T contains the components of the vector $\mathbf{t}^{(1)}$ on a plane through Q perpendicular to the x_1-axis; the second row is for a plane through Q perpendicular to the x_2-axis; and the third row is for a plane through Q perpendicular to the x_3-axis. In each row, the first subscript identifies the plane, while the second subscript identifies the component of the vector.†

We will now show how the traction vector on an arbitrary plane can be expressed in terms of $\mathbf{t}^{(1)}$, $\mathbf{t}^{(2)}$, $\mathbf{t}^{(3)}$ or in terms of the nine stress components T_{ij} displayed in the stress matrix, so that the entire state of stress at a point is determined when these nine components are given. The state of stress at a point cannot be specified completely by a single vector with three components; it requires a second-order tensor with nine components. (We shall see later that only six of the nine stress components are independently necessary to specify the state of stress at a point. This is because the symmetrically placed off-diagonal elements of the stress matrix are equal, i.e., $T_{21} = T_{12}$, $T_{31} = T_{13}$, $T_{32} = T_{23}$, when there are no distributed body or surface couples.)

Traction vector on an arbitrary plane, the Cauchy tetrahedron.†† Imagine that at a point O in a continuous medium a set of rectangular coordinate axes is drawn and a free body is chosen in the form of a tetrahedron or triangular pyramid bounded by parts of the three coordinate planes through O and a fourth plane ABC not passing through O, as shown in Fig. 3.7. The figure illustrates

†Some authors, e.g., Truesdell and Noll (1965), reverse this convention, using the first index for the vector component and the second index for the plane. Their stress tensor is then the transpose of the one defined here.

††These results are the work of A.-L. Cauchy in publications of 1823 and 1827.

the case in which the outward normal $\hat{\mathbf{n}}$ to the oblique plane points into the positive octant; the derivation given is for this case only, but the equations obtained are quite general, applying even to cases of planes parallel to one or two of the coordinate axes so that no tetrahedron is formed with the coordinate planes.

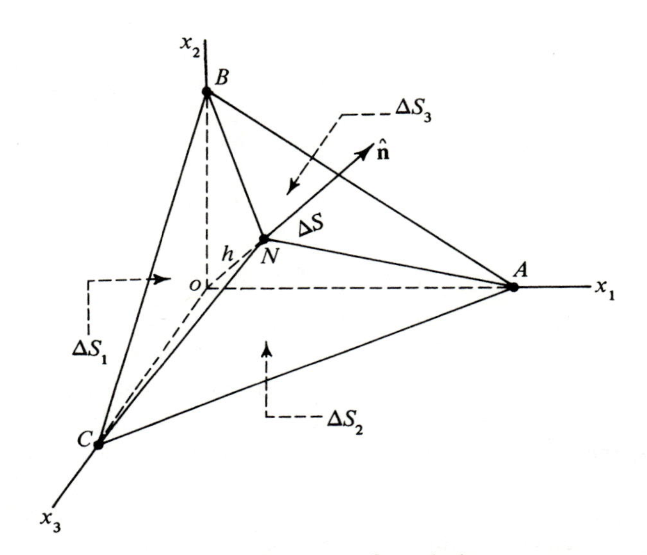

Fig. 3.7 Geometry of tetrahedron.

The components of the unit vector normal $\hat{\mathbf{n}}$ are the direction cosines of its direction. Thus in Fig. 3.7,

$$n_1 = \cos(\angle AON) \qquad n_2 = \cos(\angle BON) \qquad n_3 = \cos(\angle CON). \tag{3.2.3}$$

The altitude ON, of length h, is a leg of each of the three right triangles ANO, BNO, CNO with hypotenuses OA, OB, OC. Hence

$$h = OA\, n_1 = OB\, n_2 = OC\, n_3. \tag{3.2.4}$$

The volume of the tetrahedron is one-third of the base multiplied by the altitude. Considering each of the four bases in turn, four equivalent expressions for the volume ΔV are obtained,

$$\Delta V = \tfrac{1}{3}h\,\Delta S = \tfrac{1}{3}OA\,\Delta S_1 = \tfrac{1}{3}OB\,\Delta S_2 = \tfrac{1}{3}OC\,\Delta S_3. \tag{3.2.5}$$

Substituting for h the first expression of Eq. (4) we obtain from Eq. (5) the first equation below; the others are similarly obtained.

$$\Delta S_1 = \Delta S\, n_1, \qquad \Delta S_2 = \Delta S\, n_2, \qquad \Delta S_3 = \Delta S\, n_3$$

or in indicial notation:

$$\Delta S_i = \Delta S\, n_i. \tag{3.2.6}$$

These equations express the fact that three of the faces are projections of the oblique face onto the coordinate planes.

Figure 3.8 is a free-body diagram of the same tetrahedron, showing five vectors representing the resultant surface force on each of the four faces and

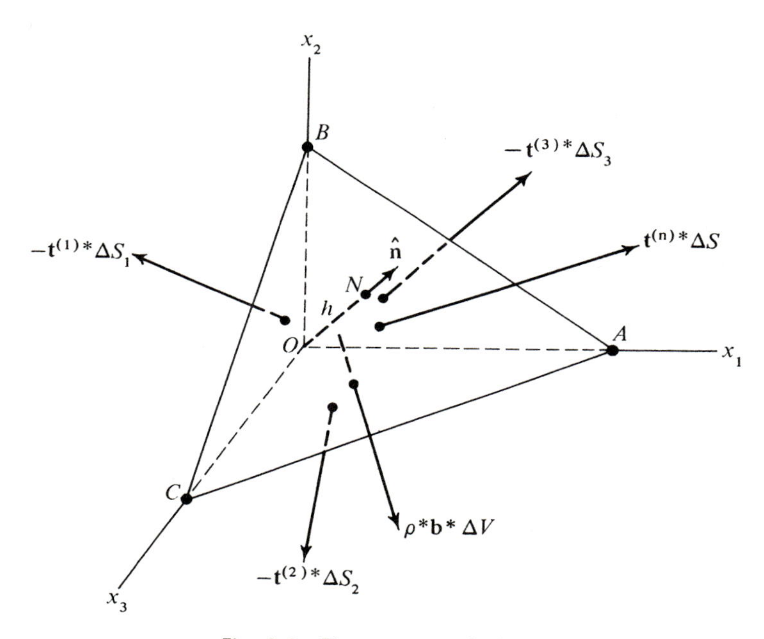

Fig. 3.8 Forces on tetrahedron.

the resultant body force $\rho^*\mathbf{b}^*\Delta V$. The asterisks indicate average values; thus $\mathbf{b}^*$ is the average value of the body force ***per unit mass*** in the tetrahedron. $\mathbf{t}^{(\mathbf{n})*}$ is the average value of the surface traction per unit area on the oblique face; $-\mathbf{t}^{(1)*}$ the average value of the surface traction per unit area on the area ΔS_1, etc. The negative sign appears because $\mathbf{t}^{(1)*}$ would denote the average traction on a surface whose outward normal pointed in the positive x_1-direction, while here the outward normal to ΔS_1 points in the negative x_1-direction. Similar comments apply to the other two faces.

An expression will be derived for $\mathbf{t}^{(\mathbf{n})*}$, and then the altitude h will be allowed to approach zero so that the volume and the four surface areas simultaneously approach zero, while the orientation of ON and the position of O do not change. We postulate the continuity of all the components of the stress vectors and the body force and the density as functions of position; it follows that the average values will approach the local values at the point O, and the result will be an expression for the traction vector $\mathbf{t}^{(\mathbf{n})}$ ***at the point O*** in terms of the three special surface stress vectors $\mathbf{t}^{(k)}$ at O.

The expression for $\mathbf{t}^{(\mathbf{n})*}$ is obtained by using the ***momentum principle of a collection of particles***, which is postulated to apply to our idealized continuous medium as well as to a collection of discrete idealized particles. This principle states that the vector sum of all the external forces acting on the free body (collection of particles) is equal to the rate of change of the total momentum. The total momentum is

$$\int_{\Delta V} \mathbf{v}\rho\, dV = \int_{\Delta m} \mathbf{v}\, dm,$$

where dm is the element of mass, and the integration applies to the mass instantaneously occupying ΔV. By the mean-value theorem of the integral calculus, this equals $\mathbf{v}^*\Delta m$, where $\mathbf{v}^*$ is the value of the velocity at some interior point. Since we are considering the momentum of a given collection of particles, Δm does not change with time; hence the time rate of change of the total momentum is $\Delta m\, d\mathbf{v}^*/dt = \rho^*\Delta V d\mathbf{v}^*/dt$, where ρ^* is the average density.

Then the equation expressing the principle of motion for the free body is

$$\mathbf{t}^{(\mathbf{n})*}\Delta S + \rho^*\mathbf{b}^*\Delta V - \mathbf{t}^{(1)*}\Delta S_1 - \mathbf{t}^{(2)*}\Delta S_2 - \mathbf{t}^{(3)*}\Delta S_3 = \rho^*\Delta V \frac{d\mathbf{v}^*}{dt}.$$

Substituting for ΔV, ΔS_1, ΔS_2, and ΔS_3 according to the first Eq. (5) and Eqs. (6), dividing through by ΔS, and rearranging terms we obtain

$$\mathbf{t}^{(\mathbf{n})*} + \frac{1}{3} h\rho^*\mathbf{b}^* = \mathbf{t}^{(1)*}n_1 + \mathbf{t}^{(2)*}n_2 + \mathbf{t}^{(3)*}n_3 + \frac{1}{3} h\rho^* \frac{d\mathbf{v}^*}{dt}.$$

We now let h approach zero. The last term in each member then approaches zero, while the vectors in the other terms approach the vectors at the point O, as is indicated by dropping the asterisks. The result is in the limit

$$\mathbf{t}^{(\mathbf{n})} = \mathbf{t}^{(1)}n_1 + \mathbf{t}^{(2)}n_2 + \mathbf{t}^{(3)}n_3 = \mathbf{t}^{(k)}n_k. \tag{3.2.7}$$

This important equation permits us to determine the traction $\mathbf{t}^{(\mathbf{n})}$ at a point, acting on an arbitrary plane through the point, when we know the tractions on only three mutually perpendicular planes through the point. Note that this result was obtained without any assumption of equilibrium. It applies just as well in fluid dynamics as in solid mechanics.

Equation (7) is a vector equation; the sum in the right member represents vector addition. The corresponding algebraic equations for the components of $\mathbf{t}^{(\mathbf{n})}$ are

$$
\begin{aligned}
t_1^{(\mathbf{n})} &= T_{11}n_1 + T_{21}n_2 + T_{31}n_3,\\
t_2^{(\mathbf{n})} &= T_{12}n_1 + T_{22}n_2 + T_{32}n_3,\\
t_3^{(\mathbf{n})} &= T_{13}n_1 + T_{23}n_2 + T_{33}n_3,\\
\text{or in indicial notation:}\quad t_i^{(\mathbf{n})} &= T_{ji}n_j,\\
\text{in matrix notation:}\quad t^{(\mathbf{n})} &= nT = T^T n,\\
\text{in dyadic notation:}\quad \mathbf{t}^{(\mathbf{n})} &= \hat{\mathbf{n}}\cdot\mathbf{T} = \mathbf{T}^T\cdot\hat{\mathbf{n}}.
\end{aligned}
\tag{3.2.8}
$$

In matrix notation, n must be regarded as a row matrix when it premultiplies the square matrix T and as a column matrix when it follows T^T.

We have thus established that the nine components T_{ij} are components of a second-order tensor, the ***stress tensor*** of Cauchy. The stress tensor is the linear vector function which associates with each argument unit normal vector $\hat{\mathbf{n}}$ the traction vector $\mathbf{t}^{(\mathbf{n})}$ acting at the point across the surface whose normal is $\hat{\mathbf{n}}$.

The components $\bar{T}_{ij}$ in any other rectangular Cartesian system, rotated with respect to the first system, are similarly defined in terms of the three traction vectors acting across the three planes perpendicular to the new coordinate directions, and will give the components $\bar{t}_i$ of the same traction vector $\mathbf{t}$ referred to the new axes by the same Eqs. (8) written with overbars—e.g., $\bar{t}_i^{(\mathbf{n})} = \bar{T}_{ji}\bar{n}_j$, where the $\bar{n}_j$ are the new components of the same $\hat{\mathbf{n}}$. Since $\mathbf{T}$ is a second-order tensor (linear vector function), it follows that the components $\bar{T}_{ij}$ are related to T_{pq} by the tensor transformation equations [Eqs. (2.4.11) and (2.4.12)].

$$
\begin{aligned}
&\bar{T}_{ij} = a_i^p a_j^q T_{pq} \qquad T_{pq} = a_i^p a_j^q \bar{T}_{ij}\\
&\text{or in matrix notation}\\
&\bar{T} = A^T T A \qquad T = A\bar{T}A^T,
\end{aligned}
\tag{3.2.9}
$$

where A is the matrix of direction cosines $a_k^r = \cos(\bar{x}_k, x_r)$ of the angles between the $\bar{x}_k$- and x_r-axes, and A^T is its transpose.

Symmetry of stress tensor, conjugate shear stresses. The stress matrix of Eq. (2) displays in each row the components of the stress vector acting on a plane perpendicular to a coordinate axis. When there are no distributed body or surface couples, the stress matrix is symmetric, i.e., the terms symmetrically placed with respect to the main diagonal of the matrix are equal, so that only six of the nine components are specified independently in order to define completely the state of stress at the point. The equality of symmetrically placed off-diagonal elements then means that

$$T_{21} = T_{12}, \qquad T_{31} = T_{13}, \qquad T_{32} = T_{23}. \tag{3.2.10}$$

The proof that the stress matrix is symmetric will be given here only for a special case of ***homogeneous*** (i.e., uniform) stress in a body in equilibrium. However, the conclusion is valid under much more general conditions, including dynamic as well as equilibrium conditions, and is not restricted to homogeneous states of stress. The proof for the general case is given in Sec. 5.3.

The nine stress components are represented in Fig. 3.6 acting on a rectangular element whose dimensions are Δx_1, Δx_2, Δx_3. If we assume that the stress state is homogeneous, the resultant force exerted on the positive x_1 face has components

$$T_{11}\Delta x_2 \Delta x_3, \qquad T_{12}\Delta x_2 \Delta x_3, \qquad T_{13}\Delta x_2 \Delta x_3$$

and acts at the center of the face, while the force on the positive x_2 side of the element has components

$$T_{21}\Delta x_3 \Delta x_1, \qquad T_{22}\Delta x_3 \Delta x_1, \qquad T_{23}\Delta x_3 \Delta x_1$$

and acts at the center of that face. Consider moment equilibrium about the x_3-axis. Since the stress is assumed to be homogeneous, the normal force on the negative side of the element is equal, opposite, and collinear to that on the positive face so that their moments cancel. The moment $(\Delta x_2/2)\,T_{31}\Delta x_1 \Delta x_2$ of the x_1-component acting on the front face is likewise balanced by the moment of an equal component on the back face, which is oppositely directed and has the same moment-arm $\Delta x_2/2$ with respect to the x_3-axis. The net moment about the x_3-axis is seen to be $\Delta x_1(T_{12}\Delta x_2 \Delta x_3) - \Delta x_2(T_{21}\Delta x_3 \Delta x_1)$, omitting any body forces acting on the element. Since we have assumed the element to be in equilibrium, this moment must be zero, whence $T_{12} = T_{21}$. The other equations (10) may be derived similarly. The omission of body forces does not invalidate the result, since the moment of the body force would vanish in the limit as Δx_1, Δx_2, Δx_3 tend to zero, because the body-force moment term involves the product of four factors of the order of the element dimensions instead of three factors as in the terms retained. However, the presence of distributed body couples or couple stresses will make the stress matrix nonsymmetric (see Sec. 5.3).

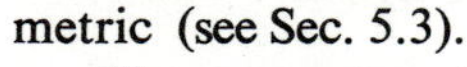

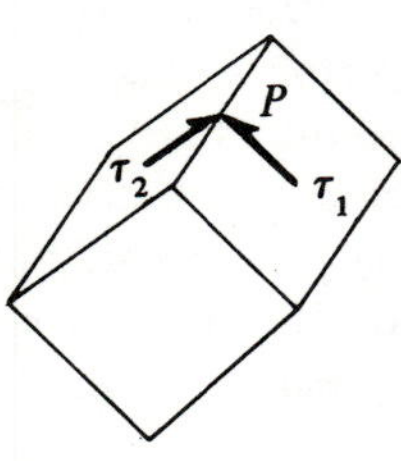

Fig. 3.9 Conjugate shear stresses.

The symmetry of the stress matrix implies what is sometimes called the ***theorem of conjugate shear stresses***, which states that when two planes intersect at right angles as in Fig. 3.9, that component of the shear stress on one of the planes which is perpendicular to the line of intersection is equal to the similar shear component on the other plane, $\tau_1 = \tau_2$. The two components either both act toward the line of intersection or both act away from it. Since coordinate axes at P may be chosen in such a way that the x_1- and x_2-axes are parallel, respectively, to the two conjugate components, the symmetry of the matrix proves the theorem.

A more general ***projection theorem***† states that if $\mathbf{t}^{(1)}$ is the stress vector on a plane with normal $\hat{\mathbf{n}}^{(1)}$, and $\mathbf{t}^{(2)}$ is the stress vector at the same point on a plane normal to $\hat{\mathbf{n}}^{(2)}$ as in Fig. 3.4, where $\hat{\mathbf{n}}^{(1)}$ and $\hat{\mathbf{n}}^{(2)}$ are in general not perpendicular, then

$$\mathbf{t}^{(1)} \cdot \hat{\mathbf{n}}^{(2)} = \mathbf{t}^{(2)} \cdot \hat{\mathbf{n}}^{(1)}. \tag{3.2.11}$$

Using matrix notation as in Eq. (8), we have

$$t^{(1)} = n^{(1)} T \quad \text{and} \quad t^{(2)} = n^{(2)} T$$

where $n^{(1)}$ and $n^{(2)}$ are the row matrices of components of the two unit vectors. The scalar products can also be represented as matrix products, e.g., $\mathbf{t}^{(1)} \cdot \hat{\mathbf{n}}^{(2)} = t^{(1)} n^{(2)}$ where now $n^{(2)}$ is the column matrix of the same components of $\hat{\mathbf{n}}^{(2)}$. The proof of Eq. (11) then requires that

$$[n^{(1)} T]\, n^{(2)} = [n^{(2)} T]\, n^{(1)}$$

which will be true if T is a symmetric matrix, as the reader can verify by writing out the components.

Other notations for stress components. The components of $\mathbf{t}^{(1)}$ are given by several writers as X_x, Y_x, Z_x with the subscripts identifying the plane on which the vector acts, while the capital letters identify the components. The vector $\mathbf{t}^{(2)}$ then has components X_y, Y_y, Z_y while $\mathbf{t}^{(3)}$ has components X_z, Y_z, Z_z.

When indicial notation is used, any symbol may be used to carry the subscripts. Commonly occurring notations are σ_{ij}, τ_{ij}, t_{ij}, P_{ij}, all representing the same thing as our T_{ij}. In curvilinear components the notations T^{ij}, T_{ij}, $T^{i}_{\cdot j}$, and $T_i^{\cdot j}$ represent different kinds of curvilinear components (see App. I) of the same tensor $\mathbf{T}$, but in rectangular Cartesian components they all mean the same thing.

Some authors use the second subscript to identify the plane, while the first subscript identifies the component. With that convention they would use $\mathbf{t} = \mathbf{T} \cdot \hat{\mathbf{n}}$ or $t_i = T_{ij} n_j$ instead of our Eq. (8), since their $\mathbf{T}$ is the transpose of ours.

A very common notation in American engineering publications, which we will use occasionally in this book, is illustrated in Fig. 3.10. We will frequently use this or the indicial notation σ_{ij} with the matrices of rectangular Cartesian components:

$$\begin{bmatrix} \sigma_x & \tau_{xy} & \tau_{xz} \\ \tau_{yx} & \sigma_y & \tau_{yz} \\ \tau_{zx} & \tau_{zy} & \sigma_z \end{bmatrix} \quad \text{or} \quad \begin{bmatrix} \sigma_{11} & \sigma_{12} & \sigma_{13} \\ \sigma_{21} & \sigma_{22} & \sigma_{23} \\ \sigma_{31} & \sigma_{32} & \sigma_{33} \end{bmatrix}. \tag{3.2.12}$$

† Also known as the reciprocal theorem of Cauchy (1829).

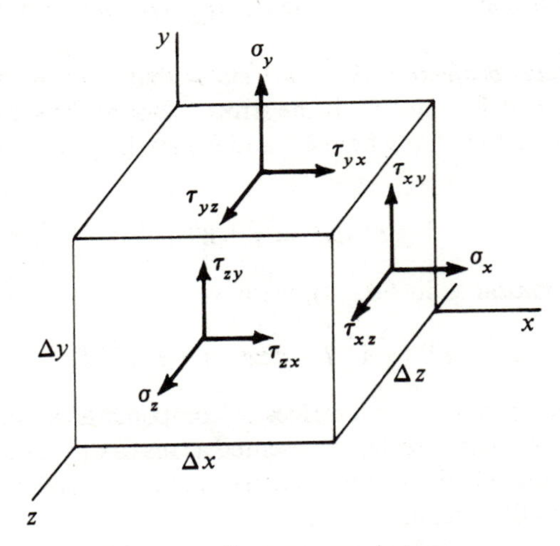

Fig. 3.10 Stress components.

The three rectangular components of the vector on the plane perpendicular to the x-axis are denoted by $(\sigma_x, \tau_{xy}, \tau_{xz})$. A normal component (in this case σ_x) is denoted by a Greek lower-case sigma, tangential or shear components (τ_{xy} and τ_{xz}) by the Greek lower-case tau. The first subscript on τ_{xy} or τ_{xz} identifies the plane on which the stress vector acts, while the second subscript identifies the component direction. A consistent notation for the normal stress would be τ_{xx}, but most American engineering writers use σ instead of τ for a normal component and do not repeat the subscript.

Several useful tensor properties will be presented in Sec. 3.3 for the stress tensor. The material there, including principal axes, invariants, and spherical and deviatoric tensors, is also applicable to any other symmetric second-order Cartesian tensor; it will be used again in Chap. 4 in the discussion of the strain tensor.

EXERCISES

1. Surface force per unit area **t** on the surface of hemisphere ABC is always directed toward the center O with magnitude Kz psi, where K is a known constant. (See diagram at top of page 81.) Find by integration the z-component of the total surface force on the hemisphere. If this is balanced by uniformly distributed body force ***in the whole sphere***, what is the magnitude ρb of the body force per unit volume? See Eqs. (3.1.1) and (3.1.2).

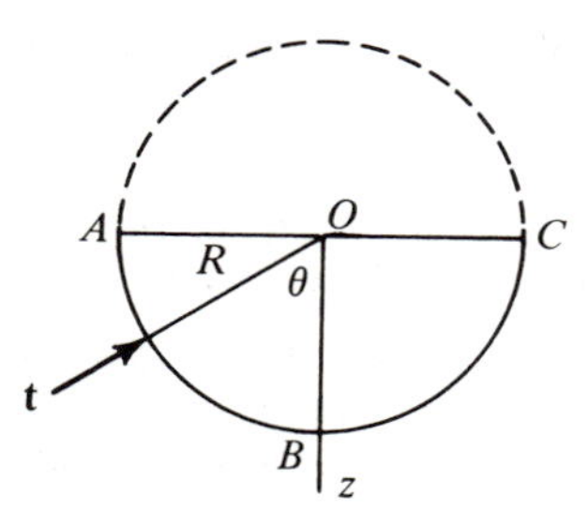

2. At a point the stress matrix, referred to xyz-axes, has the components shown below (kpsi). Find:
   (a) the three rectangular components of the traction vector acting on a plane through the point with unit normal $(2/3, -2/3, 1/3)$,
   (b) the magnitude of the traction vector of (a),
   (c) its component in the direction of the normal,
   (d) the angle between the traction vector and the normal.

$$\begin{bmatrix} 36 & 27 & 0 \\ 27 & -36 & 0 \\ 0 & 0 & 18 \end{bmatrix}$$

3. The stress distribution, according to the linear theory of elasticity, for the concentrated edge loading shown on a semi-infinite plate $x \geq 0, 0 \leq z \leq d$, is

$$T_{xx} = -\frac{2p}{\pi}\frac{\cos^3\theta}{r} \qquad T_{yy} = -\frac{2p}{\pi}\frac{\sin^2\theta\cos\theta}{r}$$

$$T_{xy} = T_{yx} = -\frac{2p}{\pi}\frac{\sin\theta\cos^2\theta}{r}.$$

All other stresses are zero under the assumption of plane stress. The total edge load is $P = pd$. Find the traction vector xyz-components on a plane perpendicular to the radius vector at a typical point (r, θ). Evaluate for $\theta = \pi/4$ and for $\theta = -\pi/6$.

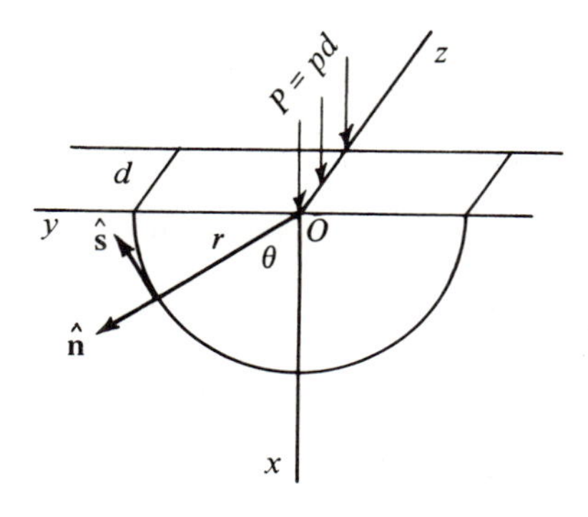

4. Find the normal component and the tangential component of the traction vector of Ex. 3, that is, the components in the directions of $\hat{\mathbf{n}}$ and $\hat{\mathbf{s}}$, and evaluate them at $\theta = \pi/4$ and at $\theta = -\pi/6$.

5. Show that the traction distribution on the half-cylinder $r = a$ of Ex. 3 is in equilibrium with the edge load, for any value of a.

6. If $\hat{\mathbf{u}}$ is a unit vector and T is a positive number with dimensions of stress, show that the stress tensor $\mathbf{T} = T\hat{\mathbf{u}}\hat{\mathbf{u}}$ represents uniaxial tension in the direction of $\hat{\mathbf{u}}$. Determine the traction vector $\mathbf{t}$ and its normal component t_n on a surface whose unit normal makes angle θ with $\hat{\mathbf{u}}$. What is the magnitude of the tangential component of the traction vector on the same surface?

7. Make the derivation of Eq. (3.2.7) for the case that $\hat{\mathbf{n}}$ points into the positive quadrant of the x_1x_2-plane by replacing the tetrahedron of Fig. 3.8 by the triangular prism shown below with altitudes h parallel to $\hat{\mathbf{n}}$ and to the x_3-axis.

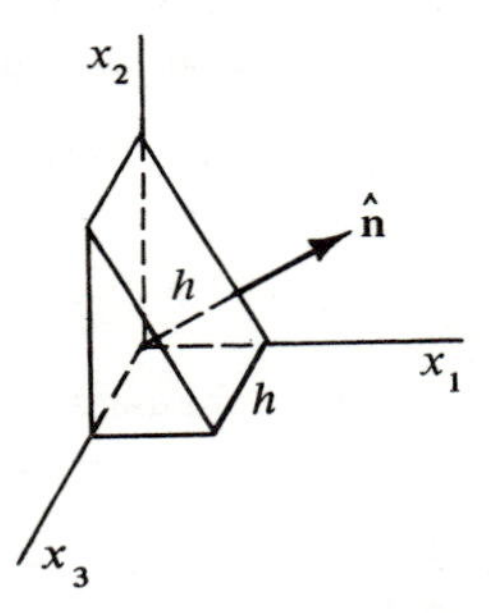

8. The matrix of rectangular Cartesian components of stress at a point is shown below, except that T_{11} is not given (units kpsi). Choose T_{11} so that there will be a traction-free plane through the point, and determine the unit normal $\hat{\mathbf{n}}$ of the traction-free plane.

$$\begin{bmatrix} T_{11} & 2 & 1 \\ 2 & 0 & 2 \\ 1 & 2 & 0 \end{bmatrix}$$

9. (a) For the curve C in the xy-plane in the diagram below, show that the unit tangent $\hat{\mathbf{s}}$ in the direction of increasing arc-length parameter s, and the unit normal forming with it a right-handed ns-system, are given by

$$\hat{\mathbf{s}} = \frac{dx}{ds}\mathbf{i} + \frac{dy}{ds}\mathbf{j} \qquad \text{and} \qquad \hat{\mathbf{n}} = \frac{dy}{ds}\mathbf{i} - \frac{dx}{ds}\mathbf{j}.$$

(b) For twisting of a prismatic bar whose cross-section is parallel to the xy-plane, it is assumed that the only stress components are $T_{zx} = T_{xz}$ and $T_{zy} = T_{yz}$ and that these are related to a scalar stress function $\phi(x, y)$ by

$$T_{zx} = \frac{\partial \phi}{\partial y} \qquad T_{zy} = -\frac{\partial \phi}{\partial x}.$$

Determine the boundary condition imposed on the function $\phi(x, y)$ on the bounding curve C of a typical cross-section by the requirement that the lateral surface of the bar be traction-free.

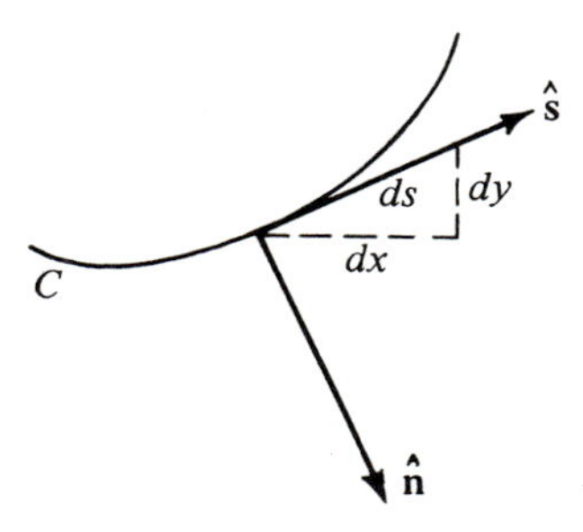

10. If the rectangular components of stress are as in the matrix below, determine the unit normal of a plane parallel to the x_3-axis on which the traction vector is tangential to the plane.

$$\begin{bmatrix} a & 0 & d \\ 0 & b & e \\ d & e & c \end{bmatrix}$$

11. If a state of plane stress parallel to the xy-plane is determined in terms of a scalar stress function $\phi(x, y)$ by

$$T_{xx} = \frac{\partial^2 \phi}{\partial y^2} \qquad T_{yy} = \frac{\partial^2 \phi}{\partial x^2} \qquad T_{xy} = T_{yx} = \frac{-\partial^2 \phi}{\partial x\, \partial y}$$

with all other components zero, express the traction components t_x and t_y on a bounding curve C [see Ex. 9(a)] in terms of the variation of $\partial\phi/\partial x$ and $\partial\phi/\partial y$ along the curve.

12. A state of plane stress has the components σ_x, σ_y, and $\tau_{xy} = \tau_{yx}$ in the notation of Fig. 3.10. Determine the normal component of the traction on a plane with $\hat{\mathbf{n}} = \cos\alpha\ \mathbf{i} + \sin\alpha\ \mathbf{j}$. What is the tangential component of the same traction vector in the direction $\hat{\mathbf{s}} = \cos(\alpha + \pi/2)\ \mathbf{i} + \sin(\alpha + \pi/2)\ \mathbf{j}$?

13. The stress matrix T referred to xyz-axes is shown below at the left (in kpsi). New $\bar{x}\bar{y}\bar{z}$-axes are chosen by rotating the axes as shown at the right.

$$T = \begin{bmatrix} 50 & 37.5 & 0 \\ 37.5 & -25 & -50 \\ 0 & -50 & 25 \end{bmatrix}$$

	$\bar{x}$	$\bar{y}$	$\bar{z}$
x	$\frac{3}{5}$	0	$\frac{4}{5}$
y	0	1	0
z	$-\frac{4}{5}$	0	$\frac{3}{5}$

(a) Determine the traction vectors on each of the new coordinate planes ***in terms of components referred to the old axes.*** For example, determine $\mathbf{t}^{(\bar{x})}$ in the form $\mathbf{t}^{(\bar{x})} = (\quad)\mathbf{i} + (\quad)\mathbf{j} + (\quad)\mathbf{k}$.

(b) Now project each of the vectors obtained in (a) onto the three new coordinate axes, and verify that the nine new components thus obtained for the stress matrix $\bar{T}$ are the same as those given by the formulas for transformation of a second-order tensor under rotation of axes.

14. Verify that the symmetry of the stress matrix implies Eq. (3.2.11).

15. A certain straight bar has elliptical cross-sections $x^2/a^2 + y^2/b^2 = 1$ with centers on the z-axis. If the stress produced by elastic bending and torsion is $T_{zz} = Ay$ plus torsional stresses given [as in Ex. 9(b)] in terms of the stress function $\phi = B(x^2/a^2 + y^2/b^2 - 1)$, where A and B are constants, what is the vector moment about the origin of the tractions on the cross-section through the origin with normal $\hat{\mathbf{n}} = \mathbf{k}$? Verify that the total force on the cross-section is zero.

16. The matrix of rectangular Cartesian stress components at a point is shown below (kpsi). Determine a direction $\hat{\mathbf{n}}$ such that the traction vector on a plane normal to $\hat{\mathbf{n}}$ has $t_x = t_y = 0$, and determine t_z on that plane.

$$\begin{bmatrix} 2 & 0 & 4 \\ 0 & 3 & 6 \\ 4 & 6 & 0 \end{bmatrix}$$

17. Cylindrical coordinates (r, θ, z) are chosen, related to the Cartesians by $x = r\cos\theta$, $y = r\sin\theta$, $z = z$. At a point P on the surface of the cylinder $r =$ constant, new local Cartesian axes are chosen with $\bar{x}$ in the radial direction, $\bar{y}$ in the transverse direction of increasing θ, and $\bar{z}$ in the z-direction.
    (a) Display the matrix A for calculating $\bar{T}$ in terms of T by Eqs. (3.2.9).
    (b) Express the components $t_{\bar{x}}, t_{\bar{y}}, t_{\bar{z}}$ of the traction vector on the cylindrical surface at the point P in terms of the elements of the original (unbarred) stress matrix T.

18. Spherical coordinates (r, θ, ϕ) are chosen, related to the Cartesians by $x = r\sin\theta\cos\phi$, $y = r\sin\theta\sin\phi$, $z = r\cos\theta$. At a point P on the surface of a sphere $r =$ constant, new local Cartesian axes are chosen with $\bar{x}$ in the radial direction, $\bar{y}$ in the downward meridional direction of increasing colatitude θ, and $\bar{z}$ in the direction of increasing azimuth angle ϕ.
    (a) Display the matrix A for calculating $\bar{T}$ in terms of T by Eqs. (3.2.9).
    (b) Express in terms of the elements of the original (unbarred) stress matrix T the components $t_{\bar{x}}, t_{\bar{y}}, t_{\bar{z}}$ of the traction vector acting at P on the material inside the sphere (exerted from the outside).
    (c) Repeat (b) for the conical surface $\theta =$ constant through P with the "outside" defined by the direction of increasing θ.

19. The rotation of axes from x, y, z to $\bar{x}, \bar{y}, \bar{z}$ is specified in terms of Euler's angular coordinates for rotation of a rigid body as follows. Suppose that the barred axes are fixed to an imaginary rigid body and rotate with it after initially coinciding with the unbarred axes: (1) Rotate about z-axis through angle ϕ to obtain posi-

tions x_1, y_1, z_1; (2) then rotate about x_1-axis through θ to obtain x_2, y_2, z_2; and finally rotate about z_2 through ψ. For the final position:

(a) Express $\bar{x}, \bar{y}, \bar{z}$ in terms of $x, y, z, \phi, \theta, \psi$.

(b) Display the matrix for calculating $\bar{T}$ in terms of T by Eqs. (3.2.9).

20. ***Plane of Symmetry of Stress.*** A stress state in a body is said to be symmetric with respect to a plane if on each pair of (imaginary internal) surfaces symmetrically placed with respect to the plane of symmetry, so that each is a mirror image of the other, the traction vectors at corresponding points are also mirror images. For example, in the figure shown, S and S^* are mirror images with respect to plane OO' (all planes are shown in edge view), and the traction vectors $\mathbf{t}$ and $\mathbf{t}^*$ are mirror images. There are no shear tractions acting on a plane of symmetry. For example, at point P in the figure, if $\mathbf{a}$ were the traction vector exerted across

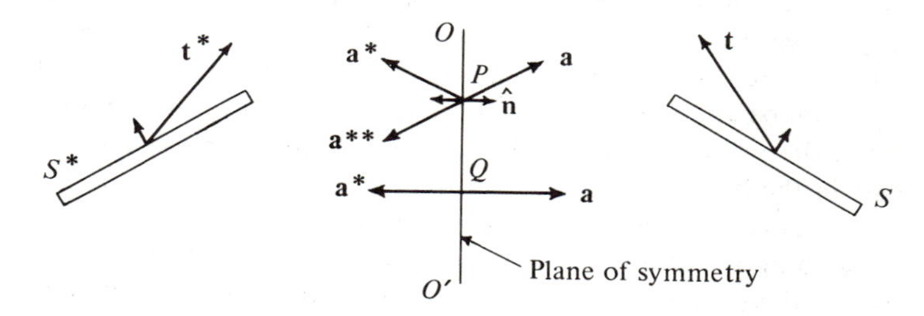

the plane by the material on the right, symmetry would require that the material on the left exert the traction $\mathbf{a}^*$, but Newton's Third Law requires instead the equal and opposite reaction $\mathbf{a}^{**}$. These two conditions can be met only if $\mathbf{a}$ and $\mathbf{a}^*$ are perpendicular to the plane of symmetry as shown at Q, and hence there is no shear traction on the plane of symmetry. If the yz-plane is a plane of symmetry of stress,

(a) What are the possible nonzero components of the stress matrix at a point in the yz-plane?

(b) Which components of the stress matrix must be even functions of x, and which components must be odd functions of x?

(c) Of which components is the partial derivative with respect to x necessarily equal to zero at points in the yz-plane of symmetry?

3.3 Principal axes of stress and principal stresses; invariants; spherical and deviatoric stress tensors

Regardless of the state of stress (so long as the stress tensor is symmetric) at a given point, it is always possible to choose a special set of axes through the point so that the shear stress components vanish when the stress components are referred to this system of axes. These special axes are called ***principal axes*** of stress or ***principal directions.*** On a plane perpendicular to a principal axis, the traction vector is entirely normal; it is therefore equal to a number multiplied by the unit normal. The three planes through the point perpendicular to the three principal axes are called ***principal planes***; the normal stress components on the three principal planes are called ***principal stresses*** and will be denoted by $\sigma_1, \sigma_2, \sigma_3$ or sometimes T_1, T_2, T_3. The traction vector magnitudes on the principal planes are equal to the absolute values of $\sigma_1, \sigma_2, \sigma_3$. The algebraically greatest of the three principal stresses is the algebraically greatest normal stress component acting on any plane through the point, and the algebraically smallest of the principal stresses is the algebraically smallest normal stress component on any plane through the point. That there is always such a set of three mutually perpendicular directions at a point can be inferred from the fact that stress is a symmetric second-order tensor, since this is a property common to all tensors of this type (see Sec. 2.4). It must be remembered that all of these statements apply to the stress at any one point; the principal directions at another point will in general be different. If the state of stress is uniform throughout a certain body, then the principal directions and the principal stresses are the same at all points in the body.

The principal directions at any point may be found as follows when the nine stress components T_{rs} are known in some Cartesian coordinate system. Let $\hat{\mathbf{n}}$ be a unit vector in one of the unknown directions sought with components n_i, and let λ represent the principal-stress component on the plane whose normal is $\hat{\mathbf{n}}$. (Both λ and $\hat{\mathbf{n}}$ are unknowns at first.) Since there is to be no shear stress component on the plane perpendicular to $\hat{\mathbf{n}}$, the stress vector on this plane must be parallel to $\hat{\mathbf{n}}$:

$$\mathbf{t}^{(\hat{\mathbf{n}})} = \lambda\hat{\mathbf{n}}. \tag{3.3.1a}$$

Using Eq. (3.2.8) and denoting the stress tensor by $\mathbf{T}$ then gives

$$\hat{\mathbf{n}}\cdot\mathbf{T} = \lambda\hat{\mathbf{n}}. \tag{3.3.1b}$$

In indicial notation, Eq. (1b) takes the form

$$n_r T_{rs} = \lambda n_s$$

or

$$(T_{rs} - \lambda\delta_{rs})n_r = 0 \tag{3.3.2a}$$

or in matrix notation

$$n(T - \lambda I) = 0, \tag{3.3.2b}$$

where I is the identity matrix (see Sec. 2.4). Equations (2a) are a set of three linear homogeneous algebraic equations for the direction cosines n_1, n_2, n_3, which must also satisfy

$$\left.\begin{aligned} n_1^2 + n_2^2 + n_3^2 &= 1 \\ \text{or indicial notation} \qquad n_i n_i &= 1. \end{aligned}\right\} \tag{3.3.3}$$

In the matrix form, Eqs. (2b), I is the unit matrix

$$I \equiv \begin{bmatrix} 1 & 0 & 0 \\ 0 & 1 & 0 \\ 0 & 0 & 1 \end{bmatrix}, \tag{3.3.4}$$

while n is the row matrix (n_1, n_2, n_3) and λ is the numerical value sought. Thus

$$\lambda I = \begin{bmatrix} \lambda & 0 & 0 \\ 0 & \lambda & 0 \\ 0 & 0 & \lambda \end{bmatrix}. \tag{3.3.5}$$

The three direction cosines cannot all be zero, since they must satisfy Eq. (3). A system of linear homogeneous equations like Eqs. (2) has solutions which are not all zero if and only if the determinant of the coefficients is equal to zero, i.e., if

$$\begin{vmatrix} T_{11} - \lambda & T_{12} & T_{13} \\ T_{21} & T_{22} - \lambda & T_{23} \\ T_{31} & T_{32} & T_{33} - \lambda \end{vmatrix} = 0 \tag{3.3.6a}$$

or, in indicial notation, if

$$|T_{rs} - \lambda \delta_{rs}| = 0 \tag{3.3.6b}$$

or, in matrix notation, if

$$|T - \lambda I| = 0. \tag{3.3.6c}$$

For given values of the nine stress components, Eq. (6) is a cubic equation for the unknown magnitude λ. Because the stress matrix is symmetric and has real elements, the three roots of the cubic equation are all real numbers† [see Eq. (2.4.57)]. The three roots may be determined by any of the standard

†The fact that the principal stresses are all real was established by Cauchy (1828, 1829).

methods of numerical analysis. We will see an alternative explicit method in Eqs. (18) to (22) of this section.

If $\sigma_1, \sigma_2, \sigma_3$ are the three principal stresses (three roots of the cubic equation for λ), then when any one of them, say σ_1, is substituted for λ in the three Eqs. (2) those equations reduce to only two independent linear equations, which may be solved together with the quadratic Eq. (3) to determine the direction cosines $n_r^{(1)}$ of the normal $\hat{\mathbf{n}}^{(1)}$ to the plane on which σ_1 acts. Since one of the equations is quadratic, two solutions will be found representing two oppositely directed normals to the same plane; the choice of which one is the positive direction is arbitrary. This process is now repeated, substituting σ_2 for λ to find the second principal direction $\hat{\mathbf{n}}^{(2)}$, and, finally, the third principal direction $\hat{\mathbf{n}}^{(3)}$ may be found by substituting σ_3. The third principal direction may more conveniently be found as the direction perpendicular to the first two, since the three directions are mutually perpendicular as will be proved below. Because it is desired that the three directions form a right-handed system, when the components of $\hat{\mathbf{n}}^{(1)}$ and $\hat{\mathbf{n}}^{(2)}$ are known, the components of $\hat{\mathbf{n}}^{(3)}$ are found by

$$\hat{\mathbf{n}}^{(3)} = \hat{\mathbf{n}}^{(1)} \times \hat{\mathbf{n}}^{(2)}. \tag{3.3.7}$$

When the three roots of the characteristic equation are distinct (no two principal stresses equal), three mutually perpendicular principal directions are found; they are uniquely determined except for the arbitrary choice of the positive sense on each axis. If they are to be used as reference axes, the three positive directions should be so chosen that they form a right-handed system. When two of the principal stresses are equal and the third one is different, the direction associated with the third one is uniquely determined (except for sense), while the other two may be arbitrarily chosen as any two perpendicular to each other and to the one uniquely determined direction. The state of stress is said in this case to be ***cylindrical***; there is no shear stress on any plane parallel to the cylinder axis (the one determined direction). If all three principal stresses are equal, the stress state is said to be ***spherical***; in this case there is no shear stress on any plane through the point, and any three mutually perpendicular directions may be selected as the principal directions. A spherical state of stress is sometimes called a ***hydrostatic*** state of stress, since it is the only kind of stress that can exist in a fluid ***at rest***. Several assertions have been made without proof here: that the algebraically greatest and least principal stresses are in fact the maximum and minimum values among the normal stresses on all the planes through the point, that the three principal directions are mutually perpendicular, and the discussion of what happens if two or three of the principal stresses are equal. Analytical proofs can be given for these assertions, but all of them except the perpendicularity will be evident from the discussion of Mohr's Circles to be given in Sec. 3.4. To prove that when two of the principal stresses, say σ_1 and

σ_2, are not equal the corresponding principal directions, $\hat{\mathbf{n}}^{(1)}$ and $\hat{\mathbf{n}}^{(2)}$, are perpendicular, proceed as follows. Write Eqs. (2) with σ_1 substituted for λ and $n_r^{(1)}$ for n_r. Multiply the first equation by $n_1^{(2)}$, the second by $n_2^{(2)}$, the third by $n_3^{(2)}$ and add the three equations. Repeat the process interchanging superscripts 1 and 2, and subtract the resulting equation from the single equation obtained from the first application of the process. Thus you can obtain (using $T_{12} = T_{21}$, etc.)

$$[\sigma_2 - \sigma_1][n_1^{(1)}n_1^{(2)} + n_2^{(1)}n_2^{(2)} + n_3^{(1)}n_3^{(2)}] = 0.$$

Since the principal stresses were assumed to be unequal, the second factor must be the one that vanishes. This proves the perpendicularity of $\hat{\mathbf{n}}^{(1)}$ and $\hat{\mathbf{n}}^{(2)}$.

Invariants. The principal stresses are physical quantities, whose values do not depend on the coordinate system in which the components of stress were initially given. They are therefore ***invariants*** of the stress state, invariant with respect to rotation of the coordinate axes to which the stresses are referred. When the determinant in the characteristic Eq. (6) is expanded, the cubic equation takes the form

$$\lambda^3 - \mathrm{I}_T\lambda^2 - \mathrm{II}_T\lambda - \mathrm{III}_T = 0, \tag{3.3.8}$$

where the symbols I_T, II_T, III_T denote the following scalar expressions in the stress components (by expanding the indicial forms, the reader may verify that they are equivalent to the expanded forms):

$$\left.\begin{aligned}
\mathrm{I}_T &= T_{11} + T_{22} + T_{33} = T_{kk} = tr\,\mathbf{T} \\
\mathrm{II}_T &= -(T_{11}T_{22} + T_{22}T_{33} + T_{33}T_{11}) + T_{23}^2 + T_{31}^2 + T_{12}^2 \\
&= \tfrac{1}{2}(T_{ij}T_{ij} - T_{ii}T_{jj}) = \tfrac{1}{2}T_{ij}T_{ij} - \tfrac{1}{2}\mathrm{I}_T^2 \\
&= \tfrac{1}{2}(\mathbf{T}:\mathbf{T} - \mathrm{I}_T^2) \\
\mathrm{III}_T &= \det\mathbf{T} = \begin{vmatrix} T_{11} & T_{12} & T_{13} \\ T_{21} & T_{22} & T_{23} \\ T_{31} & T_{32} & T_{33} \end{vmatrix} = \tfrac{1}{6}e_{ijk}e_{pqr}T_{ip}T_{jq}T_{kr}.
\end{aligned}\right\} \tag{3.3.9}$$

Since the roots of the cubic equation (i.e., the principal stresses) do not depend on the choice of coordinate axes, the coefficients of the cubic equation cannot depend on the choice of axes either. Hence, the expressions I_T, II_T, III_T are also scalar invariants with respect to rotation of the Cartesian reference axes. If the axes of reference are chosen to coincide with the principal axes of stress, the simpler algebraic forms given below are found in terms of the principal stresses.

$$\left.\begin{aligned} \mathrm{I}_T &= \sigma_1 + \sigma_2 + \sigma_3, \\ \mathrm{II}_T &= -(\sigma_1\sigma_2 + \sigma_2\sigma_3 + \sigma_3\sigma_1), \\ \mathrm{III}_T &= \begin{vmatrix} \sigma_1 & 0 & 0 \\ 0 & \sigma_2 & 0 \\ 0 & 0 & \sigma_3 \end{vmatrix} = \sigma_1\sigma_2\sigma_3. \end{aligned}\right\} \tag{3.3.10}$$

In any second system, the numerical value of any of the individual components will in general be different ($\bar{T}_{11} \neq T_{11}$, $\bar{T}_{12} \neq T_{12}, \ldots$), but the numerical value of the invariants will be the same $\bar{T}_{11} + \bar{T}_{22} + \bar{T}_{33} = T_{11} + T_{22} + T_{33} = \sigma_1 + \sigma_2 + \sigma_3$, etc. I_T is called the first invariant of the stress tensor; it is also called the ***trace*** of the stress matrix, denoted tr **T**, which is the sum of the elements on the main diagonal in the matrix of rectangular Cartesian components of **T** (the three normal stresses). The ***second invariant*** II_T is a homogeneous quadratic expression in the stress components and is often referred to as the ***quadratic invariant*** of the stress tensor; it is the negative of the sum of the three minor determinants of the three diagonal elements in the determinant of the stress matrix. The ***third invariant*** III_T is the determinant of the stress matrix and is a homogeneous cubic expression in the stress components.

Some authors define II_T as the negative of the expression given above and write Eq. (8) with a plus sign before the third term. The three invariants are frequently denoted by I_1, I_2, I_3 or by J_1, J_2, J_3, although more often J_1, J_2, J_3 denote the invariants of the stress deviator, a tensor which we will define below. These invariant expressions have played a prominent role in the constitutive equations of isotropic materials and in the formulation of yield conditions under combined stresses in theories of plasticity of isotropic materials. Of course any function of the three invariants is also an invariant.

The principal stresses can always be found by solving Eq. (8) using numerical methods, when the stress state is given so that the three coefficients I_T, II_T, III_T are numbers. An alternative procedure permitting explicit solution of the cubic will be given below in terms of the stress deviator or deviatoric stress tensor.

Spherical and deviatoric stress tensors. Let σ denote the mean normal stress (p is the mean normal pressure)

$$\sigma = -p = \tfrac{1}{3}(T_{11} + T_{22} + T_{33}) = \tfrac{1}{3}T_{kk}. \tag{3.3.11}$$

Then the stress tensor can be written as the sum of two tensors, one representing a spherical or hydrostatic state of stress in which each normal stress is equal to $-p$ and all shear stresses are zero, and the second called the ***stress deviator***†

†The first explicit study of the deviator was published by M. Kleitz in 1873.

or ***deviatoric stress tensor*** $\mathbf{T}'$. The ***spherical stress*** matrix is

$$\sigma I = -pI = \begin{bmatrix} -p & 0 & 0 \\ 0 & -p & 0 \\ 0 & 0 & -p \end{bmatrix}, \tag{3.3.12}$$

the product of the unit matrix I by the scalar $-p$. The stress deviator matrix is

$$T' = \begin{bmatrix} T_{11} - \sigma & T_{12} & T_{13} \\ T_{21} & T_{22} - \sigma & T_{23} \\ T_{31} & T_{32} & T_{33} - \sigma \end{bmatrix}, \tag{3.3.13}$$

where

$$\left. \begin{aligned} T'_{11} = T_{11} - \sigma = T_{11} + p, \quad \text{etc.,} \\ \text{but} \quad T'_{ij} = T_{ij} \quad \text{for} \quad i \neq j. \end{aligned} \right\} \tag{3.3.14}$$

Thus we have for the deviator $\mathbf{T}'$

	$\mathbf{T}' = \mathbf{T} - \sigma\mathbf{1} = \mathbf{T} + p\mathbf{1}$
with Cartesian form	$T'_{ij} = T_{ij} - \sigma\delta_{ij} = T_{ij} + p\delta_{ij}$
or in matrix notation	$T' = T - \sigma I = T + pI$

(3.3.15)

Many writers use the notation s_{ij} or σ'_{ij} instead of T'_{ij} for the deviatoric stress.

It is usually supposed that the deviatoric stress brings about change of shape, while hydrostatic stress produces volume change without change of shape in an isotropic continuum, i.e., in a material with the same material properties in all directions. Clearly a uniform all-around pressure should merely decrease the volume of a sphere of material with the same strength in all directions. However, if the sphere were weaker in one direction, that diameter would be changed more than others; thus hydrostatic pressure can produce a change of shape in anisotropic materials.

Principal values of stress deviator. The principal directions of the stress deviator tensor are the same as those of the stress tensor, since both represent directions perpendicular to planes having no shear stress. The principal deviator stresses are then

$$\sigma'_k = \sigma_k - \sigma = \sigma_k + p. \tag{3.3.16}$$

In three-dimensional problems, it is usually easier to solve for the σ'_i and then use Eq. (16) to calculate the σ_i. This is easier because the cubic Eq. (17) below, analogous to Eq. (8), lacks the quadratic term. If λ denotes any one of the principal deviator stresses, a derivation parallel to that of Eq. (8) yields

$$\lambda^3 - \mathrm{II}_{T'}\lambda - \mathrm{III}_{T'} = 0, \tag{3.3.17}$$

where $\mathrm{I}_{T'} = 0$, $\mathrm{II}_{T'}$, $\mathrm{III}_{T'}$ are the scalar invariants of the stress deviator $\mathbf{T}'$, analogous to those given for the stress tensor by Eqs. (9), but calculated with the components of the deviator. Several useful alternative expressions for $\mathrm{II}_{T'}$ are given in Ex. 6 following this section.

Equation (17) may be solved explicitly by the substitution

$$\lambda = 2(\tfrac{1}{3}\mathrm{II}_{T'})^{1/2} \cos \alpha, \tag{3.3.18}$$

which transforms it into

$$2(\tfrac{1}{3}\mathrm{II}_{T'})^{3/2} [4 \cos^3 \alpha - 3 \cos \alpha] = \mathrm{III}_{T'}. \tag{3.3.19}$$

The quantity in brackets is equal to $\cos 3\alpha$. Eq. (19) is equivalent to

$$\cos 3\alpha = \frac{\mathrm{III}_{T'}}{2}\left(\frac{3}{\mathrm{II}_{T'}}\right)^{3/2}. \tag{3.3.20}$$

Let α_1 be the angle satisfying $0 \leq 3\alpha_1 \leq \pi$ whose cosine is given by Eq. (20). Then $3\alpha_1$, $3\alpha_1 + 2\pi$, and $3\alpha_1 - 2\pi$ all have the same cosine, given in terms of the calculated invariants of the deviator by the expression above, and this furnishes three independent roots of Eq. (17), namely

$$\sigma'_k = 2(\cos \alpha_k)(\tfrac{1}{3}\mathrm{II}_{T'})^{1/2}, \tag{3.3.21}$$

where

$$\alpha_2 = \alpha_1 + \frac{2\pi}{3} \quad \text{and} \quad \alpha_3 = \alpha_1 - \frac{2\pi}{3}. \tag{3.3.22}$$

The cubic equation is thus solved explicitly for the three roots σ'_k by determining $3\alpha_1$ from a table of cosines. Since p can be calculated from the given stresses, Eqs. (16) furnish the values of σ_k. The corresponding principal directions may be determined from Eqs. (2) and (3).

We have proved that there always exists at least one set of mutually orthogonal directions at a point such that when the stress at the point is referred to new coordinate axes $(\bar{x}_1, \bar{x}_2, \bar{x}_3)$ in these directions all of the shear-stress components are zero so that the traction on each face of the element is entirely normal. We have not yet proved the second property of the principal stresses, that the algebraically greatest principal stress is the algebraically greatest normal stress on any plane through the point, while the algebraically least principal stress is the algebraically least normal stress on any plane through the point. These conclusions will be established in Sec. 3.4 by our study of Mohr's Circles, and we will also show there that the maximum shear stress magnitude equals half the difference between the greatest and least principal stresses.

EXERCISES

1. Determine the principal stresses for the stress tensors with rectangular Cartesian components given in the matrices below (units are kpsi). Determine also direction cosines for each principal axis of a right-handed system. *Hint*: The cubic equations are easily factored.

(a) $\begin{bmatrix} 18 & 0 & 24 \\ 0 & -50 & 0 \\ 24 & 0 & 32 \end{bmatrix}$ (b) $\begin{bmatrix} 3 & -10 & 0 \\ -10 & 0 & 30 \\ 0 & 30 & -27 \end{bmatrix}$

(c) $\begin{bmatrix} 1 & 0 & 0 \\ 0 & 3 & -1 \\ 0 & -1 & 3 \end{bmatrix}$ (d) $\begin{bmatrix} 2 & -1 & 1 \\ -1 & 0 & 1 \\ 1 & 1 & 2 \end{bmatrix}$

2. Evaluate the three invariants I_T, II_T, III_T for the stress tensors of Ex. 1 by using the given components and also by using the values obtained in that problem for the principal stresses, and check that the same results are obtained by the two procedures.

3. Separate each of the matrices in Ex. 1 into the sum of two matrices, representing the spherical and deviatoric stresses.

4. Verify that the indicial forms of Eq. (9) give the same results as the expanded forms given for: (a) I_T, (b) II_T, (c) III_T. *Hint for* (c): For each choice of p, q, r, such that $e_{pqr} \neq 0$, show that $e_{ijk} T_{ip} T_{jq} T_{kr} = e_{pqr} III_T$.

5. Note that for the deviatoric stress $II_{T'} = \frac{1}{2} T'_{ij} T'_{ij}$, since $I_{T'} = 0$. Evaluate $II_{T'}$ for each of the tensors of Ex. 1.

6. In the notation of Fig. 3.10, with the deviatoric normal components denoted by s_x, s_y, s_z and the three invariants of the deviator denoted by I_s, II_s, III_s, we have

$$II_s = -(s_x s_y + s_y s_z + s_z s_x) + \tau_{yz}^2 + \tau_{zx}^2 + \tau_{xy}^2. \tag{1}$$

(a) Show that $I_s = 0$. Then, by adding $I_s^2/2 = 0$ to the expression (1) above, show that

$$II_s = \tfrac{1}{2}(s_x^2 + s_y^2 + s_z^2) + \tau_{yz}^2 + \tau_{zx}^2 + \tau_{xy}^2. \tag{2}$$

(b) By adding $I_s^2/3 = 0$ to (1), show that

$$II_s = \tfrac{1}{6}[(s_x - s_y)^2 + (s_y - s_z)^2 + (s_z - s_x)^2] + \tau_{yz}^2 + \tau_{zx}^2 + \tau_{xy}^2. \tag{3}$$

(c) This may also be written as shown in (4). Why?

$$II_s = \tfrac{1}{6}[(\sigma_x - \sigma_y)^2 + (\sigma_y - \sigma_z)^2 + (\sigma_z - \sigma_x)^2] + \tau_{yz}^2 + \tau_{zx}^2 + \tau_{xy}^2 \tag{4}$$

(d) Express II_s in terms of the principal stresses $\sigma_1, \sigma_2, \sigma_3$.
(These various forms for II_s are useful in plasticity theory; see Secs. 6.6 and 6.7.)

7. If the coordinate axes are principal axes, so that the stress matrix has the form shown below:

$$\begin{bmatrix} \sigma_1 & 0 & 0 \\ 0 & \sigma_2 & 0 \\ 0 & 0 & \sigma_3 \end{bmatrix}$$

(a) Determine the normal component σ_n and the magnitude τ_n of the tangential component of the traction vector $\mathbf{t}$ on a plane with unit normal $\hat{\mathbf{n}} = n_x\mathbf{i} + n_y\mathbf{j} + n_z\mathbf{k}$.
(b) Determine also the components of a unit vector $\hat{\mathbf{e}}_\tau$ in the plane, such that $\mathbf{t} = \sigma_n\hat{\mathbf{n}} + \tau_n\hat{\mathbf{e}}_\tau$.

8. Find an expression in terms of the principal stresses for the magnitude t_0 of the traction vector on an ***octahedral plane***, that is, a plane whose normal makes equal angles with the three principal directions.

9. Show that the normal component σ_0 of the traction vector on the octahedral plane of Ex. 8 is equal to $\frac{1}{3}I_T$, where I_T is the first invariant of the stress tensor. (σ_0 is called the ***octahedral normal stress***.)

10. Show that the ***octahedral shear stress*** τ_0 (the magnitude of the tangential component of the traction vector on the octahedral plane of Ex. 8) is given by $\tau_0 = (\frac{2}{3}II_s)^{1/2}$, where II_s is the second invariant of the stress deviator. (See your results in Ex. 6.)

11. Show that the unit vector $\hat{\mathbf{e}}_{\tau_0}$ (see Ex. 7b) for the octahedral plane whose normal points into the positive octant is given by

$$\hat{\mathbf{e}}_{\tau_0} = \frac{(\sigma_1'\mathbf{i} + \sigma_2'\mathbf{j} + \sigma_3'\mathbf{k})}{(\sigma_1'^2 + \sigma_2'^2 + \sigma_3'^2)^{1/2}},$$

referred to the principal axes as reference axes. (See Exs. 8 to 10 for discussions of the octahedral plane.)

12. If two roots of the cubic equation for principal stress are equal [say that $(\lambda - \sigma_1)(\lambda - \sigma_2)^2 = 0$ and $\hat{\mathbf{n}}^{(1)}$, $\hat{\mathbf{n}}^{(2)}$ are corresponding principal axes for which $\lambda = \sigma_1$ and $\lambda = \sigma_2$, respectively],
(a) show that $\hat{\mathbf{n}}^{(1)} \times \hat{\mathbf{n}}^{(2)}$ is a principal direction for which the principal stress is also σ_2, and then
(b) show that every direction $\hat{\mathbf{n}}$ perpendicular to $\hat{\mathbf{n}}^{(1)}$ is a principal axis for which the principal stress is σ_2.

13. Evaluate the principal deviator stresses and the second invariant $II_{T'}$ for the following stress states (components not given are zero):
    (a) Uniaxial stress $T_{11} = 30{,}000$ psi.
    (b) Biaxial normal stress $T_{11} = 20{,}000$ psi, $T_{22} = 10{,}000$ psi.
    (c) Pure shear $T_{12} = T_{21} = 10{,}000$ psi.
    (d) Combined tension and shear $T_{11} = \sigma$, $T_{12} = T_{21} = \tau$.
    (e) Biaxial shear $T_{12} = T_{21} = A$, $T_{31} = T_{13} = B$.

14. Determine the direction cosines of each principal axis in Ex. 13(c, d, e).

15. Use Eqs. (3.3.20) and (21) to determine the principal stresses for the stress states given below, and determine the principal axis corresponding to the maximum principal stress.

$$\text{(a)} \begin{bmatrix} 0 & a & a \\ a & 0 & a \\ a & a & 0 \end{bmatrix} \qquad \text{(b)} \begin{bmatrix} 0 & 1 & 2 \\ 1 & 0 & 2 \\ 2 & 2 & 0 \end{bmatrix} \quad \text{(kpsi)}$$

Ans.: (a) $\sigma_1' = 2a$, $\sigma_2' = \sigma_3' = -a$; (b) $\sigma_1' = 3.37$, $\sigma_2' = -2.37$, $\sigma_3' = -1.00$.

16. For the stress matrix shown below, show that one principal stress is 2 kpsi, and find the corresponding principal direction.

$$\begin{bmatrix} 5 & 3 & -3 \\ 3 & 0 & 2 \\ -3 & 2 & 0 \end{bmatrix} \quad \text{(kpsi)}$$

17. Show that

$$\frac{\partial II_{T'}}{\partial T_{pq}} = T'_{pq},$$

where $II_{T'}$ is the second invariant of the deviatoric stress $\mathbf{T}'$. (*Warning*: The derivative is with respect to T_{pq}, not T'_{pq}.)

3.4 Mohr's Circles†

In Sec. 3.2, we saw how to express the components t_x, t_y, t_z of the stress vector $\mathbf{t}$ on an inclined plane in terms of the nine components of the stress matrix. The stress vector on the plane may alternatively be resolved into two components, one normal to the plane and one tangential. Let σ_n denote the normal component and τ_n the magnitude of the tangential or shear component. Then

$$\sigma_n = \mathbf{t}\cdot\hat{\mathbf{n}}, \tag{3.4.1}$$

†Due to O. Mohr in publications dating from 1882.

and

$$\tau_n^2 = t^2 - \sigma_n^2. \tag{3.4.2}$$

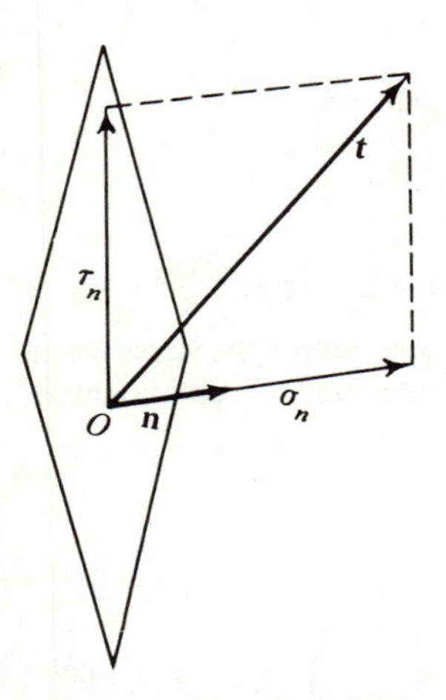

Fig. 3.11 Normal and tangential stress components.

In Mohr's ***stress representation plane***, Fig. 3.13, (σ_n, τ_n) are plotted as coordinates; the point (σ_n, τ_n) represents the state of stress on a plane through O in the physical body of Fig. 3.12. If another surface is chosen through O with a different normal, the stress vector will be different, and a different (σ_n, τ_n) point will be obtained in the stress representation σ_n, τ_n-plane. It is of interest to see what portion of the σ_n, τ_n-plane is occupied by the points plotted in this way for all possible orientations of the plane through O with a given state of stress at O. Points of special interest will be those which show the maximum and minimum values of the normal stress and the maximum shear stress. In the derivation below, we consider the general case for which no two of the principal stresses are equal, but it is easy to see from the results what happens when two of them are equal—for example, when σ_{I} is allowed to approach σ_{II}.

In order to simplify the discussion, we choose the xyz-axes in the physical body, Fig. 3.12, to coincide with the principal axes of stress at O—with the x-direction being that of the algebraically greatest principal stress σ_{I} and the z-axis that of the algebraically smallest principal stress σ_{III}. The Roman numeral subscripts here indicate that the three principal stresses have been ordered. Then

$$T_{11} = \sigma_{\text{I}}, \qquad T_{22} = \sigma_{\text{II}}, \qquad T_{33} = \sigma_{\text{III}}, \\ \sigma_{\text{I}} \geq \sigma_{\text{II}} \geq \sigma_{\text{III}}, \tag{3.4.3}$$

and all shear components are zero. Equations (3.2.8) then give the following simple expressions in terms of these matrix components for the three rectangular components of $\mathbf{t}$:

$$t_x = \sigma_{\text{I}} n_x, \qquad t_y = \sigma_{\text{II}} n_y, \qquad t_z = \sigma_{\text{III}} n_z, \tag{3.4.4}$$

and Eq. (1) becomes

$$\sigma_n = \sigma_{\text{I}} n_x^2 + \sigma_{\text{II}} n_y^2 + \sigma_{\text{III}} n_z^2. \tag{3.4.5}$$

From Eqs. (2) and (4), we also have for the square of the magnitude t of the vector $\mathbf{t}$

$$t^2 = \sigma_n^2 + \tau_n^2 = \sigma_{\mathrm{I}}^2 n_x^2 + \sigma_{\mathrm{II}}^2 n_y^2 + \sigma_{\mathrm{III}}^2 n_z^2. \tag{3.4.6}$$

Equations (5) and (6) form with

$$n_x^2 + n_y^2 + n_z^2 = 1 \tag{3.4.7}$$

a system of three linear equations for the three quantities

$$n_x^2, \qquad n_y^2, \qquad n_z^2.$$

If we eliminate n_z^2 from Eqs. (7) and (5), we obtain

$$(\sigma_{\mathrm{III}} - \sigma_{\mathrm{I}})n_x^2 + (\sigma_{\mathrm{III}} - \sigma_{\mathrm{II}})n_y^2 = \sigma_{\mathrm{III}} - \sigma_n,$$

while a similar elimination between Eqs. (5) and (6) yields

$$\sigma_{\mathrm{I}}(\sigma_{\mathrm{III}} - \sigma_{\mathrm{I}})n_x^2 + \sigma_{\mathrm{II}}(\sigma_{\mathrm{III}} - \sigma_{\mathrm{II}})n_y^2 = \sigma_n(\sigma_{\mathrm{III}} - \sigma_n) - \tau_n^2.$$

We now eliminate n_y^2 from the last two equations to get the expression below for n_x^2. The expressions for n_y^2 and n_z^2 are obtained similarly.

$$\left.\begin{aligned} n_x^2 &= \frac{\sigma_n^2 + \tau_n^2 - \sigma_n(\sigma_{\mathrm{II}} + \sigma_{\mathrm{III}}) + \sigma_{\mathrm{II}}\sigma_{\mathrm{III}}}{(\sigma_{\mathrm{II}} - \sigma_{\mathrm{I}})(\sigma_{\mathrm{III}} - \sigma_{\mathrm{I}})} \\ n_y^2 &= \frac{\sigma_n^2 + \tau_n^2 - \sigma_n(\sigma_{\mathrm{III}} + \sigma_{\mathrm{I}}) + \sigma_{\mathrm{III}}\sigma_{\mathrm{I}}}{(\sigma_{\mathrm{III}} - \sigma_{\mathrm{II}})(\sigma_{\mathrm{I}} - \sigma_{\mathrm{II}})} \\ n_z^2 &= \frac{\sigma_n^2 + \tau_n^2 - \sigma_n(\sigma_{\mathrm{I}} + \sigma_{\mathrm{II}}) + \sigma_{\mathrm{I}}\sigma_{\mathrm{II}}}{(\sigma_{\mathrm{I}} - \sigma_{\mathrm{III}})(\sigma_{\mathrm{II}} - \sigma_{\mathrm{III}})} \end{aligned}\right\} \tag{3.4.8}$$

Figure 3.12 shows a unit sphere centered at the point O of the physical body. The radius vector **ON** is the unit normal $\hat{\mathbf{n}}$ at O. If n_x is held constant, the point N moves on the circle DE on the unit sphere with constant angle $\alpha = AON$.

Since the first Eq. (8) contains only the one direction cosine n_x of $\hat{\mathbf{n}}$, for a given value of n_x it is the equation of the locus in Fig. 3.13 on which lie all the (σ_n, τ_n) points obtainable for this value of n_x.

If this equation is cleared of fractions and $\frac{1}{4}(\sigma_{\mathrm{II}} + \sigma_{\mathrm{III}})^2$ is added to both sides, it may be put into the form of the first Eq. (9).

$$\left.\begin{aligned} [\sigma_n - \tfrac{1}{2}(\sigma_{\mathrm{II}} + \sigma_{\mathrm{III}})]^2 + \tau_n^2 &= R_{\mathrm{I}}^2 \\ [\sigma_n - \tfrac{1}{2}(\sigma_{\mathrm{III}} + \sigma_{\mathrm{I}})]^2 + \tau_n^2 &= R_{\mathrm{II}}^2 \\ [\sigma_n - \tfrac{1}{2}(\sigma_{\mathrm{I}} + \sigma_{\mathrm{II}})]^2 + \tau_n^2 &= R_{\mathrm{III}}^2 \end{aligned}\right\} \tag{3.4.9}$$

where

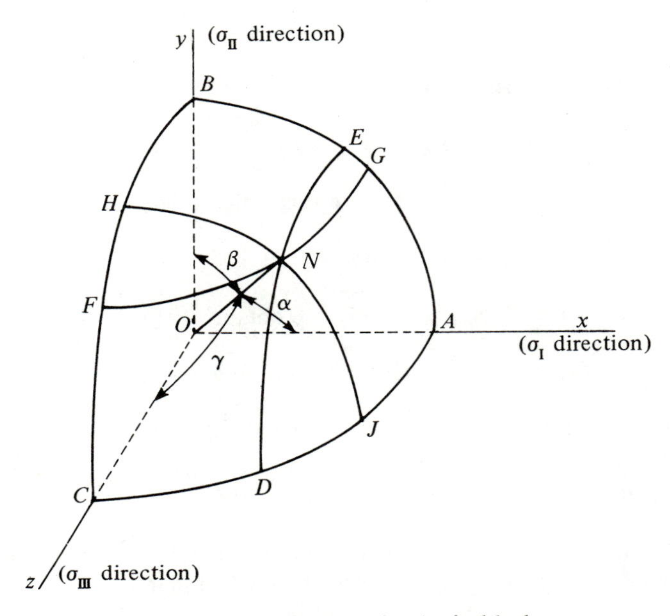

Fig. 3.12 Unit sphere in physical body.

$$\left.\begin{aligned} R_{I}^2 &= \tfrac{1}{4}(\sigma_{II} + \sigma_{III})^2 + (\sigma_I - \sigma_{II})(\sigma_I - \sigma_{III})n_x^2 - \sigma_{II}\sigma_{III} \\ R_{II}^2 &= \tfrac{1}{4}(\sigma_{III} + \sigma_I)^2 - (\sigma_I - \sigma_{II})(\sigma_{II} - \sigma_{III})n_y^2 - \sigma_{III}\sigma_I \\ R_{III}^2 &= \tfrac{1}{4}(\sigma_I + \sigma_{II})^2 + (\sigma_I - \sigma_{III})(\sigma_{II} - \sigma_{III})n_z^2 - \sigma_I\sigma_{II} \end{aligned}\right\} \tag{3.4.10}$$

The other two Eqs. (9) are obtained similarly from the last two Eqs. (8).

If n_x is held constant, the first Eq. (9) represents a circle in the σ_n, τ_n-plane with radius R_I and center at $\sigma_n = \frac{1}{2}(\sigma_{II} + \sigma_{III}), \tau_n = 0$. Similarly, the other two equations represent circles with radii R_{II} and R_{III} and centers on the σ_n-axis at $\sigma_n = \frac{1}{2}(\sigma_{III} + \sigma_I)$ and $\sigma_n = \frac{1}{2}(\sigma_I + \sigma_{II})$, respectively. The smallest R_I value is for $n_x^2 = 0$, i.e., for $\alpha = 90°$; the circle marked min R_I in Fig. 3.13 therefore contains the σ_n, τ_n values corresponding to directions ON lying in the yz-plane of Fig. 3.12. Similarly the min R_{III} circle corresponds to $\gamma = 90°$ and represents directions lying in the xy-plane of Fig. 3.12. For $\beta = 90°$, however, we obtain the largest R_{II} circle (marked max R_{II} in Fig. 3.13); it represents directions lying in the xz-plane of Fig. 3.12. The largest and smallest values for each radius are shown by the inequalities below.

$$\left.\begin{aligned} \tfrac{1}{2}(\sigma_{II} - \sigma_{III}) &\leq R_I \leq [\sigma_I - \tfrac{1}{2}(\sigma_{II} + \sigma_{III})] \\ [\sigma_{II} - \tfrac{1}{2}(\sigma_I + \sigma_{III})] &\leq R_{II} \leq \tfrac{1}{2}(\sigma_I - \sigma_{III}) \\ \tfrac{1}{2}(\sigma_I - \sigma_{II}) &\leq R_{III} \leq [\tfrac{1}{2}(\sigma_I + \sigma_{II}) - \sigma_{III}] \end{aligned}\right\} \tag{3.4.11}$$

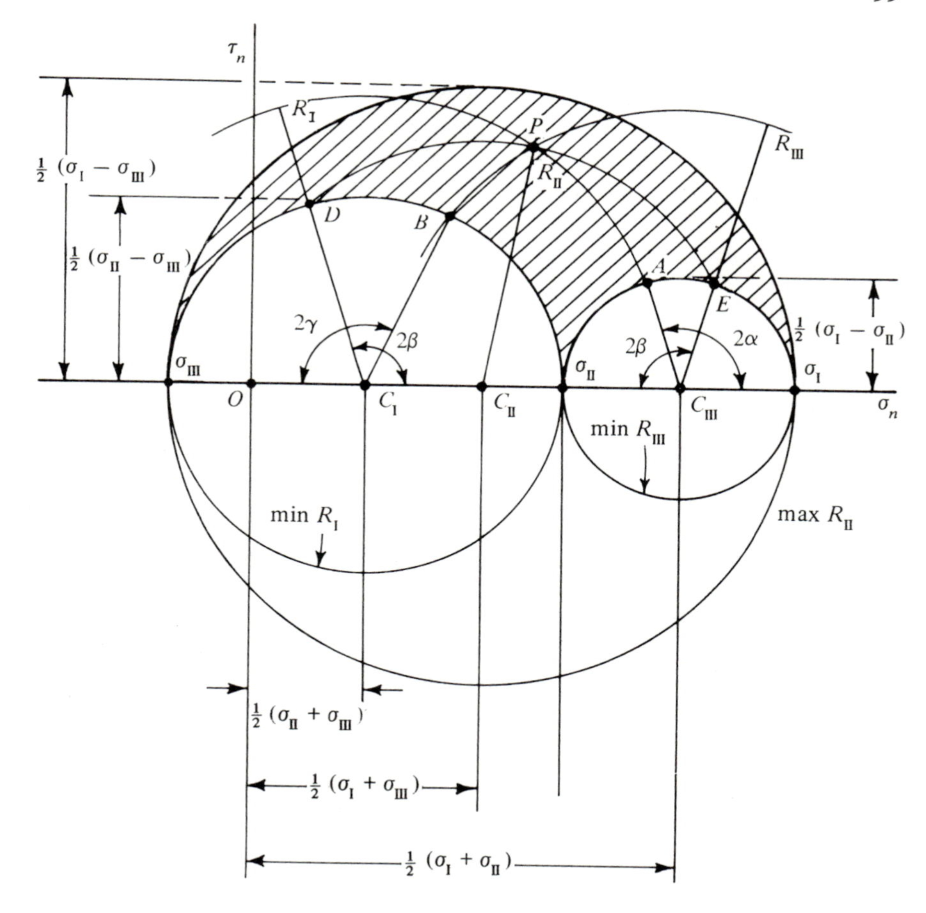

Fig. 3.13 Mohr's Circles in stress-representation plane.

Figure 3.13 shows the circles for the largest possible value of R_{II} and the smallest possible values of R_I and R_{III} for given values of $\sigma_I, \sigma_{II}, \sigma_{III}$. For any choice of the direction of $\hat{\mathbf{n}}$ the σ_n, τ_n point must be somewhere in the hatched region bounded by the three circles, e.g., at point P. (Since τ_n was defined as the ***magnitude*** of the shear component, it is always positive.) This shows the truth of the assertions made in Sec. 3.3 that the maximum value of the normal stress σ_n is equal to σ_I, the largest of the three principal stresses, while the minimum value of σ_n equals σ_{III}, the smallest of the three.

We also see that the maximum shear stress magnitude is $\tau_{max} = \frac{1}{2}(\sigma_I - \sigma_{III})$ and that the normal stress on the plane on which the maximum shear stress acts is $\frac{1}{2}(\sigma_I + \sigma_{III})$. If these values of τ_n and σ_n are substituted into Eqs. (8), it can be seen that on the maximum shear stress plane $n_x^2 = n_z^2 = \frac{1}{2}$ and $n_y = 0$;

this shows that the maximum shear stress acts on the two planes bisecting the dihedral angles between the two planes of maximum and minimum normal stress. In the present section, Roman numeral subscripts have been used to denote the ordering of the principal stresses by magnitude. If the ordering is not specified, we can say only that the maximum shearing stress is the greatest of the magnitudes of the three ***principal shear stresses***,

$$\tau_1 = \tfrac{1}{2}(\sigma_2 - \sigma_3), \qquad \tau_2 = \tfrac{1}{2}(\sigma_3 - \sigma_1), \qquad \tau_3 = \tfrac{1}{2}(\sigma_1 - \sigma_2), \tag{3.4.12}$$

which act on planes bisecting the angles between the principal planes.

The diagram of Fig. 3.13 is chiefly of theoretical interest to show graphically the range of possible values of σ_n and τ_n. It is easier to calculate accurate values by Eqs. (5) and (6) or (8) than to read them from the figure. However, a graphical solution is possible when the three principal stresses are known, as follows. Begin by marking on the σ_n-axis points with abscissas $\sigma_{\text{I}}, \sigma_{\text{II}}, \sigma_{\text{III}}$. Then locate the centers $C_{\text{I}}, C_{\text{II}}, C_{\text{III}}$, and draw the three basic circles with radii $\min R_{\text{I}} = \frac{1}{2}(\sigma_{\text{II}} - \sigma_{\text{III}})$, $\max R_{\text{II}} = \frac{1}{2}(\sigma_{\text{I}} - \sigma_{\text{III}})$, and $\min R_{\text{III}} = \frac{1}{2}(\sigma_{\text{I}} - \sigma_{\text{II}})$. To find the stresses σ_n and τ_n on an element whose normal makes angles α, β, γ with the principal directions in the body, lay off the two angles 2α and 2γ, as shown in Fig. 3.13, locating points A and B. Point P is then found as the intersection of the R_{I} circle through A with the R_{III} circle through B. The coordinates of P on the figure are the required values of σ_n and τ_n. The construction can be checked by laying off the angle 2β in the two places shown and drawing the R_{II} circle through D and E; the circle should pass through P. The converse problem is solved by reversing the process outlined, beginning with given values for σ_n, τ_n and ending with the measurement of the angles $2\alpha, 2\beta, 2\gamma$. (The three principal stresses are also part of the given data in order to permit construction of the basic circles.)

This graphical procedure for the three-dimensional case is seldom used to determine the stresses σ_n and τ_n on a given plane. But a graphical procedure for plane stress, given in Sec. 3.5, is frequently used by stress analysts with data obtained from surface-strain measurements. The plane-stress graphical procedure includes graphical determination of the principal stresses.

EXERCISES

1. (a) Sketch the three Mohr's Circles for a stress state with $\sigma_{\text{I}} = 2a, \sigma_{\text{II}} = a, \sigma_{\text{III}} = 0$.

   (b) For the stress state of (a), with coordinate axes in the principal directions, determine direction cosines of the normal to a surface on which $\sigma_n = a$, $\tau_n = 3a/4$.

(c) Determine the equation of the locus of all (σ_n, τ_n) points for surfaces such that $n_x^2 = 3/8$; sketch the locus, and determine the maximum σ_n on it.

Ans.: (b) $n_x^2 = 9/32$, $n_y^2 = 14/32$, $n_z^2 = 9/32$, (c) $(\sigma_n)_{\max} = 11a/8$.

2. For each of the following stress states (values not given are zero), sketch the three Mohr's Circles. What is the maximum shear stress in each case, and what is the normal stress on the plane of maximum shear stress? (Units are kpsi.)
   (a) Uniaxial compression $T_{11} = -60$.
   (b) Biaxial stress $T_{11} = 20$, $T_{22} = -60$.
   (c) Hydrostatic compression of magnitude 100.
   (d) $T_{12} = T_{21} = 30$, $T_{23} = T_{32} = 40$.
   (e) $T_{11} = -50$, $T_{22} = -10$, $T_{33} = 10$.

3. For each of the following stress states (values not given are zero), sketch the three Mohr's Circles, and determine the maximum shear stress and the normal stress on the maximum shear stress plane. [Units are kpsi, except for (c).]
   (a) Uniaxial tension $T_{11} = 40$.
   (b) Biaxial stress $T_{11} = -10$, $T_{22} = 30$.
   (c) Hydrostatic tension of magnitude 100 psi.
   (d) $T_{11} = -60$, $T_{22} = 100$, $T_{33} = 40$.
   (e) $T_{11} = 10$, $T_{22} = 40$, $T_{12} = T_{21} = 20$.

4. For $\sigma_{\text{I}} = t$, $\sigma_{\text{II}} = 0$, $\sigma_{\text{III}} = -t$, determine the maximum R_{II} and minimum R_{I} and R_{III}, and
   (a) Calculate the radius R_{I} for a plane whose normal makes a 45° angle with the σ_{I}-direction, and sketch the Mohr's Circle for this case ($\alpha = 45°$). Determine the maximum and minimum values of σ_n for $\alpha = 45°$.
   (b) Calculate R_{II} for $\beta = 60°$ and sketch the circle. Determine the maximum and minimum values of σ_n and τ_n for $\beta = 60°$.

5. For the stress state of Ex. 4 locate the point the Mohr's Circle plane corresponding to an octahedral plane [one such that $\alpha = \beta = \gamma$; see Exercise 10 for Sec. 3.3 (p. 94)]. Show that τ_n equals the octahedral shear stress τ_0 not only on the octahedral plane but also on a family of planes. What are the maximum and minimum values of σ_n on planes where $\tau_n = \tau_0$ for the given stress state?

6. Sketch the three Mohr's Circles for the ***deviatoric stress*** for the five stress states of Ex. 13, Sec. 3.3 [circles determined by $\sigma'_{\text{I}}, \sigma'_{\text{II}}, \sigma'_{\text{III}}$ bounding areas containing points (σ'_n, τ_n)]. Determine the maximum shear stress in each case and the normal stress on the plane of maximum shear stress.

7. (a) and (b). Sketch the three Mohr's Circles for the stress states of Exercises 15(a) and (b) of Sec. 3.3. Determine the maximum shear stress in each case and the normal stress on the plane of maximum shear stress. Determine also the octahedral shear stress τ_0 (see Exercise 10, Sec. 3.3), and show in the Mohr's Circle diagram the locus of (σ_n, τ_n) points such that $\tau_n = \tau_0$.

8. Show that the sum of the squares of the principal shear stresses defined by Eq. (3.4.12) is equal to 1.5 times the second invariant of the stress deviator. (See Ex. 6, Sec. 3.3.)

3.5 Plane stress; Mohr's Circle

A state of ***plane stress*** parallel to the xy-plane is said to exist at a point if at that point [in the notation of Eq. (3.2.12)]

$$\sigma_z = 0, \qquad \tau_{xz} = \tau_{zx} = 0, \qquad \tau_{yz} = \tau_{zy} = 0. \tag{3.5.1}$$

Then

$$T = \begin{bmatrix} \sigma_x & \tau_{xy} & 0 \\ \tau_{yx} & \sigma_y & 0 \\ 0 & 0 & 0 \end{bmatrix} \tag{3.5.2}$$

At an unloaded free surface of a body, the stress state is locally plane, since if we take the tangent plane as xy-plane all stress components with a subscript z are traction components on the unloaded surface and must therefore vanish. A state of plane stress throughout a body exists approximately in a thin plate under edge tractions parallel to the plane of the plate, if no buckling or bending occurs.†

For a plane stress state, the Mohr's Circle furnishes a convenient method of determining the stress vector on an arbitrary element in terms of $\sigma_x, \sigma_y, \tau_{xy}$. Two different sign conventions are in common use for the shear stress in connection with the plane stress Mohr's Circles, and this often leads to confusion unless the reader has ascertained carefully which convention is being used. Both conventions will be discussed here and the relation between them pointed out. We first treat the traction components $\bar{\sigma}_x$ and $\bar{\tau}_{xy}$ on a plane whose normal makes angle α with the x-axis, using the usual sign convention for shear stress in rectangular coordinates; later we will consider the other convention, according to which the shear stress on the plane with normal at angle α is $\tau_n = -\bar{\tau}_{xy}$, and that for $\alpha = 0$ is $\tau_n = -\tau_{xy}$.

Consider a new set of $\bar{x}\bar{y}$-axes in the xy-plane as shown in Fig. 3.14(a). If α denotes the angle between the $\bar{x}$-axis and the x-axis, then the direction cosines of the $\bar{x}$- and $\bar{y}$-directions are

$$\begin{aligned} a_{\bar{x}}^{x} &= \cos\alpha, & a_{\bar{x}}^{y} &= \sin\alpha, & a_{\bar{x}}^{z} &= 0 \\ a_{\bar{y}}^{x} &= -\sin\alpha, & a_{\bar{y}}^{y} &= \cos\alpha, & a_{\bar{y}}^{z} &= 0. \end{aligned}$$

Equations (3.2.9) then give

†The components (2) usually vary somewhat with z. To get a two-dimensional problem we can take their average values through the thickness of the plate as ***generalized plane stress components***. Remark: The Mohr's circle procedure can be applied to stresses in the plane at a point even if the stress state is not plane, since, for $n_z = 0$, Equations (3.2.9) still give Equations (3.5.3) and (3.5.4).

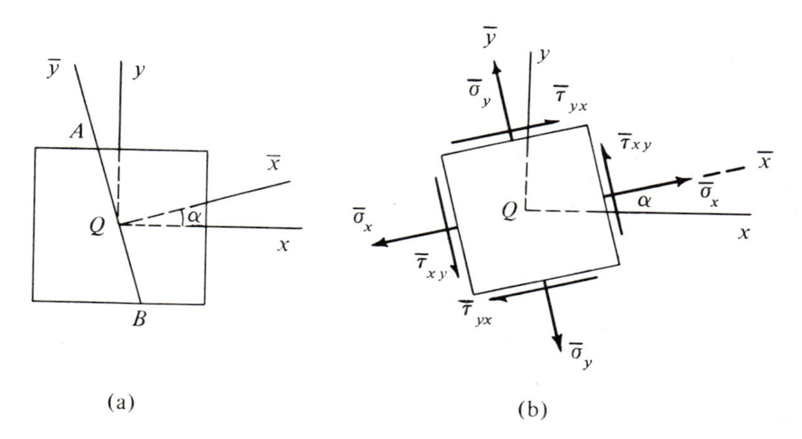

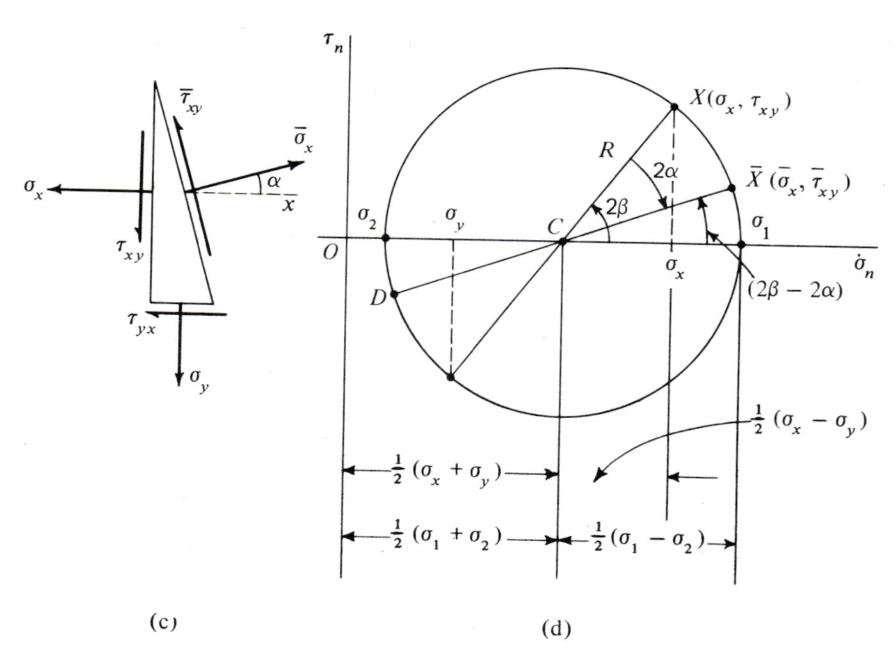

Fig. 3.14 Mohr's representation of plane stress.

$$\bar{\sigma}_x = \sigma_x \cos^2 \alpha + \sigma_y \sin^2 \alpha + 2\tau_{xy} \sin \alpha \cos \alpha, \tag{3.5.3}$$

$$\bar{\tau}_{xy} = \tau_{xy} (\cos^2 \alpha - \sin^2 \alpha) - (\sigma_x - \sigma_y) \sin \alpha \cos \alpha. \tag{3.5.4}$$

These formulas can be given a convenient form by substituting

$$\cos^2\alpha = \frac{1+\cos 2\alpha}{2}, \qquad \sin^2\alpha = \frac{1-\cos 2\alpha}{2},$$

$$\cos^2\alpha - \sin^2\alpha = \cos 2\alpha, \qquad 2\sin\alpha\cos\alpha = \sin 2\alpha.$$

After some rearrangement of terms, we obtain

$$\bar{\sigma}_x = \tfrac{1}{2}(\sigma_x + \sigma_y) + \tfrac{1}{2}(\sigma_x - \sigma_y)\cos 2\alpha + \tau_{xy}\sin 2\alpha, \tag{3.5.5}$$

$$\bar{\tau}_{xy} = \tau_{xy}\cos 2\alpha - \tfrac{1}{2}(\sigma_x - \sigma_y)\sin 2\alpha. \tag{3.5.6}$$

The points with coordinates (σ_x, τ_{xy}), $(\sigma_x, 0)$, $(\sigma_y, 0)$, and $[(\sigma_x + \sigma_y)/2, 0]$ are plotted in the stress representation σ_n, τ_n-plane of Fig. 3.14(d) with normal stress as abscissa and shear stress as ordinate. Here we permit τ_n to have negative values, thus differing from the discussion in Sec. 3.4. Let 2β be the angle shown in Fig. 3.14(d). Then

$$\boxed{\tfrac{1}{2}(\sigma_x - \sigma_y) = R\cos 2\beta, \qquad \tau_{xy} = R\sin 2\beta} \tag{3.5.7}$$

where

$$\boxed{R = [\tfrac{1}{4}(\sigma_x - \sigma_y)^2 + \tau_{xy}^2]^{1/2}} \tag{3.5.8}$$

and

$$\boxed{\tan 2\beta = \frac{2\tau_{xy}}{\sigma_x - \sigma_y}.} \tag{3.5.9}$$

Equations (5) and (6) then take the form

$$\bar{\sigma}_x = \tfrac{1}{2}(\sigma_x + \sigma_y) + R\,(\cos 2\beta\cos 2\alpha + \sin 2\beta\sin 2\alpha),$$

$$\bar{\tau}_{xy} = R\,(\sin 2\beta\cos 2\alpha - \cos 2\beta\sin 2\alpha),$$

which by use of trigonometric addition formulas are given the simple form

$$\boxed{\begin{aligned} \bar{\sigma}_x &= \tfrac{1}{2}(\sigma_x + \sigma_y) + R\cos(2\beta - 2\alpha), \\ \bar{\tau}_{xy} &= R\sin(2\beta - 2\alpha). \end{aligned}} \tag{3.5.10}$$

For a given state of stress, σ_x, σ_y, R, and β are definite numbers. The stresses $\bar{\sigma}_x$, $\bar{\tau}_{xy}$ on planes at various angles α in Fig. 3.14(a) are then given by Eqs. (10). If the points with coordinates $(\bar{\sigma}_x, \bar{\tau}_{xy})$ are plotted in the stress representation plane for various values of α, all such points will lie on the circle with center at $[\frac{1}{2}(\sigma_x + \sigma_y), 0]$ and radius R in Fig. 3.14(d). Equations (10) are parametric equations of the circle in terms of the parameter α. Its Cartesian equation is found, by eliminating the trigonometric terms, to be

$$[\bar{\sigma}_x - \tfrac{1}{2}(\sigma_x + \sigma_y)]^2 + \bar{\tau}_{xy}^2 = R^2, \tag{3.5.11}$$

where R is given by Eq. (8).

Graphical solution for stress on inclined plane. If $\sigma_x, \sigma_y, \tau_{xy}$ are known, the stresses on any inclined plane may be found graphically as follows.

1. Plot the points $(\sigma_x, 0)$, $(\sigma_y, 0)$,

$$C: \quad [\tfrac{1}{2}(\sigma_x + \sigma_y), 0], \quad \text{and} \quad X: \quad (\sigma_x, \tau_{xy})$$

   in a stress representation σ_n, τ_n-plane.
2. Draw the line CX. This is the reference line corresponding to a plane in the physical body whose normal is the positive x-direction.
3. Draw a circle with center C and radius $R = CX$.
4. To find the point in the stress representation plane that represents any plane in the physical body with normal making a counterclockwise angle α with the x-direction, lay off ***angle 2α clockwise*** from CX. The terminal side $C\bar{X}$ of this angle intersects the circle in point $\bar{X}$ whose coordinates are $(\bar{\sigma}_x, \bar{\tau}_{xy})$.
5. To find $\bar{\sigma}_y$, consider the plane whose normal makes angle $\alpha + \frac{1}{2}\pi$ with the positive x-axis in the physical plane of Fig. 3.14(a). The corresponding angle on the representation circle is $2\alpha + \pi$ measured clockwise from the reference line CX. This locates point D which is at the opposite end of the diameter through $\bar{X}$. The coordinates of D are $\bar{\sigma}_y, -\bar{\tau}_{xy}$.

Figure 3.15 shows the graphical solution with the same stresses $(\sigma_x, \sigma_y, \tau_{xy})$ as in Fig. 3.14 but for a plane whose normal makes a larger angle α with the x-axis, so that $\bar{\tau}_{xy}$ is negative. The negative value for the shear stress means that the traction on the positive-$\bar{x}$ face of the element acts in the negative $\bar{y}$-direction as is shown in Fig. 3.15(b).

Principal stresses in the plane. It may be seen in Figs. 3.14(d) and 3.15(a) that the principal stresses σ_1 and σ_2 in the xy-plane are found at angles 2α equal to 2β and $2\beta + \pi$, respectively. (In the present section, we will always suppose $\sigma_1 \geq \sigma_2$, but the third principal stress, $\sigma_3 = \sigma_z = 0$, is not necessarily ordered, since one or both of σ_1, σ_2 may be negative.) The principal directions at Q in the physical diagram of Fig. 3.14(a) are therefore at angles α equal to β and $\beta + \frac{1}{2}\pi$, respectively, measured counterclockwise from the x-axis. The

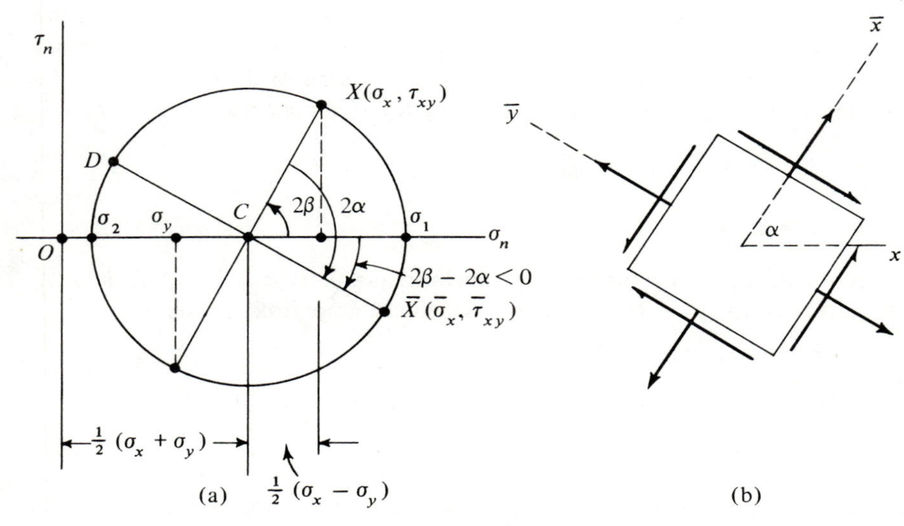

Fig. 3.15 Negative $\overline{\tau}_{xy}$.

principal planes are the planes perpendicular to these principal directions. The angle β may be determined by measurement on the diagram of the graphical solution or calculated from Eqs. (7) and (9). The third principal direction is the z-direction, corresponding to $\sigma_3 = \sigma_z = 0$. The maximum shear stress at the point is $\tau_{\max} = \frac{1}{2}(\sigma_{\max} - \sigma_{\min})$. This may not correspond to $\tau_n = R$ on the plane stress Mohr's Circle. In fact, when σ_1 and σ_2 have the same sign, as in Figs. 3.14 and 3.15, and $\sigma_z = 0$, the maximum shear stress at the point is half the maximum normal stress magnitude as may be seen by drawing all three of the Mohr's Circles as in Sec. 3.4. See Fig. 3.16.

Figure 3.17 shows a convenient graphical method of principal axis deter-

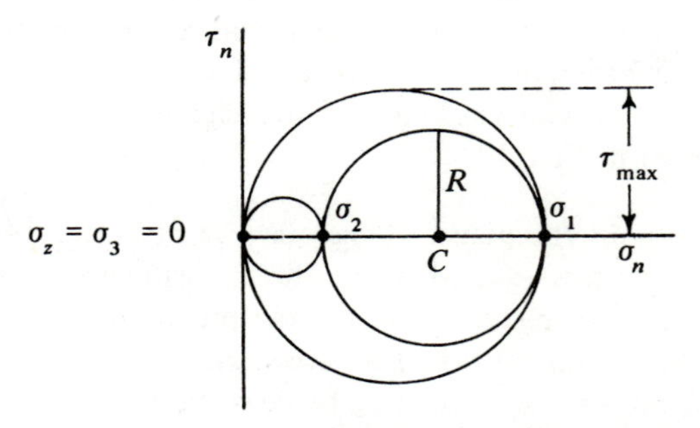

Fig. 3.16 Three Mohr's Circles for plane stress case $\sigma_1 > \sigma_2 > \sigma_3 = 0$.

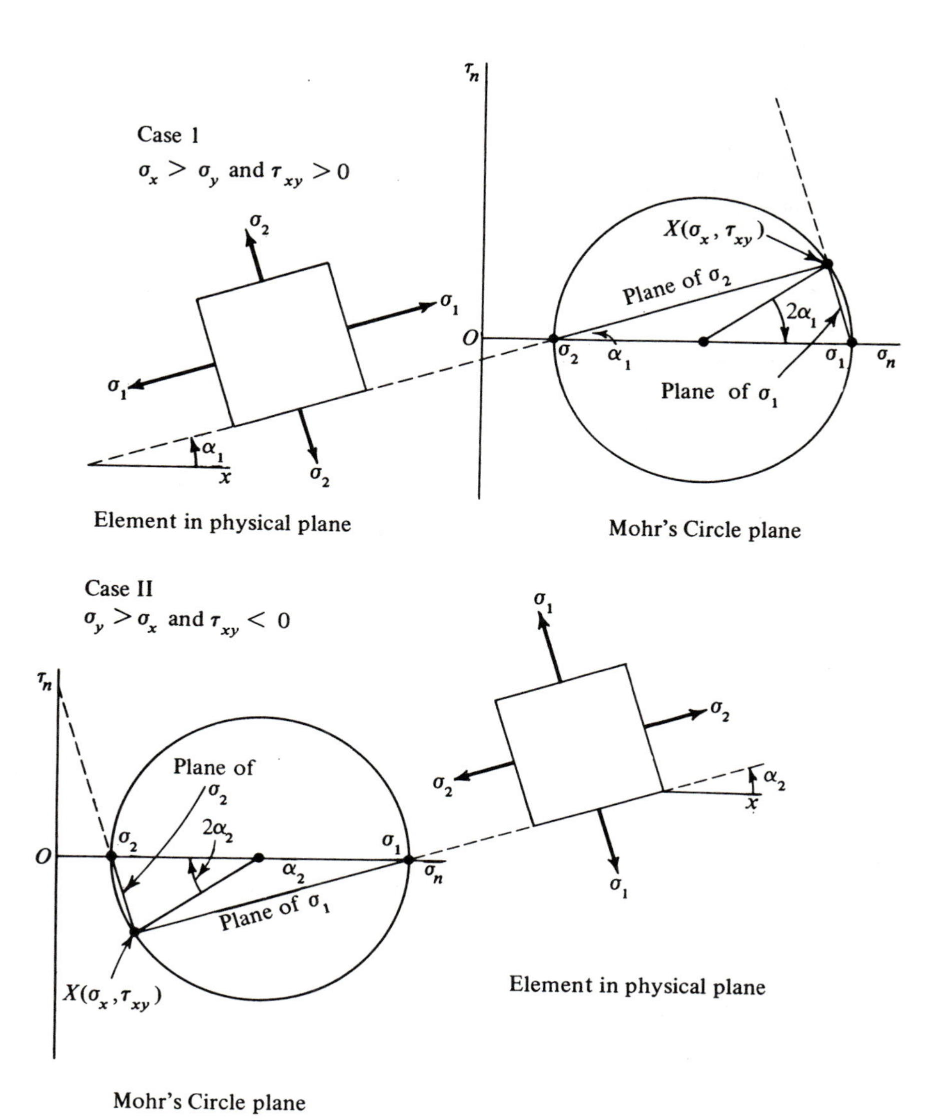

Fig. 3.17 Graphical determination of principal axes.

mination, which makes it easy to avoid the common error of laying off the angles in the wrong direction. The Mohr's Circle is constructed as before and lines drawn from the point X (sometimes called the ***origin of the planes*** or the ***pole***) to the points σ_1 and σ_2 at the ends of the horizontal diameter. The line $X\sigma_1$ is parallel to the plane in the physical body on which σ_1 acts, while $X\sigma_2$

is parallel to the plane on which σ_2 acts. This is illustrated for two cases in Fig. 3.17. Note, for example, that α_1 is the inclination of the ***normal*** of the plane on which σ_1 acts and not the inclination of the plane itself.

After the principal axes have been determined, a simpler representation for the stresses $\bar{\sigma}_x$ and $\bar{\tau}_{xy}$ on an inclined plane is obtained by taking the principal axes as reference axes. For example, let $\sigma_x = \sigma_1, \sigma_y = \sigma_2$. Then $\tau_{xy} = 0$ and $\beta = 0$, and Eqs. (10) take the simpler form:

$$\begin{aligned} \bar{\sigma}_x &= \tfrac{1}{2}(\sigma_x + \sigma_y) + R\cos 2\alpha, \\ \bar{\tau}_{xy} &= -R\sin 2\alpha. \end{aligned} \tag{3.5.12}$$

Figure 3.18 illustrates two cases. Note that the second Eq. (12) clearly requires the angle 2α to be laid off on the Mohr's Circle in the sense opposite to the

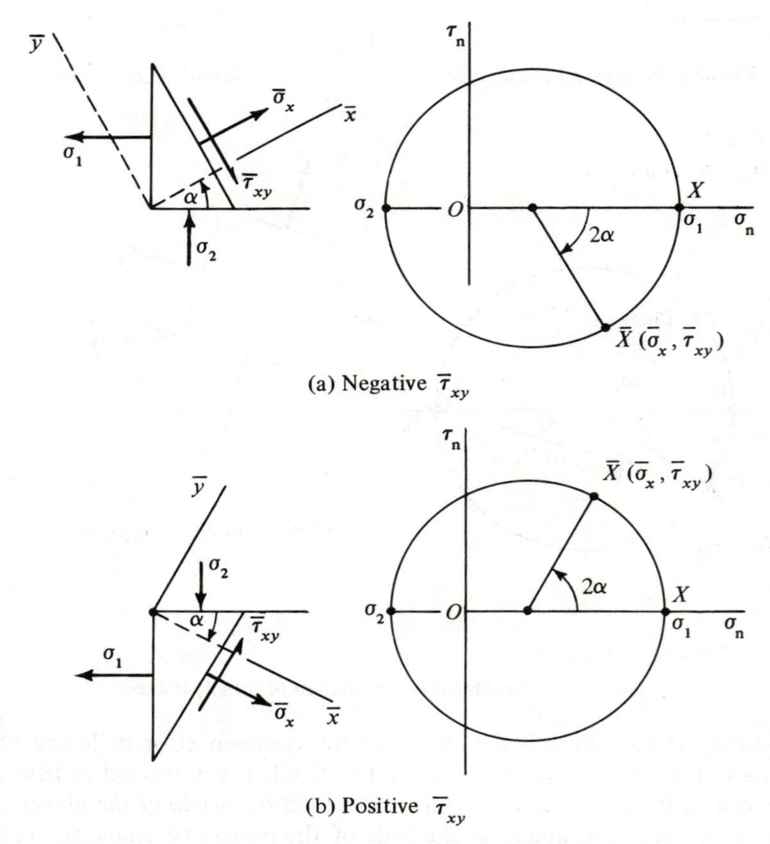

Fig. 3.18 Mohr's Circle when reference axes are principal axes.

sense of angle α in the physical plane if the shear stress sign convention of Sec. 3.2 is used for $\bar{\tau}_{xy}$ and if shear stresses are plotted positively upward in the Mohr's Circle plane.

Alternative sign convention for Mohr's Circle shear stress. Some authors use a different sign convention for τ_n from that used above where σ_n and τ_n were identified with $\bar{\sigma}_x$ and $\bar{\tau}_{xy}$ of a right-handed system of $\bar{x}\bar{y}$-axes and the sign convention for τ_n was the same as for $\bar{\tau}_{xy}$. If left-handed NT-axes are chosen (N normal and T tangential) as in Fig. 3.19(a), then the T-direction is

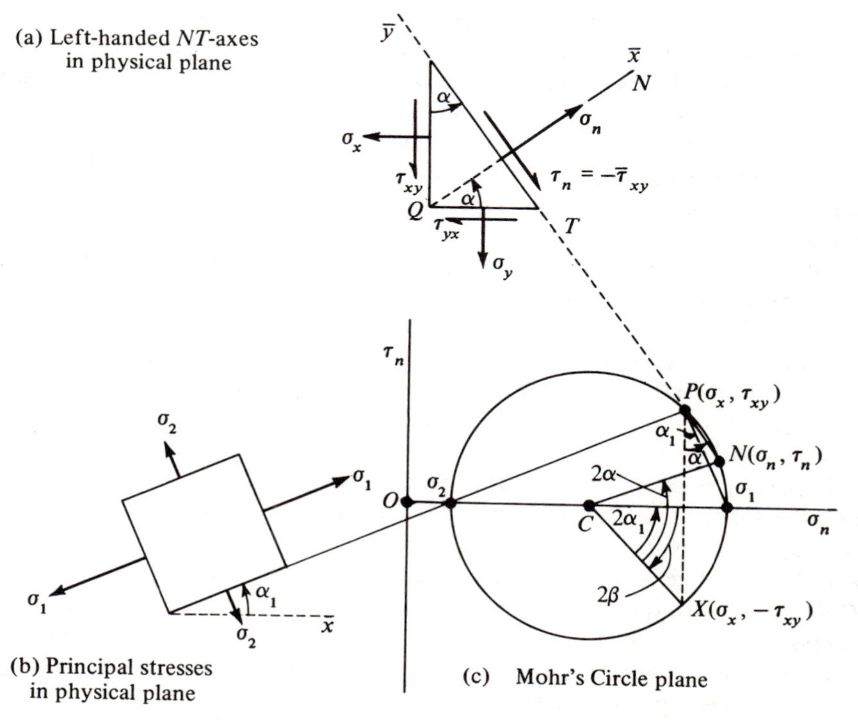

Fig. 3.19 Mohr's Circle for left-handed NT axes.

the negative $\bar{y}$-direction, and if τ_n is reckoned positive when it points in the T-direction, then $\tau_n = -\bar{\tau}_{xy}$. In Fig. 3.14(d), the angle 2β would be measured clockwise from the σ_n axis to CX, and 2α would be laid off counterclockwise from CX to CN as in Fig. 3.19. This alternative sign convention implies that a positive shear stress tends to rotate an element clockwise about Q as does the stress shown on the face perpendicular to the $\bar{x}$-direction in Fig. 3.19(a).

In the graphical determination of σ_n, τ_n for a plane with a normal at angle α to the x-direction, i.e., a plane making angle α with the negative y-direction, we now proceed as follows.

1. Plot points $(\sigma_x, 0)$, $(\sigma_y, 0)$, and $C[\frac{1}{2}(\sigma_x + \sigma_y), 0]$ as before, but plot X as $(\sigma_x, -\tau_{xy})$.
2. Draw reference line CX.
3. Draw circle with center C and radius CX.
4. Lay off 2α ***counter clockwise*** from CX to locate CN; the coordinates of N are the required values (σ_n, τ_n).

The principal directions may still be determined graphically as in Fig. 3.17, but it is necessary first to reflect the point X of Fig. 3.19(c) in the horizontal axis to locate the pole P, the origin of the planes. The present construction may also be used to find the orientation of the physical plane with any traction (σ_n, τ_n) on the Mohr's Circle. To find the plane whose traction is represented by point N, draw PN from the pole P. Since angle XPN equals α (half of the central angle XCN), measured counterclockwise from PX, and since the plane in Fig. 319(a) makes angle α measured counterclockwise from the $-y$-direction, we see that this plane in the Fig. 3.19(a) is parallel to the line PN in Fig. 3.19(c).

Because of the two different sign conventions for the shear stresses used by different writers the proper sense of the angle 2α in the representation plane may be difficult to remember. The following special case furnishes an additional check by physical considerations. Consider the case that $\sigma_1 + \sigma_2 = \sigma_x + \sigma_y = 0$ and take the coordinate axes in the principal directions so that $\sigma_x = \sigma_1 > 0, \sigma_y = \sigma_2 = -\sigma_1 < 0$, and $\tau_{xy} = 0$. Figure 3.20(a) illustrates an element $ABCD$ rotated 45° from the coordinate directions, which will have only shear stresses on its faces. The shear stresses must act in the sense shown in order for the triangles ABD and ABC to be in equilibrium. The question is which way to lay off the angle $2\alpha = 90°$ from CX on the Mohr's Circle, and the answer depends on the sign associated with the shear stress on AB. If ON is the $\bar{x}$-axis, then the shear stress by the usual Cartesian stress sign convention is negative on AB, and in Fig. 3.20(b) we should lay off the angle 2α clockwise to $\bar{X}$ (where τ is negative), as in Fig. 3.15. However, if we are using the alternative sign convention that a positive shear stress on the face AB tends to rotate the element $ABCD$ in a clockwise direction, the shear stress illustrated is a positive shear stress and in Fig. 3.20(b) the angle 2α must be laid off counterclockwise to N (where τ is positive), as in Fig. 3.19.

The procedures given in this article are for ***plane stress***. They may also be applied to the analysis of stress in ***plane strain*** (see Chap. 4), in isotropic elasticity and plasticity, where the stress-strain relations imply that the only nonzero shear stress is $\tau_{xy} = \tau_{yx}$. In general, σ_z is not zero in plane strain, but that fact does not affect the procedures given above.

Graphical determinations by means of Mohr's Circle are in fact conve-

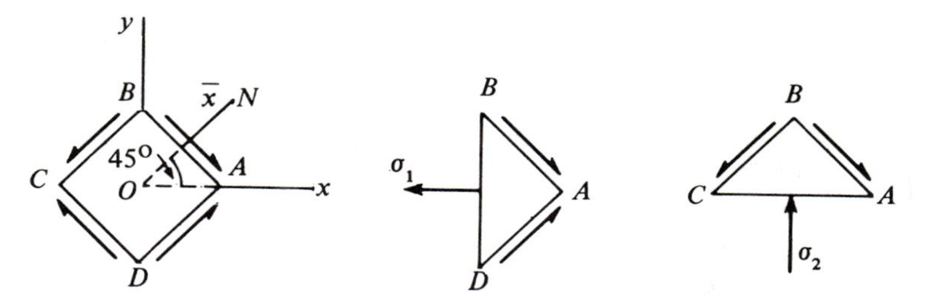

(a) Physical plane

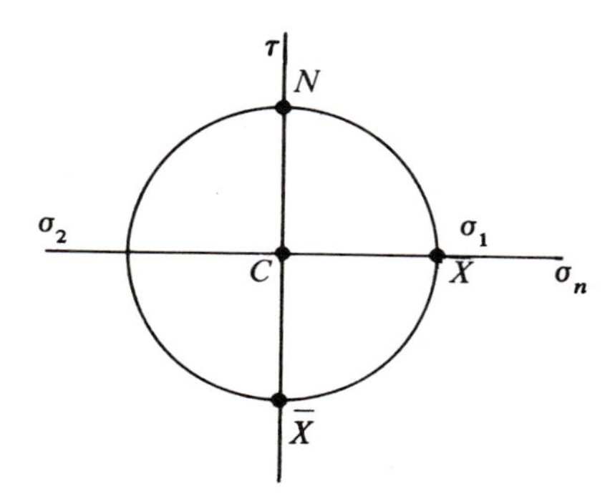

(b) Mohr's Circle

Fig. 3.20 Special case for checking sign convention.

niently made for any symmetric, two-dimensional, second-order Cartesian tensor, since the equations on which our derivation was based are simply the transformation equations for such a tensor. Additional examples of Mohr's Circles will appear in Sec. 3.6 in the discussion of stress resultants in thin plates.

EXERCISES

1. For the plane-stress state $\sigma_x = 10, \sigma_y = 40, \tau_{xy} = 20$ (kpsi), what are the stress components $\bar{\sigma}_x$ and $\bar{\tau}_{xy}$ on a plane whose normal lies in the xy-plane and makes an angle of 30° with the x-axis?

2. For the stress state of Ex. 1, what are the directions of the two principal axes in the xy-plane? Sketch the element on which the principal stresses act.

3. Repeat Ex. 2 for the following plane-stress states (units kpsi), and determine the maximum shear stress in each case.

   (a) $\sigma_x = 55, \quad \sigma_y = 15, \quad \tau_{xy} = 10.$
   (b) $\sigma_x = 15, \quad \sigma_y = 55, \quad \tau_{xy} = 0.$
   (c) $\sigma_x = -30, \quad \sigma_y = 10, \quad \tau_{xy} = 20.$
   (d) $\sigma_x = 30, \quad \sigma_y = -10, \quad \tau_{xy} = -20.$
   (e) $\sigma_x = -10, \quad \sigma_y = 30, \quad \tau_{xy} = -20.$

4. If the stress is plane stress parallel to the xy-plane, and both the yz-plane and the zx-plane are planes of symmetry of the stress state, discuss the evenness and oddness of the dependence on x and the dependence on y for the plane stress components. [See Exercise 20 of Sec. 3.2 (p. 85).]

5. If the yz-plane is a plane of symmetry of stress (not necessarily plane stress), derive a Mohr's-Circle representation for the stresses $\bar{\sigma}_y$ and $\bar{\tau}_{yz}$ at a point in the yz-plane, acting on a plane whose normal is in the direction of the $\bar{y}$-axis. Assume that the $\bar{x}$ and x axes coincide.

6. Sketch the plane-stress Mohr's Circle for $\sigma_x = \sigma, \tau_{xy} = \tau$, and $\sigma_y = 0$. Determine in terms of σ and τ the principal stresses in the plane and the maximum shear stress, and the directions of the normals to the planes on which they act.

7. Use the plane-stress Mohr's Circle to determine the principal directions in the plane for the following plane-stress states. What is the maximum shear stress? Units are kpsi.

   (a) $\sigma_x = 30, \quad \sigma_y = -10, \quad \tau_{xy} = 10.$
   (b) $\sigma_x = 10, \quad \sigma_y = 30, \quad \tau_{xy} = 10.$
   (c) $\sigma_x = 40, \quad \sigma_y = -40, \quad \tau_{xy} = -30.$
   (d) $\sigma_x = 5, \quad \sigma_y = 15, \quad \tau_{xy} = -12.$
   (e) $\sigma_x = 14, \quad \sigma_y = -10, \quad \tau_{xy} = 5.$

3.6 Stress resultants in the simplified theory of bending of thin plates

For many applications of continuum mechanics the problem of determining the three-dimensional stress distribution is too difficult to solve. In some cases, however, a useful result can be obtained by replacing the three-dimensional problem by a one- or two-dimensional simplified theory. This usually occurs in a body having one dimension small compared to the others and possessing certain symmetries of geometrical shape and load distribution. The best known of these simplified theories are the so-called "engineering theories" of beams and of plates or shells. Instead of solving for the stress components throughout the body we solve for certain ***stress resultants***. In plate theory, these stress resultants are line distributions of force or couple per unit length along curves in the middle plane. These line distributions are statically equivalent to the distributions of stress on certain planes of the three-dimensional body and are therefore resultants of the stress on such planes. In this article we show the relationship of these stress resultants to the stresses, and consider some properties of the stress resultants.

In its undeformed condition, a plate is bounded by two parallel planes; we take the xy-plane to be the middle plane between the two planes and take the z-axis downward. The plate edge boundary surface is perpendicular to the xy-plane, and we suppose the thickness h to be small compared to any dimension

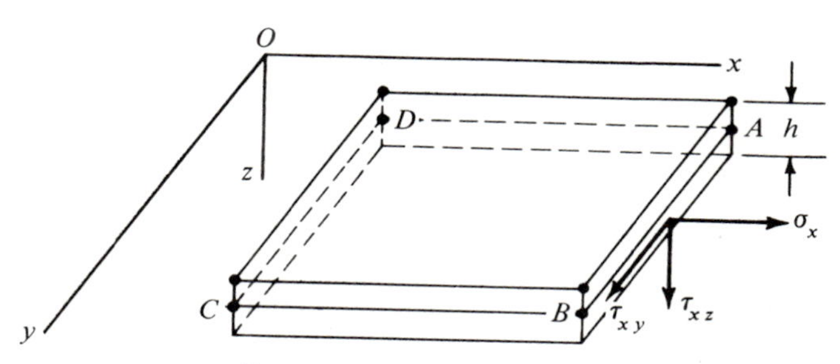

Fig. 3.21 Element of plate.

of the plate in the xy-plane. Figure 3.21 shows an element of the deformed plate cut out by planes parallel to the xz- and yz-planes. Plate theory replaces the distributed normal stress σ_x per unit of area on the positive x-face of the element by the following stress resultants: normal force N_x ***per unit length*** along AB, and ***bending moment*** M_x ***per unit length along*** AB,

$$N_x = \int_{-h/2}^{h/2} \sigma_x \, dz, \tag{3.6.1}$$

$$M_x = \int_{-h/2}^{h/2} z\sigma_x \, dz. \tag{3.6.2}$$

The line distribution of force N_x and couple M_x are statically equivalent to the surface distribution of σ_x. The distribution of vertical shear stress τ_{xz} is statically equivalent to a ***vertical shear force*** Q_x per unit length along AB.

$$Q_x = \int_{-h/2}^{h/2} \tau_{xz} \, dz. \tag{3.6.3}$$

The ***twisting moment*** M_{xy} per unit length due to horizontal shear forces is defined by

$$M_{xy} = -\int_{-h/2}^{h/2} z\tau_{xy} \, dz. \tag{3.6.4}$$

It measures the twisting couple about an axis parallel to the x-axis due to the distribution of horizontal shear stress τ_{xy} on the face. Note that $M_{xy}\,dy$ is taken as the total twisting moment on a strip of infinitesimal width dy, because the twisting moment due to the vertical shear stresses τ_{xz} on the strip is assumed negligible since its moment arm is of the order of dy. The minus sign is inserted in Eq. (4) so that a positive twisting moment turns the y-axis toward the z-axis in a counterclockwise sense as seen from the positive x-axis. The corresponding twisting moment M_{yx} per unit length along BC is, however,

$$M_{yx} = +\int_{-h/2}^{h/2} z\tau_{yx}\,dz. \tag{3.6.5}$$

A positive M_{yx} turns the z-axis toward the x-axis in a counterclockwise sense as seen from the positive y-axis.† Since $\tau_{yx} = \tau_{xy}$ at a point, it follows that at a point

$$M_{yx} = -M_{xy}. \tag{3.6.6}$$

The horizontal shear forces N_{xy} and N_{yx} per unit length are defined by

$$N_{xy} = \int_{-h/2}^{h/2} \tau_{xy}\,dz, \qquad N_{yx} = \int_{-h/2}^{h/2} \tau_{yx}\,dz. \tag{3.6.7}$$

It follows that at a point

$$N_{xy} = N_{yx}. \tag{3.6.8}$$

The distribution of τ_{xy} on the plane perpendicular to the x-axis is thus statically equivalent to the two line distributions, of twisting moment M_{xy} and of horizontal shear force N_{xy} per unit length.

Forces N_y and Q_y and bending moments M_y per unit length are defined by replacing x by y in Eqs. (1) to (3). Figure 3.22 summarizes the positive sign conventions. Note that both M_{xy} and M_{yx} are illustrated with the sense they would have if they were both positive numbers. If the element is shrunk to a point so that they are both acting at the same point they cannot actually both be positive, since at a point $M_{yx} = -M_{xy}$.

The forces (per unit length) N_x, N_y, N_{xy} acting in the plane of the plate are called ***membrane forces***. If no external loads act parallel to the plane of the plate and the edges are not restrained against motion in the plane of the plate, the membrane forces are usually supposed to be negligible when the stretching strain in the middle surface is small in comparison with the bending strain on the cross-section. This limitation requires the lateral deflection to be small in comparison with the plate thickness h.‡ On the other hand, in a very thin plate or membrane with negligible bending stiffness, the displacements may be controlled by the membrane forces.

If we choose new $\bar{x}\bar{y}$-axes in the xy-plane with the $\bar{x}$-axis inclined at angle α to the old x-axis as in Sec. 3.5, the new values of $\bar{\sigma}_x$ and $\bar{\tau}_{xy}$ are again given by Eqs. (3.5.3) and (3.5.4), because we still have $a^z_{\bar{x}} = a^z_{\bar{y}} = 0$. If we multiply

†These sign conventions are the most commonly used. See, for example, S. Timoskenko and S. Woinowsky-Krieger, *Theory of Plates and Shells*. New York: McGraw-Hill, 2nd ed., 1959. Some writers, however, define the twisting moments so that $M_{xy} = M_{yx}$, either both defined by Eq. (4), or alternatively both by Eq. (5).

‡See Timoshenko and Woinowsky-Krieger, *ibid*., p. 48.

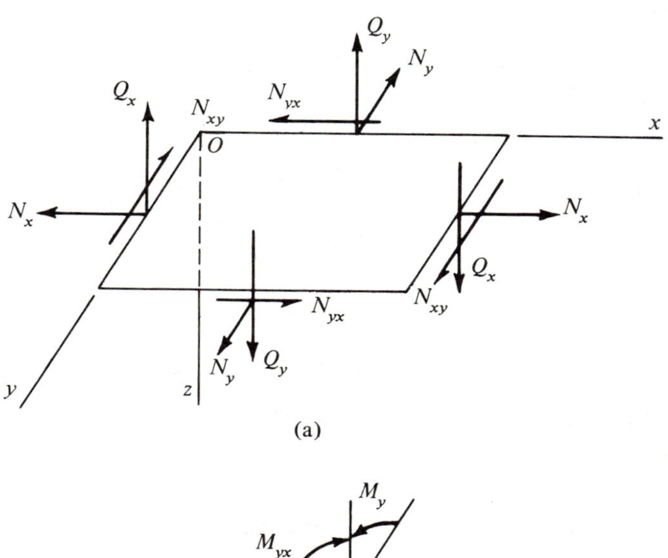

(a)

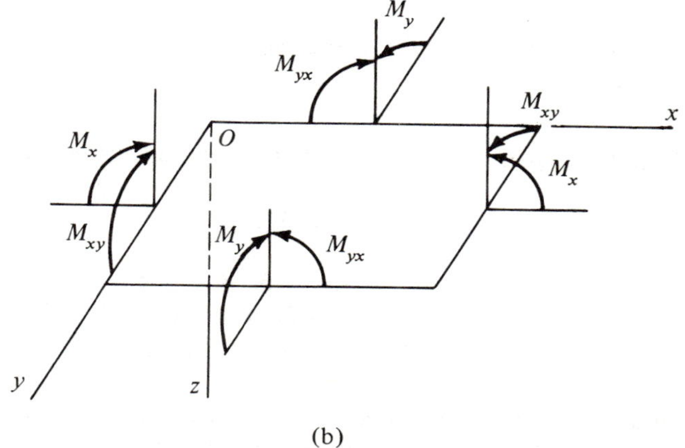

(b)

Fig. 3.22 Sign convention for stress resultants: (a) positive forces; (b) positive moments.

each of these equations through by $z\,dz$ and integrate with respect to z from $-h/2$ to $+h/2$, we obtain (denoting $\bar{M}_x$ by M_n and $\bar{M}_{xy}$ by M_{nt})

$$\left.\begin{aligned} M_n &= M_x \cos^2\alpha + M_y \sin^2\alpha - 2M_{xy}\sin\alpha\cos\alpha \\ -M_{nt} &= -M_{xy}(\cos^2\alpha - \sin^2\alpha) - (M_x - M_y)\sin\alpha\cos\alpha \end{aligned}\right\}. \tag{3.6.9}$$

Or using the trigonometric double angle identities, as in Sec. 3.5, we have

$$\left.\begin{aligned} M_n &= \tfrac{1}{2}(M_x + M_y) + \tfrac{1}{2}(M_x - M_y)\cos 2\alpha - M_{xy}\sin 2\alpha \\ M_{nt} &= M_{xy}\cos 2\alpha + \tfrac{1}{2}(M_x - M_y)\sin 2\alpha. \end{aligned}\right\} \tag{3.6.10}$$

If we now let

$$R = [\tfrac{1}{4}(M_x - M_y)^2 + M_{xy}^2]^{1/2} \tag{3.6.11}$$

and

$$\left.\begin{aligned} \tfrac{1}{2}(M_x - M_y) = R\cos 2\beta, \qquad M_{xy} = R\sin 2\beta \\ \tan 2\beta = \frac{2M_{xy}}{M_x - M_y}, \end{aligned}\right\} \tag{3.6.12}$$

Equations (10) take the form

$$\begin{aligned} M_n &= \tfrac{1}{2}(M_x + M_y) + R\,(\cos 2\beta \cos 2\alpha - \sin 2\beta \sin 2\alpha) \\ M_{nt} &= R\,(\sin 2\beta \sin 2\alpha + \cos 2\beta \cos 2\alpha) \end{aligned}$$

or

$$M_n = \tfrac{1}{2}(M_x + M_y) + R\cos(2\beta + 2\alpha) \tag{3.6.13}$$

$$M_{nt} = R\sin(2\beta + 2\alpha). \tag{3.6.14}$$

Hence M_n and M_{nt} plot on the Mohr's Circle whose Cartesian equation is

$$[M_n - \tfrac{1}{2}(M_x + M_y)]^2 + M_{nt}^2 = R^2 \tag{3.6.15}$$

in a manner similar to $\bar{\sigma}_x$ and $\bar{\tau}_{xy}$ in Sec. 3.5. But since the trigonometric functions in Eqs. (13) and (14) are of $2\beta + 2\alpha$ instead of $2\beta - 2\alpha$, the angle 2α should be laid off counterclockwise from the radius CX to locate CN; see Fig. 3.23.

The maximum and minimum values of M_n occur for values of α where M_{nt} is zero. These are the principal values M_1 and M_2 of the bending moment. The principal axes, lines across which these principal bending moments act, may be determined graphically as indicated in Fig. 3.23. The pole P is the reflection of X in the M_n-axis, and lines PM_1 and PM_2 in the representation plane are parallel to the lines in the physical plane across which act the principal bending moments M_1 and M_2, respectively. A line PN in the representation plane would also be parallel to the line in the physical plane across which act bending moment M_n and twisting moment M_{nt}. Note that, for example, α_2 is the angle from the x-axis to the normal to the line on which M_2 acts; it is also the angle to the line itself from the negative y-axis. The same double interpretation may be made of the angle α for any (M_n, M_{nt}) point on the Mohr's Circle.

In Fig. 3.23, the xy-axes and M_n, M_{nt}-axes are both right-handed. This would be the view as seen from the positive z-axis. Since the positive z-axis

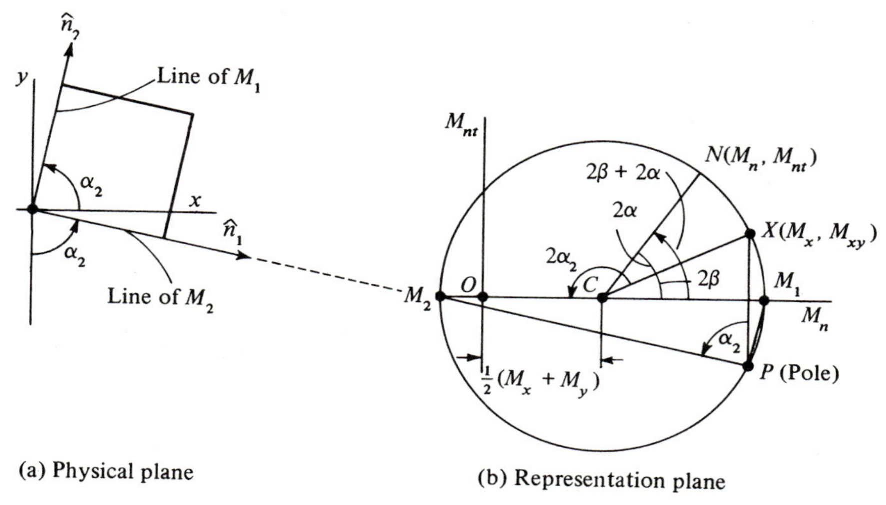

(a) Physical plane (b) Representation plane

Fig. 3.23 Mohr's Circle for bending and twisting moments.

is downward, many writers prefer to use a view of the top of the plate, where the xy-axes appear left-handed. The graphical procedure is adapted to this view by using left-handed M_n, M_{nt}-axes also and measuring all angles clockwise in both planes.

We have now seen three different versions of the two-dimensional Mohr's Circle procedure with the differences occasioned by varying sign conventions. The last version above differed from our first one for plane stress in the sign conventions for both M_{nt} and M_{xy}, while the second version (Fig. 3.19 of Sec. 3.5) differed from the first version in the sign convention for τ_n but not in that for τ_{xy}.

The proper direction for laying off the angles 2α from CX can be checked by the following physical consideration for a special case of moments in a plane. Consider the case that at a point $M_x + M_y = M_1 + M_2 = 0$ so that $M_2 = -M_1$, and take reference axes in the physical plane so that the x-axis is normal to the line on which M_1 acts. Figure 3.24(a) shows an element of the middle plane of the plate bounded by lines parallel to the x-axis and the y-axis and by a third line AB whose normal ON bisects the angle between the axes ($\alpha = 45°$). The case illustrated is such that $M_x = M_1 > 0$, and $M_{xy} = M_{yx} = 0$. According to the Mohr's Circle for this point, Fig. 3.24(b), $M_n = 0$ on AB; the question is to decide whether to rotate 90° counterclockwise to N or 90° clockwise to N' to determine M_{nt}. Consideration of static equilibrium of the element in Fig. 3.24(a) shows that M_{nt} must have the sense illustrated in order to balance the turning moments of M_1 and M_2 about ON; since this sense is a positive twisting moment we conclude that we must rotate counterclockwise to N, where M_{nt} is positive, as shown.

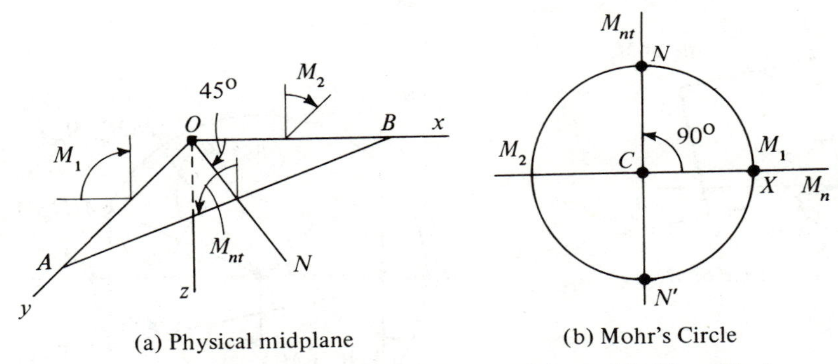

(a) Physical midplane (b) Mohr's Circle

Fig. 3.24 Check for sense of angle 2α (case $M_2 = -M_1$, M_1 positive).

This chapter has introduced the stress tensor and its properties using Cartesian coordinates. Chapter 5 will include additional properties of stress. First, however, we turn our attention to the description of strain and deformation.

EXERCISES

1. For a prismatic beam whose cross-section $x =$ const. has two axes of symmetry, assume that the stress resultants N_{xi} and M_{xi} are defined by the following relationships to the total force **f** at the center and the couple **c** which together with **f** forms a system statically equivalent to the tractions on the cross-section A.

$$\mathbf{f} = N_{xx}\mathbf{i} + N_{xy}\mathbf{j} + N_{xz}\mathbf{k} \qquad \mathbf{c} = M_{xx}\mathbf{i} + M_{xy}\mathbf{j} + M_{xz}\mathbf{k}$$

Express each resultant as an integral over the cross-section A, in which the integrand contains one or more stresses, and interpret the resultants as normal or transverse shear forces and bending or twisting moments. For example, one bending moment is $M_{xz} = -\int_A y\sigma_x \, dA$.

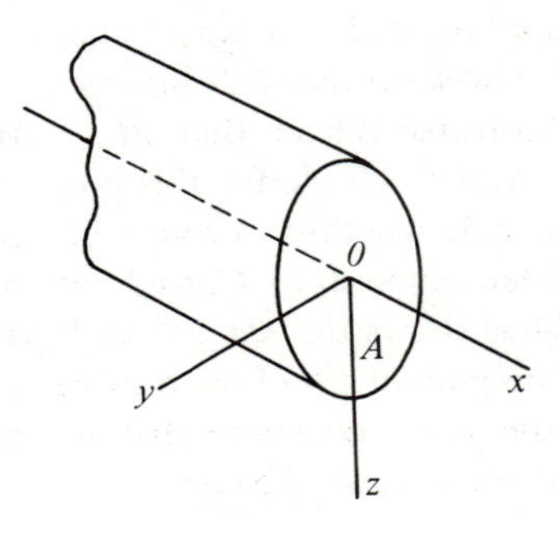

2. For each of the following conditions at a point in a plate, sketch the Mohr's Circle and determine M_1 and M_2 and the lines across which they act. What is the maximum M_{nt}? What M_n and M_{nt} act across a line at $\alpha = 64.4°$ to $-y$?

   (a) $M_x = 2M, \quad M_y = -2M, \quad M_{xy} = M.$
   (b) $M_x = -M, \quad M_y = 3M, \quad M_{xy} = M.$
   (c) $M_x = 4M, \quad M_y = 0, \quad M_{xy} = -M.$
   (d) $M_x = 0, \quad M_y = 4M, \quad M_{xy} = -M.$

3. For each of the following conditions at a point in a plate, sketch the Mohr's Circle and determine M_1 and M_2 and the lines across which they act. What is the maximum M_{nt}? What are M_n and M_{nt} acting across a line at $\alpha = 37.5°$ to $-y$?

   (a) $M_x = 3M, \quad M_y = M, \quad M_{xy} = M.$
   (b) $M_x = -2M, \quad M_y = M, \quad M_{xy} = M.$
   (c) $M_x = 2M, \quad M_y = 0, \quad M_{xy} = -M.$
   (d) $M_x = -M, \quad M_y = M, \quad M_{xy} = -M.$

4. Outline a Mohr's Circle procedure for the "membrane forces" N_x, N_y, N_{xy}, being sure to use a consistent relationship for plotting positive N_{xy} and positive α.

5. (a) Where on the Mohr's Circle for M_n and M_{nt} is the point $Y(M_y, M_{yx})$?
   (b) In polar coordinates (r, θ) the bending moments M_r and M_θ act across the circle $r =$ const. and the line $\theta =$ const., respectively. Give formulas for M_r, M_θ, $M_{r\theta}$ in terms of M_x, M_y, M_{xy} and show how they plot on Mohr's Circle.

6. Derive a relationship between the vertical shear force Q_n and the shear forces Q_x and Q_y, where Q_n acts on an element whose normal is parallel to the xy-plane and makes angle α with the x-axis.

CHAPTER 4

Strain and Deformation

4.1 Small strain and rotation in two dimensions

In a uniaxial tension test of an elastic metal, strain is ordinarily defined as change in length per unit of initial length. Thus if a two-inch gage length is stretched to 2.002 in. length, the strain is 0.001, or 0.1 per cent, and this statement is sufficient to characterize the state of extensional deformation. But if the gage length is first stretched to a length of 3 in., and the specimen is subsequently compressed back to a length of 2.002 in., a statement that the final strain is 0.001 does not adequately describe what has happened. The plastic deformation accompanying both the stretching and the subsequent compression leaves the material, in general, in quite a different state from the specimen which was merely strained elastically to a length of 2.002 in. When plastic deformation occurs, a complete characterization of the deformed state requires us, in general, even in uniaxial loading, to follow the history of straining; the final strain alone does not characterize the deformation that has occurred. Following the history can be very difficult; we may have to be content with describing a small increment of the deformation or formulating our equations in terms of the rate of deformation at a given instant.

Rate of deformation also figures prominently in the viscous behavior of fluids and viscoelastic or viscoplastic substances. The mathematical description of rate of deformation will be given in Sec. 4.4, where the rate-of-deformation tensor is introduced and defined with respect to the instantaneous state of the material. In elasticity, the strain definition is based on an ***initial state*** against which the current configuration is compared; it is assumed that the detailed process by which the material has moved from the initial state to the current configuration does not affect the final state, provided the process is elastic.

Even with this assumption the characterization of the state of strain is not simple unless the displacements and displacement gradients are small; in fact only when they are infinitesimal is the so-called small-strain theory rigorously correct. Nevertheless, in metals, where elastic strains usually are not much greater than 0.002, the infinitesimal strain theory gives good results for practical purposes. The finite-strain definitions will be discussed in Sec. 4.5. We postpone these complications and consider first the case of small strain in two dimensions. The three-dimensional case will be given in Sec. 4.2.

Elementary definitions of strain. Figure 4.1 illustrates three simple cases

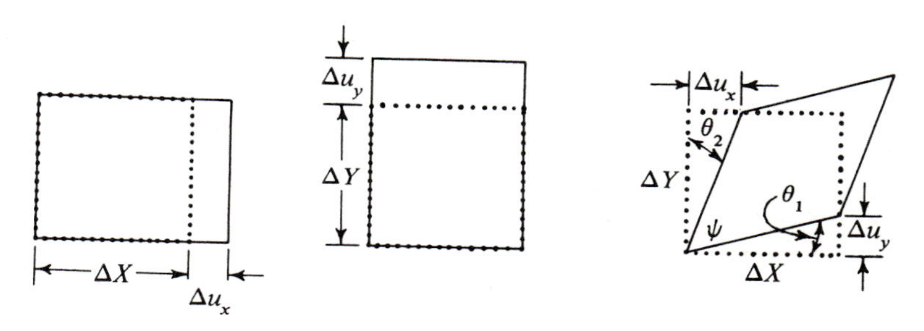

(a) Uniaxial extension in x direction

$\epsilon_x \approx \dfrac{\Delta u_x}{\Delta X}$

(b) Uniaxial extension in y direction

$\epsilon_y \approx \dfrac{\Delta u_y}{\Delta Y}$

(c) Pure shear without rotation ($\theta_1 = \theta_2$)

$\gamma_{xy} = \dfrac{\pi}{2} - \psi = \theta_2 + \theta_1$

$\epsilon_{xy} = \frac{1}{2}\gamma_{xy} \approx \dfrac{1}{2}\left(\dfrac{\Delta u_x}{\Delta Y} + \dfrac{\Delta u_y}{\Delta X}\right)$

Fig. 4.1 Pure strain in two dimensions.

of strain. In each case, the initial position of the element is outlined by the dotted lines. If the ***unit extensions*** ϵ_x and ϵ_y are defined as change in length per unit initial length, and the ***shear strain*** ϵ_{xy} is defined as half the decrease γ_{xy} in the right angle initially formed by the sides parallel to the X- and Y-axes in Fig. 4.1(c), we have in the limit as ΔX and ΔY approach zero,

$$\epsilon_x = \frac{\partial u_x}{\partial X}, \qquad \epsilon_y = \frac{\partial u_y}{\partial Y}, \qquad \epsilon_{xy} = \frac{1}{2}\gamma_{xy} = \frac{1}{2}\left(\frac{\partial u_x}{\partial Y} + \frac{\partial u_y}{\partial X}\right), \quad (4.1.1)$$

where (u_x, u_y) are the particle displacement components parallel to the X- and Y-axes. In obtaining the expression for the shear strain, the angle θ_1 has been approximated by its tangent $\Delta u_y/\Delta X$ and θ_2 by $\Delta u_x/\Delta Y$. This is a good approximation if the angles are small compared to one radian, a limitation not serious for elastic strains in metals (of the order of a thousandth of a radian), but preventing the application of this "small-strain theory" to materials in which the strains may be large. The reason for introducing the factor of $\frac{1}{2}$ in the shear strain definition is so that the strain components will be components of a second-order tensor, as will be seen later in this section and in Sec. 4.2.

We use capital letters to represent the coordinates of a particle in the initial state and lower-case letters for the final or current position coordinates of the particle. Then $x = X + u_x$, etc., as in Sec. 4.3 on kinematics. When the displacements are small, as in elasticity of metals, it is usually possible to ignore the distinction between (X, Y, Z) and (x, y, z) when evaluating some function of the coordinates; for example, we often write $\epsilon_x = \partial u_x/\partial x$, but for strict accuracy the distinction must be made. In Sec. 4.5 we will see an alternative formulation in which the final position is taken as the independent variable instead of the initial position.

We see that the strain components are quantitative measures of certain types of relative displacement between neighboring parts of the material. A solid material in general resists such relative displacement of its parts. The result is that internal forces are brought into play; these are the stresses discussed in Chap. 3. However, not all kinds of relative motion give rise to strains and stresses in a solid. If a body moves as a rigid body, the rotational part of its motion produces relative displacement. For example, if the block *ABCD* of Fig. 4.2 moves as a rigid body to position *abcd*, the relative vertical displacement component of *B* relative to *A* is *fb*, and the other particles of the body all have some displacement relative to *A*; no strain or deformation is, however, produced.

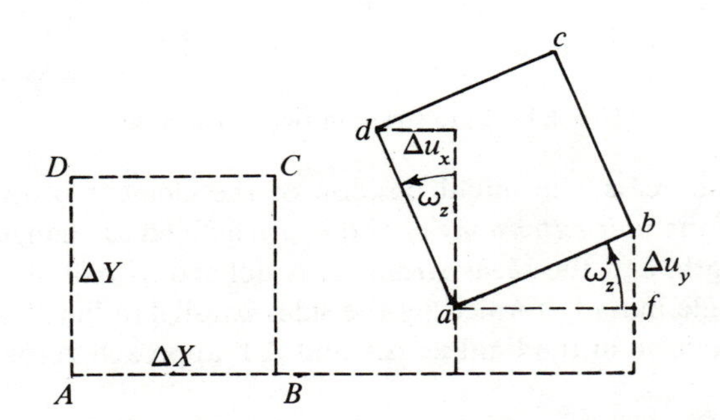

Fig. 4.2 Displacement and rigid rotation in *XY*-plane.

The analysis of displacement and strain for small deformations of a solid body is complicated by the fact that we are seldom concerned with the simple uniform strains illustrated in Fig. 4.1, but rather with cases in which all three of the kinds of strain occur simultaneously in one body, while the body is rotating. Moreover, both the strain and the rotation are variable functions of position in the body. For example, in Fig. 4.3, according to elementary beam theory, the two overhanging ends of the beam undergo essentially pure rotations about the supports, while between the two supports there is both rotation and strain.

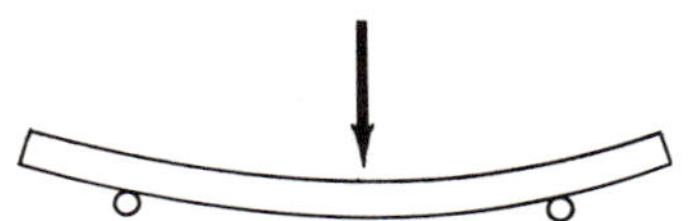

Fig. 4.3 Illustrating rotation and deformation variable with position in the body.

The general problem is to express the strain in terms of the displacements by separating off that part of the displacement distribution which does not contribute to the strain. It is a formidable task in general. When the displacement gradients are small compared to unity, the separation can be achieved, and it turns out, surprisingly enough, that the relevant measures of strain are still precisely the elementary expressions given in Eqs. (1). In order to show that this is the case, the relative displacements in a general two-dimensional motion will be analyzed in detail under the assumption that the displacement gradients are small compared to unity. The comparable results for three dimensions will be given in Sec. 4.2.

Relative displacements. Consider two neighboring particles P and Q in a continuous medium. Suppose that as the medium is displaced and deformed, they move in a plane parallel to the XY-plane until they occupy the positions p and q, as in Fig. 4.4. The ***relative displacement*** of Q relative to P is

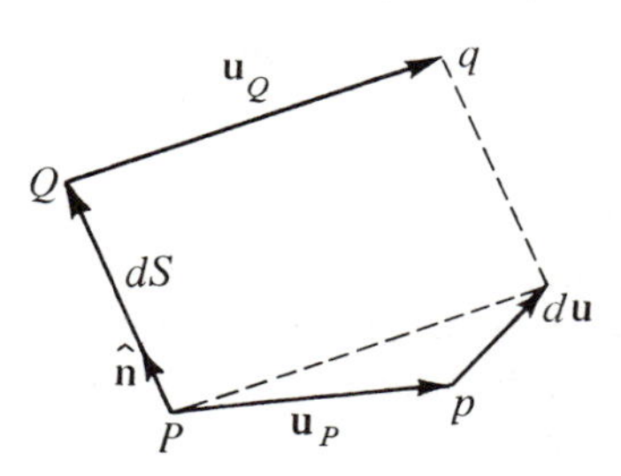

Fig. 4.4 Relative displacement $d\mathbf{u}$ of Q relative to P.

$$d\mathbf{u} = \mathbf{u}_Q - \mathbf{u}_P.$$

The ***unit relative displacement*** is $d\mathbf{u}/dS$, where the magnitude dS is the length of PQ.† The rectangular components of the unit relative displacement are (in the limit as $dS \to 0$):

$$\frac{du_x}{dS} = \frac{\partial u_x}{\partial X}\frac{dX}{dS} + \frac{\partial u_x}{\partial Y}\frac{dY}{dS}, \qquad \frac{du_y}{dS} = \frac{\partial u_y}{\partial X}\frac{dX}{dS} + \frac{\partial u_y}{\partial Y}\frac{dY}{dS}, \tag{4.1.2}$$

where the partial derivatives are evaluated at the point P and do not depend upon the components $(dX/dS,\ dY/dS)$ of the unit vector $\hat{\mathbf{n}}$ in the direction of PQ. Equations(2) may be written in matrix form as

†The symbol dS was used in Sec. 3.1 for the element of surface area. No confusion should arise, since it will almost never be used both ways in one discussion. If necessary to distinguish, we can use dA for area or ds_0 for initial length of PQ.

$$\begin{bmatrix} \frac{du_x}{dS} \\ \frac{du_y}{dS} \end{bmatrix} = \begin{bmatrix} \frac{\partial u_x}{\partial X} & \frac{\partial u_x}{\partial Y} \\ \frac{\partial u_y}{\partial X} & \frac{\partial u_y}{\partial Y} \end{bmatrix} \begin{bmatrix} \frac{dX}{dS} \\ \frac{dY}{dS} \end{bmatrix} \tag{4.1.3a}$$

or

$$\frac{du}{dS} = J_u n, \tag{4.1.3b}$$

representing the tensor equation

$$\frac{d\mathbf{u}}{dS} = \mathbf{u}\overleftarrow{\nabla}\cdot\hat{\mathbf{n}} \tag{4.1.3c}$$

where du/dS is the column matrix of unit relative displacement vector components of Q relative to P, n is the column matrix of direction cosines ($n_x = dX/dS$, $n_y = dY/dS$) of PQ, and J_u is the 2×2 matrix shown in Eq. (3a) called the ***displacement gradient matrix.*** Each row of the square matrix contains the components of the gradient of one displacement component. The square matrix is also called the ***Jacobian matrix*** or the ***unit relative displacement matrix.*** It may be regarded as an operator which operates on any direction matrix n to yield the unit relative displacement components for an infinitesimal element PQ with initial direction n. Its transpose is the matrix of $\overrightarrow{\nabla}\mathbf{u}$, such that $d\mathbf{u}/dS = \hat{\mathbf{n}}\cdot\overrightarrow{\nabla}\mathbf{u}$. See Eqs. (2.5.32).

We now seek to separate J_u into the sum of two matrices, one of which accounts for the strain or deformation in the neighborhood of P and the other for those rigid-body relative displacements not giving rise to any deformation.† It turns out that what we need is to separate J_u into the sum of a symmetric matrix and a skew-symmetric matrix as shown in Eq. (4).

$$\begin{bmatrix} \frac{\partial u_x}{\partial X} & \frac{\partial u_x}{\partial Y} \\ \frac{\partial u_y}{\partial X} & \frac{\partial u_y}{\partial Y} \end{bmatrix} = \begin{bmatrix} \frac{\partial u_x}{\partial X} & \frac{1}{2}\left(\frac{\partial u_x}{\partial Y} + \frac{\partial u_y}{\partial X}\right) \\ \frac{1}{2}\left(\frac{\partial u_y}{\partial X} + \frac{\partial u_x}{\partial Y}\right) & \frac{\partial u_y}{\partial Y} \end{bmatrix} + \begin{bmatrix} 0 & \frac{1}{2}\left(\frac{\partial u_x}{\partial Y} - \frac{\partial u_y}{\partial X}\right) \\ \frac{1}{2}\left(\frac{\partial u_y}{\partial X} - \frac{\partial u_x}{\partial Y}\right) & 0 \end{bmatrix} \tag{4.1.4}$$

And we recognize the elements in the symmetric matrix as being, with the

†The principle that, in the case of small displacement gradients, the three-dimensional displacement gradient can be resolved into the sum of a strain and a rotation was implicit in the work of Cauchy (1789–1857).

factors of $\frac{1}{2}$, the strain components defined by Eqs. (1). Let

$$\left.\begin{aligned} \Omega_{xy} &= \frac{1}{2}\left(\frac{\partial u_x}{\partial Y} - \frac{\partial u_y}{\partial X}\right), \qquad \Omega_{yx} = -\,\Omega_{xy} = \frac{1}{2}\left(\frac{\partial u_y}{\partial X} - \frac{\partial u_x}{\partial Y}\right) \\ \epsilon_{xy} &= \frac{1}{2}\left(\frac{\partial u_x}{\partial Y} + \frac{\partial u_y}{\partial X}\right), \qquad \epsilon_{yx} = \epsilon_{xy} = \frac{1}{2}\left(\frac{\partial u_y}{\partial X} + \frac{\partial u_x}{\partial Y}\right) \end{aligned}\right\} \tag{4.1.5}$$

The matrix equation (4.1.4) becomes

$$J_u = E + \Omega, \tag{4.1.6}$$

where E is the ***strain matrix*** and Ω is the ***rotation matrix,*** defined by

$$E = \begin{bmatrix} \epsilon_x & \epsilon_{xy} \\ \epsilon_{yx} & \epsilon_y \end{bmatrix}, \qquad \Omega = \begin{bmatrix} 0 & \Omega_{xy} \\ \Omega_{yx} & 0 \end{bmatrix}. \tag{4.1.7}$$

The first row in each matrix is designated the x row, while the second is the y row (based on the corresponding displacement components in J_u); the first column is designated the x column and the second column is the y column (based on the differentiation variable in J_u). The first subscript on an element with two subscripts identifies the row, while the second identifies the column; this agrees with the way the elements were placed in the stress matrix. If these conditions are observed, then the proper algebraic sign for the definition of Ω_{xy} or Ω_{yx} can always be verified by checking the matrix addition Eq. (4). We still have to show that the names we have given the two matrices are appropriate by considering the two separately to see that one does indeed produce deformation while the other produces rotation.

Pure strain without rotation. If the elements of the rotation matrix Ω are all zero, the unit relative displacements can be found by the matrix product En. For example, in Fig. 4.5 we may find the unit relative displacement of B relative to A as in the equation below. The direction matrix n has components 1 and 0 if n represents the direction AB. Hence

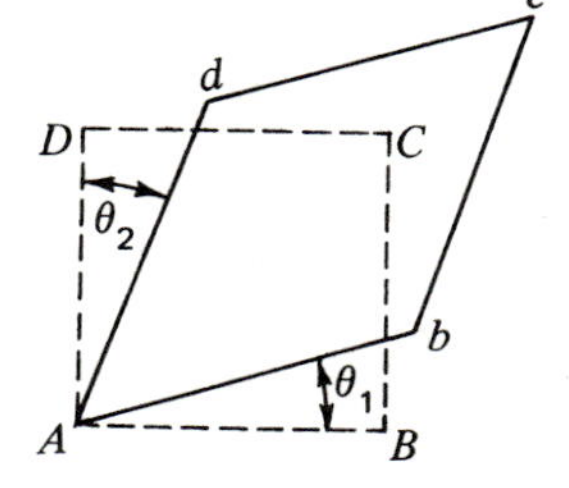

Fig. 4.5 Strain without rotation.

$$\begin{bmatrix} \dfrac{du_x}{dS} \\[2ex] \dfrac{du_y}{dS} \end{bmatrix} = \begin{bmatrix} \epsilon_x & \epsilon_{xy} \\ \epsilon_{yx} & \epsilon_y \end{bmatrix} \begin{bmatrix} 1 \\ 0 \end{bmatrix} = \begin{bmatrix} \epsilon_x \\ \epsilon_{yx} \end{bmatrix}.$$

Thus for AB,

$$\frac{du_x}{dS} = \epsilon_x, \qquad \frac{du_y}{dS} = \epsilon_{yx}.$$

Similarly for AD,

$$\frac{du_x}{dS} = \epsilon_{xy}, \qquad \frac{du_y}{dS} = \epsilon_y.$$

Since $\epsilon_{yx} = \epsilon_{xy}$, we have $\theta_1 = \theta_2$, and there is no net rotation of the block $ABCD$.

Evidently the columns in the strain matrix individually represent the unit relative displacement vectors for two elements AB and AD, the first column for the element in the X-direction and the second for the element in the Y-direction. Since the strain matrix is symmetric, we could just as well suppose that the first row corresponds to the element in the X-direction, and the second row to the element in the Y-direction and have a closer analogy to the rows of the stress matrix; but this will not work with the matrix J_u or the rotation matrix Ω, where the columns must be used. This example illustrates another physical interpretation of the shear components $\epsilon_{xy} = \epsilon_{yx}$. When there is no rotation, ϵ_{xy} is equal to the unit lateral relative displacement of B relative to A for an infinitesimal material vector AB initially parallel to the X-axis, i.e., it is the displacement in the Y-direction of B relative to A, divided by the length AB. It is also equal to the unit relative displacement of D relative to A for another element AD initially pointing in the Y-direction. Of course, when there is both rotation and shear, the shear represents only part of each unit lateral relative displacement.

Pure rotation. Before examining the alleged rotation matrix Ω, we consider geometrically the pure rotation of a vector PQ through an angle ω_z about an axis parallel to the Z-axis. The relative displacements of Q relative to P are shown in Fig. 4.6. Since only the relative displacements are considered, the point P has been shown fixed. (Otherwise, we can imagine the final position pq to be translated back parallel to itself until p coincides with P. This does not affect the relative displacements.)

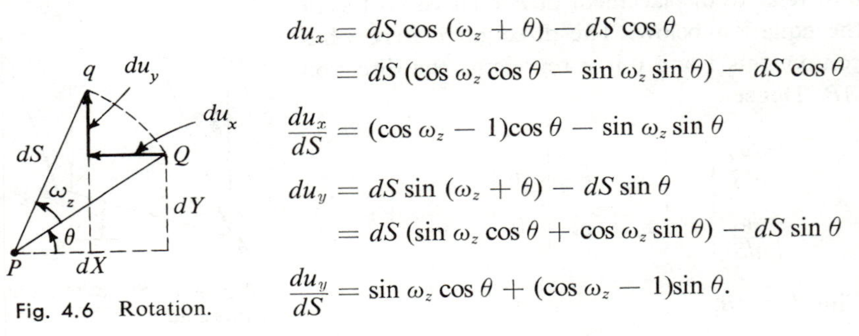

Fig. 4.6 Rotation.

$$du_x = dS\cos(\omega_z + \theta) - dS\cos\theta$$
$$= dS(\cos\omega_z\cos\theta - \sin\omega_z\sin\theta) - dS\cos\theta$$
$$\frac{du_x}{dS} = (\cos\omega_z - 1)\cos\theta - \sin\omega_z\sin\theta$$
$$du_y = dS\sin(\omega_z + \theta) - dS\sin\theta$$
$$= dS(\sin\omega_z\cos\theta + \cos\omega_z\sin\theta) - dS\sin\theta$$
$$\frac{du_y}{dS} = \sin\omega_z\cos\theta + (\cos\omega_z - 1)\sin\theta.$$

The equations for the unit relative displacements produced by rotation may be put in the following matrix form.

$$\begin{bmatrix} \dfrac{du_x}{dS} \\ \dfrac{du_y}{dS} \end{bmatrix} = \begin{bmatrix} \cos\omega_z - 1 & -\sin\omega_z \\ \sin\omega_z & \cos\omega_z - 1 \end{bmatrix} \begin{bmatrix} \cos\theta \\ \sin\theta \end{bmatrix}.$$

We now introduce the assumption that ω_z is small compared to one radian, and accordingly approximate $\sin\omega_z$ by ω_z in radians, and $\cos\omega_z$ by 1. The matrix operation takes the form below. (Note that $\cos\theta = dX/dS$, $\sin\theta = dY/dS$.)

$$\begin{bmatrix} \dfrac{du_x}{dS} \\ \dfrac{du_y}{dS} \end{bmatrix} = \begin{bmatrix} 0 & -\omega_z \\ \omega_z & 0 \end{bmatrix} \begin{bmatrix} \dfrac{dX}{dS} \\ \dfrac{dY}{dS} \end{bmatrix}. \tag{4.1.8}$$

The square matrix in Eq. (8) has the form of the rotation matrix Ω of Eq. (7), and will be identical with it if we set

$$\omega_z = \Omega_{yx} = -\Omega_{xy} \tag{4.1.9}$$

We see that if all the elements of the strain matrix are zero, ***and if moreover*** Ω_{xy} ***is small compared to one radian,*** then the unit relative displacements of Q relative to P given by the matrix product Ωn are the same as would be produced by a rigid-body rotation through the small angle $\omega_z = -\Omega_{xy}$. This is true for every infinitesimal element PQ at P; we say therefore that when the strain matrix components are zero, the local motion of the material in the neighborhood of P is an infinitesimal rigid-body motion.

Unit extention of an inclined element. We now show that the unit extension of an element initially inclined at an arbitrary angle to the x-axis can be calculated in terms of the angle of inclination of the element and the components of the strain matrix E even when a small rotation occurs simultaneously. We will also find that when all of the components of the strain matrix are zero, the unit extension of every such element is zero; hence the strain matrix components are a general measure of the strain in the neighborhood of a point. We assume in this derivation that the displacement derivatives are small compared to unity, as we have throughout the development in this section. We further assume that the unit extension sought is small compared to unity, and denote it by ϵ_n. Let the direction cosines of PQ be dX/dS and dY/dS, and let its initial length be dS. Its final length pq is then $(1 + \epsilon_n)dS$. Hence

$$\frac{|pq|^2 - |PQ|^2}{|PQ|^2} = \frac{(1+\epsilon_n)^2(dS)^2 - (dS)^2}{(dS)^2} = 2\epsilon_n + \epsilon_n^2.$$

Another expression for the same thing is

$$\frac{|pq|^2 - |PQ|^2}{|PQ|^2} = \frac{[(dX + du_x)^2 + (dY + du_y)^2] - [(dX)^2 + (dY)^2]}{(dS)^2}$$
$$= 2\left(\frac{dX}{dS}\frac{du_x}{dS} + \frac{dY}{dS}\frac{du_y}{dS}\right) + \left(\frac{du_x}{dS}\right)^2 + \left(\frac{du_y}{dS}\right)^2.$$

We equate the right-hand sides of these two expressions, neglect ϵ_n^2 in comparison to $2\epsilon_n$ and neglect the squares of the space derivatives in comparison to the first powers of them (note that dX/dS and dY/dS are not necessarily small). We obtain, after dividing by 2,

$$\epsilon_n = \frac{dX}{dS}\frac{du_x}{dS} + \frac{dY}{dS}\frac{du_y}{dS}. \tag{4.1.10}$$

In this equation, we substitute Eqs. (2) and rearrange to obtain

$$\begin{aligned}\epsilon_n &= \frac{\partial u_x}{\partial X}\left(\frac{dX}{dS}\right)^2 + \left(\frac{\partial u_x}{\partial Y} + \frac{\partial u_y}{\partial X}\right)\frac{dX}{dS}\frac{dY}{dS} + \frac{\partial u_y}{\partial Y}\left(\frac{dY}{dS}\right)^2 \\ &= \epsilon_x \cos^2\theta + (\tfrac{1}{2}\gamma_{xy})2\cos\theta\sin\theta + \epsilon_y \sin^2\theta,\end{aligned} \tag{4.1.11}$$

where θ is the angle of inclination of the element to the X-axis in the initial state. Compare this expression for the unit extension of an arbitrary element with Eq. (3.5.3) for the normal stress on an arbitrarily inclined element in plane stress. The two formulas are analogous if $\frac{1}{2}\gamma_{xy}$ is taken to be the analogue of the shear stress instead of γ_{xy}. If the unit extension is to be zero for all possible directions of the element, the coefficients of $\cos^2\theta$, $2\cos\theta\sin\theta$, and $\sin^2\theta$ must all be zero. Hence the condition for local rigid-body motion of the element is again that all of the elements of the strain matrix be zero.

We observe by Eq. (10) that the unit extension of an arbitrary element is obtained by projecting the unit relative displacement vector onto the initial direction of the element

$$\epsilon_n = \frac{d\mathbf{u}}{dS} \cdot \hat{\mathbf{n}}. \tag{4.1.12}$$

The unit relative displacement vector involves displacement associated with rotation as well as displacement associated with strain, but for small rotations the rotational part of $d\mathbf{u}/dS$ is perpendicular to $\hat{\mathbf{n}}$ and does not contribute to ϵ_n. We could introduce a strain vector $\boldsymbol{\epsilon}$ analogous to the traction vector $\mathbf{t}$ associated with the direction $\hat{\mathbf{n}}$. Its rectangular components would be given by the matrix product En and then its normal component could be found as $\epsilon_n = \boldsymbol{\epsilon} \cdot \hat{\mathbf{n}}$ in analogy with $\sigma_n = \mathbf{t} \cdot \hat{\mathbf{n}}$. Thus the strain is a linear operator, which associates with an arbitrary direction vector $\hat{\mathbf{n}}$ another vector $\boldsymbol{\epsilon}$. Such a linear operator is a tensor. In Sec. 4.2, we consider the three-dimensional strain and rotation tensors.

4.2 Small strain and rotation in three dimensions

The development of the two-dimensional small strain and rotation definitions in Sec. 4.1 emphasized the geometric interpretation, which is especially easy to visualize in two dimensions. A similar development could be made in three dimensions, but we shall make a more formal development here guided by the geometrical ideas already developed in two dimensions.

Consider an arbitrary infinitesimal line element vector $d\mathbf{X} = \mathbf{PQ}$, Fig. 4.7.

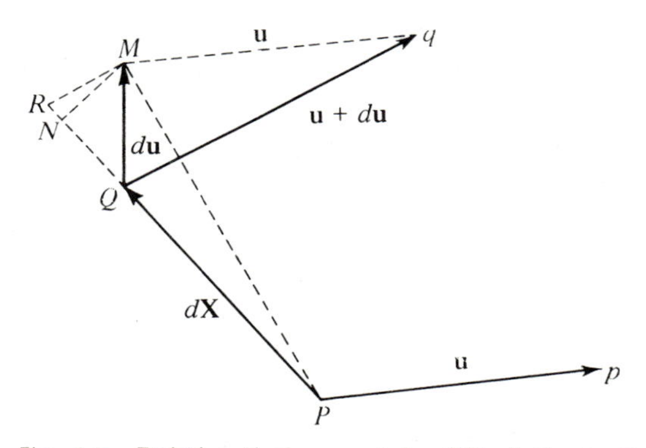

Fig. 4.7 Relative displacement $d\mathbf{u}$ of Q relative to P.

We seek an expression for the components of the unit relative displacement vector $d\mathbf{u}/dS$ similar to the one found in the preceding section for the two-dimensional motion. The three-dimensional counterpart of Eqs. (4.1.2) and (4.1.3) is in indicial notation

$$\frac{du_i}{dS} = \frac{\partial u_i}{\partial X_j}\frac{dX_j}{dS} \tag{4.2.1a}$$

(summed on the suffix j), or in x-y-z notation

$$\begin{bmatrix} \dfrac{du_x}{dS} \\ \dfrac{du_y}{dS} \\ \dfrac{du_z}{dS} \end{bmatrix} = \begin{bmatrix} \dfrac{\partial u_x}{\partial X} & \dfrac{\partial u_x}{\partial Y} & \dfrac{\partial u_x}{\partial Z} \\ \dfrac{\partial u_y}{\partial X} & \dfrac{\partial u_y}{\partial Y} & \dfrac{\partial u_y}{\partial Z} \\ \dfrac{\partial u_z}{\partial X} & \dfrac{\partial u_z}{\partial Y} & \dfrac{\partial u_z}{\partial Z} \end{bmatrix} \begin{bmatrix} \dfrac{dX}{dS} \\ \dfrac{dY}{dS} \\ \dfrac{dZ}{dS} \end{bmatrix}, \tag{4.2.1b}$$

which may be written briefly as

$$\frac{du}{dS} = J_u n \quad \text{or} \quad d\mathbf{u}/dS = \mathbf{u}\overleftarrow{\nabla}\cdot\hat{\mathbf{n}}, \tag{4.2.1c}$$

where du/dS and n are the column matrix representations of the vector $d\mathbf{u}/dS$ and of the unit vector $\hat{\mathbf{n}}$ in the direction of $d\mathbf{X}$, i.e., the unit vector whose components are dX/dS, dY/dS, dZ/dS. By transposing the square matrix J_u to obtain the matrix of $\overrightarrow{\nabla}\mathbf{u}$, we obtain the alternative form

$$\frac{d\mathbf{u}}{dS} = \hat{\mathbf{n}}\cdot\overrightarrow{\nabla}\mathbf{u} \tag{4.2.1d}$$

whose matrix equivalent uses row matrices for the rectangular Cartesian components of the two vectors. See Eq. (2.5.32). [Some authors use the symbol $\nabla\mathbf{u}$ for the operator we have denoted by $\mathbf{u}\overleftarrow{\nabla}$, but operate with it as in Eq.(1c).] J_u is the 3×3 Jacobian matrix or displacement gradient matrix with element $\partial u_i/\partial X_j$ in row i and column j.

The gradient of a vector is a second-order tensor, and we can conclude therefore that the quantities $\partial u_i/\partial X_j$ transform according to the transformation formula for a second-order Cartesian tensor. Note that formally each row of the displacement gradient matrix contains the vector components of the gradient of one scalar component of the displacement vector. The Jacobian matrix is also called the ***unit relative displacement*** matrix, since it is an operator which associates with each vector $d\mathbf{X} = \mathbf{PQ}$ the unit relative displacement $d\mathbf{u}/dS$ of Q relative to P, by the operation of matrix multiplication shown in Eq. (1b).

Guided by our two-dimensional development, we write J_u as the sum of a symmetric matrix E and a skew-symmetric matrix Ω (Greek capital omega). We form E by placing in each position half the sum of two symmetrically placed elements of J_u with respect to the main diagonal. Here we use ϵ_{ij} to denote the components of the small-strain tensor $\mathbf{E}$. For example, in the first row and second column, we place

$$\epsilon_{xy} = \frac{1}{2}\left(\frac{\partial u_x}{\partial Y} + \frac{\partial u_y}{\partial X}\right) = \frac{1}{2}\gamma_{xy},$$

while in the second row and first column we place

$$\epsilon_{yx} = \frac{1}{2}\left(\frac{\partial u_y}{\partial X} + \frac{\partial u_x}{\partial Y}\right) = \frac{1}{2}\gamma_{yx}.$$

(Note that $\epsilon_{yx} = \epsilon_{xy}$. The order of the two terms in each element is selected to agree with the conventional definitions of shear strain and to have the first subscript of ϵ_{xy} or ϵ_{yx} identify the row in the matrix, while the second subscript identifies the column. Each tensor shear strain component ϵ_{ij} equals $\frac{1}{2}\gamma_{ij}$, half the angle change of Sec. 4.1.) The elements of the skew-symmetric matrix Ω are then just the quantities which must be added to the elements of E to give the elements of J_u. Thus

$$E = \begin{bmatrix} \dfrac{\partial u_x}{\partial X} & \dfrac{1}{2}\left(\dfrac{\partial u_x}{\partial Y} + \dfrac{\partial u_y}{\partial X}\right) & \dfrac{1}{2}\left(\dfrac{\partial u_x}{\partial Z} + \dfrac{\partial u_z}{\partial X}\right) \\ \dfrac{1}{2}\left(\dfrac{\partial u_y}{\partial X} + \dfrac{\partial u_x}{\partial Y}\right) & \dfrac{\partial u_y}{\partial Y} & \dfrac{1}{2}\left(\dfrac{\partial u_y}{\partial Z} + \dfrac{\partial u_z}{\partial Y}\right) \\ \dfrac{1}{2}\left(\dfrac{\partial u_z}{\partial X} + \dfrac{\partial u_x}{\partial Z}\right) & \dfrac{1}{2}\left(\dfrac{\partial u_z}{\partial Y} + \dfrac{\partial u_y}{\partial Z}\right) & \dfrac{\partial u_z}{\partial Z} \end{bmatrix}$$

$$\mathbf{E} = \tfrac{1}{2}(\mathbf{u}\overleftarrow{\nabla} + \overrightarrow{\nabla}\mathbf{u}) \tag{4.2.2}$$

and

$$\Omega = \begin{bmatrix} 0 & \dfrac{1}{2}\left(\dfrac{\partial u_x}{\partial Y} - \dfrac{\partial u_y}{\partial X}\right) & \dfrac{1}{2}\left(\dfrac{\partial u_x}{\partial Z} - \dfrac{\partial u_z}{\partial X}\right) \\ -\dfrac{1}{2}\left(\dfrac{\partial u_x}{\partial Y} - \dfrac{\partial u_y}{\partial X}\right) & 0 & \dfrac{1}{2}\left(\dfrac{\partial u_y}{\partial Z} - \dfrac{\partial u_z}{\partial Y}\right) \\ -\dfrac{1}{2}\left(\dfrac{\partial u_x}{\partial Z} - \dfrac{\partial u_z}{\partial X}\right) & -\dfrac{1}{2}\left(\dfrac{\partial u_y}{\partial Z} - \dfrac{\partial u_z}{\partial Y}\right) & 0 \end{bmatrix}$$

$$\mathbf{\Omega} = \tfrac{1}{2}(\mathbf{u}\overleftarrow{\nabla} - \overrightarrow{\nabla}\mathbf{u}) \tag{4.2.3}$$

It can be shown that the components of the column matrix Ωn produced by the rotation matrix Ω operating on the column matrix n are just the unit relative displacements given by the vector product $\boldsymbol{\omega} \times \hat{\mathbf{n}}$, where $\boldsymbol{\omega}$ is the vector

$$\boldsymbol{\omega} = -\Omega_{yz}\mathbf{i} - \Omega_{zx}\mathbf{j} - \Omega_{xy}\mathbf{k} \qquad \text{or} \qquad \omega_i = -\tfrac{1}{2}e_{ijk}\,\Omega_{jk}.$$

Also

$$\boldsymbol{\omega} = \tfrac{1}{2}\nabla \times \mathbf{u} \qquad \text{or} \qquad \omega_i = \tfrac{1}{2}\,e_{ijk}\,\partial_j u_k. \tag{4.2.4}$$

These are therefore the unit relative displacements produced by small rotation through angle $|\boldsymbol{\omega}|$ about an axis in the direction of $\boldsymbol{\omega}$, provided that the components

$$\omega_x = -\Omega_{yz} \qquad \omega_y = -\Omega_{zx} \qquad \omega_z = -\Omega_{xy} \tag{4.2.5}$$

are all small compared to one radian. The separation of J_u into symmetric E and skew-symmetric Ω can always be made, and Ωn is always equivalent to $\boldsymbol{\omega} \times \hat{\mathbf{n}}$, but only when the components are small compared to one radian does this represent a rigid-body rotation. For this it is evidently sufficient that all components of the displacement gradient be small compared to unity.

When all of the $\partial u_i/\partial X_j$ are small compared to unity, the strain vector $\boldsymbol{\epsilon}$ with components given by $En = J_u n - \Omega n$ represents that part of the unit relative displacement vector not attributable to local rigid-body rotation of the element initially at P.

$$\boldsymbol{\epsilon} = \mathbf{E}\cdot\hat{\mathbf{n}} \tag{4.2.6}$$

The unit extension of PQ is

$$\epsilon_n = \boldsymbol{\epsilon}\cdot\hat{\mathbf{n}}\,, \tag{4.2.7}$$

while the magnitude of that part of the unit lateral displacement due to shear (of Q relative to P) is

$$|\tfrac{1}{2}\gamma_n| = [|\boldsymbol{\epsilon}|^2 - \epsilon_n^2]^{1/2}. \tag{4.2.8}$$

When there is rotation, the rotation also contributes to the unit lateral displacement of Q relative to P, but the unit extension of PQ is still given by Eq. (7), since the rotation does not contribute to the unit extension. If there were no rotation, the unit relative displacement of Q relative to P in Fig. 4.7 would be $\boldsymbol{\epsilon}$; the total relative displacement would be $d\mathbf{u} = \boldsymbol{\epsilon}\, dS$; and the total elongation would be $QN = (\boldsymbol{\epsilon}\, dS)\cdot\hat{\mathbf{n}}$, neglecting the second-order error NR in Fig. 4.7. The argument given at the close of Sec. 4.1 can be generalized to three dimensions to show that when the strain matrix elements are all zero, the unit elongation of every line element at P will be zero.

Strain and rotation tensors. We have several times referred to the strain and rotation components as tensor components. That they are in fact tensors may be seen as follows. Both $\mathbf{u}\overleftarrow{\nabla}$ and $\overrightarrow{\nabla}\mathbf{u}$ are second-order tensors because they are linear vector functions associating with antecedent vectors $\mathbf{n}$ the image vectors $d\mathbf{u}/dS$. Hence

$$\mathbf{E} = \tfrac{1}{2}(\mathbf{u}\overleftarrow{\nabla} + \overrightarrow{\nabla}\mathbf{u}) \quad \text{and} \quad \boldsymbol{\Omega} = \tfrac{1}{2}(\mathbf{u}\overleftarrow{\nabla} - \overrightarrow{\nabla}\mathbf{u}), \tag{4.2.9}$$

with components†

$$\epsilon_{ij} = \frac{1}{2}\left(\frac{\partial u_i}{\partial X_j} + \frac{\partial u_j}{\partial X_i}\right) \quad \text{and} \quad \Omega_{ij} = \frac{1}{2}\left(\frac{\partial u_i}{\partial X_j} - \frac{\partial u_j}{\partial X_i}\right), \tag{4.2.10}$$

are tensors because the sum or difference of two second-order tensors is a second-order tensor. It follows that under a change of rectangular Cartesian axes from

†The components ϵ_{ij} will frequently be denoted E_{ij} when there is no danger of confusing them with the finite strains E_{ij} of Sec. 4.5.

X_p to $\bar{X}_r$ the components transform according to Eqs. (2.4.11) and (2.4.12). For example,

$$\bar{\epsilon}_{rs} = a_r^p a_s^q \epsilon_{pq} \quad \text{or} \quad \bar{E} = A^T E A, \tag{4.2.11}$$

where A is the matrix of direction cosines

$$a_r^p = \cos(X_p, \bar{X}_r) \tag{4.2.12}$$

and inversely

$$\epsilon_{pq} = a_r^p a_s^q \bar{\epsilon}_{rs} \quad \text{or} \quad E = A\bar{E}A^T. \tag{4.2.13}$$

Recall that the shear component of the tensor ϵ_{ij} is half the angle change —for example, if x_1 and x_2 correspond to x and y, $\epsilon_{12} = \frac{1}{2}\gamma_{xy}$. In order to make it transform as a tensor, we must write the matrix as

$$E = \begin{bmatrix} \epsilon_x & \frac{1}{2}\gamma_{xy} & \frac{1}{2}\gamma_{xz} \\ \frac{1}{2}\gamma_{yx} & \epsilon_y & \frac{1}{2}\gamma_{yz} \\ \frac{1}{2}\gamma_{zx} & \frac{1}{2}\gamma_{zy} & \epsilon_z \end{bmatrix}, \tag{4.2.14}$$

where γ_{xy}, γ_{yz}, γ_{zx} are the (small) changes in right angles whose sides were initially parallel to the x-and y-, y- and z-, z- and x-axes, respectively. Or of course the components in (14) can be written as ϵ_{xx}, ϵ_{xy}, etc.

Because strain is a symmetric second-order tensor, the whole development of ***principal strains*** $\epsilon_1, \epsilon_2, \epsilon_3$, invariants $\mathrm{I}_E, \mathrm{II}_E, \mathrm{III}_E$, ***spherical and deviator strains***, and Mohr's Circles may be made for strain in precisely the same way as was done in Chap. 3 for stress. (Note that the shear component plotted as ordinate in the Mohr's Circle plane must be $\frac{1}{2}\gamma$, not γ.)

We denote the mean normal strain by $e/3$:

$$\begin{aligned} \tfrac{1}{3}e &= \tfrac{1}{3}(\epsilon_x + \epsilon_y + \epsilon_z) = \tfrac{1}{3}\mathrm{I}_E = \tfrac{1}{3}\,\mathrm{tr}\,\mathbf{E} \\ &= \tfrac{1}{3}\epsilon_{ii} \quad \text{in indicial notation.} \end{aligned} \tag{4.2.15}$$

Then the components ϵ'_{ij} of the ***strain deviator*** $\mathbf{E}'$ are given by

$$\left.\begin{aligned} &\epsilon'_{ij} = \epsilon_{ij} - \tfrac{1}{3}e\delta_{ij} \qquad \text{where} \quad \mathbf{E}' = \mathbf{E} - \tfrac{1}{3}e\,\mathbf{1} \\ &\text{or in matrix notation,} \qquad E' = E - \tfrac{1}{3}e\,I, \end{aligned}\right\} \tag{4.2.16}$$

where the E' denotes the deviator matrix, E denotes the strain matrix, and I is the identity matrix. Invariants of the strain deviator will be denoted by $\mathrm{I}_{E'}$, $\mathrm{II}_{E'}$, $\mathrm{III}_{E'}$.

The strain deviator $\mathbf{E}'$ measures the change in shape of an element, while the ***spherical*** or ***hydrostatic*** strain $\frac{1}{3}e\,\mathbf{1}$ represents the volume change. Indeed for small strains, the ***volume strain*** (change in volume per unit initial volume), also called the ***dilatation***, is equal to e, neglecting second order effects. This is most easily seen if the strains are referred to principal axes. If $\bar{X}, \bar{Y}, \bar{Z}$ are principal axes of strain, a rectangular element with initial dimensions $\Delta\bar{X}, \Delta\bar{Y}, \Delta\bar{Z}$ has, after straining, edge lengths $(1 + \bar{\epsilon}_x)\,\Delta\bar{X}$, $(1 + \bar{\epsilon}_y)\,\Delta\bar{Y}$, $(1 + \bar{\epsilon}_z)\,\Delta\bar{Z}$. Since these are principal axes, the edges of the element are still mutually perpendicular (no shear strain), and the volume strain is therefore $\bar{\epsilon}_x + \bar{\epsilon}_y + \bar{\epsilon}_z$, neglecting terms involving products of the strains. Since this is an invariant it is equal to $e = \epsilon_x + \epsilon_y + \epsilon_z$ measured in any system of axes. Thus

$$\boxed{e \equiv \mathrm{I}_E \equiv \operatorname{tr}\mathbf{E} = (V - V_0)/V_0 \qquad \text{Dilatation}} \tag{4.2.17}$$

Plane strain and surface strains. A state of ***plane strain*** parallel to the XY-plane is said to exist ***at a point*** in a body if at that point

$$\epsilon_z = 0, \qquad \gamma_{xz} = \gamma_{zx} = 0, \qquad \gamma_{yz} = \gamma_{zy} = 0. \tag{4.2.18a}$$

Plane strain exists throughout the body if these equations are satisfied at each point, and if in addition the remaining strain components,

$$\epsilon_x, \epsilon_y, \quad \text{and} \quad \tfrac{1}{2}\gamma_{xy} = \tfrac{1}{2}\gamma_{yx} \tag{4.2.18b}$$

are independent of Z. The Mohr's Circle for plane strain is exactly analogous to that for plane stress given in Sec. 3.5. We plot $\bar{\epsilon}_x$ and $\frac{1}{2}\bar{\gamma}_{xy}$ for $\bar{X}$ inclined at various angles to the X-axis.

The Mohr's Circle also furnishes a convenient representation of surface strains. We choose the XY-plane tangent to the bounding surface at the point in question. Strains in the Z-direction will not affect the result. Since shear strains are not easy to measure experimentally, the Mohr's Circle must usually be constructed from a knowledge of the unit extensions in three directions, measured by electric strain gages mounted at angles α_1, α_2, α_3 to the x-axis, forming a strain-gage rosette. We write the equation

$$\bar{\epsilon}_x = \tfrac{1}{2}(\epsilon_x + \epsilon_y) + \tfrac{1}{2}(\epsilon_x - \epsilon_y)\cos 2\alpha + \tfrac{1}{2}\gamma_{xy}\sin 2\alpha \tag{4.2.19}$$

for three different choices of the $\bar{x}$-direction inclined at angles α_1, α_2, α_3 to the x-axis and solve the three simultaneous equations for ϵ_x, ϵ_y, γ_{xy} in terms of the three measured strains. If two gages are perpendicular (rectangular rosette), their directions can be taken as axes and the third equation solved for γ_{xy}. In any case, one gage can be taken on the x-axis and there will be only two equations to solve for ϵ_y and γ_{xy}.

The small-strain definitions introduced in this section depend on our being able to neglect squares and products of the displacement gradient components $\partial u_i/\partial X_j$ in comparison to linear terms. For this reason it is always an approximate representation, a very good one in the usual elasticity theory of metals, but less good and possibly not good at all in polymer materials where elastic strains may be large. Some of the finite-strain definitions that have been proposed will be discussed in Sec. 4.5, but first we will consider the kinematics of the motion and deformation of a continuous medium, leading up to the definition of rate of deformation in Sec. 4.4, a concept useful in viscous flow and plasticity theories.

EXERCISES

1. For each of the following displacement gradient matrices sketch the deformed position of an element which was initially a square in the XY-plane with sides parallel to the axes.

$$\begin{bmatrix} 0 & 0.01 & 0 \\ 0.01 & 0 & 0 \\ 0 & 0 & 0 \end{bmatrix} \quad \begin{bmatrix} 0 & -0.01 & 0 \\ 0.01 & 0 & 0 \\ 0 & 0 & 0 \end{bmatrix} \quad \begin{bmatrix} 0 & 0 & 0 \\ 0.02 & 0 & 0 \\ 0 & 0 & 0 \end{bmatrix}$$

2. For each of the following displacement gradient matrices J_u, determine the strain matrix, the rotation matrix, the volume strain, and the deviatoric strain matrix.

$$\text{(a) } 10^{-4}\begin{bmatrix} 9 & 10 & -4 \\ -10 & 18 & -18 \\ -14 & -18 & 27 \end{bmatrix} \qquad \text{(b) } 10^{-4}\begin{bmatrix} 4 & 1 & 4 \\ -1 & -4 & 0 \\ 0 & 2 & 6 \end{bmatrix}$$

3. For each matrix J_u of Ex. 2, evaluate:
   (1) the three invariants I_E, II_E, III_E of the strain
   (2) the three invariants $\mathrm{I}_{E'}$, $\mathrm{II}_{E'}$, $\mathrm{III}_{E'}$ of the deviatoric strain.

4. Show that the matrix operation Ωn gives the same unit relative displacements as the vector product $\boldsymbol{\omega} \times \mathbf{n}$, where $\boldsymbol{\omega}$ is as defined in Eq. (4.2.4).

5. A strain-gage rosette measures unit elongations of 0.6, -1.1, -1.8 milli-inches per inch at directions making angles of $0°$, $45°$, $90°$, respectively, with the X-axis. Find the principal strains in the plane of the measurement, and show on a sketch the inclinations of the principal directions to the X-axis.

6. If at a point $\epsilon_x = -0.005$, $\epsilon_y = -0.001$, $\epsilon_z = 0.001$, and $\gamma_{xy} = \gamma_{yz} = \gamma_{zx} = 0$, find:
   (a) the maximum shear strain at the point (angle change),

(b) the normal strain (unit extension) in the direction normal to the plane of the maximum shear strain.

7. If the Jacobian matrix of unit relative displacements in a certain small deformation is as in Ex. 2(a), find the unit extension of a material element whose direction cosines were initially $\frac{2}{3}$, $-\frac{2}{3}$, $\frac{1}{3}$.

8. If the displacement field is given by

$$u_1 = kX_1X_2 \qquad u_2 = kX_1X_2 \qquad u_3 = 2k(X_1 + X_2)X_3$$

where k is a constant small enough to ensure applicability of small deformation theory:

(a) Express the small strain tensor components and rotation components as functions of X_1, X_2, X_3 and display them in matrices.

(b) At the point (1, 1, 0) determine the principal strains and principal axes of strain for the strain matrix of (a).

9. If Eq. (4.2.10) defines ϵ_{ij} in any rectangular Cartesian coordinate system, show that the vector component transformation formulas for $\bar{u}_p$ imply the tensor transformation formulas for $\bar{\epsilon}_{pq}$ in a new rectangular Cartesian system such that

$$\bar{X}_r = a_r^s(X_s - b_s) \qquad X_s = b_s + a_r^s\bar{X}_r.$$

10. The three-dimensional theory of elasticity solution for the displacement distribution in a rectangular beam bent by end couples is shown below. The beam initially occupied the rectangular region $0 \leq X \leq L$ (length), $-c \leq Y \leq c$ (depth), $-b \leq Z \leq b$ (width), and the centerline has been bent to radius of curvature R.

$$u_x = \frac{1}{R}XY + MY + NZ + P \qquad u_z = -\frac{1}{R}\nu YZ - AY - NX + B$$

$$u_y = -\frac{1}{2R}[X^2 + \nu(Y^2 - Z^2)] - MX + AZ + C$$

(ν is Poisson's ratio; A, B, C, M, N, P are arbitrary constants).

(a) What are the components of the small strain matrix and the small rotation matrix at a general point (X, Y, Z)?

(b) If the end $X = 0$ is constrained in such a way that all displacements vanish at the origin, what arbitrary constants are thereby determined?

(c) If the end restraint is such that all rotation components vanish at the origin as well as all displacement components, what are the rotation components at the point $(L, 0, 0)$ on the other end of the centerline?

(d) Determine the centerline slope at the point $(L, 0, 0)$ from the rotation matrix, and verify that it agrees with the slope calculated from the equation for u_y.

11. If the motion of a continuum from initial positions $\mathbf{X}$ to current positions $\mathbf{x}$ is defined by $\mathbf{x} = (\mathbf{1} + \mathbf{B})\cdot\mathbf{X}$, where $\mathbf{1}$ is the unit tensor and $\mathbf{B}$ is a second-order tensor whose components are all constants small in comparison to unity, display in matrices the components of the displacement $\mathbf{u}$, small strain $\mathbf{E}$, rotation $\mathbf{\Omega}$, and strain deviator $\mathbf{E}'$. What is the volume strain?

12. If two successive motions of a continuum from general points $\mathbf{X}$ to $\mathbf{x}$ and then to $\mathbf{y}$ are such that the relative motions in the neighborhood of a certain particle are given by

$$d\mathbf{x} = (\mathbf{1} + \mathbf{A})\cdot d\mathbf{X} \quad \text{and} \quad d\mathbf{y} = (\mathbf{1} + \mathbf{B})\cdot d\mathbf{x} \tag{1}$$

where the rectangular components of **A** and **B** are constants small compared to unity, display in a matrix the components of the single tensor giving the motion from $d\mathbf{X}$ to $d\mathbf{y}$. Show that to the accuracy implicit in the small strain and rotation theory the strain and rotation tensors in the motion from $d\mathbf{X}$ to $d\mathbf{y}$ are the sums of the two strain tensors and the two rotation tensors, respectively, implied by (1).

13. If the displacement gradient $\mathbf{u}\overleftarrow{\nabla}$ has the matrix of rectangular Cartesian components shown in Ex. 2(b), what elements initially perpendicular to the Z-axis have undergone unit extension $\epsilon_n = 2(10^{-4})$?

14. If the small strain tensor at a point in a body made of a moldable plastic could be determined experimentally by measuring the unit extensions in six directions at the point (the coordinate directions and the three directions bisecting the angles between the axes: ϵ_m for the direction between X_1 and X_2, ϵ_n between X_2 and X_3, and ϵ_t between X_3 and X_1), determine the shear strain components ϵ_{12}, ϵ_{23}, and ϵ_{31} in terms of the six measured unit extensions.

15. In a simple cubic crystal, if plastic slip takes place only on planes parallel to the cube faces and in the directions parallel to the cube edges to give the matrix of small strains shown below, referred to the cubic axes, obtain an equation that could be used to determine the maximum unit extension and explain how you would determine the magnitude and direction of the maximum unit extension if given numerical values for the three matrix components.

$$\begin{bmatrix} 0 & \epsilon_{12} & \epsilon_{31} \\ \epsilon_{21} & 0 & \epsilon_{23} \\ \epsilon_{31} & \epsilon_{23} & 0 \end{bmatrix}$$

16. In a face-centered cubic crystal, plastic slip takes place parallel to the most closely packed crystal planes, which are planes whose normals make equal angles with the cubic axes of the crystals. There are only four families of such planes, since the eight such normals occur in equal and opposite pairs. Slip occurs in one of the three closely packed directions in each plane. For example, express in terms of the small-strain tensor referred to the cubic axes the shear strain component in the direction $(\mathbf{i} - \mathbf{j})/\sqrt{2}$ parallel to the plane whose normal makes equal angles with the positive axes.

17. If the YZ-plane is a plane of symmetry of the displacement distribution, so that $\mathbf{u}$ at any point $(-a, b, c)$ is a mirror image of $\mathbf{u}$ at (a, b, c), which rectangular components of small strain are even functions of X and which are odd? *Hint*: Differentiation changes an even function into an odd one and vice-versa.

18. For the rectangular component small-strain matrix shown below, determine the principal strains and the direction of the maximum unit extension.

$$(10^{-4})\begin{bmatrix} 1 & 0 & 0 \\ 0 & 0 & -2 \\ 0 & -2 & 3 \end{bmatrix}$$

4.3 Kinematics of a continuous medium; material derivatives

Four types of description of the motion of a continuum are in common use, all based on classical nonrelativistic kinematics. Truesdell (1965) calls them the:

1. ***Material description***, whose independent variables are the particle X and the time t,
2. ***Referential description***, whose independent variables are the position $\mathbf{X}$ of the particle in an arbitrarily chosen reference configuration, and the time t. (In elasticity, the reference configuration is usually chosen to be the natural or unstressed state.) When the reference configuration is chosen to be the actual initial configuration at $t = 0$, the referential description is often called the ***Lagrangian description***, although many authors call it the ***material description***, using the reference position $\mathbf{X}$ as a label for the material particle X of the first type of description listed above.
3. ***Spatial description***, whose independent variables are the present position $\mathbf{x}$ occupied by the particle at time t and the present time t. (The spatial description fixes attention on a given region of space instead of on a given body of matter. It is the description most used in fluid mechanics, often called the ***Eulerian description***.)
4. ***Relative description***, whose independent variables are the present position $\mathbf{x}$ of the particle and a variable time τ. The variable time τ is the time when the particle occupied another position, say $\boldsymbol{\xi}$, and the motion is described with $\boldsymbol{\xi}$ as dependent variable, say $\boldsymbol{\xi} = \chi_t\ (\mathbf{x}, \tau)$ where the t on the function symbol χ_t emphasizes that the reference configuration is the configuration occupied at time t. [See the discussion of Eqs. (12) and (13) in this section.]

We shall use mostly the referential and spatial descriptions in this book. The other two, prominent in some recent studies, will be mentioned only briefly.

The ***material description*** is in terms of the material particles X, the undefined primitive elements of the medium. Thus the ***motion*** $\mathbf{x} = \mathbf{x}\ (X, t)$ gives symbolically the position $\mathbf{x}$ occupied by the particle X at time t. The material description is frequently confused with the referential description because of the custom of using the particle's position $\mathbf{X}$ in the reference configuration as a label for the particle and often indeed referring to it as "the particle $\mathbf{X}$." The notational distinction here is that the reference position $\mathbf{X}$ is printed in boldface while the particle X is not. Now a reference position is, of course, not the same thing as a particle, but no harm will come from the usage as long as we understand the meaning of the conventional language.

The referential description refers the motion to a reference configuration in which the particle X occupies position $\mathbf{X}$. In elasticity the reference configuration is usually chosen as the initial unstressed state, the configuration to which the body will return when it is unloaded. In fluid flow and in continuum mechanics generally the reference configuration may be chosen arbitrarily and need not even be a configuration actually assumed during a particular motion.

For most of our purposes we will suppose that the ***reference configuration*** is the configuration occupied at time $t = 0$ (***Lagrangian description***), and we will take the reference position $\mathbf{X}$ of the particle X as an identifying label for the particle, using the language "$\mathbf{x}$ is the ***place*** occupied at time t by the ***particle*** $\mathbf{X}$" to mean "$\mathbf{x}$ is the position occupied at time t by the particle which occupied position $\mathbf{X}$ in the reference configuration"; and writing symbolically

$$\mathbf{x} = \mathbf{x}(\mathbf{X}, t) \tag{4.3.1a}$$

with rectangular Cartesian components

$$x_r = x_r(X_1, X_2, X_3, t) \tag{4.3.1b}$$

where X_1, X_2, X_3 are the ***material coordinates*** of the particle, i.e., the coordinates of its position in the reference configuration, while x_1, x_2, x_3 are its ***spatial coordinates***, giving its place $\mathbf{x}$ at time t.

The material coordinates and spatial coordinates are usually measured with respect to the same coordinate axes, although the axes could be different.† The notations $\mathbf{x}$ and $\mathbf{p}$ are interchangeable if we use only one reference origin for the ***position vector*** $\mathbf{p}$, but both "place" and "particle" have meaning independent of the choice of origin. The same place $\mathbf{x}$ may have many different position vectors $\mathbf{p}$ corresponding to different choices of origin, but the relative position vectors $d\mathbf{p} = d\mathbf{x}$ of neighboring positions will be the same for all origins. The functional forms for a given motion as described by Eqs. (1) depend upon the choice of the reference configuration of the body as well as upon the choice of origin. If more than one reference configuration is used in the discussion, it is necessary to label the functions accordingly. For example, relative to two configurations κ_1 and κ_2 the same motion could be represented symbolically by the two equations $\mathbf{x} = \mathbf{x}_{\kappa_1}(\mathbf{X}, t)$ and $\mathbf{x} = \mathbf{x}_{\kappa_2}(\mathbf{X}, t)$. For simplicity we will suppose that only one reference configuration is used in each discussion throughout most of this book and will write the motion equation $\mathbf{x} = \mathbf{x}(\mathbf{X}, t)$ without any subscript to identify the reference configuration. In the motion equation the symbol $\mathbf{x}$ appears twice with two different meanings. On the right-hand side $\mathbf{x}$ represents the function whose arguments are $\mathbf{X}$ and t, while on the left-hand side it represents the ***value*** of the function, i.e., the point $\mathbf{x}$. This double usage,

†According to Truesdell (1952) the material coordinates were introduced by Euler in 1762, although they are now widely called Lagrangian coordinates, while the spatial coordinates, often called Eulerian coordinates, were introduced by d'Alembert in 1752.

common in the mechanics literature, may occasionally cause confusion, which may be avoided by using a different symbol for the function, e.g., χ (Greek chi), so that Eq. (1) is written as

$$\mathbf{x} = \chi(\mathbf{X}, t). \tag{4.3.1c}$$

The spatial description fixes attention on a given region of space and takes the spatial position $\mathbf{x}$ and t as the independent variables. The spatial description is especially useful in fluid mechanics, where we may observe a flow in a wind tunnel or in a channel, a fixed region in space. We can imagine a perfect instrumentation with which an observer records the fluid velocity at a fixed point in space as a function of time. If this is imagined done for every point in the region, we have the spatial description of the velocity $\mathbf{v}$ as a function of position $\mathbf{x}$ and time t, say $\mathbf{v} = \mathbf{g}(\mathbf{x}, t)$. In a general nonsteady flow we usually do not know the particle paths, and at any time t we do not know what was the reference position $\mathbf{X}$ occupied at $t = 0$ by the fluid particle now at $\mathbf{x}$. If we did know the complete motion for all particles, say $\mathbf{x} = \chi(\mathbf{X}, t)$, we could calculate any particle's velocity simply by taking the partial time derivative with $\mathbf{X}$ held constant. Some common notations are illustrated by

$$\mathbf{v} = \mathbf{G}(\mathbf{X}, t) \qquad \text{where} \quad \mathbf{G}(\mathbf{X}, t) \equiv \frac{\partial}{\partial t}\chi(\mathbf{X}, t) \equiv \left(\frac{\partial \mathbf{x}}{\partial t}\right)_{\mathbf{X}} \tag{4.3.2a}$$

The last notation uses $\mathbf{x}$ for the function name instead of χ, and the subscript $\mathbf{X}$ identifies the variable held constant in taking the partial time derivative. Similarly the acceleration is

$$\mathbf{a} = \left(\frac{\partial \mathbf{v}}{\partial t}\right)_{\mathbf{X}} \equiv \frac{\partial}{\partial t}\mathbf{G}(\mathbf{X}, t) = \frac{\partial^2}{\partial t^2}\chi(\mathbf{X}, t). \tag{4.3.2b}$$

The partial time derivative with the material coordinates $\mathbf{X}$ held constant should be carefully distinguished from the partial time derivative with spatial coordinates $\mathbf{x}$ held constant. For example, in the spatial description $\mathbf{v} = \mathbf{g}(\mathbf{x}, t)$,

$$\left(\frac{\partial \mathbf{v}}{\partial t}\right)_{\mathbf{x}} \equiv \frac{\partial}{\partial t}\mathbf{g}(\mathbf{x}, t), \tag{4.3.3}$$

and this is different from the acceleration $(\partial \mathbf{v}/\partial t)_{\mathbf{X}}$ of Eq. (2b). The derivative $(\partial \mathbf{v}/\partial t)_{\mathbf{x}}$ of Eq. (3) is called the ***local rate of change*** of $\mathbf{v}$. It is the rate of change of the reading of an ideal velocity meter located at the fixed place $\mathbf{x}$, which is not the same thing as the acceleration of the particle just now passing the place $\mathbf{x}$. For example, in a steady flow the local rate of change is everywhere zero, but this does not imply that the acceleration is zero, since even in a steady flow the velocity varies in general from point to point, and a particle changes its velocity as it moves from one place of constant velocity to another place having a different constant velocity. Since the laws of dynamics deal with particle accelerations and not with local rates of change, we need to be able to calculate the particle accelerations from a knowledge of the spatial description $\mathbf{v} = \mathbf{g}(\mathbf{x}, t)$. This can be accomplished by calculating the material derivative as follows.

Material time derivative in spatial coordinates. The material time derivative is the time derivative with the material coordinates held constant as in Eqs. (2). But if we know only the spatial description $\mathbf{v} = \mathbf{g}(\mathbf{x}, t)$ and not the referential description $\mathbf{v} = \mathbf{G}(\mathbf{X}, t)$ we cannot use Eqs. (2) to calculate the material derivative. We assume, however, that there always exists a sufficiently differentiable and single-valued function $\mathbf{x}(\mathbf{X}, t)$ defining the motion, even though it may be unknown to us. Then by imagining it substituted into the spatial description of the velocity $\mathbf{v} = \mathbf{g}(\mathbf{x}, t)$, we obtain

$$\mathbf{v} = \mathbf{G}(\mathbf{X}, t) \equiv \mathbf{g}[\mathbf{x}(\mathbf{X}, t), t]$$

or in rectangular Cartesians

$$v_m = g_m[x_1(X_1, X_2, X_3, t), \quad x_2(X_1, X_2, X_3, t), \quad x_3(X_1, X_2, X_3, t), \quad t]$$

and, by the chain rule of calculus,

$$\left(\frac{\partial v_m}{\partial t}\right)_{\mathbf{X}} = \left(\frac{\partial v_m}{\partial t}\right)_{\mathbf{x}} + \frac{\partial v_m}{\partial x_k}\left(\frac{\partial x_k}{\partial t}\right)_{\mathbf{X}} \qquad \text{(sum on } k\text{)}. \tag{4.3.4}$$

Now, by Eq. (2a),

$$\left(\frac{\partial \mathbf{x}}{\partial t}\right)_{\mathbf{X}} = \mathbf{v} \quad \text{with Cartesian components} \quad \left(\frac{\partial x_k}{\partial t}\right)_{\mathbf{X}} = v_k. \tag{4.3.5}$$

Although we may not know the description $\mathbf{x} = \mathbf{x}(\mathbf{X}, t)$, we do know the distribution of the derivatives $(\partial x_k/\partial t)_{\mathbf{X}}$ when we know the spatial description $\mathbf{v} = \mathbf{g}(\mathbf{x}, t)$. Hence, by substituting v_k for $(\partial x_k/\partial t)_{\mathbf{X}}$ on the right side of Eq. (4) we obtain an expression for the acceleration component a_m entirely in terms of quantities calculable from the spatial description $\mathbf{v} = \mathbf{g}(\mathbf{x}, t)$:

$$a_m = \left(\frac{\partial v_m}{\partial t}\right)_{\mathbf{X}} = \left(\frac{\partial v_m}{\partial t}\right)_{\mathbf{x}} + v_k \frac{\partial v_m}{\partial x_k} \qquad \text{(sum on } k\text{)}. \tag{4.3.6a}$$

The notations dv_m/dt or $\dot{v}_m$ will usually be used in this book to represent the material derivative instead of the more awkward $(\partial v_m/\partial t)_{\mathbf{X}}$, and $\partial v_m/\partial t$ will denote the local time derivative $(\partial v_m/\partial t)_{\mathbf{x}}$. Equation (6a) can then be written in the matrix form

$$[a_m] = \left[\frac{dv_m}{dt}\right] \equiv [\dot{v}_m] = \left[\frac{\partial v_m}{\partial t}\right] + [v_k][\partial_k v_m]. \tag{4.3.6b}$$

The notation $\partial_k v_m$ in Eq. (6b) means $\partial v_m/\partial x_k$ as in Eq. (2.5.20b). The matrix equation represents the vector equation

$$\mathbf{a} = \frac{d\mathbf{v}}{dt} \equiv \dot{\mathbf{v}} = \frac{\partial \mathbf{v}}{\partial t} + \mathbf{v}\cdot\text{grad } \mathbf{v}. \tag{4.3.6c}$$

In the last form grad $\mathbf{v}$ is the gradient with respect to the spatial coordinates, sometimes denoted $\nabla_{\mathbf{x}}\mathbf{v}$ to distinguish it from the gradient $\nabla_{\mathbf{X}}\mathbf{v}$ with respect to the material coordinates. When only spatial coordinates appear in a discussion it may be denoted simply $\nabla\mathbf{v}$, but when any possibility of confusion arises we use the convention that

$$\begin{aligned} \nabla\mathbf{v} \quad &\text{means} \quad \nabla_{\mathbf{X}}\mathbf{v}, \quad \text{while} \\ \text{grad } \mathbf{v} \quad &\text{means} \quad \nabla_{\mathbf{x}}\mathbf{v}. \end{aligned} \tag{4.3.7}$$

In Eqs. (6) we have used the material derivative of the velocity to calculate the acceleration from the spatial description of the velocity. The rate of change of any other material property may also be calculated if its spatial description is known. For example, if the spatial description of density is given by the scalar function $\rho(\mathbf{x}, t)$

$$\rho = \rho(\mathbf{x}, t),$$

the rate of change of the density in the neighborhood of the particle instantaneously at $\mathbf{x}$ is given by

$$\begin{aligned} \frac{d\rho}{dt} &= \frac{\partial \rho}{\partial t} + \mathbf{v}\cdot\text{grad } \rho \\ &= \frac{\partial \rho}{\partial t} + v_k \partial_k \rho. \end{aligned} \tag{4.3.8}$$

The first term $\partial\rho/\partial t$ gives the local rate of change of the density in the neighborhood of the place $\mathbf{x}$, while the second term gives the ***convective rate of change*** of the density in the neighborhood of a particle as it moves to a place having a different density. The first term vanishes in a steady flow, while the second term vanishes in a uniform flow.

The material derivative operator

$$\frac{d}{dt} = \left(\frac{\partial}{\partial t} + \mathbf{v}\cdot\text{grad}\right) \tag{4.3.9}$$

can be applied to a scalar, a vector, or a tensor function of the spatial position

x and t. We have already seen as examples the vector velocity and the scalar density. For any scalar f, vector **u**, or tensor **T** the formulas and Cartesian component representations are summarized in Eqs. (10).

$$\dot{f} \equiv \frac{df}{dt} = \frac{\partial f}{\partial t} + \mathbf{v} \cdot \operatorname{grad} f = \frac{\partial f}{\partial t} + v_k \partial_k f \tag{4.3.10a}$$

$$\left.\begin{aligned} \dot{\mathbf{u}} &\equiv \frac{d\mathbf{u}}{dt} = \frac{\partial \mathbf{u}}{\partial t} + \mathbf{v} \cdot \operatorname{grad} \mathbf{u} \\ \dot{u}_m &\equiv \frac{du_m}{dt} = \frac{\partial u_m}{\partial t} + v_k \partial_k u_m \end{aligned}\right\} \tag{4.3.10b}$$

$$\left.\begin{aligned} \dot{\mathbf{T}} &\equiv \frac{d\mathbf{T}}{dt} = \frac{\partial \mathbf{T}}{\partial t} + \mathbf{v} \cdot \operatorname{grad} \mathbf{T} \\ \dot{T}_{km} &\equiv \frac{dT_{km}}{dt} = \frac{\partial T_{km}}{\partial t} + v_r \partial_r T_{km} \end{aligned}\right\} \tag{4.3.10c}$$

Another name for the material derivative† is the ***substantial derivative***. Another notation sometimes used for it is D/Dt. Then

$$\frac{Df}{Dt} \equiv \frac{df}{dt} \equiv \dot{f} \equiv \left(\frac{\partial f}{\partial t}\right)_{\mathbf{X}}.$$

In Sec. 5.1 we shall see how to calculate the material derivative of a quantity defined as an integral over a given body in terms of an integral over the region of space instantaneously occupied by the body.

Streamlines and particle paths. A curve such that at each of its points the tangent line has the direction of the velocity vector of the particle instantaneously at the point is called a ***streamline***. In ***steady flow***, all flow properties are constant in time at each point. Hence, in steady flow the streamline pattern does not change with time, and each ***particle path*** is along one of the unchanging streamlines. In ***unsteady flow***, on the other hand, the streamline pattern changes with time, and a particle path may not coincide with any streamline, although at any given instant the particle path is tangent to the streamline which at that instant passes through the place occupied by the particle. The streamlines on the surface of a fluid can be approximately located by floating some reflecting particles in the fluid and making a short-time exposure photograph. Each reflecting particle then produces a short line on the photograph, approximating the tangent line to a streamline. To photograph a path line it is necessary to use a long-time exposure, and in the case of unsteady flow a

†The concept of the material derivative and formulas for it in Cartesian spatial coordinates go back to work by Euler in 1770 and Lagrange in 1783.

small number of reflecting particles must be used because the paths may cross over each other.

It may happen that the time-average of the fluid properties over a short time does not change, although the instantaneous values fluctuate in a random manner about the average value; this is the case of ***turbulence*** superposed onto a steady flow, and in this case the streamlines are defined relative to the average flow.

The material derivative takes an especially simple form in terms of the directional derivative along a streamline. For example, if $d\mathbf{u}/ds$ is the derivative of $\mathbf{u}$ with respect to arc length s along a streamline in the direction of motion defined by the unit tangent $\hat{\mathbf{e}}_t$, then, along the streamline, $\mathbf{u} = \mathbf{u}(s, t)$ and

$$\frac{d\mathbf{u}}{dt} = \frac{\partial \mathbf{u}}{\partial t} + \frac{d\mathbf{u}}{ds}\frac{ds}{dt}. \tag{4.3.11}$$

Since $d\mathbf{u}/ds = \hat{\mathbf{e}}_t \cdot \operatorname{grad} \mathbf{u}$ and $\hat{\mathbf{e}}_t(ds/dt) = \mathbf{v}$, Eq. (11) is the same as Eq. (10b). If $\mathbf{u}$ is the velocity vector $\mathbf{v}$ with magnitude v, Eq. (11) yields

$$\mathbf{a} = \frac{d\mathbf{v}}{dt} = \frac{\partial \mathbf{v}}{\partial t} + v\frac{d\mathbf{v}}{ds} \tag{4.3.12}$$

The relative description,† the fourth type of description listed at the beginning of this article, expresses the position $\boldsymbol{\xi}$ occupied at time τ by the particle which occupies the spatial position $\mathbf{x}$ at time t in terms of the independent variables $\mathbf{x}$ and τ. Thus we write

$$\boldsymbol{\xi} = \boldsymbol{\chi}_t(\mathbf{x}, \tau), \tag{4.3.13a}$$

where the subscript t on the function symbol is used to recall that the reference configuration is the configuration at time t. This is actually a special case of the referential description, differing from the Lagrangian description in that the reference position is now denoted by $\mathbf{x}$ at time t instead of $\mathbf{X}$ at time $t = 0$. Since this is a referential description, the velocity and acceleration in terms of the reference position and the variable time are given in analogy with Eqs. (2) by

$$\left.\begin{aligned} \mathbf{v}(\mathbf{x}, \tau) &= \left(\frac{\partial \boldsymbol{\xi}}{\partial \tau}\right)_{\mathbf{x}} \equiv \frac{\partial \boldsymbol{\chi}_t}{\partial \tau} \\ \mathbf{a}(\mathbf{x}, \tau) &= \left(\frac{\partial \mathbf{v}}{\partial \tau}\right)_{\mathbf{x}} \equiv \frac{\partial^2 \boldsymbol{\chi}_t}{\partial \tau^2}. \end{aligned}\right\} \tag{4.3.13b}$$

Change of reference frame is discussed in Sec. 6.7 in connection with the

†See Truesdell and Noll (1965), Sec. 21.

principle of material frame indifference, one of the important postulates of the modern theory of constitutive equations.

We continue our study of kinematics in Sec. 4.4 with the rate-of-deformation and spin tensors.

4.4 Rate-of-deformation tensor (stretching); spin tensor (vorticity); natural strain increment

Our definition of small strain in Sec. 4.2 was made in terms of displacement components u_i from an initial position P to a final position p. This displacement vector is a straight line from the initial position to the final position, whether or not the actual motion from P to p was along a straight line. In elasticity theory the actual process by which a continuum goes from one configuration to another is usually considered irrelevant, the initial and final states alone being sufficient to define the strain. In describing the effects of viscosity in a fluid or the deformation of a plastic solid, however, it is necessary to specify the process and follow the history of the deformation. The deformation can in principle be represented by giving the trajectory of every particle $\mathbf{X}$,

$$\mathbf{x} = \mathbf{x}(\mathbf{X}, t) \quad \text{or} \quad x_i = x_i(X_1, X_2, X_3, t), \tag{4.4.1}$$

as in Eq. (4.3.1), though the explicit writing down of such equations for even quite simple deformation processes is a difficult problem.

The important kinematic tensors in linear viscosity theory and in most plasticity theories are the ***rate-of-deformation tensor***† (also called the ***stretching tensor*** or occasionally ***velocity strain***) and the ***spin tensor*** (also called the ***vorticity tensor***), whose rectangular Cartesian components will be defined in terms of the velocity components $v_i = dx_i/dt$ expressed in terms of the spatial coordinates and the time,

$$v_i = v_i(x_1, x_2, x_3, t). \tag{4.4.2}$$

These velocity components describe what is sometimes called the ***tangent motion.*** In Fig. 4.8, the dashed lines from P to p and from Q to q represent the trajectories of two particles P and Q, each describable in principle by an equation like Eq. (1). The velocity vectors $\mathbf{v}$ at p and $\mathbf{v} + d\mathbf{v}$ at q are tangent to the two trajectories. We suppose that an observer cannot see the trajectories or the initial points P and Q; he can only record the tangent velocities as a function of the spatial coordinates x_i and the time. The instantaneous rate-of-deformation and spin tensors at p are defined in terms of the relative velocity of the

†The rate-of-deformation tensor was introduced by Euler in 1770.

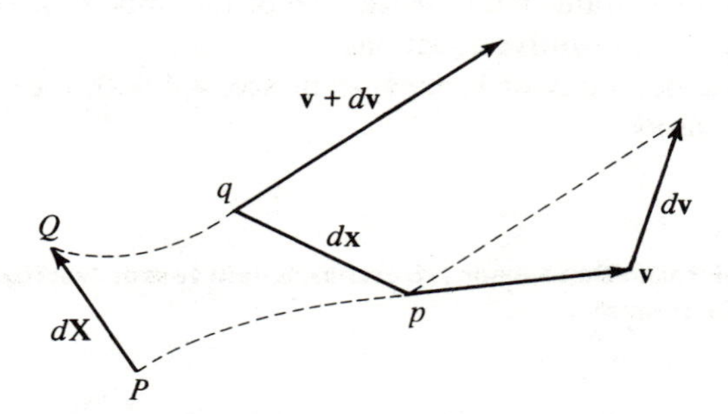

Fig. 4.8 Relative velocity $d\mathbf{v}$ of particle Q at point q relative to particle P at point p.

particles instantaneously at neighboring points q relative to the particle instantaneously at p.

The relative velocity components dv_k of the particle at q relative to the particle at p are given by

$$dv_k = \frac{\partial v_k}{\partial x_m} dx_m \tag{4.4.3a}$$

or in matrix form

$$[dv_k] = [v_{k,m}]\,[dx_m] = [dx_m]\,[\partial_m v_k] \tag{4.4.3b}$$

representing the tensor equations

$$d\mathbf{v} = \mathbf{L}\cdot d\mathbf{x} = d\mathbf{x}\cdot\mathbf{L}^T, \tag{4.4.3c}$$

where

$$\left.\begin{array}{l} \mathbf{L} = \mathbf{v}\overleftarrow{\nabla}_{\mathbf{x}} \quad \text{and} \quad \mathbf{L}^T = \overrightarrow{\nabla}_{\mathbf{x}}\mathbf{v}, \\ \text{with Cartesian components} \\ L_{km} = v_{k,m} \qquad (L^T)_{km} = \partial_k v_m = v_{m,k} \end{array}\right\} \tag{4.4.4}$$

are the ***spatial gradients of the velocity***. (Compare with the similar development in Sec. 4.2 in terms of small displacement components u_i instead of velocities v_i; there all equations were divided through by dS to give ***unit*** relative displacements, while here we have considered actual relative velocities.) We write $\mathbf{L}$ as the sum of a symmetric tensor $\mathbf{D}$ called the ***rate-of-deformation tensor*** or the ***stretching tensor*** and a skew-symmetric tensor $\mathbf{W}$ called the ***spin tensor*** or the ***vorticity tensor*** as follows. Let

$$\mathbf{D} = \tfrac{1}{2}(\mathbf{L} + \mathbf{L}^T) \quad \text{and} \quad \mathbf{W} = \tfrac{1}{2}(\mathbf{L} - \mathbf{L}^T)$$

with rectangular Cartesian components (4.4.5a)

$$D_{km} = \tfrac{1}{2}(v_{k,m} + v_{m,k}) \quad \text{and} \quad W_{km} = \tfrac{1}{2}(v_{k,m} - v_{m,k}).$$

Then

$$\mathbf{L} = \mathbf{D} + \mathbf{W}, \tag{4.4.5b}$$

and

$$D_{km} = D_{mk} \quad \text{while} \quad W_{km} = -W_{mk}. \tag{4.4.6}$$

Equation (3a) can then be written as

$$dv_k = D_{km}\,dx_m + W_{km}\,dx_m \tag{4.4.7a}$$

representing the tensor equation

$$d\mathbf{v} = \mathbf{D}\cdot d\mathbf{x} + \mathbf{W}\cdot d\mathbf{x} \tag{4.4.7b}$$

or in matrix notation

$$dv = D\,dx + W\,dx, \tag{4.4.7c}$$

where D is the rate-of-deformation matrix with elements D_{km} and W is the spin matrix with elements W_{km}. Many authors use Ω and ω_{km} for the spin matrix and its components and also use the same symbols for the small rotation matrix of Secs. 4.1 and 4.2. This will not often lead to confusion since the small rotation matrix and the spin matrix are seldom used in the same discussion; however, we will denote the spin matrix by W to distinguish it from the small rotation matrix.

The physical significance of the spin tensor may be seen by considering the case that all of the rate-of-deformation components D_{km} are zero; the instantaneous motion is then a rigid-body rotation, and the relative-velocity components of the particle at q relative to the particle at p are the elements of the column matrix given by the matrix product $W\,dx$. Note that the components of the column matrix $W\,dx$ are the same as the velocity vector components $\mathbf{w} \times d\mathbf{x}$ due to rigid-body rotation about p with angular velocity $\mathbf{w}$, where

$$\mathbf{w} = -W_{23}\mathbf{i}_1 - W_{31}\mathbf{i}_2 - W_{12}\mathbf{i}_3 \quad \text{or} \quad w_i = -\tfrac{1}{2}\,e_{ijk}W_{jk}.$$

Also (4.4.8)

$$\mathbf{w} = \tfrac{1}{2}\nabla \times \mathbf{v} \quad \text{or} \quad w_i = \tfrac{1}{2}\,e_{ijk}\partial_j v_k.$$

If, on the other hand, the spin-tensor components W_{km} are all zero in a region, the velocity field is said to be ***irrotational*** in the region.

The tensor **D** is called the ***rate-of-deformation tensor***.† When the velocity field is irrotational, the relative velocity of the particle at q relative to the particle at p is given by $d\mathbf{v} = \mathbf{D} \cdot d\mathbf{x}$ for each infinitesimal vector $d\mathbf{x}$ originating at p. Note that since we are interested in quantities evaluated at p, the vector $d\mathbf{x}$ represented by the column matrix dx must be infinitesimal, but the present development differs fundamentally from that of Sec. 4.2 because we do not require the velocity components or the velocity gradients to be small. The physical interpretation of **D** and **W** as rates of deformation and rotation‡ is strictly applicable even when their components are all finite. The rate-of-deformation tensor is a symmetric tensor; hence all of the development of principal axes, invariants, and Mohr's Circles, given in Chap. 3, may be applied to rate of deformation.

***The suitability of the name* "rate of deformation"** for the tensor **D** may be seen as follows. Fix attention on a given material vector $d\mathbf{X}$. We choose the instantaneous rate of change at time t of the squared length $(ds)^2$ of this vector as a measure of the rate of deformation. Note that $d/dt\,[(ds)^2] = 2\,ds(d/dt)(ds)$. At time t, the material of $d\mathbf{X}$ has moved to position $d\mathbf{x}$, and

$$(ds)^2 = d\mathbf{x} \cdot d\mathbf{x} = dx_k\, dx_k. \tag{4.4.9a}$$

Hence

$$\frac{d}{dt}[(ds)^2] = 2\,d\mathbf{x} \cdot \frac{d}{dt}(d\mathbf{x}) = 2\,dx_k \frac{d}{dt}(dx_k). \tag{4.4.9b}$$

But from Eq. (1)

$$d\mathbf{x} = \mathbf{x}\overleftarrow{\nabla} \cdot d\mathbf{X} \quad \text{or} \quad dx_k = \frac{\partial x_k}{\partial X_m} dX_m$$

whence

$$\frac{d}{dt}(d\mathbf{x}) = \left[\frac{d}{dt}(\mathbf{x}\overleftarrow{\nabla})\right] \cdot d\mathbf{X} + \mathbf{x}\overleftarrow{\nabla} \cdot \frac{d}{dt}(d\mathbf{X}).$$

The last term vanishes because $d\mathbf{X}$ does not change with time, being the ***initial***

†Other notations sometimes used for its components are d_{km}, V_{km} and $\dot{\epsilon}_{km}$; the latter unfortunately seems to imply that D_{km} is the time derivative of the small strain ϵ_{km}, which is not the case as will be shown later in this section. An earlier name proposed for **D** was ***velocity strain.***

‡Cauchy showed in 1841 that the vector of Eq. (8) represents a local instantaneous rate of rotation.

relative-position vector of the two particles. Interchanging the order of differentiation in the first term then yields

$$\frac{d}{dt}(d\mathbf{x}) = \left(\frac{d\mathbf{x}}{dt}\overleftarrow{\nabla}\right)\cdot d\mathbf{X} = \mathbf{v}\overleftarrow{\nabla}\cdot d\mathbf{X}. \tag{4.4.10a}$$

Note that here $\mathbf{v}\overleftarrow{\nabla}$ is the gradient of $\mathbf{v}$ with respect to the material coordinates $\mathbf{X}$ and is not the same thing as $\mathbf{L} = \mathbf{v}\overleftarrow{\nabla}_{\mathbf{x}}$. But both

$$\mathbf{v}\overleftarrow{\nabla}\cdot d\mathbf{X} = d\mathbf{v}$$

and

$$\mathbf{L}\cdot d\mathbf{x} = d\mathbf{v}$$

whence $\mathbf{v}\overleftarrow{\nabla}\cdot d\mathbf{X} = \mathbf{L}\cdot d\mathbf{x}$, so that

$$\frac{d}{dt}(d\mathbf{x}) = \mathbf{L}\cdot d\mathbf{x} \quad \text{or} \quad \frac{d}{dt}(dx_k) = v_{k,m}\,dx_m. \tag{4.4.10b}$$

Equation (9b) then becomes

$$\begin{aligned}\frac{d}{dt}[(ds)^2] &= 2dx_k\,v_{k,m}\,dx_m = 2d\mathbf{x}\cdot\mathbf{L}\cdot d\mathbf{x}\\ &= 2dx_k\,D_{km}\,dx_m + 2dx_k\,W_{km}\,dx_m\end{aligned}$$

where the last line follows by use of Eq. (5b). But the sum in the last term vanishes,

$$dx_k\,W_{km}\,dx_m = 0,$$

because $W_{km} = -W_{mk}$ while $dx_k dx_m = dx_m dx_k$. Hence

$$\boxed{\frac{d}{dt}[(ds)^2] = 2dx_k\,D_{km}\,dx_m = 2d\mathbf{x}\cdot\mathbf{D}\cdot d\mathbf{x}.} \tag{4.4.11}$$

It follows that the rate of change of the squared length $(ds)^2$ of the material instantaneously occupying any infinitesimal relative position $d\mathbf{x}$ at p is determined by the tensor $\mathbf{D}$ at p. This is why it is called the rate-of-deformation tensor.

Comparison of D and $\dot{\mathbf{E}}$. The rate-of-deformation tensor $\mathbf{D}$ is sometimes confused with the time derivative of the strain tensor. The small strain tensor of Sec. 4.2 is defined in terms of the material or initial coordinates so that

$$u_k = u_k(X_1, X_2, X_3, t)$$

and

$$\epsilon_{km} = \frac{1}{2}\left(\frac{\partial u_k}{\partial X_m} + \frac{\partial u_m}{\partial X_k}\right).$$

Since

$$\frac{\partial v_k}{\partial X_m} = \frac{\partial}{\partial X_m}\left(\frac{dx_k}{dt}\right) = \frac{d}{dt}\left(\frac{\partial x_k}{\partial X_m}\right) = \frac{d}{dt}\left(\frac{\partial(X_k + u_k)}{\partial X_m}\right)$$
$$= \frac{d}{dt}\left(\delta_{km} + \frac{\partial u_k}{\partial X_m}\right) = \frac{d}{dt}\left(\frac{\partial u_k}{\partial X_m}\right),$$

we have formally

$$\frac{d\epsilon_{km}}{dt} = \frac{1}{2}\left(\frac{\partial v_k}{\partial X_m} + \frac{\partial v_m}{\partial X_k}\right). \tag{4.4.12a}$$

But

$$D_{km} = \frac{1}{2}\left(\frac{\partial v_k}{\partial x_m} + \frac{\partial v_m}{\partial x_k}\right). \tag{4.4.12b}$$

We see that the two expressions differ in that in Eq. (12a) the derivatives are with respect to the material coordinates X_i, while in Eq. (12b) they are with respect to the spatial coordinates x_i. When the displacements and displacement gradients are very small the two tensors are approximately equal. But the rate-of-deformation tensor **D** is usable in situations where the displacements are not small (e.g., fluid flow, drawing, extrusion, etc.).†

Natural-strain increments. In the analysis of tensile tests, the conventional strain e is defined as the change in length per unit initial length.

$$e = \frac{L - L_0}{L_0} \quad \text{or} \quad e = \frac{1}{L_0}\int_{L_0}^{L} dL, \qquad \text{where} \quad de = \frac{dL}{L_0}. \tag{4.4.13}$$

Instead of defining the strain increment de as dL/L_0 in the conventional manner, we can define a new measure $d\epsilon$ of the deformation called the ***natural-strain increment*** by writing

$$d\epsilon = \frac{dL}{L}. \tag{4.4.14}$$

†See Sec. 4.5 for a relationship between D_{km} and the rate of change dE_{km}/dt of finite strain E_{km}.

The increment $d\epsilon$ is the change in length per unit contemporary length. In a tensile test, we can integrate to find an expression for the ϵ attained during a finite amount of stretching.

$$\epsilon = \int_{L_0}^{L} \frac{dL}{L} = \ln \frac{L}{L_0} = \ln (1 + e), \tag{4.4.15}$$

where ln denotes the natural logarithm. This is the logarithmic strain† or natural strain.

A three-dimensional generalization of the infinitesimal natural-strain increment in time dt is

$$d\epsilon_{km} = D_{km} dt. \tag{4.4.16}$$

Let

$$\overline{du}_k = v_k \, dt. \tag{4.4.17}$$

(Note that $\overline{du}_k$ is the infinitesimal change in position of one particle in time dt and is not the same thing as du_k which is the relative displacement of one particle with respect to another one.) The infinitesimal increments $d\epsilon_{km}$ can then be written as

$$d\epsilon_{km} = \frac{1}{2}\left[\frac{\partial(\overline{du}_k)}{\partial x_m} + \frac{\partial(\overline{du}_m)}{\partial x_k}\right]. \tag{4.4.18}$$

We remark that we have defined $d\epsilon_{km}$, but have no meaning as yet for ϵ_{km}. We could define ϵ_{km} as

$$\int_{t_0}^{t} d\epsilon_{km} = \int_{t_0}^{t} D_{km} \, dt,$$

but there is no physical significance to such an integration at one space point even in a steady flow situation. We may choose to interpret the integral as a material time integral (analogous to the material time derivative) and evaluate it by following a particle through its motion, evaluating the integrand for the given particle at each time t. This may be more meaningful physically, but it poses a formidable mathematical problem.

R. Hill‡ uses the natural-strain increments defined above denoted by $d\epsilon_{ij}$. In his equations, only the increments appear, not the ϵ_{ij}; this is essentially equivalent to using the rate of deformation D_{ij}, since $d\epsilon_{ij}$ equals $D_{ij}\,dt$ and

†The logarithmic-strain measure was introduced by P. Ludwik in 1909. For a discussion of the theory of the tensile test see Sec. 6.5 and the references given there.

‡R. Hill, *The Mathematical Theory of Plasticity* (Oxford, Clarendon Press, 1950).

all of his equations are homogeneous in the time and may be expressed in terms of the D_{ij} by dividing through by dt.

The natural strain increments $d\epsilon_{ij}$ are components of a Cartesian tensor, and the transformation formulas, principal axis theory, etc. may be applied. The quantities ϵ_{ij} defined by integration are not components of a Cartesian tensor. The integration can be performed in special cases when the increment principal axes do not rotate during the deformation, but the results appear to be of little use.

The rate-of-deformation tensor has been especially useful in theories of viscosity and viscoplasticity whose constitutive equations express the stress in terms of the instantaneous rate of deformation. Some new approach may still be needed in work-hardening metal plasticity in order to keep track of the previous deformation, but this is yet to be worked out—see Sec. 6.6. The small-strain tensor discussed in Sec. 4.2, on the other hand, is more suitable for elasticity of metals. When we deal with polymer materials, rubber or any of the plastics which we attempt to characterize as viscoelastic materials, we are faced with the necessity of somehow preserving a reference to the initial state but nevertheless simultaneously dealing with large displacements and also with time or rate effects. Just how this is to be accomplished in general is not yet settled, but a great deal of effort is being invested in the problem. In Secs. 4.5 and 4.6, we examine some of the finite-strain measures that have been proposed.

EXERCISES

1. In a certain region the flow velocity components are

$$v_x = -A(x^3 + xy^2)e^{-kt} \qquad v_y = A(x^2y + y^3)e^{-kt} \qquad v_z = 0$$

where A and k are given constants, x, y, z are spatial coordinates and t is time. Find the acceleration components at the point (1, 1, 0) at time $t = 0$.

2. If the intensity of illumination of a fluid particle at (x, y, z) at time t is given by $I = Ae^{-3t}/(x^2 + y^2 + z^2)$ and the fluid velocity field is given by

$$v_x = B(y + 2z), \qquad v_y = B(y + 3z), \qquad v_z = B(2x + 3y + 2z),$$

where A and B are known constants, determine the rate of change of the illumination experienced at time t by the fluid particle which is at point (1, 2, −2) at time t.

3. For the flow of Ex. 1, display the matrices of rectangular Cartesian components of the spatial gradient **L**, the rate of deformation **D** and the spin **W** at point (1, 1, 0) at time $t = 0$. What are the principal values of the rate of deformation, and what is the maximum shear rate of deformation, i.e., maximum D_{nt}?

4. For the flow of Ex. 3 calculate the unit rate of change $(1/ds)(d/dt)(ds)$ for the arc length of the element $d\mathbf{x} = ds(\mathbf{i} + \sqrt{3}\,\mathbf{j})/2$ at the point (1, 1, 0) at $t = 0$.
5. The motion of a certain continuous medium is defined by the equations

$$x_1 = \tfrac{1}{2}(X_1 + X_2)e^t + \tfrac{1}{2}(X_1 - X_2)e^{-t}, \qquad x_2 = \tfrac{1}{2}(X_1 + X_2)e^t - \tfrac{1}{2}(X_1 - X_2)e^{-t},$$

$$x_3 = X_3.$$

   (a) Express the velocity components in terms of the material coordinates and time.
   (b) Express the velocity components in terms of spatial coordinates and time.
   (c) Calculate the components of the rate-of-deformation matrix.
   (d) Express the displacement components u_1, u_2, u_3 in terms of the material coordinates and time.
   (e) Express the small-strain components in terms of material coordinates and time. Express them in terms of hyperbolic functions. Evaluate them at $t = 0$ and at $t = 0.05$.
   (f) Calculate the rate of change $d\epsilon_{ij}/dt$ of the small-strain components and compare with D_{ij} at $t = 0$ and at $t = 0.05$.
6. For the steady motion defined by $v_x = my$, $v_y = v_z = 0$, where m is a constant:
   (a) Display the matrices of rectangular Cartesian components of the velocity gradient **L**, the rate of deformation **D**, and spin **W**.
   (b) Express the relative motion components ξ, η, ζ of $\boldsymbol{\xi}$ in terms of x, y, z, and m for $\tau < t$. (See Eqs. 4.3.13.)
7. If the velocity components are given by

$$v_x = v_y = 0, \qquad v_z = v_1 z \frac{[1 - (x^2 + y^2)/a^2]}{L}$$

   where v_1, L, and a are constants, calculate and display component matrices of (a) rate of deformation, (b) spin, (c) rate-of-deformation deviator. (d) What is the volumetric rate of deformation per unit current volume?
8. If at a point the rate-of-deformation components are $D_{xx} = -6$, $D_{yy} = -2$, $D_{zz} = 4$, $D_{xy} = D_{yz} = D_{zx} = 0$ (each times 10^{-2} sec^{-1}), determine at that point the maximum shear deformation rate (max D_{nt}), and the extensional deformation rate of an element perpendicular to the surface of maximum D_{nt}.
9. If at a point the rectangular Cartesian rate-of-deformation components are as shown below (units sec^{-1}), determine the maximum shear deformation rate (max D_{nt}) at the point, and the extensional deformation rate of an element perpendicular to the surface of max D_{nt}.

$$\begin{bmatrix} 14 & -8 & 0 \\ -8 & 26 & 0 \\ 0 & 0 & -10 \end{bmatrix}$$

10. By differentiating with respect to time the expression $d\mathbf{x}^{(1)} \cdot d\mathbf{x}^{(2)} = ds^{(1)} ds^{(2)} \cos\theta_{12}$, where $d\mathbf{x}^{(1)}$ and $d\mathbf{x}^{(2)}$ are the current positions of two infinitesimal material vectors from the same point, and then evaluating for the special case that at time

t the two vectors are instantaneously parallel to the x_1 and x_2 axes, show that for this special case $\dot{\theta}_{12} = -2D_{12}$.

11. In a certain region of a steady fluid flow the velocity components are

$$v_x = -wy - \frac{Ay}{x^2 + y^2}, \qquad v_y = wx + \frac{Ax}{x^2 + y^2}, \qquad v_z = 0,$$

where w and A are constants. Display the matrices of components of rate of deformation **D** and spin **W** at a typical point (x, y, z).

12. If the electrical potential P at a point in space is given by $P = (P_0/r)\sin qt$, where P_0 and q are constants, r is distance from the origin, and t is time, write an expression for the dP/dt experienced by the fluid particle at position (x, y, z) at time t in the flow of Ex. 11.

13. Show that $d[\ln(ds)]/dt = \mathbf{n}\cdot\mathbf{D}\cdot\mathbf{n}$, where $\mathbf{n} = d\mathbf{x}/ds$.

14. Show that in general a skew-symmetric tensor such as the spin tensor **W** has only one real principal value λ. Determine λ and the corresponding direction $\hat{\mathbf{n}}$ in terms of the components w_1, w_2, w_3 of the vector **w** of Eq. (4.4.8).

15. For certain steady flows of ideal frictionless fluid the velocity vector is the gradient of a scalar ***velocity potential*** function $\phi(x, y, z)$, $\mathbf{v} = \nabla\phi$.
    (a) Show that such a flow is irrotational, i.e., that its spin components vanish.
    (b) What boundary conditions must be satisfied by ϕ at a fixed boundary surface through which there is no flow? (Note that the stated boundary condition for a frictionless fluid does not forbid flow tangential to the boundary.)

16. A particle of fluid is flowing in a wind tunnel where the velocity components are given by

$$v_x = U\left[1 + \left(\frac{a^2}{r^2}\right) - \left(\frac{2a^2x^2}{r^4}\right)\right], \qquad v_y = -\frac{2Ua^2xy}{r^4}, \qquad v_z = 0,$$

where $r^2 = x^2 + y^2$ and where U and a are given constants. A fixed electrical conductor on the z-axis creates the electrostatic potential field P given by $P = A\cos qt\ln(x^2 + y^2)$, where A and q are constants. If the moving particle is at $(a, a, 0)$ at $t = t_1$, express the rate of change of P as seen by the particle at t_1 in terms of the given quantities.

17. Display the rectangular component matrices of the rate of deformation and spin at $(a, a, 0)$ for the flow of Ex. 16.

18. For the flow of Ex. 2 calculate the unit rate of change $(1/ds)(d/dt)(ds)$ for the arc length of element bisecting the angle between the x- and y-axes.

4.5 Finite strain and deformation; Eulerian and Lagrangian formulations; geometric measures of strain; relative deformation gradient

When the displacement-gradient components are not small compared to unity, the problem of characterizing the strain from the initial state is more difficult than in the small-strain case of Sec. 4.2. Several different kinds of finite-strain definitions have been proposed, which fall into two classes: definitions in terms of the undeformed configuration, and definitions in terms of the deformed configuration. When coordinates are introduced, the first class uses material coordinates in the undeformed configuration, while the second class uses spatial coordinates in the deformed configuration. The formulation in terms of the undeformed configuration in usually called the ***Lagrangian formulation***, while the formulation in terms of the deformed configuration is called the ***Eulerian formulation***. In this section, we will develop the two formulations together, and in each we will consider four tensors: ***a deformation-gradient tensor***, ***its transpose***, ***a deformation tensor***, and ***a strain tensor***.

Some other possible measures of finite strain will be discussed in Sec. 4.6, especially the stretch tensors which combine multiplicatively with the rotation tensor, so that the product of a stretch tensor and a rotation tensor is equal to the deformation-gradient tensor.

The deformation-gradient tensor is the simplest to define in terms of the deformation equations, Eqs. (4.3.1), which are repeated below as Eqs. (1) of this section; this tensor also includes more information about the motion than do the other two, but this can be a disadvantage. Because the deformation-gradient tensor includes the rotation as well as the deformation, constitutive equations employing it will have to be so constructed that they will not predict a stress due to rigid-body rotation. The strain tensor, on the other hand, vanishes for a rigid-body rotation; hence, for example, a simple linear homogeneous stress-strain equation will satisfy the requirement that no stress will be predicted to arise from rigid-body rotation.

The Lagrangian formulation seems the more suitable one in elasticity, since in elasticity there is usually assumed to be a natural undeformed state, to which the body would return when it is unloaded. But the stress equations of motion or equilibrium must be satisfied in the deformed or contemporary configuration, and stress is therefore defined in the deformed configuration. If a stress-strain equation is to be written, either the stresses must be referred back to the undeformed configuration, or else the strains must be referred to the deformed configuration in order to use the same reference for all the tensors appearing in the equation. In this section we consider only the second of the two alternatives, leaving until later (see Sec. 5.3) the question of how to refer the stress to the undeformed configuration in order to use it in constitutive equations with the strains of the Lagrangian formulation.

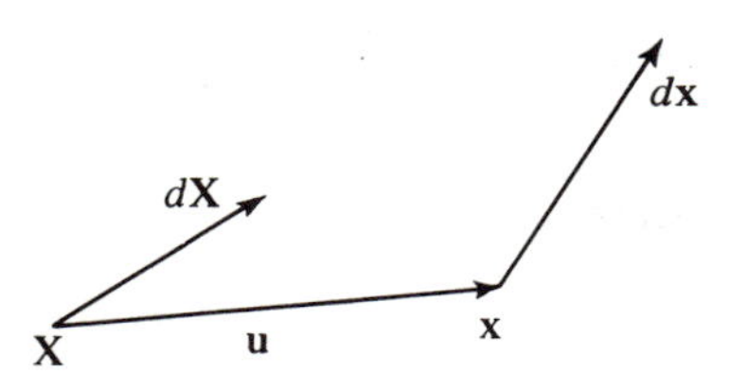

Fig. 4.9 Displacement, stretch, and rotation of material vector $d\mathbf{X}$ to new position $d\mathbf{x}$.

Figure 4.9 shows the displacement of a particle from its initial position $\mathbf{X}$ to the current position $\mathbf{x}$, defined by the deformation equations

$$\mathbf{x} = \mathbf{x}(\mathbf{X}, t) \quad \text{or} \quad x_i = x_i(X_1, X_2, X_3, t) \tag{4.5.1a}$$

and inversely by

$$\mathbf{X} = \mathbf{X}(\mathbf{x}, t) \quad \text{or} \quad X_i = X_i(x_1, x_2, x_3, t). \tag{4.5.1b}$$

The Eulerian formulation of Eq. (1b) gives the initial position of the particle now (at time t) occupying the position $\mathbf{x}$. The ***deformation gradients*** are the gradients of the functions on the right-hand side of these equations. Note that the rectangular Cartesian components of the gradients are derivatives with respect to material coordinates in the Lagrangian formulation of Eq. (1a). To further emphasize the distinction between the two sets of coordinates, we can use capital letters also for the indices on coordinates and tensors referred to the undeformed configuration, e.g., E_{KM} for the strain and X_K for the coordinates. A gradient component could then be written as $\partial x_k/\partial X_K$, where no summation is implied because k and K are different letters. When we limit ourselves to rectangular Cartesian coordinates, we will usually refer both the x_k and the X_K to the same reference axes, but this is not necessary. When rotations of coordinate axes are permitted independently for the two, this is an example of a ***two-point tensor field***, whose components transform like those of a vector under rotations of only one of the two reference axes and like a two-point tensor when the two sets of axes are rotated independently. (If T_{Km} are the two-point tensor components referred to X_K- and x_m- axes, and we change to two different sets of axes, $\bar{X}_J$ and $\bar{x}_p$, then $\bar{T}_{Jp} = A_J^K a_p^m T_{Km}$, where $A_J^K = \cos(\bar{X}_J, X_K)$ and $a_p^m = \cos(\bar{x}_p, \bar{x}_m)$.) The distinction becomes more important in curvilinear coordinates, because even with the same reference axes, the r-direction in polar coordinates, for example, varies from point to point (see Sec. I.4).

The deformation gradient referred to the undeformed configuration is denoted by $\mathbf{F}$ or by $\mathbf{x}\overleftarrow{\nabla}$; its transpose is $\mathbf{F}^T = \overrightarrow{\nabla}\mathbf{x}$. The deformation gradient $\mathbf{F}$ is defined as the tensor whose rectangular Cartesian components are the partial derivatives $\partial x_k/\partial X_K$ and which operates on an arbitrary infinitesimal material vector $d\mathbf{X}$ at $\mathbf{X}$ to associate with it a vector $d\mathbf{x}$ at $\mathbf{x}$ as follows

$$d\mathbf{x} = \mathbf{F} \cdot d\mathbf{X} \quad \text{or} \quad d\mathbf{x} = d\mathbf{X} \cdot \mathbf{F}^T,$$

since

$$d\mathbf{x} = (\mathbf{x}\overleftarrow{\nabla}) \cdot d\mathbf{X} = d\mathbf{X} \cdot \overrightarrow{\nabla}\mathbf{x}.$$

The rectangular Cartesian component matrix forms are

$$[dx_k] = [x_{k,M}][dX_M] \quad \text{or} \quad [dx_k] = [dX_M][\partial_M x_k] \tag{4.5.2a}$$

both representing the same indicial form

$$dx_k = \frac{\partial x_k}{\partial X_M} dX_M.$$

Here the notation $x_{k,M}$ has been used for $F_{kM} = \partial x_k/\partial X_M$ and $\partial_M x_k$ for $(F^T)_{Mk} = (\partial/\partial X_M)(x_k)$ to keep the indices in the proper order so that the first index identifies the row in the matrix, while the second index identifies the column.

The spatial deformation gradients are the comparable tensors referred to the deformed configuration (Eulerian formulation),† denoted by $\mathbf{F}^{-1} = \mathbf{X}\overleftarrow{\nabla}_{\mathbf{x}}$ and $(\mathbf{F}^{-1})^T = \overrightarrow{\nabla}_{\mathbf{x}}\mathbf{X}$ and defined by

$$d\mathbf{X} = \mathbf{F}^{-1}\cdot d\mathbf{x} \quad \text{or} \quad d\mathbf{X} = d\mathbf{x}\cdot(\mathbf{F}^{-1})^T$$
$$d\mathbf{X} = (\mathbf{X}\overleftarrow{\nabla}_{\mathbf{x}})\cdot d\mathbf{x} = d\mathbf{x}\cdot\overrightarrow{\nabla}_{\mathbf{x}}\mathbf{X}.$$

The rectangular Cartesian component matrix forms are

$$[dX_M] = [X_{M,k}][dx_k] \quad \text{or} \quad [dX_M] = [dx_k][\partial_k X_M] \tag{4.5.2b}$$

both representing the same indicial form

$$dX_M = \frac{\partial X_M}{\partial x_k} dx_k.$$

The spatial deformation gradient at $\mathbf{x}$ is inverse to the two-point tensor $\mathbf{F}$ at $\mathbf{X}$, since it operates on $d\mathbf{x}$ at $\mathbf{x}$ to yield $d\mathbf{X}$ at $\mathbf{X}$. The concept of the inverse of a tensor is discussed further in Sec. 4.6. In terms of the components, the two inverse relationships $\mathbf{F}\cdot\mathbf{F}^{-1} = \mathbf{1}$ and $\mathbf{F}^{-1}\cdot\mathbf{F} = \mathbf{1}$ are expressed by two sets of nine simultaneous equations,

$$x_{k,M} X_{M,m} = \delta_{km} \quad \text{and} \quad X_{K,m} x_{m,M} = \delta_{KM}. \tag{4.5.2c}$$

Although the components of the deformation gradient are finite, in Eqs. (2) we are considering the deformation of an infinitesimal neighborhood of the particle. In the following we shall usually interpret $d\mathbf{x}$ as the position occupied by the deformed material vector $d\mathbf{X}$ in the same sense as $dy = f'(x)\,dx$ is interpreted in the usual applications of elementary calculus as the change in y accompanying an infinitesimal change in x. When the material vector $d\mathbf{X}$ is

†The Cartesian-component forms of Eqs. (2) were introduced in papers from 1762 to 1770 by Euler, who also gave the solution by Cramer's rule of the nine component equations (2c), expressing each of the nine $X_{K,m}$ as a rational function of the $x_{k,m}$.

small but finite, the associated $d\mathbf{x}$ for our purposes is defined by the first Eq. (2a) and does not exactly coincide with the deformed position of the material $d\mathbf{X}$, which now lies along a curve in the deformed material. A careful discussion of this point is given by Truesdell and Toupin (1960), Sec. 20.

The strain tensors ($\mathbf{E}$ and $\mathbf{E}^*$) are defined so that they give the change in the squared length of the material vector $d\mathbf{X}$ as follows. For the Lagrangian formulation, we write (component forms are for rectangular Cartesians)

$$(ds)^2 - (dS)^2 = 2d\mathbf{X}\cdot\mathbf{E}\cdot d\mathbf{X}$$
$$(ds)^2 - (dS)^2 = 2dX_I E_{IJ}\, dX_J \tag{4.5.3a}$$

while the Eulerian formulation is

$$(ds)^2 - (dS)^2 = 2d\mathbf{x}\cdot\mathbf{E}^*\cdot d\mathbf{x}$$
$$(ds)^2 - (dS)^2 = 2dx_k\, E^*_{km}\, dx_m \tag{4.5.3b}$$

Another common notation for the Eulerian strain component is e_{km}.

The deformation tensors $\mathbf{C}$ and $\mathbf{B}^{-1}$ are closely related to the classical strain tensors; instead of giving the change in the squared length the Green deformation tensor $\mathbf{C}$, referred to the undeformed configuration, gives the new squared length $(ds)^2$ of the element into which the given element $d\mathbf{X}$ is deformed, while the Cauchy deformation tensor† $\mathbf{B}^{-1}$ (alternatively denoted by $\mathbf{c}$) gives the initial squared length $(dS)^2$ of an element $d\mathbf{x}$ identified in the deformed configuration:

$$(ds)^2 = d\mathbf{X}\cdot\mathbf{C}\cdot d\mathbf{X} \qquad (ds)^2 = dX_I C_{IJ} dX_J \tag{4.5.4a}$$
$$(dS)^2 = d\mathbf{x}\cdot\mathbf{B}^{-1}\cdot d\mathbf{x} \qquad (dS)^2 = dx_k (B^{-1})_{km}\, dx_m \tag{4.5.4b}$$

Comparing Eqs. (3) and (4) we obtain the following relationships between the deformation tensors and strain tensors

$$2\mathbf{E} = \mathbf{C} - \mathbf{1} \qquad 2E_{IJ} = C_{IJ} - \delta_{IJ} \tag{4.5.5a}$$
$$2\mathbf{E}^* = \mathbf{1} - \mathbf{B}^{-1} \qquad 2E^*_{km} = \delta_{km} - (B^{-1})_{km} \tag{4.5.5b}$$

†The notation $\mathbf{B}^{-1}$ is used here, because it is the inverse of a tensor commonly denoted by $\mathbf{B}$ as will be seen in Sec. 4.6. Another notation for $\mathbf{B}^{-1}$ is $\mathbf{c}$. This tensor was introduced by Cauchy in 1827, while $\mathbf{C}$ was used by Green in 1841. In Truesdell and Noll (1965), Sec. 24, $\mathbf{C}$ is called the right Cauchy-Green tensor, while $\mathbf{B}$ is called the left Cauchy-Green tensor.

We note that the only exception to the formal interchange of symbols in going from the Lagrangian formulation to the Eulerian formulation is that the place of **E** in the interchange would have to be taken by $-\mathbf{E}^*$ instead of $\mathbf{E}^*$. A completely symmetric interchange would have required the left-hand side of Eq. (3b) to be $(dS)^2 - (ds)^2$, the negative of the customary form. The deformation tensors reduce to the unit tensor **1** when there is no strain, instead of to zero.

Expressions for the strain and deformation tensors in terms of the deformation gradient. Using the two forms of the first line of Eq. (2a) we can express $(ds)^2$ as follows:

$$(ds)^2 = d\mathbf{x}\cdot d\mathbf{x} = (d\mathbf{X}\cdot\mathbf{F}^T)\cdot(\mathbf{F}\cdot d\mathbf{X}) = d\mathbf{X}\cdot[(\mathbf{F}^T\cdot\mathbf{F})]\cdot d\mathbf{X} \tag{4.5.6a}$$

Starting with Eq. (2b) we similarly obtain

$$(dS)^2 = d\mathbf{X}\cdot d\mathbf{X} = [d\mathbf{x}\cdot(\mathbf{F}^{-1})^T]\cdot[\mathbf{F}^{-1}\cdot d\mathbf{x}] = d\mathbf{x}\cdot[(\mathbf{F}^{-1})^T\cdot\mathbf{F}^{-1}]\cdot d\mathbf{x}. \tag{4.5.6b}$$

Comparison with the defining equations for the deformation tensors, Eqs. (4), then shows that

$$\boxed{\begin{gathered}\mathbf{C} = \mathbf{F}^T\cdot\mathbf{F} \quad \text{and} \quad \mathbf{B}^{-1} = (\mathbf{F}^{-1})^T\cdot\mathbf{F}^{-1} \\ \text{with rectangular Cartesian components} \\ C_{IJ} = \frac{\partial x_k}{\partial X_I}\frac{\partial x_k}{\partial X_J} \quad \text{and} \quad B^{-1}_{km} = \frac{\partial X_J}{\partial x_k}\frac{\partial X_J}{\partial x_m}.\end{gathered}} \tag{4.5.7}$$

summed on k and J, respectively.

The expressions for the strain tensors can then be obtained from Eqs. (5).

$$\boxed{\begin{gathered}\mathbf{E} = \tfrac{1}{2}[\mathbf{F}^T\cdot\mathbf{F} - \mathbf{1}] \quad \text{and} \quad \mathbf{E}^* = \tfrac{1}{2}[\mathbf{1} - (\mathbf{F}^{-1})^T\cdot\mathbf{F}^{-1}] \\ \text{with rectangular Cartesian components} \\ E_{IJ} = \frac{1}{2}\left[\frac{\partial x_k}{\partial X_I}\frac{\partial x_k}{\partial X_J} - \delta_{IJ}\right] \quad \text{and} \quad E^*_{km} = \frac{1}{2}\left[\delta_{ij} - \frac{\partial X_J}{\partial x_k}\frac{\partial X_J}{\partial x_m}\right]\end{gathered}} \tag{4.5.8}$$

summed on k and J, respectively. From Eqs. (7) and (8) we see that the strain and deformation matrices are symmetric. That they are tensors may then be inferred from the tensor test, Eq. (2.4.27), and the defining Eqs. (3) and (4) above. Alternatively, Eqs. (7) establish **C** and $\mathbf{B}^{-1}$ as tensors, since the product of two tensors is a tensor. The tensor character of **E** and $\mathbf{E}^*$ then follows from Eq. (5).

Since both **C** and **E** are symmetric tensors, they each have three real principal values, and the entire discussion of principal directions, invariants, and Mohr's Circles developed in Chap. 3 for the stress tensor applies also to **C** and **E**. Equation (5a) shows that the principal axes of the two tensors **C** and **E** coincide, since if the coordinate axes are so chosen that the off-diagonal elements C_{IJ} are zero, then the off-diagonal elements E_{IJ} will be zero, referred to the same axes.

A similar discussion applies to the two tensors $\mathbf{B}^{-1}$ and $\mathbf{E}^*$, whose principal axes coincide at the spatial point **x**, but which are in general not parallel to the principal axes of **C** and **E** at **X**. In Sec. 4.6, the rotation tensor **R** will be defined as the tensor rotating the principal axes of **C** at **X** to the directions of the principal axes of $\mathbf{B}^{-1}$ at **x**.

Since $(ds)^2$ is always positive, for arbitrary nonzero choice of the initial vector $d\mathbf{X}$, and $(dS)^2$ is positive for arbitrary nonzero $d\mathbf{x}$, the quadratic forms of Eqs. (4) are positive-definite. Hence the matrices of Cartesian components are positive-definite matrices, which implies that the principal values of **C** and $\mathbf{B}^{-1}$ are all positive, a conclusion also apparent from the geometrical meanings of the diagonal components to be discussed later in this section. First, however, we compare the finite strains to the small-strain definitions.

For comparison with the small-strain components, we write out E_{11} and E_{12}. Using the same reference axes for both x_i and X_i, and using lower-case subscripts for both, we have

$$\left.\begin{aligned} x_i &= X_i + u_i, \\ &\text{with displacement components} \\ u_i &= u_i(X_1, X_2, X_3, t). \end{aligned}\right\} \tag{4.5.9}$$

We obtain

$$E_{11} = \frac{1}{2}\left[\left(1 + \frac{\partial u_1}{\partial X_1}\right)\left(1 + \frac{\partial u_1}{\partial X_1}\right) + \frac{\partial u_2}{\partial X_1}\frac{\partial u_2}{\partial X_1} + \frac{\partial u_3}{\partial X_1}\frac{\partial u_3}{\partial X_1} - 1\right]$$

$$E_{12} = \frac{1}{2}\left[\left(1 + \frac{\partial u_1}{\partial X_1}\right)\frac{\partial u_1}{\partial X_2} + \frac{\partial u_2}{\partial X_1}\left(1 + \frac{\partial u_2}{\partial X_2}\right) + \frac{\partial u_3}{\partial X_1}\frac{\partial u_3}{\partial X_2} - 0\right]$$

or

$$\left.\begin{aligned} E_{11} &= \frac{\partial u_1}{\partial X_1} + \frac{1}{2}\left[\left(\frac{\partial u_1}{\partial X_1}\right)^2 + \left(\frac{\partial u_2}{\partial X_1}\right)^2 + \left(\frac{\partial u_3}{\partial X_1}\right)^2\right], \\ E_{12} &= \frac{1}{2}\left(\frac{\partial u_1}{\partial X_2} + \frac{\partial u_2}{\partial X_1}\right) + \frac{1}{2}\left[\frac{\partial u_1}{\partial X_1}\frac{\partial u_1}{\partial X_2} + \frac{\partial u_2}{\partial X_1}\frac{\partial u_2}{\partial X_2} + \frac{\partial u_3}{\partial X_1}\frac{\partial u_3}{\partial X_2}\right]. \end{aligned}\right\} \tag{4.5.10}$$

In terms of the displacements, the general expression for E_{ij} in Eq. (8) takes the form

$$E_{ij} = \frac{1}{2}\left[\frac{\partial u_i}{\partial X_j} + \frac{\partial u_j}{\partial X_i} + \frac{\partial u_k}{\partial X_i}\frac{\partial u_k}{\partial X_j}\right]. \tag{4.5.11a}$$

We see that if the partial derivatives of the displacements u_1, u_2, u_3 with respect to the material coordinates X_1, X_2, X_3 are all small compared to unity, the squares and products of these derivatives may be neglected in comparison to the linear terms. The remaining terms are the small-strain components defined in Sec. 4.2. In the elastic deformation of metals, the displacement gradients are often of the order of a few thousandths of an inch per inch, and the small-strain theory is then quite adequate. In rubber-like materials and some other synthetic plastics, recoverable elastic deformation may be of much larger magnitude, requiring the use of the finite strains. The plastic deformation of metals after yielding may also lead to large strains, but there is usually a considerable range of deformation beyond yield in which small-strain theories may still be used; in metal forming, where large deformations occur, the small-strain components are of doubtful physical significance. Incremental theories of plasticity in effect analyze these operations in terms of rate of deformation rather than strain. The components of $\mathbf{E}^*$ can also be expressed in terms of the displacement derivatives with respect to the spatial coordinates, if Eq. (9) is re-written with the u_i expressed as functions of the x_j.

$$X_i = x_i - u_i(x_1, x_2, x_3, t).$$

The result is

$$E^*_{ij} = \frac{1}{2}\left[\frac{\partial u_i}{\partial x_j} + \frac{\partial u_j}{\partial x_i} - \frac{\partial u_k}{\partial x_i}\frac{\partial u_k}{\partial x_j}\right]. \tag{4.5.11b}$$

When the squares and products of the derivatives can be neglected in comparison with the linear terms, these components also reduce to the form of the small-strain components, except that the derivatives are with respect to spatial coordinates instead of material coordinates. The E^*_{ij} are usually called ***Eulerian strain components***, although, according to Truesdell (1952), they were introduced for infinitesimal strains by Cauchy (1827) and for finite strain by Almansi (1911) and Hamel (1912). The E_{ij} are usually called ***Lagrangian strain components***, although, again according to Truesdell (1952), they were introduced by Green (1841) and St.-Venant (1844). ***When the displacements and displacement gradients are sufficiently small, as in the classical linearized theory of elasticity, the distinction between the two small-strain definitions is usually ignored.***

Note that the finite-strain components involve only linear and quadratic terms in the components of the displacement gradient. This is the complete finite-strain tensor and not merely a second-order approximation to it.

Strain rate compared to rate of deformation. By taking the material derivative of both sides of Eq. (3a), we see that

$$\frac{d}{dt}(ds)^2 = 2\,d\mathbf{X}\cdot\frac{d\mathbf{E}}{dt}\cdot d\mathbf{X} = 2\,dX_I\frac{dE_{IJ}}{dt}\,dX_J, \tag{4.5.12}$$

since $d\mathbf{X}$ and dS are constants. But Eq. (4.4.11) shows that

$$\frac{d}{dt}(ds)^2 = 2\,d\mathbf{x}\cdot\mathbf{D}\cdot d\mathbf{x},$$

whence by Eq. (2)

$$\begin{aligned}\frac{d}{dt}(ds)^2 &= 2(d\mathbf{X}\cdot\mathbf{F}^T)\cdot\mathbf{D}\cdot(\mathbf{F}\cdot d\mathbf{X})\\ &= 2\,d\mathbf{X}\cdot(\mathbf{F}^T\cdot\mathbf{D}\cdot\mathbf{F})\cdot d\mathbf{X}.\end{aligned}$$

Comparing this with Eq. (12), we see that

$$d\mathbf{X}\cdot\left[\frac{d\mathbf{E}}{dt} - \mathbf{F}^T\cdot\mathbf{D}\cdot\mathbf{F}\right]\cdot d\mathbf{X} = 0$$

for arbitrary $d\mathbf{X}$. Hence the quantity in square brackets must be identically zero, which implies that the ***strain rate*** $d\mathbf{E}/dt$ is related to the rate of deformation $\mathbf{D}$ by†

$$\boxed{\frac{d\mathbf{E}}{dt} = \mathbf{F}^T\cdot\mathbf{D}\cdot\mathbf{F} \quad\text{or}\quad \frac{dE_{IJ}}{dt} = \frac{\partial x_m}{\partial X_I}D_{mn}\frac{\partial x_n}{\partial X_J}} \tag{4.5.13a}$$

where the second form applies only to rectangular Cartesian coordinates. From Eq. (5) we also see that

$$\frac{d\mathbf{C}}{dt} = 2\frac{d\mathbf{E}}{dt}. \tag{4.5.13b}$$

Note that, since $\mathbf{x} = \mathbf{X} + \mathbf{u}$, $\mathbf{F}^T = \mathbf{1} + \overleftarrow{\nabla}\mathbf{u}$ and $\mathbf{F} = \mathbf{1} + \mathbf{u}\overleftarrow{\nabla}$. Hence, when the displacement-gradient components are small compared to unity, Eq. (13a) reduces to

$$\left.\begin{aligned}\frac{d\mathbf{E}}{dt} \approx \mathbf{1}\cdot\mathbf{D}\cdot\mathbf{1} = \mathbf{D} \quad\text{or}\\ \frac{dE_{ij}}{dt} \approx D_{ij} \quad\text{for}\quad \left|\frac{\partial u_i}{\partial X_j}\right| \ll 1,\end{aligned}\right\} \tag{4.5.13c}$$

†The results of Eqs. (13a) are due to E. and F. Cosserat in 1896.

with similar expressions for $\Lambda^2_{(2)}$ and $\Lambda^2_{(3)}$, while the stretch of the element currently parallel to the x_1-axis, for which $\hat{\mathbf{n}}$ has components (1, 0, 0), is denoted by $\lambda_{(1)}$ and given by

$$\frac{1}{\lambda^2_{(1)}} = (B^{-1})_{11} = 1 - 2\,E^*_{11} \tag{4.5.17b}$$

with similar expressions for $1/\lambda^2_{(2)}$ and $1/\lambda^2_{(3)}$. As a rule, $\lambda_{(1)}$ is not equal to $\Lambda_{(1)}$, since the element now in the x_1-direction is not the same material element as was initially in the X_1-direction. Equations (17) show that in the matrices of rectangular Cartesian components the diagonal elements for $\mathbf{C}$ and $\mathbf{B}^{-1}$ must be positive, while the diagonal elements for $\mathbf{E}$ must be greater than $-\frac{1}{2}$ and those for $\mathbf{E}^*$ must be less than $+\frac{1}{2}$.

The unit extension of the element, $(ds-dS)/dS$ is equal to the stretch minus one, $(ds/dS)-1$, and can be expressed in terms either of $\hat{\mathbf{N}}$ or of $\hat{\mathbf{n}}$, denoted $E_{(\hat{\mathbf{N}})}$ and $E^*_{(\hat{\mathbf{n}})}$, respectively. (Do not confuse these with the principal values E_K of $\mathbf{E}$ and E^*_k of $\mathbf{E}^*$.) For example, the unit extension of the element initially parallel to the X_1-axis is given by

$$\left.\begin{aligned} E_{(1)} &= \Lambda_{(1)} - 1 = \sqrt{C_{11}} - 1 = \sqrt{1 + 2\,E_{11}} - 1 \\ \text{whence} \qquad & \\ E_{11} &= E_{(1)} + \tfrac{1}{2}\,E^2_{(1)} \end{aligned}\right\} \tag{4.5.18}$$

with similar expressions for $E_{(2)}$ and $E_{(3)}$. When the unit extension is small compared to unity, the quadratic term in the last equation may be neglected in comparison to the linear term; then the corresponding strain component is approximately equal to the unit extension, as it was with small strain in Sec. 4.2.

Angle change. The elementary definition of shear strain was half the change in the angle between two of the edges of a rectangular element with edges initially parallel to the coordinate axes. The cosines of the angles between any two material elements in both the initial and the deformed configuration can be calculated by using the scalar product formula, Eq. (2.3.8). The angle change can then be found by subtracting the two angles. If cos $(\hat{\mathbf{N}}_1, \hat{\mathbf{N}}_2)$ denotes the cosine of the initial angle between the two elements $d\mathbf{X}_1$ and $d\mathbf{X}_2$ and cos $(\hat{\mathbf{n}}_1, \hat{\mathbf{n}}_2)$ denotes the cosine of the angle between the ***same two material elements,*** which in the deformed configuration, occupy positions $d\mathbf{x}_1$ and $d\mathbf{x}_2$, then

$$\cos\,(\hat{\mathbf{n}}_1, \hat{\mathbf{n}}_2) = \hat{\mathbf{n}}_1 \cdot \hat{\mathbf{n}}_2 = \frac{d\mathbf{x}_1 \cdot d\mathbf{x}_2}{|d\mathbf{x}_1||d\mathbf{x}_2|} = \frac{[d\mathbf{X}_1 \cdot \mathbf{F}^T] \cdot [\mathbf{F} \cdot d\mathbf{X}_2]}{\sqrt{d\mathbf{X}_1 \cdot \mathbf{C} \cdot d\mathbf{X}_1}\,\sqrt{d\mathbf{X}_2 \cdot \mathbf{C} \cdot d\mathbf{X}_2}},$$

where we have used Eq. (2a) in the numerator and Eq. (4a) in the denominator. Now divide numerator and denominator by $|d\mathbf{X}_1|$ and $|d\mathbf{X}_2|$, and substitute $\hat{\mathbf{N}}_1 = d\mathbf{X}_1/|d\mathbf{X}_1|$, $\hat{\mathbf{N}}_2 = d\mathbf{X}_2/|d\mathbf{X}_2|$, and $\mathbf{C} = \mathbf{F}^T \cdot \mathbf{F}$ by Eq. (7), to get

$$\cos(\hat{\mathbf{n}}_1, \hat{\mathbf{n}}_2) = \frac{\hat{\mathbf{N}}_1 \cdot \mathbf{C} \cdot \hat{\mathbf{N}}_2}{\sqrt{\hat{\mathbf{N}}_1 \cdot \mathbf{C} \cdot \hat{\mathbf{N}}_1}\,\sqrt{\hat{\mathbf{N}}_2 \cdot \mathbf{C} \cdot \hat{\mathbf{N}}_2}} = \frac{\hat{\mathbf{N}}_1 \cdot \mathbf{C} \cdot \hat{\mathbf{N}}_2}{\Lambda_{(\hat{\mathbf{N}}_1)} \Lambda_{(\hat{\mathbf{N}}_2)}}. \tag{4.5.19a}$$

Since $\cos(\hat{\mathbf{N}}_1, \hat{\mathbf{N}}_2) = \hat{\mathbf{N}}_1 \cdot \hat{\mathbf{N}}_2$ can be calculated immediately when the unit vectors are given in the undeformed configuration, the two angles determined from the two cosines can be subtracted to give the angle change. If the two unit vectors $\hat{\mathbf{n}}_1$ and $\hat{\mathbf{n}}_2$ are specified in the deformed configuration, on the other hand, we obtain $\cos(\hat{\mathbf{n}}_1, \hat{\mathbf{n}}_2) = \hat{\mathbf{n}}_1 \cdot \hat{\mathbf{n}}_2$ immediately, and calculate $\cos(\hat{\mathbf{N}}_1, \hat{\mathbf{N}}_2)$ by

$$\cos(\hat{\mathbf{N}}_1, \hat{\mathbf{N}}_2) = \frac{\hat{\mathbf{n}}_1 \cdot \mathbf{B}^{-1} \cdot \hat{\mathbf{n}}_2}{\sqrt{\hat{\mathbf{n}}_1 \cdot \mathbf{B}^{-1} \cdot \hat{\mathbf{n}}_1}\,\sqrt{\hat{\mathbf{n}}_2 \cdot \mathbf{B}^{-1} \cdot \hat{\mathbf{n}}_2}} = \lambda_{(\hat{\mathbf{n}}_1)} \lambda_{(\hat{\mathbf{n}}_2)} (\hat{\mathbf{n}}_1 \cdot \mathbf{B}^{-1} \cdot \hat{\mathbf{n}}_2), \tag{4.5.19b}$$

which is derived similarly to Eq. (19a), the last form following from Eq. (17b). For example, let θ_{12} denote the angle between the deformed elements which were initially parallel to the X_1-and X_2-Cartesian axes. Then $\hat{\mathbf{N}}_1$ has components (1, 0, 0), $\hat{\mathbf{N}}_2$ has components (0, 1, 0), and

$$\cos\theta_{12} = \frac{C_{12}}{\Lambda_{(1)}\,\Lambda_{(2)}} = \frac{C_{12}}{\sqrt{C_{11}\,C_{22}}} = \frac{2E_{12}}{\sqrt{(1 + 2E_{11})(1 + 2E_{22})}}. \tag{4.5.20}$$

Similarly, if we specify $\hat{\mathbf{n}}_1$ and $\hat{\mathbf{n}}_2$ to have the directions of the x_1-and x_2-axes in the deformed configuration, the cosine of the angle between the same two material elements in the undeformed configuration is

$$\lambda_{(1)}\lambda_{(2)}(B^{-1})_{12} = \frac{(B^{-1})_{12}}{\sqrt{(B^{-1})_{11}(B^{-1})_{22}}}.$$

We see that in general the shear deformation (as defined by angle change) is determined not only by the "shear components" of the finite strain or deformation tensors, but depends also on the stretches of the elements involved, thus differing from the small-strain case.

Equation (20) shows that when the X_1-and X_2-axes are principal axes of $\mathbf{E}$ (and $\mathbf{C}$) the new angle θ_{12} is a right angle, and there has been no angle change. Thus any three mutually perpendicular principal axes of $\mathbf{C}$ at $\mathbf{X}$ are deformed into three mutually perpendicular elements at $\mathbf{x}$. A derivation similar to that of Eq. (20) shows that any three mutually perpendicular principal axes of $\mathbf{B}^{-1}$ at $\mathbf{x}$ also have mutually perpendicular antecedent material elements at $\mathbf{X}$. When the principal values are distinct, there is a unique set of three mutually perpendicular principal directions for each tensor. Hence, ***when no two principal***

***strains are equal, the principal axes of* C *at* X *are rotated and translated by the deformation into the principal axes of* $\mathbf{B}^{-1}$ *at* x.**

When the initial elements are not parallel to principal axes, but the unit extensions and the angle changes are small compared to unity, Eq. (20) reduces to

$$\frac{\pi}{2} - \theta_{12} \approx \sin\left(\frac{\pi}{2} - \theta_{12}\right) = \cos\theta_{12} \approx 2E_{12}\,, \qquad (4.5.21)$$

so that E_{12} is then approximately half the angle change as it was in Sec. 4.2 for small strains.

The volume ratio dV/dV_0 of a rectangular element with edges initially parallel to coordinate axes in the principal directions of **E** and **C** at the point can be calculated by using Eq. (18) and the similar equations for the other three elements forming the edges of the rectangular volume element. Since there has been no shear of this element, the new element is still rectangular, and

$$\frac{dV}{dV_0} = \Lambda_{(1)}\Lambda_{(2)}\Lambda_{(3)} = \sqrt{(1+2E_1)(1+2E_2)(1+2E_3)}$$

$$= \sqrt{C_1C_2C_3} = \sqrt{\mathrm{III}_C} \qquad (4.5.22\text{a})$$

where III_C is the cubic invariant of **C**. Since this invariant can be calculated without reference to principal axes, the last form gives a general expression† for dV/dV_0. The ratio dV/dV_0 can also be expressed in terms of the invariants I_E, II_E, III_E, of **E**. The volume ratio may alternatively be calculated for a rectangular element whose final position has edges parallel to the principal axes of $\mathbf{B}^{-1}$ in the deformed configuration. Thus

$$\frac{dV}{dV_0} = \lambda_{(1)}\lambda_{(2)}\lambda_{(3)} = \frac{1}{\sqrt{(1-2E_1^*)(1-2E_2^*)(1-2E_3^*)}}$$

$$= \frac{1}{\sqrt{B_1^{-1}B_2^{-1}B_3^{-1}}} = \frac{1}{\sqrt{\mathrm{III}_{B^{-1}}}},$$

which can also be expressed in terms of the invariants of $\mathbf{E}^*$. The volume ratio is the reciprocal of the ratios of the densities. It is a basic hypothesis of continuum mechanics that the density is a scalar point function in each configuration. Hence the volume ratio will be the same for an infinitesimal element of arbitrary shape at the point (assuming that all its dimensions are infinitesimal), and our calculation, based on the particular volume element with edges parallel to the principal axes of strain, will be valid for a volume element of any shape.

†The last expression for dV/dV_0 in each of Eqs. (22a) and (22b) is due to Cauchy (1827).

The ratio of volume elements can also be expressed in terms of the determinant J of the deformation-gradient matrix. If the material coordinates are regarded as curvilinear coordinates in the deformed configuration (a curved coordinate surface $X_1 = \text{const.}$, for example, being the surface occupied in the deformed position by the particles initially on the plane $X_1 = \text{const.}$), then the deformation equations, Eqs. (1), represent a change of variables in the deformed state between curvilinear coordinates X_N and rectangular Cartesian coordinates x_m. As is shown in advanced calculus textbooks, any integral over the deformed volume V can be expressed in terms of the curvilinear coordinates as follows:

$$\int f(\mathbf{x})\, dx_1 dx_2 dx_3 = \int f[\mathbf{x}(\mathbf{X})]|J| dX_1 dX_2 dX_3\,, \tag{4.5.23}$$

where $|J|$ is the absolute value of the Jacobian determinant whose elements are $\partial x_m / \partial X_N$. When the change of variables represents a continuous deformation, the determinant will always be positive and the absolute value signs can be omitted. If we now interpret the integral on the right as an integral over the volume V_0 in the undeformed configuration, with rectangular coordinate element of volume $dV_0 = dX_1 dX_2 dX_3$, we see that the ratio of the two volume elements is equal to J†.

$$\left.\begin{aligned} \frac{\rho_0}{\rho} &= \frac{dV}{dV_0} = J \\ \text{and inversely} \qquad & \\ \frac{\rho}{\rho_0} &= \frac{dV_0}{dV} = J^{-1} \end{aligned}\right\} \tag{4.5.24}$$

where ρ_0 and ρ are the densities in the undeformed and deformed configurations, respectively, and where J^{-1} is the Jacobian determinant of the inverse deformation, equal to the reciprocal of J. [In Sec. 5.2 Eq. (24) will reappear as the continuity equation in the Lagrangian formulation.] Comparison with Eqs. (22) then shows that we must have

$$\mathrm{III}_C = J^2 \quad \text{and} \quad \mathrm{III}_{B^{-1}} = J^{-2} \quad \text{where} \quad J = \det \mathbf{F} = \det|x_{m,N}|, \tag{4.5.25}$$

which also follows directly from Eqs. (7). Any third-order determinant, with elements a_{mn} can be expanded as follows using the permutation symbol e_{rst} defined by Eq. (2.3.22)

$$e_{rst} \det|a_{mn}| = e_{ijk} a_{ir} a_{js} a_{kt} \tag{4.5.26}$$

[see Eq. (I.3.17)]. Using this expansion for J and J^{-1}, we obtain the two fol-

†The relationship $dV = J\, dV_0$ is due to Euler (1762).

lowing expressions for the density ratios

$$e_{rst}\frac{\rho}{\rho_0} = e_{rst}J^{-1} = e_{IJK}\frac{\partial X_I}{\partial x_r}\frac{\partial X_J}{\partial x_s}\frac{\partial X_K}{\partial x_t}$$

$$e_{IJK}\frac{\rho_0}{\rho} = e_{IJK}J = e_{rst}\frac{\partial x_r}{\partial X_I}\frac{\partial x_s}{\partial X_J}\frac{\partial x_t}{\partial X_K}. \tag{4.5.27}$$

The area change of an infinitesimal area dS_0 in the initial configuration into an area dS in the deformed configuration will be useful for referring the stress to the initial configuration in order to use it in constitutive equations with the strain and deformation tensors of the Lagrangian formulation. Do not confuse this area dS with the initial arc length, which does not appear here. The area change may be calculated as follows. Let $\hat{\mathbf{N}}$ be the unit normal to the area dS_0 of the parallelogram with two edges $d\mathbf{X}$ and $\delta\mathbf{X}$ in the undeformed configuration, and let $\hat{\mathbf{n}}$ be the unit normal in the deformed configuration to the area dS of the parallelogram with edges $d\mathbf{x}$ and $\delta\mathbf{x}$ given in terms of $d\mathbf{X}$ and $\delta\mathbf{X}$ by Eqs. (2a). Then

$$\hat{\mathbf{N}}\,dS_0 = d\mathbf{X}\times\delta\mathbf{X} \quad \text{and} \quad \hat{\mathbf{n}}\,dS = d\mathbf{x}\times\delta\mathbf{x}$$

with rectangular Cartesian components: (4.5.28)

$$N_I dS_0 = e_{IJK}dX_J\delta X_K \quad \text{and} \quad n_r dS = e_{rst}dx_s\delta x_t.$$

Substituting for dX_J and δX_K the expressions given by Eq. (2b), we obtain

$$N_I dS_0 = e_{IJK}\frac{\partial X_J}{\partial x_s}\frac{\partial X_K}{\partial x_t}dx_s\delta x_t.$$

Multiplying both sides of this equation by $\partial X_I/\partial x_r$(with summation on I) and using the first Eq. (27), we get

$$\frac{\partial X_I}{\partial x_r}N_I dS_0 = \frac{\rho}{\rho_0}e_{rst}dx_s\delta x_t.$$

Hence, using Eq. (28), we have finally†

$$n_r dS = \frac{\rho_0}{\rho}\frac{\partial X_I}{\partial x_r}N_I dS_0 = J\,N_I X_{I,r}\,dS_0 \tag{4.5.29a}$$

or, since $X_{I,r}$ are the components of $\mathbf{F}^{-1} \equiv \mathbf{X}\overleftarrow{\nabla}_{\mathbf{x}}$,

$$\hat{\mathbf{n}}\,dS = \frac{\rho_0}{\rho}\hat{\mathbf{N}}\cdot\mathbf{F}^{-1}\,dS_0 = J\hat{\mathbf{N}}\cdot\mathbf{F}^{-1}\,dS_0 \tag{4.5.29b}$$

†Equation (29a) is due to Nanson, 1878.

which is sometimes written by introducing vector areas $\mathbf{dA}_0 = \hat{\mathbf{N}}\, dS_0$ and $\mathbf{dA} = \hat{\mathbf{n}}\, dS$ as

$$\mathbf{dA} = \frac{\rho_0}{\rho}\mathbf{dA}_0 \cdot \mathbf{F}^{-1} = J\,\mathbf{dA}_0 \cdot \mathbf{F}^{-1}. \tag{4.5.30}$$

These area-deformation formulas will be used in Sec. 5.3 to introduce the Piola-Kirchhoff stress tensors referred to the initial configuration.

The relative deformation gradient $\mathbf{F}_t$ is the gradient of the relative motion function of Eq. (4.3.12)

$$\boldsymbol{\xi} = \boldsymbol{\chi}_t(\mathbf{x}, \tau) \tag{4.5.31}$$

Thus

$$\mathbf{F}_t(\mathbf{x}, \tau) = \boldsymbol{\chi}_t \overleftarrow{\nabla}_{\mathbf{x}}$$

with components (4.5.32)

$$(\mathbf{F}_t)_{km} = \xi_{k,m} \equiv \frac{\partial \xi_k}{\partial x_m}.$$

If the fixed reference position $\mathbf{X}$, the current position $\mathbf{x}$, and the variable position $\boldsymbol{\xi}$ are all referred to rectangular Cartesian axes, the chain rule of calculus establishes the component relations

$$\frac{\partial \xi_k}{\partial X_J} = \frac{\partial \xi_k}{\partial x_m}\frac{\partial x_m}{\partial X_J} \quad \text{or} \quad [\xi_{k,J}] = [\xi_{k,m}][x_{m,J}],$$

which represent the tensor equation (4.5.33)

$$\mathbf{F}(\tau) = \mathbf{F}_t(\tau) \cdot \mathbf{F}(t).$$

In Sec. 4.6 we will consider some other measures of deformation, the stretch tensors, which combine multiplicatively with a rigid rotation to give the deformation gradient as the product of two tensors.

EXERCISES

1. (a) Sketch the initial and final positions of an element which was initially a square $ABCD$ in the XY-plane, of side length dL, with AB parallel to the X-axis and AD to the Y-axis, which undergoes the following simple shear.

$$x = X + kY \qquad y = Y \qquad z = Z$$

(b) Calculate and display in three matrices the components of the deformation gradients **F** and $\mathbf{F}^T$ and the deformation tensor **C**. *Partial answer*:

$$[C_{IJ}] = \begin{bmatrix} 1 & k & 0 \\ k & 1 + k^2 & 0 \\ 0 & 0 & 1 \end{bmatrix}$$

2. Compare the matrices of rectangular Cartesian components of the finite strain E_{IJ} and the small strain ϵ_{ij} for Ex. 1. Use the finite strain tensor to calculate the change in squared length of edges AB and AD and diagonals AC and DB. *Partial answer*: For AC

$$(ds)^2 - (dS)^2 = (2k + k^2)(dL)^2.$$

3. Calculate the stretch $\Lambda_{(\hat{\mathbf{N}})}$ and the unit elongation $E_{(\hat{\mathbf{N}})}$ for AB, AD, AC, and DB. Use Eq. (4.5.20) to calculate the angle change at A, and check your result by trigonometry on the sketch of Ex. 1. *Partial ans.*: $\Lambda_{AC} = [1 + k + \frac{1}{2}k^2]^{1/2}$.

4. Note that $C_3 = 1$ is a principal value of **C** in Ex. 1, corresponding to the principal direction $\hat{\mathbf{N}}_3$ in the Z-direction, since $C_{ZX} = C_{ZY} = 0$. Determine C_1 and C_2. *Ans.*:

$$C_1 = 1 + \tfrac{1}{2}k^2 + k\sqrt{1 + \tfrac{1}{4}k^2} \qquad C_2 = 1 + \tfrac{1}{2}k^2 - k\sqrt{1 + \tfrac{1}{4}k^2}.$$

Compare C_1 numerically with the square of the stretch of AC for $k = 1$.

5. Show that two principal directions of **C** in Ex. 4 are

$$\hat{\mathbf{N}}_1 = \frac{\mathbf{I}_1 + (\frac{1}{2}k + \sqrt{1 + \frac{1}{4}k^2})\mathbf{I}_2}{[2 + \frac{1}{2}k^2 + k\sqrt{1 + \frac{1}{4}k^2}]^{1/2}} \qquad \hat{\mathbf{N}}_2 = \frac{-\mathbf{I}_1 - (\frac{1}{2}k - \sqrt{1 + \frac{1}{4}k^2})\mathbf{I}_2}{[2 + \frac{1}{2}k^2 - k\sqrt{1 + \frac{1}{4}k^2}]^{1/2}}.$$

6. Solve the equations of Ex. 1 for X, Y, Z, in terms of x, y, z, and evaluate the components of $\mathbf{B}^{-1}$ and $\mathbf{E}^*$. *Partial answer*:

$$[B_{ij}^{-1}] = \begin{bmatrix} 1 & -k & 0 \\ -k & 1 + k^2 & 0 \\ 0 & 0 & 1 \end{bmatrix}$$

7. Find the principal directions of $\mathbf{B}^{-1}$ in Ex. 6. *Hint*: You would expect to get the answer simply by replacing k by $-k$ in the solution of Ex. 5, and $\mathbf{I}_1$ by $\mathbf{i}_1$, etc.

8. The most general two-dimensional homogeneous finite-strain distribution is defined by giving the spatial coordinates as linear homogeneous functions, say

$$x_1 = X_1 + AX_1 + BX_2 \qquad x_2 = X_2 + CX_1 + DX_2.$$

(a) Express the components of the deformation tensor **C** and the finite strain tensor **E** in terms of the given constants $A, B, C,$ and D. Display your answers in two matrices.

(b) Calculate $(ds)^2$ and $(ds)^2 - (dS)^2$ for the $d\mathbf{X}$ with components $(dL, dL, 0)$.

9. If the Lagrangian finite-strain components at a point are as shown in the matrix below, find (a) the principal strains, (b) the direction cosines of the principal axis having the largest principal strain, and (c) the maximum shear strain component (max E_{nt}).

$$E = \begin{bmatrix} -1 & 0 & 0 \\ 0 & 1.64 & -0.48 \\ 0 & -0.48 & 1.36 \end{bmatrix}$$

10. For the deformation of Ex. 9, calculate $\Lambda^2_{(\hat{\mathbf{N}})}$ for $\hat{\mathbf{N}}$ making angles $[\pi/2, \beta, (\pi/2) - \beta]$ with the X_1, X_2, X_3 directions, respectively through the point. Determine β for maximum Λ^2.

11. For the deformation of Ex. 9, find the new angle θ_{23} between elements initially parallel to the X_2 and X_3 directions at the point.

12. When the displacement gradients are small compared to unity, the finite strain components E_{ij} may be replaced by the small-strain components ϵ_{ij}, so that

$$(ds)^2 - (dS)^2 = 2\epsilon_{ij}\, dX_i\, dX_j.$$

Show that when this equation is used to calculate the unit extension ϵ_n of an element initially having the direction of unit vector $\hat{\mathbf{n}}$, such that $ds = (1 + \epsilon_n)dS$, the resulting value for ϵ_n is the same as would be given by the tensor transformation formula for $\bar{\epsilon}_{11}$, if the $\bar{x}_1$-direction is taken in the direction of $\hat{\mathbf{n}}$, and higher-order terms in ϵ_n are neglected.

13. The principal axes $\hat{\mathbf{N}}_1, \hat{\mathbf{N}}_2, \hat{\mathbf{N}}_3$ of the deformation tensor $\mathbf{C}$ at a point $\mathbf{X}$ and the stretch ratios of elements initially in the principal directions are

$$\hat{\mathbf{N}}_1 = \tfrac{4}{5}\mathbf{i}_1 + \tfrac{3}{5}\mathbf{i}_2 \qquad \hat{\mathbf{N}}_2 = -\tfrac{3}{5}\mathbf{i}_1 + \tfrac{4}{5}\mathbf{i}_2 \qquad \hat{\mathbf{N}}_3 = \mathbf{i}_3$$

$$\Lambda_{(\hat{\mathbf{N}}_1)} = 1.5 \qquad \Lambda_{(\hat{\mathbf{N}}_2)} = 1.0 \qquad \Lambda_{(\hat{\mathbf{N}}_3)} = 0.5,$$

where the $\mathbf{i}_k$ are unit vectors in a ***fixed*** rectangular Cartesian system. Determine:
(a) the principal values C_1, C_2, C_3 of $\mathbf{C}$
(b) the matrix of components of $\mathbf{C}$ in the fixed system
(c) the Lagrangian strain matrix, referred to the fixed system of axes.

14. If in finite, plane strain an experimental measurement gives stretch ratios Λ of 0.8 and 0.6 in the X_1 and X_2 directions, respectively, and 0.5 in the direction bisecting the angle between X_1 and X_2 directions, determine the components of $\mathbf{C}$ and $\mathbf{E}$ at the point.

15. Use your results of Ex. 14 to determine the new angle between:
(a) elements initially parallel to the axes;
(b) the element initially in the X_1 direction and the element bisecting the angle between the axes.

16. Use Eq. (4.5.14) to show that $\dot{\mathbf{F}}^{-1} = -\mathbf{F}^{-1}\boldsymbol{\cdot}\mathbf{L}$. Write a component form of the result. *Hint*: Recall $\mathbf{F}\boldsymbol{\cdot}\mathbf{F}^{-1} = \mathbf{F}^{-1}\boldsymbol{\cdot}\mathbf{F} = \mathbf{1}$.

4.6 Rotation and stretch tensors

When the displacement gradients are finite, the symmetric and skew parts of the displacement-gradient matrix $[\partial u_i/\partial X_j]$ of Sec. 4.2 no longer provide an additive decomposition of the displacement gradient into the sum of a pure strain plus a pure rotation, since when the displacement-gradient components are not small compared to unity the two matrices no longer represent pure strain and pure rotation, respectively. But a multiplicative decomposition of the deformation gradient $\mathbf{F} = \mathbf{x}\overleftarrow{\nabla}$ into the product of two tensors, one of which represents a rigid-body rotation, while the other is a symmetric positive-definite tensor is always possible. If $\mathbf{R}$ denotes the orthogonal ***rotation tensor***, which rotates the principal axes of $\mathbf{C}$ at $\mathbf{X}$ into the directions of the principal axes of $\mathbf{B}^{-1}$ at $\mathbf{x}$, then we will see in Eqs. (14) to (21) below that there exist two tensors $\mathbf{U}$ and $\mathbf{V}$ satisfying†

$$\mathbf{F} = \mathbf{R}\cdot\mathbf{U} = \mathbf{V}\cdot\mathbf{R} \tag{4.6.1a}$$

so that

$$d\mathbf{x} = (\mathbf{R}\cdot\mathbf{U})\cdot d\mathbf{X} = (\mathbf{V}\cdot\mathbf{R})\cdot d\mathbf{X}$$

with rectangular Cartesian component forms (referring x_i and X_i to the same reference axes and using lower-case indices for both)

$$\begin{aligned} x_{k,p} &= R_{kq}U_{qp} = V_{kq}R_{qp} \quad \text{and} \\ dx_k &= R_{kq}U_{qp}\,dX_p = V_{kq}R_{qp}\,dX_p \end{aligned} \tag{4.6.1b}$$

such that

$$\left.\begin{aligned} R_{im}R_{jm} = \delta_{ij} \quad &\text{and} \quad R_{kp}R_{kq} = \delta_{pq} \quad \text{or} \\ \mathbf{R}\cdot\mathbf{R}^T = \mathbf{1} \quad &\text{and} \quad \mathbf{R}^T\cdot\mathbf{R} = \mathbf{1}. \end{aligned}\right\} \tag{4.6.1c}$$

$\mathbf{U}$ is called the ***right stretch tensor***, and $\mathbf{V}$ is called the ***left stretch tensor***. Either stretch tensor operating as $\mathbf{U}\cdot$ or as $\mathbf{V}\cdot$ on the set of all vectors at a point produces length changes (stretch) in the vectors and also produces additional rotation of all vectors except those in the principal directions of the stretch tensor, in addition to the rigid-body rotation $\mathbf{R}$ of the whole set of vectors. Both $\mathbf{U}$ and $\mathbf{V}$ are symmetric and positive-definite.

Equations (1) show that we may consider the motion and deformation of an infinitesimal volume element at $\mathbf{X}$ to consist of the successive application‡ of:

†The notation here is that of Truesdell and Noll (1965), following Noll (1958), except that they omit the dots, giving $\mathbf{F} = \mathbf{RU} = \mathbf{VR}$.

‡The fact that the deformation at a point may be considered as the result of a translation followed by a rotation of the principal axes of strain, and stretches along the principal axes, was apparently recognized by Thomson and Tait in 1867, but first explicitly stated by Love in 1892.

$$\left.\begin{array}{l}\text{1. a stretch by the operator } \mathbf{U}\cdot \\ \text{2. a rigid-body rotation by the operator } \mathbf{R}\cdot \\ \text{3. and finally a translation to } \mathbf{x}\end{array}\right\} \tag{4.6.2a}$$

Alternatively the same deformation can be produced by the successive application of:

$$\left.\begin{array}{l}\text{1. a translation to } \mathbf{x} \\ \text{2. a rigid-body rotation by } \mathbf{R}\cdot \\ \text{3. and finally a stretch by } \mathbf{V}\cdot\end{array}\right\} \tag{4.6.2b}$$

The translation does not change a material vector or its rectangular components with respect to the common reference axes, but would change curvilinear components (see Sec. I.7). We shall see below that $\mathbf{U}$ is related to $\mathbf{C}$ at $\mathbf{X}$, while $\mathbf{V}$ is related to $\mathbf{B}$ at $\mathbf{x}$, where $\mathbf{B}$ is the inverse of the deformation tensor denoted by $\mathbf{B}^{-1}$ in Sec. 4.5. This accounts for the placement of the translation in the sequences outlined above, so that $\mathbf{U}$ may be considered to act at $\mathbf{X}$ while $\mathbf{V}$ acts at $\mathbf{x}$. Before proceeding with the derivation of Eqs. (1), we will consider the concepts of the inverse of a tensor and of positive, negative, and fractional powers of a tensor, since we will find that

$$\mathbf{U} = \mathbf{C}^{1/2} \quad \text{and} \quad \mathbf{V} = \mathbf{B}^{1/2} \tag{4.6.3a}$$

where $\mathbf{B}$ is the inverse of $\mathbf{B}^{-1}$. It can be shown† that for Cartesian components

$$B_{ij} = x_{i,K}\partial_K x_j \quad \text{or} \quad \mathbf{B} = \mathbf{F}\cdot\mathbf{F}^T. \tag{4.6.3b}$$

The inverse of a second-order tensor T (Linear Vector Function). If the second-order tensor $\mathbf{T}$ has an inverse, it is denoted by $\mathbf{T}^{-1}$ (sometimes written ${}^{-1}\mathbf{T}$) and is the tensor such that, for arbitrary vectors $\mathbf{v}$, the pairings

$$\mathbf{u} = \mathbf{T}\cdot\mathbf{v} \quad \text{and} \quad \mathbf{v} = \mathbf{T}^{-1}\cdot\mathbf{u} \tag{4.6.4a}$$

are equivalent, whence

$$\mathbf{T}\cdot\mathbf{T}^{-1} = \mathbf{T}^{-1}\cdot\mathbf{T} = \mathbf{1}. \tag{4.6.4b}$$

A sufficient condition for the existence of an inverse is that the matrix of coefficients of the linear transformation of rectangular Cartesian components be nonsingular [see Eqs. (2.4.47) and (2.4.48)].

†See, for example, Truesdell (1952) or Eringen (1967). The tensor $\mathbf{B}^{-1}$ is frequently denoted by $\mathbf{c}$. Its inverse or reciprocal $\mathbf{B}$ or $\mathbf{c}^{-1}$ was introduced by Finger in 1894. Truesdell and Noll (1965) call $\mathbf{B}$ the left Cauchy-Green tensor, and $\mathbf{C}$ the right Cauchy-Green tensor of the deformation.

Since both $\mathbf{C}$ and $\mathbf{B}^{-1}$ are positive-definite (see Sec. 4.5), their principal values are all positive, and the determinants $III_C = C_1C_2C_3$ and $III_{B^{-1}} = B_1^{-1}B_2^{-1}B_3^{-1}$ of their respective component matrices are positive. Hence the inverses $\mathbf{C}^{-1}$ and $\mathbf{B}$ both exist and are also positive-definite. ***The principal values of the inverse tensor are the reciprocals of the principal values of the direct tensor, and the principal directions coincide.*** We shall have particular use for $\mathbf{B}$, since the left stretch tensor will be $\mathbf{V} = \mathbf{B}^{1/2}$.

Powers of a second-order tensor. Integral powers of a second-order tensor $\mathbf{T}$ are defined inductively by

$$\mathbf{T}^n = \mathbf{T}^{n-1}\cdot\mathbf{T} = \mathbf{T}\cdot\mathbf{T}^{n-1} \quad \text{with} \quad \mathbf{T}^0 = \mathbf{1}, \tag{4.6.5}$$

where n is restricted to be positive unless $\mathbf{T}$ has an inverse. When $\mathbf{T}$ has an inverse, the power for $n = -1$ is the inverse tensor $\mathbf{T}^{-1}$, and other negative integral powers follow inductively by successive multiplication by $\mathbf{T}^{-1}$, in agreement with Eq. (5). We will see at the end of this subsection that any principal direction of $\mathbf{T}$ associated with principal value T_α is also a principal direction of $\mathbf{T}^n$ with associated principal value $(T_\alpha)^n$. It is then consistent to define for a positive-definite tensor such as $\mathbf{C}$ the power $\mathbf{C}^m$ for any real number m as that tensor with the same principal directions as $\mathbf{C}$ but with corresponding principal values equal to the mth powers of the principal values of $\mathbf{C}$. For example, each principal value of $\mathbf{U} = \mathbf{C}^{1/2}$ is the positive square root of the corresponding (positive) principal value of $\mathbf{C}$, $U_\alpha = \sqrt{C_\alpha}$ and the principal directions of $\mathbf{U}$ coincide with those of $\mathbf{C}$. Similarly the principal values of $\mathbf{V} = \mathbf{B}^{1/2}$ are $V_\alpha = \sqrt{B_\alpha}$ and the principal axes of $\mathbf{V}$ coincide with those of $\mathbf{B}$ (and $\mathbf{B}^{-1}$). If the unit vectors $\hat{\mathbf{N}}_\alpha$ are in the directions of the principal axes of $\mathbf{C}$ at $\mathbf{X}$, with rectangular Cartesian components $N_{(\alpha)m}$, then $a_\alpha^m = N_{(\alpha)m}$ are the direction cosines for transformation of the tensor components from barred Cartesian axes in the principal directions to the unbarred axes of the common reference.† Hence, by Eq. (2.4.12),

$$U_{qp} = (C^{1/2})_{qp} = \sum_\alpha \sqrt{C_\alpha}\, N_{(\alpha)q}N_{(\alpha)p} \tag{4.6.6a}$$

and similarly

$$V_{kq} = (B^{1/2})_{kq} = \sum_\alpha \sqrt{B_\alpha}\, n_{(\alpha)k}n_{(\alpha)q} = \sum_\alpha \frac{1}{\sqrt{B_\alpha^{-1}}}\, n_{(\alpha)k}n_{(\alpha)q}. \tag{4.6.6b}$$

To show that these definitions for arbitrary real powers of a positive-definite tensor are consistent with the inductive definitions for integral powers, we still

†As in Eqs. (1), we are using lower-case indices for both x_i and X_i and using a single set of Cartesian axes for the common reference.

need to show that for integral n, the principal axes of $\mathbf{T}$ are also principal axes of $\mathbf{T}^n$ when $\mathbf{T}^n$ is defined by Eq. (5). We must also show that the nth powers of the principal values T_α of $\mathbf{T}$ are the corresponding principal values of $\mathbf{T}^n$ as asserted above. If $\hat{\mathbf{N}}_\alpha$ is a unit vector in the principal direction of $\mathbf{T}$ corresponding to principal value T_α, then

$$\mathbf{T}\cdot\hat{\mathbf{N}}_\alpha = T_\alpha\hat{\mathbf{N}}_\alpha = T_\alpha(\mathbf{1}\cdot\hat{\mathbf{N}}_\alpha) \qquad \text{(no sum)}.$$

Multiplication of both sides by $\mathbf{T}$ yields

$$\mathbf{T}^2\cdot\hat{\mathbf{N}}_\alpha = T_\alpha(\mathbf{T}\cdot\hat{\mathbf{N}}_\alpha) \qquad \text{(no sum)}. \tag{4.6.7}$$

Since $\mathbf{T}\cdot\hat{\mathbf{N}}_\alpha = T_\alpha\hat{\mathbf{N}}_\alpha$, this reduces to

$$\mathbf{T}^2\cdot\hat{\mathbf{N}}_\alpha = (T_\alpha)^2\hat{\mathbf{N}}_\alpha \qquad \text{(no sum)},$$

which shows that $\hat{\mathbf{N}}_\alpha$ is a principal direction of $\mathbf{T}^2$ with principal value $(T_\alpha)^2$. By further successive multiplications by $\mathbf{T}$ we can establish inductively that

$$\mathbf{T}^n\cdot\hat{\mathbf{N}}_\alpha = (T_\alpha)^n\hat{\mathbf{N}}_\alpha \tag{4.6.8}$$

for any positive integer n. For negative integers, assuming the inverse $\mathbf{T}^{-1}$ exists with principal value $1/T_\alpha$ corresponding to direction $\hat{\mathbf{N}}_\alpha$, beginning with

$$\mathbf{T}^{-1}\cdot\hat{\mathbf{N}}_\alpha = \frac{1}{T_\alpha}\hat{\mathbf{N}}_\alpha = \frac{1}{T_\alpha}(\mathbf{1}\cdot\hat{\mathbf{N}}_\alpha), \tag{4.6.9}$$

we can establish Eq. (8) inductively by successive multiplications by $\mathbf{T}^{-1}$. Furthermore any direction is a principal direction of $\mathbf{T}^0 = \mathbf{1}$, with principal value $T_\alpha^0 = 1$. Therefore the integral powers as defined inductively at the beginning of this subsection also satisfy the conditions of the definition for any real power of a positive-definite tensor.

The rotation tensor $\mathbf{R}$ was defined at the beginning of this section as the orthogonal tensor which rotates the principal axes of $\mathbf{C}$ at $\mathbf{X}$ into the directions of the principal axes of $\mathbf{B}^{-1}$ at $\mathbf{x}$. When the principal values C_α are distinct, there is a unique set of three orthogonal principal axes at $\mathbf{X}$, which we denote by the unit vectors $\hat{\mathbf{N}}_\alpha$ in their directions, and the deformation carries them into a unique set of mutually orthogonal directions, designated by unit vectors $\hat{\mathbf{n}}_\alpha$, which are the principal directions of $\mathbf{B}^{-1}$ at $\mathbf{x}$ [see Eq. (4.5.20)]. If the C_α are not distinct, the $\hat{\mathbf{N}}_\alpha$ are not uniquely determined, but it is always possible to select an orthogonal set of three $\hat{\mathbf{N}}_\alpha$ which are principal directions of $\mathbf{C}$ at $\mathbf{X}$ and which are carried by the deformation into an orthogonal set of three principal directions of $\mathbf{B}^{-1}$ at $\mathbf{x}$, which we will denote by the unit vectors $\hat{\mathbf{n}}_\alpha$. In the following we will assume that this selection has already been made at any point under discussion; then the tensor $\mathbf{R}$ which rotates the $\hat{\mathbf{N}}_\alpha$ into the $\hat{\mathbf{n}}_\alpha$ is completely determined in terms of the $\hat{\mathbf{N}}_\alpha$ and the $\hat{\mathbf{n}}_\alpha$.

Figure 4.10 illustrates two such associated orthogonal triads of unit vectors.

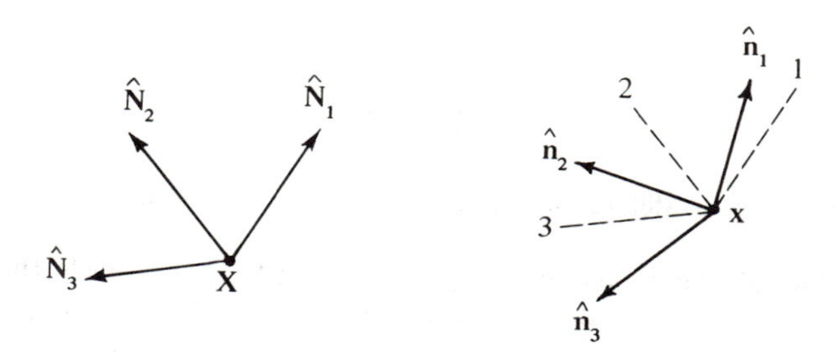

Fig. 4.10 Rotation of principal direction unit vectors $\hat{\mathbf{N}}_\alpha$ of $\mathbf{C}$ at $\mathbf{X}$ into unit vectors $\hat{\mathbf{n}}_\alpha$ of $\mathbf{B}^{-1}$ at $\mathbf{x}$.

The dashed lines marked 1, 2, 3 at $\mathbf{x}$ are parallel to the corresponding $\hat{\mathbf{N}}_\alpha$ at $\mathbf{X}$. Then $\mathbf{R}$ is the orthogonal tensor such that

$$\hat{\mathbf{n}}_\alpha = \mathbf{R}\cdot\hat{\mathbf{N}}_\alpha \tag{4.6.10a}$$

with rectangular Cartesian components in a common set of axes,

$$n_{\alpha i} = R_{im} N_{\alpha m}. \tag{4.6.10b}$$

Multiplying Eq. (10b) by $N_{\alpha j}$, we obtain

$$R_{ij} = n_{\alpha i} N_{\alpha j} \qquad \text{(summed on } \alpha) \tag{4.6.11}$$

since $N_{\alpha m} N_{\alpha j} = \delta_{mj}$ by Eq. (2.4.8). (The nine $N_{\alpha p}$ are direction cosines for rotation to new Cartesian axes in the direction of the $\hat{\mathbf{N}}_\alpha$.) The inverse tensor $\mathbf{R}^{-1}$ which rotates the $\hat{\mathbf{n}}_\alpha$ into the directions of the $\hat{\mathbf{N}}_\alpha$

$$\hat{\mathbf{N}}_\alpha = \mathbf{R}^{-1}\cdot\hat{\mathbf{n}}_\alpha$$

with rectangular Cartesian component form

$$N_{\alpha p} = R^{-1}_{pi} n_{\alpha i} \tag{4.6.12}$$

is similarly seen to have rectangular Cartesian components

$$R^{-1}_{pq} = N_{\alpha p} n_{\alpha q} \qquad \text{(summed on } \alpha). \tag{4.6.13}$$

The orthogonality conditions of Eqs. (1c) must be satisfied because $\mathbf{R}$ was defined as an orthogonal tensor. We can check that the components given by Eq. (11) do satisfy Eq. (1c) as follows. $R_{im} = n_{\alpha i}N_{\alpha m}$ and $R_{jm} = n_{\beta j}N_{\beta m}$. Hence

$$R_{im}R_{jm} = (n_{\alpha i}n_{\beta j})(N_{\alpha m}N_{\beta m}) = (n_{\alpha i}n_{\beta j})\delta_{\alpha\beta} = n_{\alpha i}n_{\alpha j} = \delta_{ij}$$

or

$$\mathbf{R}\cdot\mathbf{R}^T = \mathbf{1}.$$

We are now ready to derive the main result of this section.

Derivation of fundamental rotation theorem.† To establish the theorem implied by Eqs. (1), that the deformation gradient can be decomposed into the product of the rotation tensor **R** and a stretch tensor we begin by considering the elements

$$d\mathbf{x}_\alpha = ds_{(\alpha)}\hat{\mathbf{n}}_{(\alpha)} \qquad \text{(no sum on } \alpha)$$

in the principal direction $\hat{\mathbf{n}}_\alpha$ of $\mathbf{B}^{-1}$ at $\mathbf{x}$ and (4.6.14)

$$d\mathbf{X}_\alpha = dS_{(\alpha)}\hat{\mathbf{N}}_{(\alpha)} \qquad \text{(no sum on } \alpha)$$

in the corresponding principal direction of **C** at **X**. (If the principal values are not distinct, the choice of $\hat{\mathbf{n}}_\alpha$ is not uniquely determined, but we suppose that a selection has been made as at the beginning of the preceding subsection on the rotation tensor.) Since

$$ds_{(\alpha)} = \sqrt{C_{(\alpha)}}\, dS_{(\alpha)} \qquad \text{(no sum)} \tag{4.6.15}$$

by Eq. (4.5.17a), the first Eq. (14) takes the form

$$d\mathbf{x}_\alpha = \sqrt{C_{(\alpha)}}\, dS_{(\alpha)}\hat{\mathbf{n}}_{(\alpha)} \qquad \text{(no sum)}$$

with rectangular Cartesian components (referring x_i and X_i to the same axes)

$$x_{k,m}dX_{(\alpha)m} = dx_{(\alpha)k} = \sqrt{C_{(\alpha)}}\, dS_{(\alpha)}n_{(\alpha)k} \qquad \text{(no sum on } \alpha), \tag{4.6.16}$$

where $x_{k,m}$ denotes $\partial x_k/\partial X_m$. Now substitute

$$dX_{(\alpha)m} = dS_{(\alpha)}N_{(\alpha)m} \qquad \text{(no sum)}$$

and divide by $dS_{(\alpha)}$ to obtain

†The derivation given here in Cartesian components is similar to one given in curvilinear components by Truesdell and Toupin (1960), Sec. 37, following Truesdell (1953) and Toupin (1956).

$$x_{k,m}N_{(\alpha)m} = \sqrt{C_{(\alpha)}}\, n_{(\alpha)k} \qquad \text{(no sum on } \alpha\text{)}.$$

This can be solved for $x_{k,p}$ by multiplying both sides by $N_{(\alpha)p}$ and summing on α, since $\sum_{\alpha} N_{(\alpha)m}N_{(\alpha)p} = \delta_{mp}$ [see above following Eq. (11)]. Because the index appears more than twice on the right, we indicate the summation explicitly by $\sum_{\alpha}$. Thus

$$x_{k,p} = \sum_{\alpha}\sqrt{C_{(\alpha)}}R_{kq}N_{(\alpha)q}N_{(\alpha)p} \tag{4.6.17a}$$

(summed also on q). Equation (6a) then reduces this to

$$x_{k,p} = R_{kq}(\mathbf{C}^{1/2})_{qp} \tag{4.6.17b}$$

whence, by Eq. (3)

$$x_{k,p} = R_{kq}U_{qp} \quad \text{or} \quad \mathbf{F} = \mathbf{R}\cdot\mathbf{U} \tag{4.6.17c}$$

which establishes the first Eq. (1) involving the right stretch tensor $\mathbf{U}$. Thus the vectors $d\mathbf{X}$ at $\mathbf{X}$ are changed into the corresponding vectors $d\mathbf{x}$ by successive application of the right stretch tensor $\mathbf{U}$ and the rigid-body rotation $\mathbf{R}$. (Do not forget that the stretch tensor itself produces some rotation of all vectors except those in the principal directions, in addition to the rigid-body rotation $\mathbf{R}$ of the whole set of vectors.)

The sequence of operations (2a) can be represented by the three matrix equations (18) below for the matrices of rectangular Cartesian components, all referred to one set of reference axes. Let $[dX_q]_S$ denote the column matrix of vector components produced by the stretch of $[dX_p]$ at $\mathbf{X}$, and let $[dX_k]_R$ be the new components after rigid-body rotation of all the stretched vectors at $\mathbf{X}$, and finally let $[dx_k]$ represent the same rotated vector after translation to $\mathbf{x}$. Then

$$\left.\begin{array}{lll} \text{1. Stretch:} & [dX_q]_S & = [U_{qp}][dX_p] \\ \text{2. Rotation:} & [dX_k]_R & = [R_{kq}][dX_q]_S \\ \text{3. Translation to } \mathbf{x}\text{:} & [dx_k] & = [dX_k]_R. \end{array}\right\} \tag{4.6.18}$$

Besides the stretch and rotation, the deformation also involves the third step, translation of the whole set of vectors from $\mathbf{X}$ to $\mathbf{x}$, but this does not change any vector or its Cartesian components. (For the change in curvilinear components, see Sec. I.7.)

The second form of Eqs. (1) involving the left stretch $\mathbf{V}$ is derived similarly. Substitute $N_{(\alpha)m}\, dS_{(\alpha)}$ for $dX_{(\alpha)m}$ on the left side of the first Eq. (16) to obtain

$$x_{k,m}dX_{(\alpha)m} = dx_{(\alpha)k} = ds_{(\alpha)}n_{(\alpha)k} \qquad \text{(no sum on } \alpha\text{)}. \tag{4.6.19}$$

By Eq. (4.5.17b),

$$\frac{ds_{(\alpha)}}{dS_{(\alpha)}} = \lambda_{(\hat{\mathbf{n}}_\alpha)} = \frac{1}{\sqrt{B_\alpha^{-1}}}. \tag{4.6.20}$$

Also $B_\alpha = 1/B_\alpha^{-1}$. Hence $ds_{(\alpha)} = \sqrt{B_\alpha}\, dS_{(\alpha)}$. Substituting this on the right side of Eq. (19) and $dX_{(\alpha)m} = dS_{(\alpha)}N_{(\alpha)m}$ on the left, we obtain after dividing by $dS_{(\alpha)}$

$$x_{k,m}N_{(\alpha)m} = \sqrt{B_\alpha}\, n_{(\alpha)k}.$$

In order to get the component $V_{kq} = (B^{1/2})_{kq}$ on the right side of this equation, we multiply by $n_{(\alpha)q}$ and sum on α

$$x_{k,m}\sum_\alpha N_{(\alpha)m}n_{(\alpha)q} = \sum_\alpha \sqrt{B_\alpha}\, n_{(\alpha)k}\, n_{(\alpha)q}.$$

Now use Eq. (6b) on the right side, and $n_{(\alpha)q} = R_{qs}N_{(\alpha)s}$ on the left to obtain

$$x_{k,m}R_{qs}\sum_\alpha N_{(\alpha)s}N_{(\alpha)m} = (B^{1/2})_{kq}.$$

But $\sum_\alpha N_{(\alpha)s}N_{(\alpha)m} = \delta_{sm}$ and $(B^{1/2})_{kq} = V_{kq}$. Hence

$$x_{k,m}R_{qm} = V_{kq}.$$

To get rid of the R_{qm} on the left, multiply by R_{qp} and use $R_{qm}R_{qp} = \delta_{mp}$ by Eq. (1c). The result is

$$x_{k,p} = V_{kq}R_{qp} \quad \text{or} \quad \mathbf{F} = \mathbf{V}\cdot\mathbf{R} \tag{4.6.21a}$$

the second form of Eq. (1).

The principal values of $\mathbf{B}^{-1}$ were used in Eq. (20) instead of those of $\mathbf{C}$, because in this second decomposition we are thinking of the stretch being performed at $\mathbf{x}$ after the translation. The tensor $\mathbf{B}$ is also thought of in this decomposition as acting at point $\mathbf{x}$, and the same is true of $\mathbf{V}$ and $\mathbf{R}$. The sequence of operations (2b) can now be represented by the three matrix equations (21b) below for the rectangular components all referred to one set of reference axes. Since the translation is performed first, the rotation now produces $[dx_q]_R$ by acting on the $[dx_p]_T$ obtained by translating the $[dx_p]$ to $\mathbf{x}$.

$$\left.\begin{aligned} &\text{1. Translation:} && [dx_p]_T = [dX_p] \\ &\text{2. Rotation:} && [dx_q]_R = [R_{qp}][dx_p]_T \\ &\text{3. Stretch:} && [dx_k] \;= [V_{kq}][dx_q]_R. \end{aligned}\right\} \tag{4.6.21b}$$

The decomposition of $\mathbf{F}$ into the product of an orthogonal tensor and a symmetric positive-definite tensor, Eq. (1), is frequently called the ***polar decomposition*** theorem.

The decomposition is of great theoretical importance, but the stretch tensors themselves are not convenient for specific problems because their components are complicated irrational functions of the deformation gradients. The deformation tensors $\mathbf{C}$ and $\mathbf{B}^{-1}$ (or its inverse $\mathbf{B}$) are more useful for specific problems.

Relative stretch and rotation.† The relative deformation gradient $\mathbf{F}_t$ of Eq. (4.5.32) can similarly be separated into the product of a relative rotation $\mathbf{R}_t$ and stretch $\mathbf{U}_t$ or $\mathbf{V}_t$. Thus

$$\mathbf{F}_t\,(\mathbf{x}, \tau) = \mathbf{R}_t \cdot \mathbf{U}_t = \mathbf{V}_t \cdot \mathbf{R}_t \tag{4.6.22}$$

and the relative deformation tensors are

$$\mathbf{C}_t = \mathbf{U}_t^2 \qquad \mathbf{B}_t = \mathbf{V}_t^2.$$

The rotation $\mathbf{R}_t$ rotates the principal axes of $\mathbf{C}_t = \mathbf{F}_t^T \cdot \mathbf{F}_t$ at the reference position $\mathbf{x}$ at time t to the directions occupied by the same infinitesimal material elements at $\boldsymbol{\xi}$ at time τ, and the right stretch tensor also operates on material elements at $\mathbf{x}$.

The rates of change of the relative stretch and rotation tensors, evaluated at $\tau = t$, are equal to the rate of deformation $\mathbf{D}$ and spin $\mathbf{W}$ at t, as the following derivation shows.‡ By Eq. (4.5.33)

$$\mathbf{F}(\tau) = \mathbf{F}_t(\tau) \cdot \mathbf{F}(t)$$

whence

$$\frac{d}{d\tau}\,\mathbf{F}(\tau) = \left[\frac{d}{d\tau}\,F_t(\tau)\right] \cdot \mathbf{F}(t). \tag{4.6.23}$$

Let

$$\dot{\mathbf{F}}(t) \quad \text{denote} \quad \frac{d}{d\tau}\,\mathbf{F}(\tau)\bigg|_{\tau=t} = \frac{d}{dt}\,\mathbf{F}(t)$$

and

$$\dot{\mathbf{F}}_t(t) \quad \text{denote} \quad \frac{d}{d\tau}\,\mathbf{F}_t(\tau)\bigg|_{\tau=t}.$$

Then Eq. (23) becomes

$$\dot{\mathbf{F}}(t) = \dot{\mathbf{F}}_t(t) \cdot \mathbf{F}(t)$$

†See Truesdell and Noll (1965), Sec. 23.
‡See Truesdell and Noll (1965), Sec. 24.

whence

$$\dot{\mathbf{F}}_t(t) = \dot{\mathbf{F}}(t) \cdot \mathbf{F}^{-1}(t) = \mathbf{L}(t) \tag{4.6.24}$$

by Eq. (4.5.14).

Now

$$\mathbf{F}_t(\tau) = \mathbf{R}_t(\tau) \cdot \mathbf{U}_t(\tau)$$

by Eq. (22), and therefore

$$\dot{\mathbf{F}}_t(\tau) = \dot{\mathbf{R}}_t(\tau) \cdot \mathbf{U}_t(\tau) + \mathbf{R}_t(\tau) \cdot \dot{\mathbf{U}}_t(\tau). \tag{4.6.25}$$

Since

$$\mathbf{R}_t = \mathbf{1} \quad \text{and} \quad \mathbf{U}_t = \mathbf{1} \qquad \text{at} \qquad \tau = t,$$

Eq. (25) evaluated at $\tau = t$ yields

$$\dot{\mathbf{F}}_t(t) = \dot{\mathbf{R}}_t(t) + \dot{\mathbf{U}}_t(t). \tag{4.6.26}$$

Because $\mathbf{R}_t \cdot \mathbf{R}_t^T = \mathbf{1}$, we have $(d/d\tau)(\mathbf{R}_t \cdot \mathbf{R}_t^T) = 0$. When this is evaluated at $t = \tau$, the result is $\dot{\mathbf{R}}_t + \dot{\mathbf{R}}_t^T = 0$ at $t = \tau$. Hence $\dot{\mathbf{R}}_t(t)$ is skew-symmetric. Since $\dot{\mathbf{R}}_t$ is skew and $\dot{\mathbf{U}}_t$ is symmetric, Eq. (26) is the additive decomposition of $\dot{\mathbf{F}}_t(t)$ into the sum of a skew tensor and a symmetric tensor. But, by Eq. (24), $\dot{\mathbf{F}}_t(t) = \mathbf{L}(t)$, and we know from Eq. (4.4.5) that the symmetric part of $\mathbf{L}$ is $\mathbf{D}$ while the skew part is $\mathbf{W}$. A similar argument can be made beginning with $\mathbf{F}_t = \mathbf{V}_t \cdot \mathbf{R}_t$. Hence

$$\mathbf{D} = \dot{\mathbf{U}}_t(t) = \dot{\mathbf{V}}_t(t) \quad \text{and} \quad \mathbf{W} = \dot{\mathbf{R}}_t(t). \tag{4.6.27}$$

We conclude our study of the strain and deformation tensors with a study of the conditions under which a proposed set of functions can actually be the strain tensor components for a possible deformation. These conditions, called ***compatibility conditions***, will be developed in Sec. 4.7.

EXERCISES

1. What are the principal directions and principal values of the right stretch tensor **U** in the deformation of Ex. 1 of Sec. 4.5? *Partial answer*:

$$U_1 = [1 + \tfrac{1}{2}k^2 + k\sqrt{1 + \tfrac{1}{4}k^2}]^{1/2}$$

2. By using Eq. (4.6.6a) get an expression for the component U_{12} of the right stretch tensor $\mathbf{U}$ of Ex. 1. (It is not necessary to simplify the expression.)

3. Use the results of Exercises 5 and 7 of Sec. 4.5 to show that the components of the rotation tensor $\mathbf{R}$ for that deformation are

$$[R_{ij}] = \begin{bmatrix} (1+\frac{1}{4}k^2)^{-1/2} & \frac{1}{2}k(1+\frac{1}{4}k^2)^{-1/2} & 0 \\ -\frac{1}{2}k(1+\frac{1}{4}k^2)^{-1/2} & (1+\frac{1}{4}k^2)^{-1/2} & 0 \\ 0 & 0 & 1 \end{bmatrix}.$$

4. If the deformation of Ex. 13, Sec. 4.5, takes particle $\mathbf{X}$ to position $\mathbf{x}$ where the principal axes of $\mathbf{B}^{-1}$ are:

$$\hat{\mathbf{n}}_1 = \tfrac{4}{5}\mathbf{i}_1 - \tfrac{3}{5}\mathbf{i}_3 \qquad \hat{\mathbf{n}}_2 = \mathbf{i}_2 \qquad \hat{\mathbf{n}}_3 = \tfrac{3}{5}\mathbf{i}_1 + \tfrac{4}{5}\mathbf{i}_3,$$

determine the component matrix of the rotation tensor $\mathbf{R}$.

5. For the deformation of Ex. 4, determine, at the given point:
   (a) the principal values of the right stretch tensor $\mathbf{U}$
   (b) the matrix of components of $\mathbf{U}$ in the fixed axes
   (c) the components of the deformation gradient $\mathbf{F}$.

6. For the steady motion of Ex. 6, Sec. 4.4 ($v_x = my$, $v_y = v_z = 0$):
   (a) Display the matrices of the relative deformation gradient $\mathbf{F}_t$ and the relative deformation tensor $\mathbf{C}_t$ at time τ [see Eq. (4.5.32)]
   (b) Verify that $\dot{\mathbf{F}}_t(t) = \mathbf{D} + \mathbf{W}$, where $\dot{\mathbf{F}}_t(t)$ means

$$\left.\frac{d\mathbf{F}_t}{d\tau}\right|_{\tau = t}.$$

   (c) What are $\dot{\mathbf{R}}_t(t)$, $\dot{\mathbf{U}}_t(t)$, and $\dot{\mathbf{V}}_t(t)$ for this motion? [See Eqs. (4.6.27).]

7. Carry out the details of showing that $\dot{\mathbf{R}}_t(t)$ is skew-symmetric as suggested below Eq. (4.6.26).

8. Show that:
   (a) $\mathbf{V} = \mathbf{F}\cdot\mathbf{R}^T = \mathbf{R}\cdot\mathbf{U}\cdot\mathbf{R}^T$
   (b) $\mathbf{B} = \mathbf{V}\cdot\mathbf{V} = \mathbf{R}\cdot\mathbf{C}\cdot\mathbf{R}^T = \mathbf{F}\cdot\mathbf{F}^T$.

4.7 Compatibility conditions; determination of displacements when strains are known

If the (small) strain distribution in a body is known, then Eqs. (4.2.10) (strain definitions in terms of the displacement) are six independent† partial differential equations for the three unknown displacement functions u_i, namely

$$\frac{1}{2}(\mathbf{u}\overleftarrow{\nabla} + \overrightarrow{\nabla}\mathbf{u}) = \mathbf{E} \quad \text{or} \quad \frac{1}{2}\left(\frac{\partial u_i}{\partial X_j} + \frac{\partial u_j}{\partial X_i}\right) = \epsilon_{ij}, \tag{4.7.1}$$

†There are actually nine equations, but since both sides are symmetric in i and j, this yields only six independent equations.

where the small strains ϵ_{ij} are thought of as known functions of the X_i. Equations (1) offered no difficulty when the displacements were considered given and the strains were to be calculated; any differentiable functions whatever could be used for the u_i to calculate the strains. However, when we attempt to reverse the process by arbitrarily assuming some well-behaved functions of position as a strain distribution, we find that there may not exist a displacement solution which would produce our assumed strain distribution; this means that our assumed strain distribution is not a possible strain distribution. The mathematical difficulty is that Eqs. (1) are six partial differential equations for only three unknown functions, which would in general overdetermine the three unknown functions u_i. We will find some conditions, known as ***St.-Venant's compatibility equations***†, that must be satisfied by the strain distribution in order that Eqs. (1) may be integrable.

The compatibility equations are six partial differential equations, which together with the three equilibrium differential equations, and the six stress-strain relations form a system of 15 equations to solve for 12 unknown functions (six independent stresses and six independent strains) in a boundary-value problem of classical theory of elasticity. This would appear to overdetermine the elasticity problem; but the six compatibility equations are not in fact independent as will be shown below. ***If the displacement functions are included explicitly as unknowns in formulating the elasticity boundary-value problem, then the compatibility equations are not needed.*** There are then 15 unknown functions (three displacements, six stresses, six strains) and 15 equations [six Eqs. (1), six stress-strain equations, and three equilibrium equations]. For practicable solutions, however, it is necessary to reduce the number of equations and unknowns; this is frequently done by introducing stress functions as the only unknown functions, and it is then necessary to use the compatibiliy conditions along with the equilibrium conditions to solve for the stress functions.

It should not be thought that the compatibility conditions are limited to elasticity theory, however. We limit ourselves to small strains in most of the development below, but no restriction is placed on the material properties. The physical meaning of the compatibility conditions may be seen by imagining that the body is cut up into small cubes before it is strained, and then each cube is given a certain strain. In general the strained pieces cannot be fitted back together to form a continuous body without further deformation, but if the strain in each part is related to the strain in its neighbors according to the compatibility conditions, i. e., if its strain is ***compatible*** with that of its neighbors, then they can be fitted back together to form a continuous body. If the displacement components are retained as unknown functions, and these are required to be continuous and single-valued functions of the coordinates, the compatibility requirement is thereby fulfilled. But when the displacements are not

†B. de Saint-Venant, 1864. See A. E. H. Love (1944), p. 17 and pp. 48–50.

explicitly retained as unknowns the compatibility conditions must be imposed on the strain-distribution functions to ensure that there exists a continuous single-valued displacement distribution corresponding to the strain distribution.

Derivation of the compatibility equations as necessary conditions for existence of single-valued displacements. In x-y-z notation Eqs. (1) are

$$\left.\begin{aligned} &\frac{\partial u_x}{\partial X} = \epsilon_{xx}, \qquad \frac{\partial u_y}{\partial Y} = \epsilon_{yy}, \qquad \frac{\partial u_z}{\partial Z} = \epsilon_{zz}, \\ &\frac{\partial u_x}{\partial Y} + \frac{\partial u_y}{\partial X} = 2\epsilon_{xy}, \qquad \frac{\partial u_y}{\partial Z} + \frac{\partial u_z}{\partial Y} = 2\epsilon_{yz}, \qquad \frac{\partial u_z}{\partial X} + \frac{\partial u_x}{\partial Z} = 2\epsilon_{zx}. \end{aligned}\right\} \tag{4.7.2}$$

We begin by assuming that there does exist a solution of Eqs. (2), single-valued and possessing continuous third partial derivatives. The mixed third partial derivatives of the displacements are then equal when the order of differentiation is changed. Differentiating ϵ_{xy} with respect to X and Y in the fourth Eq. (2), we obtain

$$2\frac{\partial^2 \epsilon_{xy}}{\partial X \partial Y} = \frac{\partial^3 u_x}{\partial Y^2 \partial X} + \frac{\partial^3 u_y}{\partial X^2 \partial Y}.$$

From the first two Eqs. (2) we obtain

$$\frac{\partial^2 \epsilon_{xx}}{\partial Y^2} + \frac{\partial^2 \epsilon_{yy}}{\partial X^2} = \frac{\partial^3 u_x}{\partial Y^2 \partial X} + \frac{\partial^3 u_y}{\partial X^2 \partial Y}.$$

Hence

$$\frac{\partial^2 \epsilon_{xx}}{\partial Y^2} + \frac{\partial^2 \epsilon_{yy}}{\partial X^2} = 2\frac{\partial^2 \epsilon_{xy}}{\partial X \partial Y}. \tag{4.7.3}$$

Two similar equations are found by beginning with the other two shear components.

Now we differentiate ϵ_{xy} with respect to X and Z and ϵ_{xz} with respect to Y and X and add. By Eqs. (2), we have then

$$\begin{aligned} 2\left[\frac{\partial^2 \epsilon_{xy}}{\partial X \partial Z} + \frac{\partial^2 \epsilon_{zx}}{\partial Y \partial X}\right] &= \frac{\partial^2}{\partial X \partial Z}\left(\frac{\partial u_x}{\partial Y} + \frac{\partial u_y}{\partial X}\right) + \frac{\partial^2}{\partial Y \partial X}\left(\frac{\partial u_z}{\partial X} + \frac{\partial u_x}{\partial Z}\right) \\ &= 2\frac{\partial^2}{\partial Y \partial Z}\left(\frac{\partial u_x}{\partial X}\right) + \frac{\partial^2}{\partial X^2}\left(\frac{\partial u_y}{\partial Z} + \frac{\partial u_z}{\partial Y}\right) \\ &= 2\left[\frac{\partial^2 \epsilon_{xx}}{\partial Y \partial Z} + \frac{\partial^2 \epsilon_{yz}}{\partial X^2}\right], \end{aligned}$$

or

$$\frac{\partial^2 \epsilon_{xx}}{\partial Y \partial Z} = \frac{\partial}{\partial X}\left(-\frac{\partial \epsilon_{yz}}{\partial X} + \frac{\partial \epsilon_{zx}}{\partial Y} + \frac{\partial \epsilon_{xy}}{\partial Z}\right). \tag{4.7.4}$$

Two similar equations can be found beginning first with ϵ_{yz} and ϵ_{xy}, and then with ϵ_{yz} and ϵ_{xz}. The entire system of six compatibility equations can be written down by cyclic permutation of the subscripts in the two equations already found. The derivation above shows that these are necessary conditions for integrability, i.e., if Eqs. (2) are integrable, then it is necessary that the conditions be satisfied. The complete set of six equations, known as ***St.-Venant's compatibility equations*** is given below, where we also define functions S_{ij} or $R_x, R_y, R_z, U_x, U_y, U_z$ as the functions appearing on the left-hand sides of the equations.

Compatibility Equations

$$\left.\begin{aligned}
-S_{33} &\equiv R_z \equiv \frac{\partial^2 \epsilon_{xx}}{\partial Y^2} + \frac{\partial^2 \epsilon_{yy}}{\partial X^2} - 2\frac{\partial^2 \epsilon_{xy}}{\partial X \partial Y} = 0 \\
-S_{11} &\equiv R_x \equiv \frac{\partial^2 \epsilon_{yy}}{\partial Z^2} + \frac{\partial^2 \epsilon_{zz}}{\partial Y^2} - 2\frac{\partial^2 \epsilon_{yz}}{\partial Y \partial Z} = 0 \\
-S_{22} &\equiv R_y \equiv \frac{\partial^2 \epsilon_{zz}}{\partial X^2} + \frac{\partial^2 \epsilon_{xx}}{\partial Z^2} - 2\frac{\partial^2 \epsilon_{zx}}{\partial Z \partial X} = 0 \\
-S_{23} &\equiv U_x \equiv -\frac{\partial^2 \epsilon_{xx}}{\partial Y \partial Z} + \frac{\partial}{\partial X}\left(-\frac{\partial \epsilon_{yz}}{\partial X} + \frac{\partial \epsilon_{zx}}{\partial Y} + \frac{\partial \epsilon_{xy}}{\partial Z}\right) = 0 \\
-S_{31} &\equiv U_y \equiv -\frac{\partial^2 \epsilon_{yy}}{\partial Z \partial X} + \frac{\partial}{\partial Y}\left(\frac{\partial \epsilon_{yz}}{\partial X} - \frac{\partial \epsilon_{zx}}{\partial Y} + \frac{\partial \epsilon_{xy}}{\partial Z}\right) = 0 \\
-S_{12} &\equiv U_z \equiv -\frac{\partial^2 \epsilon_{zz}}{\partial X \partial Y} + \frac{\partial}{\partial Z}\left(\frac{\partial \epsilon_{yz}}{\partial X} + \frac{\partial \epsilon_{zx}}{\partial Y} - \frac{\partial \epsilon_{xy}}{\partial Z}\right) = 0.
\end{aligned}\right\} \tag{4.7.5a}$$

These equations are conveniently remembered in the form of Eq. (5b) below using the curl operators $(\vec{\nabla} \times)$ and $(\times \overleftarrow{\nabla})$.

Compatibility Equation

$$\vec{\nabla} \times \mathbf{E} \times \overleftarrow{\nabla} = 0 \tag{4.7.5b}$$

Equation (5b) is also useful for obtaining the form of the compatibility equations in orthogonal curvilinear coordinates; see App. II. Since in general the cross product is not associative it might be thought that parentheses would be needed in Eq. (5b) to indicate the order of the operations. But it turns out that for a symmetric tensor $\mathbf{E}$ possessing continuous second derivatives, we have

$$(\vec{\nabla} \times \mathbf{E}) \times \overleftarrow{\nabla} \equiv \vec{\nabla} \times (\mathbf{E} \times \overleftarrow{\nabla})$$

so that the parentheses may be omitted. Moreover the second-order tensor $\mathbf{S}$

defined by

$$\mathbf{S} = \overleftarrow{\nabla} \times \mathbf{E} \times \overleftarrow{\nabla}$$

is symmetric so that there are only six distinct component equations as in Eqs. (5a). It is left as an exercise for the reader to obtain the component equations of Eq. (5b) by using the formulas of Sec. 2.5 and to show that these component equations are equivalent to Eqs. (5a). See Exercises 12 to 14 of Sec. 2.5.

Demonstration that the compatibility equations are not independent conditions. We have proved that the six Eqs. (5) are necessary for the existence of a compatible strain distribution and will show below that they are also sufficient to ensure the existence of a single-valued displacement distribution in a simply connected body. First, however, we note that the six equations do not represent six independent conditions, since the "incompatibility components" R_x, R_y, R_z, U_x, U_y, U_z (which vanish when the strains are compatible) satisfy the following three identities,

$$\left.\begin{aligned} \frac{\partial R_x}{\partial X} + \frac{\partial U_z}{\partial Y} + \frac{\partial U_y}{\partial Z} &= 0 \\ \frac{\partial U_z}{\partial X} + \frac{\partial R_y}{\partial Y} + \frac{\partial U_x}{\partial Z} &= 0 \\ \frac{\partial U_y}{\partial X} + \frac{\partial U_x}{\partial Y} + \frac{\partial R_z}{\partial Z} &= 0. \end{aligned}\right\} \tag{4.7.6}$$

These identities are known as the ***Bianchi formulas*** in Riemannian geometry. The discussion given here follows K. Washizu (1958), who attributes the form of the Bianchi equations for small strains [Eqs. (6)] to S. Moriguti (1947). It is not sufficient,† however, to use only three of Eqs. (5). For example, neither

$$R_x = 0, \qquad R_y = 0, \qquad R_z = 0 \tag{4.7.7a}$$

alone, nor

$$U_x = 0, \qquad U_y = 0, \qquad U_z = 0 \tag{4.7.7b}$$

alone are sufficient to ensure the existence of single-valued displacements. Washizu showed that if either Eqs. (7a) or (7b) were satisfied in the interior of a simply connected body, while the other set was satisfied on the boundary, then the other set was also satisfied in the interior. But one set alone in the interior is not sufficient, and the application of the other set on the boundary may not be convenient in formulating a boundary-value problem. The usual procedure is

†See R. V. Southwell (1938)

to include all six equations in the interior problem formulation, but to remember that they represent only three independent conditions.

Proof that compatibility equations are necessary and sufficient conditions for single-valued displacements in a simply connected body.† We begin by expressing the displacements u_j^P at a typical point P of the region in terms of the displacements u_j^0 at an arbitrary point P_0 and the integral $\int du_j$ of the relative displacements du_j of particles initially lying along some curve C in the body connecting P_0 to P:

$$u_j^P = u_j^0 + \int_C du_j = u_j^0 + \int_C \frac{\partial u_j}{\partial X_k}\, dX_k$$
$$= u_j^0 + \int_C \epsilon_{jk}\, dX_k + \int_C \Omega_{jk} dX_k. \tag{4.7.8}$$

The last form follows from the fact that $J_u = E + \Omega$ as in Sec. 4.2. In the last integral we replace dX_k by $d(X_k - X_k^P)$, where X_k^P are the coordinates of P (constant in the integration). Then, integrating by parts, we get

$$\underset{C}{\int_{P_0}^{P}} \Omega_{jk} dX_k \equiv \underset{C}{\int_{P_0}^{P}} \Omega_{jk} d(X_k - X_k^P)$$
$$= (X_k^P - X_k^0)\Omega_{jk}^0 - \underset{C}{\int_{P_0}^{P}} (X_k - X_k^P) \frac{\partial \Omega_{jk}}{\partial X_m}\, dX_m, \tag{4.7.9}$$

since the integrated part vanishes at the upper limit. We express the partial derivatives of the rotation components in terms of partial derivatives of the strain components as follows:

$$\frac{\partial \Omega_{jk}}{\partial X_m} = \frac{\partial}{\partial X_m}\left[\frac{1}{2}\left(\frac{\partial u_j}{\partial X_k} - \frac{\partial u_k}{\partial X_j}\right)\right].$$

By adding and subtracting $\frac{1}{2}(\partial^2 u_m/\partial X_j \partial X_k)$ we bring this into the form

$$\frac{\partial \Omega_{jk}}{\partial X_m} = \frac{\partial}{\partial X_k}\left[\frac{1}{2}\left(\frac{\partial u_m}{\partial X_j} + \frac{\partial u_j}{\partial X_m}\right)\right] - \frac{\partial}{\partial X_j}\left[\frac{1}{2}\left(\frac{\partial u_k}{\partial X_m} + \frac{\partial u_m}{\partial X_k}\right)\right]$$
$$= \frac{\partial \epsilon_{mj}}{\partial X_k} - \frac{\partial \epsilon_{km}}{\partial X_j}. \tag{4.7.10}$$

Let

$$F_{jm} = \epsilon_{jm} - (X_k - X_k^P)\left(\frac{\partial \epsilon_{mj}}{\partial X_k} - \frac{\partial \epsilon_{km}}{\partial X_j}\right). \tag{4.7.11}$$

†This presentation of the proof follows Sokolnikoff (1956). The proof is originally due to E. Cesaro (1906).

Then, using Eqs. (9) to (11) we can write Eq. (8) in the form

$$u_j^P = u_j^0 + (X_k^P - X_k^0)\Omega_{jk}^0 + \int_{P_0 \atop C}^{P} F_{jm}\, dX_m, \tag{4.7.12}$$

in which the second term on the right gives the relative displacement of P relative to P_0 due to the rigid rotation of the element at P_0, while the integral in the last term gives the additional relative displacements due to the strain and rotation in the material elements initially located along the path C of the line integral. If the displacements are to be single-valued functions of the coordinates of point P, this means that the line integral in the last term must be independent of the path of integration between P_0 and P so that the quantities $F_{jm}\,dX_m$ are exact differentials of certain functions Φ_j of position. Note that Eqs. (12) are really three equations, one for each choice of the free index j, with one line integral in each equation. The necessary and sufficient conditions that these line integrals be independent of the path in a simply connected region are

$$\frac{\partial F_{jn}}{\partial X_m} - \frac{\partial F_{jm}}{\partial X_n} = 0. \tag{4.7.13}$$

Substituting for F_{jn} and F_{jm} according to Eq. (11), we obtain from Eq. (13)

$$\frac{\partial \epsilon_{jn}}{\partial X_m} - \delta_{km}\left[\frac{\partial \epsilon_{nj}}{\partial X_k} - \frac{\partial \epsilon_{kn}}{\partial X_j}\right] - \frac{\partial \epsilon_{jm}}{\partial X_n} + \delta_{kn}\left[\frac{\partial \epsilon_{mj}}{\partial X_k} - \frac{\partial \epsilon_{km}}{\partial X_j}\right]$$
$$- (X_k - X_k^P)\left[\frac{\partial^2 \epsilon_{nj}}{\partial X_k \partial X_m} - \frac{\partial^2 \epsilon_{kn}}{\partial X_j \partial X_m} - \frac{\partial^2 \epsilon_{mj}}{\partial X_k \partial X_n} + \frac{\partial^2 \epsilon_{km}}{\partial X_j \partial X_n}\right] = 0.$$

The first line of this equation vanishes identically, and, since the equation must hold for all choices of $X_k - X_k^P$, it follows that the coefficients of $X_k - X_k^P$ must each vanish; this yields the following equations:

$$\frac{\partial^2 \epsilon_{nj}}{\partial X_k \partial X_m} - \frac{\partial^2 \epsilon_{kn}}{\partial X_j \partial X_m} - \frac{\partial^2 \epsilon_{mj}}{\partial X_k \partial X_n} + \frac{\partial^2 \epsilon_{km}}{\partial X_j \partial X_n} = 0. \tag{4.7.14}$$

These are the compatibility equations in indicial notation. Because there are four free indices, the system of Eqs. (14) consists of $3^4 = 81$ equations, but some of them are trivial identities and some are repetitions because of the symmetry in the indices nj and km. It can be shown that there are really only six distinct equations, and that when written out they are the same as Eqs. (5). These do not represent six independent conditions, however, because of the three dependency identities of Eqs. (6).

We have proved that the compatibility equations are necessary and sufficient conditions for the ***existence*** of single-valued displacements in a simply connected body, but ***they do not ensure uniqueness*** of the displacement distribution for given strains. Indeed the displacements are not unique, since we can always superpose a rigid-body motion, which changes the displacement distribution but not the strains. If we specify further the displacement and rotation of the element at P_0, then Eq. (12) gives the displacements uniquely for a simply connected body. Actual determination of displacements is frequently done more conveniently without specific use of Eq. (12), as illustrated in the example below.

Example of displacement determination. Pure bending of an elastic beam. Consider a uniform beam with its centerline along the horizontal X-axis, bent by end couples in the XY-plane. The usual assumptions of beam theory yield an axial strain distribution $\epsilon_{xx} = Y/R$, where R is the radius of the circular arc which is assumed to be the deformed centerline and Y is the initial coordinate of the element. The only stress is assumed to be $\sigma_{xx} = E\epsilon_{xx}$, where E is the modulus of elasticity, and the only other nonzero strains are therefore $\epsilon_{yy} = \epsilon_{zz} = -\nu Y/R$, according to these assumptions (ν is Poisson's ratio). This strain distribution satisfies the compatibiliy conditions, Eqs. (5), ensuring the integrability of Eqs. (2), which now take the form

$$\frac{\partial u_x}{\partial X} = \frac{Y}{R} \qquad \frac{\partial u_y}{\partial Y} = -\frac{\nu Y}{R} \qquad \frac{\partial u_z}{\partial Z} = -\frac{\nu Y}{R} \tag{4.7.15a}$$

$$\frac{\partial u_x}{\partial Y} + \frac{\partial u_y}{\partial X} = 0 \qquad \frac{\partial u_y}{\partial Z} + \frac{\partial u_z}{\partial Y} = 0 \qquad \frac{\partial u_z}{\partial X} + \frac{\partial u_x}{\partial Z} = 0. \tag{4.7.15b}$$

The first Eq. (15a) can be integrated explicitly with respect to X if we add an arbitrary function F of Y and Z as a "constant of integration." Thus

$$u_x = \frac{XY}{R} + F(Y, Z). \tag{4.7.16}$$

We substitute this result into the first and third Eqs. (15b) to obtain

$$\frac{\partial u_y}{\partial X} = -\frac{X}{R} - \frac{\partial F}{\partial Y} \quad \text{and} \quad \frac{\partial u_z}{\partial X} = -\frac{\partial F}{\partial Z},$$

which can also be integrated with respect to X (remember that F does not depend on X) to yield

$$u_y = -\frac{X^2}{2R} - X\frac{\partial F}{\partial Y} + G(Y, Z) \qquad u_z = -X\frac{\partial F}{\partial Z} + H(Y, Z), \tag{4.7.17}$$

where G and H are arbitrary functions of Y and Z. We now substitute these results into the second and third Eqs. (15a) to obtain

$$-X\frac{\partial^2 F}{\partial Y^2} + \frac{\partial G}{\partial Y} = -\frac{\nu Y}{R} \quad \text{and} \quad -X\frac{\partial^2 F}{\partial Z^2} + \frac{\partial H}{\partial Z} = -\frac{\nu Y}{R}. \tag{4.7.18}$$

Equations (18) must be identities, since they were obtained by substituting the solutions into the differential equations. Therefore, since X appears only in the first term of each, and F is independent of X, we must have

$$\left.\begin{aligned} \frac{\partial^2 F}{\partial Y^2} &= 0 \qquad \frac{\partial^2 F}{\partial Z^2} = 0 \\ \frac{\partial G}{\partial Y} &= -\frac{\nu Y}{R} \qquad \frac{\partial H}{\partial Z} = -\frac{\nu Y}{R}. \end{aligned}\right\} \tag{4.7.19}$$

The last two of these can be integrated to yield

$$G(Y, Z) = -\frac{\nu Y^2}{2R} + g(Z) \qquad H(Y, Z) = -\frac{\nu YZ}{R} + h(Y), \tag{4.7.20}$$

where g is an arbitrary function of Z only, and h is an arbitrary function of Y only. Hence Eqs. (17) take the form

$$u_y = -\frac{X^2}{2R} - X\frac{\partial F}{\partial Y} - \frac{\nu Y^2}{2R} + g(Z) \qquad u_z = -X\frac{\partial F}{\partial Z} - \frac{\nu YZ}{R} + h(Y). \tag{4.7.21}$$

We now substitute these results in the remaining equation, the second Eq. (15b), to obtain

$$-2X\frac{\partial^2 F}{\partial Y \partial Z} + g'(Z) - \frac{\nu Z}{R} + h'(Y) = 0. \tag{4.7.22}$$

This equation must also be an identity, whence

$$\frac{\partial^2 F}{\partial Y \partial Z} = 0 \quad \text{and} \quad g'(Z) - \frac{\nu Z}{R} + h'(Y) = 0. \tag{4.7.23}$$

Equations (19) and (23) show that F can only be a linear function, since all of its second derivatives vanish, say

$$F(Y, Z) = MY + NZ + P, \tag{4.7.24}$$

where M, N, and P are arbitrary constants. The second Eq. (23), when written as

$$g'(Z) - \frac{\nu Z}{R} = -h'(Y),$$

implies that both sides of the equation must be equal to a constant, say A, since the left side is independent of Y and the right side is independent of Z, and the two are equal for all values of Y and Z. Thus

$$h'(Y) = -A \quad \text{and} \quad g'(Z) = \frac{\nu Z}{R} + A$$

whence

$$h(Y) = -AY + B \quad \text{and} \quad g(Z) = \frac{\nu Z^2}{2R} + AZ + C,$$

where A, B, and C are arbitrary constants. The solution, Eqs. (16) and (21), therefore finally takes the form†

$$\left.\begin{aligned} u_x &= \frac{XY}{R} + MY + NZ + P \\ u_y &= -\frac{1}{2R}[X^2 + \nu(Y^2 - Z^2)] - MX + AZ + C \\ u_z &= -\frac{\nu YZ}{R} - NX - AY + B. \end{aligned}\right\} \tag{4.7.25}$$

The linear terms in Eqs. (25) represent a rigid-body motion, which does not affect the strain distribution. The six arbitrary constants may be determined by support conditions, which must be sufficiently simple that they can be satisfied by appropriate choices of the constants; otherwise, the support constraints will give rise to additional strains.

Displacements for plane strain. When the displacement distribution is one of plane motion parallel to the XY-plane, a much simpler procedure suffices. If $u_z = 0$ identically, and u_x and u_y are independent of Z, then in the first integration with respect to X the arbitrary function to be added is a function of Y only, while in an integration with respect to Y an arbitrary function of X only is added. The two extensional strain expressions can be integrated in this way, and when the results are substituted into the one shear-strain equation, the requirement that this be an identity suffices to yield two equations to be integrated to determine the two arbitrary functions.

Compatibility conditions for rate of deformation. The whole derivation given for small strains can be repeated for the components D_{ij} of the rate-of-deformation tensor. This yields six distinct equations, like Eqs. (5) or (14), with ϵ replaced by D, as the necessary and sufficient conditions that the distribution of D_{ij} can be associated with a single-valued velocity distribution in a simply connected region. (The partial derivatives are now with respect to the spatial

†See Timoshenko and Goodier (1951), Sec. 88.

coordinates x_i.) These conditions again represent only three independent conditions.

Finite-strain compatibility conditions using general-tensor components in curvilinear coordinates. [This subsection may be omitted by readers not familiar with the general-tensor concepts.] The problem of integrating the finite strains to determine the displacement field or the deformation field $x_k(X_1, X_2, X_3, t)$ for a given t is much more difficult than for the infinitesimal strain case discussed above, since Eqs. (1) must be replaced by nonlinear partial differential equations, e. g., Eq. (26) below. Indeed the task is so difficult that it is almost never attempted. The compatibility equations ensuring integrability can, however, be stated. The question may be avoided completely if we retain the x_k as basic unknown functions in our formulation of the problem, since any continuous solution of the field equations in terms of these unknowns will automatically satisfy compatibility. It is only when these unknowns are eliminated and strain measures, for example the deformation-tensor components C_{KM} of Sec. 4.5, are taken as the basic unknowns of the problem that it is necessary to satisfy compatibility conditions.

We may ask what are necessary and sufficient conditions to ensure that a proposed set of functions C_{KM} (symmetric in K, M) can be the deformation-tensor components of some deformation field. A direct approach would be to attempt to integrate the set of (six independent) nonlinear partial differential equations

$$\frac{\partial x_k}{\partial X_K}\frac{\partial x_k}{\partial X_M} = C_{KM} \tag{4.7.26}$$

with right-hand sides known as functions of the X_K. This is too difficult to discuss in general terms. But some results from general-tensor calculus will furnish the required compatibility conditions.† From Eqs. (4.5.4a) and (4.5.7), we have

$$(ds)^2 = C_{KM}\, dX^K\, dX^M \tag{4.7.27}$$

and

$$C_{KM} = \frac{\partial z^k}{\partial X^K}\frac{\partial z^m}{\partial X^M}\delta_{km} \tag{4.7.28}$$

if the X^K are interpreted as curvilinear coordinates in the deformed body. That is, for example, a plane $X^1 =$ const. in the undeformed body is deformed into a curved surface $X^1 =$ const. in the deformed body, while the line ($X^1 =$ const., $X^2 =$ const.) in the undeformed body becomes the curve ($X^1 =$ const.,

†See Truesdell and Toupin (1960), Sec. 34.

$X^2 =$ const.) in the deformed body, etc. Thus, giving the material coordinates X^1, X^2, X^3 of a particle locates its deformed position as the intersection of three curved coordinate surfaces in the deformed body. But this same point in space is also identified by the spatial coordinates, which were denoted z^k in Eq. (28) to signify that they are rectangular Cartesian coordinates (see App. I). The deformation equations, Eqs. (4.5.1),

$$z^k = z^k(X^1, X^2, X^3, t), \qquad X^K = X^K(z^1, z^2, z^3, t) \tag{4.7.29}$$

for fixed t may be regarded as a coordinate transformation in the deformed body between rectangular Cartesian coordinates z^k and curvilinear coordinates X^K.

Equation (27) then shows that the C_{KM} are the (covariant) components of the metric tensor with respect to the curvilinear coordinates X^K, i.e., the C_{KM} are the coefficients of $dX^K dX^M$ in the quadratic form for $(ds)^2$ (see Sec. I.4). It is necessary to make this interpretation in the deformed state, because ds is the arc length in the deformed state. A theorem of Riemann states that for a symmetric tensor to be a metric tensor for a Euclidean space, it is necessary and sufficient that it be a nonsingular positive-definite tensor and that the Riemann-Christoffel tensor formed from it vanishes identically.† [See Eq. (I.6.26).] Thus for the proposed C_{KM} to be covariant components of a metric tensor, we must have

$$R^{(\mathbf{C})}_{KRMN} = 0. \tag{4.7.30}$$

The superscript (**C**) is added as a reminder that R_{KRMN} is formed with **C**. It can be shown that†

$$R_{KRMN} = R_{MNKR} \tag{4.7.31a}$$

$$R_{KRMN} = -R_{KRNM} = -R_{RKMN} \tag{4.7.31b}$$

$$R_{KRMN} + R_{KMNR} + R_{KNRM} = 0. \tag{4.7.31c}$$

Equations (31) imply that of the $3^4 = 81$ Eqs. (30) there are only six distinct equations:

$$\left.\begin{aligned} R^{(\mathbf{C})}_{3131} &= 0 \qquad R^{(\mathbf{C})}_{3232} = 0 \qquad R^{(\mathbf{C})}_{1212} = 0 \\ R^{(\mathbf{C})}_{3132} &= 0 \qquad R^{(\mathbf{C})}_{3212} = 0 \qquad R^{(\mathbf{C})}_{3112} = 0. \end{aligned}\right\} \tag{4.7.32}$$

These are the six compatibility conditions for the C_{KM}. but they are not all independent, since they satisfy the Bianchi identities†

†For a proof of the theorem (30) and the identities (31) and (33), see, e.g., Veblen (1927).

$$R^{(C)K}{}_{.RMN}|_P + R^{(C)K}{}_{.RNP}|_M + R^{(C)K}{}_{.RPM}|_N = 0, \tag{4.7.33}$$

where the vertical bars denote the covariant derivative of Sec. I.6.

The compatibility Eqs. (32) can be expressed in terms of the Christoffel symbols (see Sec. I.6) and their derivatives, which in turn can be expressed in terms of the covariant metric tensor components C_{KM} and the associated contravariant components C^{KM} [see Eq. (I.4.24)] and their derivatives. The result is (note summation on S)

$$R^{(C)}_{KRMN} = C_{KS}\left[\frac{\partial}{\partial X^M}\begin{Bmatrix} S \\ R \quad N \end{Bmatrix} - \frac{\partial}{\partial X^N}\begin{Bmatrix} S \\ R \quad M \end{Bmatrix} + \begin{Bmatrix} P \\ R \quad N \end{Bmatrix}\begin{Bmatrix} S \\ P \quad M \end{Bmatrix} - \begin{Bmatrix} T \\ R \quad M \end{Bmatrix}\begin{Bmatrix} S \\ T \quad N \end{Bmatrix}\right] \tag{4.7.34}$$

where the Christoffel symbols $\begin{Bmatrix} S \\ M \quad N \end{Bmatrix}$ and $[MN, R]$ are defined by

$$\begin{Bmatrix} S \\ M \quad N \end{Bmatrix} = C^{SR}[MN, R] \tag{4.7.35}$$

and

$$[MN, R] = \frac{1}{2}\left(\frac{\partial C_{NR}}{\partial X^M} + \frac{\partial C_{RM}}{\partial X^N} - \frac{\partial C_{MN}}{\partial X^R}\right). \tag{4.7.36}$$

The finite-strain compatibility equations (32) can also be expressed in terms of the components of $\mathbf{E}$. Comparable equations may be written for the Eulerian formulation in terms of $\mathbf{B}^{-1}$ or $\mathbf{E}^*$. See, e.g., Truesdell and Toupin (1960), Sec. 34 or Eringen (1962), Sec. 13.

We have now completed our formal study of the stress tensor of Chap. 3 and of the strain and deformation tensors in this chapter. In Chap. 5 we will examine a number of general principles of continuum mechanics, which are valid for all continuous materials. Then in Chap. 6 we will take up the constitutive equations characterizing different materials.

EXERCISES

1. Show that one of the following strain distributions is a possible state of small strain, while the other is not. (a is a constant with dimensions of length.) For the possible strain distribution find the displacements as functions of X and Y, if the element at the origin has zero displacement and zero rotation.

   (a) Plane strain with $u_z = 0$ and with $\epsilon_{xx} = (X^2 + Y^2)/a^2$, $\epsilon_{yy} = Y^2/a^2$, $\epsilon_{xy} = XY/a^2$

(b) $\epsilon_{xx} = Z(X^2 + Y^2)/a^3 \qquad \epsilon_{yy} = Y^2Z/a^3, \qquad \epsilon_{xy} = XYZ/a^3,$
$\epsilon_{zz} = \epsilon_{yz} = \epsilon_{zx} = 0.$

2. In the theory of small-strain elasticity the isotropic stress-strain relations for plane strain may be written as follows:

$$\epsilon_{xx} = \frac{1+\nu}{E}[(1-\nu)\sigma_x - \nu\sigma_y], \quad \epsilon_{xy} = \frac{1+\nu}{E}\tau_{xy},$$

$$\epsilon_{yy} = \frac{1+\nu}{E}[-\nu\sigma_x + (1-\nu)\sigma_y]$$

where E is the modulus of elasticity and ν is Poisson's ratio. Use these equations to obtain a compatibility equation in terms of the stresses σ_x, σ_y, and τ_{xy}.

3. (a) Repeat Ex. 2 for plane stress, where the stress-strain relation is

$$\epsilon_{xx} = \frac{1}{E}[\sigma_x - \nu\sigma_y], \qquad \epsilon_{yy} = \frac{1}{E}[-\nu\sigma_x + \sigma_y], \qquad \epsilon_{xy} = \frac{1+\nu}{E}\tau_{xy}.$$

(b) For the plane-stress case the strain ϵ_{zz} is not usually zero, but is given by $\epsilon_{zz} = -\nu(\sigma_x + \sigma_y)/E$. If the strains are all assumed to be independent of z, what additional restrictions are placed on the plane-stress field by the requirement that it satisfy the full three-dimensional compatibility conditions, in addition to the one equation determined in (a)?

4. (a) Show that the plane-strain field

$$\epsilon_{xx} = kXY, \qquad \epsilon_{yy} = -\nu kXY, \qquad \epsilon_{xy} = \frac{k}{2}(1+\nu)(c^2 - Y^2),$$

where k, ν, and c are constants, is a possible field.

(b) Assuming that $u_z \equiv 0$ and that u_x and u_y are independent of Z, determine the displacements corresponding to this strain field.

5. Verify that the expansion of Eq. (4.7.5b) gives the component Eqs. (5a). [See Exs. 12 to 14 of Sec. 2.5.]

6. (a) If the only strains are ϵ_{zx} and ϵ_{zy} (and of course ϵ_{xz} and ϵ_{yz}), and the strains are independent of z, show that the two compatibility equations not identically satisfied can be integrated to give a single first-order compatibility equation for ϵ_{zx} and ϵ_{zy}, containing an arbitrary constant.

(b) In the St.-Venant theory of torsion of a prismatic bar, the displacements (for small z) are assumed given by $u_x = -\theta yz$, $u_y = \theta xz$, $u_z = \theta f(x,y)$, where θ is the given angle of twist per unit length (assumed independent of x, y, and z) and $f(x, y)$ is the warping function. (The axial displacements are assumed proportional to θ in the linear elastic case. In plasticity one usually assumes instead $u_z = f(x, y, \theta)$ not proportional to θ.) Show that the assumed displacements lead to the strains of (a), and determine the arbitrary constant for the compatibility equation obtained in (a), for the linear elastic case.

7. Hooke's law gives $\tau_{zx} = 2G\epsilon_{zx}$, $\tau_{zy} = 2G\epsilon_{zy}$, and the shear stresses are also given by $\tau_{zx} = \partial\phi/\partial y$, $\tau_{zy} = -\partial\phi/\partial x$ in terms of the torsion stress function $\phi(x, y)$ of Ex. 9, Sec. 3.2. Hence, derive the second-order partial differential equation satisfied by $\phi(x, y)$ from the compatibility equation found in Ex. 6.

CHAPTER 5

General Principles

5.1 Introduction; integral transformations; flux

We have studied the stress tensor and its properties, and several tensors describing strain and deformation at a point. In general the tensors vary from point to point and represent a ***tensor field***, i.e., a tensor function of position. In Sec. 4.7 we have already obtained compatibility conditions, partial differential equations governing the way the strain can vary in the neighborhood of a point. In this chapter, we will derive additional differential equations governing the way the stress and deformation vary in the neighborhood of a point and with time. The equations derived in this chapter, which apply to any continuous medium, will not be sufficient in number to determine the unknown tensor functions, because they will not include the constitutive equations characterizing the material, which will be considered in Chap. 6. The constitutive equations (e.g., stress-strain equations) and boundary and initial conditions must be added to obtain a well-defined mathematical problem to solve for the stress and deformation distributions or the displacement or velocity fields.

In the next three sections we will derive differential equations expressing locally the conservation of mass, momentum, and energy. These differential equations of balance will be derived from integral forms of the equations of balance expressing fundamental postulates of continuum mechanics. The derivations will make use of certain integral transformation formulas, especially the divergence theorem (Gauss's theorem). In the invariant form of Eqs. (4b) or (5) below, this theorem is independent of any choice of coordinates, but it is simpler to present the derivation of Gauss's theorem in terms of rectangular Cartesian components than in terms of curvilinear components.

Integral transformations in rectangular Cartesian coordinates. The derivation of Gauss's theorem makes use of the following relationship between a volume integral and a surface integral over the bounding surface of the volume.

If $f(x_1, x_2, x_3)$ has continuous first partial derivatives with respect to the rectangular Cartesian coordinates x_i, then

$$\boxed{\int_S f n_i \, dS = \int_V \frac{\partial f}{\partial x_i} \, dV} \tag{5.1.1}$$

This is often called ***Green's theorem***, but since several different theorems go by that name, the present one will be referred to in this book as "***the transformation from a surface integral to a volume integral.***" Here the unit vector components n_i are the direction cosines of the outward normal to the surface S at a point in the surface area element dS; the n_i are functions of position on the surface. The restrictions on the surface regularity are not great; it will be sufficient if the surface is piecewise smooth and is topologically such that it clearly defines an inside and an outside. The volume need not be simply connected, provided that the surface integral is extended over all the bounding surfaces, and the outer normal on each is taken to point away from the enclosed volume. It should be possible by inserting a finite number of interior boundaries to display the total volume as the sum of a finite number of volumes each bounded by simple closed surfaces. For the proof of the theorem, see a textbook on advanced calculus or vector analysis. The basic ideas of the proof will be illustrated below for a special case of the analogous two-dimensional theorem.

The analogous theorem in the plane for a function $f(x, y)$ on an area A bounded by a curve C is

$$\int_C f n_x \, ds = \int_A \frac{\partial f}{\partial x} \, dA \qquad \int_C f n_y \, ds = \int_A \frac{\partial f}{\partial y} \, dA \tag{5.1.2a}$$

where

$$n_x = \frac{dy}{ds} \qquad n_y = -\frac{dx}{ds}. \tag{5.1.2b}$$

Consider the special case of this plane theorem for which C is a smooth simple closed curve cut by any line parallel to the x- or y-axis at most twice as in Fig. 5.1. For this case the normal direction cosines are as given in Eq. (2), since $\hat{\mathbf{n}}$ is perpendicular to the unit tangent $\hat{\mathbf{t}} = (dx/ds)\,\mathbf{i} + (dy/ds)\,\mathbf{j}$. The bounding curve C is composed of two parts; the lower boundary C_1 may be represented as $y = y_1(x)$, while the upper curve C_2 is represented by $y = y_2(x)$. Consider the area integral of $\partial f/\partial y$.

$$\begin{aligned}
\int_A \frac{\partial f}{\partial y} \, dA &= \int_a^b \left[\int_{y_1(x)}^{y_2(x)} \frac{\partial f}{\partial y} \, dy \right] dx \\
&= \int_a^b \{ f[x, y_2(x)] - f[x, y_1(x)] \} \, dx \\
&= \underset{C_2}{\int_a^b} f \, dx - \underset{C_1}{\int_a^b} f \, dx.
\end{aligned}$$

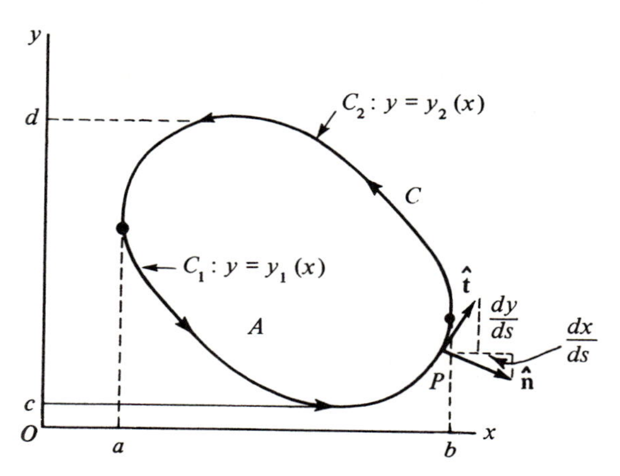

Fig. 5.1 Simple closed plane curve cut by a line parallel to a coordinate axis at most twice.

The term $\int_{a\atop C_2}^{b} f\,dx$ represents a line integral along C_2 in the negative sense and therefore equals $-\int_{b\atop C_2}^{a} f\,dx$. Thus, dropping the limits, we have

$$\begin{aligned}
\int_A \frac{\partial f}{\partial y}\,dA &= -\left[\int_{C_2} f\,dx + \int_{C_1} f\,dx\right] \\
&= -\int_C f\frac{dx}{ds}\,ds \\
&= +\int_C f n_y\,ds,
\end{aligned}$$

which establishes the second of Eqs. (2a). The first Eq. (2a) is proved similarly by integrating the area integral first with respect to x.

The transformation formulas are very easy to remember, since each boundary integral involves the function times a unit normal component, while the integrand of the integral over the region is the partial derivative with respect to the corresponding Cartesian coordinate.

These integral transformations are frequently encountered in the form of integration by parts. For example, we may need to evaluate $\int_V u(\partial v/\partial x_i)\,dV$. Since, according to Eq. (1),

$$\int_V \frac{\partial}{\partial x_i}(uv)\,dV = \int_S (uv)n_i\,dS,$$

it follows from expanding the derivative of the product uv that

$$\int_V u \frac{\partial v}{\partial x_i} \, dV = \int_S uvn_i \, dS - \int_V \frac{\partial u}{\partial x_i} v \, dV, \tag{5.1.3}$$

an integration by parts analogous to the familiar equation

$$\int_a^b u \, dv = [uv]_a^b - \int_a^b v \, du.$$

For the plane case, based on Eqs. (2), we can in the resulting line integral $\int_C uvn_i \, ds$ replace $n_x \, ds$ by dy or $n_y \, ds$ by $-dx$.

Vector integral transformations: divergence theorem. If Eq. (1) is written with the three components v_i of a vector $\mathbf{v}$ successively substituted for f and the three resulting equations are added, the result is

$$\int_S v_i n_i \, dS = \int_V \frac{\partial v_i}{\partial x_i} \, dV \qquad \text{(Cartesian form)} \tag{5.1.4a}$$

or in vector notation [see Eq. (2.5.25)]

$$\int_S \mathbf{v} \cdot \hat{\mathbf{n}} \, dS = \int_V \nabla \cdot \mathbf{v} \, dV. \tag{5.1.4b}$$

This is the ***divergence theorem*** or Gauss's theorem which states that the integral of the outer normal component of a vector over a closed surface is equal to the integral of the divergence of the vector over the volume bounded by the closed surface (subject, of course, to the same kind of restrictions on the continuity of derivatives, etc. as were stated above for the component equations). The indicial form of the equation is especially easy to remember and use in Cartesian coordinates. The vector form has the advantage that it may be applied in curvilinear coordinates, if the expressions $\mathbf{v} \cdot \hat{\mathbf{n}}$ and $\nabla \cdot \mathbf{v}$ as well as dS and dV are expressed in terms of the curvilinear coordinates.

A generalized Gauss's theorem can be stated by introducing a notation called the "star product."† The star product $\mathbf{a} * \mathbf{b}$ represents either the dot product or the cross product, and can be used to state those general propositions common to both, e.g.,

$$\frac{d}{dt}(\mathbf{a} * \mathbf{b}) = a * \frac{d\mathbf{b}}{dt} + \frac{d\mathbf{a}}{dt} * \mathbf{b}.$$

More generally still, it can include products like $a\mathbf{b}$ of a scalar by a vector or

†See, for example, Nathaniel Coburn, *Vector and Tensor Analysis*, Macmillan, New York, 1955, p. 23 and p. 74.

ab the tensor product [see Eq. (2.4.21)] of two vectors leading to a second-order tensor with Cartesian components $T_{ij} = a_i b_j$. To obtain the generalized results we must require the star products to be distributive and to satisfy theorems on the algebra of limits, and we must be careful to keep the factors in the proper order. Then the generalized Gauss's theorem takes the form

$$\int_S \hat{\mathbf{n}} * \mathscr{A}\, dS = \int_V \nabla * \mathscr{A}\, dV, \tag{5.1.5}$$

where the script $\mathscr{A}$ may denote a scalar, vector, or tensor. Some special cases are written below in indicial notation for rectangular Cartesian components:

$$\mathscr{A} = f, \quad \text{scalar} \qquad \int_S n_i f\, dS = \int_V \frac{\partial f}{\partial x_i}\, dV \tag{5.1.6}$$

$$\mathscr{A} = \mathbf{v}, \quad \text{vector (dot product)} \qquad \int_S n_i v_i\, dS = \int_V \frac{\partial v_i}{\partial x_i}\, dV \tag{5.1.7}$$

$$\mathscr{A} = \mathbf{T}, \quad \text{tensor} \qquad \int_S n_i T_{ij}\, dS = \int_V \frac{\partial T_{ij}}{\partial x_i}\, dV \tag{5.1.8}$$

Another special case for $\mathscr{A} = \mathbf{v}$ is obtained by using the cross product:

$$\int_S e_{ijk} n_j v_k\, dS = \int_V e_{ijk} \frac{\partial}{\partial x_j}(v_k)\, dV \tag{5.1.9a}$$

or [see Eq. (2.5.26b)]

$$\int_S \hat{\mathbf{n}} \times \mathbf{v}\, dS = \int_V \nabla \times \mathbf{v}\, dV. \tag{5.1.9b}$$

Stokes theorem is another important vector integral transformation theorem. It is not concerned with a volume at all, but instead it relates an integral around a closed curve in space to an integral over a portion of an orientable surface in space bounded by the curve. Figure 5.2 illustrates the meaning of oriented surface and curve. If we arbitrarily assign a positive side to the surface S and take the normal $\hat{\mathbf{n}}$ toward the positive side, then the positive sense on C is implied by the right-hand screw rule applied to a screw lying along a normal $\hat{\mathbf{n}}$, as is illustrated by the unit tangent $\hat{\mathbf{t}}$.

A generalized version of Stokes theorem then states that if $\mathscr{A}$ is a continuous field (scalar, vector, or tensor) with continuous partial derivatives on S, then

$$\int_S (\hat{\mathbf{n}} \times \nabla) * \mathscr{A}\, dS = \int_C \hat{\mathbf{t}} * \mathscr{A}\, ds. \tag{5.1.10}$$

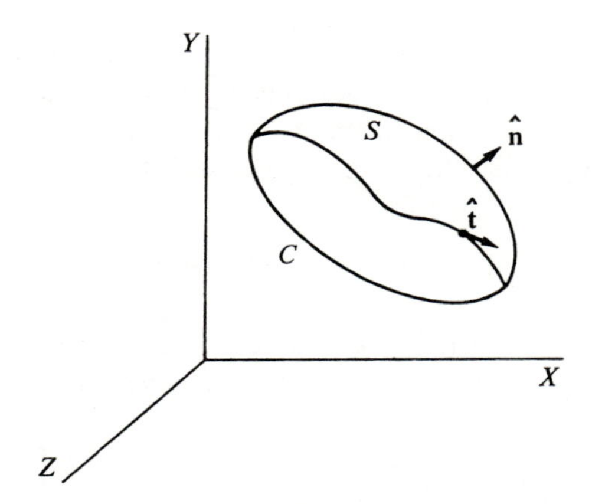

Fig. 5.2 Oriented surface S and bounding curve C.

The proof is omitted.† The most commonly used version is for $\mathscr{A} = \mathbf{v}$, a vector, and with the star product representing the dot product. Thus, since, by Eq. (2.3.20), $\mathbf{a \cdot b \times c = a \times b \cdot c}$, we have

$$\boxed{\int_S \hat{\mathbf{n}} \cdot (\nabla \times \mathbf{v})\, dS = \int_C \hat{\mathbf{t}} \cdot \mathbf{v}\, ds,} \tag{5.1.11}$$

stating that the integral of the normal component of curl $\mathbf{v}$ over the surface is equal to the integral of the tangential component of $\mathbf{v}$ around the curve. This is the usual form of ***Stokes theorem***.

Figure 5.2 illustrated a simply connected surface S, but the theorem applies to multiply connected surfaces, provided the line integral is taken in the correct sense around the interior boundaries. For example, in the case illustrated in Fig. 5.3 the positive direction on C_2 is such that "a man walking along the curve in the positive sense will find the interior of S to the left, provided his head is in the positive $\hat{\mathbf{n}}$ direction." The correct sense can be defined more precisely by introducing suitable cuts to make the region simply connected, traversing each of them twice, and maintaining the same sense as that defined for the outer boundary, e.g., cuts AB and CD in Fig. 5.3.

If curl $\mathbf{v}$ is identically zero in a simply connected region of space, then Stokes theorem implies that $\int_C \hat{\mathbf{t}} \cdot \mathbf{v}\, ds$ vanishes for every closed curve C in the region, and this implies that the line-integral is independent of the path between any two points in the region and that therefore the integrand

†See, for example, Coburn (1955), p. 83.

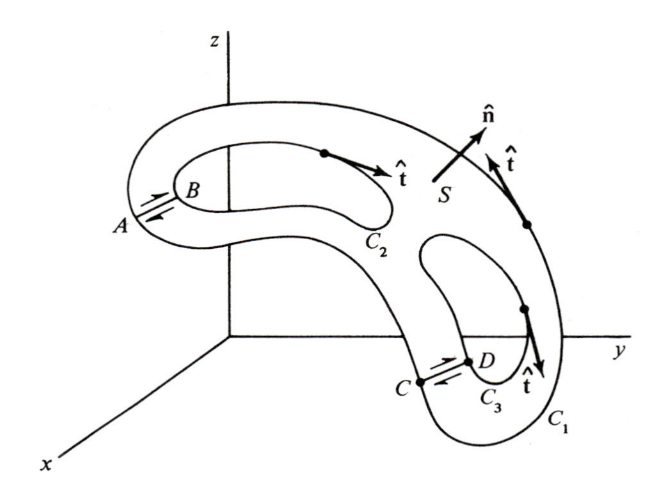

Fig. 5.3 Multiply connected surface S in space.

$\hat{\mathbf{t}} \cdot \mathbf{v}\, ds = v_x\, dx + v_y\, dy + v_z\, dz$ is an exact differential of a scalar function, as was pointed out in Sec. 2.5.

If $\mathbf{v}$ is the velocity vector of a flow then $\int_C \hat{\mathbf{t}} \cdot \mathbf{v}\, ds$ is called the ***circulation*** around the curve C, and the theorem shows that the circulation around a boundary C is equal to the integral of the normal component of the ***vorticity vector*** $\nabla \times \mathbf{v}$ over any surface bounded by C.

Flux across a surface. Consider an imaginary surface fixed in space with continuous medium flowing through it. If we assign a positive side to the surface (outward in the case of a closed surface, otherwise arbitrarily) and take the normal $\hat{\mathbf{n}}$ in the positive sense, then the volume of material flowing through the infinitesimal surface area dS in time dt is equal to the volume of the cylinder with base dS and slant height $v\, dt$ parallel to the velocity vector $\mathbf{v}$. See Fig. 5.4. The altitude of the cylinder is $v_n\, dt = \mathbf{v} \cdot \hat{\mathbf{n}}\, dt$. Hence the volume in time dt is $v_n\, dt\, dS = \mathbf{v} \cdot \hat{\mathbf{n}}\, dt\, dS$ and the ***flux*† *of volume*** or volume per unit time flowing through dS is $v_n\, dS = \mathbf{v} \cdot \hat{\mathbf{n}}\, dS$. (If $\mathbf{v} \cdot \hat{\mathbf{n}}$ is negative, the flow is in the negative direction.) The total flux of volume through a finite surface S is given by the surface integral.

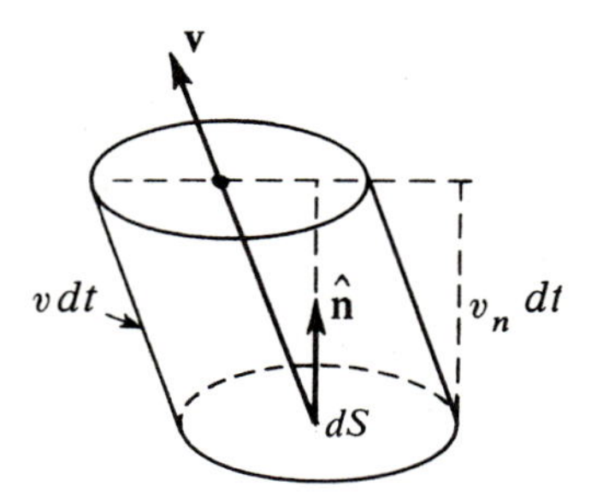

Fig. 5.4 Flux through dS.

†The name *flux* derives from Maxwell (1873).

$$\text{Volume Flux} = \int_S \mathbf{v} \cdot \hat{\mathbf{n}} \, dS = \int_S v_j n_j \, dS \tag{5.1.12}$$

where the last form is for rectangular Cartesian components.

The ***mass flux*** per unit area through dS is obtained by multiplying the volume flux by the local density. Hence

$$\text{Mass Flux} = \int_S \rho \mathbf{v} \cdot \hat{\mathbf{n}} \, dS = \int_S \rho v_j n_j \, dS. \tag{5.1.13}$$

The rate of flow of any material property carried by mass transport through the surface can be similarly defined. The momentum per unit volume is $\rho \mathbf{v}$ and the kinetic energy per unit volume is $\frac{1}{2}\rho v^2$. Hence

$$\underset{\text{(a vector)}}{\text{Momentum Flux}} = \int_S \rho \mathbf{v}(\mathbf{v} \cdot \hat{\mathbf{n}}) \, dS = \mathbf{i}_k \int_S \rho v_k v_j n_j \, dS \tag{5.1.14}$$

and

$$\underset{\text{(a scalar)}}{\text{Kinetic Energy Flux}} = \int_S \tfrac{1}{2}\rho v^2(\mathbf{v} \cdot \hat{\mathbf{n}}) \, dS = \int_S \tfrac{1}{2}\rho v_i v_i v_j n_j \, dS. \tag{5.1.15}$$

In general, if $\mathscr{A}$ denotes any quantity reckoned per unit mass of the medium ($\mathscr{A}$ may be a scalar, vector, or tensor), then its ***flux carried by mass transport*** through the surface is given by

$$\text{Flux of } \mathscr{A} = \int_S \rho \mathscr{A} \mathbf{v} \cdot \hat{\mathbf{n}} \, dS = \int_S \rho \mathscr{A} v_j n_j \, dS. \tag{5.1.16}$$

Many authors use a "vector element of area," combining $\hat{\mathbf{n}} \, dS$ into $\mathbf{dA}$, or $dA_j = n_j \, dS$ to shorten the notation. In this notation

$$\text{Flux of } \mathscr{A} = \int_S \rho \mathscr{A} \mathbf{v} \cdot \mathbf{dA} = \int_S \rho \mathscr{A} v_j \, dA_j. \tag{5.1.17}$$

When any of these flux integrals is taken over a closed surface it gives the net outward flux through the surface, i.e., the net amount of the quantity flowing out through the surface per unit time. These integral expressions for the flux across a surface will be used in the following sections together with the integral transformations of Eqs. (4) to derive some important differential equations from the assumption of conservation of mass and the momentum principles.

Flux of nonmechanical quantities may also occur through a surface whether or not there is any mass flux through the surface. An important example is heat transfer by conduction. If $\mathbf{q}$ denotes the ***heat flux vector***, the flux through

dS is $\mathbf{q} \cdot \hat{\mathbf{n}}\, dS$. [Note the analogy with Eq. (12) for volume flux.] Another example is electric current flow; if $\mathbf{J}$ denotes the current density vector, then $\mathbf{J} \cdot \hat{\mathbf{n}}\, dS$ is the flux of positive charge through S. Flux of nonmechanical quantities may occur through a ***material boundary***, say S_M, not fixed in space but separating two material regions of the continuous medium and moving with the material so that there is no mass flux through S. In fact, electrical current flow or heat conduction is most naturally defined relative to the conducting medium rather than relative to "fixed" space. In Sec. 5.4, we will assume that the nonmechanical energy flux $\mathbf{q} \cdot \hat{\mathbf{n}}\, dS$ is relative to the material, but we write dS instead of dS_M, since the integral will be taken over the spatial surface S instantaneously occupied by S_M.

Material volume integrals (over a given material region instead of a given spatial region) are especially useful for formulating the balance equations, e.g., momentum balance and energy balance treated in Secs. 5.3 and 5.4. In Sec. 5.2, we will see how the time rate of change of an integral over a given mass system $\mathscr{M}$ can be expressed in terms of integrals over the spatial volume V instantaneously occupied by $\mathscr{M}$ and over its boundary surface S. First, however, we will derive the continuity equation in spatial coordinates from the condition of conservation of mass in an arbitrary spatial volume.

5.2 Conservation of mass; the continuity equation

We begin our study of the differential equations giving local expressions to the conservation laws by deriving the continuity equation from the conservation of mass. Consider an arbitrary volume V fixed in space, bounded by surface S, as shown in Fig. 5.5. If continuous medium of density ρ fills the volume at time t, the total mass in V is

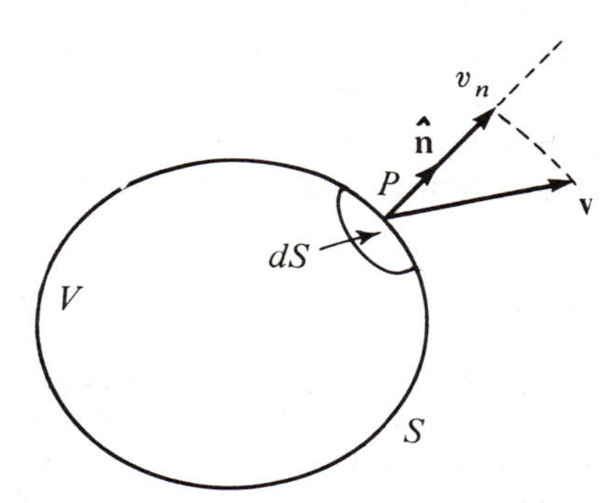

Fig. 5.5 Reference volume in space.

$$M = \int_V \rho \, dV. \tag{5.2.1}$$

The density ρ depends on position and time

$$\rho = \rho(x, y, z, t). \tag{5.2.2}$$

The existence of a piecewise continuous† density function giving the total mass by Eq. (1) is postulated as part of the definition of a ***continuous*** medium. The rate of increase of the total mass in the volume is

$$\frac{\partial M}{\partial t} = \int_V \frac{\partial \rho}{\partial t} \, dV. \tag{5.2.3}$$

If no mass is created or destroyed inside V, this must also be equal to the rate of inflow of mass through the surface. The flux or rate of mass ***outflow*** through the element dS of surface at P is $(\rho v_n \, dS)$, where $v_n = \mathbf{v} \cdot \hat{\mathbf{n}}$ is the outward normal component of the velocity [see Eq. (5.1.13)]. Hence the rate of ***inflow*** through S is given by

$$\begin{aligned} \int_S (-\rho v_n) \, dS &= -\int_S \rho \mathbf{v} \cdot \hat{\mathbf{n}} \, dS \\ &= -\int_V \nabla \cdot (\rho \mathbf{v}) \, dV \end{aligned}$$

where the last line follows from the divergence theorem, Eq. (5.1.4b). Equating this to $\partial M / \partial t$, given by Eq. (3), we obtain

$$\int_V \left[\frac{\partial \rho}{\partial t} + \nabla \cdot (\rho \mathbf{v}) \right] dV = 0. \tag{5.2.4}$$

Since the integral of Eq. (4) vanishes for arbitrary choice of the volume V, it follows that the integrand must vanish‡ at each point of a region in which no mass is created or destroyed, assuming as usual that the integrand is a continuous function. The resulting equation, a consequence of the conservation of mass, is known as the ***continuity equation***, well known in fluid dynamics. We have

†Isolated singularities at which ρ becomes infinite could be allowed provided that the integral of Eq. (1) exists.

‡This is demonstrated by the following argument. Suppose the integrand differs from zero at an interior point, say it is positive at P. Then, since it is continuous, it must be positive in a neighborhood of P and the integral over this neighborhood would be positive, contradicting the hypothesis that the integral vanishes for an arbitrary choice of the volume V, and thus proving the falsity of our supposition that there could be an interior point where the integrand was positive.

$$\frac{\partial \rho}{\partial t} + \nabla \cdot (\rho \mathbf{v}) = 0 \quad \text{or} \quad \frac{\partial \rho}{\partial t} + \frac{\partial (\rho v_i)}{\partial x_i} = 0 \quad \text{in rectangular Cartesian coordinates.} \tag{5.2.5}$$

The equation takes a slightly different form when the derivatives of products are written out. We have

$$\frac{\partial (\rho v_i)}{\partial x_i} = \frac{\partial \rho}{\partial x_i} v_i + \rho \frac{\partial v_i}{\partial x_i}.$$

Then by using the notation for the material derivative of Sec. 4.3, namely

$$\frac{d\rho}{dt} = \frac{\partial \rho}{\partial t} + v_i \frac{\partial \rho}{\partial x_i}, \tag{5.2.6}$$

we obtain the following form for the continuity equation. [The equations marked (b) and (c) are the same equation in x-y-z notation and vector notation, respectively.]

Continuity Equation

$$\frac{d\rho}{dt} + \rho \frac{\partial v_i}{\partial x_i} = 0, \quad \text{i.e.,} \tag{5.2.7a}$$

$$\frac{d\rho}{dt} + \rho \left[\frac{\partial v_x}{\partial x} + \frac{\partial v_y}{\partial y} + \frac{\partial v_z}{\partial z} \right] = 0, \quad \text{or} \tag{5.2.7b}$$

$$\frac{d\rho}{dt} + \rho \operatorname{div} \mathbf{v} = 0. \tag{5.2.7c}$$

The ***continuity equation***† in the vector form of Eq. (7c) is independent of any choice of coordinates. Equation (7) shows that the divergence of the velocity vector field equals $-(1/\rho)(d\rho/dt)$. Thus the divergence of the velocity vector measures the rate of flow of material away from the particle and is equal to the unit rate of decrease of density ρ in the neighborhood of the particle.

If the material is incompressible so that the density in the neighborhood of each material particle remains constant as it moves, the continuity equation takes the simpler form

$$\frac{\partial v_i}{\partial x_i} = 0, \tag{5.2.8a}$$

i.e.,

†The general equation is due to Euler (1757), although d'Alembert had obtained it in 1752 for the special case of steady rotationally symmetric motion.

$$\frac{\partial v_x}{\partial x} + \frac{\partial v_y}{\partial y} + \frac{\partial v_z}{\partial z} = 0, \tag{5.2.8b}$$

or

$$\text{div } \mathbf{v} = 0. \tag{5.2.8c}$$

This is the ***condition of incompressiblity***, which has been important in classical hydrodynamics and in plasticity theories.

Material form of the continuity conditions. The continuity equation derived above is for velocity components and density expressed in terms of the spatial coordinates, the formulation most common in fluid mechanics. If material coordinates (X, Y, Z) are used, the conservation of mass implies

$$\int_{V_0} \rho(X, Y, Z, t_0)\, dV_0 = \int_V \rho(x, y, z, t)\, dV, \tag{5.2.9}$$

where V is the volume occupied at time t by the material which occupied V_0 at time t_0. The second integral can be evaluated by a change of the variables of integration† from x, y, z to X, Y, Z. Then

$$\int_V \rho(x, y, z, t)\, dV = \int_{V_0} \rho[x(X, Y, Z, t), y(X, Y, Z, t), z(X, Y, Z, t), t]\, |J|\, dV_0 \tag{5.2.10}$$

where $|J|$ is the absolute value of the ***Jacobian determinant*** J.‡ (We will see that J is in fact positive.)

$$J \equiv |\mathbf{x}/\mathbf{X}| \equiv \begin{vmatrix} \frac{\partial x}{\partial X} & \frac{\partial x}{\partial Y} & \frac{\partial x}{\partial Z} \\ \frac{\partial y}{\partial X} & \frac{\partial y}{\partial Y} & \frac{\partial y}{\partial Z} \\ \frac{\partial z}{\partial X} & \frac{\partial z}{\partial Y} & \frac{\partial z}{\partial Z} \end{vmatrix}. \tag{5.2.11}$$

For brevity we denote the first integrand of Eq. (9) by ρ_0 and the second integrand of Eq. (10) by $\rho|J|$. The conservation of mass expressed by Eq. (9) then becomes

†For information on change of integration variables in a volume integral, see any textbook on advanced calculus.

‡Note that J is the determinant of the matrix of rectangular Cartesian components of $\mathbf{x}\overleftarrow{\nabla}$. $J = \sqrt{III_C}$—see Eq. (4.5.22a). Also $J = \det \mathbf{F}$. See Eq. (4.5.25).

$$\int_{V_0} [\rho_0 - \rho|J|]\, dV_0 = 0$$

for an arbitrary initial volume V_0. The integrand must therefore vanish, whence we obtain the continuity equation in terms of material coodinates, namely

$$\rho|J| = \rho_0. \tag{5.2.12}$$

We suppose that the initial density ρ_0 is everywhere positive in V_0, i.e., there are no empty spaces in V_0. It follows then from Eq. (12) that J cannot vanish during the motion. Since at time $t = t_0$, $J = 1$, we conclude that J is positive throughout the motion so that we can drop the absolute value signs and write

$$\rho J = \rho_0, \tag{5.2.13}$$

constant in time for each particle, the material form of the continuity equation, due to Euler (1762), or

$$\frac{d}{dt}(\rho J) = 0. \tag{5.2.14}$$

This is sometimes interpreted by saying that in the integrals of Eq. (9) $\rho_0\, dV_0 = \rho\, dV$, since $dV = J\, dV_0$ (for positive J), when the integrals are taken over the same mass system. Thus

$$\rho\, dV = \rho_0\, dV_0 \quad \text{or} \quad \frac{d}{dt}(\rho\, dV) = 0 \tag{5.2.15}$$

for a given material element $d\mathscr{M}$. For example, if u is internal energy per unit mass, then the material derivative $d(\rho u\, dV)/dt$ equals simply $(du/dt)\rho\, dV$ and does not require a term with the derivative of $\rho\, dV$.

Equation (14) in material coordinates bears no obvious resemblance to Eq. (7) in spatial coordinates. They must be equivalent, however, since they both express the same physical assumption, the conservation of mass. The equivalence can be seen explicitly by realizing that any time may be chosen as time t_0. By evaluating the derivative of Eq. (14) at t_0, we can show that it reduces to Eq. (7). The details of the rather long computation will be omitted. (Differentiate first with respect to t; then put $t = t_0$, $x = X$, etc., and $\partial x/\partial X = 1$, $\partial x/\partial Y = 0$, etc.)

The more commonly used form of the continuity equation is that of Eq. (7) in terms of spatial coordinates. It is an important fundamental partial differential equation in all branches of continuum mechanics. In Sec. 5.3, the partial differential equations of motion, also valid in all branches of continuum mechanics, will be derived from the momentum principles of a collection of particles. For this it will be easier to use integrals over a given mass of material, instead of integrals over a given spatial volume.

Reynolds transport theorem, material time derivative of volume integral. If the closed surface S_M is defined as the boundary of a given mass system, then it will no longer be fixed in space, but will move with the material. There is no mass transport through such a surface; hence the flux integrals of Eqs. (5.1.12) to (5.1.15) extended over the material boundary will all vanish. Flux of nonmechanical quantities may still occur across the material boundaries.

The time rate of change of an integral is important in the expressions for mass, momentum, and energy balance. For a surface fixed in space, this offers no problem, since for example,

$$\frac{\partial}{\partial t}\int_V \rho\, dV = \int_V \frac{\partial \rho}{\partial t}\, dV,$$

as in Eq. (3). However, if the surface is taken as the boundary of a given mass system, then not only does the integrand change with time, but so does the volume over which the integral is taken. We wish to define a ***material time derivative of a volume integral*** in such a way that it measures the rate of change of the total amount of the quantity carried by a given mass system in space. It is clear that, if d/dt denotes such a material differentiation, then

$$\frac{d}{dt}\int_V \rho\, dV = 0 \qquad \text{by the conservation of mass.}$$

We have not used the notation $\mathscr{M}$ here for the material volume, as in Sec. 5.1, since, just as in Sec. 4.3, we want to calculate the material derivative with the integrand expressed in spatial coordinates. This is done by integrating over the spatial volume V instantaneously occupied by the material of $\mathscr{M}$. If V is the spatial volume bounded by a ***control surface*** S fixed in space, and $\mathscr{A}$ denotes any property of the material, reckoned per unit mass, then

$$\begin{bmatrix}\text{Rate of increase of}\\ \text{the total amount}\\ \text{of } \mathscr{A} \text{ inside the}\\ \text{control surface } S\end{bmatrix} = \begin{bmatrix}\text{Rate of increase of}\\ \text{the amount of } \mathscr{A}\\ \text{possessed by the}\\ \text{material instantan-}\\ \text{eously inside the}\\ \text{control surface}\end{bmatrix} - \begin{bmatrix}\text{Net rate of outward}\\ \text{flux of } \mathscr{A} \text{ carried by}\\ \text{mass transport}\\ \text{through the control}\\ \text{surface}\end{bmatrix}$$

or, if $\mathscr{A}\rho$ and $(\partial/\partial t)(\mathscr{A}\rho)$ are continuous functions,

$$\int_V \frac{\partial}{\partial t}(\mathscr{A}\rho)\, dV = \frac{d}{dt}\int_V \mathscr{A}\rho\, dV - \int_S \mathscr{A}\rho \mathbf{v}\cdot\hat{\mathbf{n}}\, dS. \tag{5.2.16}$$

(Here $\mathscr{A}$ may be a scalar, vector, or tensor.) The first term on the right is the desired material time derivative of $\int_V \mathscr{A}\rho\, dV$. Hence, after solving for it, we have the following expression for the ***material time derivative of the volume integral*** over a spatial volume V bounded by S

$$\frac{d}{dt}\int_V \mathscr{A}\rho\, dV = \int_V \frac{\partial}{\partial t}(\mathscr{A}\rho)\, dV + \int_S \mathscr{A}\rho v_j n_j\, dS. \tag{5.2.17}$$

This is Reynolds transport theorem.† A more convenient form is obtained by transforming the surface integral to a volume integral, using the divergence theorem, Eq. (5.1.4). In rectangular Cartesians $\mathbf{v}\cdot\hat{\mathbf{n}} = v_j n_j$. Hence we obtain

$$\frac{d}{dt}\int_V \mathscr{A}\rho\, dV = \int_V \left[\frac{\partial}{\partial t}(\mathscr{A}\rho) + \frac{\partial}{\partial x_j}(\mathscr{A}\rho v_j)\right] dV.$$

By regrouping the terms in the integral on the right side we may write it as

$$\int_V \left\{\rho\left[\frac{\partial \mathscr{A}}{\partial t} + v_j\frac{\partial \mathscr{A}}{\partial x_j}\right] + \mathscr{A}\left[\frac{\partial \rho}{\partial t} + \frac{\partial(\rho v_j)}{\partial x_j}\right]\right\} dV.$$

The quantity in the last bracket vanishes by Eq. (5), while that in the first bracket is $d\mathscr{A}/dt$. Hence we get the simple result

$$\boxed{\frac{d}{dt}\int_V \rho\mathscr{A}\, dV = \int_V \rho\frac{d\mathscr{A}}{dt}\, dV.} \tag{5.2.18}$$

This may be remembered by noting that in taking the material time derivative of $\rho\mathscr{A}\, dV$, $\rho\, dV = d\mathscr{M}$ is a constant by Eq. (15).‡ Although we used rectangular Cartesian coordinates in deriving Eq. (18) from Eq. (16) the result is independent of any choice of coordinates. Recall that $\mathscr{A}$ may be a scalar point function, a vector, or a tensor.

The material derivative of a material line integral $\int_C f\, dx_k$ can be obtained from Eq. (4.4.10b), which gives $d(dx_k)/dt = v_{k,m}\, dx_m$. Thus§

$$\frac{d}{dt}\int_C f\, dx_k = \int_C \left[\frac{df}{dt}\, dx_k + f v_{k,m}\, dx_m\right]. \tag{5.2.19a}$$

†Asserted by Reynolds in 1903, proved by Spielrein in 1916. See Truesdell and Toupin (1960), Sec. 81.

‡If we were taking the partial time derivative as in Eq. (3) we would need to write

$$\frac{\partial}{\partial t}\int_V \rho\mathscr{A}\, dV = \int_V \frac{\partial}{\partial t}(\rho\mathscr{A})\, dV.$$

§See Truesdell and Toupin (1960), Eq. (79.2). The results are implicit in the work of Kelvin, 1869.

(In terms of material coordinates the curve C is a fixed curve, so that d/dt may be brought inside the integral. But we actually evaluate the material derivative in spatial coordinates and interpret C as the spatial curve instantaneously occupied by the material curve.) A vector form of Eq. (19a) (where $d\mathbf{x} = \hat{\mathbf{t}}\, ds$ and $\hat{\mathbf{t}}$ is the unit tangent) is

$$\frac{d}{dt}\int_C f\,d\mathbf{x} = \int_C \left[\frac{df}{dt}\,d\mathbf{x} + f\mathbf{L}\cdot d\mathbf{x}\right]. \tag{5.2.19b}$$

In Secs. 5.3 and 5.4, Eq. (18) will be used to derive the momentum and energy equations from the momentum and energy balance principles expressed as integrals over a given mass of the material.

EXERCISES

1. Express the surface integral over a closed surface S as a volume integral over the enclosed volume V, if the integrand of the surface integral is: (a) $(x^2 + y^2)n_x$, (b) $\mathbf{w}\cdot\hat{\mathbf{n}}$, (c) $\hat{\mathbf{n}}\cdot\mathbf{T}$, (d) $\hat{\mathbf{n}}\cdot\mathbf{T}\cdot\mathbf{v}$, (e) $e_{rms}\, x_m\, T_{js}\, n_j$.
2. Verify Stokes theorem for the plane area bounded by the square with corners $(0, 0)$, $(b, 0)$, (b, b), $(0, b)$ in the xy-plane, if $v_x = Ay$ and $v_y = v_z = 0$, where A is a constant.
3. If the torsion stress function $\phi\,(x, y)$ of Ex. 9, Sec. 3.2, satisfies the boundary condition $\phi = 0$ on the boundary curve C of a simply connected cross-section, show that the total torque (moment) about the origin is equal to $\int_A 2\phi\, dA$ over the area bounded by C.
4. For the case of ***incompressible*** irrotational flow with $\mathbf{v} = \nabla\phi$, as in Ex. 15, Sec. 4.4, write the continuity equation as a partial differential equation for the velocity potential function ϕ.
5. Use the principle of ***conservation of electric charge*** in a stationary conducting solid to derive an equation of continuity in spatial coordinates involving the following two functions: Scalar q is charge per unit volume, while vector $\mathbf{J} = J_1\mathbf{i}_1 + J_2\mathbf{i}_2 + J_3\mathbf{i}_3$ is current density whose magnitude is the flux of charge per unit area across any surface perpendicular to $\mathbf{J}$.
6. Show that the spatial form, Eq. (5.2.7), follows from the material form, Eq. (5.2.14), of the continuity equation. See the remarks in the paragraph following Eq. (5.2.15).
7. A vector field $\mathbf{v}$ satisfying $\nabla\cdot\mathbf{v} = 0$ is called ***solenoidal.*** A volume-preserving motion is called ***isochoric,*** i.e., a motion for which the density in the neighborhood of any particle remains constant as the particle moves. (The flow of an incompressible fluid is necessarily isochoric, but there may be isochoric flows of compressible fluids.)

(a) Show that in an isochoric motion the velocity field is solenoidal, and conversely.

(b) Show that any velocity field $\mathbf{v}$ given in terms of a vector potential function $\boldsymbol{\psi}$ by $\mathbf{v} = \nabla \times \boldsymbol{\psi}$ is solenoidal and the flow isochoric.

8. For incompressible (or isochoric) plane flow parallel to the xy-plane, the vector potential of Ex. 7(b) has only one component, the z-component, which is independent of z. Thus $\boldsymbol{\psi} = \psi(x, y)\mathbf{k}$, where $\psi(x, y)$ is called the ***stream function.*** Show that the volume flux Q across any plane curve C from point A to point B equals $\psi(B) - \psi(A)$, if the positive normal is from left to right as we look along the curve from A toward B. (By flux across the plane curve we mean, of course, flux through a cylindrical surface of unit height above the plane curve.)

9. Show that if $\Gamma = \oint_C \mathbf{v} \cdot d\mathbf{x}$ is the circulation around a closed curve C, then $d\Gamma/dt = \oint_C \mathbf{a} \cdot d\mathbf{x} + \oint_C \mathbf{v} \cdot d(d\mathbf{x})/dt$, where $\mathbf{a}$ is the acceleration. Here $d\mathbf{x} = \hat{\mathbf{t}}\, ds$ and $\hat{\mathbf{t}}$ is the unit tangent. C is the spatial curve instantaneously occupied by the material curve. Since $d(d\mathbf{x})/dt = d[d\mathbf{x}/dt] = d\mathbf{v}$, show that the second integral vanishes around a closed curve.

10. Combine Eqs. (5.2.7) and (5.2.14) to obtain an expression for dJ/dt in terms of $\mathbf{v}$.

11. Use the result of Ex. 10 and the result of Ex. 16, Sec. 4.5, to show that the material derivative of the vector area element $\hat{\mathbf{n}}\, dS$ of Eq. (4.5.29) is given by $d(\hat{\mathbf{n}}\, dS)/dt = (\nabla \cdot \mathbf{v})\hat{\mathbf{n}}\, dS - \mathbf{L} \cdot \mathbf{n}\, dS$, where $\mathbf{L}$ is the velocity-gradient tensor.

12. Use the result of Ex. 11 to show that

$$\frac{d}{dt}\int_S f\hat{\mathbf{n}}\, dS = \int_S \left[\frac{df}{dt}\hat{\mathbf{n}} + f\{(\nabla \cdot \mathbf{v})\hat{\mathbf{n}} - \mathbf{L} \cdot \hat{\mathbf{n}}\}\right] dS. \qquad (f \text{ is a scalar function.})$$

Write the Cartesian component form of this equation, introducing the components of $\mathbf{L}$ in terms of those of $\mathbf{v}$.

5.3 Momentum principles; equations of motion and equilibrium; couple stresses

The momentum principle for a collection of particles states that the time rate of change of the total momentum of a ***given set of particles*** equals the vector sum of all the ***external*** forces acting on the particles of the set, provided Newton's Third Law of action and reaction governs the internal forces. The continuum form of this principle is a basic ***postulate*** of continuum mechanics. Consider a given mass of the medium, instantaneously occupying a volume V bounded by surface S (as in Fig. 5.6) and acted upon by external surface and body forces $\mathbf{t}$ per unit area and $\mathbf{b}$ per unit mass. The rate of change of the total momentum of the given mass is $(d/dt)\int \rho\mathbf{v}\, dV$, where d/dt denotes the material derivative of the integral. Then the momentum balance expressed by

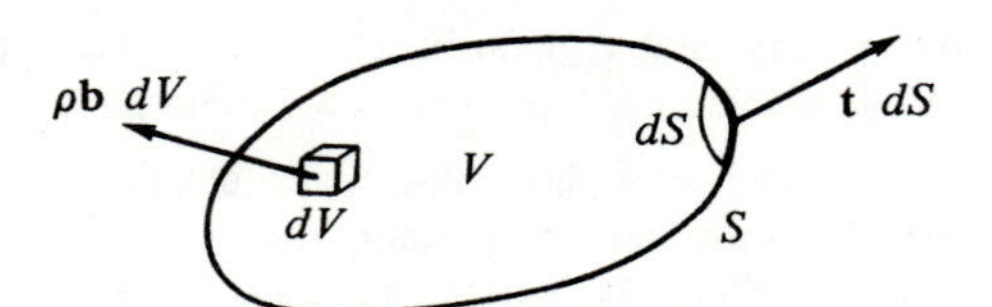

Fig. 5.6 Momentum balance.

the postulate is

$$\int_S \mathbf{t}\, dS + \int_V \rho \mathbf{b}\, dV = \frac{d}{dt}\int_V \rho \mathbf{v}\, dV \tag{5.3.1}$$

or in rectangular coordinates

$$\int_S t_i\, dS + \int_V \rho b_i\, dV = \frac{d}{dt}\int_V \rho v_i\, dV. \tag{5.3.2}$$

Substitute $t_i = T_{ji} n_j$ by Eq. (3.2.8), transform the surface integral by using the divergence theorem, Eq. (5.1.4), and use Eq. (5.2.18) on the right to obtain

$$\int_V \left(\frac{\partial T_{ji}}{\partial x_j} + \rho b_i\right) dV = \int_V \rho \frac{dv_i}{dt}\, dV. \tag{5.3.3}$$

Hence the momentum balance takes the form

$$\int_V \left[\frac{\partial T_{ji}}{\partial x_j} + \rho b_i - \rho \frac{dv_i}{dt}\right] dV = 0$$

for an arbitrary volume V, whence at each point we have

Cauchy's Equations of Motion†

$$\frac{\partial T_{ji}}{\partial x_j} + \rho b_i = \rho \frac{dv_i}{dt} \quad \text{or} \quad \nabla \cdot \mathbf{T} + \rho \mathbf{b} = \rho \frac{d\mathbf{v}}{dt}. \tag{5.3.4}$$

This derivation based on a given material domain is simpler than a possible one based on a given spatial domain, since the basic laws of mechanics apply to a given collection of particles instead of to a given volume of space. A derivation based on a given control surface in space may, however, be convenient for generalizing to momentum balance for a mixture of two or more materials and possibly for inclusion of the nonmaterial momentum of an electromagnetic

†This is Cauchy's first law of motion (1827).

field. For simplicity, we will use only the material-domain derivation for the derivations of the local equations representing the balance of momentum and energy.

Equilibrium equations. In the special case of static equilibrium of the medium, important in solid mechanics, the acceleration $d\mathbf{v}/dt$ is zero and Eqs. (4) reduce to the partial differential equations of equilibrium, where $\mathbf{f} = \rho\mathbf{b}$ is body force per unit volume:

$$\frac{\partial T_{ji}}{\partial x_j} + f_i = 0 \quad \text{or} \quad \nabla \cdot \mathbf{T} + \mathbf{f} = 0. \tag{5.3.5}$$

These equations of equilibrium do not contain any kinematic variables, but they do not in general suffice to determine the stress distribution, even when the boundary tractions and body forces are given, since there are only three first-order partial differential equations for six independent unknown stress components. Additional equations must be found, and these in general include the constitutive equations or stress-strain relations defining the nature of the particular material under consideration. Since the constitutive equations do include kinematic variables, e.g., displacements or strains, the problem of determining the stress distribution cannot in general be separated from the consideration of strains and displacements; it is a statically indeterminate problem except in a few special cases. We have still to prove that in general $T_{ij} = T_{ji}$ so that there really are only six independent stress components instead of nine. This will appear as a consequence of the moment of momentum principle.

The moment of momentum principle. In a collection of particles whose interactions are equal, opposite, and collinear forces, the time rate of change of the total moment of momentum for the given collection of particles is equal to the vector sum of the moments of the external forces acting on the system. In the absence of distributed couples, we postulate the same principle for a continuum. Thus

$$\int_S (\mathbf{r} \times \mathbf{t})\, dS + \int_V (\mathbf{r} \times \rho\mathbf{b})\, dV = \frac{d}{dt}\int_V (\mathbf{r} \times \rho\mathbf{v})\, dV \tag{5.3.6a}$$

or using Eq. (2.3.23) to express the cross products in indicial notation,

$$\int_S e_{rmn} x_m t_n \, dS + \int_V e_{rmn} x_m b_n \rho \, dV = \frac{d}{dt}\int_V e_{rmn} x_m \rho v_n \, dV. \tag{5.3.6b}$$

Substitute $t_n = T_{jn}n_j$. Then transform the surface integral to a volume integral and use Eq. (5.2.18) for the material derivative of a volume integral to obtain

$$\int_V e_{rmn}\left[\frac{\partial(x_m T_{jn})}{\partial x_j} + x_m \rho b_n\right] dV = \int_V e_{rmn} \frac{d}{dt}(x_m v_n)\rho \, dV.$$

Since $dx_m/dt = v_m$, this becomes

$$\int_V e_{rmn}\left[x_m\left(\frac{\partial T_{jn}}{\partial x_j} + \rho b_n\right) + \delta_{mj}T_{jn}\right] dV = \int_V e_{rmn}\left(v_m v_n + x_m \frac{dv_n}{dt}\right)\rho\, dV.$$

Now $e_{rmn}v_m v_n = 0$, since $v_m v_n$ is symmetric in the indices mn while e_{rmn} is antisymmetric, and the last term on the right cancels with the first term on the left by Eq. (4), leaving only (note: $\delta_{mj}T_{jn} = T_{mn}$)

$$\int_V e_{rmn}T_{mn}\, dV = 0$$

for an arbitrary volume V, whence

$$e_{rmn}T_{mn} = 0 \tag{5.3.7a}$$

at each point. This yields

$$\begin{aligned} &\text{for } r = 1 \qquad T_{23} - T_{32} = 0,\\ &\text{for } r = 2 \qquad T_{31} - T_{13} = 0,\\ &\text{for } r = 3 \qquad T_{12} - T_{21} = 0, \end{aligned} \tag{5.3.7b}$$

establishing the symmetry of the stress matrix in general without any assumption of equilibrium or of uniformity of the stress distribution.† The symmetry of the stress matrix is related to the moment of momentum principle in the general nonpolar case (i.e., in the case where there are no assigned traction couples or body couples and no couple stresses) and related to the moment equilibrium condition in the static case.

In the theory of elasticity, it is convenient to identify a material particle of the continuous body by giving its initial coordinates, i.e., its position coordinates before deformation. The position coordinates x, y, z with respect to which the partial derivatives and the integrals are taken in the foregoing derivations are, however, the instantaneous position coordinates. For an elastic body in equilibrium they represent the coordinates of a particle in its new position in the deformed body. When the strains and displacements are small, it may be possible to suppose that Eqs. (5) and (7) apply with sufficient accuracy when x, y, z are interpreted as initial coordinates of the particle, i.e., to suppose that "equilibrium conditions are satisfied in the undeformed configuration of the body." But the equilibrium differential equations are strictly applicable and the stress matrix is strictly symmetric for the nonpolar case only when defined in the instantaneous deformed position, and even in small-strain theory of elasticity it is necessary to take account of this in applications where instability

†The symmetry of the stress matrix is Cauchy's second law of motion (1827).

may occur, as in the buckling of a column or a shell. Asymmetry of the stress matrix also occurs when there is distributed couple stress.

Couple stress.† Up to this point we have assumed the ***nonpolar case*** with no distributed assigned couples per unit area on the boundary surfaces and no distributed assigned body couples (per unit volume or per unit mass), but only external surface traction forces and body forces; and with no couple stresses, i.e., no distributed couples per unit area across internal surfaces. Assigned external couple distributions could result from the action of an external magnetic field on magnetized particles of the material or the action of an electric field on polarized matter. Even without assigned external couples, couple stress can arise from interactions between adjacent parts of the material other than central-force interactions. Couple stresses were included in continuum mechanics by Voigt (1887, 1894) and the Cosserats (1909), but these formulations were not much applied. More recently Truesdell and Toupin (1960) have discussed them; Aero and Kuvshinskii (1960), Toupin (1962), and Mindlin and Tiersten (1962) have considered elastic materials with couple stress; and several authors have considered possible effects on stress concentration.

We assume that the interaction across any internal surface consists of equal, opposite, and collinear forces plus equal and opposite couples. Consider again the Cauchy tetrahedron of Fig. 3.8 (Sec. 3.2) reproduced here in Fig. 5.7. The average couple tractions $-\mathbf{m}^{(1)*}$, $-\mathbf{m}^{(2)*}$, $-\mathbf{m}^{(3)*}$, $\mathbf{m}^*$ per unit area and the average total body couple $\mathbf{c}^*$ (per unit mass) are shown separately in the

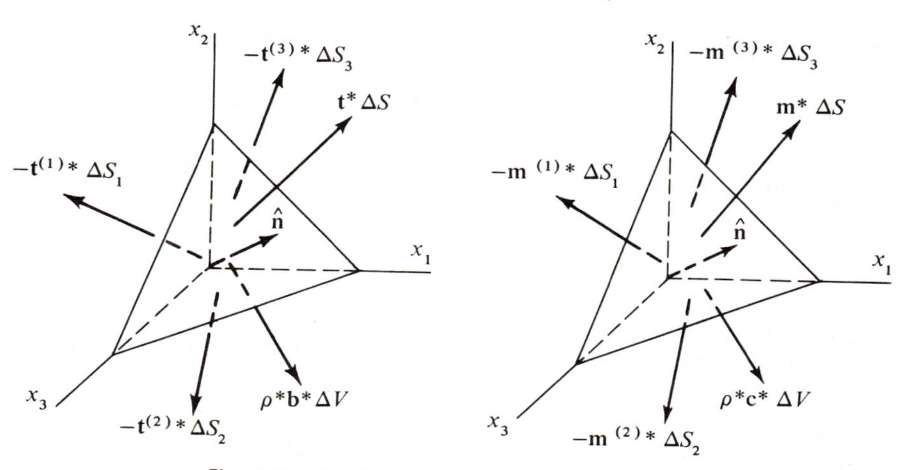

Fig. 5.7 Cauchy tetrahedrons with couple stress.

†This subsection on couple stress may be omitted without interrupting the main line of development.

right-hand figure to avoid confusion, although they act simultaneously on the same body as the tractions and body forces.

The presence of the couples does not affect the consideration of linear momentum, and the limiting process of Sec. 3.2 still yields

$$\mathbf{t} = \mathbf{t}^{(1)}n_1 + \mathbf{t}^{(2)}n_2 + \mathbf{t}^{(3)}n_3 \quad \text{or} \quad t_i = T_{ji}n_j, \tag{5.3.8}$$

where T_{ji} is the ith component of $\mathbf{t}^{(j)}$. We now apply the moment of momentum principle to the mass inside the tetrahedron. We suppose that the particles may possess ***spin angular momentum*** $\mathbf{l}$ (per unit mass) in addition to the moment of linear momentum. (Spin angular momentum is also sometimes called ***intrinsic angular momentum.***) Then the moment of momentum Eq. (6a) becomes

$$\int_S [(\mathbf{r} \times \mathbf{t}) + \mathbf{m}]\, dS + \int_V [(\mathbf{r} \times \mathbf{b}) + \mathbf{c}]\rho\, dV = \frac{d}{dt}\int_V [(\mathbf{r} \times \rho\mathbf{v}) + \mathbf{l}\rho]\, dV, \tag{5.3.9}$$

which we apply to the material contained in the tetrahedron of Fig. 5.7. With $\mathbf{r}$ measured from the vertex of the tetrahedron, it is apparent that for nonvanishing $\mathbf{m}$, $\mathbf{c}$, and $\mathbf{l}$ and for finite $\mathbf{t}$, $\mathbf{b}$, and $\mathbf{v}$ the contributions of the terms containing the cross products are of higher order than the other terms as the altitude h of the tetrahedron tends to zero. We accordingly omit them now and apply the mean-value theorem of the integral calculus to the integrals over the faces and the volume of the tetrahedron. We thus obtain

$$-\mathbf{m}^{(1)*}\,\Delta S_1 - \mathbf{m}^{(2)*}\,\Delta S_2 - \mathbf{m}^{(3)*}\,\Delta S_3 + \mathbf{m}^*\,\Delta S + \mathbf{c}^*\rho^*\,\Delta V = \rho^*\frac{d\mathbf{l}^*}{dt}\,\Delta V.$$

Now substituting $\Delta S_i = n_i\,\Delta S$, $\Delta V = \frac{1}{3}h\,\Delta S$, dividing by ΔS, and letting $h \to 0$, we obtain

$$\mathbf{m} = \mathbf{m}^{(1)}n_1 + \mathbf{m}^{(2)}n_2 + \mathbf{m}^{(3)}n_3, \tag{5.3.10}$$

where the couple traction vector $\mathbf{m}$ on an arbitrary plane is expressed in terms of the couple traction on the coordinate planes through the point by an equation of the same form as Eq. (8) for the traction vector.

We display the nine components of the three couple tractions $\mathbf{m}^{(i)}$ in a square matrix M with components μ_{ij}. (The components of $\mathbf{m}^{(1)}$ are in the first row, etc.) Then

$$m_i = \mu_{ji}n_j \quad \text{or in matrix notation} \quad m = nM \tag{5.3.11a}$$

or

$$\mathbf{m} = \hat{\mathbf{n}} \cdot \mathbf{M} \tag{5.3.11b}$$

in strict analogy with Eq. (8) for **t**. Since the m_i are components of a vector given by Eq. (11) for arbitrary choice of the direction $\hat{\mathbf{n}}$, it follows that **M** is a linear vector operator and hence that μ_{ij} are Cartesian components of a tensor and transform accordingly under changes of the coordinate system (see Sec. 2.4).

The presence of assigned couples and couple stresses does not affect the linear momentum principle; hence the equations of motion, Eqs. (4), still apply. We must, however, reconsider the moment of momentum equation, Eq. (6a), which now takes the form

$$\int_S [(\mathbf{r} \times \mathbf{t}) + \mathbf{m}]\, dS + \int_V [(\mathbf{r} \times \rho\mathbf{b}) + \rho\mathbf{c}]\, dV = \frac{d}{dt}\int_V [(\mathbf{r} \times \rho\mathbf{v}) + \rho\mathbf{l}]\, dV.$$

Note that

$$\frac{d}{dt}\int_V \rho\mathbf{l}\, dV = \int_V \rho \frac{d\mathbf{l}}{dt}\, dV \qquad \text{by Eq. (5.2.18).}$$

Introducing indicial notation as before, with $m_r = \mu_{jr}n_j$, and transforming surface integrals to volume integrals, we find the additional terms $(\partial\mu_{jr}/\partial x_j) + \rho c_r$ in the integrand on the left and $\rho(dl_r/dt)$ in the integrand on the right. Hence Eq. (7) must be replaced by

$$\frac{\partial\mu_{jr}}{\partial x_j} + \rho c_r + e_{rmn}T_{mn} = \rho\frac{dl_r}{dt}, \tag{5.3.12}$$

a set of three partial differential equations expressing the rotational momentum principle in the presence of couple stresses μ_{jr}, body couples c_r per unit mass, and spin angular momentum l_r per unit mass.

The stress tensor is no longer symmetric, but may be expressed as the sum of a symmetric part $T_{(mn)}$ and a skew-symmetric part $T_{[mn]}$:

$$\left.\begin{aligned} T_{(mn)} &= \tfrac{1}{2}(T_{mn} + T_{nm}), \\ T_{[mn]} &= \tfrac{1}{2}(T_{mn} - T_{nm}). \end{aligned}\right\} \tag{5.3.13}$$

The quantity $e_{rmn}T_{mn}$ no longer vanishes, but instead we have

$$\begin{aligned} e_{1mn}T_{mn} &= 2T_{[23]}, \\ e_{2mn}T_{mn} &= 2T_{[31]}, \\ e_{3mn}T_{mn} &= 2T_{[12]}. \end{aligned} \tag{5.3.14}$$

If there is no spin angular momentum in addition to the moment of linear momentum, Eq. (12) takes, even in the dynamic case, the form†

$$\frac{\partial \mu_{jr}}{\partial x_j} + \rho c_r + e_{rmn} T_{mn} = 0. \tag{5.3.15}$$

Some authors express assigned couples as an equivalent skew-symmetric second-order tensor m_{ij} or c_{ij} instead of as a vector m_i or c_i. [Compare this statement with the discussion of Eq. (4.4.8).] For example,

$$m_{ij} = \tfrac{1}{2} e_{ijk} m_k, \tag{5.3.16}$$

or

$$[m_{ij}] = \begin{bmatrix} 0 & \tfrac{1}{2}m_3 & -\tfrac{1}{2}m_2 \\ -\tfrac{1}{2}m_3 & 0 & \tfrac{1}{2}m_1 \\ \tfrac{1}{2}m_2 & -\tfrac{1}{2}m_1 & 0 \end{bmatrix}. \tag{5.3.17}$$

Thus

$$\mathbf{m} = -m_{yz}\mathbf{i} - m_{zx}\mathbf{j} - m_{xy}\mathbf{k}. \tag{5.3.18}$$

The couple stress can also be represented in terms of a skew-symmetric third-order relative tensor $\mu_{kpr} = \tfrac{1}{2} e_{ipr} \mu_{ki}$. See, for example, Eringen (1962), p. 101, Truesdell and Toupin (1960), and Toupin (1962).

Multipolar media. In addition to the couple stresses, other more general multipolar forces have recently been introduced in the rapidly expanding theoretical study of multipolar continuum mechanics, a topic beyond the scope of this book. Special cases of multipolar media have been variously called oriented media, continuous media with microstructure, materials of grade N, Cosserat continua or structured continua. See Jaunzemis (1967), Sec. 11, for an account of the theory.

Equations of motion in the reference state, Piola-Kirchhoff stress tensors. We noted earlier that the Cauchy equations of motion, Eqs. (4), apply to the current deformed configuration. The Cauchy stress tensor field $\mathbf{T}$ is defined as a function of the spatial point $\mathbf{x}$, and it is a symmetric tensor in the nonpolar case (no couple stress or assigned couples). A suitable strain measure to use with the Cauchy stress tensor would therefore be one of the strain or deformation tensors of the Eulerian formulation in terms of the spatial position in the deformed configuration. But, as was pointed out at the beginning

†This is equivalent to Eq. (11) of Aero and Kuvshinskii (1960). Their last term appears to be equivalent to $e_{rmn} T_{nm}$, the negative of our last term, because their stress component convention differs—with $t_n = T_{nj} n_j$ instead of $t_n = T_{jn} n_j$.

of Sec. 4.5, the Lagrangian or material formulation is often preferred in elasticity, where it is assumed that there exists a natural state to which the body would return when it is unloaded. When we define strain as a function of the material point $\mathbf{X}$ in a reference state other than the current configuration (not necessarily the natural state just mentioned, though this is the usual reference state) we need also to express the stress as a function of the material point $\mathbf{X}$ and derive equations of motion in the reference state. The two Piola-Kirchhoff stress tensors† discussed below are two alternative definitions of stress in the reference state. The first Piola-Kirchhoff stress tensor (sometimes called the ***Lagrangian stress tensor***) is the simpler one to define and leads to a simple form of the equations of motion, Eqs. (27) below, quite similar to Eq. (4), except that the derivatives are now with respect to material coordinates instead of spatial coordinates. But this tensor has the disadvantage of being not symmetric and therefore awkward to use in constitutive equations with a symmetric strain tensor. The second Piola-Kirchhoff tensor is symmetric whenever the Cauchy stress tensor is symmetric (nonpolar case) and has been preferred in finite-strain elasticity formulations even though it leads to a more complex form for the equations of motion or equilibrium.

It is not merely a question of a change of variables in the equations of motion, expressing the Cauchy stress component T_{ij} as a function of $\mathbf{X}$ by substituting $\mathbf{x} = \mathbf{x}(\mathbf{X}, t)$, and writing, for example,

$$\frac{\partial T_{ji}}{\partial x_j} = \frac{\partial T_{ji}}{\partial X_P}\frac{\partial X_P}{\partial x_j}$$

but rather ***we seek expressions for a force per unit undeformed area.***

The basic ideas of the two Piola-Kirchhoff stress tensors can be indicated in terms of the force vectors illustrated in Fig. 5.8; precise definitions will be given below.

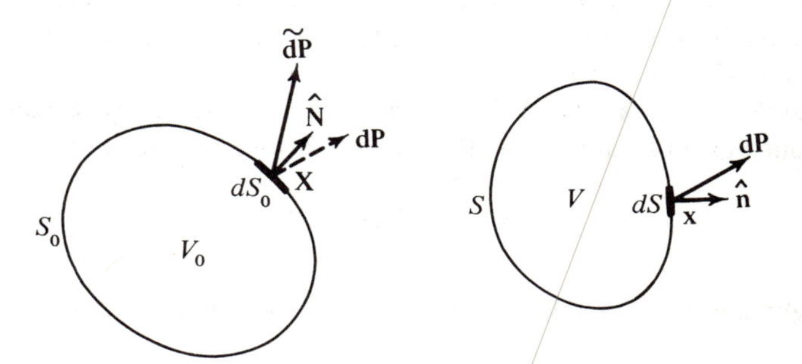

Fig. 5.8 Force vectors for Piola-Kirchhoff stress definitions.

†Due to Piola in 1833 and Kirchhoff in 1852. The equations of motion in terms of these tensors and material coordinates [Eqs. (27) and (28), this section] were given by Piola in 1833. See Truesdell and Toupin (1960), Sec. 210.

The first Piola-Kirchhoff stress tensor $\mathbf{T}^0$ gives the actual force $\mathbf{dP}$ on the deformed dS, but it is reckoned per unit area of the undeformed dS_0 and expresses the force in terms of the normal $\hat{\mathbf{N}}$ to dS_0 at $\mathbf{X}$. Thus

$$(\hat{\mathbf{N}}\cdot\mathbf{T}^0)\,dS_0 = \mathbf{dP} = (\hat{\mathbf{n}}\cdot\mathbf{T})\,dS. \tag{5.3.19}$$

The second Piola-Kirchhoff stress tensor $\tilde{\mathbf{T}}$ is formulated somewhat differently. Instead of the actual force $\mathbf{dP}$ on dS it gives a ***force*** $\widetilde{\mathbf{dP}}$ ***related to the force*** $\mathbf{dP}$ ***in the same way that a material vector*** $d\mathbf{X}$ ***at*** $\mathbf{X}$ ***is related by the deformation to the corresponding spatial vector*** $d\mathbf{x}$ ***at*** $\mathbf{x}$. That is,

$$\widetilde{\mathbf{dP}} = \mathbf{F}^{-1}\cdot\mathbf{dP} \quad \text{just as} \quad d\mathbf{X} = \mathbf{F}^{-1}\cdot d\mathbf{x}, \quad \text{where} \quad \mathbf{F}^{-1} = \mathbf{X}\overleftarrow{\nabla}_{\mathbf{x}}. \tag{5.3.20a}$$

$\mathbf{F}^{-1}$ is the spatial or inverse deformation gradient of Eq. (4.5.2b). This relationship has been illustrated schematically in the two-dimensional Fig. 5.8 where $\widetilde{\mathbf{dP}}$ and dS_0 have been stretched and rotated the same amount relative to the final positions $\mathbf{dP}$ and dS, respectively. Then

$$(\hat{\mathbf{N}}\cdot\tilde{\mathbf{T}})\,dS_0 = \widetilde{\mathbf{dP}} = \mathbf{F}^{-1}\cdot\mathbf{dP} = \mathbf{F}^{-1}\cdot(\hat{\mathbf{n}}\cdot\mathbf{T})\,dS = (\hat{\mathbf{n}}\cdot\mathbf{T})\cdot(\mathbf{F}^{-1})^T dS. \tag{5.3.20b}$$

The Piola-Kirchhoff stress tensors are sometimes called ***pseudo-stress*** tensors, and we can define ***pseudo-traction vectors*** $\mathbf{t}^0$ and $\tilde{\mathbf{t}}$ such that

$$\left.\begin{aligned} &\mathbf{t}^0\,dS_0 = \mathbf{dP} = \mathbf{t}\,dS \qquad \tilde{\mathbf{t}}\,dS_0 = \widetilde{\mathbf{dP}} = \mathbf{F}^{-1}\cdot\mathbf{t}\,dS, \\ &\text{where} \\ &\mathbf{t}^0 = \hat{\mathbf{N}}\cdot\mathbf{T}^0 = \mathbf{F}\cdot\tilde{\mathbf{t}} \qquad \tilde{\mathbf{t}} = \hat{\mathbf{N}}\cdot\tilde{\mathbf{T}} = \mathbf{F}^{-1}\cdot\mathbf{t}^0. \end{aligned}\right\} \tag{5.3.21}$$

Thus $\mathbf{t}^0$ is the force acting on the deformed element ***per unit undeformed area,*** while $\tilde{\mathbf{t}}$ is the dot product of $\mathbf{F}^{-1}$ into the force acting on the deformed element, per unit of undeformed area.

Expressions for the Piola-Kirchhoff stresses in terms of Cauchy stress follow from Eqs. (19) and (20b). By Eq. (4.5.29) we can substitute

$$\hat{\mathbf{n}}\,dS = \frac{\rho_0}{\rho}\hat{\mathbf{N}}\cdot\mathbf{F}^{-1}\,dS_0$$

on the right side of Eq. (19) and divide by dS_0 to obtain, for arbitrary unit vector $\hat{\mathbf{N}}$,

$$\hat{\mathbf{N}}\cdot\left[\mathbf{T}^0 - \frac{\rho_0}{\rho}\mathbf{F}^{-1}\cdot\mathbf{T}\right] = 0, \quad \text{whence}$$

$$\mathbf{T}^0 = \frac{\rho_0}{\rho}\mathbf{F}^{-1}\cdot\mathbf{T} \quad \text{with rectangular components} \quad T^0_{Ji} = \frac{\rho_0}{\rho}X_{J,r}T_{ri}. \tag{5.3.22}$$

$\mathbf{T}^0$ is a two-point tensor (see appendix Sec. I.7) with components T^0_{Ji} associating force vector components dP_i at $\mathbf{x}$ with vector area components $N_J dS_0$ at $\mathbf{X}$. Similarly, for the second Piola-Kirchhoff tensor $\tilde{\mathbf{T}}$, starting with Eq. (20b), we obtain

$$(\hat{\mathbf{N}}\cdot\tilde{\mathbf{T}})\,dS_0 = \mathbf{F}^{-1}\cdot\left[\frac{\rho_0}{\rho}\hat{\mathbf{N}}\cdot dS_0\mathbf{F}^{-1}\cdot\mathbf{T}\right] = \left[\hat{\mathbf{N}}\cdot\frac{\rho_0}{\rho}dS_0\mathbf{F}^{-1}\cdot\mathbf{T}\right]\cdot(\mathbf{F}^{-1})^T$$

$$\text{or}\quad N_J\tilde{T}_{JI} = X_{I,i}\frac{\rho_0}{\rho}N_J X_{J,s}T_{si}$$

for arbitrary unit vectors $\hat{\mathbf{N}}$, whence

$$\tilde{T}_{JI} = \frac{\rho_0}{\rho}X_{J,s}T_{si}\partial_i X_I \quad\text{or}\quad \tilde{T}_{JI} = X_{I,i}T^0_{Ji} = T^0_{Ji}\partial_i X_I$$

by Eq. (22). Thus

$$\tilde{\mathbf{T}} = \frac{\rho_0}{\rho}\mathbf{F}^{-1}\cdot\mathbf{T}\cdot(\mathbf{F}^{-1})^T \quad\text{or}\quad \tilde{\mathbf{T}} = \mathbf{T}^0\cdot(\mathbf{F}^{-1})^T. \tag{5.3.23}$$

The first Eq. (23) shows that $\tilde{\mathbf{T}}$ is symmetric whenever $\mathbf{T}$ is symmetric (nonpolar case), but Eq. (22) shows that $\mathbf{T}^0$ is in general not symmetric.

The ***inverse relationship*** to Eq. (22) can be obtained by multiplying the component form by $\partial x_j/\partial X_J$ and noting that $(\partial x_j/\partial X_J)(\partial X_J/\partial x_r) = \delta_{jr}$. Thus

$$T_{ji} = \frac{\rho}{\rho_0}x_{j,J}T^0_{Ji} \quad\text{or}\quad \mathbf{T} = \frac{\rho}{\rho_0}\mathbf{F}\cdot\mathbf{T}^0. \tag{5.3.24}$$

A similar inversion of the second Eq. (23) yields, after a change of indices,

$$T^0_{Ji} = \tilde{\mathbf{T}}_{JI}\partial_I x_i \quad\text{or}\quad \mathbf{T}^0 = \tilde{\mathbf{T}}\cdot\mathbf{F}^T. \tag{5.3.25}$$

Substitution of this into Eq. (24) gives finally an expression for the Cauchy stress in terms of the second Piola-Kirchhoff stress tensor $\tilde{\mathbf{T}}$:

$$T_{ji} = \frac{\rho}{\rho_0}x_{j,J}\tilde{T}_{JI}\partial_I x_i \quad\text{or}\quad \mathbf{T} = \frac{\rho}{\rho_0}\mathbf{F}\cdot\tilde{\mathbf{T}}\cdot\mathbf{F}^T. \tag{5.3.26}$$

The equations of motion in the reference state are derived from the condition that the vector sum of the external forces on the material in V, which initially occupied V_0, is equal to the rate of change of the momentum. (Note that, although the integrals are over V_0, the forces act on the material in V and the momentum is the momentum in V at time t.) Here $\mathbf{b}_0 = \mathbf{b}[\mathbf{x}(\mathbf{X})]$ is denoted $\mathbf{b}_0$ for brevity:

$$\int_{S_0} \hat{\mathbf{N}} \cdot \mathbf{T}^0 \, dS_0 + \int_{V_0} \rho_0 \mathbf{b}_0 \, dV_0 = \int_{V_0} \rho_0 \frac{d^2\mathbf{x}}{dt^2} \, dV_0.$$

Transformation of the surface integral to a volume integral by the divergence theorem, and the requirement that the equation hold for arbitrary volumes V_0, lead to the equation of motion for $\mathbf{T}^0$.

$$\vec{\nabla}_{\mathbf{X}} \cdot \mathbf{T}^0 + \rho_0 \mathbf{b}_0 = \rho_0 \frac{d^2\mathbf{x}}{dt^2} \quad \text{or} \quad \frac{\partial T^0_{Ji}}{\partial X_J} + \rho_0 b_{0i} = \rho_0 \frac{d^2 x_i}{dt^2}, \tag{5.3.27}$$

which is similar in form to Eq. (4) for the Cauchy stress tensor in spatial coordinates. The equation of motion for the second Piola-Kirchhoff tensor $\tilde{\mathbf{T}}$ results when Eq. (25) is substituted into Eq. (27).

$$\vec{\nabla} \cdot [\tilde{\mathbf{T}} \cdot \mathbf{F}^T] + \rho_0 \mathbf{b}_0 = \rho_0 \frac{d^2\mathbf{x}}{dt^2} \quad \text{or} \quad \partial_J [\tilde{T}_{JI} \partial_I x_i] + \rho_0 b_{0i} = \rho_0 \frac{d^2 x_i}{dt^2}. \tag{5.3.28}$$

The transpose of $\mathbf{T}^0$ is used by some writers instead of $\mathbf{T}^0$. Truesdell and Noll (1965) use the notation $\mathbf{T}_R$ for the transpose of our $\mathbf{T}^0$:

$$\mathbf{T}_R = (\mathbf{T}^0)^T. \tag{5.3.29}$$

They are led naturally to this choice because their Cauchy stress is also the transpose of ours. The first subscript on a component of their T_{ij} identifies the component, while the second identifies the plane on which it acts, reversing our convention of Fig. 3.6.

When one of the strain or deformation tensors of the Lagrangian formulation in terms of the reference state is used in finite-deformation elasticity, one of the two Piola-Kirchhoff tensors is appropriate to use as the stress variable in the stress-strain law. The appropriate one to use will be the one which appears as a conjugate variable, as in Eqs. (5.4.8).

In Sec. 5.4 the energy balance of the first law of thermodynamics will lead to another field equation, the energy equation.

EXERCISES

1. (a) Write out the component equations of Eq. (5.3.4) without using the summation convention.
   (b) Write out the plane stress equilibrium equations using the notation σ_x, σ_y, and τ_{xy}.
   (c) For the plane stress state $\sigma_x = 12\,Ax^2y$, $\sigma_y = By^3$, $\tau_{xy} = -Cxy^2$, determine the constants B and C in terms of A for equilibrium with negligible body forces.

2. Derive the equilibrium partial differential equations, Eqs. (5.3.5), by beginning with the balance of forces in integral form, instead of as a special case of the dynamic equations.

3. If in plane stress with no body forces it is assumed that

$$\sigma_x = \frac{\partial^2 \phi}{\partial y^2} \qquad \sigma_y = \frac{\partial^2 \phi}{\partial x^2} \qquad \tau_{xy} = -\frac{\partial^2 \phi}{\partial x\, \partial y}$$

where $\phi(x, y)$ is the ***Airy stress function***, an unknown function to be determined, show that the equilibrium equations are identically satisfied no matter what sufficiently differentiable function is used for $\phi(x, y)$.

4. In an ideal nonviscous fluid there can be no shear stress. Hence, the stress tensor is entirely hydrostatic, $T_{ij} = -p\delta_{ij}$. Show that this leads to the following form, known as Euler's equation of motion for a frictionless fluid:

$$-\frac{1}{\rho}\nabla p + \mathbf{b} = \frac{\partial \mathbf{v}}{\partial t} + \mathbf{v} \cdot \vec{\nabla} \mathbf{v}.$$

5. For the case of uniform body forces show that the plane stress equilibrium equations imply that

$$2\frac{\partial^2 \tau_{xy}}{\partial x\, \partial y} = -\frac{\partial^2 \sigma_x}{\partial x^2} - \frac{\partial^2 \sigma_y}{\partial y^2}.$$

6. Use the result of Ex. 5 to show that the stress compatibility equation obtained in Ex. 3, Sec. 4.7, can be brought into the form $[(\partial^2/\partial x^2) + (\partial^2/\partial y^2)](\sigma_x + \sigma_y) = 0$, if, in small-displacement elasticity, we ignore the distinction between the spatial coordinates x, y and material coordinates X, Y.

7. Use the result of Ex. 6 to obtain a partial differential equation for the Airy stress function of Ex. 3.

8. Show that the stresses given by the torsion stress function $\phi(x, y)$ of Ex. 9(b), Sec. 3.2, namely $T_{zx} = \partial\phi/\partial y$ and $T_{zy} = -\partial\phi/\partial x$, satisfy identically the equilibrium differential equations for no body forces.

9. If Hooke's law gives $T_{zx} = 2G\epsilon_{zx}$ and $T_{zy} = 2G\epsilon_{zy}$, determine the partial differential equation that must be satisfied by the torsional warping function $f(x, y)$ of Ex. 6(b), Sec. 4.7, in order that these stresses satisfy the equilibrium differential equations for no body forces.

10. Derive the partial differential equations of motion [Eqs. (5.3.4)], by a momentum balance for a fixed volume V bounded by a fixed control surface S. *Hint*: The rate of change of the ith component of momentum inside S is $(\partial/\partial t)\int_V \rho v_i\, dV$, which must be equal to the inward flux of ith component momentum carried by the mass flux plus the total ith component of external forces.

11. When the displacements and displacement gradients are small, the second Piola-Kirchhoff stress $\tilde{\mathbf{T}}$ is approximately equal to the Cauchy stress $\mathbf{T}$, since, by Eq. (5.3.23),

$$\tilde{\mathbf{T}} = \frac{\rho_0}{\rho}\mathbf{F}^{-1} \cdot \mathbf{T} \cdot (\mathbf{F}^{-1})^T, \quad \text{while} \quad \frac{\rho_0}{\rho} \approx 1, \quad \mathbf{F}^{-1} \approx \mathbf{1} \approx (\mathbf{F}^{-1})^T$$

under the small-displacement assumption. Thus for small vibrations of an elastic system Eq. (5.3.28) reduces to

$$\frac{\partial T_{ji}}{\partial X_j} + \rho_0 b_{0i} = \rho_0 \frac{\partial^2 u_i}{\partial t^2} .$$

Assume the Hooke's law $T_{ij} = \lambda e \delta_{ij} + 2G\epsilon_{ij}$, where $e = \epsilon_{kk}$ and λ and G are the Lamé elastic moduli. Show that the first equation of motion then leads to the following displacement equation of motion:

$$(\lambda + G)\frac{\partial e}{\partial X} + G\nabla^2 u_x + \rho_0 b_{0x} = \rho_0 \frac{\partial^2 u_x}{\partial t^2},$$

and write the vector equation of which this is the first component.

12. Show that in an ***incompressible*** Newtonian fluid whose constitutive equation is $T_{ij} = -p\delta_{ij} + 2\mu D_{ij}$, where μ is the coefficient of viscosity, the x_k-component of the total surface force is given by

$$\int_S t_k \, dS = \int_V \left[-\frac{\partial p}{\partial x_k} + \mu \frac{\partial^2 v_k}{\partial x_j \, \partial x_j} \right] dV.$$

Write a vector form of this rectangular Cartesian component result.

13. Write out the ***three*** components of Eqs. (5.3.28) for the case of ***equilibrium*** under a plane deformation defined by $x_3 = X_3$ and

$$x_1 = X_1 + \tfrac{1}{2}A_{11}X_1^2 + A_{12}X_1X_2 + \tfrac{1}{2}A_{22}X_2^2,$$
$$x_2 = X_2 + \tfrac{1}{2}B_{11}X_1^2 + B_{12}X_1X_2 + \tfrac{1}{2}B_{22}X_2^2$$

where the A_{ij} and B_{ij} are known constants.

14. Modify Ex. 13 by supposing that the A_{ij} and B_{ij} are given functions of the time instead of constants. How do the resulting equations of motion differ from the equations of equilibrium found in Ex. 13?

15. Verify the correctness of Eqs. (5.3.14), and write out the three component Eqs. (5.3.12), using the results of Eqs. (5.3.14).

16. (a) Define couple-stress tensors $\mathbf{M}^0$ and $\tilde{\mathbf{M}}$ as functions of the material point $\mathbf{X}$ in the reference state analogous to the Piola-Kirchhoff stresses $\mathbf{T}^0$ and $\tilde{\mathbf{T}}$.
    (b) Develop expressions relating $\mathbf{M}, \mathbf{M}^0$, and $\tilde{\mathbf{M}}$ analogous to the equations relating $\mathbf{T}, \mathbf{T}^0$, and $\tilde{\mathbf{T}}$.

17. Develop the version of Eq. (5.3.12) valid in the reference state using the tensors: (a) $\mathbf{M}^0$, and (b) $\tilde{\mathbf{M}}$, defined in Ex. 16.

5.4 Energy balance; first law of thermodynamics; energy equation

An additional field equation, the energy equation, is a consequence of the energy-balance postulate of the first law of thermodynamics. Since the energy equation involves an additional unknown quantity, the ***internal energy***, the

where $\mathbf{q}$ is the heat flux vector. The negative sign is needed because $\int_S \mathbf{q}\cdot\hat{\mathbf{n}}\,dS$ is the outward heat flux. When this is combined with the input power it is necessary to write the total input as $P_{\text{input}} + JQ_{\text{input}}$, where J is the mechanical equivalent of heat if Q is expressed in thermal units. It is customary, however, to omit the J in the equations with the understanding that Q_{input} and P_{input} have been expressed in the same units—for example, by multiplying the thermal values of $\mathbf{q}$ and r by J.

First law of thermodynamics. Generalizing from many experimental observations it is found that when a system is carried through a cycle and returned to its initial state

$$\oint P_{\text{input}}\,dt \neq 0 \qquad \oint Q_{\text{input}}\,dt \neq 0$$

where $\oint dt$ denotes the integral throughout the cycle. Hence, there is no work-function of which $(P_{\text{input}}\,dt)$ is an exact differential and likewise no heat-function of which $(Q_{\text{input}}\,dt)$ is an exact differential. It is not possible to speak of the work content or the heat content of the system at any one time; work and heat are not state functions or system properties; they exist only in the form of energy being transferred and have no individual identities in the system. On the other hand, it is found that in any cycle

$$\oint [P_{\text{input}} + Q_{\text{input}}]\,dt = 0, \tag{5.4.4a}$$

which shows that there does exist a function, called the ***total energy of the system***, E_{total}, such that

$$\dot{E}_{\text{total}} = P_{\text{input}} + Q_{\text{input}}, \tag{5.4.4b}$$

and

$$dE_{\text{total}} = (P_{\text{input}} + Q_{\text{input}})\,dt \tag{5.4.4c}$$

is an exact differential. Then in any change from state 1 to state 2 the change in the total energy of the system is

$$\Delta E_{\text{total}} \equiv (E_{\text{total}})_2 - (E_{\text{total}})_1 = \int_{t_1}^{t_2} [P_{\text{input}} + Q_{\text{input}}]\,dt. \tag{5.4.4d}$$

Equations (4) are alternative expressions of the first law of thermodynamics.†

†The interconvertibility of heat and mechanical work was known to Carnot by 1832. It was unequivocally stated and verified experimentally by Joule in 1843 and 1845. A very general statement of the first law was given by Duhem in 1892.

(In most thermodynamics textbooks the principle is stated in terms of power output instead of input, so that instead of $\dot{E} = Q + P$, the equation is $\dot{E} = Q - P$, a more convenient form for dealing with heat engines.)

The total energy of the system will be considered as the sum of two parts, the ***kinetic energy*** K and ***internal energy*** U. By kinetic energy, we mean the macroscopic kinetic energy associated with the usual macroscopically observable velocity of the continuum. The kinetic energy of the random thermal motions of molecules, associated with temperature measurements instead of velocity measurements, is considered a part of the internal energy. The internal energy also includes stored elastic energy and possibly other forms of energy not specified explicitly.

The internal energy per unit mass or ***specific internal energy*** is denoted by u; then ρu is internal energy per unit volume. The first law, Eq. (4b), with use of Eqs. (2) and (3) now takes the form

$$\frac{d}{dt}\int_V \left[\frac{1}{2}\rho\mathbf{v}\cdot\mathbf{v} + \rho u\right] dV = \frac{d}{dt}\int_V \left[\frac{1}{2}\rho\mathbf{v}\cdot\mathbf{v}\right] dV + \int_V \mathbf{T}:\mathbf{D}\, dV$$
$$- \int_S \mathbf{q}\cdot\hat{\mathbf{n}}\, dS + \int_V \rho r\, dV. \tag{5.4.5}$$

When the surface integral is transformed by the divergence theorem, and all terms are collected on the left side of the equation, the result is

$$\int_V \left[\nabla\cdot\mathbf{q} + \rho\frac{du}{dt} - \mathbf{T}:\mathbf{D} - \rho r\right] dV = 0 \tag{5.4.6}$$

for arbitrary choice of the volume V within the continuous medium. Hence, the integrand must vanish at each point of the medium, establishing the energy equation, Eq. (7), as the field equation expressing at each point the conservation of energy implied by the first law of thermodynamics.

Energy Equation (nonpolar case)†

$$\rho\frac{du}{dt} = \mathbf{T}:\mathbf{D} + \rho r - \nabla\cdot\mathbf{q} \quad \text{or}$$
$$\rho\frac{du}{dt} = T_{ij}D_{ij} + \rho r - \frac{\partial q_j}{\partial x_j} \quad \text{Cartesian components.} \tag{5.4.7}$$

†Unfortunately notations are not at all standardized. Truesdell and Toupin (1960) and Truesdell and Noll (1965) use q for internal heat supply strength instead of our r and denote by the vector $\mathbf{h}$ the negative of the usual heat flux vector $\mathbf{q}$, giving a heat flux vector $\mathbf{h}$ pointing from low temperature toward higher temperature. Their divergence term would then be $+\nabla\cdot\mathbf{h}$ on the right side of Eq. (7). Eringen (1962) uses $\mathbf{q}$ for what they call $\mathbf{h}$, the negative of our $\mathbf{q}$, and h for the internal supply strength, our r. Fung (1965) uses $\mathbf{h}$ for our $\mathbf{q}$, and does not include a supply term.

The energy equation and the method of derivation are due to Kirchhoff (1868). Although his original treatment was for the infinitesimal motion of a perfect gas, he generalized the result in 1894.

The energy equation is the internal energy balance for the volume element. The terms on the right represent that part of the input power (per unit volume) not converted into kinetic energy plus ρr, the internal supply of heat per unit volume, and the negative divergence $-\nabla \cdot \mathbf{q}$, the inflow per unit volume of heat through the boundaries of the element. The left side $\rho\, du/dt$ is the rate of increase of the internal energy ρu per unit volume. [Recall that $\rho\, dV$ is constant by the continuity equation, Eq. (5.2.15); hence, $d(\rho u\, dV)/dt = \rho(du/dt)\, dV$.] {In many elementary textbooks the equation is stated only for a nonviscous fluid in which $T_{ij} = -p\delta_{ij}$ and $T_{ij}D_{ij} = -pD_{ii} = -p\rho\dot{v}$, where† v is the specific volume or volume per unit mass, leading to the usual form $du = đq - p\, dv$, per unit mass, where $đq$ denotes the net heat input to the element per unit mass, equal to $[r - (\nabla \cdot \mathbf{q}/\rho)]$ in the notation of Eq. (7). The symbol $đ$ is used for $đq$ to emphasize that it is not the differential of any function. Even in the absence of viscosity this form is inadequate for solids exhibiting elasticity in shear.}

In ideal elasticity, heat transfer is considered insignificant, and all of the input work is assumed converted into internal energy in the form of recoverable stored elastic strain energy, which can be recovered as work when the body is unloaded. In general, however, the major part of the input work into a deforming material is not recoverably stored, but dissipated by the deformation process, causing an increase in the body's temperature and eventually being conducted away as heat.

Generalization to include electrical and other energies. The first law in the forms of Eqs. (4) is quite general if E_{total} is interpreted to include all kinds of energy content and P_{input} includes all kinds of energy input except heat transfer. But Eq. (2) accounts only for mechanical power input, and hence, when other kinds of power input are present additional terms will appear in the energy equation. For some purposes these may be avoided by not including the other energies in the balance; for example, if the only pertinent effect of an electric current flowing through a solid body is the internal heating produced, it may be possible to account for that input through the internal heat supply term ρr. Another approach sometimes used is to call $\mathbf{q}$ the ***nonmechanical energy flux*** instead of the heat flux and thus to include all other energy inputs with the term $-\int_S \mathbf{q} \cdot \hat{\mathbf{n}}\, dS$ and include all other energy content in the term $\int_V \rho u\, dV$ in Eq. (5), making that equation and Eq. (7) quite general (possibly even without the internal supply term ρr). But we will want to isolate the heat flux in order to relate it to the temperature gradient by the constitutive equation of heat conduction and in order to express the entropy flux in terms of the heat flux in Sec. 5.6. For the time being, we do not formulate the extra terms needed when there are electromagnetic inputs. See, for example, Truesdell and Toupin (1960), Secs. 284 to 700.

†The last equality follows from the continuity equation.

Stress power in the reference configuration (nonpolar case). The stress power $\int_V \mathbf{T} : \mathbf{D}\, dV$ in a volume V of the current deformed configuration can be referred back to the volume V_0 occupied by the same material in the reference configuration, e.g., the natural state of an elastic body. This takes a particularly simple form in terms of the second Piola-Kirchhoff stress tensor $\tilde{\mathbf{T}}$ and the finite-strain rate $d\mathbf{E}/dt$. We treat here only the nonpolar case (no couple stresses or assigned couples). In the nonpolar case, $\mathbf{T}$ is a symmetric tensor. We recall that

$$dV = J\, dV_0 = \frac{\rho_0}{\rho}\, dV_0, \qquad \text{see Eq. (5.2.12),}$$

and that, by Eq. (5.3.26),

$$T_{ij} = \frac{\rho}{\rho_0} \frac{\partial x_i}{\partial X_I} \frac{\partial x_j}{\partial X_J} \tilde{T}_{IJ}.$$

Hence

$$\int_V T_{ij} D_{ij}\, dV = \int_{V_0} \tilde{T}_{IJ} \frac{\partial x_i}{\partial X_I} \frac{\partial x_j}{\partial X_J} D_{ij}\, dV_0.$$

But, by Eq. (4.5.13),

$$\frac{\partial x_i}{\partial X_I} \frac{\partial x_j}{\partial X_J} D_{ij} = \frac{dE_{IJ}}{dt} = \frac{1}{2} \frac{dC_{IJ}}{dt}.$$

Thus

$$\int_V T_{ij} D_{ij}\, dV = \int_{V_0} \tilde{T}_{IJ} \frac{dE_{IJ}}{dt}\, dV_0 = \int_{V_0} \tilde{\mathbf{T}} : \frac{d\mathbf{E}}{dt}\, dV_0 = \frac{1}{2} \int_{V_0} \tilde{\mathbf{T}} : \frac{d\mathbf{C}}{dt}\, dV_0, \tag{5.4.8a}$$

and the ***stress power per unit undeformed volume*** is $\tilde{\mathbf{T}} : d\mathbf{E}/dt$, which shows that $\tilde{\mathbf{T}}$ and $d\mathbf{E}/dt$ are conjugate variables in the reference state, as are $\mathbf{T}$ and $\mathbf{D}$ in the deformed configuration.†

By beginning with the stress power per unit deformed volume as $T_{ji}(\partial v_i/\partial x_j)$ as in Eq. (2a), and introducing the first Piola-Kirchhoff tensor $\mathbf{T}^0$ instead of $\tilde{\mathbf{T}}$, it can be shown that

$$\int_V T_{ji} v_{i,j}\, dV = \int_{V_0} T^0_{Ji} \frac{d}{dt}(x_{i,J})\, dV_0 = \int_{V_0} \mathbf{T}^0 \cdot\cdot \dot{\mathbf{F}}\, dV_0 = \int_{V_0} \mathbf{T}_R : \dot{\mathbf{F}}\, dV_0, \tag{5.4.8b}$$

where $\mathbf{T}_R$ is the transpose of $\mathbf{T}^0$. Details are left as an exercise.

†Expressions for the stress power in the reference state were given by Kirchhoff in 1852.

Remark on internal energy as a state function. The conclusion in Eqs. (4) that there exists a state function E_{total} such that $(P_{\text{input}} + Q_{\text{input}})\, dt$ is its exact differential, being based on the assumed vanishing of the integral in an arbitrary cycle, tacitly assumes that the material is capable of being carried through a cycle and restored to its initial state. The postulate is a generalization from experimental observations on a gas whose state (for a given gas) can be characterized by one mechanical variable and one thermal variable, say by the specific volume $v = 1/\rho$ and the temperature. One may question, for example, whether a plastically cold-worked metal can ever be restored to its virgin state. The assumption implicit in the postulate of energy balance is that, even if it were necessary to melt it down before recrystallizing it and working it back to the initial state, a precise record of all the energy exchanges during the physical and chemical transformations would verify the energy balance. Experiments by Taylor and Quinney (1934, 1937) on several metals cold-worked in torsion indicated that from 5 to 15 per cent of the input work remained in the metal as latent energy while the remainder appeared as heat. In some cases, subsequent heating to a temperature above the annealing but below the melting point resulted in the evolution of an additional amount of heat approximately equal to the estimated latent energy introduced by the cold work, although the measurements were so difficult that there was not a precise check.

Evidently though, if the only state variables considered are the deformation gradient $\mathbf{F} \equiv \mathbf{x}\overleftarrow{\nabla}$ for the mechanical state and the temperature θ for the thermal state, it is too much to expect that the specific internal energy u will be a function only of these variables, say

$$u = u(\mathbf{F}, \theta).$$

Such an equation would in fact be one of the constitutive equations for a particular kind of material, an ideal thermoelastic solid. Clearly a bar which is first stretched inelastically to twice its initial length and then pushed back to its initial length and cooled to the initial temperature has not thereby been restored to its initial state, and we would not expect such a simple constitutive equation to give its final internal energy. The equation would have to be replaced by some kind of a constitutive equation in which u is not merely a function of the present values of the variables $\mathbf{F}$ and θ, but is a functional or path-dependent function of the history of $\mathbf{F}$ and θ. This does not invalidate the concept of the internal energy as a path-independent function of all the state variables as implied by Eq. (4d). However it does point out the limitations in applying the concept to materials for which we may find it difficult to enumerate all the relevant state variables from the macroscopic point of view, to say nothing of recording their values experimentally.

Power of couple stresses.† When assigned couples and couple stresses are included, as in a subsection of Sec. 5.3, the power input of Eq. (2) and the application of the first law of thermodynamics in Eq. (5) must both be modified. We consider only the case of vanishing spin angular momentum **l**; then the total angular momentum is the total moment of the linear momentum.‡ The power of the external forces and couples is (using rectangular Cartesian components)

$$P_{\text{input}} = \int_S (t_i v_i + m_i w_i)\, dS + \int_V (\rho b_i v_i + \rho c_i w_i)\, dV \tag{5.4.9}$$

where

$$\mathbf{w} = \tfrac{1}{2}\,\text{curl}\,\mathbf{v} \quad \text{or} \quad w_i = -\tfrac{1}{2} e_{mni} W_{mn} \tag{5.4.10}$$

is the angular-velocity vector [see Eq. (4.4.8)].

If we substitute $t_i = T_{ji} n_j$ and $m_i = \mu_{ji} n_j$ and transform the surface integral as before, we obtain

$$P_{\text{input}} = \int_V \left[\left(\frac{\partial T_{ji}}{\partial x_j} + \rho b_i \right) v_i + T_{ji} \frac{\partial v_i}{\partial x_j} + \left(\frac{\partial \mu_{ji}}{\partial x_j} + \rho c_i \right) w_i + \mu_{ji} \frac{\partial w_i}{\partial x_j} \right] dV$$

or, using Eqs. (5.3.4b) and (5.3.12) with $l_i = 0$,

$$P_{\text{input}} = \int_V \left[\rho v_i \frac{dv_i}{dt} + T_{ji} \frac{\partial v_i}{\partial x_j} - e_{mni} T_{mn} w_i + \mu_{ji} \frac{\partial w_i}{\partial x_j} \right] dV.$$

We recall that, in the presence of couple stress, T_{ji} is not symmetric, and write it as the sum of its symmetric and antisymmetric parts:

$$T_{ji} = T_{(ji)} + T_{[ji]} \quad \text{and also} \quad \frac{\partial v_i}{\partial x_j} = D_{ij} + W_{ij}, \tag{5.4.11}$$

where in both cases the first term on the right is the symmetric part and the second term is skew-symmetric. Then§

$$T_{ji} \frac{\partial v_i}{\partial x_j} = T_{(ji)} D_{ij} + T_{[ji]} W_{ij}. \tag{5.4.12}$$

Moreover,

$$T_{[ji]} W_{ij} = e_{mni} T_{mn} w_i,$$

†This subsection may be omitted without interrupting the main line of development.

‡Toupin (1962) includes the spin angular momentum in his energy formulation.

§Equation (12) shows that the symmetric part of the stress does work in stretching the body, while the skew-symmetric part does work in spinning the elements of the body. See Truesdell and Toupin (1960), Sec. 217, who attribute this result to Voigt (1887) and Combebiac (1902).

as can be verified by using Eq. (10) and writing out the two products. Hence, the external power takes the form

$$P_{\text{input}} = \int_V \left[\rho v_i \frac{dv_i}{dt} + T_{(ji)} D_{ij} + \mu_{ji} \frac{\partial w_i}{\partial x_j} \right] dV. \qquad (5.4.13)$$

The last term in the integrand shows that the appropriate deformation variable to associate with the couple stress is the ***angular-velocity gradient*** $\partial w_i/\partial x_j$. This is a deviatoric second-order tensor. (Since $\mathbf{w} = \frac{1}{2}$ curl $\mathbf{v}$ and the divergence of a curl is zero, the trace $\partial w_i/\partial x_i$ of the matrix $\partial w_i/\partial x_j$ vanishes.) In elasticity, if $\mathbf{w}$ is replaced by $\boldsymbol{\omega}$, representing a small rotation instead of an angular velocity, then the gradient $\partial \omega_i/\partial x_j$ represents bending and twist associated with the bending and twisting couple stress μ_{ij}. In Eq. (13) the $\partial w_i/\partial x_j$ represent bending and twisting deformation rates, while the D_{ij} represent extension and shear deformation rates of the volume element.

Energy equation with couple stresses. In the assumed absence of spin angular momentum, the assigned body couples and surface couples modify the energy balance of Eq. (5) only through the terms representing the power input. Hence, the only change in the energy equation, Eq. (7), is the adding of the one extra term $\mu_{ji}\, \partial w_i/\partial x_j$ appearing in the integrand of Eq. (13), and replacing T_{ij} by $T_{(ij)}$ [the latter term is equal to $T_{(ji)}$ the symmetric part of T_{ji}]. The differential equation expressing the balance of energy, replacing the second Eq. (7), is then†

$$\rho \frac{du}{dt} = T_{(ij)} D_{ij} + \mu_{ji} \frac{\partial w_i}{\partial x_j} + \rho r - \frac{\partial q_j}{\partial x_j}. \qquad (5.4.14)$$

Before discussing, in Sec. 5.6, the second law of thermodynamics and the concept of entropy, we consider the principle of virtual displacements, involving an equation similar in form to Eq. (2), but differing from it fundamentally in that the displacements are not actual displacements of a real motion but merely hypothetical, kinematically possible displacements, called ***virtual displacements***, from an equilibrium configuration.

EXERCISES

1. In an ideal frictionless fluid there are no shear tractions on any plane.
   (a) Show that this implies that $T_{ij} = -p\delta_{ij}$, and

†When the intrinsic angular momentum is not zero, and the angular velocity of a particle is not assumed equal to the mean angular velocity of material lines as given by Eq. (10), this equation must be modified. See, for example, the discussion of multipolar media in Jaunzemis (1967), Sec. 11.

(b) that for this case the stress power $T_{ij}D_{ij}$ is given by

$$T_{ij}D_{ij} = \frac{p}{\rho}\frac{d\rho}{dt}.$$

Express this result in terms of the specific volume $v = 1/\rho$.

2. Show that in the frictionless fluid of Ex. 1, if there is no internal heat source and if the heat conduction is governed by Fourier's law $\mathbf{q} = -k\,\nabla\theta$ where θ is the temperature, then the energy equation becomes

$$\rho\frac{du}{dt} = \nabla\cdot(k\,\nabla\theta) - p\,\nabla\cdot\mathbf{v}.$$

3. Show that in a fluid which is frictionless and also incompressible the external mechanical power input is equal to the rate of change of the kinetic energy of the system. Show also that for such a fluid the rate of change of the internal energy is equal to the heat input rate.

4. For the special case where no motion occurs, assume that the specific internal energy u is given by $u = c_v\theta$ where c_v is the specific heat at constant volume and θ is the absolute temperature. Assume further that the only energy transfer is heat conduction given by $\mathbf{q} = -k\,\nabla\theta$, where k is thermal conductivity. If there are no heat sources or sinks inside an arbitrary control surface fixed in space, derive from the integral form of the energy balance a partial differential equation for θ as a function of the spatial coordinates and the time. Consider both the case where c_v and k are functions of θ and the case where they are constant.

5. Assume an adiabatic process with r and q_j everywhere zero, and assume that the internal energy density u is entirely strain energy W such that $\rho\,du \equiv dW$ is an identity for arbitrary $d\epsilon_{ij}$ (within small-strain ϵ_{ij} assumptions), where $W = \frac{1}{2}\lambda e^2 + G\epsilon_{ij}\epsilon_{ij}$ and where $e = \epsilon_{kk}$ (λ and G are the adiabatic Lamé constants of elasticity). For the small-strain case assume $d\epsilon_{ij} = D_{ij}\,dt$, and show that the energy equation then implies the stress-strain relation $T_{ij} = \lambda e\delta_{ij} + 2G\epsilon_{ij}$.

6. In a Newtonian viscous fluid the constitutive equation is $T_{ij} = -p\delta_{ij} + \kappa D_{kk}\delta_{ij} + 2\,\mu D'_{ij}$, where κ is the bulk viscosity, μ is the shear viscosity, and p is the thermodynamic pressure (not in general equal to $-T_{kk}/3$). Show that in such a fluid the stress power consists of two parts, one of which is always nonnegative and therefore represents a dissipation of mechanical power (not convertible into kinetic energy). If this dissipation power is denoted by $2W_D$ write the forms taken by Eqs. (5.4.2) and (5.4.7).

7. (a) Carry out the details of deriving Eq. (5.4.8b) by making the substitutions suggested in the text.
   (b) Alternatively derive the reference configuration version of Eq. (5.4.2) by transforming Eq. (1) to the reference configuration before transforming the surface integral to a volume integral.

8. Show that for the irrotational incompressible flow of Ex. 4, Sec. 5.2, where $\mathbf{v} = \nabla\phi$, the total kinetic energy in a volume V bounded by S is equal to $(\rho/2)\int_S \phi(\partial\phi/\partial n)\,dS$, where ρ is the density (assumed constant in time and uniform throughout the volume).

9. Carry out the details of deriving Eqs. (12) and (13).

5.5 Principle of virtual displacements†

Variational methods have played a prominent role in continuum mechanics. The fundamental variational principle is the principle of virtual displacements, sometimes called the ***principle of virtual work.*** Although it takes a form similar to Eq. (5.4.2), this is not really an energy principle, because the work computed is a fictitious work computed with a set of ***statically admissible forces and stresses*** assumed to remain constant while they do work on a set of infinitesimal, ***kinematically admissible displacements.*** The stresses need not be the actual stresses occurring in the real physical material, and the displacements need not be the actual displacements. Moreover the stress distribution and the displacement distribution may be independently prescribed; this differs from the case of a real motion in a real material, since in the real motion the displacements in general give rise to strains and deformations which are related to the stresses by the constitutive equations of the material.

In the statics of particles and rigid bodies the principle of virtual displacements is an alternative way of expressing the equilibrium conditions, and we shall see that in a deformable medium also it is equivalent to the equilibrium conditions.

When certain external (body and surface) forces are prescribed as acting on a deformable body, a ***statically admissible*** stress distribution is defined as one satisfying the equilibrium partial differential equations, Eqs. (5.3.5), in the interior of the body,‡ and boundary conditions $T_{ji}n_j = t_i$ wherever any boundary tractions t_i are prescribed. It is important to remember that the proposed equilibrium stress distribution used in connection with the principle need not be the actual stress distribution in the deformed body. Even when all the boundary conditions are in terms of stress, the stress distribution is not completely determined by the equilibrium conditions and the boundary conditions but depends in general upon material properties. Usually many different possible statically admissible stress distributions exist—all satisfying equilibrium requirements. Any one of them may be the distribution referred to in the principle of virtual displacements. We restrict our attention to the nonpolar case; hence the Cauchy stress is symmetric, $T_{ij} = T_{ji}$.

A ***kinematically admissible*** displacement distribution is one satisfying any prescribed displacement boundary conditions and possessing continuous first partial derivatives in the interior of the body. Since the virtual displacements to be considered are ***additional*** displacements from the equilibrium configura-

†This section may be omitted without interruping the main line of development.

‡For the time being we assume the continuity of all derivatives appearing in the discussion. The principle can be modified to permit a finite number of interior surfaces across which certain kinds of discontinuities occur. See the discussion in connection with Fig. 5.9 and Eq. (8) below.

tion, a virtual displacement component must be zero wherever the actual displacement is prescribed by the boundary conditions.

Virtual displacements. Suppose that a body is in a certain equilibrium configuration and that each point of the body is given an infinitesimal virtual displacement δu_i from the equilibrium configuration. Each component of the virtual displacement is a function of position in the body. We suppose that the three functions δu_i have continuous first partial derivatives with respect to x_1, x_2, x_3 and that $\delta u_i = 0$ on those parts of the boundary surface where the actual displacement u_i is prescribed. Thus, "the virtual displacements satisfy displacement boundary conditions where prescribed" means that no additional virtual displacement δu_i is permitted where the boundary condition fixes the value of u_i. The displacements are called virtual because they are not actual physical displacements which would occur under the given loads, but merely hypothetical, kinematically possible displacements. The boundary conditions may be stated more generally as follows.

Boundary conditions. At each boundary point we choose a local Cartesian axis system $\bar{x}_i$ (usually with one axis along the normal) and require:

$$\text{either } \bar{T}_{j1}\bar{n}_j = \bar{t}_1 \quad \text{or} \quad \delta\bar{u}_1 = 0, \qquad \text{but not both;} \tag{5.5.1a}$$

$$\text{either } \bar{T}_{j2}\bar{n}_j = \bar{t}_2 \quad \text{or} \quad \delta\bar{u}_2 = 0, \qquad \text{but not both;} \tag{5.5.1b}$$

$$\text{either } \bar{T}_{j3}\bar{n}_j = \bar{t}_3 \quad \text{or} \quad \delta\bar{u}_3 = 0, \qquad \text{but not both.} \tag{5.5.1c}$$

In Eqs. (1) the barred symbols $\bar{t}_i$ are given functions of position on those parts of the boundary where the traction components are prescribed. The traction vector components t_i in a fixed system x_i can then be expressed in terms of the $\bar{t}_j$, and $t_i\delta u_i\, dS = \bar{t}_i\delta\bar{u}_i\, dS$ represents the virtual work of the traction forces on dS.

Equation (2) gives the ***virtual work*** δW_{ext} of the external surface tractions t_i and body forces f_i (per unit volume) if the forces are assumed to remain unchanged during the virtual displacement. With $t_i = T_{ji}n_j$,

$$\delta W_{\text{ext}} = \int_S T_{ji}n_j\delta u_i\, dS + \int_V f_i\,\delta u_i\, dV. \tag{5.5.2}$$

We transform the first integral to a volume integral by using Eq. (5.1.1) to obtain

$$\delta W_{\text{ext}} = \int_V \left[T_{ji}\frac{\partial(\delta u_i)}{\partial x_j} + \delta u_i\left(\frac{\partial T_{ji}}{\partial x_j} + f_i\right)\right] dV. \tag{5.5.3}$$

The term in parentheses vanishes by the equations of equilibrium, Eqs. (5.3.5). We write

$$\frac{\partial(\delta u_i)}{\partial x_j} = \delta\epsilon_{ij} + \delta\omega_{ij}$$

where

$$\delta\epsilon_{ij} = \frac{1}{2}\left(\frac{\partial\delta u_i}{\partial x_j} + \frac{\partial\delta u_j}{\partial x_i}\right) \quad \text{and} \quad \delta\omega_{ij} = \frac{1}{2}\left(\frac{\partial\delta u_i}{\partial x_j} - \frac{\partial\delta u_j}{\partial x_i}\right) \tag{5.5.4}$$

are the virtual strains and rotations associated with the infinitesimal virtual displacement distribution.† Hence

$$T_{ji}\frac{\partial(\delta u_i)}{\partial x_j} = T_{ij}\delta\epsilon_{ij} + T_{ij}\delta\omega_{ij}.$$

Moreover $T_{ij}\delta\omega_{ij} = 0$, because T_{ij} is symmetric and $\delta\omega_{ij}$ is skew-symmetric. Thus Eq. (3) takes the form

$$\delta W_{\text{ext}} = \int_V T_{ij}\delta\epsilon_{ij}\,dV. \tag{5.5.5}$$

This shows that if the stress field is statically admissible, the total virtual work of the external forces on any kinematically admissible virtual displacement field is equal to $\int_V T_{ij}\delta\epsilon_{ij}\,dV$. This and its converse, together, are the ***principle of virtual displacements*** for a deformable body. The converse proposition states that if the virtual work of the prescribed external forces is equal to $\int_V T_{ij}\delta\epsilon_{ij}\,dV$ for a certain assumed stress field T_{ij} ***for every kinematically admissible virtual displacement field***, then the stress field is statically admissible, i.e., it satisfies equilibrium conditions in the interior and traction boundary conditions wherever traction boundary conditions are prescribed on the surface.

In proving the converse proposition, we have by hypothesis

$$\int_S \bar{t}_i\,\delta\bar{u}_i\,dS + \int_V f_i\,\delta u_i\,dV = \int_V T_{ij}\delta\epsilon_{ij}\,dV.$$

We add and subtract $\bar{T}_{ji}\bar{n}_j\delta\bar{u}_i$ in the surface integral, and add $T_{ij}\delta\omega_{ij} = 0$ in the volume integral on the right. Then (note: $\bar{T}_{ji}\bar{n}_j\delta\bar{u}_i = T_{ji}n_j\delta u_i$)

$$\int_S [\bar{t}_i - \bar{T}_{ji}\bar{n}_j]\delta\bar{u}_i\,dS + \int_S T_{ji}n_j\delta u_i\,dS + \int_V f_i\,\delta u_i\,dV = \int_V T_{ji}\frac{\partial\delta u_i}{\partial x_j}\,dV.$$

†Note that, while the virtual displacements and associated strains and rotations are infinitesimal, no restrictions are placed on the magnitudes of the actual displacements from any reference configuration to the equilibrium configuration. The principle can therefore be used in finite-displacement problems.

If we transform the second surface integral as in the derivation of (3), the above equation takes the form

$$\int_S [\bar{t}_i - \bar{T}_{ji}\bar{n}_j]\delta\bar{u}_i\, dS + \int_V \left[\frac{\partial(T_{ji}\,\delta u_i)}{\partial x_j} + f_i\,\delta u_i\right] dV = \int_V T_{ji}\frac{\partial \delta u_i}{\partial x_j}\, dV,$$

whence we obtain

$$\int_S [\bar{t}_i - \bar{T}_{ji}\bar{n}_j]\delta u_i\, dS + \int_V \delta u_i\left[\frac{\partial T_{ji}}{\partial x_j} + f_i\right] dV = 0.$$

Since this equation is satisfied for arbitrary δu_i, the terms in brackets in the volume integral must vanish at every interior point of the body. Otherwise, if there is a point where, for example, the term for $i = 1$ does not vanish, say it is positive there, then since the term is a continuous function it will be positive in some neighborhood of the point. We choose δu_1 so that it is positive in a small sphere around the point in which the bracketed term is also positive: outside of the sphere, δu_1 is chosen to be zero. Also we choose δu_2 and δu_3 to be zero everywhere. The preceding equation then requires

$$\int_{\text{sphere}} \delta u_1\left[\frac{\partial T_{j1}}{\partial x_j} + f_1\right] dV = 0,$$

while the fact that both the bracketed sum and δu_1 are positive everywhere inside the sphere requires the integral over the sphere to be positive. This contradiction proves that there is no interior point where the bracketed sum fails to vanish, and that therefore the assumed stress distribution satisfies the first equilibrium differential equation at every interior point of the body. That the other equilibrium differential equations are also satisfied is proved in a similar fashion. (Note that after we have arrived at our contradiction, we may choose another arbitrary distribution of virtual displacements for the next part of the proof.) Since all of the equilibrium equations are satisfied throughout the body, the volume integral vanishes. We can now prove in similar fashion that the bracketed terms in the surface integral must also vanish at every point of the surface where displacements are not prescribed. Thus we show that the assumed stress distribution satisfies the given traction boundary conditions, and we complete the proof that it is statically admissible.

We remark that this principle has nothing whatsoever to do with conservation of energy. It applies even when mechanical energy is not conserved, e.g., in plastic deformation.†

†Because the principle has been widely applied in the theory of elasticity and elastic structural analysis, the principle is often stated in a form valid only when an elastic strain-energy function exists. This special form obscures the principle's great generality. Such elasticity principles grew out of the work of Green, 1839. The strain-energy function or elastic potential function will be discussed in Sec. 6.2.

The variables used in this development are spatial coordinates rather than material coordinates; the infinitesimal virtual strains are defined with reference to the equilibrium configuration as initial state and not with reference to the unstressed state. This is comparable to the reference state for rate of deformation and natural strain in Sec. 4.4. If the infinitesimal virtual particle displacements δu_i are imagined to take place in an infinitesimal time interval δt, then the virtual velocity is $\delta u_i/\delta t$. By dividing through by δt, we obtain an alternative statement of Eq. (5), namely

$$\frac{\delta W_{\text{ext}}}{\delta t} \equiv \int_S \bar{t}_i \frac{\delta \bar{u}_i}{\delta t} dS + \int_V f_i \frac{\delta u_i}{\delta t} dV = \int_V T_{ij} \frac{\delta \epsilon_{ij}}{\delta t} dV,$$

where $\delta W_{\text{ext}}/\delta t$ is the rate at which the external forces are doing work on the virtual velocities, and where $\delta\epsilon_{ij}/\delta t$ is the virtual rate of deformation. This alternative statement is called the ***principle of virtual velocities.*** It is not really a dynamical principle, since the stresses still satisfy equilibrium equations instead of equations of motion. The only advantage of this formulation is that the virtual velocities $\delta u_i/\delta t$ and virtual rates of deformation $\delta\epsilon_{ij}/\delta t$ can be considered as arbitrary finite quantities, without invoking the imprecise notion of "infinitesimal" virtual displacements.

By including the inertia forces $-\rho d^2 u_i/dt^2$ as body forces, the principle of virtual displacements can be extended to the fictitious equilibrium state obtained. But the u_i appearing in the accelerations are not varied in the virtual displacements. To demonstrate this we make use of a theorem of stress means, which further emphasizes that this whole argument is concerned with the conditions of equilibrium (or perhaps fictitious "dynamic equilibrium" with the inertia forces) and has nothing to do with energy principles. It is simply a mathematical identity, which can be obtained by integration of the equations of equilibrium (or motion) over the body.

The theorem of stress means states that, if f is any continuously differentiable function whatever, while the Cauchy stress T_{ij} satisfies the Cauchy equations of motion, Eqs. (5.3.4), then

$$\int_S f T_{ji} n_j \, dS = \int_V \left[T_{ji} \frac{\partial f}{\partial x_j} + f\rho(\dot{v}_i - b_i) \right] dV. \tag{5.5.6}$$

To derive this we note that

$$(f T_{ji})_{,j} = T_{ji} f_{,j} + f T_{ji,j} = T_{ji} f_{,j} + f\rho(\dot{v}_i - b_i)$$

by the Cauchy equation of motion. When this is integrated over an arbitrary volume V, and the integral $\int_V (f T_{ji})_{,j} \, dV$ transformed to an integral over the bounding surface S, the result is Eq. (6), the theorem of stress means. Since f

is merely an arbitrary continuously differentiable function, no physical significance is to be attached to it.

But if we now substitute successively for f the three virtual displacement components δu_i for $i = 1, 2, 3$, and add the three resulting equations, we can write the result as

$$\int_S n_j T_{ji} \delta u_i \, dS + \int_V \rho(b_i - \dot{v}_i) \delta u_i \, dV = \int_V T_{ji} \delta u_{i,j} \, dV. \qquad (5.5.7)$$

By virtue of Eqs. (4) this is equivalent to the result of Eq. (5) provided we include in the virtual work of the external forces the virtual work of the inertia forces $-\rho \dot{v}_i$. The satisfaction of Eq. (7) for arbitrary virtual displacements then is an expression of the fact that the given stresses ***and the given accelerations*** satisfy the equations of motion.

The theorem of stress means is due to Signorini (1933), but the result of Eq. (7) is much older. For further generalizations of the principle of virtual work, see Truesdell and Toupin (1960), Sec. 232.

Discontinuity surfaces. If some of the stress or velocity components are discontinuous or have discontinuous derivatives across an interior surface of the body, the principle of virtual displacements cannot be applied to the whole body, but can be written for each subregion in which the derivatives are continuous. In plasticity theory, for example, the tangential components of displacement are sometimes assumed to be discontinuous across an internal surface while the normal components must be continuous in order that no gaps can open up in the interior.† The traction components acting on any interior surface must be continuous by Newton's Third Law, but a stress component which is not a traction component may be discontinuous, e.g., σ_x, σ_y, and τ_{xy} could be discontinuous across a plane parallel to the xy-plane, but σ_z, τ_{zx}, and τ_{zy} would have to be continuous,

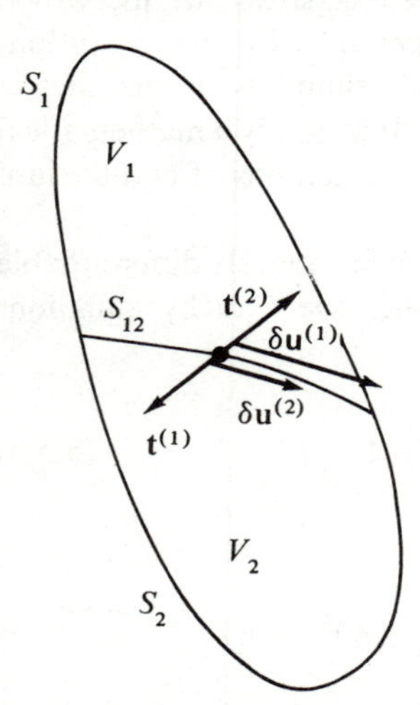

Fig. 5.9 Discontinuity surface.

We illustrate the procedure for the case of a single discontinuity surface S_{12} separating two parts, V_1 and V_2, of the volume V in Fig. 5.9. The case illustrated has a discontinuity in tangential displacement, and the normal component is zero on both sides. The traction vector $\mathbf{t}^{(1)}$ acts on the material in V_1, while $\mathbf{t}^{(2)}$ acts on V_2 (equal and opposite). S_1 denotes that part of S which

†See, for example, Chaps. 7 and 8 in *Theory of Perfectly Plastic Solids*, by W. Prager and P. G. Hodge, New York, John Wiley & Sons, Inc., 1951.

along with S_{12} forms the boundary of V_1, while S_2 is part of the boundary of V_2. We write Eq. (5) for each subregion.

$$\int_{S_1} \bar{t}_i \, \delta \bar{u}_i \, dS + \int_{S_{12}} t_i^{(1)} \delta u_i^{(1)} \, dS + \int_{V_1} f_i \, \delta u_i \, dV = \int_{V_1} T_{ij} \delta \epsilon_{ij} \, dV$$

$$\int_{S_2} \bar{t}_i \, \delta \bar{u}_i \, dS + \int_{S_{12}} t_i^{(2)} \delta u_i^{(2)} \, dS + \int_{V_2} f_i \, \delta u_i \, dV = \int_{V_2} T_{ij} \delta \epsilon_{ij} \, dV.$$

Adding these two equations and substituting $-t_i^{(2)}$ for $t_i^{(1)}$, we obtain

$$\int_{S} \bar{t}_i \, \delta \bar{u}_i \, dS + \int_{V} f_i \, \delta u_i \, dV + \int_{S_{12}} t_i^{(2)} (\delta u_i^{(2)} - \delta u_i^{(1)}) \, dS = \int_{V} T_{ij} \delta \epsilon_{ij} \, dV. \tag{5.5.8}$$

We see that when there is an internal sliding surface, the frictional work at such a surface must be added to the work of the external forces in applying the principle of virtual displacements. If the displacement field were an actual displacement field, this frictional work would have to be negative, since frictional forces would oppose the motion.

Virtual work of stress resultants in finite plane bending of a beam. The principle of virtual displacements is basic to all variational principles in structural analysis. When simplified theories of beams, plates, or shells are used in the analysis, the principle must be formulated in terms of the variables used in the simplified theory. The statically admissible quantities are then certain generalized stresses (i.e., stress resultants such as bending moments and shear forces instead of the three-dimensional continuum stresses T_{ij}), and the virtual displacements lead to virtual changes in certain generalized strain measures appropriately conjugate to the generalized stresses. Here "appropriately conjugate" means defined in such a way that if Q_i are the generalized stresses and δq_i are the generalized strains, then the virtual work of the external forces (per unit generalized volume) is $Q_i \delta q_i$ (summed over all values of i). "Generalized volume" would mean area in the case of a two-dimensional continuum theory such as the theory of plates or shells; it would mean arc length in the case of a one-dimensional beam or arch theory.

Does the three-dimensional principle proved above imply that the corresponding principle holds in the one-dimensional and two-dimensional theories? It is not obvious that it always will. But if the generalized strains are defined to be appropriately conjugate to the generalized stresses, then the principle will hold for the simplified theory. In fact, the definition of "appropriately conjugate" given above would make it hold. In practice, one sure way of selecting appropriately conjugate, generalized strain definitions is to carry out the derivation of the principle of virtual displacements for the theory in question. The procedure is illustrated here for finite plane bending of a beam.

Figure 5.10 shows a portion of the deformed centerline of the beam loaded by a distributed vertical load w per unit length. Stress resultants are the axial forces T, transverse shears Q, and bending moments M at sections 1 and 2. We adopt sign conventions so that the six resultants as shown are positive.

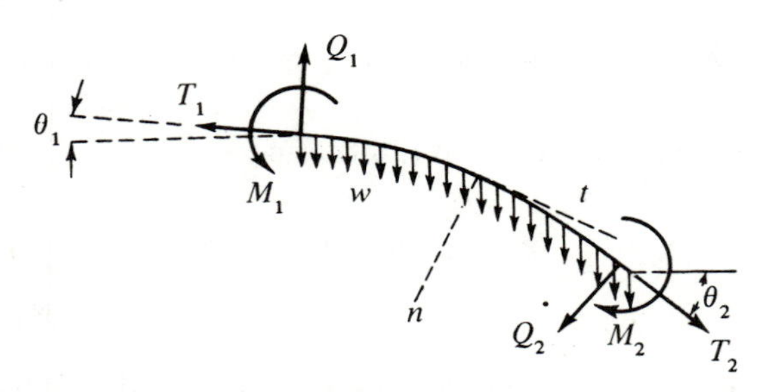

Fig. 5.10 Portion of deformed beam in finite plane bending.

Consider infinitesimal virtual displacements from the equilibrium configuration, with normal and tangential components δu_n and δu_t as functions of arc length s. If the beam were straight, according to elementary beam theory with plane normal sections remaining plane and normal (neglecting shear deformation), the incremental rotation would be $\delta\theta = d(\delta u_n)/ds$ and the incremental axial strain would be $\delta\epsilon = d(\delta u_t)/ds$. The curvature $d\theta/ds$ adds $(d\theta/ds)\delta u_t$ to the rotation and $-(d\theta/ds)\delta u_n$ to the strain. Thus

$$\delta\theta = \frac{d(\delta u_n)}{ds} + \frac{d\theta}{ds}\delta u_t, \qquad \delta\epsilon = \frac{d(\delta u_t)}{ds} - \frac{d\theta}{ds}\delta u_n \tag{5.5.9}$$

The virtual work of the external forces (with the forces assumed to remain constant during the infinitesimal virtual displacement) is

$$\delta W_{\text{ext}} = [M\delta\theta + T\delta u_t + Q\delta u_n]_1^2 + \int_1^2 (w \sin\theta\delta u_t + w\cos\theta\delta u_n)\,ds, \tag{5.5.10}$$

where the first term in brackets is to be evaluated at the two limits and prefixed by a minus sign at the lower limit. This first term is equal to

$$\int_1^2 \left[\frac{d(M\delta\theta)}{ds} + \frac{d(T\delta u_t)}{ds} + \frac{d(Q\delta u_n)}{ds}\right] ds =$$

$$\int_1^2 \left[\frac{dM}{ds}\delta\theta + M\delta\left(\frac{d\theta}{ds}\right) + \frac{dT}{ds}\delta u_t + T\delta\frac{(du_t)}{ds} + \frac{dQ}{ds}\delta u_n + Q\delta\frac{(du_n)}{ds}\right] ds. \tag{5.5.11}$$

We use Eqs. (9), and, for infinitesimal motion, $d(\delta\theta)/ds = \delta(d\theta/ds)$ to obtain

$$\delta W_{\text{ext}} = \int_1^2 \left[M\delta\left(\frac{d\theta}{ds}\right) + T\delta\epsilon \right] ds + \int_1^2 \left[\left(\frac{dM}{ds} + Q\right)\delta\theta + \left(\frac{dT}{ds} - Q\frac{d\theta}{ds} + w\sin\theta\right)\delta u_t + \left(\frac{dQ}{ds} + T\frac{d\theta}{ds} + w\cos\theta\right)\delta u_n \right] ds \tag{5.5.12}$$

The second integral in Eq. (12) vanishes if the equilibrium equations

$$\frac{dM}{ds} + Q = 0 \qquad \frac{dT}{ds} - Q\frac{d\theta}{ds} + w\sin\theta = 0 \qquad \frac{dQ}{ds} + T\frac{d\theta}{ds} + w\cos\theta = 0 \tag{5.5.13}$$

are satisfied. Then

$$\delta W_{\text{ext}} = \int_1^2 \left[M\delta\left(\frac{d\theta}{ds}\right) + T\delta\epsilon \right] ds \tag{5.5.14}$$

and the kinematic variables conjugate to M and T are the curvature $d\theta/ds$ and strain ϵ, respectively. There is no generalized strain conjugate to Q. Because the imposed constraints of the simple beam theory permit no shear deformation, the shear force Q should be regarded as a reaction to the internal constraints instead of as a generalized stress.

For the inverse part of the derivation, we assume that Eq. (14) holds for arbitrary virtual displacements. This implies that the second integral in Eq. (12) vanishes for arbitrary virtual displacements δu_n and δu_t. But we cannot immediately conclude that the three expressions in parentheses vanish, since $\delta\theta$ is not independent of δu_n and δu_t. We replace $\delta\theta$ by using Eq. (9) and integrate the first term by parts to obtain

$$\left[\left(\frac{dM}{ds} + Q\right)\delta u_n\right]_1^2 + \int_1^2 \left[\left(\frac{dT}{ds} + \frac{dM}{ds}\frac{d\theta}{ds} + w\sin\theta\right)\delta u_t + \left(-\frac{d^2M}{ds^2} + T\frac{d\theta}{ds} + w\cos\theta\right)\delta u_n \right] ds = 0$$

for arbitrary δu_t and δu_n. This implies

$$\frac{dT}{ds} + \frac{dM}{ds}\frac{d\theta}{ds} + w\sin\theta = 0 \qquad -\frac{d^2M}{ds^2} + T\frac{d\theta}{ds} + w\cos\theta = 0 \tag{5.5.15}$$

in the interior, plus the result

$$\frac{dM}{ds} + Q = 0 \tag{5.5.16}$$

at sections 1 and 2. Since sections 1 and 2 could have been chosen arbitrarily in the beam, Eq. (16) holds at any point. Use of Eq. (16) then transforms Eqs. (15) to the desired last two Eqs. (13). This awkward derivation was a result of the internal restriction preventing shear deformation. See Exercise 3 of this section, where this restriction is dropped and Q becomes a generalized stress.

A principle of virtual displacements in terms of the Piola-Kirchhoff stress tensors in the reference state can be derived as follows.† The virtual work of the external forces in an infinitesimal virtual displacement $\delta\mathbf{u}$ from the current deformed state is, by Eq. (2),

$$\delta W_{\text{ext}} = \int_S \mathbf{t}\cdot\delta\mathbf{u}\, dS + \int_V \mathbf{f}\cdot\delta\mathbf{u}\, dV. \tag{5.5.17}$$

Now $\mathbf{u} = \mathbf{x} - \mathbf{X}$ and therefore $\delta\mathbf{u} = \delta\mathbf{x}$. Moreover $\mathbf{t}\, dS = \mathbf{t}^0\, dS_0$, where $\mathbf{t}^0$ is the pseudo-traction vector introduced in Eq. (5.3.21), giving the force on the deformed element dS per unit undeformed area. Hence

$$\delta W_{\text{ext}} = \int_{S_0} \mathbf{t}^0\cdot\delta\mathbf{x}\, dS_0 + \int_{V_0} f[\mathbf{x}(\mathbf{X})]\cdot\delta\mathbf{x}\, J\, dV_0,$$

where $J = \det \mathbf{F} = \rho_0/\rho$ is the Jacobian determinant introduced by the change to the material coordinates to obtain an integral over the undeformed volume. Let

$$\mathbf{f}_0 = \rho_0\mathbf{b}_0 = \frac{\rho_0}{\rho} f[\mathbf{x}(\mathbf{X})]$$

denote the body force in the deformed configuration per unit undeformed volume, where $\mathbf{b}_0 = \mathbf{b}[\mathbf{x}(\mathbf{X})]$ is the body force per unit mass. Then, since by Eq. (5.3.21) $\mathbf{t}^0 = \hat{\mathbf{N}}\cdot\mathbf{T}^0$, we obtain

$$\delta W_{\text{ext}} = \int_{S_0} \hat{\mathbf{N}}\cdot\mathbf{T}^0\cdot\delta\mathbf{x}\, dS_0 + \int_{V_0} \mathbf{f}_0\cdot\delta\mathbf{x}\, dV_0, \tag{5.5.18}$$

where $\mathbf{T}^0$ is the nonsymmetric first Piola-Kirchhoff stress tensor. Introducing rectangular Cartesian components and transforming the surface integral over S_0 to a volume integral over V_0, we obtain

†See Truesdell and Toupin (1960), Sec. 232, for a discussion of early formulations by Piola (1848) of the principle of virtual work, using material coordinates. We use a different procedure.

$$\delta W_{\text{ext}} = \int_{V_0} \left[T^0_{Ji} \frac{\partial \delta x_i}{\partial X_J} + \left(\frac{\partial T^0_{Ji}}{\partial X_J} + f_{0i} \right) \delta x_i \right] dV_0. \tag{5.5.19}$$

The expression in parentheses vanishes by the equilibrium version of Eq. (5.3.27), with $\mathbf{f}_0 = \rho_0 \mathbf{b}_0$ and $d^2\mathbf{x}/dt^2 = 0$. Moreover $\partial(\delta x_i)/\partial X_J = \delta(\partial x_i/\partial X_J) = \delta F_{iJ}$ is the component of the virtual change $\delta \mathbf{F}$ in the deformation gradient. Thus

$$\delta W_{\text{ext}} = \int_{V_0} T^0_{Ji} \delta F_{iJ} \, dV_0 = \int_{V_0} (\mathbf{T}^0)^T : \delta \mathbf{F} \, dV_0 = \int_{V_0} \mathbf{T}^0 : \delta \mathbf{F}^T \, dV_0. \tag{5.5.20}$$

Since neither $\mathbf{F}$ nor $\mathbf{T}^0$ is symmetric, it is necessary to use some care in evaluating $T^0_{Ji} \delta F_{iJ} \equiv T^0_{Ji}(\partial \delta x_i/\partial X_J)$. Equation (20) indicates that "appropriately conjugate" variables are $\mathbf{F}$ and $(\mathbf{T}^0)^T$. [Truesdell and Noll (1965) denote $(\mathbf{T}^0)^T$ by $\mathbf{T}_R$.] Another possible choice of conjugate variables would be $\mathbf{T}^0$ and $\mathbf{F}^T$. Equation (20) is the form taken by the virtual displacement Eq. (5) in terms of the first Piola-Kirchhoff tensor, with an integration over the undeformed volume.

To get a form involving the symmetric second Piola-Kirchhoff tensor $\tilde{\mathbf{T}}$ we can substitute into Eq. (20) the expression

$$\mathbf{T}^0 = \tilde{\mathbf{T}} \cdot \mathbf{F}^T \quad \text{or} \quad T^0_{Ji} = \tilde{T}_{JI} \partial_I x_i$$

given by Eq. (5.3.25). Thus

$$\delta W_{\text{ext}} = \int_{V_0} \tilde{T}_{JI} \frac{\partial x_i}{\partial X_I} \frac{\partial \delta x_i}{\partial X_J} dV_0. \tag{5.5.21}$$

Now, by Eq. (4.5.7)

$$C_{IJ} = \frac{\partial x_i}{\partial X_I} \frac{\partial x_i}{\partial X_J}, \quad \text{whence} \quad \delta C_{IJ} = \frac{\partial \delta x_i}{\partial X_I} \frac{\partial x_i}{\partial X_J} + \frac{\partial x_i}{\partial X_I} \frac{\partial \delta x_i}{\partial X_J}.$$

It can be shown that $\tilde{T}_{IJ} \delta C_{IJ}$ is equal to twice the integrand of Eq. (21). For example, $\tilde{T}_{12} \delta C_{12} + \tilde{T}_{21} \delta C_{21}$ is equal to

$$\begin{aligned} &\tilde{T}_{12} \frac{\partial \delta x_i}{\partial X_1} \frac{\partial x_i}{\partial X_2} + \tilde{T}_{12} \frac{\partial x_i}{\partial X_1} \frac{\partial \delta x_i}{\partial X_2} + \tilde{T}_{21} \frac{\partial \delta x_i}{\partial X_2} \frac{\partial x_i}{\partial X_1} + \tilde{T}_{21} \frac{\partial x_i}{\partial X_2} \frac{\partial \delta x_i}{\partial X_1} \\ &= \tilde{T}_{12} \frac{\partial x_i}{\partial X_2} \frac{\partial \delta x_i}{\partial X_1} + \tilde{T}_{21} \frac{\partial x_i}{\partial X_1} \frac{\partial \delta x_i}{\partial X_2} + \tilde{T}_{21} \frac{\partial x_i}{\partial X_1} \frac{\partial \delta x_i}{\partial X_2} + \tilde{T}_{12} \frac{\partial x_i}{\partial X_2} \frac{\partial \delta x_i}{\partial X_1}. \end{aligned}$$

To get the second form we interchange the order of the derivative factors in the first and third terms and substitute $\tilde{T}_{21}$ for $\tilde{T}_{12}$ in the second term and $\tilde{T}_{12}$ for $\tilde{T}_{21}$ in the fourth term. Thus

$$\tilde{T}_{12} \delta C_{12} + \tilde{T}_{21} \delta C_{21} = 2 \left[\tilde{T}_{21} \frac{\partial x_i}{\partial X_1} \frac{\partial \delta x_i}{\partial X_2} + \tilde{T}_{12} \frac{\partial x_i}{\partial X_2} \frac{\partial \delta x_i}{\partial X_1} \right].$$

The other pairs of terms may be treated similarly. Hence Eq. (21) may be written as

$$\delta W_{\text{ext}} = \int_{V_0} \tfrac{1}{2}\tilde{\mathbf{T}} : \delta\mathbf{C}\, dV_0. \tag{5.5.22}$$

Alternatively, since $2\mathbf{E} = \mathbf{C} - \mathbf{1}$, by Eq. (4.5.5a), we have $\delta\mathbf{E} = \delta\mathbf{C}/2$, whence

$$\delta W_{\text{ext}} = \int_{V_0} \tilde{\mathbf{T}} : \delta\mathbf{E}\, dV_0, \tag{5.5.23}$$

showing that $\tilde{\mathbf{T}}$ and $\mathbf{E}$ are conjugate variables as in Eq. (5.4.8a).

EXERCISES

1. Assuming that the isotropic elastic Hooke's law $T_{ij} = \lambda\epsilon_{kk}\delta_{ij} + 2G\epsilon_{ij}$ applies to the stresses in the principle of virtual displacements, find a function W of the strains, such that the virtual work of the external forces is $\delta W_{\text{ext}} = \int_V (\partial W/\partial\epsilon_{ij})\delta\epsilon_{ij}\, dV$. For this case $\delta W = (\partial W/\partial\epsilon_{ij})\delta\epsilon_{ij}$ is called the ***first variation of the strain-energy density*** W. (Do not confuse δW with δW_{ext}.) This, however, is only a very special case of the principle of virtual displacements, since the principle applies to inelastic media, while this does not.
2. Derive Eq. (5.5.23) by transforming the volume integral of Eq. (5.5.5) to an integral over the undeformed volume V_0. *Hint*: See Eq. (5.4.8a), and note that $\delta\epsilon_{ij} = D_{ij}\delta t$, where D_{ij} is the virtual rate of deformation.
3. When plane normal sections of a beam are not constrained to remain plane and normal, Eq. (5.5.9) no longer holds. If $\delta\gamma$ denotes the amount by which $\delta\theta$ exceeds the rotation $\delta\phi$ of the normal, derive the resulting principle of virtual displacements for the beam. *Remark*: Note that the variable conjugate to M turns out to be $d\phi/ds$ and not the centerline curvature $d\theta/ds$.
4. For the case of small deflections of a beam initially parallel to the horizontal X-direction, which deflects in the Y-direction, express in terms of u_x and u_y (evaluated at $Y = 0$) the deformation variables of (a) Eq. (14), (b) Ex. 3.
5. Derive an equation for δW_{ext} replacing Eq. (5.5.5) when there are couple stresses and body couples. *Hint*: See the derivation of Eq. (5.4.13), and note that the present context concerns equilibrium.

5.6 Entropy and the second law of thermodynamics; the Clausius-Duhem inequality†

The first law of thermodynamics can be regarded as an expression of the interconvertibility of heat and work, maintaining an energy balance; as such

†This section and Sec. 5.7 can be omitted, although some of the results are cited in Chap. 6 on constitutive equations.

it places no restriction on the direction of the process. In the classical mechanics of particles and rigid bodies, kinetic energy and potential energy may be fully transformed from one to the other in the absence of friction or other dissipative mechanisms; the transformation can equally well proceed in either direction, for example, in a swinging pendulum or a vibrating spring-supported mass.

The situation is quite different when thermal phenomena are involved. By means of a friction brake, the kinetic energy of a flywheel can all be converted into internal energy; if the whole system is insulated, the internal energy remains in the system causing its temperature to rise. As far as the first law is concerned, the process could equally well be reversed; the flywheel could be set in motion by converting internal energy into kinetic energy, while the temperature of the system decreases. Such a reversal never occurs; the frictional dissipation is an ***irreversible process***. The second law of thermodynamics puts limits on the direction of such processes.

Another example of the preferred direction is in the flow of heat between two systems at different temperatures; the heat never flows spontaneously from the colder system to the warmer.† This does not mean that heat can never be transferred from a cold system to a hot one; every mechanical refrigerator accomplishes this, but work must be input to do it, and the transfer does not occur spontaneously. The friction-brake example does not prove that heat can never be converted into work either. Thermodynamics textbooks discuss in detail the operation of a heat engine, a cyclically operating system across whose boundaries flow only heat and work, utilizing two heat reservoirs at different temperatures. Heat is extracted from the hot reservoir and partly converted into work by the heat engine, but never entirely so, since a considerable portion of the energy must always be rejected in the form of heat transfer to the cold reservoir. If this were not the case, the cold reservoir would not be needed and, for example, an ocean liner could be propelled across the sea by drawing heat energy from the ocean and converting it into work. One of the classical statements of the second law of thermodynamics, the Kelvin-Planck statement, specifies that it is impossible to construct an engine which, operating continuously in a cycle, will produce no other effect than the extraction of heat from a single heat reservoir and the performance of an equivalent amount of work. These examples not relevant to continuum mechanics have been included only to give some feeling for the historical importance of the second law as a law of nature permitting energy transfer to occur spontaneously only in certain preferred directions. This limitation will be expressed mathematically as an inequality stating that the internal entropy production is always nonnegative and is positive for an irreversible process. This inequality, known as the Clausius-Duhem inequality, is presented in the next-to-last subdivision of this

†This is essentially the statement of the second law of thermodynamics given by Clausius in 1850.

section, where the continuum-mechanics formulation of the second law is given. The lengthy digression preceding it is included in an attempt to give some feeling for the physical meaning of the concept of entropy. The specific entropy (per unit mass) will be denoted by s and the total entropy of the system by S. Then in continuum mechanics $S = \int_V s\rho\, dV$.

Entropy and probability, ordered and disordered states. Just as the first law is intimately related to the concept of internal energy, the second law deals with another new concept not included in classical mechanics, that of ***entropy***.† Unfortunately the concept of entropy is not so readily accessible to intuitions accustomed to mechanical concepts. To get a real feeling for its physical significance, we must temporarily put aside the continuum theory and think of matter as an aggregate of molecules. In statistical mechanics the entropy of a state is related to the probability of the occurrence of that state among all the possible states that could occur. And it is found that changes of state are more likely to occur in the direction of greater disorder when a system is left to itself. Thus, increasing entropy is associated with increasing disorder; the second law seems to imply an almost metaphysical principle of preference for disorder. But the tendency appears quite natural when the combinatorial method is used, as a simple example will illustrate.

Consider a system consisting of a box separated into two chambers by a wall with a hole in it, and containing two identical balls A and B. We can shake the box and then look to see what the state of the system is. When both balls are in the same chamber, the system is more ordered than when there is one in each chamber. There are four possible configurations, as shown in Fig. 5.11, but the two configurations on the right are indistinguishable from each other because the two balls are identical. There are only three observable states (macrostates) for the four configurations (microstates). The disordered macrostate is twice as likely to occur as either of the ordered macrostates, assuming that the microstates are equally probable. When larger numbers are involved the effect is even more striking; with 20 balls the chance of finding them all in one chamber is only about one in a million. Molecular aggregates of

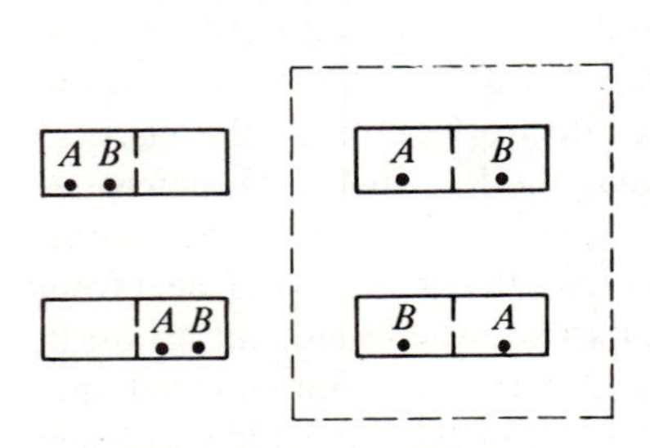

Fig. 5.11 Illustrating ordered and disordered states.

†The name "entropy" is due to Clausius (1865). The concept as he used it is similar to the "calorique" of Carnot (1824), though different from the "caloric" of some other early writers on thermal effects. The gist of the second law of thermodynamics was perceived by Carnot, but the first precise statements of it were made by Clausius.

macroscopic interest contain millions of molecules; this and the probabilistic concepts sketched above should make it clear why, for example, any two equal volume elements of a confined gas always appear to contain the same number of molecules.

These combinatorial ideas suggest the Boltzmann principle postulate that entropy of a state is proportional to the logarithm of its probability. Elementary statistical mechanics based on considering a gas as a collection of rigid molecules leads to an expression of the following form for the total entropy S of a sample of the gas containing N molecules.†

$$S = kN[\ln V + \tfrac{3}{2}\ln\theta] + C \tag{5.6.1}$$

where V is volume, θ is absolute temperature, k is Boltzmann's constant, and C is a constant for which a definite value is obtained only by means of quantum theory. We shall not make explicit use of this formula except to note that it has the same form as an equation developed earlier in classical thermodynamics without use of statistical mechanics, but using certain assumptions about a perfect gas, as follows.

Entropy in classical thermodynamics. The entropy concept first appeared in thermodynamics as a state function related to the heat transfer. We have seen in Sec. 5.4 that the heat input $đq$ per unit mass is not a perfect differential of any function, but it is assumed in classical thermodynamics that a function s, the specific entropy (or entropy per unit mass) exists, such that in any reversible process

$$ds = \left(\frac{đq}{\theta}\right)_{\text{rev}} \tag{5.6.2}$$

is a perfect differential. To illustrate this by a simple example given in most thermodynamics textbooks, consider an ideal gas defined by the gas law

$$pv = R\theta, \tag{5.6.3a}$$

where R is the gas constant for the particular gas, and by the additional assumption‡ that the specific internal energy u is a function only of the temperature θ,

$$u = u(\theta), \tag{5.6.3b}$$

†See, for example, Sommerfeld (1964), p. 229.

‡Some writers call a gas "perfect" or "ideal" if it satisfies only Eq. (3a), at least approximately. Without the additional assumption of Eq. (3b), irreversible processes are possible in a gas described by Eq. (3a).

as suggested by experiments of Joule in 1843. Any process in such an ideal gas is reversible because whenever the independent variables, say specific volume v and temperature θ, return to their initial values, so do the dependent variables, u and pressure p. The first law of thermodynamics takes the form

$$du = đq - p\,dv. \tag{5.6.4}$$

For a constant-volume process this gives

$$du = đq = c_v\,d\theta,$$

where c_v is the specific heat at constant volume. Then the assumption of Eq. (3b) that u is a function of θ only implies that c_v is a function of θ only and that

$$du = c_v(\theta)\,d\theta$$

in any process (volume not constant) for an ideal gas. Hence, the first law, Eq. (4), can be written

$$đq = c_v(\theta)\,d\theta + R\theta\frac{dv}{v}$$

by using Eq. (3a). Division by θ then shows that $đq/\theta$ is a perfect differential; thus the integral of Eq. (2),

$$s - s_0 = \int_{p_0,\,v_0}^{p,\,v}\frac{đq}{\theta} = \int_{\theta_0}^{\theta}c_v(\theta)\frac{d\theta}{\theta} + R\ln\frac{v}{v_0}, \tag{5.6.5}$$

gives the change in entropy for any process (necessarily reversible) in an ideal gas. For this case, entropy is evidently a state function, returning to its initial value whenever the temperature returns to its initial value, as it does according to Eq. (3a) whenever p and v return to their initial values. If we further assume that c_v is independent of θ, Eq. (5) takes the form of Eq. (1), except that it now gives the specific entropy s of unit mass instead of the total entropy $S = \int_V s\rho\,dV$ for N molecules.

The identity in form of Eqs. (1) and (5) then suggests that entropy in statistical physics and thermodynamic entropy are really the same thing. The statistical physics interpretation in terms of probability and tendency toward disordered macrostates furnishes a physical significance for the otherwise rather abstract thermodynamic concept. Of course, the identity of the two forms has been displayed here only for a very special case, the ideal gas. For more general materials, explicit formulas for the entropy are not easy to come by either in thermodynamics or in statistical physics. But it has been possible to make calculations of the entropy of solids, for example. And it should come as no surprise that carbon has a lower entropy in the form of diamond, a hard crystal with atoms closely bound in a highly ordered array, than it has as graphite. See, for example, Kubaschewski and Evans (1958) for numerical values.

The Gibbs relation; entropy as variable conjugate to temperature. Since $\bar{d}q = \theta\, ds$ in a reversible process, the first law, Eq. (4), can be written for an ideal gas in the following form known as a ***Gibbs relation***:

$$du = \theta\, ds - p\, dv \tag{5.6.6a}$$

or, for a homogeneous system in equilibrium,

$$dU = \theta\, dS - p\, dV. \tag{5.6.6b}$$

This shows that entropy S is the extensive variable conjugate to the intensive variable temperature θ for calculating the thermal energy input in the same way that volume V is the extensive variable conjugate to the stress, $-p$, for calculating the mechanical work input. An ***extensive variable*** is one that in a homogeneous system is proportional to the total mass; in general the total amount of it in the system is the sum of the amounts in all its parts, e.g., $S = \int_V s\rho\, dV$. An intensive variable, on the other hand, has the same value at all points in a homogeneous system in equilibrium; in general the point value of an intensive variable does not depend on the size of the system. The densities of extensive variables (e.g., u, s, v) are intensive variables. Since in continuum mechanics we work mainly with the densities, the distinction between extensive and intensive variables tends to be obscured.

Entropy change in an irreversible process. Only reversible processes are possible in the ideal gas of Eqs. (3), and Eq.(5) shows that in the ideal gas the entropy is a state function, determined by the current value of the state variables θ and v up to an arbitrary constant s_0 not determinable in macroscopic thermodynamics. Classical thermodynamics postulates that the entropy of an equilibrium state is a state function, determined (up to an arbitrary constant) by the equilibrium values of the independent state variables. Although explicit formulas for either u or s in terms of the independent state variables are rare, the first law still furnishes a method for calculating the change in u by recording the work input and heat input for a particular process. And Eq. (2) furnishes a method for computing the change in s if we know the heat-input and temperature history for any (possibly hypothetical) reversible process between the same two states

$$\Delta s \equiv s_2 - s_1 = \int_1^2 \left(\frac{\bar{d}q}{\theta}\right)_{\text{rev}}. \tag{5.6.7}$$

Unfortunately, this may not be a useful equation for inelastic deformation processes in materials if the requisite hypothetical reversible process is not known. In a hypothetical, cyclic reversible process returning to the initial state, we have

$$\oint ds = \oint \left(\frac{đq}{\theta}\right)_{\text{rev}} = 0, \tag{5.6.8a}$$

while in an irreversible cyclic process returning to the same state we still have $\oint ds = 0$, but $\oint (đq/\theta)_{\text{actual}} \neq 0$. In fact, experience indicates that in general

$$\oint \left(\frac{đq}{\theta}\right)_{\text{irrev}} < 0. \tag{5.6.8b}$$

We interpret $đq/\theta$ as the entropy input from outside carried by the heat input $đq$; thus in an irreversible cycle the net entropy input is negative. Since the entropy has been assumed to be a state function, returning to its initial value at the end of the cycle, the negative value for net entropy input implies that entropy has been created inside the system; the ***internal entropy production*** is a result of dissipative irreversible processes, e.g., internal friction.

In an irreversible change from state 1 to state 2, the entropy increase is greater than the entropy input by heat transfer

$$\Delta s > \int_1^2 \left(\frac{đq}{\theta}\right)_{\text{irrev}} \tag{5.6.9}$$

because of the internal entropy production, which is always positive in an irreversible process.

In an isolated system, i.e., one with no heat transfer, reversible processes do not change the entropy, while irreversible processes always increase the entropy. All real processes are irreversible, but in some cases the dissipation is small enough to be negligible; the reversible process, even though it is hypothetical, is a useful conceptual model, playing a role in thermodynamics analogous to the frictionless motions of classical mechanics.

The second law of thermodynamics postulates the existence of entropy as a state function, satisfying Eq. (7) for a reversible process and the inequality (9) for an irreversible process. It can be shown to imply the limitations on physical processes mentioned in the introduction to this section. We are now ready to formulate it as an inequality in a form more suitable for use in continuum mechanics.

The Clausius-Duhem inequality. The content of the second law will be rephrased in the notation of continuum mechanics as used in the energy balance and energy equation of Sec. 5.4. The entropy input rate, carried by heat transfer into the mass system instantaneously occupying the volume V bounded by the surface S, is defined by

$$\text{Entropy Input Rate} = \int_V \frac{\rho r}{\theta}\, dV + \int_S -\frac{\mathbf{q}}{\theta}\cdot\hat{\mathbf{n}}\, dS,$$

where r is the internal heat supply per unit mass and unit time (possibly from a radiation field), and $\mathbf{q}$ is the outward heat flux vector. Recall that we are dealing with a closed system not gaining or losing mass; an open system in a control surface fixed in space would have additional entropy input $\int_S -\rho s\mathbf{v}\cdot\hat{\mathbf{n}}\, ds$ carried by the mass flux. According to the second law,

$$\text{Rate of Entropy Increase} \geq \text{Entropy Input Rate}$$

or

$$\frac{d}{dt}\int_V s\rho\, dV \geq \int_V \frac{r}{\theta}\rho\, dV + \int_S -\frac{\mathbf{q}}{\theta}\cdot\hat{\mathbf{n}}\, dS. \tag{5.6.10a}$$

This is the integral form of the ***Clausius-Duhem inequality.*** The inequality implies internal entropy production in an irreversible process; the equality holds for a reversible process. The inequality is postulated to hold for arbitrary choice of the volume V. Then, after transforming the surface integral to a volume integral, we conclude that the following local version of the Clausius-Duhem inequality holds at each point:

Clausius-Duhem Inequality

$$\frac{ds}{dt} \geq \frac{r}{\theta} - \frac{1}{\rho}\operatorname{div}\frac{\mathbf{q}}{\theta} \tag{5.6.10b}$$

or

$$\gamma \equiv \frac{ds}{dt} - \frac{r}{\theta} + \frac{1}{\rho\theta}\operatorname{div}\mathbf{q} - \frac{\mathbf{q}}{\rho\theta^2}\cdot\operatorname{grad}\theta \geq 0, \tag{5.6.10c}$$

where γ is the ***internal entropy production*** rate per unit mass.

[*Remark*: Even when there is no heat input to a given finite body, so that the right side of Eq. (10a) is zero, it does not follow that the right side of Eq. (10b) is zero, because of the possibility of heat conduction from one part of the body to another.]

Truesdell and Noll (1965), Sec. 79, propose a stronger assumption, requiring separately†

$$\gamma_{\text{loc}} \geq 0 \quad \text{and} \quad \gamma_{\text{con}} \geq 0, \quad \text{where} \tag{5.6.10d}$$

$$\gamma_{\text{loc}} = \frac{ds}{dt} - \frac{r}{\theta} + \frac{1}{\rho\theta} \operatorname{div} \mathbf{q} \qquad \gamma_{\text{con}} = -\frac{1}{\rho\theta^2} \mathbf{q} \cdot \operatorname{grad} \theta \tag{5.6.11}$$

(and requiring both $\gamma_{\text{loc}} = 0$ and $\gamma_{\text{con}} = 0$ for a reversible process), where γ_{loc} is the ***local entropy production*** and γ_{con} is the ***entropy production by heat conduction.***

In continuum mechanics, the Clausius-Duhem inequality, which must be satisfied for every possible process, imposes restrictions on constitutive equations. For example, if the constitutive equation for heat conduction is proposed as $\mathbf{q} = \kappa \operatorname{grad} \theta$, the second inequality (10d) requires the constant κ to be negative; restrictions are also imposed on the signs and magnitudes of the elastic constants in linearized elasticity (see Chap. 6).

Entropy as a functional in continuum mechanics. The Clausius-Duhem inequality as used in continuum mechanics differs from the usual second law of thermodynamics only in not postulating explicitly that entropy is a state function determined by the instantaneous values of the other state variables. Many writers on continuum thermodynamics have in fact postulated it as a state function of ***all*** the state variables, including possibly some ***hidden variables*** not available for macroscopic observation. If, however, the "state" of a continuum is taken to be defined by a limited number of explicitly enumerated macroscopic state variables, observable at least in principle, then the entropy must in general depend on the history of this limited number of state variables and not merely on their current values. We have already seen in Sec. 5.4 that the internal energy as a function of the deformation gradient $\mathbf{F}$ and the temperature θ must be history-dependent for inelastic deformation of a solid. The same is true of the entropy. Coleman (1964) at the outset of his thermodynamics of simple materials with memory defines a simple material as one in which the present values of specific entropy s, specific internal energy u, stress $\mathbf{T}$, and heat flux $\mathbf{q}$ are determined by the ***histories*** of $\mathbf{F}$ and θ and by the present value of grad θ for the particle. Such a function of the history of the variables is called a ***functional*** of the variables to distinguish it from an ordinary function depending only on the instantaneous values of the variables. Only in very special thermoelastic materials are u, $\mathbf{T}$ and $\mathbf{q}$ determined by the instantaneous values of θ, grad θ, and $\mathbf{F}$.

†The second inequality (10d) implies that heat does not flow spontaneously toward a hotter part of the body. The first inequality (10d) has been shown to follow from an apparently less specific assumption in Coleman's (1964) thermodynamics of simple materials. See, for example, Truesdell and Noll (1965), Eq. (96b.103).

Coleman and Mizel (1964) considered the class of materials in which history appeared only to the extent of determining the derivative $\dot{\mathbf{F}}$, postulating constitutive equations of the form

$$\begin{aligned} s &= s(\theta, \text{grad } \theta, \mathbf{F}, \dot{\mathbf{F}}) \\ u &= u(\theta, \text{grad } \theta, \mathbf{F}, \dot{\mathbf{F}}) \\ \mathbf{T} &= \mathbf{T}(\theta, \text{grad } \theta, \mathbf{F}, \dot{\mathbf{F}}) \\ \mathbf{q} &= \mathbf{q}(\theta, \text{grad } \theta, \mathbf{F}, \dot{\mathbf{F}}). \end{aligned} \tag{5.6.12}$$

Assuming that the second Eq. (12) was invertible to solve for θ, they then expressed the first equation in the form

$$s = f(u, \text{grad } \theta, \mathbf{F}, \dot{\mathbf{F}})$$

and were able to show by the Clausius-Duhem inequality that the function f in this last equation must be independent of both grad θ and $\dot{\mathbf{F}}$, so that a ***caloric equation of state*** of the form

$$s = f(u, \mathbf{F}) \tag{5.6.13}$$

must hold in any material of the class defined by Eqs. (12). Earlier treatments of continuum thermodynamics [see, for example, Truesdell and Toupin (1960)] have ***postulated*** that a caloric equation of state of the form

$$u = u(s, \boldsymbol{\nu}) \tag{5.6.14}$$

[or inversely $s = s(u, \boldsymbol{\nu})$] always holds. Here the ***thermodynamic substate*** $\boldsymbol{\nu}$ represents a set of n variables including ***all*** the mechanical and electrical state variables; it is assumed that just one more state variable of a thermal character (here chosen as s) is needed to define completely the thermodynamic state. The cited works of Coleman and Mizel (1964) and Coleman (1964) do not invalidate the earlier assumption; they merely ignore it; the entropy may or may not be a state function of all the state variables, but it is in general not a state function of a limited number of them. Coleman's approach promises to be useful, since it deals with a limited number of explicitly enumerated macroscopic state variables and not a vague collection of unspecified substate variables.

When a caloric equation of state is assumed, it permits the generalization of the Gibbs relation, given in Eq. (6a) for an ideal gas, and indicates appropriate thermodynamic tensions conjugate to chosen thermodynamic substate variables. These implications of an assumed caloric equation of state and the use of thermodynamic potentials such as Helmholtz free energy, enthalpy, and free enthalpy will be discussed in Sec. 5.7, which also includes the concept of a dissipation function to describe the internal entropy production.

EXERCISES

A ***pure substance*** is one which is chemically homogeneous and remains invariant in chemical composition during the process. Experience indicates that in the absence of motion, gravity, capillarity, electricity, and magnetism the state of a pure substance in the form of a fluid is completely determined by any two independent properties. Assuming that this is strictly correct, then the various other properties are in principle expressible as functions of the two chosen independent properties by ***equations of state.*** For example, if temperature θ and specific volume $v = 1/\rho$ are chosen as independent properties, then

$$u = u(\theta, v) \quad \text{and} \quad p = p(\theta, v).$$

Alternatively, we may invert the second equation to obtain $v = v(\theta, p)$, whence the first equation becomes $u = u[\theta, v(\theta, p)]$, expressing the specific internal energy u as a function of θ and the thermodynamic pressure p. Exercises 1–8 are concerned with pure substances which are assumed to obey such equations of state. Some exercises assume either the ideal gas or a possibly better approximation to a real gas called the ***van der Waals' gas.*** For more refined approximations to real fluids see books on thermodynamics for engineers.

1. A ***specific heat*** is defined by the ratio $\bar{d}q/d\theta$, which in a fluid pure substance is equal to $(du + p\,dv)/d\theta$ by the first law. Show that the ***specific heat at constant pressure*** is $c_p = (\partial u/\partial\theta)_p + p(\partial v/\partial\theta)_p$. The subscripts indicate the variable held constant. Show also that $(\partial u/\partial\theta)_p = (\partial u/\partial\theta)_v + (\partial u/\partial v)_\theta(\partial v/\partial\theta)_p$ and hence that $c_p - c_v = [p + (\partial u/\partial v)_\theta](\partial v/\partial\theta)_p$. Show that this can be brought to the form

$$c_p - c_v = pv\beta + v\beta\left(\frac{\partial u}{\partial v}\right)_\theta \quad \text{where} \quad \beta = \frac{(\partial v/\partial\theta)_p}{v}$$

   is the coefficient of volume expansion at constant pressure.

2. Evaluate $c_p - c_v$ for the ideal gas of Eqs. (5.6.3). See Ex. 1.

3. In any ideal-gas process $p\,dv + v\,dp = R\,d\theta$.
   (a) Show that in an ***adiabatic process*** where $\bar{d}q = 0$, this equation can be brought to the form $(c_v + R)p\,dv + c_v v\,dp = 0$ by using the first law of thermodynamics and $du = c_v\,d\theta$.
   (b) Now use the result of Ex. 2 to derive $pv^\gamma = \text{const.}$, where $\gamma = c_p/c_v$, for an adiabatic process in an ideal gas, if γ is assumed to remain constant.

4. Show that in a fluid pure substance where $u = u(\theta, v)$, the total differential du is given by

$$du = c_v d\theta + \left(\frac{c_p - c_v}{\beta v} - p\right) dv.$$

   See Ex. 1.

5. Use the results of Ex. 4 to show that

$$u = u_0 + \int_{\theta_0}^{\theta} c_v d\theta + a\left(\frac{1}{v_0} - \frac{1}{v}\right)$$

in a ***van der Waals' gas*** defined by

$$\left(p + \frac{a}{v^2}\right)(v - b) = R\theta \quad \text{and} \quad \left(\frac{\partial u}{\partial v}\right)_\theta = \frac{a}{v^2}$$

where a, b and R are constants.

6. When the equations of state relating three variables such as p, v, θ are invertible some important relationships among the partial derivatives can be derived. For example, from $v = v[\theta, p(\theta, v)]$ we get

$$1 \equiv \left(\frac{\partial v}{\partial v}\right)_\theta = \left(\frac{\partial v}{\partial p}\right)_\theta \left(\frac{\partial p}{\partial v}\right)_\theta \quad \text{and} \quad 0 \equiv \left(\frac{\partial v}{\partial \theta}\right)_v = \left(\frac{\partial v}{\partial \theta}\right)_p + \left(\frac{\partial v}{\partial p}\right)_\theta \left(\frac{\partial p}{\partial \theta}\right)_v.$$

Show that these imply the cyclic relation

$$\left(\frac{\partial v}{\partial p}\right)_\theta \left(\frac{\partial p}{\partial \theta}\right)_v \left(\frac{\partial \theta}{\partial v}\right)_p = -1.$$

7. The Gibbs relation $du = \theta\, ds - p\, dv$ of Eq. (5.6.6a) is assumed valid for any process in a pure fluid substance, since it is a relationship between state variables, even though it was obtained by using $\bar{d}q = \theta\, ds$ which applies only to a reversible process. If $u = u(\theta, v)$, show that the Gibbs relation leads to

$$ds = \frac{1}{\theta}\left(\frac{\partial u}{\partial \theta}\right)_v d\theta + \frac{1}{\theta}\left[p + \left(\frac{\partial u}{\partial v}\right)_\theta\right] dv$$

and that this implies

$$\left(\frac{\partial s}{\partial \theta}\right)_v = \frac{1}{\theta}\left(\frac{\partial u}{\partial \theta}\right)_v \quad \text{and} \quad \left(\frac{\partial s}{\partial v}\right)_\theta = \frac{1}{\theta}\left[p + \left(\frac{\partial u}{\partial v}\right)_\theta\right].$$

From these last two equations calculate $\partial^2 s/\partial\theta\, \partial v$ and $\partial^2 s/\partial v\, \partial\theta$, and from the equality of the two mixed derivatives show that $p + (\partial u/\partial v)_\theta = \theta(\partial p/\partial\theta)_v$. Now use the results of Ex. 1 and the first result of Ex. 6, and the definitions of c_v and of $\kappa = -(\partial v/\partial p)_\theta/v$, the isothermal compressibility, to obtain

$$ds = \frac{c_v}{\theta} d\theta + \frac{\beta}{\kappa} dv.$$

8. For the van der Waals' gas defined in Ex. 5, show that

$$\beta = \frac{Rv^2(v - b)}{R\theta v^3 - 2a(v - b)^2} \qquad \kappa = \frac{v^2(v - b)^2}{R\theta v^3 - 2a(v - b)^2}$$

where β and κ are defined in Ex. 1 and Ex. 7. (*Hint*: See Ex. 6.) Hence use the result of Ex. 7 to show that

$$s - s_0 = \int_{\theta_0}^{\theta} \frac{c_v}{\theta} d\theta + R \ln\left(\frac{v - b}{v_0 - b}\right).$$

9. In the derivation of the Clausius-Duhem inequality, show that Eq. (5.6.10b) follows from (10a), and that (10c) follows from (10b) with appropriate definition of γ. Write the forms of Eqs. (11) when $\mathbf{q} = -k \operatorname{grad} \theta$.

10. By using the energy Eq. (5.4.7) show that the Clausius-Duhem inequality (10c) takes the form $\rho(\theta\dot{s} - \dot{u}) + \mathbf{T} : \mathbf{D} - (1/\theta)\, \mathbf{q} \cdot \operatorname{grad} \theta \geq 0$.

11. As an example of how the Clausius-Duhem inequality places restrictions on possible constitutive equations, consider a class of viscous fluids assumed defined

by four constitutive equations of the form

$$\mathbf{T} = \mathbf{T}\left(s, \frac{1}{\rho}, \mathbf{D}\right) \quad \mathbf{q} = \mathbf{q}\left(s, \frac{1}{\rho}, \mathbf{D}\right) \quad u = u\left(s, \frac{1}{\rho}, \mathbf{D}\right) \quad \theta = \theta\left(s, \frac{1}{\rho}, \mathbf{D}\right)$$

giving the stress, heat flux vector, internal energy, and temperature in terms of entropy, density, and rate of deformation. Show that the result of Ex. 10 takes the form

$$\rho\left(\theta - \frac{\partial u}{\partial s}\right)\dot{s} - \rho\frac{\partial u}{\partial D_{km}}\dot{D}_{km} + \frac{\partial u}{\partial(1/\rho)}\frac{\dot{\rho}}{\rho} + T_{km}\,D_{km} - \frac{1}{\theta}q_k\frac{\partial \theta}{\partial x_k} \geq 0.$$

If this is to hold for arbitrary independent prescriptions of $\dot{s}$ and $\dot{D}_{km}$ we must have $\theta = \partial u/\partial s$ and $\partial u/\partial D_{km} = 0$. Hence u is independent of $\mathbf{D}$, and therefore $\theta = \partial u/\partial s$ is also independent of $\mathbf{D}$. If we define the thermodynamic pressure p by $p = -\partial u/\partial(1/\rho)$ and use the equation of continuity, show that the inequality then reduces to $(T_{km} + p\delta_{km})D_{km} - (1/\theta)q_k\,(\partial\theta/\partial x_k) \geq 0$. Since $\mathbf{T}$, p, and $\mathbf{q}$ are independent of $\partial\theta/\partial x_k$ this requires $q_k = 0$. Hence this class of fluids must have $\mathbf{q} \equiv 0$ and can only respond adiabatically. Evidently a more general assumption is needed for a realistic description of a fluid, including grad θ among the independent variables. With $q_k \equiv 0$, the inequality reduces to $(T_{km} + p\delta_{km})D_{km} \geq 0$.

5.7 The caloric equation of state; Gibbs relation; thermodynamic tensions; thermodynamic potentials; dissipation function

Continuum thermodynamics based on a caloric equation of state assumes that the local internal energy u per unit mass is determined by the ***thermodynamic state***, specified by $n + 1$ state variables $\nu_1, \nu_2, \cdots, \nu_n$, and s, where the ν_j are the ***thermodynamic substate variables*** and s is the specific entropy.† The substate variables have mechanical or electromagnetic dimensions, but are otherwise left arbitrary in the general formulation. In the simplest case of a fluid pure substance there is just one substate variable, the specific volume v. In ideal elasticity we will have nine substate variables, the components of one of the strain or deformation tensors. In other cases, the identification of suitable substate variables may be difficult. The ***basic assumption of thermodynamics*** has been that in addition to the n substate variables ν_j just one additional dimensionally independent scalar parameter suffices to determine the specific internal energy u. This assumes that there exists a ***caloric equation of state***

$$u = u(s, \nu, \mathbf{X}). \tag{5.7.1}$$

†See Truesdell and Toupin (1960). In this formulation entropy is taken as the primitive concept for the construction of thermodynamics, as it was by Gibbs in 1873, instead of temperature.

In any particular motion $\mathbf{x}=\mathbf{x}(\mathbf{X}, t)$ of the continuum, the thermodynamic state variables ν_j and s as well as the internal energy u will be functions of position $\mathbf{x}$ and time t, but the basic assumption is that we do not need to know the current position or the history of the motion to determine u for a particle, if we know the current values of ν_j and s. The appearance of $\mathbf{X}$ in the equation permits the possibility that the functional dependence on the ν_j and s can be different for different particles in inhomogeneous media; we suppose the dependence on $\mathbf{X}$, as on the other variables, to be continuous, but for most purposes we do not explicitly include $\mathbf{X}$ in the equations.

Thermodynamic temperature θ and ***thermodynamic tensions*** τ_j are defined by

$$\theta \equiv \left(\frac{\partial u}{\partial s}\right)_{\mathbf{v}} \quad \tau_j \equiv \left(\frac{\partial u}{\partial \nu_j}\right)_s \qquad j = 1, 2, \cdots, n. \tag{5.7.2}$$

(Subscripts outside the parentheses indicate variables held constant.) Thus, in any real or hypothetical change in the thermodynamic state of a given particle $\mathbf{X}$

$$du = \theta\, ds + \tau_j\, d\nu_j \qquad \text{sum } j = 1 \text{ to } n. \tag{5.7.3}$$

This is known as a ***Gibbs relation***, although Gibbs gave the equation only for the case of a fluid, whose only substate variable is the specific volume v. For the fluid the Gibbs relation takes the form

$$du = \theta\, ds - p\, dv, \quad \text{so that} \quad \theta = \left(\frac{\partial u}{\partial s}\right)_v \quad \text{and} \quad -p = \left(\frac{\partial u}{\partial v}\right)_s \tag{5.7.4}$$

where p is the thermodynamic pressure; the thermodynamic tension ***conjugate to*** the specific volume v is $-p$, as θ is conjugate to s. This p is not in general equal to the mean pressure† $\bar{p}$, and its relation to any measured pressure for nonideal fluids in motion must be established. [In Sec. 6.3 it is shown that p will be equal to $\bar{p}$ in a Newtonian fluid only in the two special cases: (1) when there is no volume viscosity, or (2) when the flow is at constant density.]

From the caloric equation of state, Eq. (1), and the definitions, Eq. (2), it follows that the temperature and the thermodynamic tensions are functions of the thermodynamic state: For a given particle

$$\theta = \theta(s, \boldsymbol{\nu}) \qquad \tau_j = \tau_j(s, \boldsymbol{\nu}). \tag{5.7.5}$$

We assume that the first Eq. (5) is invertible to yield

$$s = s(\theta, \boldsymbol{\nu}) \tag{5.7.6}$$

†In Chap. 3 the mean pressure was denoted by p. Here we use $\bar{p} = T_{kk}/3$.

and we substitute this into Eq. (1) to obtain an alternative form for the caloric equation of state:

$$u = u(\theta, \boldsymbol{\nu}, \mathbf{X}). \tag{5.7.7}$$

Substituting Eq. (6) into the second Eq. (5), we obtain

$$\tau_j = \tau_j(\theta, \boldsymbol{\nu}, \mathbf{X}) \tag{5.7.8}$$

or, assuming invertibility,

$$\nu_j = \nu_j(\theta, \boldsymbol{\tau}, \mathbf{X}). \tag{5.7.9}$$

These last two equations are called ***thermal equations of state***. (In most cases, explicit dependence on the particle **X** will not be indicated.) The thermal equations of state resemble stress-strain relations, but some caution is necessary in interpreting the tensions as stresses and the ν_j as strains. Even if the ν_j are elastic strains, the τ_j may differ from the usual components of the stress tensor **T**, as the remarks about the difference between $\bar{p}$ and p in a fluid should warn us. The simplest example of a thermal equation of state is the equation of state of a ***perfect gas***, Eq. (5.6.3a) where $\bar{p} = p$.

Thermodynamic potentials. Based on the assumed existence of a caloric equation of state, four thermodynamic potentials are introduced, each useful for a certain choice of the independent state variables, as tabulated in Table 5-1.

TABLE 5-1

THERMODYNAMIC POTENTIALS

Potential		*Relation to u*	*Independent Variables*
Internal energy	u	u	s, ν_j
Helmholtz free energy	ψ	$\psi = u - s\theta$	θ, ν_j
Enthalpy	h	$h = u - \tau_j \nu_j$	s, τ_j
Free enthalpy, or Gibbs function	g	$g = u - s\theta - \tau_j \nu_j$ $= h - s\theta$	θ, τ_j

The enthalpy definition here, following Truesdell and Toupin (1960), differs from Gibbs (1875), who put $h = u + pv$, whence $g = h - s\theta = u + pv - s\theta$, when there is more than one substate variable. The Gibbs procedure is useful when the temperature is controlled and a uniform hydrostatic pressure is maintained. By means of Eqs. (1) to (9) any one of the potentials can be expressed in terms of any one of the four choices of state variables listed in Table 5-1, but when the choices are made as listed, especially convenient formulations result. In any actual or hypothetical change obeying the equations of state we have

$$\left.\begin{aligned} du &= \theta\, ds + \tau_j\, dv_j, \\ d\psi &= -s\, d\theta + \tau_j\, dv_j, \\ dh &= \theta\, ds - v_j\, d\tau_j, \\ dg &= -s\, d\theta - v_j\, d\tau_j. \end{aligned}\right\} \tag{5.7.10}$$

[The first of these is the Gibbs relation of Eq. (3).] From these differential expressions we obtain the following partial derivative expressions. (Subscripts indicate what state variables are held constant. Of course, when the partial derivative is with respect to v_j, for example, the other v_k for $k \neq j$ are held constant too.)

$$\begin{aligned} \theta &= \left(\frac{\partial u}{\partial s}\right)_{\mathbf{v}} & \tau_j &= \left(\frac{\partial u}{\partial v_j}\right)_s \\ s &= -\left(\frac{\partial \psi}{\partial \theta}\right)_{\mathbf{v}} & \tau_j &= \left(\frac{\partial \psi}{\partial v_j}\right)_\theta \\ \theta &= \left(\frac{\partial h}{\partial s}\right)_{\boldsymbol{\tau}} & v_j &= -\left(\frac{\partial h}{\partial \tau_j}\right)_s \\ s &= -\left(\frac{\partial g}{\partial \theta}\right)_{\boldsymbol{\tau}} & v_j &= -\left(\frac{\partial g}{\partial \tau_j}\right)_\theta. \end{aligned} \tag{5.7.11}$$

The ***free energy*** ψ is the portion of the internal energy available for doing work at constant temperature. The ***enthalpy*** h as defined here is the portion of the internal energy that can be released as heat when the thermodynamic tensions are held constant. Thermodynamic functions related to the potentials defined here were introduced by Massieu (1869, 1876) The theory was largely developed by Gibbs (1875); see also Helmholtz (1882).

From the first two Eqs. (11) it is evident that the internal energy density u is a potential for the thermodynamic tensions in an isentropic process ($s =$ const.), while the Helmholtz free energy density ψ is a potential for the tensions in an isothermal process ($\theta =$ const.). With certain additional assumptions implying full recoverability and for certain specific choices of the substate variables $\mathbf{v}$ the thermodynamic tensions may be identified with the Piola-Kirchhoff stress tensors $\tilde{\mathbf{T}}$ and $\mathbf{T}^0$ introduced in Sec. 5.3, as is shown below. ***Two fully recoverable cases*** are the cases of†

†These idealized cases are sufficient conditions for recoverability, but not necessary. For a set of necessary and sufficient conditions, see Truesdell and Toupin (1960), Sec. 256A, who also give references to earlier formulations in elasticity.

Case 1. adiabatic and isentropic deformation, or
Case 2. isothermal deformation with reversible heat transfer.

In these two cases, we can show that the work of the thermodynamic tensions is recoverable and that the external stress power equals the rate of work of the thermodynamic tensions, i.e.

$$T_{ij}D_{ij} = \rho \sum_{j=1}^{n} \tau_j \dot{v}_j, \tag{5.7.12}$$

so that the external power input is recoverable.

Case 1. The first Eq. (10) shows that $\rho\, du = \rho\theta\, ds + \rho\tau_j dv_j$. If the deformation is isentropic, $ds = 0$ and $\tau_j\, dv_j = du$. Since du is the differential of a state function, $\oint \tau_j\, dv_j = 0$, which shows that the work of the τ_j is recoverable. If we further assume adiabatic conditions, then $\rho r - q_{j,j} = 0$ and the energy equation, Eq. (5.4.7), $\rho\dot{u} = T_{ij}D_{ij} + \rho r - q_{j,j}$ becomes $\rho\dot{u} = T_{ij}D_{ij}$ implying Eq. (12). Note that it is not sufficient to have no heat input into a finite body; there must be no heat conduction between parts of the body. This is not ensured by insulating the body, but is believed to be approximated by a rapid deformation, e.g., in one cycle of a vibration where not enough time is allowed for significant heat transfer.

Case 2. The second Eq. (10) shows that $\rho d\psi = -\rho s d\theta + \rho\tau_j dv_j$. Hence, in an isothermal deformation where $d\theta = 0$, we have $\tau_j dv_j = d\psi$, the differential of the Helmholtz free energy, again implying recoverability of the work of the thermodynamic tensions. By means of the relation $\psi = u - s\theta$, the energy equation, Eq. (5.4.7), can be written in the form

$$\rho\dot{\psi} = T_{ij}D_{ij} + \rho r - q_{j,j} - \rho\dot{s}\theta - \rho s\dot{\theta}. \tag{5.7.13}$$

In the isothermal deformation, the last term vanishes, and the further assumption of reversible heat transfer is made, so that

$$\dot{s} = \frac{1}{\theta}\left[r - \frac{1}{\rho} q_{j,j}\right]$$

reduces Eq. (13) to $\rho\dot{\psi} = T_{ij}D_{ij}$, again implying Eq. (12), since $\rho\dot{\psi} = \rho\tau_j\dot{v}_j$ in this case.†

Identification of thermodynamic tensions with stress components. This is possible for the two fully recoverable cases for certain choices of the substate

†The concept of recoverable work is bound to the existence of a caloric equation of state, and only thermodynamic tensions derived from a potential defined by such an equation do work not contributing to entropy production. The assumed existence of a caloric equation of state does not imply our knowledge of a formula for it.

variables ν_j. We consider here two possible choices of substate variables: the Lagrangian finite-strain tensor **E** (which reduces to the small-strain tensor when displacement gradients are small) and the deformation-gradient tensor **F**, assuming that the substate is completely defined by the nine components of **E** or **F**.

When the nine rectangular Cartesian components E_{IJ} of the Lagrangian strain tensor are chosen as the substate variables and the tensions τ_j are relabeled τ_{IJ}, Eq. (12) can be brought to the following form by use of Eq. (5.4.8) for arbitrary $\dot{E}_{IJ}$:

$$\frac{\rho}{\rho_0}\tilde{T}_{IJ}\dot{E}_{IJ} = \rho\tau_{IJ}\dot{E}_{IJ} \tag{5.7.14}$$

per unit deformed volume (recall that $\rho\, dV = \rho_0\, dV_0$), where the $\tilde{T}_{IJ}$ are the rectangular Cartesian components of the second Piola-Kirchhoff stress tensor introduced in Sec. 5.3, a symmetric tensor in the nonpolar case assumed here. Unless the internal energy function u (or ψ in Case 2) has been symmetrized in its arguments E_{IJ} and E_{JI} [e.g., by replacing each of them by $(E_{IJ} + E_{JI})/2$], we cannot conclude immediately that $\tau_{IJ} = (1/\rho_0)\tilde{T}_{IJ}$. For example, since $\dot{E}_{IJ}$ is symmetric, we cannot choose the $\dot{E}_{IJ}$ so that only $\dot{E}_{12}$ is nonzero but only so that $\dot{E}_{12}$ and $\dot{E}_{21}$ are nonzero; then Eq. (14) becomes

$$\frac{\rho}{\rho_0}[\tilde{T}_{12}\dot{E}_{12} + \tilde{T}_{21}\dot{E}_{21}] = \rho[\tau_{12}\dot{E}_{12} + \tau_{21}\dot{E}_{21}]$$

whence, using the symmetry of $\tilde{\mathbf{T}}$ and **E**,

$$\frac{1}{\rho_0}\tilde{T}_{12} = \frac{1}{2}(\tau_{12} + \tau_{21}) = \frac{1}{2}\left(\frac{\partial u}{\partial E_{12}} + \frac{\partial u}{\partial E_{21}}\right)_s$$

or, in general (Case 1),

$$\frac{1}{\rho_0}\tilde{T}_{IJ} = \frac{1}{2}(\tau_{IJ} + \tau_{JI}) = \frac{1}{2}\left(\frac{\partial u}{\partial E_{IJ}} + \frac{\partial u}{\partial E_{JI}}\right)_s. \tag{5.7.15}$$

If the function u (or ψ in Case 2) has been symmetrized in its arguments, then $\partial u/\partial E_{IJ} = \partial u/\partial E_{JI}$, and it follows that

$$\frac{1}{\rho_0}\tilde{T}_{IJ} = \tau_{IJ}, \tag{5.7.16}$$

showing that ***the thermodynamic tension conjugate to*** **E** is $(1/\rho_0)\tilde{\mathbf{T}}$.

The small-strain case follows immediately under the assumption that all displacement-gradient components are small compared to unity. Then E_{IJ} reduces to the small strain ϵ_{ij}, $\tilde{\mathbf{T}}$ reduces to **T**, and Eq. (12) becomes

$T_{ij}D_{ij} = T_{ij}\dot{e}_{ij}$. (Note that $\rho/\rho_0 = 1$ to the same degree of approximation.) The thermodynamic tensions are then the Cauchy stress components T_{ij}.

If the substate variables are chosen as the rectangular Cartesian components $x_{i,J}$ of the deformation gradient **F**, then the ***thermodynamic tension conjugate to the deformation gradient*** $x_{i,J}$ is

$$\tau_{iJ} = \frac{1}{\rho_0} T^0_{Ji}, \tag{5.7.17}$$

so that $\rho_0 \tau_{iJ}$ are the rectangular components of ***the transpose of the first Piola-Kirchhoff stress tensor*** defined in Sec. 5.3. {Some writers [e.g., Truesdell and Toupin (1960) and Truesdell and Noll (1965)] define the Cauchy stress so that the second index on its rectangular Cartesian component T_{ij} identifies the plane, while the first one identifies the component of the traction vector on the plane; then their resulting Piola-Kirchhoff stress $\mathbf{T}_R$ is the transpose of $\mathbf{T}^0$ and is equal to ρ_0 times the thermodynamic tension conjugate to **F**.} To derive Eq. (17) we begin with the expression $T_{ji} v_{i,j}$ for the stress power per unit volume [see Eq. (5.4.2a)], and substitute

$$T_{ji} = \frac{\rho}{\rho_0} x_{j,J} T^0_{Ji},$$

by Eq. (5.3.24), to obtain

$$\left.\begin{aligned} T_{ij}D_{ij} = T_{ji}v_{i,j} = \frac{\rho}{\rho_0} x_{j,J} T^0_{Ji} v_{i,j} = \frac{\rho}{\rho_0} T^0_{Ji} \frac{d}{dt}(x_{i,J}) \quad \text{or} \\ \mathbf{T}:\mathbf{D} = \frac{\rho}{\rho_0}(\mathbf{T}^0)^T : \dot{\mathbf{F}} = \frac{\rho}{\rho_0}\mathbf{T}_R : \dot{\mathbf{F}} \end{aligned}\right\} \tag{5.7.18}$$

for the stress power per unit deformed volume. The last step in the first Eq. (18) follows from the fact that

$$\left.\begin{aligned} \frac{d}{dt}(x_{i,J}) = v_{i,j}x_{j,J} \quad \text{or} \\ \dot{\mathbf{F}} = \mathbf{L}\cdot\mathbf{F} \qquad \text{[see Eq. (4.5.14)].} \end{aligned}\right\} \tag{5.7.19}$$

In Eq. (19), **L** denotes the spatial gradient of **v** with rectangular Cartesian components $v_{i,j}$. Equation (17) follows from the requirement that Eqs. (18) hold for arbitrary choices of the $x_{i,J}$. Hence,

$$\begin{aligned} &\text{in Case 1:} \quad \frac{1}{\rho_0} T^0_{Ji} = \left(\frac{\partial u}{\partial x_{i,J}}\right)_s \\ &\text{in Case 2:} \quad \frac{1}{\rho_0} T^0_{Ji} = \left(\frac{\partial \psi}{\partial x_{i,J}}\right)_\theta. \end{aligned} \tag{5.7.20}$$

[Alternatively if the substate variable is chosen as $\mathbf{F}^T$ with components $\partial_J x_i$, the thermodynamic tensions are T^0_{Ji}, since $\dot{\mathbf{F}}^T = \mathbf{F}^T \cdot \mathbf{L}^T$, where the components of $\mathbf{L}^T$ are $\partial_j v_i$. Then

$$\mathbf{T}:\mathbf{D} = \frac{\rho}{\rho_0}\mathbf{T}^0 : \dot{\mathbf{F}}^T \quad \text{and}$$

$$\text{in Case 1:} \quad \frac{1}{\rho_0} T^0_{Ji} = \left(\frac{\partial u}{\partial(\partial_J x_i)_s}\right) \tag{5.7.21}$$

$$\text{in Case 2:} \quad \frac{1}{\rho_0} T^0_{Ji} = \left(\frac{\partial \psi}{\partial(\partial_J x_i)_\theta}\right)$$

Since neither $\mathbf{T}^0$ nor $\mathbf{F}$ is symmetric, it is necessary to be consistent in the choice.]

Dissipative power and internal entropy production. Dissipation function. When the stress power is not fully recoverable, it is sometimes assumed that the stress tensor $\mathbf{T}$ can be separated into the sum of two parts, say

$$\mathbf{T} = {}_R\mathbf{T} + {}_D\mathbf{T}, \tag{5.7.22}$$

where ${}_R\mathbf{T}$ is the recoverable part for which the internal energy is assumed to be a potential, related to ${}_R\mathbf{T}$ in the same way it is related to $\mathbf{T}$ in the fully recoverable cases, and ${}_D\mathbf{T}$ is the dissipative part of the Cauchy stress. It is not at all clear that such a separation can always be made,† especially for an explicit choice of a limited number of substate variables. Truesdell and Toupin (1960) and Eringen (1962) have given general formulations without explicit identification of the substate variables. See Ziegler (1963) for a motivation for the assumption based on a modified statistical mechanics for irreversible processes. To illustrate the idea of dissipation we will consider only a few simple cases.

For example, in a fluid we assume ${}_RT_{ij} = -p\delta_{ij}$ [see Eq. (4)]. With the substate choices above [see Eqs. (15), (20), (5.3.26), and (5.3.24)], we assume

$${}_RT_{ji} = \rho \frac{\partial x_j}{\partial X_J}\frac{\partial x_i}{\partial X_I}\left(\frac{\partial u}{\partial E_{IJ}}\right)_s = \rho \frac{\partial x_j}{\partial X_J}\left(\frac{\partial u}{\partial(x_{i,J})}\right)_s, \tag{5.7.23}$$

while the dissipative part is the remainder of the Cauchy stress,

$${}_DT_{ji} = T_{ji} - {}_RT_{ji}. \tag{5.7.24}$$

It is supposed that only ${}_D\mathbf{T}$ contributes to the internal entropy production. By

†Alternatively, we could deal only with the total Cauchy stress, and suppose the strain to be separated into a recoverable part and a dissipative part, as in plasticity (see Secs. 6.5 and 6.6) where, for small strains, it is assumed that the strain is the sum of an elastic strain and a plastic strain (permanent set).

eliminating u between Eq. (3) and the energy equation, Eq. (5.4.7), we obtain the following differential equation for the specific entropy s (with $T_{ij}D_{ij} = T_{ji}v_{i,j}$ in the nonpolar case)

$$\rho\theta\dot{s} = T_{ji}v_{i,j} + \rho r - q_{j,j} - \rho\sum_{j=1}^{n}\tau_j\dot{\nu}_j. \tag{5.7.25}$$

With the ν_j selected as the nine $x_{j,J}$, disregarding any possible additional substate variables, and using $\dot{x}_{i,J} = v_{i,j}x_{j,J}$ by Eq. (19), we obtain [see Eqs. (23)]

$$\rho\theta\dot{s} = (T_{ji} - {}_RT_{ji})v_{i,j} - q_{j,j} + \rho r$$

or, using Eq. (24),

$$\rho\theta\dot{s} = {}_DT_{ji}v_{i,j} - q_{j,j} + \rho r. \tag{5.7.26}$$

Comparing this with the definition of the specific internal entropy production rate γ, Eq. (5.6.10c), which may be written

$$\rho\theta\gamma = \rho\theta\dot{s} - \rho r + q_{j,j} - \frac{1}{\theta}q_p\theta_{,p},$$

we conclude that

$$\rho\theta\gamma = {}_DT_{ji}v_{i,j} - \frac{1}{\theta}q_p\theta_{,p}. \tag{5.7.27}$$

Equation (23) shows that ${}_RT_{ji}$ is symmetric, whence so is ${}_DT_{ji}$, and

$$\left.\begin{aligned} \rho\theta\gamma &= {}_DT_{ij}D_{ij} - \frac{1}{\theta}q_p\theta_{,p} \quad \text{or} \\ \rho\theta\gamma &= {}_D\mathbf{T}:\mathbf{D} - \frac{1}{\theta}\mathbf{q}\cdot\operatorname{grad}\theta. \end{aligned}\right\} \tag{5.7.28}$$

The internal entropy production rate γ is thus separated into two parts, that due to the dissipative power ${}_D\mathbf{T}:\mathbf{D}$ and that due to irreversible heat conduction in the presence of a thermal gradient. The strong form, Eq. (5.6.10d), of the Clausius-Duhem inequality would then require separately:

$$\rho\theta\gamma_{\text{loc}} \equiv {}_D\mathbf{T}:\mathbf{D} \geq 0 \tag{5.7.29}$$

and

$$\rho\theta\gamma_{\text{con}} \equiv -\frac{1}{\theta}\mathbf{q}\cdot\operatorname{grad}\theta \geq 0. \tag{5.7.30}$$

The quantities entering the expression for internal entropy production are often designated as ***generalized irreversible forces*** $\mathbf{X}^{(i)}$ and ***fluxes*** $\mathbf{J}$. For example in the formulation above, we may choose

$$\left.\begin{array}{l}\text{Generalized Irreversible Forces: } \dfrac{1}{\rho}\,{}_D\mathbf{T} \quad \text{and} \quad -\dfrac{1}{\rho\theta}\operatorname{grad}\theta \\ \text{and} \quad \text{Generalized Fluxes: } \mathbf{D} \quad \text{and} \quad \mathbf{q}.\end{array}\right\} \tag{5.7.31}$$

The choice of which factor is flux and which is force is somewhat arbitrary, but we will suppose that they are always chosen so that the dot product of the generalized irreversible force vector $\mathbf{X}^{(i)}$ into the generalized flux vector $\mathbf{J}$ gives the dissipation power (mechanical and thermal) per unit mass

$$\theta\gamma = \mathbf{X}^{(i)}\boldsymbol{\cdot}\mathbf{J} \equiv X_m^{(i)} J_m. \tag{5.7.32}$$

In irreversible thermodynamics [see, for example, Fitts (1962) or de Groot (1952)], it is usually assumed that the constitutive equations give the fluxes as functions of the forces (or inversely), and that at least "in the neighborhood of equilibrium" the constitutive equations (called ***phenomenological equations***) are equations of the form

$$J_m = L_{mk} X_k^{(i)} \qquad X_k^{(i)} = a_{km} J_m, \tag{5.7.33}$$

where it is assumed that the coefficients satisfy the ***Onsager reciprocal relations***

$$L_{mk} = L_{km} \quad \text{or} \quad a_{km} = a_{mk}, \tag{5.7.34}$$

making the coefficient matrix symmetric. [It has been pointed out that in certain cases the Onsager relations (34) must be replaced by $a_{km}(\mathbf{b}) = a_{mk}(-\mathbf{b})$ or $a_{km} = -a_{mk}$, for example in systems affected by a magnetic field $\mathbf{b}$; see de Groot (1952).] Onsager (1931) proposed the reciprocal relations on the basis of statistical mechanics, and they have since been extensively used. The wide use of the procedure has been criticized as inadequately founded for application to field theory until some rule can be given for selecting the forces and the fluxes in a way that will guarantee the applicability of the Onsager relations. [See, for example, Truesdell (1966).]

Substitution of the phenomenological relations into the entropy production, Eq. (32), yields two quadratic forms

$$\left.\begin{array}{l}\theta\gamma = L_{mk} X_m^{(i)} X_k^{(i)} \geq 0 \quad \text{or} \\ \theta\gamma = a_{km} J_k J_m \geq 0.\end{array}\right\} \tag{5.7.35}$$

These quadratic forms must be positive-definite. [A necessary and sufficient condition for the positive-definiteness of a quadratic form having a symmetric coefficient matrix with real elements is that all the eigenvalues of the matrix be positive, i.e., that all the roots λ of the determinantal equations $|L_{mk} - \lambda\delta_{mk}| = 0$ or $|a_{km} - \lambda\delta_{km}| = 0$ be positive.] The second quadratic form of Eq. (35) is called a ***dissipation function***, say D,

$$D(\mathbf{J}) = a_{km} J_k J_m \tag{5.7.36}$$

and the second form of the phenomenological equation (33) is equivalent to the assumption that

$$X_k^{(i)} = \frac{1}{2} \frac{\partial D}{\partial J_k}. \tag{5.7.37}$$

Alternatively postulating (36) and (37), with the dissipation function in (36) symmetrized [for example, replacing both a_{12} and a_{21} by $\frac{1}{2}(a_{12} + a_{21})$], will ensure the satisfaction of the Onsager relations (34). In general D does not separate into the sum of two parts as it did in Eq. (28) where thermal and mechanical generalized forces were uncoupled.

Ziegler (1963) has shown that the phenomenological relations (37) [and hence (33), and also the Onsager relations (34) if D has been symmetrized in its arguments] follow from a principle of maximum rate of entropy production or maximum dissipation power, which he deduces for "quasistatic processes" by a statistical mechanics approach suitably modified to include irreversible processes. [It should not be thought that all quasistatic processes are reversible. This widespread erroneous belief stems from the fact that most elementary thermodynamics books emphasize applications to gases. The plastic deformation of a metal is always irreversible, no matter how slow.] Ziegler's principle offers no advantage over the Onsager approach when the dissipation function $D(\mathbf{J})$ is a quadratic form as in Eq. (36) that leads to a linear constitutive relation like (33). But he proposed it as applying also to a more general form for the dissipation function, leading to nonlinear constitutive equations for quasistatic behavior, and discussed applications to viscoplasticity and plasticity.

This completes our introduction to the thermodynamics of continuous media. For a more complete development of the thermodynamic formulation based on the caloric equation of state, see Truesdell and Toupin (1960). For more recent theoretical developments based on functionals of the state variables instead of functions of their current values, see Coleman (1964) and Truesdell and Noll (1965).

An equation of state is actually one of the constitutive equations of the particular material it is assumed to describe. Hence this Sec. 5.7, based largely on the assumed existence of the caloric equation of state, is less general in its application than the general principles discussed in the other sections of this chapter. It was included here because of its close relationship to Secs. 5.4 and 5.6.

We now turn our attention in Chap. 6 to the constitutive equations defining particular materials.

EXERCISES

1. (a) For the ideal gas of Eqs. (5.6.3), with c_v assumed constant, what equations express the specific forms of the caloric and thermal equations of state: (1), (7), (8), and (9) of Sec. 5.7?
   (b) For the ideal gas express the specific enthalpy h as a function of s and p, and express the Helmholtz free-energy density ψ and the Gibbs function g (free-enthalpy density) as functions of θ and v.
2. The second Eq. (5.7.10) may be verified as follows. From Table 5-1 on page 262, $\psi = u - s\theta$. Hence $d\psi = (\partial u/\partial s)_v\, ds + (\partial u/\partial v_j)_s\, dv_j - \theta\, ds - s\, d\theta = \tau_j\, dv_j - s\, d\theta$, since $\theta = (\partial u/\partial s)_v$ and $\tau_j = (\partial u/\partial v_j)_s$. Verify similarly the third and fourth Eqs. (5.7.10).
3. A pure substance in the fluid state is characterized by any two of its properties. [See remarks before Ex. 1, Sec. 5.6, and see also Eq. (5.7.4).] Write the forms of Eqs. (5.7.10) for a fluid pure substance.
4. (a) In a fluid $h = u + pv$. Combine this with the results of Exs. 5 and 8 of Sec. 5.6 for the van der Waals' gas to express h and g as functions of θ and v. How could h be expressed in principle in terms of s and p?
   (b) Express the free energy ψ of the van der Waals' gas as a function of θ and v.
5. (a) Verify Eq. (5.7.13), and derive analogous expressions for ρh and ρg. These will be somewhat more complicated because of the presence of $\tau_j v_j$.
   (b) Write the forms assumed by these equations when $T_{ij} D_{ij} = \rho \tau_j \dot{v}_j$.
6. How are Eqs. (15) and (16) modified if the substate variables are chosen as C_{IJ} instead of E_{IJ}? See Eq. (4.5.5a).
7. If in linear uncoupled isotropic thermoelasticity the free energy is given by

$$\rho\psi = G\epsilon_{ij}\epsilon_{ij} + \frac{1}{2}\lambda e^2 - \frac{E\alpha}{1 - 2\nu}(\theta - \theta_0)e,$$

   where $e = \epsilon_{kk}$ and where $\theta - \theta_0$ is temperature change from the unstrained state, and if the thermodynamic tension τ_{ij} is assumed equal to $(1/\rho)T_{ij}$, derive an expression for T_{rs} in terms of the strains and the temperature change. (G, λ, E, α, ν, and θ_0 are constants, and, for a linear theory, we consider ρ also constant.)
8. (a) Repeat Ex. 7 for the general anisotropic linear case obtained by truncating an assumed power series $\rho\psi = C_0 - \frac{1}{2}\beta_{ij}(\theta - \theta_0)\epsilon_{ij} + \frac{1}{2}C_{ijkm}\epsilon_{ij}\epsilon_{km} + \cdots$. (Assume the quadratic terms symmetrized so that $C_{ijkm} = C_{kmij}$.)
   (b) Derive Clapeyron's formula $W \equiv \rho\psi = \frac{1}{2}T_{ij}\epsilon_{ij}$ for the elastic strain energy (per unit volume) for the case of an ***isothermal*** linear-elastic deformation, assuming that the strain energy vanishes in the unstrained state.

9. Verify the details of the derivation of Eqs. (23) to (28).

10. In a Newtonian viscous fluid the recoverable part of the Cauchy stress is assumed to be ${}_RT_{ij} = -p\delta_{ij}$, where p is the thermodynamic pressure, while the remainder of the stress is given by ${}_DT_{km} = \lambda D_{ii}\delta_{km} + 2\mu D_{km}$, where λ and μ may depend on the temperature. (This last is the form of Eqs. (33) giving the irreversible part of the Cauchy stress.)
    (a) Write out the quadratic form of Eq. (36) for the Newtonian fluid, and verify that the dissipative stresses are given by Eq. (37). Assume separation of the thermal and mechanical dissipation as in Eq. (28).
    (b) If the free energy $\psi(\theta, v)$ is the potential used, express p and the specific entropy s in terms of ψ, and hence express the internal energy u in terms of θ, ψ, and derivatives of ψ.

CHAPTER 6

Constitutive Equations

6.1 Introduction; ideal materials

Up to this point, we have considered the mathematical descriptions of stress, strain, and rate of deformation. We have also developed in Chap. 5 a number of general theorems applicable to all continuous media. We now consider equations characterizing the individual material and its reaction to applied loads; such equations are called ***constitutive equations***, since they describe the macroscopic behavior resulting from the internal constitution of the material. But materials, especially in the solid state, behave in such complex ways when the entire range of possible temperatures and deformations is considered that it is not feasible to write down one equation or set of equations to describe accurately a real material over its entire range of behavior. Instead, we formulate separately equations describing various kinds of ***ideal material response***, each of which is a mathematical formulation designed to approximate physical observations of a real material's response over a suitably restricted range.

Some of the ideas involved in formulating simple equations for such ideal materials will be illustrated below in two examples, the ideal elastic (Hookean) solid and the ideal viscous (Newtonian) fluid. These two especially simple ideal materials will be considered in some detail in two separate sections; subsequent sections present some of the classical constitutive equations of viscoelasticity, viscoplasticity, and plasticity. These classical equations were introduced separately to meet specific needs, and made as simple as possible, oversimplifying many physical situations.

The modern continuum theory of constitutive equations, which has flourished in the last 15 years, is guided by a different philosophy; it begins with very general functional constitutive equations, seeks to determine the limits imposed on the forms of the equations by certain general principles, and specializes the equations as late as possible and as little as possible. This approach has the advantage of not overlooking coupling effects between different kinds of behavior (e.g., thermal and mechanical) and also provides many general results applicable to all of the possible specializations. An extensive treatment of the modern theory is beyond the scope of this introductory book, but some of its ideas will be presented in Secs. 6.7 and 6.8. A comprehensive account of the theory up to 1965 is given by Truesdell and Noll (1965).

The ideal elastic solid, or Hookean solid, is the ideal material most commonly assumed for stress analysis in structures and machine parts. It is assumed to obey Hooke's law, which in a uniaxial stress situation takes the form $T_{xx} = E\epsilon_{xx}$ expressing a linear relation between the axial stress and strain, where E is the modulus of elasticity. For pure shear, closely approximated by the twisting of a thin-walled circular tube, where the shear stress is assumed uniformly distributed over the cross-section, the comparable equation is $\tau = G\gamma$, where G is the shear modulus, τ the shear stress in the circumferential direction, and γ the angle change for elements initially in the axial and circumferential directions. Under triaxial loading, classical elasticity theory assumes a generalized Hooke's law expressing each stress component as a linear combination of all the strains, $T_{ij} = c_{ijrs}\epsilon_{rs}$. These nine equations contain a total of 81 coefficients c_{ijrs}, but not all the coefficients are independent. The symmetry of T_{ij} and ϵ_{ij} reduces the number of independent coefficients to 36, and for elastically isotropic material, it will be shown in Sec. 6.2 that all the coefficients can be expressed in terms of just two elastic moduli. The equations characterizing isotropic Hookean response can then be written most simply in terms of the deviatoric components $T'_{ij} = T_{ij} + p\delta_{ij}$ and $\epsilon'_{ij} = \epsilon_{ij} - \frac{1}{3}e\delta_{ij}$ as follows:

$$T'_{ij} = 2G\epsilon'_{ij} \quad \text{or} \quad \mathbf{T}' = 2G\mathbf{E}' \quad \text{and} \quad p = -Ke, \tag{6.1.1}$$

where p is the mean normal pressure, $e = \epsilon_{kk}$ is the volume strain,† and K is the bulk modulus. For the common structural metals these equations quite accurately describe the material behavior in a single loading cycle if the strain components do not exceed a magnitude of about 0.001 and the temperature does not change appreciably during the loading cycle.

The ideal Newtonian fluid is assumed for many applications in viscous fluid analysis. A fluid differs from a solid in that it cannot support a shear stress; in a fluid, shear deformation will continue as long as any shear stress is applied. The constitutive equation for a fluid relates the rate of deformation $\mathbf{D}$ to the applied stress. For a parallel flow between flat plate boundaries it is found that an equation of the form $\tau = \mu\, dv/dy$ gives a good account of the relation, where y is the coordinate perpendicular to the direction of the flow at speed $v(y)$, and μ is the coefficient of viscosity. For a three-dimensional flow of ***incompressible***, isotropic Newtonian fluid, the generalization comparable to Eq. (1) is

†Many authors use e for the mean normal strain (one-third of the volume strain) and write the second Eq. (1) as $p = -3Ke$.

$$T'_{ij} = 2\mu D_{ij} \quad \text{or} \quad \mathbf{T}' = 2\mu \mathbf{D} \tag{6.1.2}$$

(where $\mathbf{D} = \mathbf{D}'$ under the assumption of incompressibility). The viscosity coefficient μ for a given liquid depends strongly on temperature but is almost independent of pressure at moderate pressures.

These two classical linear equations, Eq. (1) for classical elasticity and Eq. (2) for classical viscosity, are especially simple examples of two fundamental types of ideal constitutive equations. In a more general ideal elastic body we say merely that stress depends upon strain or deformation from a certain ***natural state***; symbolically

$$\text{Stress Tensor} = f(\text{Deformation tensor}). \tag{6.1.3}$$

The deformation tensor used as an argument for the function of Eq. (3) may be one of the finite-strain tensors of Sec. 4.5 or 4.6 or possibly the deformation gradient $\mathbf{F} \equiv x\overleftarrow{\nabla}$; if $\mathbf{F}$ is used, the constitutive equations must be so formulated that a rigid-body rotation does not change the stress predicted by the equation. The functional relationship between the two tensors in Eq. (3) may be nonlinear, but must express a one-to-one relationship between stress and strain such that, for example, the Lagrangian finite strain $\mathbf{E}$ vanishes whenever the stress does. A comparable generalization of Eq. (2) is the general ***Stokesian fluid*** in which

$$\mathbf{T}' = f(\mathbf{D}). \tag{6.1.4}$$

Unlike the elastic material, the Stokesian fluid carries no memory of its initial state; the stress at a point depends exclusively on the instantaneous rate of deformation at that point. Of course, in either type of material the equations may contain coefficients dependent not only on the particular material involved but also upon temperature and possibly upon still other factors. However, the idealization involved in Eq. (3) implies that either the deformation takes place at constant temperature, or else that temperature changes accompanying loading have an effect on the stress negligible in comparison to the effect of the deformation. Thermoelastic equations in which the linear thermal expansions are incorporated in classical linearized elasticity, but which still do not couple the elastic constants with the temperature will be discussed in Eqs. (6.2.27) to (6.2.33), but the coupled thermoelastic equations are not considered in this book. See, for example, Boley and Weiner (1960) for a discussion of them. In the viscous fluid, mechanical energy is certainly dissipated, but if the heat is conducted away sufficiently rapidly there may be no significant coupling, so that the viscosity coefficients may be considered constant during the motion, even though the viscosity coefficients are much more highly dependent on temperature than are the elastic constants of a metal. The general case, including coupling, offers great difficulties.

The examples so far proposed by no means exhaust the possibilities. There is no reason to suppose that all materials can be separated into two categories: materials in which stress depends upon strain, and materials in which stress depends upon rate of deformation. The elastic solid and the viscous fluid are merely the two extreme cases of a wide range of materials, including ***visco-elastic solids***, elastic solids that have some rate dependence but still possess a well-defined natural state to which they eventually return when the loads are removed, and ***viscoplastic materials***, more like a thick fluid but still exhibiting some aspects of a solid. Even the most nearly ideally elastic real materials exhibit some time dependence under cyclic loading or long-time static loading. The best tuning fork will not continue to vibrate forever, even in a vacuum, and the dissipation causing its vibrations to die out is associated with the dependence of stress on the rate of deformation. The phenomena of creep under constant load even below the conventional elastic limit and of stress relaxation under constant extension are well-known. Metals at temperatures approaching their melting points exhibit these phenomena more markedly, and at room temperature various plastics and high polymers show significant time dependence even in a single cycle of loading and unloading. Classical viscoelasticity and viscoplasticity will be considered in later sections. We begin with simpler examples.

In Secs. 6.2 and 6.3 on the constitutive equations of classical linearized elasticity and of Newtonian fluids, certain restrictions on the possible form of the constitutive equations will be developed. For example, it will be shown that in classical linearized elasticity there are only two independent elastic constants for an isotropic material. An almost identical argument produces similar results for linear viscosity and for viscoplasticity. The key point in these arguments is that one tensor, say stress, is assumed to depend linearly on ***only one*** other tensor (strain in the elastic case). The coefficients of the linear terms relating the two second-order tensors are themselves components of a fourth-order tensor. The result is then simply the conclusion that all the components of a fourth-order isotropic tensor relating two symmetric second-order tensors can be expressed in terms of two independent components, which are constants in the elasticity and viscosity equations, but are functions of the invariants of the rate-of-deformation tensor in the viscoplasticity equations. No such simple results follow when the stress depends on both strain and rate of deformation.

Isotropic tensors. An isotropic tensor is one whose rectangular Cartesian components are unchanged by any orthogonal transformation of the coordinate axes. A trivial example is the zero tensor of any order. All tensors of order zero (scalars) are isotropic, but there are no isotropic first-order tensors (vectors) except the zero vector. The unit tensor **1**, whose components are given in any rectangular Cartesian system by the Kronecker delta δ_{ij}, is isotropic, and it can be proved that this and scalar multiples of it are the only nontrivial second-order isotropic tensors. See, for example, Aris (1962), pp. 30–34. The only nontrivial isotropic third-order tensors are the one whose rectangular Cartesian components are given by the permutation symbols e_{mnr} of Eq. (2.3.22) and scalar multiples of it.

The most general fourth-order isotropic tensor c_{ijrs} has rectangular Cartesian components of the form

$$c_{ijrs} = \lambda\delta_{ij}\delta_{rs} + \mu(\delta_{ir}\delta_{js} + \delta_{is}\delta_{jr}) + \nu(\delta_{ir}\delta_{js} - \delta_{is}\delta_{jr}) \tag{6.1.5}$$

where λ, μ, and ν have the same value in all coordinate systems. For a proof, see, for example, Aris (1962), p. 33. We can make a further simplification, given in Eq. (6b), when the isotropic c_{ijrs} are coefficients in a linear homogeneous equation such as

$$T_{ij} = c_{ijrs}E_{rs} \tag{6.1.6a}$$

where at least one of the two tensors **T** or **E** is symmetric. If $E_{rs} = E_{sr}$, there is no loss of generality in choosing the coefficients c_{ijrs} in Eq. (6a) to be symmetric in the indices r, s, since the same sum results if we replace both c_{ijrs} and c_{ijsr} by $(c_{ijrs} + c_{ijsr})/2$ for each numerical choice of rs. Also, if $T_{ij} = T_{ji}$, we must have $c_{ijrs} = c_{jirs}$. Thus:

If $T_{ij} = T_{ji}$ we must have $c_{ijrs} = c_{jirs}$.

If $E_{rs} = E_{sr}$ we may choose $c_{ijrs} = c_{ijsr}$.

In either case, if c_{ijrs} is required to be symmetric in either ij or rs, we must have $\nu = 0$.

Hence, the most general fourth-order isotropic tensor with rectangular Cartesian components c_{ijrs} symmetric in either ij or rs has rectangular Cartesian components of the form

$$c_{ijrs} = \lambda\delta_{ij}\delta_{rs} + \mu(\delta_{ir}\delta_{js} + \delta_{is}\delta_{jr}). \tag{6.1.6b}$$

Eq. (6b) will be used in Secs. 6.2 and 6.3 to represent the coefficients for isotropic linearized elasticity and Newtonian viscosity. A second derivation will be given in Sec. 6.2, not using Eq. (6b) but examining the consequences of the existence of planes of elastic symmetry before treating the fully isotropic case and in effect deriving Eq. (6b) for the elastic case.

EXERCISES

1. Show that if a tensor **T** satisfies $T_{rs} = \delta_{rs}$ in one rectangular Cartesian system, it satisfies $\bar{T}_{pq} = \delta_{pq}$ in any other rectangular Cartesian system.

2. It is often said that in an isotropic material the principal axes of stress coincide with the principal axes of strain. Show that this is true for an elastic solid governed by $T'_{ij} = 2G\epsilon'_{ij}$ but not necessarily in a viscoelastic material governed by $T'_{ij} = 2G\epsilon'_{ij} + 2\eta\, d\epsilon'_{ij}/dt$.
3. If the reference axes are principal axes of stress in an isotropic elastic solid governed by $\sigma'_i = 2G\epsilon'_i$ and $p = -Ke$, combine these equations to obtain three equations relating $\sigma_1, \sigma_2, \sigma_3$ to $\epsilon_1, \epsilon_2, \epsilon_3$.
4. Combine the first six Eqs. (6.1.1), namely $T'_{ij} = 2G\epsilon'_{ij}$, with $p = -Ke$ to obtain equations for $T_{11}, T_{12}, \ldots T_{33}$.
5. Show that if c_{ijrs} has the form given in Eq. (6.1.6b), then Eq. (6a) reduces to $T_{ij} = \lambda E_{kk}\delta_{ij} + 2\mu E_{ij}$.

6.2 Classical elasticity, generalized Hooke's law, isotropy; hyperelasticity, the strain-energy function or elastic potential function; elastic symmetry; thermal stresses

A material is called ***ideally elastic*** when a body formed of the material recovers its original form completely upon removal of the forces causing the deformation, and there is a one-to-one relationship between the state of stress and the state of strain, for a given temperature. Structural metals approximate this behavior if the deformations are sufficiently small. Since there is a one-to-one relationship between stress and strain, no creep at constant load or stress relaxation at constant strain are included in the theory. The coefficients of the constitutive equation specifying the relationship between stress and strain for the material in general depend on the temperature, but we usually assume that the dependence is sufficiently slight or else that the temperature variation is sufficiently small that the coefficients may be treated as constants during the deformation.

Even though we neglect the variation of the elastic constants with temperature, we may be compelled to take account of the thermal expansion of the material, which often produces dimensional changes as large as those produced by the applied forces; or, if the dimensional change is prevented by support constraints or surrounding material, ***thermal stresses*** are induced in addition to the stresses related to the strains according to the elastic constitutive equations. When the strains and temperature variations are sufficiently small and the geometry change is negligible (buckling situations are excluded by this assumption), the equations are all linear and permit superposition of thermal stresses and stresses due to loads. We consider first only the elastic constitutive equations and neglect thermal effects. The superposition of the thermal expansion is included in Eqs. (27) to (34).

The classical elastic constitutive equations, often called the generalized Hooke's law, are nine equations expressing the stress components as linear homogeneous functions of the nine strain components, e.g.,

$$\tilde{T}_{IJ} = c_{IJRS} E_{RS} \tag{6.2.1}$$

if the Lagrangian finite-strain tensor $\mathbf{E}$ and the second Piola-Kirchhoff stress tensor $\tilde{\mathbf{T}}$ are taken as the constitutive field variables referred to the natural state of the material. When displacement gradients are everywhere small compared to unity,† the displacements in a finite body will be sufficiently small (apart from a rigid translation of the whole body, which may be imagined removed) that no distinction need be made between the initial coordinates X_I and the deformed current spatial position x_i of the same particle in writing the field equations; under these conditions also we may replace Eq. (1) with sufficient accuracy by

$$T_{ij} = c_{ijrs} E_{rs} \tag{6.2.2}$$

using the small-strain components $E_{rs}(= \epsilon_{rs})$ and the Cauchy stress T_{ij}. In the following, E_{rs} will frequently be used for small strains instead of ϵ_{rs} when the Lagrangian finite strains do not appear in the discussion. The development given below based on Eq. (2) can be rephrased for Eq. (1).

The nine Eqs. (2) contain 81 constants, but since T_{ij} is symmetric, we must have $c_{ijrs} = c_{jirs}$. Also, since E_{rs} is symmetric in the dummy indices rs, there is no loss of generality in supposing the coefficients c_{ijrs} symmetrized in rs. Thus

$$c_{ijrs} = c_{jirs} \quad \text{and} \quad c_{ijrs} = c_{ijsr}. \tag{6.2.3}$$

When the material is ***elastically isotropic***, i.e., when there are no preferred directions in the material, the elastic constants must be the same at a given particle for all possible choices of rectangular Cartesian coordinates in which to evaluate the components T_{ij} and E_{rs} (including left-handed systems). This means that the c_{ijrs} are components of a fourth-order isotropic tensor symmetric in the indices ij and also in rs. Hence, Eq. (2) takes the following form since, according to Eq. (6.1.6), the most general fourth-order isotropic tensor c_{ijrs} symmetric with respect to either ij or rs has the form of the bracketed terms in

$$T_{ij} = [\lambda\delta_{ij}\delta_{rs} + \mu(\delta_{ir}\delta_{js} + \delta_{is}\delta_{jr})]E_{rs}$$

†Note that this is a stronger condition than merely requiring small strains, since it also implies small rotations, thereby excluding certain cases of large deflections of slender structural members.

whence we obtain the isotropic Hooke's law†

$$T_{ij} = \lambda E_{kk}\delta_{ij} + 2\mu E_{ij}, \tag{6.2.4}$$

where λ and μ are the Lamé elastic constants and δ_{ij} is the Kronecker delta.

If we put $i = j$ in Eq. (4), implying summation, we obtain

$$T_{ii} = (3\lambda + 2\mu)E_{ii}. \tag{6.2.5}$$

Substituting $E_{kk} = T_{kk}/(3\lambda + 2\mu)$ into Eq. (4), and solving for E_{ij} we obtain

$$E_{ij} = -\frac{\lambda\delta_{ij}}{2\mu(3\lambda + 2\mu)} T_{kk} + \frac{1}{2\mu} T_{ij} \tag{6.2.6}$$

as the inverse of Eq. (4), expressing strain in terms of stress.

The two Lamé elastic constants λ and μ, introduced by Lamé in 1852, are related to the more familiar shear modulus G, Young's modulus E, and Poisson's ratio ν as follows:

$$\mu = G = \frac{E}{2(1 + \nu)} \quad \text{and} \quad \lambda = \frac{\nu E}{(1 + \nu)(1 - 2\nu)}. \tag{6.2.7}$$

By means of the identities of Eqs. (7) the constants in Eq. (6) can be expressed in terms of E and ν, so that the inverse equation takes the form

$$E_{ij} = -\frac{\nu}{E} T_{kk}\delta_{ij} + \frac{1 + \nu}{E} T_{ij}. \tag{6.2.8}$$

The reader can verify that the identities of Eqs. (7) make the isotropic Hooke's law Eq. (8) equivalent to the six equations

$$\left.\begin{aligned} \epsilon_x &= \frac{1}{E}[\sigma_x - \nu(\sigma_y + \sigma_z)] & \gamma_{yz} &= \frac{1}{G}\tau_{yz} \\ \epsilon_y &= \frac{1}{E}[\sigma_y - \nu(\sigma_z + \sigma_x)] & \gamma_{zx} &= \frac{1}{G}\tau_{zx} \\ \epsilon_z &= \frac{1}{E}[\sigma_z - \nu(\sigma_x + \sigma_y)] & \gamma_{xy} &= \frac{1}{G}\tau_{xy} \end{aligned}\right\} \tag{6.2.9}$$

in the stress notation of Eq. (3.2.12) and the strain notation of Eq. (4.2.14). (Recall $\gamma_{xy} = 2\epsilon_{xy}$, etc.)

†This law (with $\lambda = \mu$ because of an inadequate molecular model) was proposed to the Paris Academy by Navier in 1821. The two-constant version was presented to the Academy in 1823 by Cauchy, who formulated the continuum theory of linearized elasticity in virtually the same form it has today. This law, of course, greatly generalizes the observations of Hooke (around 1660) and Thomas Young (1807).

The isotropic Hooke's law takes an especially simple form if the deviatoric stress T'_{ij} and strain E'_{ij} are used. Recall that

$$T'_{ij} = T_{ij} - \tfrac{1}{3}\, T_{kk}\delta_{ij} \qquad E'_{ij} = E_{ij} - \tfrac{1}{3}\, E_{kk}\delta_{ij}. \tag{6.2.10}$$

Then the whole relationship can be expressed by the two equations given in Sec. 6.1:

$$T'_{ij} = 2GE'_{ij} \quad \text{or} \quad \mathbf{T}' = 2G\mathbf{E}' \quad \text{and} \quad p = -Ke, \tag{6.2.11}$$

where $p = -T_{kk}/3$ is the mean pressure, $e = E_{kk}$ is the volume strain,† and K is the bulk modulus, related to the other elastic constants by the equations

$$K = \lambda + \frac{2}{3}G = \frac{E}{3(1-2\nu)} \tag{6.2.12}$$

For the case of an isotropic material, all the elastic constants can be expressed in terms of just two independent constants. The equations are often written using three constants as in Eqs. (9), but only two of the three can be specified independently. For the small strains and displacements contemplated, the distinction between material coordinates and spatial coordinates is ignored and both the stresses and the strains appearing in Hooke's law are taken to be defined with respect to the same independent position variables. This procedure is correct only in the limit of infinitesimal displacement gradients, but gives satisfactory results for the small (but finite) displacements occurring in structural metals below the elastic limit, except for certain cases of large deflections of slender members.

Matrix form of generalized Hooke's law. Since both T_{ij} and E_{rs} are symmetric, the Eqs. (2) really represent only six independent equations, each containing six terms, for a total of only 36 elastic constants or moduli instead of 81. This is conveniently represented as a matrix equation expressing a six-element column matrix of stresses in terms of a six-element column matrix of strains as follows:

$$\begin{bmatrix} T_{11} \\ T_{22} \\ T_{33} \\ T_{23} \\ T_{31} \\ T_{12} \end{bmatrix} = \begin{bmatrix} c_{11} & c_{12} & c_{13} & c_{14} & c_{15} & c_{16} \\ c_{21} & c_{22} & c_{23} & c_{24} & c_{25} & c_{26} \\ c_{31} & c_{32} & c_{33} & c_{34} & c_{35} & c_{36} \\ c_{41} & c_{42} & c_{43} & c_{44} & c_{45} & c_{46} \\ c_{51} & c_{52} & c_{53} & c_{54} & c_{55} & c_{56} \\ c_{61} & c_{62} & c_{63} & c_{64} & c_{65} & c_{66} \end{bmatrix} \begin{bmatrix} E_{11} \\ E_{22} \\ E_{33} \\ 2E_{23} \\ 2E_{31} \\ 2E_{12} \end{bmatrix}. \tag{6.2.13}$$

†Some writers use e for mean strain $\frac{1}{3} E_{kk}$ and write $p = -3Ke$.

With the following enumeration for the elements of the two column matrices T and E,

$$\begin{array}{llllll} T_1 = T_{11} & T_2 = T_{22} & T_3 = T_{33} & T_4 = T_{23} & T_5 = T_{31} & T_6 = T_{12} \\ E_1 = E_{11} & E_2 = E_{22} & E_3 = E_{33} & E_4 = 2E_{23} & E_5 = 2E_{31} & E_6 = 2E_{12} \end{array} \tag{6.2.14}$$

the matrix Eq. (13) becomes

$$T = CE \tag{6.2.15}$$

or

$$T_m = c_{mn} E_n \tag{6.2.16}$$

in indicial notation, where C is the 6×6 square matrix of Eq. (13) whose elements are the elastic moduli† c_{mn}. (Note that $E_6 = \gamma_{12}$, where $\gamma_{12} = 2E_{12}$, etc.) Comparing Eqs. (16) to Eqs. (2) we see, for example, that $c_{3112} E_{12} + c_{3121} E_{21} = c_{56}(2E_{12})$, whence

$$c_{3112} = c_{3121} = c_{56}.$$

The number of independent constants is further reduced from 36 to 21 when a strain-energy function exists.

Elastic-potential or strain-energy function; Green-elastic or hyperelastic material. Green (1839, 1841) defined an elastic material as one for which a strain-energy function exists; this has been the basic notion of most of the work on "perfect elasticity" since that time. More recently the name ***hyperelastic*** has been used for the Green-elastic material [see, e.g., Truesdell and Toupin (1960) and Truesdell and Noll (1965)]. We will call a material ***Green-elastic*** or ***hyperelastic*** if there exists an ***elastic potential function W*** (or ***strain-energy function***), a scalar function of one of the strain or deformation tensors, whose derivative with respect to a strain component determines the corresponding stress component.

For the fully recoverable case of isothermal deformation with reversible heat conduction, Eqs. (5.7.11) and (5.7.16) show that (when the free energy ψ has been symmetrized in the finite strains E_{IJ} and E_{JI}) we have

$$\tilde{T}_{IJ} = \rho_0 \left(\frac{\partial \psi}{\partial E_{IJ}} \right)_\theta ;$$

†A capital letter C will be used in writing the components C_{pq} of the inverse relation $E_p = C_{pq} T_q$. Do not confuse C_{pq} with c_{pq}.

hence, $W = \rho_0 \psi$ is an elastic potential function for this case, while $W = \rho_0 u$ is one for the adiabatic isentropic case. Hyperelasticity ignores the thermal effects and assumes that the elastic potential function always exists, a function of the strains alone; it is a purely mechanical theory. We assume†

$$\tilde{T}_{IJ} = \frac{\partial W(\mathbf{E})}{\partial E_{IJ}} \tag{6.2.17a}$$

(with W symmetrized in the components E_{IJ} and E_{JI}). $W(\mathbf{E})$ is the ***strain energy per unit undeformed volume.***

Many recent writers take the deformation gradient $\mathbf{F}$ with components $x_{i,J}$ as the independent variable in W. For example, Truesdell and Noll (1965) define a hyperelastic material as one for which

$$T_{km} = \rho x_{k,J} \frac{\partial \sigma(\mathbf{F})}{\partial (x_{m,J})} = \rho x_{m,J} \frac{\partial \sigma(\mathbf{F})}{\partial (x_{k,J})} \tag{6.2.17b}$$

[see their Eqs. (82.11) and (84.8)], where $\sigma(\mathbf{F})$ is called the ***strain-energy function.*** Note that the second equality in Eq. (17b) requires some symmetry in the structure of σ as a function of the deformation-gradient components. The requirement of objectivity or frame-indifference [see Eqs. (6.7.53) to (6.7.55)] for the hyperelastic constitutive equations then implies that the strain energy can depend on $\mathbf{F}$ only through the right stretch tensor $\mathbf{U}$ (see our Sec. 4.6) or alternatively through the Cauchy-Green deformation tensor $\mathbf{C} = \mathbf{U}^2$, for example as

$$\tilde{T}_{IJ} = 2\rho_0 \frac{\partial \bar{\sigma}(\mathbf{C})}{\partial C_{IJ}}, \tag{6.2.17c}$$

where $\bar{\sigma}$ denotes the strain energy as a function of $\mathbf{C}$. Since $\mathbf{C} = 2\mathbf{E} + \mathbf{1}$, this may be rewritten as

$$\tilde{T}_{IJ} = \rho_0 \frac{\partial \hat{\sigma}(\mathbf{E})}{\partial E_{IJ}}, \tag{6.2.17d}$$

where $\hat{\sigma}(\mathbf{E}) = \bar{\sigma}(\mathbf{C})$ and $2\partial\bar{\sigma}/\partial C_{IJ} = \partial\hat{\sigma}/\partial E_{IJ}$. This is equivalent to our assumption in Eq. (17a), if we take $W(\mathbf{E}) = \rho_0\hat{\sigma}(\mathbf{E})$.

The form (17c) was the starting point for Green (1839, 1841), who introduced the method. We have used the equivalent form (17a), because it reduces conveniently to the usual small-strain form, which is our primary concern here.

When displacement gradients are small compared to unity, Eq. (17a) may be replaced by

†Eq. (17a) is essentially the form used by the Cosserats (1896).

$$T_{ij} = \frac{\partial W}{\partial E_{ij}} \tag{6.2.18}$$

in terms of the Cauchy stress T_{ij} and small strain E_{ij}; the following argument is limited to this case. We will always suppose W symmetrized in the variables E_{ij} and E_{ji}. If these symmetries are not required, the symmetry of T_{ij} can be maintained by writing

$$T_{ij} = \frac{1}{2}\left(\frac{\partial W}{\partial E_{ij}} + \frac{\partial W}{\partial E_{ji}}\right)$$

in place of Eq. (18), a procedure used by some writers.† We assume that the elastic potential function is represented as a power series in the small-strain components. We can then deduce the six linear equations of the generalized Hooke's law with a symmetric coefficient matrix as follows. For small strains, neglect terms of higher than the second degree in the series expansion. Then W is a quadratic function of the strains:

$$W = c_0 + c_q E_q + \tfrac{1}{2}\bar{c}_{rs} E_r E_s \tag{6.2.19a}$$

in the notation of Eqs. (14) or, combining terms like $\frac{1}{2}[\bar{c}_{12}E_1E_2 + \bar{c}_{21}E_2E_1]$ into a single term $c_{12}E_1E_2$, where $c_{12} = \frac{1}{2}(\bar{c}_{12} + \bar{c}_{21})$,

$$\begin{aligned} W = c_0 &+ c_1E_1 + c_2E_2 + c_3E_3 + c_4E_4 + c_5E_5 + c_6E_6 \\ &+ \tfrac{1}{2}c_{11}E_1^2 + c_{12}E_1E_2 + c_{13}E_1E_3 + c_{14}E_1E_4 + c_{15}E_1E_5 + c_{16}E_1E_6 \\ &\qquad + \tfrac{1}{2}c_{22}E_2^2 + c_{23}E_2E_3 + c_{24}E_2E_4 + c_{25}E_2E_5 + c_{26}E_2E_6 \\ &\qquad\qquad + \tfrac{1}{2}c_{33}E_3^2 + c_{34}E_3E_4 + c_{35}E_3E_5 + c_{36}E_3E_6 \\ &\qquad\qquad\qquad + \tfrac{1}{2}c_{44}E_4^2 + c_{45}E_4E_5 + c_{46}E_4E_6 \\ &\qquad\qquad\qquad\qquad + \tfrac{1}{2}c_{55}E_5^2 + c_{56}E_5E_6 \\ &\qquad\qquad\qquad\qquad\qquad + \tfrac{1}{2}c_{66}E_6^2. \end{aligned} \tag{6.2.19b}$$

We require that W vanish in the unstrained state. Then $c_0 = 0$. We now apply Eq. (18) to the quadratic expression for W and obtain, for example,

$$T_5 = \frac{\partial W}{\partial E_5} = c_5 + c_{15}E_1 + c_{25}E_2 + c_{35}E_3 + c_{45}E_4 + c_{55}E_5 + c_{56}E_6.$$

If the stress is also to be zero in the unstrained state, we must have $c_5 = 0$ and similarly all coefficients in the first row of Eq. (19b) must be zero, and the elastic potential function is a ***homogeneous quadratic*** function of the strains. The

†See, for example, Green and Zerna (1954), Eq. (5.3.1).

generalized Hooke's law expressing the stress matrix T in terms of the strain matrix E, each represented as a six element column matrix then takes the form of Eq. (20) with a symmetric coefficient matrix C containing only 21 independent elastic constants.

$$T = CE \quad \text{or} \quad \begin{bmatrix} T_{11} \\ T_{22} \\ T_{33} \\ T_{23} \\ T_{31} \\ T_{12} \end{bmatrix} = \begin{bmatrix} c_{11} & c_{12} & c_{13} & c_{14} & c_{15} & c_{16} \\ c_{12} & c_{22} & c_{23} & c_{24} & c_{25} & c_{26} \\ c_{13} & c_{23} & c_{33} & c_{34} & c_{35} & c_{36} \\ c_{14} & c_{24} & c_{34} & c_{44} & c_{45} & c_{46} \\ c_{15} & c_{25} & c_{35} & c_{45} & c_{55} & c_{56} \\ c_{16} & c_{26} & c_{36} & c_{46} & c_{56} & c_{66} \end{bmatrix} \begin{bmatrix} E_{11} \\ E_{22} \\ E_{33} \\ 2E_{23} \\ 2E_{31} \\ 2E_{12} \end{bmatrix} \tag{6.2.20}$$

By use of Eqs. (20) the reader can show that, for symmetric c_{ij} and with $c_0 = c_q = 0$, Eq. (19) is equivalent to the ***Clapeyron formula***†

$$W = \tfrac{1}{2}T_k E_k \quad \text{or} \quad W = \tfrac{1}{2}T_{ij}E_{ij} \tag{6.2.21}$$

when the strain-energy function is a homogeneous quadratic function of the strains.

Elastic symmetry. In general the elastic moduli c_{ij} relating the Cartesian components of stress and strain depend on the orientation of the coordinate system with respect to the body. If the form of the elastic potential function W and the values of the c_{ij} are independent of the orientation, the material is said to be ***isotropic***. If it is not isotropic, it is called ***anisotropic***.

We have seen how the existence of the strain-energy function implies the symmetry of c_{ij}, reducing the number of constants from 36 to 21. We will now show how the number of constants is further reduced from 21 to 13 when there is one plane of elastic symmetry. In the case of orthotropic symmetry (three mutually perpendicular planes of elastic symmetry) the number of independent constants is reduced to nine, and finally when the material is isotropic there are just two independent moduli.

The ***material symmetry group*** or ***isotropy group*** of a material is defined as the group of transformations of the material coordinates which leave the constitutive equations invariant; see Noll (1958). The symmetry group depends in general on the choice of the reference state. An ***isotropic material*** is then one possessing at least one reference state (called the ***undistorted state***) for which its symmetry group contains the full orthogonal group of transformations of the material coordinates. Thus, its constitutive equation is unchanged in form if rectangular Cartesian reference axes are given any rigid rotation, reflection in a plane, or central inversion (reflection in a point); for the elastic

†Attributed to B. P. E. Clapeyron (1799–1864) by G. Lamé in 1852.

material under discussion any such change of coordinates would leave the elastic coefficient matrix $[c_{ij}]$ unaltered.

A material possessing some symmetry, but whose symmetry group does not contain the full orthogonal group is called ***aelotropic***. For a discussion of the effects of various kinds of crystal symmetry see, for example, Green and Adkins (1960), Chap. 1, or Voigt (1910).

When the elastic constants at a point have the same values for every pair of coordinate systems which are mirror images of each other in a certain plane, that plane is called a ***plane of elastic symmetry*** for the material at the point. Note that the plane of elastic symmetry is ***not*** in general either a plane of geometric symmetry of the body or a plane of symmetry of the stress state. The symmetry in question is a directional symmetry at a point in the plane; it has nothing whatever to do with two points not in the plane. A plane of crystal symmetry in a metal single crystal will be a plane of elastic symmetry. In a rolled plate of polycrystalline material the preferred orientations are usually so distributed that macroscopically the material has ***orthotropic symmetry*** (three mutually perpendicular planes of elastic symmetry through each point). In an isotropic material every plane is a plane of elastic symmetry. A polycrystalline metal with randomly oriented crystal grains exhibits macroscopic isotropy, but when the sample becomes of the order of the grain size it will be anisotropic. We examine first the consequences of the existence of a single plane of elastic symmetry when one coordinate plane is parallel to the plane of elastic symmetry.

Consider two coordinate systems X_K and $\bar{X}_J$ with $\bar{X}_1$, $\bar{X}_2$-axes coinciding with the X_1, X_2-axes parallel to the plane of elastic symmetry. However, choose the $\bar{X}_3$-axis with $\bar{X}_3 = -X_3$, so that one system is the mirror image of the other in the plane of elastic symmetry. (One system must be left-handed, but that will not affect the discussion.) Consideration of the tensor transformation formulas or, more simply, consideration of the definitions and sign conventions of the components shows that

$$\bar{T}_{23} = -T_{23} \quad \text{and} \quad \bar{T}_{31} = -T_{31}, \tag{6.2.22a}$$

while all of the other independent stress components are unchanged in value by the change from one coordinate system to the other. Similarly

$$\bar{E}_{23} = -E_{23} \quad \text{and} \quad \bar{E}_{31} = -E_{31}, \tag{6.2.22b}$$

while all other independent strain components are unchanged. The first Eq. (13) gives for $\bar{T}_{11}$ in terms of the components of the strain $\bar{\mathbf{E}}$

$$\bar{T}_{11} = c_{11}\bar{E}_{11} + c_{12}\bar{E}_{22} + c_{13}\bar{E}_{33} + 2c_{14}\bar{E}_{23} + 2c_{15}\bar{E}_{31} + 2c_{16}\bar{E}_{12}.$$

We substitute into this equation the results of Eqs. (22), noting that $\bar{T}_{11} = T_{11}$

and $\bar{E}_{ij} = E_{ij}$ except for those of the strains E_{ij} appearing in Eqs. (22b), to obtain

$$T_{11} = c_{11}E_{11} + c_{12}E_{22} + c_{13}E_{33} - 2c_{14}E_{23} - 2c_{15}E_{31} + 2c_{16}E_{12}.$$

But the first Eq. (13) also gives T_{11} in terms of $\mathbf{E}$ as

$$T_{11} = c_{11}E_{11} + c_{12}E_{22} + c_{13}E_{33} + 2c_{14}E_{23} + 2c_{15}E_{31} + 2c_{16}E_{12}.$$

(The elastic constants are the same for the two coordinate systems, since they are mirror images in the plane of elastic symmetry.) Subtracting the two expressions for T_{11} yields the identity:

$$0 \equiv -4c_{14}E_{23} - 4c_{15}E_{31} \qquad \text{for all values of } E_{23} \text{ and } E_{31}.$$

Hence, $c_{14} = 0$ and $c_{15} = 0$. Considering similarly two alternative expressions for each of the other stresses we find other zero coefficients so that the coefficient matrix takes the form in (23).†

Coefficient Matrix for One Plane of Elastic Symmetry (X_1X_2*-plane*)
(13 independent constants, when $c_{ij} = c_{ji}$, otherwise 20)

$$\begin{bmatrix} c_{11} & c_{12} & c_{13} & 0 & 0 & c_{16} \\ c_{21} & c_{22} & c_{23} & 0 & 0 & c_{26} \\ c_{31} & c_{32} & c_{33} & 0 & 0 & c_{36} \\ 0 & 0 & 0 & c_{44} & c_{45} & 0 \\ 0 & 0 & 0 & c_{54} & c_{55} & 0 \\ c_{61} & c_{62} & c_{63} & 0 & 0 & c_{66} \end{bmatrix} \tag{6.2.23}$$

Observe that in this case as in the general anisotropic case a pure shear strain can give rise to normal stress. For example,

$$T_{11} = 2c_{16}E_{12}$$

if all other strains are zero. Also note that the principal axes of stress do not in general coincide with those of strain, since, for example,

$$T_{12} = c_{61}E_{11} + c_{62}E_{22} + c_{63}E_{33} + 2c_{66}E_{12}$$

and T_{12} may therefore differ from zero even when all the shear strains are zero.

We have seen that when a coordinate plane is a plane of elastic symmetry there are only 13 independent elastic constants when a strain-energy function exists. If the plane of symmetry is not a coordinate plane there will be more

†See, for example, Sokolnikoff (1956), Eq. (21.5).

nonzero constants, but there will be enough algebraic relations among them so that there will still be only 13 independent constants. These relations could be determined, if necessary, by observing that the coefficients c_{mnrs} are components of a fourth-order Cartesian tensor. The components in any coordinate system may be expressed in terms of the components in a system having one coordinate plane parallel to the plane of symmetry by using the tensor transformation formulas. In most cases, it will be much more convenient to choose a coordinate plane parallel to the plane of symmetry.

Orthotropic symmetry† (3 orthogonal planes of symmetry) is a very important case of elastic symmetry, exhibited, for example, by a rolled plate, which has symmetry planes parallel to the plane of the plate and parallel and perpendicular to the direction of rolling, forming a mutually orthogonal set of symmetry planes. Remember that the symmetry under discussion is a ***directional*** property and not a ***positional*** property. Even when the material has a certain elastic symmetry at each point, the properties may vary from point to point in a manner not possessing any symmetry with respect to the shape of the body.

Arguments similar to those given for a single elastic symmetry plane lead to the conclusion that when the coordinate planes are chosen parallel to the planes of orthotropic symmetry there are only nine independent constants if a strain-energy function exists as shown in the matrix (24). In this case, if principal axes of strain coincide with the symmetry axes, then so do the principal axes of stress. However, in general, if the loads are such that the strain principal axes do not coincide with the axes of symmetry, then the principal axes of stress do not coincide with the principal axes of strain.

Coefficient Matrix for Orthotropic Symmetry
with Respect to Coordinate Planes
(9 independent constants when $c_{ij} = c_{ji}$, otherwise 12)

$$\begin{bmatrix} c_{11} & c_{12} & c_{13} & 0 & 0 & 0 \\ c_{21} & c_{22} & c_{23} & 0 & 0 & 0 \\ c_{31} & c_{32} & c_{33} & 0 & 0 & 0 \\ 0 & 0 & 0 & c_{44} & 0 & 0 \\ 0 & 0 & 0 & 0 & c_{55} & 0 \\ 0 & 0 & 0 & 0 & 0 & c_{66} \end{bmatrix} \tag{6.2.24}$$

Isotropy. In an isotropic material, the elastic constants are independent of the orientation of the coordinate axes, and we will see that this implies that there are only two independent elastic constants. For the isotropic case, the coefficient matrix can be shown to be symmetric without assuming the existence

†This case was discussed by St.-Venant; see Love (1944), Sec. 111.

of the elastic potential function. The existence of such a function can then be derived as a consequence of the assumed isotropy and the assumed generalized Hooke's law.

Consider an $\bar{X}_J$-system obtained by a 90-degree rotation around the X_3-axis from the X_K-system position. The $\bar{X}_1$-direction then coincides with X_2, the $\bar{X}_2$ with $-X_1$, and $\bar{X}_3$ with X_3. We have the following relations:

$$\left.\begin{aligned} \bar{T}_{11} &= T_{22} & \bar{T}_{22} &= T_{11} \\ \bar{T}_{33} &= T_{33} & \bar{E}_{11} &= E_{22} \\ \bar{E}_{22} &= E_{11} & \bar{E}_{33} &= E_{33}. \end{aligned}\right\} \tag{6.2.25}$$

Since an isotropic material is orthotropic, we may apply matrix (24) to obtain, for example

$$\bar{T}_{22} = c_{21}\bar{E}_{11} + c_{22}\bar{E}_{22} + c_{23}\bar{E}_{33}.$$

We substitute in this equation for the barred quantities in terms of unbarred quantities, using Eqs. (25), to obtain

$$T_{11} = c_{21}E_{22} + c_{22}E_{11} + c_{23}E_{33}.$$

But we also obtain, by applying (24) directly to the unbarred system,

$$T_{11} = c_{11}E_{11} + c_{12}E_{22} + c_{13}E_{33}.$$

Subtracting the two expressions for T_{11}, we get the identity:

$$0 \equiv E_{11}(c_{22} - c_{11}) + E_{22}(c_{21} - c_{12}) + E_{33}(c_{23} - c_{13})$$

for arbitrary values of E_{11}, E_{22}, and E_{33}, whence

$$c_{22} = c_{11}, \qquad c_{21} = c_{12}, \quad \text{and} \quad c_{23} = c_{13}.$$

Similar arguments with the other stresses show finally that

$$c_{22} = c_{33}, \qquad c_{12} = c_{21} = c_{13} = c_{31} = c_{23} = c_{32} = \lambda, \text{ say,}$$

and

$$c_{44} = c_{55} = c_{66} = \mu, \text{ say.}$$

Consideration of 45-degree rotations shows further that

$$c_{11} = \lambda + 2\mu.$$

Details of this will be left as an exercise. Since we already have $c_{22} = c_{33} = c_{11}$, the isotropic coefficient matrix has the form (26) with two independent elastic constants λ and μ.

Isotropic Coefficient Matrix

$$\begin{bmatrix} \lambda + 2\mu & \lambda & \lambda & 0 & 0 & 0 \\ \lambda & \lambda + 2\mu & \lambda & 0 & 0 & 0 \\ \lambda & \lambda & \lambda + 2\mu & 0 & 0 & 0 \\ 0 & 0 & 0 & \mu & 0 & 0 \\ 0 & 0 & 0 & 0 & \mu & 0 \\ 0 & 0 & 0 & 0 & 0 & \mu \end{bmatrix} \tag{6.2.26}$$

With this form for the coefficient matrix, the matrix Eq. (13) is equivalent to Eq. (4). Hence, in a rather roundabout manner, we have given a derivation of Eqs. (6.1.6). Note that the derivation made no use of the existence of a strain-energy function. We now consider the implications of temperature variation.

Linear thermoelastic equations. In deriving the Green-elastic or hyperelastic equations, we required the constants c_1 to c_6 to be zero in order that the stress vanish in the unstrained state. When a constrained body is subjected to a temperature change $\theta - \theta_0$ from the uniform reference state temperature, it may develop stresses without strains. For a linear theory we put $c_k = -\beta_k(\theta - \theta_0)$ for $k = 1$ to 6, and add terms like $c_7(\theta - \theta_0) + c_8(\theta - \theta_0)^2$ to the quadratic expression of Eq. (19), where we now interpret $(1/\rho_0)W$ as the free energy ψ. The stress-strain matrix equations (15) become

$$[T_k] = [-\beta_k(\theta - \theta_0)] + [c_{km}][E_m] \tag{6.2.27}$$

or, returning to the Cartesian tensor notation with $\beta_{11} = \beta_1$, $\beta_{22} = \beta_2$, $\beta_{33} = \beta_3$, $\beta_{23} = \beta_{32} = \beta_4$, $\beta_{31} = \beta_{13} = \beta_5$, $\beta_{12} = \beta_{21} = \beta_6$, we obtain

$$T_{ij} = -\beta_{ij}(\theta - \theta_0) + c_{ijrs}E_{rs} \tag{6.2.28}$$

the generalized Duhamel-Neumann form of Hooke's law.† For a linear theory we suppose the β_{ij} independent of strain and the c_{ijrs} independent of temperature change from the natural state. Then for isotropic behavior with $\theta = \theta_0$, c_{ijrs} must have the form of Eq. (6.1.6), while for isotropic behavior with $E_{rs} = 0$ we must have $\beta_{ij} = \beta\delta_{ij}$, where δ_{ij} is the Kronecker delta. The isotropic equations are then

$$T_{ij} = \lambda E_{kk}\delta_{ij} + 2GE_{ij} - \beta(\theta - \theta_0)\delta_{ij} \tag{6.2.29}$$

†Attributed to Duhamel (1838) and Neumann (1841) by Sokolnikoff (1956), Sec. 99.

or

$$\left. \begin{aligned} T'_{ij} &= 2GE'_{ij} \\ p &= -Ke + \beta(\theta - \theta_0). \end{aligned} \right\} \tag{6.2.30}$$

Inversely

$$\left. \begin{aligned} E'_{ij} &= \frac{T'_{ij}}{2G} \\ e &= -\frac{p}{K} + 3\alpha(\theta - \theta_0) \end{aligned} \right\} \tag{6.2.31}$$

or

$$E_{ij} = -\frac{\nu}{E} T_{kk}\delta_{ij} + \frac{1+\nu}{E} T_{ij} + \alpha(\theta - \theta_0)\delta_{ij} \tag{6.2.32}$$

where

$$3\alpha = \frac{\beta}{K} \quad \text{or} \quad \beta = \frac{E\alpha}{1 - 2\nu} \tag{6.2.33}$$

and α is the usual ***linear coefficient of thermal expansion***, assumed constant in the linear theory though it may depend on the reference state temperature θ_0. Comparison with Eqs. (4) and (8) shows that the only effect in the linear isotropic case is to add an expansion $\alpha(\theta - \theta_0)$ to each unit extension caused by stress in Eq. (8) and to subtract $\beta(\theta - \theta_0)$ from each normal stress required for the strains of Eq. (4).

If the temperature field is taken as given, incorporation of these thermal-expansion or ***thermal-stress*** terms into the linear elastic equations does not present much difficulty. But if part of the temperature change $\theta - \theta_0$ is brought about by the elastic straining, so that the heat equation contains terms dependent on strain (the coupled heat equation), solution is considerably more complicated. Fortunately, for most applications the coupling can be neglected. Boley and Weiner (1960), Sec. 2.2, consider numerical examples in aluminum and steel indicating negligible coupling when

$$\frac{\dot{E}_{kk}}{3\alpha\dot{\theta}} \ll 20. \tag{6.2.34}$$

We conclude this article on constitutive equations of linear elasticity by considering some thermodynamic limitations on the values of the isotropic moduli.

Positive-definiteness of the strain-energy function. We assume that the strain-energy function W is positive-definite as a function of $\mathbf{E}$, vanishing only for $\mathbf{E} = 0$. This assumption is basic to the proof of uniqueness of solution of boundary-value problems in elastic equilibrium. It seems reasonable to require W to be positive-definite, since this means that in any small strain from an unstressed state the stress must do positive work. This is related to the Gibbs (1875) concept of (stable) thermodynamic equilibrium, based on the second law of thermodynamics, but a rigorous derivation for nonuniform systems is lacking. The Gibbs equilibrium concept will be sketched briefly here as motivation for our assumption. Consider a system in equilibrium state A with entropy S and internal energy U, and a neighboring state B (in the sense of variational calculus) with the same energy but with entropy $S + \Delta S$. If any spontaneous change can occur at all, the second law requires that it be in the direction of increasing entropy. Hence, if $\Delta S < 0$, a spontaneous change from A to B could not occur, while if $\Delta S > 0$ it could. Hence, ***the equilibrium state is a state of maximum entropy among all states with the same energy.*** Gibbs gave another statement of the principle more directly applicable to our assumption: ***For an isolated system in equilibrium the internal energy is minimum among all states with the same entropy.***

Now the unstrained state of our ideal elastic material, the ***natural state***, must be a state of stable thermodynamic equilibrium, since experience indicates that it will not spontaneously deform from the state. Hence, we assume that the internal energy is a strict minimum among neighboring deformed states with the same entropy. A quite similar argument can be made for the isothermal case using the free energy. We therefore postulate for the Green-elastic or hyperelastic material that in general the strain-energy function W is a strict minimum in the natural state where we have arbitrarily taken its value to be zero. In any neighboring state it must therefore be positive.

Restrictions imposed on the isotropic elastic moduli by the positive-definiteness of the strain energy may be seen as follows. By the definition of W implied by Eq. (18) for small displacement gradients, we have

$$dW = T_{ij}\, dE_{ij} \tag{6.2.35}$$

for arbitrary symmetric dE_{ij}, whence by Eqs. (10)

$$dW = -p\, de + T'_{ij}\, dE'_{ij}. \tag{6.2.36}$$

Use of Eqs. (11) then leads to the integrated form

$$W = \tfrac{1}{2}Ke^2 + GE'_{ij}E'_{ij} \tag{6.2.37}$$

if W vanishes in the natural state. Since e and the quantities entering the sum of squares $E'_{ij}E'_{ij}$ are independently variable, a necessary and sufficient condition for positive-definiteness is

$$\lambda + \tfrac{2}{3}G \equiv K > 0 \quad \text{and} \quad G > 0 \tag{6.2.38}$$

if we rule out the cases $K = 0$ or $G = 0$. Use of Eqs. (7) and (12) furnishes the equivalent set of conditions

$$E > 0 \quad \text{and} \quad -1 < \nu < \tfrac{1}{2}. \tag{6.2.39}$$

The isotropic strain-energy function can be put into the alternative form

$$W = \tfrac{1}{2}\lambda e^2 + GE_{ij}E_{ij}. \tag{6.2.40}$$

Experience does not exhibit any materials with negative values of ν, although the value for beryllium (0.01 to 0.06) is so near to zero that imprecise measurements might give a negative value. The value $\nu = \frac{1}{2}$ implies $G = E/3$ and $1/K = 0$, or elastic incompressibility, approximated by some rubber, while lead has ν of approximately 0.45. The value $\nu = \frac{1}{4}$, predicted by Poisson in 1829 on the basis of an inadequate molecular model, makes $\lambda = G$, facilitating the solution of problems.

Typical experimental values of E, G, and ν are given in Table 6-1. Relations among the different isotropic constants are summarized in Eq. (41).

TABLE 6-1

ELASTIC CONSTANTS AND LINEAR COEFFICIENT OF EXPANSION (TYPICAL VALUES)

Material	E (10^6 psi)	G (10^6 psi)	K (10^6 psi)	ν	α ($10^{-6}/°F$)
Steels	28.5–30	10.6–11.9	20.2–24.0	0.265–0.305	5.6–7.3
Copper	15.6	5.8	17.9	0.355	9.2
Aluminum alloys	9.9–10.3	3.7–3.9	9.9–10.2	0.330–0.334	10.5–13.3

The following useful relationships among the isotropic elastic constants can be established by using Eqs. (7) and (12).

$$\lambda = \frac{2G\nu}{1-2\nu} = \frac{G(E-2G)}{3G-E} = K - \frac{2}{3}G = \frac{E\nu}{(1+\nu)(1-2\nu)}$$

$$= \frac{3K\nu}{1+\nu} = \frac{3K(3K-E)}{9K-E},$$

$$\mu \equiv G = \frac{\lambda(1-2\nu)}{2\nu} = \frac{3}{2}(K-\lambda) = \frac{E}{2(1+\nu)} = \frac{3K(1-2\nu)}{2(1+\nu)}$$

$$= \frac{3KE}{9K-E},$$

$$\left.\begin{aligned}
\nu &= \frac{\lambda}{2(\lambda + G)} = \frac{\lambda}{(3K - \lambda)} = \frac{E}{2G} - 1 = \frac{3K - 2G}{2(3K + G)} \\
&= \frac{3K - E}{6K}, \\
E &= \frac{G(3\lambda + 2G)}{\lambda + G} = \frac{\lambda(1 + \nu)(1 - 2\nu)}{\nu} = \frac{9K(K - \lambda)}{3K - \lambda} \\
&= 2G(1 + \nu) = \frac{9KG}{3K + G} = 3K(1 - 2\nu), \\
K &= \lambda + \frac{2}{3}G = \frac{\lambda(1 + \nu)}{3\nu} = \frac{2G(1 + \nu)}{3(1 - 2\nu)} = \frac{GE}{3(3G - E)} \\
&= \frac{E}{3(1 - 2\nu)}.
\end{aligned}\right\} \quad (6.2.41)$$

Also useful are

$$\frac{G}{\lambda + G} = 1 - 2\nu, \qquad \frac{\lambda}{\lambda + 2G} = \frac{\nu}{1 - \nu},$$

$$\frac{\lambda + 2G}{E} = \frac{1 - \nu}{(1 + \nu)(1 - 2\nu)}, \qquad \frac{E}{1 - \nu^2} = \frac{4G(\lambda + G)}{\lambda + 2G}.$$

EXERCISES†

1. Verify that the identities of Eqs. (6.2.7) reduce Eq. (6) to Eq. (8).
2. Verify the identities $\lambda = 2G\nu/(1 - 2\nu)$ and $E = G(3\lambda + 2G)/(\lambda + G)$ of Eq. (6.2.41).
3. Solve Eqs. (6.2.9) for the stresses in terms of strains, and verify that the identities of Eqs. (7) reduce the results to the results of the matrix (26).
4. Show that the result of using the matrix (6.2.26) is the same as the indicial Eq. (4).
5. Carry out the details of obtaining Eq. (6.2.20) from Eqs. (18) and (19b).
6. Show that Eqs. (6.2.19) and (20), for symmetric c_{ij} and with $c_0 = c_q = 0$, imply (21).
7. In the derivation of the form of matrix (6.2.23), what additional zero elements are required to make $\bar{T}_{22} = T_{22}$ and $\bar{T}_{23} = -T_{23}$ for arbitrary strains?
8. Carry out the details of showing that $c_{11} = \lambda + 2\mu$ in the isotropic elastic coefficient matrix as in (6.2.26), assuming it already is known that $c_{22} = c_{33} = c_{11}$ and that all other nonzero coefficients are either equal to λ or to μ. *Hint*: Con-

†See also Exercises 11 of Sec. 5.3; 5 of Sec. 5.4; 1 of Sec. 5.5; 7 and 8 of Sec. 5.7; and the exercises in Chap. 8.

sider how the equation $\bar{T}_{12} = 2G\bar{E}_{12}$ transforms under a 45° rotation of axes with the $\bar{X}_3$-axis coinciding with the X_3-axis.

9. If α and K have the smallest values suggested for steel in Table 6-1, what stresses are induced by a temperature rise of 50°F with no strains permitted ?

10. If free expansion is allowed in the X_2 and X_3 directions, so that $T_{22} = T_{33} = 0$, while extension in the X_1-direction is not permitted, express the stress T_{11} in terms of the temperature rise, according to Eqs. (6.2.29) to (33).

11. (a) Show that Eq. (6.2.35) leads to Eq. (37) and that this is equivalent to Eq. (40). Show also that Eq. (40) can alternatively be obtained from Eq. (21) and the Hooke's Law.
    (b) How is the distortion-energy term $GE'_{ij}E'_{ij}$ in Eq. (6.2.37) related to the second invariant $\mathrm{II}_{T'}$ of the stress deviator ?

12. Write W of Eqs. (6.2.21) and (40) as a function of the stresses alone for the isotropic case.

13. Show that the isotropic Hooke's Law may be written as

$$T_{ij} = 2G\left[E_{ij} + \frac{\nu}{1 - 2\nu} e\,\delta_{ij}\right]$$

and hence derive the following form for the equilibrium equations of an elastic body in terms of the displacements u_i, neglecting the distinction between spatial and material coodinates:

$$G\left[\nabla^2 u_i + \frac{1}{1 - 2\nu}\frac{\partial^2 u_j}{\partial x_j \partial x_i}\right] + \rho b_i = 0.$$

14. (a) Write the quadratic part of Eq. (6.2.19b) for the case of orthotropic symmetry with respect to the coordinate axes.
    (b) Introduce the deviatoric and spherical strains, and determine some restrictions on the coefficients which are sufficient to ensure that W is positive-definite for the following cases: (1) $E'_i \equiv 0$, and e arbitrary; (2) $e = 0$ and any one arbitrarily chosen E'_i different from zero. (*Remark*: This still does not ensure that W is always positive-definite.)

15. If the inverse of Eq. (6.2.16) is $E_p = C_{pq}T_q$ [with capital C used for the C_{pq} of the inverse relation to distinguish it from the c_{mn} of Eq. (16)], show from Clapeyron' s formula, Eq. (21), that when W is expressed in terms of stresses alone, the equation $E_i = \partial W/\partial T_i$ results (due to Castigliano in 1875).

6.3 Fluids; ideal frictionless fluid; linearly viscous (Newtonian) fluid; Stokes condition of vanishing bulk viscosity; laminar and turbulent flow

Experience indicates that a fluid at rest or in uniform flow cannot sustain a shear stress. Hence, in a fluid at rest or in uniform flow the maximum shear stress magnitude is zero, and the stress is a purely hydrostatic state of stress,

$$T_{ij} = -\bar{p}_0\delta_{ij} \quad \text{or} \quad \mathbf{T} = -\bar{p}_0\mathbf{1} \tag{6.3.1}$$

where **1** is the unit tensor, $\bar{p}_0$ is the ***static pressure***, the mean normal pressure $\bar{p} = -T_{kk}/3$ of Eq. (3.3.11) with the overbar added to distinguish it from the thermodynamic pressure p and with the subscript zero added to indicate that it pertains to the condition of rest or uniform flow. In thermodynamics the static pressure in a fluid pure substance is assumed to be related to the density ρ and the absolute temperature θ by an equation of state $F(\bar{p}_0, \rho, \theta) = 0$ in a state of thermodynamic equilibrium. In fluid mechanics we will define the ***thermodynamic pressure*** p to be the quantity given by the same functional relation to ρ and θ that gives the static pressure $\bar{p}_0$ in an equilibrium state, that is,

$$F(p, \rho, \theta) = 0. \tag{6.3.2a}$$

Equation (2a) is an ***equation of state***, sometimes called the ***kinetic equation of state*** for the fluid to distinguish it from the ***caloric equation of state***

$$u = u(\theta, \rho) \tag{6.3.2b}$$

that expresses the dependence of the internal energy u on the state variables. [See Eq. (5.7.7).] The kinetic equation of state is an example of what we called a thermal equation of state in Sec. 5.7; it expresses the dependence of the thermodynamic tension $-p$ on the state variables as in Eq. (5.7.8). With this definition of p there is no guarantee that p will be equal to $-T_{kk}/3$, and we will see that it often is not. In this section and in Chap. 7 the overbar will be used to distinguish the mean pressure $\bar{p} = -T_{kk}/3$ from the thermodynamic pressure p. (Some authors use p for mean pressure and denote the thermodynamic pressure by π.) The reader is cautioned that the reading of a "pressure gauge" in a fluid in motion may respond to some normal component of stress, which may not be equal to either p or $\bar{p}$.

An example of an equation of state is the perfect-gas law

$$p = \rho R\theta, \tag{6.3.3}$$

where R is the gas constant for the particular gas.

Barotropic flows are defined to be flows satisfying a functional relation

$$f(p, \rho) = 0 \tag{6.3.4}$$

independent of the temperature. For such flows a formulation in terms of mechanical variables alone is possible. Under special conditions a fluid having an equation of state of the form of Eq. (2) may satisfy a barotropic equation like Eq. (4), but the functional form of f depends on the special conditions of the flow. One example is an ***isothermal flow*** with $\theta =$ const. in Eq. (2). Another example is the reversible adiabatic flow (isentropic flow) of a gas satisfying Eq.

(3). It can be shown that reversible adiabatic changes in such a perfect gas are governed by the particular barotropic relation (see Ex. 3 of Sec. 5.6).

$$\frac{p}{\rho^\gamma} = \text{const.}, \tag{6.3.5}$$

where $\gamma = c_p/c_v = 1 + (R/c_v)$ and where c_p and c_v are the specific heats at constant pressure and constant volume. This equation is approximately satisfied in adiabatic expansion of dry air with $\gamma = 1.4$.

An ideal incompressible fluid is governed by the special barotropic equation

$$\rho = \text{const.} \tag{6.3.6}$$

This special case of Eq. (4) does not determine p as a function of ρ. We will see in Eq. (17a) that in an incompressible fluid we will have $p = \bar{p}$, but this remains an unknown function of position not determined in terms of the other field variables by the field equations at a single point.

An ideal frictionless (nonviscous) fluid is defined as one such that it can sustain no shear stresses even when it is in motion. No real fluids are actually nonviscous, but it turns out that in many flow fields the pressure and body-force effects predominate over the viscous effects and the flow can be analyzed quite satisfactorily by assuming the fluid to be nonviscous everywhere except in the ***boundary layer***, a thin transition layer near a solid object in the flow field or near a container wall, where viscous effects are of the same order of magnitude as other effects. Outside the boundary layer the flow is usually analyzed as though it extended right up to the solid boundary, the boundary condition imposed being that the flow can have no relative velocity normal to the boundary: $v_n = 0$, if the boundary is stationary, while the frictionless-flow boundary condition places no limitation on the tangential component v_t. When the streamline pattern around an airfoil, for example, is calculated under these assumptions, it is found that the analytical prediction agrees quite well with the measured values if the fluid flows smoothly around the airfoil, that is with no large regions of separation (stall). In the boundary layer, the viscosity of the fluid retards the flow, so that it brings the relative tangential velocity also to zero. Some examples of boundary-layer analysis will be given in Sec. 7.5.

The constitutive equation for the ideal nonviscous fluid is then simply

$$\boxed{\text{Frictionless fluid} \quad \mathbf{T} = -p\mathbf{1} \quad \text{or} \quad T_{ij} = -p\delta_{ij}} \tag{6.3.7}$$

where p is the thermodynamic pressure, satisfying the equation of state, Eq. (2). When the equation of state is $p = p(\rho)$ the frictionless fluid is called an ***elastic fluid***. Hence, in an ideal frictionless fluid we have $\bar{p} = p$, where p is the thermodynamic pressure given by the equation of state, Eq. (2).

We shall see that in a viscous fluid in motion not only are shear stresses developed but the mean pressure $\bar{p} = -\frac{1}{3}T_{kk}$ may differ from the thermodynamic pressure p calculated according to the equation of state. Following Stokes (1845) we assume that the difference between the stress in a deforming fluid and the static equilibrium stress given by Eq. (1) is a function of the rate-of-deformation tensor $\mathbf{D}$, so that the total stress is

$$\mathbf{T} = -p\mathbf{1} + \mathbf{F}(\mathbf{D}), \tag{6.3.8}$$

where p is the thermodynamic pressure given by the equation of state, Eq. (2); the extra stress in addition to the stress $-p\mathbf{1}$ which would be present at the same density and temperature if the fluid were at rest is called the ***viscous stress*** and given by the tensor-valued function $\mathbf{F}$ of the rate of deformation $\mathbf{D}$. The viscous stress is the total stress $\mathbf{T}$ minus the equilibrium stress $(-p\mathbf{1})$ or $\mathbf{T} + p\mathbf{1}$. Stokes considered only the case of linear viscosity, although a fluid in which $\mathbf{F}$ is nonlinear is often called a ***Stokesian fluid***.

When the function $\mathbf{F}$ is linear, the fluid is called ***Newtonian*** and the constitutive equation in rectangular Cartesian components takes the form

$$T_{ij} = -p\delta_{ij} + c_{ijrs}D_{rs}, \tag{6.3.9}$$

where the c_{ijrs} are constants. Since T_{ij} is symmetric, we require that c_{ijrs} be symmetric in i and j. Moreover, since D_{rs} is symmetric, we can take c_{ijrs} to be symmetric also in r and s without loss of generality. This reduces the number of independent constants to 36. Indeed the dependence of the viscous stress $\mathbf{T} + p\mathbf{1}$ on $\mathbf{D}$ can be represented by a matrix equation similar to the equation $T = CE$ used in linearized elasticity [Eq. (6.2.13)]. We will consider only isotropic fluids,† however, and use Eq. (6.1.6b), which states that any isotropic fourth-order tensor c_{ijrs} symmetric in either ij or rs, has the form

$$c_{ijrs} = \lambda\delta_{ij}\delta_{rs} + \mu(\delta_{ir}\delta_{js} + \delta_{is}\delta_{jr}).$$

When this is substituted into Eq. (9), the constitutive equation takes the form

Navier-Poisson Law‡ of a Newtonian Fluid

$$T_{ij} = -p\delta_{ij} + \lambda D_{kk}\delta_{ij} + 2\mu D_{ij} \quad \text{or}$$
$$\mathbf{T} = -p\mathbf{1} + \lambda(\operatorname{tr}\mathbf{D})\mathbf{1} + 2\mu\mathbf{D}, \tag{6.3.10}$$

where

$$\operatorname{tr}\mathbf{D} = I_D \tag{6.3.11}$$

†It can be shown that any fluid defined by Eq. (8) and satisfying the principle of material frame-indifference (see Sec. 6.7) is necessarily isotropic and moreover is not able to support couple stress. See, for example, Truesdell and Toupin (1960).

‡The simplest case is due to Newton (1687). The three-dimensional cases were obtained from molecular models by Navier (1821) for incompressible fluid and by Poisson (1831) in general. The continuum theory is due to St.-Venant (1843) and Stokes (1845).

is the trace of **D**, equal to the first invariant of **D** (D_{kk} in rectangular Cartesian components), and λ and μ are two independent parameters characterizing the viscosity of the fluid.

The Newtonian fluid equations take an especially interesting form in terms of the deviators $T'_{ij} = T_{ij} + \bar{p}\delta_{ij}$ and $D'_{ij} = D_{ij} - \frac{1}{3}D_{kk}\delta_{ij}$. Equation (10) becomes

$$T'_{ij} = (\bar{p} - p)\delta_{ij} + (\lambda + \tfrac{2}{3}\mu)D_{kk}\delta_{ij} + 2\mu D'_{ij} \tag{6.3.12}$$

whence, putting $i = j$ and summing we find, since $T'_{ii} = 0$, $D'_{ii} = 0$ and $\delta_{ii} = 3$,

$$(\bar{p} - p) + (\lambda + \tfrac{2}{3}\mu)D_{kk} = 0. \tag{6.3.13}$$

Hence the first two terms on the right in Eq. (12) cancel, and Eq. (10) is equivalent to the two relations

Navier-Poisson Law of a Newtonian Fluid

$$T'_{ij} = 2\mu D'_{ij} \quad \text{or} \quad \mathbf{T}' = 2\mu\mathbf{D}' \quad \text{and}$$
$$\bar{p} = p - \kappa D_{kk} = p + \kappa\frac{1}{\rho}\frac{d\rho}{dt}, \tag{6.3.14}$$

where κ is the ***bulk viscosity***

$$\kappa = \lambda + \tfrac{2}{3}\mu, \tag{6.3.15}$$

and the last form of the second Eq. (14) follows from

$$D_{kk} \equiv \operatorname{div} \mathbf{v} = -\frac{1}{\rho}\frac{d\rho}{dt} \tag{6.3.16}$$

by the equation of continuity [Eq. (5.2.7)]. Equation (14) shows that the mean pressure $\bar{p}$ equals the thermodynamic pressure p if and only if one of the following two conditions is satisfied:

$$D_{kk} = 0 \qquad \left(\text{i. e., } \frac{d\rho}{dt} = 0\right) \tag{6.3.17a}$$

or

$$\text{Stokes condition:†} \quad \kappa \equiv \lambda + \tfrac{2}{3}\mu = 0. \tag{6.3.17b}$$

†Stokes (1845). See Truesdell (1952), Sec. 61A, for a history of attempts to justify this hypothesis and for references to evidence that the hypothesis is not in general correct.

The first condition is assured in an incompressible fluid. Hence, in an ***incompressible*** Newtonian fluid $p = \bar{p}$ at all times. (For nonlinear viscosity, incompressibility may not imply $p = \bar{p}$.) While no real fluids are strictly incompressible, in most flow situations the effect of compressibility is negligible in liquids; it is also negligible in gases in flow around a body when the speed relative to the body is small compared to sound speed. The second condition under which $p = \bar{p}$ is called the ***Stokes condition***, namely $\lambda + \frac{2}{3}\mu = 0$. To see its significance we note that in Eq. (14) $-\kappa D_{kk}$ is the contribution to the mean pressure due to volume viscosity (or bulk viscosity), in addition to the thermodynamic pressure p which would be there at the same temperature and density if the unit rate of change of volume D_{kk} were zero.

The significance of the bulk viscosity appears further when we calculate the dissipation power per unit volume. Introducing $D_{ij} = D'_{ij} + \frac{1}{3}D_{kk}\delta_{ij}$ in Eq. (10), we obtain the stress power per unit volume as [see Sec. 5.4, following Eq. (5.4.2)]

$$T_{ij}D_{ij} = -pD_{kk} + \kappa(D_{kk})^2 + 2\mu D'_{ij}D'_{ij}$$

or (6.3.18)

$$\mathbf{T}:\mathbf{D} = -p\,\mathrm{tr}\,\mathbf{D} + \kappa(\mathrm{tr}\,\mathbf{D})^2 + 2\mu\mathbf{D}':\mathbf{D}'.$$

The term $-pD_{kk}$ equals $(+p/\rho)\,d\rho/dt$ by the equation of continuity, or $(-p/v)\,dv/dt$ where $v = 1/\rho$. It may be positive or negative and can therefore represent a recoverable contribution to the elastic internal energy. In an isothermal cycle, for example, it represents a recoverable contribution to the free energy. The assumption that it is always recoverable and that the recoverable part of the stress ${}_RT_{ij}$ [see Eq. (5.7.26)] is given by ${}_RT_{ij} = -p\delta_{ij}$ in effect assumes the existence of an elastic strain energy function $W(v)$ such that $-p = \partial W/\partial v$, which may not be justified in all cases. But at any rate the term $-pD_{kk} = (-p/v)\,dv/dt$ ***may*** represent recoverable power not contributing to the internal entropy production, while the remaining terms in Eq. (18) are assumed to be always dissipative.

The dissipation power $2W_D$ (per unit volume) is therefore defined as

$$2W_D = \kappa(D_{kk})^2 + 2\mu D'_{ij}D'_{ij}. \qquad (6.3.19)$$

Nonnegative dissipation for arbitrary D_{ij} is required by the second law of thermodynamics (see Exercise 11, Sec. 5.6). Hence

$$\mu \geqslant 0, \qquad \text{and} \qquad \kappa \geqslant 0 \quad \text{or} \quad \lambda \geqslant -\tfrac{2}{3}\mu. \qquad (6.3.20)$$

Since $2W_D$ is nonnegative this part of the stress power never contributes to an increase of the kinetic energy of the system. [See Eq. (5.4.2).]

The dissipation potential W_D (equal to half the dissipation power)† as a function of the D_{ij} may be used to express the constitutive equation in the form

$$_DT_{ij} = \frac{\partial W_D}{\partial D_{ij}}, \tag{6.3.21}$$

where

$$_RT_{ij} = -p\delta_{ij} \quad \text{and} \quad _DT_{ij} = T_{ij} + p\delta_{ij} \tag{6.3.22}$$

are the recoverable stress $_RT_{ij}$ and the dissipative stress $_DT_{ij}$ of Eq. (5.7.26). If the bulk viscosity vanishes ($\kappa \equiv \lambda + \frac{2}{3}\mu = 0$), the first term on the right in Eq. (19) vanishes and the volume change is nondissipative, the entire dissipation being given by the last term $2\mu D'_{ij}D'_{ij}$ involving the deviatoric or shape-change rate of deformation. The dilatational dissipation is important in the damping of sound waves [see Lindsay (1960), for example, or Landau and Lifshitz (1959)], but in most fluid-flow situations the major part of the dissipation is associated with the deviator D'_{ij}.

For flow analyses we usually make the Stokes-condition assumption, Eq. (17); then $\lambda = -\frac{2}{3}\mu$ and the constitutive Eq. (10) reduces to

Navier-Poisson Law of a Newtonian Fluid with No Bulk Viscosity

$$T_{ij} = -p\delta_{ij} - \tfrac{2}{3}\mu D_{kk}\delta_{ij} + 2\mu D_{ij}$$

$$\mathbf{T} = -p\mathbf{1} - \tfrac{2}{3}\mu(\text{tr}\,\mathbf{D})\mathbf{1} + 2\mu\mathbf{D} \quad \text{or} \tag{6.3.23}$$

$$T_{ij} = -p\delta_{ij} + 2\mu D'_{ij} \qquad \mathbf{T} = -p\mathbf{1} + 2\mu\mathbf{D}'$$

equivalent to the two equations

$$\mathbf{T}' = 2\mu\mathbf{D}' \qquad \bar{p} = p. \tag{6.3.24}$$

These equations also apply, of course, to any incompressible Newtonian fluid or to a flow in which the density is constant.

In most engineering applications only the one viscosity parameter μ is considered. It is called the ***dynamic viscosity*** and has dimensions of stress multiplied by time or

$$[\mu] = \left[\frac{FT}{L^2}\right] = \left[\frac{M}{LT}\right]. \tag{6.3.25}$$

Because the ratio μ/ρ occurs so frequently in the field equations of fluid dy-

†In Sec. 5.7 the "dissipation function" D was the dissipation power per unit mass; thus $\rho D = 2W_D$, and, in Eqs. (5.7.32) to (5.7.37), the "force" corresponding to the "flux" D_{ij} is $_DT_{ij}/\rho$.

namics obtained by using the constitutive equations to eliminate the stress from the equations of motion, Eqs. (5.3.4), the parameter ν is introduced, and is defined by

$$\nu = \frac{\mu}{\rho} \qquad [\nu] = \left[\frac{L^2}{T}\right]. \tag{6.3.26}$$

This parameter is called the ***kinematic viscosity*** because its dimensions L^2/T are expressed entirely in terms of kinematic dimensions.

Experimentally determined values for the dynamic viscosity μ for particular fluids at various temperatures are given in handbooks. Water has a viscosity μ of about 3×10^{-5} lbf-sec/ft^2 at a temperature just above the melting point, dropping to about half this value as the boiling point is approached, with $\mu = 2.1 \times 10^{-5}$ lbf-sec/ft^2 at 68°F. Air has a value of about 0.0377×10^{-5} lbf-sec/ft^2 at 68°F.

Laminar flow and turbulent flow. As was pointed out in Sec. 6.1, the assumption that the shear stress τ is proportional to the velocity gradient dv/dy is well verified in parallel flow situations. The dynamic viscosity is sometimes measured by viscometers in which flow of liquid occurs between concentric cylinders rotating relative to each other; another simple type of viscometer has a heavy spherical ball falling through liquid contained in a vertical tube of radius slightly larger than the ball. Viscometers are designed to produce an essentially ***laminar flow*** with very little turbulent mixing between layers moving in parallel directions. The Navier-Poisson constitutive law, Eqs. (23) or (24), is found to apply in a satisfactory manner only to values of stress and velocity obtained from measurements in such laminar flows.

Turbulent flow describes a situation where the stress and velocity at a point fluctuate in a random fashion with time. In turbulent flow it is not possible to make measurements from which simultaneous values of the instantaneous stress and velocity at a point can be deduced and checked against the Navier-Poisson law. Many investigators believe that in most flow situations the instantaneous values do in fact satisfy the Navier-Poisson law, because, in a few cases where the structure of the turbulence was believed known, it has been found possible to make predictions about the observable mean flow which are verified within the experimental accuracy. See, for example, Townsend (1956), Secs. 2.3 and 3.4.

It is sometimes possible to make measurements from which the ***mean flow***, represented by time averages of the stress and velocity, can be deduced, and it is found that ***these mean values do not satisfy the Navier-Poisson law*** in general when the flow is turbulent. For example, in parallel mean flow in the x-direction, the mean shear stress $\bar{\tau}$ and velocity x-component $\bar{u}(y)$ do not satisfy $\bar{\tau} = \mu\, d\bar{u}/dy$ as they would in laminar flow, because the shear stress is modified by momentum transfer between parallel layers due to turbulent eddies carrying material from one layer into the other. Boussinesq expressed the additional shear stress due to turbulent eddies in terms of an ***eddy viscosity*** η that depends on the state of the turbulent motion, writing

$$\bar{\tau} = \mu\frac{d\bar{u}}{dy} + \eta\frac{d\bar{u}}{dy} \tag{6.3.27}$$

for the parallel-flow case, where $\mu\, d\bar{u}/dy$ is the apparent stress computed from the mean flow, using the dynamic viscosity μ characteristic of laminar flow in that fluid at that temperature, while $\eta\, d\bar{u}/dy$ is an additional apparent stress due to turbulent momentum transfer. The eddy viscosity η is not a fluid property but rather a characteristic of the flow. In many cases of fully developed turbulence the second term is dominant, so that it is not a question of making a small correction to the laminar equations.

A physical model for the additional apparent stress was introduced by Reynolds around 1880. Consider the parallel mean flow in adjacent layers where the x-components of mean velocity are $\bar{u}_1$ at y_1 and $\bar{u}_2 > \bar{u}_1$ at $y_1 + \Delta y$ for $\Delta y > 0$. Let the instantaneous velocity components v_x and v_y be expressed by

$$v_x = \bar{u} + u' \quad \text{and} \quad v_y = v'$$

where $\bar{u}$ is the mean value of v_x, while u' and v' are turbulent fluctuations. If the lower-velocity fluid in layer 1 were to fluctuate with velocity v' into layer 2, its velocity in the flow direction would be less than $\bar{u}_2$ by the amount u' (in the limit as $\Delta y \to 0$). The mass flux across a horizontal surface is $\rho v'$ (per unit area): hence, the upward momentum transfer is $-\rho u'v'$ or on the average $-\rho\overline{u'v'}$, where $\overline{u'v'}$ is the time average of $u'v'$. Thus

$$\bar{\tau} = \mu\frac{d\bar{u}}{dy} - \rho\overline{u'v'}. \tag{6.3.28}$$

Comparing this with Eq. (27), we see that

$$\eta = -\frac{\rho\overline{u'v'}}{d\bar{u}/dy}, \tag{6.3.29}$$

clearly indicating that η is a characteristic of the state of turbulence instead of a material parameter. For three-dimensional flows, ***Reynolds stresses***, equal to $-\rho\overline{v_i'v_j'}$ are added to the expressions for $\bar{T}_{ij}$. See, for example, Daily and Harleman (1966), Sec. 11–4, for an elementary presentation of the ***Reynolds equations***, which replace the Navier-Stokes equations of Sec. 8.1 when the modified constitutive equations including the Reynolds stresses are substituted into the equations of motion. Unfortunately the Reynolds equations contain too many unknowns, and solutions can only be obtained by making special assumptions about the nature of the turbulence.

At the present time no really satisfactory theory of turbulent flow exists, despite its great importance. It is not surprising therefore that a great deal of current research activity in fluid dynamics is concerned with turbulent flow, especially with its initiation in boundary layers. A substantial body of information is available for engineering designers, but the results do not lend themselves readily to complete mathematical formulation of turbulent flow. The study of this information is outside the scope of this book.

Temperature dependence of the viscosity has already been mentioned. It is well-known that liquid viscosity decreases with increasing temperature; a demonstration of the high viscosity at low temperatures is furnished each time we start an automobile engine on a winter morning. In gases the temperature dependence is reversed; the viscosity increases with increasing temperature. The explanation of the different behaviors requires us to leave the domain of continuum mechanics and consider fluids as collections of particles. In gases the intermolecular forces have little effect except for particle collisions. The apparent shear stress is the effect of momentum transfer in collisions of faster molecules diffusing into slower moving layers and vice versa and depends on a sort of microscopic turbulence even when the macroscopic flow appears to be laminar. Increasing temperature increases this interchange and thus increases the apparent shear stress. In liquids the spacing is much closer, and the shear stress in laminar flow is largely due to intermolecular forces, which are decreased by increasing temperature.

The fluids discussed in the present section possessed volume elasticity, or thermoelasticity specified by the equation of state, Eq. (2), but the deviator relationships of Eqs. (14) and (24) were entirely viscous. We turn our attention now to viscoelastic materials exhibiting both viscosity and elasticity in the deviator relationships. After a look at viscoelasticity in Sec. 6.4, we consider plasticity and viscoplasticity in Sec. 6.5.

EXERCISES†

1. (a) Use Eq. (6.3.1) and the partial differential equations of equilibrium, Eqs. (5.3.5) to derive differential equations expressing the variation of the hydrostatic pressure $\bar{p}_0$ with position in a uniform gravitational field acting in the negative z-direction.
   (b) Hence, show that in a homogeneous incompressible liquid, $\bar{p}_0$ varies linearly with depth.

†See also Exercises 4 and 12 of Sec. 5.3; 1, 2, 3, and 6 of Sec. 5.4; 1 to 8, 11, and 12 of Sec. 5.6; and 10 of Sec. 5.7 and Chap. 7.

(c) What additional information would be needed to make it possible to integrate the differential equations of (a) in order to determine the variation of $\bar{p}_0$ with height in a compressible fluid?

2. What is the barotropic form of the perfect-gas law for the case of isothermal motion?

3. Introduce the deviator D'_{ij}, and show that Eq. (6.3.12) follows from Eq. (10), and that the stress power may be put into the form of Eq. (18).

4. Verify that if the dissipation potential W_D is defined as half the dissipation power of Eq. (6.3.19), then the dissipative part of the stress ${}_DT_{pq} = T_{pq} + p\delta_{pq}$ is given by ${}_DT_{pq} = \partial W_D/\partial D_{pq}$. *Hint:* Express the D'_{ij} factors as functions of D_{ij} before differentiating with respect to D_{pq}. Avoid repeating any letter index more than twice in one term.

5. Show that the term $-pD_{kk}$ in Eq. (6.3.18) represents recoverable work in an isothermal flow.

6. Substitute Eq. (6.3.23) into the equations of motion, Eq. (5.3.4), and express the D_{ij} in terms of the velocities, to obtain the component forms of the Navier-Stokes equation:

$$-\frac{1}{\rho}\nabla p + \frac{\nu}{3}\nabla(\nabla\cdot\mathbf{v}) + \nu\nabla^2\mathbf{v} + \mathbf{b} = \frac{d\mathbf{v}}{dt},$$

where $\nu = \mu/\rho$ is the kinematic viscosity of Eq. (26). *Hint:* Show that $D_{rm,r} = \frac{1}{2}[(\nabla\cdot\mathbf{v})_{,m} + \nabla^2 v_m]$.

7. Show that the total traction on any surface S enclosing volume V of fluid governed by Eq. (6.3.10) is equal to the volume integral

$$\int_V [-\nabla p + (\lambda + \mu)\nabla(\nabla\cdot\mathbf{v}) + \mu\nabla^2\mathbf{v}]\,dV.$$

By applying this to smaller and smaller volumes about a point, interpret the terms of the integrand as a pressure-gradient force and viscous force (per unit volume) at the point.

8. Fluid of viscosity μ in laminar flow through a circular pipe of radius a is observed to have the axially symmetric flow $v_x = v_y = 0$, and $v_z = v_0[1 - (r^2/a^2)]$, where v_0 is a positive constant and $r = (x^2 + y^2)^{1/2}$ is the distance from the axis. Determine:
   (a) the shear stress at $r = a/2$ and at the wall $r = a$, e.g., at $x = 0$, $y = a/2$ and $x = 0$, $y = a$, and
   (b) the total drag on length L of the pipe.
   (c) Sketch the velocity profile.

9. For the flow of Ex. 8, evaluate the dissipation power $2W_D$ per unit volume at a typical point and the total dissipation power in a length L of the pipe.

10. Show that the energy equation, Eq. (5.4.7), takes the form

$$\rho\frac{du}{dt} = \nabla\cdot(k\nabla\theta) - p\nabla\cdot\mathbf{v} + \rho r + 2W_D,$$

if the stresses are given by Eq. (6.3.10), $2W_D$ by Eq. (19), and $\mathbf{q}$ by $\mathbf{q} = -k\nabla\theta$.

6.4 Linear viscoelastic response

This section is divided into four parts, as follows:

Part 1 discusses the nature of the creep and relaxation behavior observed in polymers and the way their response to periodic loading is characterized.

Part 2 introduces constitutive equations by way of analogies to spring and dashpot models, and concludes by passing to the limit of continuous creep and relaxation spectra.

Part 3 uses the Boltzmann superposition principle and the continuous spectra to formulate constitutive equations in terms of superposition integrals for a linear hereditary material and concludes by expressing the periodic response functions in terms of the creep and relaxation spectra.

Part 4 concludes the section with some references to authors who formulate boundary-value problems or discuss nonlinear viscoelasticity or thermoviscoelasticity. This part is followed by exercises and a bibliography giving a few references for collateral reading.†

Part 1. Response of Polymers

A viscoelastic material combines elastic and viscous characteristics. This type of behavior is especially noticeable in the organic high polymers, whose mechanical properties have been extensively studied in recent years because of their increasing technological importance. There are many different kinds of such materials, including various hard and soft plastics, natural and synthetic rubbers, and liquid-like molten plastics, with widely varying viscoelastic properties. Not only do the properties vary greatly from one type of polymer to another, but they depend strongly on temperature, and some properties change almost discontinuously in the neighborhood of a critical temperature called the ***glass-transition temperature*** T_g.

Polymers are broadly classified as

1. amorphous polymers
   (a) crosslinked (network)
   (b) uncrosslinked (linear)
2. crystalline.

The uncrosslinked amorphous polymer is usually called a ***linear amorphous polymer*** in the chemical and rheological literature, but the term uncrosslinked

†Additional bibliography at the end of the book contains references to twentieth-century authors mentioned in the text. For references to earlier work, see Truesdell and Toupin (1960) or Freudenthal and Geiringer (1958).

is used in this section to avoid confusion with the concept of linear response. An example of an ***uncrosslinked*** polymer is unvulcanized natural rubber. The individual long-chain molecules are randomly intertwined with each other but not chemically bonded together. Because of their thermal motions a sample of the material has been likened to "a box packed full of very long thin active earthworms" by Alfrey and Gurnee (1956) in discussing the kinetic theory of rubber elasticity. The vulcanization process introduces a bonding agent (e.g., sulfur), which randomly attaches a chain (possibly at several places along its length) to neighboring chains by means of covalent bonds to form a three-dimensional ***network*** or ***crosslinked polymer***, which, because of its relatively permanent network structure, resists large deformation. A crystalline structure is found in many polymers, e.g., in polyethylene and in many textile fibers [see, for example, Meredith (1958)]. Polycrystalline material is composed of crystallites whose linear dimensions are small compared to the chain length, so that a single chain could extend through several crystallites, or crystallites could be composed of folded chains forming a layered structure, as recent studies indicate. [See, for example, the discussions in Nielsen (1962) and Meares (1965).] The degree of crystallinity (fraction of the whole polymer which is in the crystalline state) may vary from a few per cent to over 80 per cent. Several methods exist for determining the degree of crystallinity, but unfortunately, according to Nielsen (1962), these methods often give somewhat different results. The exact nature of the region between crystallites is not quite clear, but it is often considered to be amorphous, perhaps also composed of folded chains less regularly ordered than in the crystallites. The mechanical properties of amorphous polymers are usually simpler to correlate than those of polycrystalline polymers. This is because the amorphous polymers seem to obey a time-temperature superposition principle permitting use of data obtained at different temperatures to extend the time scale at any one temperature. However, this is not generally the case for crystalline polymers, which also frequently fail to obey the laws of linear viscoelasticity even at small deformations [Tobolsky (1958)]. For this reason, most of the discussion in this section is more readily applicable to amorphous polymers. Although crystalline polymers can often be treated by the linear theory some caution is in order; it is necessary in any case to verify that the material used actually has an approximately linear response under the conditions of interest, and the amorphous polymers are more likely to meet this requirement.

The viscoelastic response characteristics of a material are most often determined by (1) creep tests, (2) stress relaxation tests, or (3) dynamic response to loads varying sinusoidally with time. Before taking up the analytical description of viscoelastic behavior, we will consider some of the different kinds of response observed in creep and relaxation studies and more briefly the method of describing the sinusoidal response. But first we digress to consider the glass transition, which affects markedly the mechanical response.

Glass-transition temperature. Some liquids can be supercooled to form glasses without crystallization. See, for example, Alfrey and Gurnee (1967), Sec. 1.5. Glass-transition temperatures are usually determined from the volume-temperature cooling curve, which shows a discontinuity in slope at the glass-transition temperature T_g. Mechanical properties are also different above and below T_g. A polymer well below T_g is an organic glass, an elastic solid with a high modulus of the order of 1 GN/m^2 (compare this with steel, E = 220 GN/m^2, and copper with E = 120 GN/m^2). There is a transition range around T_g in which viscoelastic phenomena predominate. Above the transition region (but below the melting point), a crosslinked amorphous polymer is a rubbery elastic solid with a modulus of the order of 10^6 or 10^7 dynes/cm^2, lower by a factor of about 1000 to 10,000 than the glassy modulus. See, for example, Tobolsky (1958). An uncrosslinked polymer above the transition region behaves like a viscous liquid. These different regions (glassy, rubbery, and transition) will be illustrated first in relation to creep and retarded elasticity.

Creep testing is most often done under uniaxial tensile stress because of experimental simplicity. But separate measurements of shear response under shear loading and volume response under hydrostatic compression are more useful as a basis for combined-stress behavior. Shear response may be measured by twisting a thin-walled cylinder, but for soft materials a direct shear measurement is often made, as in the loading arrangement indicated schematically in Fig. 6.1, which has been used for testing rubber [see Leaderman (1958)].

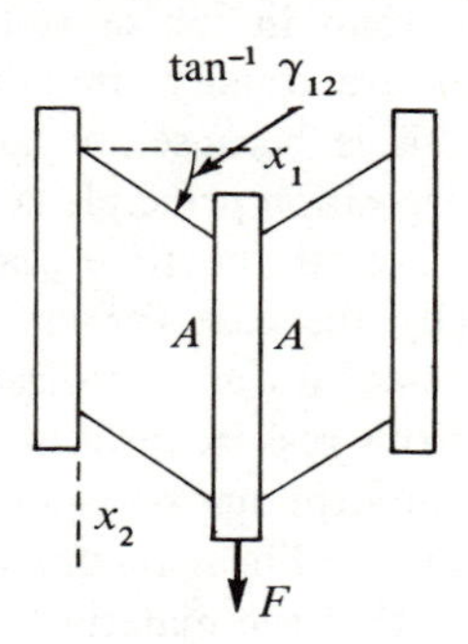

Fig. 6.1 Simple shear loading arrangement.

Two identical specimens are bonded between three parallel rigid plates. The outer plates are fixed while the inner plate moves under a vertical load F. When shear traction $T_{12} = F/A$ is applied to each specimen at $t = t_0$ and maintained constant until $t = t_1$ ($t_1 - t_0$ of the order of a few minutes to a few hours, say), and the load is removed at $t = t_1$, the resulting plot of γ_{12} versus t is called a

creep curve or a ***retarded elastic response*** curve if the crosslinking is sufficient to prevent indefinitely continued flow.

Retarded elastic response of an idealized crosslinked polymer is illustrated in Fig. 6.2. If the response is linear, plots of γ_{12}/T_{12} versus time will be inde-

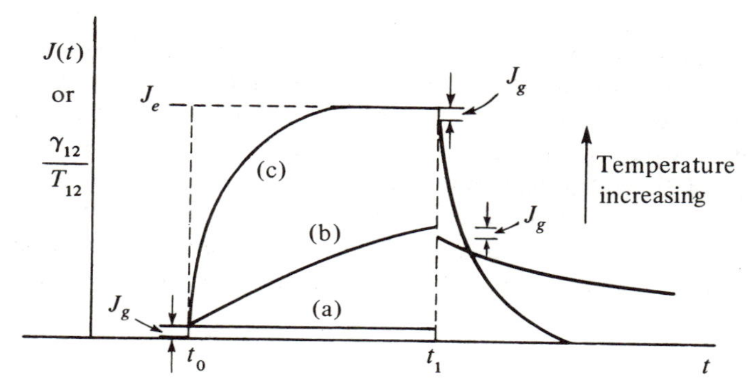

Fig. 6.2 Retarded elastic response of an idealized crosslinked polymer: (a) glassy response at temperature well below T_g; (b) transition region viscoelastic response; (c) transition to rubbery equilibrium.

pendent of the magnitude chosen for the constant T_{12}, and data obtained with various values of T_{12} all plot on the same curve, which represents the time-dependent ***creep compliance*** $J(t) = \gamma_{12}/T_{12}$. Curve (a) shows the ideal elastic response in the glassy region for temperature well below T_g, with ***glass compliance*** J_g independent of time. Curve (b), in the transition temperature region above T_g, shows an initial elastic compliance J_g followed by creep or steadily increasing strain until t_1; after unloading there is an immediate elastic recovery $\Delta\gamma_{12} = -J_g|T_{12}|$ followed by creeping recovery, which ideally would lead eventually back to the initial shape. Curve (c) for a still higher temperature in the transition range is believed to begin the same way with elastic compliance J_g, but the creep sets in so rapidly that no slope discontinuity is observed, and a ***rubbery elastic plateau*** is reached at ***equilibrium compliance*** J_e, which is not in general independent of the temperature. After unloading, the initial recovery J_g is followed by creep recovery back to the initial configuration in about the same time as was required to reach equilibrium under load. At still higher temperature, the rubbery elastic response may be almost instantaneous as indicated by the dashed curve; but J_e would not in general be the same as for curve (c), and the rubbery elastic response is likely to be nonlinear except for very small deformations. Actual materials often show a very slow creep at the so-called rubbery plateau, which eventually accelerates and leads to failure if the load is maintained indefinitely instead of being limited to a few hours.

Creep response of an uncrosslinked amorphous polymer begins in a way qualitatively similar to the retarded elastic response of a crosslinked polymer. But, instead of rubbery equilibrium, steady viscous flow results eventually, and upon unloading the recovery is not complete, a permanent set remains. The testing might be done in a viscometer with relatively rotating coaxial cylinders. Below T_g, the viscous flow is inhibited and a glassy elasticity results. Figure 6.3 shows transition region behavior for various durations of constant load (all at the same temperature) in an uncrosslinked polymer. After about point B on the loading curve the creep is steady with $d\gamma_{12} = (1/\eta)T_{12}\,dt$, where η is the coefficient of viscosity. If the response is linear, data obtained for different magnitudes of T_{12} will all plot on the same curve. Curve C_1D_1 is for unloading at time t_1, leaving a permanent set $\gamma_{12}/T_{12} = t_1/\eta$, while after t_2 the permanent set is t_2/η. Since in experiments a finite time is required to apply or remove the load, the steps of magnitude J_g may be smoothed out in both Figs. 6.2 and 6.3.

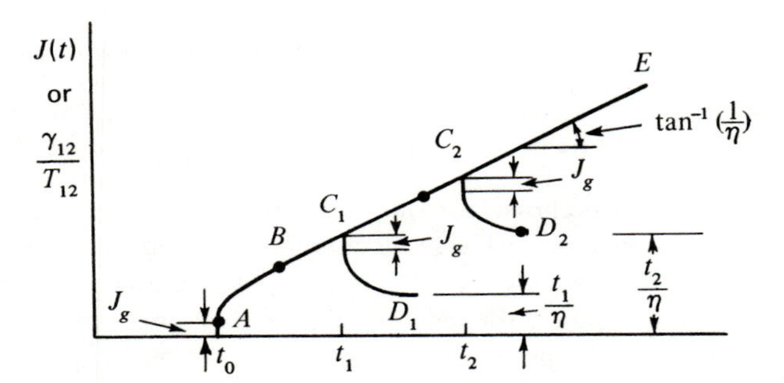

Fig. 6.3 Creep response of uncrosslinked polymer, showing effect of time at constant load.

Stress-relaxation response at constant strain is illustrated in Fig. 6.4 for an idealized crosslinked polymer at a temperature in the transition region; both scales are logarithmic to give some idea of the typical magnitudes involved as the relaxation modulus $G(t) = T_{12}/\gamma_{12}$ relaxes from the initial glassy modulus $G_g = 1/J_g$ to the equilibrium modulus G_e at the rubbery plateau. Since the relaxation covers some eight decades on the logarithmic time scale, it is not feasible to determine the whole curve by a constant-strain test at one temperature. Instead this relaxation master curve represents hypothetical behavior at one temperature obtained from several curves (possibly 10 or more) at different temperatures, each for a much shorter interval on the log t scale, and suitably translated parallel to the log t scale, so that segments from different curves join to form a continuous curve. This ***time-temperature superposition principle*** was first established empirically and has since been given a theoretical basis by statistical theories. For further details see, for example, Tobolsky (1960) or

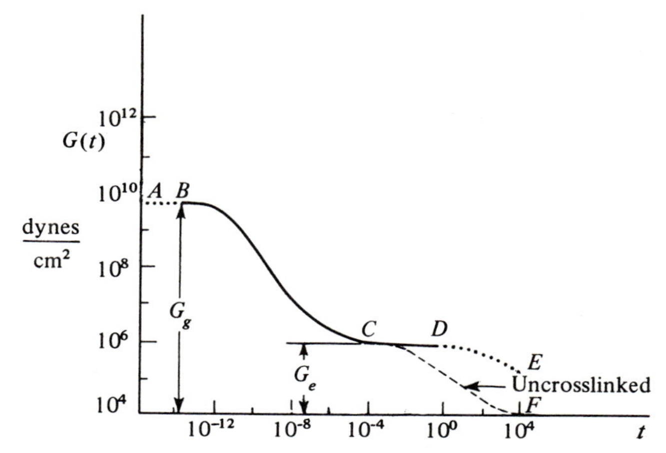

Fig. 6.4 Relaxation master curve for idealized crosslinked polymer (strain and temperature constant).

Ferry (1961). An uncrosslinked polymer shows stress relaxation to zero stress as is indicated by the dashed curve CF. And real crosslinked polymers tend to fall off the equilibrium plateau eventually as is indicated by the dotted continuation DE of the solid curve. Since a finite time is required to impose the constant strain, a rapidly relaxing material may never reach the glassy modulus in a test above T_g. Well below T_g, a constant strain yields a constant stress as in ideal elasticity, when the load is maintained for a moderate period. Actually the stress relaxes slowly even below T_g, although it might require a period of years to fall to zero. A similar behavior is ideally found (with much smaller modulus) in the rubbery elastic region, but here there is a tendency for relaxation to set in sooner if the strain is maintained; the rubbery elastic range is not as nearly ideal as the glassy range. Because of the time-temperature superposition principle, it is possible to speak of a ***transition region in time*** for any given temperature. For example, in the relaxation master curve of Fig. 6.4, representing behavior of a crosslinked polymer at a certain fixed reference temperature, the dotted segment AB is the glassy elastic region of time (possibly inaccessible in actual relaxation tests because the strain cannot be applied so rapidly), while BC is the transition region to the rubbery elastic plateau CD, followed by the rubbery flow DE.

Periodic response to periodic loading is a valuable supplement to the transient experiments (creep and relaxation) and provides information corresponding to short times. A sinusoidal stress experiment at frequency $\omega/2\pi$ is qualitatively equivalent to a transient experiment at $t = 1/\omega$; hence short-time properties are inferred from high-frequency sinusoidal testing, but this is still limited by the fact that t must still be long compared to the transit time of elastic waves through the specimen, so that stress and strain can be considered uniform throughout the test section. Under these conditions the response to a steady-state sinusoidal stress is a steady-state sinusoidal strain at the same frequency, but out of phase, e.g.,

$$T_{12} = T_0 \sin \omega t \qquad \gamma_{12} = \gamma_0 \sin(\omega t - \delta). \tag{6.4.1}$$

Both the response amplitude and the phase-shift δ are frequency-dependent, but in the linear range γ_0 is proportional to T_0. The phase relationships are conveniently shown in the rotating-vector representation of simple harmonic motion, as in Fig. 6.5. The rotating strain vector OB lags behind the stress

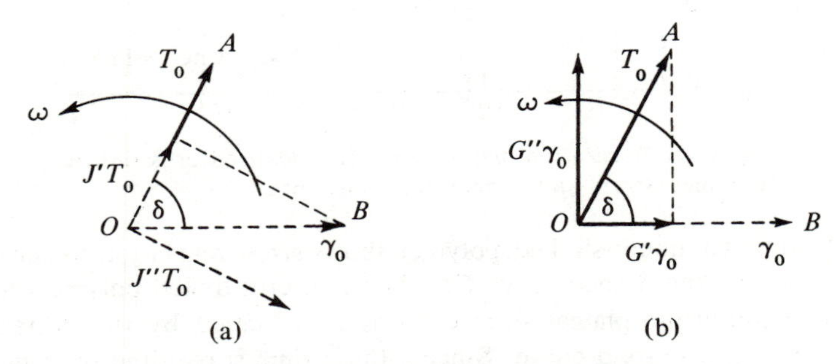

Fig. 6.5 Storage and loss compliances and moduli.

OA by δ radians; it may be resolved into two components, one in phase with and one $\pi/2$ radians out of phase with the stress, as in Fig. 6.5(a), where the in-phase component $\gamma_0 \cos \delta$ is labeled $J'T_0$ and the out-of-phase component $\gamma_0 \sin \delta$ is labeled $J''T_0$, where J' is the ***storage compliance*** and J'' is the ***loss compliance***—both of which depend on frequency. (The primes do not imply derivatives.) Alternatively, as in Fig. 6.5(b), we can resolve the stress OA into components $G'\gamma_0$, in phase with the strain, and $G''\gamma_0$ out of phase. Here G' is the ***storage modulus*** and G'' the ***loss modulus***, and both are frequency-dependent. The ***complex shear modulus*** G^* and ***complex compliance*** J^* are defined by

$$G^* = G' + iG'' \qquad J^* = 1/G^* = J' - iJ'' \tag{6.4.2}$$

with magnitudes

$$G^* = (G'^2 + G''^2)^{1/2} \qquad J^* = (J'^2 + J''^2)^{1/2}.$$

Thus

$$G' = |G^*| \cos \delta \quad G'' = |G^*| \sin \delta$$
$$J' = |J^*| \cos \delta \quad J'' = |J^*| \sin \delta.$$

Therefore

$$\frac{G''}{G'} = \frac{J''}{J'} = \tan \delta. \tag{6.4.3}$$

The ratio $G''/G' = J''/J' = \tan \delta$ is called the ***loss tangent***. If the simple harmonic stress and strain are represented by complex amplitudes T_0^* and γ_0^* times $e^{i\omega t}$, that is

$$T_{12} = T_0^* e^{i\omega t} \qquad \gamma_{12} = \gamma_0^* e^{i\omega t},$$

then

$$G^* = \frac{T_0^*}{\gamma_0^*} = \frac{T_0}{\gamma_0} e^{i\delta} \quad \text{and} \quad J^* = \frac{\gamma_0^*}{T_0^*}. \tag{6.4.4}$$

$$\mathbf{T}' = 2G^*\mathbf{E}'$$

then gives the periodic response at any circular frequency ω.

All five quantities G', G'', J', J'' and $\tan \delta$ are functions of ω. See, for example, Ferry (1961) for data on many polymers and numerous references to original investigations.

Volumetric response (also called ***dilatational response***) and ***tensile response*** can be represented qualitatively in creep, relaxation, and periodic response in the same way as the shear response, although the magnitudes of the volumetric creep compliance and relaxation modulus differ considerably from the shear values, and even an uncrosslinked polymer does not show indefinitely continued volume decrease under hydrostatic pressure. For many purposes the volume change may be considered elastic, and in some cases the material may be considered incompressible, but there will always be some dissipation in periodic hydrostatic compression, leading to a complex bulk modulus $K^* = K' + iK''$ defined in the same way as G^*.

Now that we have some acquaintance with the nature of the response of real viscoelastic materials, we turn our attention to attempts to represent the response by phenomenological theories generalizing the representation of the time-dependent response in one-dimensional stress situations like the simple shear of Fig. 6.1 to the three-dimensional, combined-stress formulation.

Part 2. Constitutive Equations Based on Analogies to Spring-and-Dashpot Models

Analogies to spring-and-dashpot models give a qualitative representation of viscoelastic behavior. Two especially simple models are illustrated in Fig. 6.6, the ***Kelvin-Voigt Element*** consisting of a spring and dashpot in parallel, and the ***Maxwell element*** consisting of a spring and dashpot in series. These simple models display qualitatively retarded-elastic, creep and relaxation phenomena, but, unlike the Hookean solid and the Newtonian fluid, neither the Voigt solid nor the Maxwell fluid gives a satisfactory quantitative representation of any real viscoelastic material.† Axial force in the model represents stress in the continuum, while axial elongation and velocity represent strain and strain rate. For the time being we suppose the continuum sample subjected to uniform strain without rotation; then for small strains the rate of deformation may be identified with the time rate of small strain. For later generalization we can suppose the stress and strain referred to moving axes fixed to the particle and rotating with the principal axes of strain, or we can use convected coordinates, or some other method to make the constitutive equations frame-indifferent (see Sec. 6.7), so that, for example, a rigid-body rotation would not predict a change in stress, according to the constitutive equation. The stress and strain taken as analogous to the force and displacement of the model can be shear, tensile, or volumetric. However, as will be seen in Ex. 17 at the end of this section, a material with a Kelvin-Voigt deviatoric response and elastic dilatational response is not governed by the same differential equation under uniaxial tensile stress as is a Kelvin-Voigt tensile element. We discuss first the deviatoric relationships, supposing that in an isotropic material each deviator component obeys the same equations as a shear component in a pure-shear deformation; hence in the models of Fig. 6.6 the force is labeled $\mathbf{T}'$ to represent the deviator tensor of the continuum, the spring constant is labeled $2G$, and the viscous element has coefficient 2η.

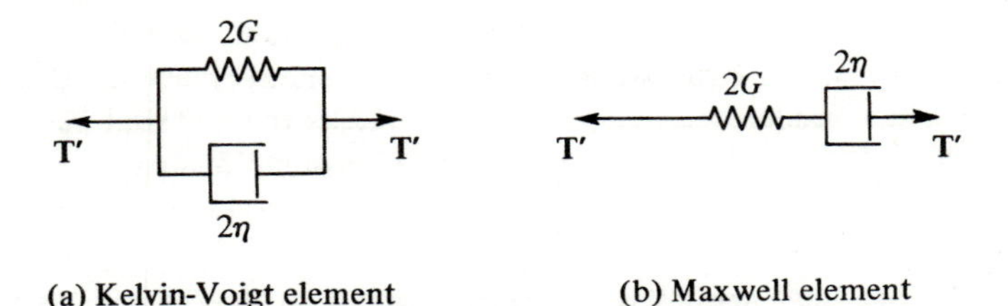

Fig. 6.6 Simple viscoelastic models.

In the Kelvin-Voigt element model of Fig. 6.6(a) the total axial force is the sum of the spring force and the dashpot force, the displacement being the same for both spring and dashpot; however, ***in the Maxwell element*** the force is the same in both spring and dashpot, but the displacements (or velocities) add. The analogous deviator equations for the continuum are therefore

$$\text{Kelvin-Voigt:} \quad \mathbf{T}' = 2G\mathbf{E}' + 2\eta\dot{\mathbf{E}}' = 2G(\mathbf{E}' + \tau\dot{\mathbf{E}}') \tag{6.4.5}$$

$$\text{Maxwell:} \quad \dot{\mathbf{E}}' = \frac{1}{2G}\dot{\mathbf{T}}' + \frac{1}{2\eta}\mathbf{T}' \quad \text{or} \quad 2G\dot{\mathbf{E}}' = \dot{\mathbf{T}}' + \frac{1}{\tau}\mathbf{T}', \tag{6.4.6}$$

†The Maxwell material was introduced by Maxwell (1868), the Kelvin-Voigt material by Voigt (1889), Kelvin (1875), and Meyer (1874).

where the ***retardation time*** or ***relaxation time*** τ is given by

$$\tau = \frac{\eta}{G}. \tag{6.4.7}$$

Retarded elastic behavior is shown by the Kelvin-Voigt element. There is no instantaneous elastic response (glassy response), but if stress $\mathbf{T}'_0$ is applied at $t = 0$ and held constant, the solution of Eq. (5) for $\mathbf{E}'$ is

$$\text{Kelvin-Voigt retarded elasticity:} \quad \mathbf{E}' = \frac{\mathbf{T}'_0}{2G}(1 - e^{-t/\tau}) \tag{6.4.8}$$

so that $1/G$ represents the equilibrium compliance J_e of Fig. 6.2, while the retardation time τ is the time required for the difference $(\mathbf{T}'_0/2G) - \mathbf{E}'$ to be reduced by a factor $1/e$. Equation (5) shows that the initial strain rate is $\mathbf{T}'_0/2G\tau$; hence τ is the time it would have taken to reach the equilibrium strain if the initial rate had been maintained; actually Eq. (8) shows that the equilibrium strain is only asymptotically approached in the Kelvin-Voigt element. ***Steady creep in the Maxwell element*** takes place under constant stress $\mathbf{T}'_0$ after an instantaneous glassy response $\mathbf{T}'_0/2G$:

$$\text{Maxwell creep:} \quad \mathbf{E}' = \frac{1}{2G}\mathbf{T}'_0 + \frac{1}{2\eta}\mathbf{T}'_0 t. \tag{6.4.9}$$

Stress relaxation is illustrated by the Maxwell element. If strain $\mathbf{E}'_0$ is suddenly applied at $t = 0$, and then held constant, the instantaneous glassy response is $\mathbf{T}'_0 = 2G\mathbf{E}'_0$; the subsequent behavior is given by Eq. (6) with $\dot{\mathbf{E}}' = 0$, namely

$$\text{Maxwell relaxation:} \quad \mathbf{T}' = 2G\mathbf{E}'_0 e^{-t/\tau}. \tag{6.4.10}$$

The relaxation time τ is the time required for the stress to relax to $1/e$ times its initial value $2G\mathbf{E}'_0$. The Kelvin-Voigt element does not permit instantaneous application of strain without infinite stress, but if a retarded elastic test at constant stress (or any other loading program) is interrupted before equilibrium is reached and the strain held constant at $\mathbf{E}'_1$, say, from that point on, the stress will relax exponentially from whatever value it had at the time of the interruption to the equilibrium value $\mathbf{E}'_1/2G$.

The three-element model has an extra spring added to represent the ***standard linear solid***. The extra spring is added in series with a Kelvin-Voigt element in Fig. 6.7(a) to provide an instantaneous glassy response with modulus G_1, while the equilibrium modulus G_e is given by $1/G_e = 1/G_1 + 1/G_2$. Alternatively, the extra spring is put in parallel with a Maxwell element in Fig. 6.7(b) to provide an equilibrium modulus G_1, while the glassy modulus is $G_g = G_1 + G_2$, since the dashpot behaves like a rigid connection for instantaneous response.

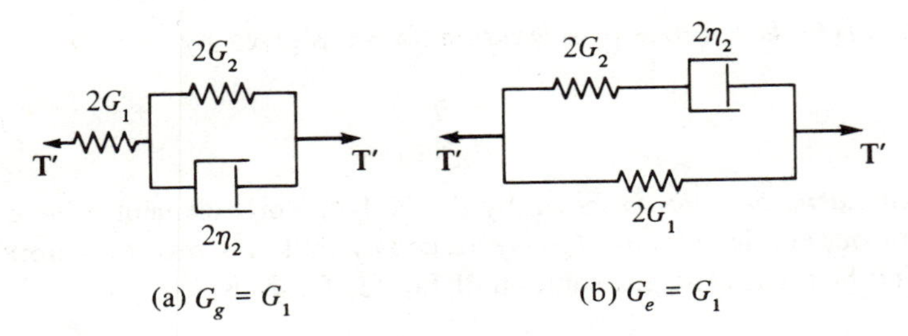

Fig. 6.7 Three-parameter models.

The constitutive equations for the three-parameter models are

$$\text{Fig. 6.7(a):} \quad \dot{\mathbf{T}}' + \frac{G_1 + G_2}{\eta_2}\mathbf{T}' = 2G_1\dot{\mathbf{E}}' + \frac{2G_1G_2}{\eta_2}\mathbf{E}' \tag{6.4.11a}$$

$$\text{Fig. 6.7(b):} \quad \dot{\mathbf{T}}' + \frac{G_2}{\eta_2}\mathbf{T}' = 2(G_1 + G_2)\dot{\mathbf{E}}' + \frac{2G_1G_2}{\eta_2}\mathbf{E}' \tag{6.4.11b}$$

so that, with different choices for the constants G_1, G_2, η_2 in the two cases, either model may be used to fit a constitutive equation of the form

$$\text{Standard linear solid:} \quad \dot{\mathbf{T}}' + p_0\mathbf{T}' = q_1\dot{\mathbf{E}}' + q_0\mathbf{E}', \tag{6.4.11c}$$

which qualitatively represents an ideal crosslinked polymer. For an uncross-linked polymer a dashpot of viscosity η_3 may be added in series to permit unrestricted creep. If the dashpot is added in series with Fig. 6.7(a), then in the resulting four-element model

$$\mathbf{E}' = \mathbf{E}'_1 + \mathbf{E}'_2 + \mathbf{E}'_3 \tag{6.4.12a}$$

$$\mathbf{T}' = 2G_1\mathbf{E}'_1 = 2G_2\mathbf{E}'_2 + 2\eta_2\dot{\mathbf{E}}'_2 = 2\eta_3\dot{\mathbf{E}}'_3 \tag{6.4.12b}$$

whence, by eliminating $\mathbf{E}'_1$, $\mathbf{E}'_2$, $\mathbf{E}'_3$, we obtain the constitutive equation

$$\ddot{\mathbf{T}}' + \left(\frac{G_1}{\eta_2} + \frac{G_1}{\eta_3} + \frac{G_2}{\eta_2}\right)\dot{\mathbf{T}}' + \frac{G_1G_2}{\eta_2\eta_3}\mathbf{T}' = 2G_1\ddot{\mathbf{E}}' + \frac{2G_1G_2}{\eta_2}\dot{\mathbf{E}}' \tag{6.4.12c}$$

or

$$\ddot{\mathbf{T}}' + p_1\dot{\mathbf{T}}' + p_0\mathbf{T}' = q_2\ddot{\mathbf{E}}' + q_1\dot{\mathbf{E}}'. \tag{6.4.12d}$$

Alternative configurations for a four-element model are given, for example, by Bland (1960).

The three-and four-element models give a good qualitative representation of viscoelastic behavior, but do not give a good quantitative fit to data over any considerable range. More complicated models can be devised leading to higher-order differential equations, represented by ***operator equations*** of the form

$$P\{\mathbf{T}'\} = Q\{\mathbf{E}'\}, \tag{6.4.13a}$$

where P and Q are differential operators

$$P\{\mathbf{T}'\} = \sum_{r=0}^{n} p_r \frac{d^r}{dt^r}\mathbf{T}' \qquad Q\{\mathbf{E}'\} = \sum_{r=0}^{m} q_r \frac{d^r}{dt^r}\mathbf{E}', \tag{6.4.13b}$$

and the p_r and q_r are constants.

The quantitative shortcoming of the single Kelvin-Voigt element or the single Maxwell element is that it has a single relaxation or retardation time τ and therefore is inadequate to fit a response like Fig. 6.4, where the relaxation extends across several decades of logarithmic time.

A generalized Kelvin-Voigt model consisting of N Kelvin-Voigt elements in series can be used to fit creep data to a high degree of accuracy. Each element has a time-dependent creep compliance given by Eq. (8), and since the total strain is the sum of the strains in the N elements, the creep compliance at any time t is

$$J(t) = J_g + \frac{1}{\eta}t + \sum_{n=1}^{N} J_n(1 - e^{-t/\tau_n}), \tag{6.4.14a}$$

where $J_n = 1/G_n$, if steady-state creep is represented by a free dashpot in series with the Kelvin-Voigt elements, with coefficient η, and instantaneous response is provided by a series spring with glassy compliance $J_g = 1/G_g$. Then, for example pure shear with constant stress T_{12}^0 gives

$$\gamma_{12} = T_{12}^0\left[J_g + \frac{1}{\eta}t + \sum_{n=1}^{N} J_n(1 - e^{-t/\tau_n})\right]. \tag{6.4.14b}$$

Although we were able to write down the creep response for such a model without even setting up the differential equation, response to any other kind of loading, e. g., for stress relaxation, would require integration of the differential equation. ***A generalized Maxwell model*** with N Maxwell elements in parallel with a free spring of modulus G_e gives the following relaxation response to constant strain γ_{12}^0

$$T_{12} = \gamma_{12}^0\left[G_e + \sum_{n=1}^{N} G_n e^{-t/\tau_n}\right] \tag{6.4.15a}$$

which can be made to fit relaxation data quite accurately. These generalized models with a discrete spectrum of N different relaxation times or retardation times have been especially interesting to materials scientists, who have sought to identify the different relaxation times with different molecular processes. See, for example, Alfrey (1948). It is necessary, however, to know the modulus G_n to associate with each relaxation or retardation time τ_n. For empirical determination of the spectrum of relaxation times, a ***continuous spectrum*** is more useful, replacing the relaxation modulus $G(t)$ in the brackets of Eq. (15a), namely

$$G(t) = G_e + \sum_{n=1}^{N} G_n e^{-t/\tau_n} \tag{6.4.15b}$$

by an integral where each G_n is replaced by $f_2(\tau)\, d\tau$ for the relaxation times between τ and $\tau + d\tau$, giving

$$G(t) = G_e + \int_0^\infty f_2(\tau) e^{-t/\tau}\, d\tau \tag{6.4.16a}$$

or, more conveniently for use on a logarithmic time scale, with $H(\tau) = \tau f_2(\tau)$,

$$G(t) = G_e + \int_{-\infty}^{\infty} H(\tau) e^{-t/\tau}\, d(\ln \tau). \tag{6.4.16b}$$

The last equation can be taken as a definition of the distribution function H, usually called the ***relaxation spectrum***, without making any use of mechanical models. A continuous spectrum of retardation times in creep also replaces Eq. (14a) by

$$J(t) = J_g + \frac{1}{\eta} t + \int_0^\infty f_1(\tau)[1 - e^{-t/\tau}] d\tau \tag{6.4.17a}$$

or, with $L(\tau) = \tau f_1(\tau)$

$$J(t) = J_g + \frac{1}{\eta} t + \int_{-\infty}^{\infty} L(\tau)(1 - e^{-t/\tau}) d(\ln \tau). \tag{6.4.17b}$$

Relationships between the ***retardation spectrum*** L and the relaxation spectrum H are given by Gross (1953). See also Leaderman (1958) and Ferry (1961); Ferry gives examples of determining the spectra from experimental data.

The superposition principle, introduced by Boltzmann (1874), may be used to determine response to loading or deformation histories other than constant stress or constant strain, ***provided the response is linear***. Consider first the shear response to a sequence of loads $T_{12}^{(k)}$ (possibly not all of the same sign) applied at times t'_k and each held constant thereafter. Then the creep response at time t will be the sum of the creep responses given by the compliance $J(t)$ of Eq. (17) for each load maintained for time $t - t'_k$

$$\gamma_{12} = \sum_{k=1}^{N} T_{12}^{(k)} J(t - t'_k). \tag{6.4.18a}$$

A similar superposition for strain increments $\gamma_{12}^{(k)}$ applied at times t'_k yields

$$T_{12} = \sum_{k=1}^{N} \gamma_{12}^{(k)} G(t - t'_k), \tag{6.4.18b}$$

where $G(t)$ is given by Eq. (16). For a continuously varying load, the superposition leads to a hereditary integral in place of the sums (18).

Part 3. Constitutive Equations of a Linear Hereditary Material Defined by Superposition Integrals

For continuously varying load, with increments $dT_{12} = \dot{T}_{12}\, dt'$ applied at time t', or with increments $d\gamma_{12}$ applied at time t', the superposition of Eqs. (18) yields

$$\left.\begin{aligned} \gamma_{12}(t) &= \int_{-\infty}^{t} \frac{dT_{12}(t')}{dt'} J(t - t')\, dt' \\ T_{12}(t) &= \int_{-\infty}^{t} \frac{d\gamma_{12}(t')}{dt'} G(t - t')\, dt' \end{aligned}\right\} \tag{6.4.19}$$

These constitutive relations based on superposition integrals [and the alternative forms below through Eq. (25)] are examples of linear ***hereditary stress-strain laws***, in which stress depends on the history of the strain instead of only on its present value. History-dependent materials are often referred to as ***materials with memory;*** the linear hereditary laws define special cases of such materials.

The instantaneous elastic response is sometimes separated out by writing

$$\left.\begin{aligned} J(t) &= J_g[1 + \psi(t)] \\ G(t) &= G_g[1 - \phi(t)], \end{aligned}\right\} \tag{6.4.20a}$$

where $\psi(t)$ is the ***dimensionless creep function*** or ***retarded elastic function*** and $\phi(t)$ is the ***dimensionless relaxation function*** [Leaderman (1958)]. Then Eqs. (19) take the form

$$\left.\begin{aligned} \gamma_{12}(t) &= J_g\left[T_{12}(t) + \int_{-\infty}^{t} \psi(t - t') \frac{dT_{12}}{dt'}\, dt'\right] \\ T_{12}(t) &= G_g\left[\gamma_{12}(t) - \int_{-\infty}^{t} \phi(t - t') \frac{d\gamma_{12}}{dt'}\, dt'\right] \end{aligned}\right\} \tag{6.4.20b}$$

assuming that stress and strain vanish at $t' = -\infty$. These forms are especially useful when the time-dependent creep or relaxation may be considered as a small perturbation on the glassy-elastic behavior. Alternatively, a change of integration variable in Eqs. (19) to the ***time lapse*** s, where

$$s = t - t' \qquad ds = -dt', \tag{6.4.21}$$

yields the forms

$$\gamma_{12}(t) = \int_{\infty}^{0} \frac{dT_{12}(t-s)}{ds} J(s)\,ds \qquad T_{12}(t) = \int_{\infty}^{0} \frac{d\gamma_{12}(t-s)}{ds} G(s)\,ds, \tag{6.4.22}$$

which, after an integration by parts, become

$$\left.\begin{aligned} \gamma_{12}(t) &= J_g T_{12}(t) + \int_0^{\infty} T_{12}(t-s) \frac{dJ(s)}{ds}\,ds \\ T_{12}(t) &= G_g \gamma_{12}(t) + \int_0^{\infty} \gamma_{12}(t-s) \frac{dG(s)}{ds}\,ds, \end{aligned}\right\} \tag{6.4.23}$$

assuming that stress and strain were zero at $t' = -\infty$, and writing J_g and G_g for J and G at $s = 0$.

The three-dimensional deviator relationships for an isotropic material follow the pattern of any one of the Eqs. (18) through (23) with $\gamma_{12} = 2E'_{12}$. For example, Eqs. (20) become

$$\left.\begin{aligned} 2\mathbf{E}' &= J_g\left[\mathbf{T}' + \int_{-\infty}^{t} \psi(t-t') \frac{d\mathbf{T}'}{dt'}\,dt'\right] \\ \mathbf{T}' &= 2G_g\left[\mathbf{E}' - \int_{-\infty}^{t} \phi(t-t') \frac{d\mathbf{E}'}{dt'}\,dt'\right]. \end{aligned}\right\} \tag{6.4.24}$$

The volumetric relationships would be of the same form, but with different functions in general. For example, if $\operatorname{tr}\mathbf{T} = T_{kk}$ denotes the trace of $\mathbf{T}$ and $e = \operatorname{tr}\mathbf{E}$ signifies the volume strain, the last Eq. (24) would have a counterpart of the form

$$\operatorname{tr}\mathbf{T} = 3K_g\left[e - \int_{-\infty}^{t} \phi_v(t-t') \frac{de}{dt'}\,dt'\right], \tag{6.4.25a}$$

where ϕ_v is the dimensionless volumetric stress-relaxation function. For some purposes, it may be possible to neglect the time-dependence of volumetric strain and use the elastic relationship

$$e = \frac{1}{3K}\operatorname{tr}\mathbf{T} + 3\alpha(\theta - \theta_0), \tag{6.4.25b}$$

where the last term is the thermal expansion due to temperature change $\theta - \theta_0$ from the reference state. The deviator relationships so far presented have tacitly assumed isothermal deformation. Considerable caution is required in using them, since a relatively small change in temperature may produce quite large changes in properties in the transition region, where viscoelastic behavior is most significant.

Periodic response for materials defined by the constitutive equations (19) may be determined as follows. Consider first a single Kelvin-Voigt element with compliance $J_k = 1/G_k$ and retardation time τ_k. It will be left as an exercise to show that the response to a sinusoidal load $T_0 \sin \omega t$ consists of an in-phase component $J'T_0 \sin \omega t$ and an out-of-phase component $J''T_0 \cos \omega t$ with

$$\text{Kelvin-Voigt element:} \quad J'(\omega) = \frac{J_k}{1 + \omega^2\tau_k^2} \qquad J''(\omega) = \frac{J_k\,\omega\tau_k}{1 + \omega^2\tau_k^2}\,. \tag{6.4.26}$$

A series array of such (massless) elements has the same force in each element so that the displacement is the sum of the displacements in the elements. Hence the dynamic compliances are obtained by summing the compliances of Eqs. (26) for $k = 1, 2, \ldots, N$. If we pass to the limit of a continuous spectrum of retardation times as in Eq. (17a), replacing τ_k by τ and J_k by $f_1(\tau)\,d\tau$, we obtain

$$\text{Kelvin-Voigt continuum:} \quad J'(\omega) = \int_0^\infty \frac{f_1(\tau)\,d\tau}{1 + \omega^2\tau^2}$$

$$J''(\omega) = \int_0^\infty \frac{f_1(\tau)\omega\tau\,d\tau}{1 + \omega^2\tau^2}\,. \tag{6.4.27}$$

To account for the in-phase compliance of the instantaneous elastic response J_g should be added to the $J'(\omega)$ of the continuous Kelvin-Voigt model (corresponding to a free spring in series with the model), and to account for the out-of-phase response of a free dashpot $1/\omega\eta$ should be added to $J''(\omega)$. Then substituting $L = \tau f_1(\tau)$ as in Eq. (17b) we get the following expressions for the dynamic compliances:

$$\left.\begin{aligned} J'(\omega) &= J_g + \int_{-\infty}^{\infty} \frac{L(\tau)}{1 + \omega^2\tau^2}\,d(\ln \tau) \\ J''(\omega) &= \frac{1}{\omega\eta} + \int_{-\infty}^{\infty} \frac{L(\tau)\omega\tau}{1 + \omega^2\tau^2}\,d(\ln \tau). \end{aligned}\right\} \tag{6.4.28}$$

For high-frequency response, the term $1/\omega\eta$ is negligible. A similar argument based on a parallel set of Maxwell elements with a spring of modulus G_e in parallel, yields

$$\left.\begin{aligned} G'(\omega) &= G_e + \int_{-\infty}^{\infty} \frac{H(\tau)\omega^2\tau^2}{1+\omega^2\tau^2}\,d(\ln\tau) \\ G''(\omega) &= \int_{-\infty}^{\infty} \frac{H(\tau)\omega\tau}{1+\omega^2\tau^2}\,d(\ln\tau). \end{aligned}\right\} \tag{6.4.29}$$

Equations (28) and (29) give the periodic response in terms of known creep and relaxation spectra. The inverse problem, to determine $L(\tau)$ or $H(\tau)$ from known dynamic compliance or modulus spectra (i.e., known as functions of ω) is not so easy. That and the problem of calculating $L(\tau)$ from $H(\tau)$ or vice versa can in principle be solved by Laplace or Fourier transform methods, although numerical inversion of the transforms may be required. See, for example, Ferry (1961), Chaps. 3 and 4.

The operator forms of Eqs. (13), particularly the special cases corresponding to models of not more than three or four elements, have been popular in engineering structural applications, because most engineers feel more at home with differential equations than with superposition integrals like those in Eqs. (18) through (25). But the superposition integrals have been gaining favor in recent years.

Part 4. Discussion and References

Lenoe and Martin (1966) discuss computer use in formulating and applying the hereditary laws in uniform strain situations. Lee and Rogers (1963) discuss the solution of viscoelastic stress analysis problems using measured creep and relaxation properties. Earlier accounts of viscoelastic stress analysis, largely based on the differential equation formulation and a correspondence principle using Laplace transforms to reduce the viscoelastic problem to a fictitious elastostatic problem where the elastic constants depend on the transform variable as a parameter, are given by Bland (1960) and by Lee (1960). See also Fung (1965), Chap. 15, and Flügge (1967).

Gurtin and Sternberg (1962) have given a comprehensive account of the mathematical theory, based on Stieltjes integrals to include discontinuous time histories without use of Dirac delta-functions. Their work includes formulation of the mixed boundary-value problem in the quasi-static linear theory, a study of the properties of solutions, work and energy identities, uniqueness questions, and theorems on integration by displacement potentials (including extensions to viscoelasticity of the Galerkin and Papkovich-Neuber solutions of elasticity).

Coleman and Noll (1961) have examined the foundations of the linear theory from the viewpoint of the modern theory of constitutive equations. See also Truesdell and Noll (1965). The second Eq. (23) is an example of a hereditary law of the type

$$\text{Stress} = \mathop{\mathscr{F}}_{s=0}^{\infty} \{\text{Strain}\}, \tag{6.4.30}$$

where the script $\mathscr{F}$ denotes a ***functional*** whose present value is determined by the history for all past time $t' = t - s$. Truesdell and Noll define a simple material† as one where the present stress is determined by the history of the deformation gradient **F**. By the requirement of frame indifference (see Sec. 6.7) they are able to show that every simple material has a constitutive equation of the form

$$\mathbf{R}^T \cdot \mathbf{T} \cdot \mathbf{R} = \mathbf{f}(\mathbf{C}) + \mathop{\mathscr{F}}_{s=0}^{\infty} \{\mathbf{G}^*(s); \mathbf{C}\} \tag{6.4.31}$$

where **R** is the rotation tensor (Sec. 4.6), **C** is the Green deformation tensor of the Lagrangian formulation, Eq. (4.5.4a), and

$$\mathbf{G}^*(s) = \mathbf{R}(t)^T \cdot \mathbf{B}^{-1}(t - s) \cdot \mathbf{R}(t) - \mathbf{1} \tag{6.4.32}$$

describes the history of the Cauchy deformation tensor $\mathbf{B}^{-1}$ of Eq. (4.5.4b). The equilibrium response is given by $\mathbf{f}(\mathbf{C})$ while the functional $\mathscr{F}$ vanishes if $\mathbf{G}^*(s) \equiv 0$, i.e, if the material has always been at rest. Coleman and Noll discuss materials with ***fading memory*** (for which they give a precise definition) and show that the constitutive functional $\mathscr{F}$ for such materials can be approximated by a sum of n linear functionals. The special case $n = 1$ can be represented by an integral, so that Eq. (31) becomes

$$\mathbf{R}^T \cdot \mathbf{T} \cdot \mathbf{R} = \mathbf{f}(\mathbf{C}) + \int_0^{\infty} \mathbf{K}(\mathbf{C}; s)[\mathbf{G}^*(s)]\, ds \tag{6.4.33}$$

where $\mathbf{K}(\mathbf{C}; s)[\mathbf{G}^*(s)]$ is for each choice of s and **C** a tensor-valued linear function of the tensor variable $\mathbf{G}^*(s)$, i.e., $\mathbf{K}[\ ]$ denotes a fourth-order tensor of the argument in brackets. Truesdell and Noll call the theory of such materials ***finite linear viscoelasticity***.

For small displacement gradients, the following approximations are valid, to higher-order terms

$$\left.\begin{aligned} G^*(s) &= 2[\mathbf{E}(t - s) - \mathbf{E}(t)] \\ \mathbf{C}(t) &= \mathbf{1} + 2\mathbf{E}(t) \\ \mathbf{R}(t) &= \mathbf{1} + \boldsymbol{\Omega}(t) \end{aligned}\right\} \tag{6.4.34}$$

where $\mathbf{E}(t)$ and $\boldsymbol{\Omega}(t)$ are the small-strain and rotation tensors, and for ***isotropic infinitesimal viscoelasticity*** the equation

†The application of such an ideal nonlinear material to real polymers is limited by the fact that the temperature history is excluded from the formulation. The same limitation applies of course to the linear hereditary materials of Part 3. Coleman (1964) has included dependence on the history of θ and grad θ.

$$\mathbf{T}(t) = \left\{[\lambda + \bar{\lambda}(0)]e(t) + \int_0^\infty \dot{\bar{\lambda}}(s)e(t-s)\,ds\right\}\mathbf{1}$$

$$+ 2[\mu + \bar{\mu}(0)]\mathbf{E}(t) + \int_0^\infty \dot{\bar{\mu}}(s)\mathbf{E}(t-s)\,ds \qquad (6.4.35)$$

results, where λ and μ are material constants and $\bar{\lambda}(s)$ and $\bar{\mu}(s)$ are scalar material functions. This last equation is equivalent to a combination of a deviator relation like the second Eq. (23) with a similar equation for the volume response and shows how the hereditary integral formulations fit into the modern nonlinear theory as presented by Truesdell and Noll (1965). Note that the limitation to small displacement gradients implied small rotations; Eq. (35) does not apply correctly to situations of small strain with finite rotation, which may be important in thin-shell structures. There it may be useful to resort to convected curvilinear coordinates [see Fredrickson (1964)].

Nonlinear viscoelasticity is beyond the scope of this book. See Truesdell and Noll (1965) for a summary of the nonlinear theory. Another recent summary has been given by Rivlin (1965). Schapery (1966) has proposed a theory of ***nonlinear thermoviscoelasticity*** based on the principles of irreversible thermodynamics, extending his earlier work on linear thermoviscoelasticity [Schapery (1964)]; his approach is quite different from the general thermodynamic studies of Coleman (1964). Nonlinear analogue models have been proposed to represent one-dimensional deformation, using nonlinear springs and dashpots. See, e.g., Schapery (1966), Alfrey (1948), and Alfrey and Gurnee (1967). Biot (1965) formulates the incremental deformation of a viscoelastic medium under initial stress, using irreversible thermodynamics.

The plasticity theories of Secs. 6.5 and 6.6 incorporate another kind of nonlinear inelastic response not concerned with time dependence.

EXERCISES

1. Derive Eqs. (6.4.8), (6.4.9), (6.4.10) from the appropriate Eq. (5) or (6).
2. Show that Eqs. (6.4.11) are the appropriate equations for the models of Fig. 6.7.
3. Solve Eq. (6.4.11a) for the retarded elastic response $E_{12}(t)$ of the three-element model of Fig. 6.7(a) to a pure shear loading T_{12}^0 applied at $t = 0$ and then held constant.
4. Solve Eq. (6.4.11b) for the stress-relaxation behavior of the three-element model of Fig. 6.7(b) to a pure shear strain E_{12}^0 applied at $t = 0$ and then held constant.
5. Determine $J(t)$ according to Eq. (6.4.14a): (a) for the three-element model of Fig. 6.7(a), and (b) for the four-element model obtained by putting a dashpot of viscosity η_3 in series with the model of Fig. 6.7(a).
6. Use the superposition principle of Eq. (6.4.18a) to obtain the response $\gamma_{12}(t)$ for $t > t_1$ of the three-element model of Fig. 6.7(a) to two loads T_{12}^0 applied at

$t = 0$ and T^1_{12} applied at $t = t_1$ and each held constant after application. (After $t = t_1$ both loads act.) See your result for Ex. 5(a).

7. Determine $G(t)$ according to Eq. (6.4.15b) for the three-element model of Fig. 6.7(b).

8. Use the superposition principle of Eq. (6.4.18b) to determine the stress-relaxation response $T_{12}(t)$ for $t > t_1$ of the three-element model of Fig. 6.7(b) to two shear-strain increments E^0_{12} applied at $t = 0$ and E^1_{12} applied at $t = t_1$ and each held constant after application. (After $t = t_1$ the total strain is $E^0_{12} + E^1_{12}$.) See your result of Ex. 7.

9. Determine the steady-state response of a Kelvin-Voigt element with time constant τ and equilibrium compliance $J = 1/G$ to the sinusoidal pure shear loading $T_{12}(t) = T_0 \sin \omega t$.

10. For the single Kelvin-Voigt element where $J(t) = (1 - e^{-t/\tau})/G$, use the superposition integral of Eq. (19) to determine $\gamma_{12}(t)$ in a pure shear loading in which $dT_{12}/dt = k$, constant for $t \geq 0$, if both T_{12} and γ_{12} were zero for $t < 0$, and show that the constant stress-rate test asymptotically approaches constant strain rate.

11. Determine the steady-state $T_{12}(t)$ response of a Kelvin-Voigt element to a sinusoidal shear strain $\gamma_{12}(t) = \gamma_0 \sin \omega t$, and hence determine G' and G'' for the element.

12. With reference to Eqs. (6.4.2) to (6.4.4), show that $J' = (1/G')/(1 + \tan^2 \delta)$, and check that this relation is satisfied by the values appropriate to a Kelvin-Voigt element. (See your results of Ex. 9 and Ex. 11.)

13. For a generalized Maxwell model having two Maxwell elements in parallel with a spring of modulus G_e, use a superposition integral to determine the stress $T_{12}(t)$ for the strain history: $\gamma_{12} = 0$ for $t \leq 0$, $\gamma_{12} = At^2$ for $t > 0$, in terms of the constants G_e, G_1, G_2, τ_1, τ_2 and A.

14. If $\gamma_{12}(t) \equiv 0$ for $t < 0$, what can be taken as the lower limit of the integral in the second Eq. (6.4.19)? Show that in this case the upper limit for the integral in the second Eq. (23) may be taken as t. Make comparable statements about the first Eq. (19) and the first Eq. (23).

15. The empirical expression $G(t) = G_e + \{(G_g - G_e)/[1 + (t/t_0)]^\beta\}$, where β and t_0 are constants, is sometimes used to approximate the transition-region relaxation modulus. Use Eq. (6.4.19) to calculate $T_{12}(t)$ with this approximation for a shear-strain history: $\gamma_{12} = 0$ for $t < 0$, and constant strain rate $d\gamma_{12}/dt = C$ for $t \geq 0$.

16. Use Eq. (6.4.19) to calculate $\gamma_{12}(t)$ for the generalized Kelvin-Voigt material of Eq. (14a) for the shear-stress history: $T_{12} = 0$ for $t < 0$, and $dT_{12}/dt = B$, constant for $t \geq 0$.

17. (a) If the dilatation is elastic according to Eq. (6.4.25b), determine the differential equations for isothermal uniaxial tensile stress $T_{11}(t)$ [with $T_{22} \equiv T_{33} \equiv 0$] versus tensile strain $E_{11}(t)$ of a material whose deviatoric response is that of the Kelvin-Voigt element of Eq. (5).

(b) Determine also the differential equation relating $E_{22}(t)$ to $T_{11}(t)$. How does your result in (a) differ from the equations you would obtain if you assumed a Kelvin-Voigt element representing the tensile response [$T_{11}(t)$ versus $E_{11}(t)$ with a spring of stiffness E and "tensile viscosity" 2η]? Note: In elasticity $E = 9KG/(3K + G)$ by Eq. (6.2.41).

18. Repeat Ex. 17(a) and (b) for the case that the deviatoric response is governed by the three-element model of Eq. (11a).

19. Write the expression for the $\phi(t)$ in Eq. (6.4.20a), if $G(t)$ is given by the empirical expression introduced in Ex. 15.

20. For volumetric response, write the counterpart of the first Eq. (6.4.24).

21. (a) In Eq. (6.4.11c) for the standard linear solid, substitute $T'_{11} = T_{11} - \frac{1}{3}T_{kk}$ and $E'_{11} = E_{11} - \frac{1}{3}E_{kk}$ etc., to obtain a set of three coupled differential equations relating T_{11}, T_{22}, T_{33} and E_{11}, E_{22}, E_{33}.
    (b) For the case of plane strain, with $E_{33} = 0$, assume elastic volumetric response and eliminate T_{33} from the first two equations obtained in (a) to obtain coupled plane-strain differential equations relating T_{11} and T_{22} to E_{11} and E_{22}.
    (c) For the case of plane stress with $T_{33} = 0$, make a similar elimination of E_{33} to obtain plane-stress differential equations.

Selected references on polymers and on linear viscoelasticity

Alfrey, T., *Mechanical Behavior of High Polymers.* New York: Interscience, 1948.

Alfrey, T. and E. F. Gurnee, *Organic Polymers.* Englewood Cliffs, N. J.: Prentice-Hall, 1967.

Biot, M. A., *Mechanics of Incremental Deformations.* New York: Wiley, 1965.

Bland, D. R., *The Theory of Linear Viscoelasticity.* New York: Pergamon, 1960.

Eirich, F. R., editor, *Rheology, Theory and Applications.* New York: Academic Press; Vol. 1, 1956; Vol. 2, 1958; Vol. 3, 1960.

Ferry, J. D., *Viscoelastic Properties of Polymers.* New York: Wiley, 1961.

Flügge, W., *Viscoelasticity.* Waltham, Mass.: Blaisdell, 1967.

Fredrickson, A. G., *Principles and Applications of Rheology.* Englewood Cliffs, N. J.: Prentice-Hall, 1964.

Freudenthal, A. M. and H. Geiringer, "The Mathematical Theories of the Inelastic Continuum," in Vol. 6 of *Encyclopedia of Physics,* ed. S. Flügge. Berlin: Springer-Verlag, 1958.

Fung, Y. C., *Foundations of Solid Mechanics.* Englewood Cliffs, N. J.: Prentice-Hall, 1965.

Meares, P., *Polymers, Structure and Bulk Properties*. Princeton, N. J.: Van Nostrand, 1965.

Nielsen, L. E., *Mechanical Properties of Polymers*. New York: Reinhold, 1962.

Tobolsky, A. V., *Properties and Structure of Polymers*. New York: Wiley, 1960.

6.5 Plasticity I. Plastic behavior of metals; examples of theories neglecting work-hardening: Levy-Mises perfectly plastic; Prandtl-Reuss elastic, perfectly plastic; and viscoplastic materials

The deformation of metals beyond the elastic limit is known as ***metal plasticity***. Constitutive equations for plasticity are not well established. Several different idealizations have been proposed, all failing to account completely for the phenomena observed, but some nevertheless are quite useful. Most of the applications made have used the perfectly plastic theories discussed in this section. Before introducing these theories, we review in Part 1 some of the known features of metal plasticity. Part 2 then presents examples of three-dimensional theories of rate-independent, perfectly plastic (nonhardening) solids, while Part 3 considers viscoplasticity. This section concludes with a list of books for collateral reading on plasticity.† Attempts to provide a more adequate theory, including work-hardening but neglecting rate and temperature effects, are discussed in Sec. 6.6. Also mentioned at the end of Sec. 6.6 is the older total-strain theory (deformation theory) of Hencky.

Part 1. Metal Plasticity

Metal plasticity is illustrated in the hypothetical uniaxial stress-strain curve of Fig. 6.8 and the sample curves of Fig. 6.9. Three features are apparent:

1. an initial steeply rising linear ***elastic range***, followed by
2. ***yield***, and then a
3. ***work-hardening plastic range*** in which the stress increases much more slowly with strain than in the elastic range, and where a ***permanent set*** ϵ_p remains after unloading.

The ***proportional limit*** P is the stress at the end of the linear range, while the ***elastic limit*** stress σ_{EL} is the highest stress that can be reached before any permanent set (plastic deformation) occurs. Both of these points are important for theoretical formulations, but they are both difficult to pinpoint precisely in ex-

†References to twentieth-century periodical literature mentioned in the text are given in the bibliography at the end of the book. References to earlier work may be found in Hill (1950) and in Freudenthal and Geiringer (1958).

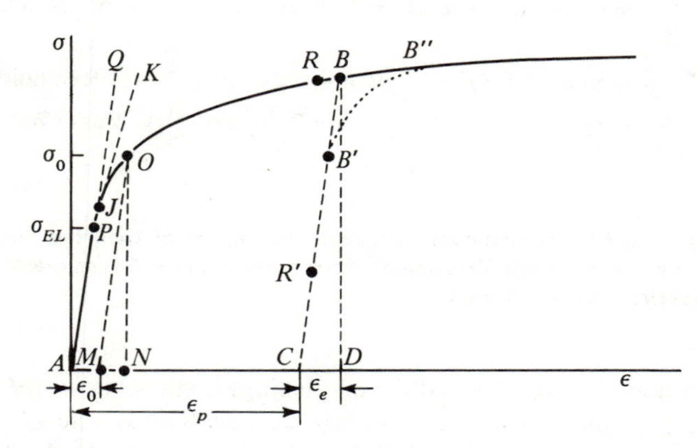

Fig. 6.8 Hypothetical tensile stress-strain curve.

periments; indeed if observations are sufficiently precise it becomes questionable whether either an exactly linear range or a completely recoverable range exists. For an idealized theory, the elastic limit and the proportional limit are usually assumed to coincide. In experimental determinations, in order to avoid ambiguity, the point is sometimes chosen as the point where the slope of the curve has decreased to an assigned percentage of its initial value. For example, ***Johnson's apparent elastic limit*** is defined as the stress where the slope drops to 50% of its initial value. This point is marked J in Fig. 6.8 (the tangent JK has half the slope of the tangent PQ).

However it is measured, the elastic limit stress is called the ***yield stress*** in theoretical discussions. For practical measurements to compare different materials, the ***yield strength*** is usually defined by an ***offset method*** as illustrated by the point O in Fig. 6.8: line OM is drawn ***parallel to the initial elastic slope***, offset to the right a distance ϵ_o. Usually $\epsilon_o = 0.002$ (0.2%). Where this line crosses the curve determines the offset ***yield strength*** σ_o. While this is more reproducible and therefore more useful for characterizing the material than is the elastic limit, it is not as satisfactory a criterion for a single specimen loaded just to the yield stress several times, because there may be enough plastic deformation on each loading to ϵ_o to cold-work the material and change its properties, especially when a new zero point (e.g., M in Fig. 6.8) is taken to start the second loading.

Postyield behavior is also illustrated in Fig. 6.8. Beyond the yield stress, the curve rises slowly, so that greater stress is required for further deformation, a phenomenon known as ***work-hardening*** or ***strain-hardening***. If the load is removed after yield has occurred, the unloading stress-strain path is different from the loading path, following an approximately straightline path very nearly parallel to the original elastic slope if the plot is true stress versus natural strain (logarithmic strain), as illustrated by the unloading lines OM and BC in Fig.

6.8. For unloading from B, if E denotes the elastic modulus the elastic strain recovered is $\epsilon_e = \sigma_B/E$, represented by the length CD on the strain axis, while the ***plastic strain*** or ***permanent set*** ϵ_p is represented by the length AC. Note that in a work-hardening material, the elastic part of the strain also continues to increase after yield, since the stress does, so that the elastic strain CD at B is greater than the elastic strain MN at O. Notice also that the unloaded reference state for the elastic part of the strain changes when plastic strain occurs. Even though $\epsilon_e \ll 1$, the changing unloaded reference state makes the meaning of ϵ_e not quite the same as it is in ordinary small-strain elasticity, where the entire motion of a particle is referred to one reference configuration. If the unloading process is stopped anywhere along BC and the load is increased again, ideally the loading returns elastically to point B, where yield occurs again followed by plastic deformation along the same curve BB'' which would have been followed if the loading had never been interrupted. In actual experiments the loading curve deviates from the elastic reloading line CB and follows a path something like the dotted curve $B'\,B''$ in Fig. 6.8. Another deviation from the ideal behavior described here, observed in careful measurements, is that the reloading path CB' does not quite coincide with the unloading path $B'C$. The unloading path is not quite straight, curving toward the origin near C; the reloading path CB' then rises slightly to the left of $B'C$, forming a ***hysteresis loop***, which becomes significant in cyclic loading because of the energy dissipation.

Because the unloading and reloading paths do not follow the initial loading path AOB, when plastic deformation occurs the ***stress is a history-dependent function of the strain***. For example, points R and R' in Fig. 6.8 have the same strain (same abscissa) but different stresses. When we later consider combined-stress loading, we will see that the stress state may depend on the history of the deformation even when there has been no unloading. ***History-dependence is the most characteristic difference between elastic and plastic deformation***. What it amounts to is that the cold-working involved in plastic deformation actually changes the material properties. A second loading therefore shows a different response, because it is a different material responding, and the change in material properties depends on the amount and nature of the prior cold work. Because of the extreme complication of the nonlinear processes involved in the hardening range, it should be no surprise that a satisfactory theory taking account of repeated loadings and load reversals after hardening does not exist.

Figure 6.9 gives some idea of the variety of metal responses in the plastic range in tension. Actually the variety is even greater than the sample indicates. High-strength steels exist with yield strengths around 250,000 psi. And the strain scale in the figure does not extend far enough to give an idea of the remarkable ductility to be expected from the annealed, low-carbon structural steel (mild steel); at room temperature a total strain of 0.25 before fracture would not be unusual. The dimensional changes in plastic deformation can be

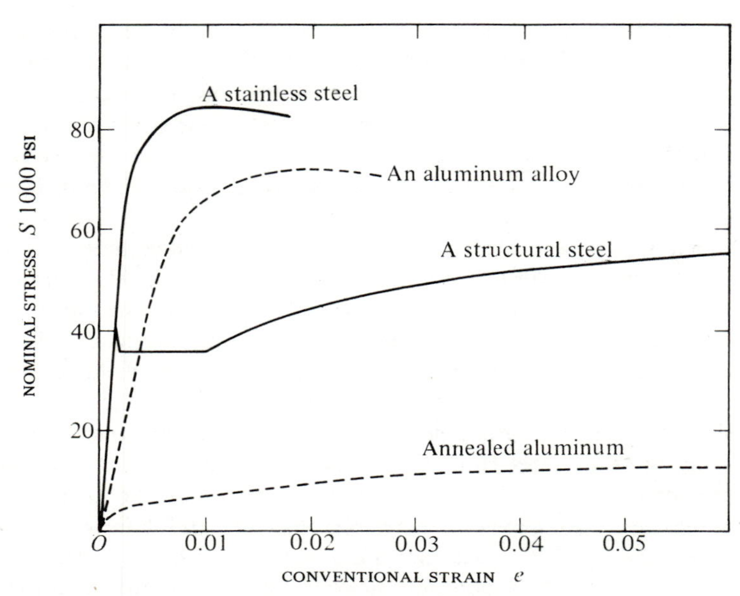

Fig. 6.9 Selected tensile stress-strain curves.

even greater in other types of loading such as compression, rolling, or extrusion. On the other hand, there are metals with apparently no plastic range at all, which fracture in a brittle fashion at the end of the elastic range.

The stress-strain curve for a given metal depends on the rate of strain and also on the temperature. These may affect the initial yield stress, the ductility, and the level of stress corresponding to any given strain. A rate-sensitivity parameter n is sometimes defined, it is determined empirically by

$$n = \log_{10} \frac{\sigma_2}{\sigma_1} \Big/ \log_{10} \frac{\dot{\varepsilon}_2}{\dot{\varepsilon}_1},$$

where σ_2 and σ_1 are stresses at rates $\dot{\varepsilon}_2$ and $\dot{\varepsilon}_1$ for the same strain. Lubahn and Felgar (1961), p. 190, cite room-temperature values of n in the neighborhood of 0.01 for several metals. With $n = 0.01$, doubling the strain rate would give less than a one per cent increase in stress, while a thousandfold increase in the rate gives only about a seven per cent increase in stress. Lowering the temperature increases the stress at any strain, and so does raising the rate of deformation. Indeed there have been attempts to formulate a mechanical equation of state in which stress is a function of the instantaneous strain, strain rate, and temperature, but these have had only limited success, because the stress appears to depend on the history of these parameters and not just on their current values. See Lubahn and Felgar (1961), Chap. 7.

One noteworthy feature of the temperature-sensitivity of the deformation of some metals is the ***ductile-to-brittle*** transition temperature. Above a certain temperature a mild steel, for example, exhibits the great ductility already mentioned in discussing Fig. 6.9, but below a transition region (possibly around −160°F) the ductility is greatly reduced and brittle fracture is likely to occur. Transition temperatures are usually determined by notched-bar impact tests. See, for example, Polakowski and Ripling (1966).

Most of the mathematical theory of combined-stress plasticity ignores the thermal phenomena and tacitly assumes isothermal deformation, although the temperature rise during metal-forming processes may be considerable. Green and Naghdi (1965) include the temperature in their thermodynamically based theory of the large deformations of an elastic-plastic continuum. See also Lee (1966, 1968). Naghdi (1960) has reviewed some earlier attempts at thermoplasticity. Most of the theories also neglect rate effects, although viscoplasticity is considered briefly at the end of this section.

The plots of Fig. 6.9 are for ***nominal stress*** (sometimes called ***engineering stress***), defined as the axial force per unit initial area A_0 of the cross section. Plastic deformation in tension takes place at approximately constant volume; hence actual cross-sectional area decreases as the specimen extends so that the ***true stress*** (axial force per unit current area A) is greater than the nominal stress in tension. The ultimate strength is defined as maximum nominal stress at the relative maximum point on the nominal stress-strain curve; it is the stress corresponding to the maximum axial load. The true-stress curve would continue to increase even beyond the maximun-load point, but this is an unstable region of the tensile test, since further deformation can continue with decreasing load; the instability leads to localization of the strain at a weak point in the test specimen, manifested by local ***necking down*** of the specimen. To estimate the true stress after necking begins it is necessary to observe the local diameter. But the problem is even more complicated. Because of the rounded notch formed by the necking, stress-concentration occurs so that the stress is not uniformly distributed over the necked-down cross section, and moreover a triaxial state of combined stress is developed. See, for example, Lubahn and Felgar (1961), p. 114 or Polakowski and Ripling (1966), p. 278. Lubahn and Felgar give additional details on complete stress-strain curves with many references. For definitions and methods of mesurement of standard mechanical properties, see, for example, Polakowski and Ripling (1966), Chap. 10, or Marin (1962).

True stress in tension at loads below the ultimate load (before necking occurs) is easily calculated in terms of the nominal stress and the conventional strain when the strain is large enough that the elastic strain is negligible in comparison to the plastic strain, if the plastic deformation is assumed to occur at constant volume. For then the current values of test-section length L and cross-sectional area A are related to the initial values L_0 and A_0 by the constant-volume condition

$$AL = A_0L_0.$$

Hence if the total axial force is F, the true stress σ is given by

$$\sigma = \frac{F}{A} = \frac{F}{A_0}\frac{A_0}{A} = \frac{F}{A_0}\frac{L}{L_0} = S(1 + e), \tag{6.5.1}$$

where

$$S = \frac{F}{A_0} \quad \text{and} \quad e = \frac{L - L_0}{L_0}, \tag{6.5.2}$$

are the nominal stress and strain. The ***natural strain*** ϵ is defined by

$$\epsilon = \ln \frac{L}{L_0} = \ln(1 + e) \qquad \text{(see sec 4. 4).} \tag{6.5.3}$$

The differential increments are

$$de = \frac{dL}{L_0} \qquad d\epsilon = \frac{dL}{L}. \tag{6.5.4}$$

The two ways of plotting the tensile curve (σ versus ϵ or S versus e) are compared in Fig. 6.10 for a hypothetical material in which the true-stress versus natural-strain curve for $\epsilon > 0.01$ is defined by $\sigma = 100{,}000\ \epsilon^{0.2}$. The two curves are almost coincident at a strain of 0.01. Natural strain is approximately equal to conventional strain up to about $e = 0.05$ where $\epsilon = 0.049$ (about two per cent below ϵ), but the difference increases so that $e = 0.30$ corresponds to

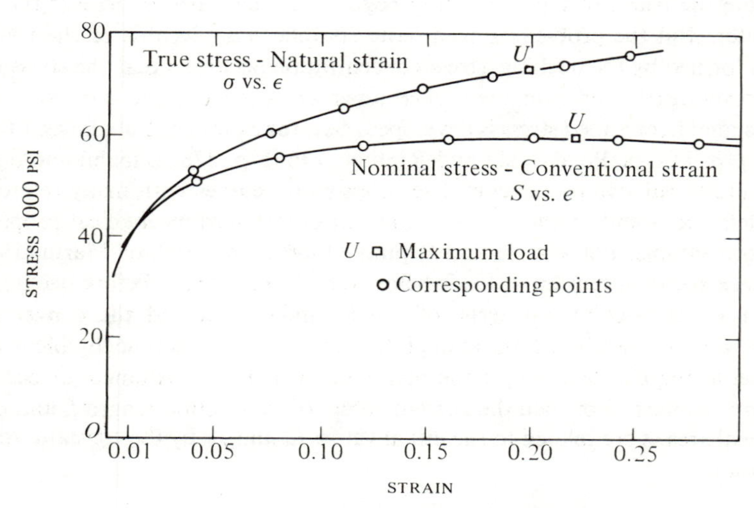

Fig. 6.10 Comparison of true-stress versus natural-strain plot with nominal-stress versus conventional-strain plot, if $\sigma = 100{,}000\ \epsilon^{0.2}$ for $\epsilon > 0.01$.

$\epsilon = 0.26$. The true stress σ diverges more rapidly from the nominal stress S, being 5 per cent above S already at the point corresponding to $e = 0.05$. The ultimate strength is easily read from the nominal-stress curve as the maximum ordinate (marked U in Fig. 6.10). The corresponding point on the true-stress versus natural-strain curve may be located as follows. Since

$$S = \frac{\sigma}{1 + e},$$

the maximum value of S occurs where

$$0 = \frac{dS}{de} = \frac{(d\sigma/d\epsilon) - \sigma}{(1 + e)^2}$$

or where the slope is numerically equal to the ordinate

$$\frac{d\sigma}{d\epsilon} = \sigma \qquad \text{(at maximum load).} \tag{6.5.5}$$

If true stress is plotted against conventional strain,

$$\frac{d\sigma}{de} = \frac{\sigma}{1 + e} \qquad \text{(at maximum load).} \tag{6.5.6}$$

For the special case of a power law,

$$\sigma = K\epsilon^n, \quad \text{we find} \quad \epsilon = n \qquad \text{(at maximum load),} \tag{6.5.7}$$

as may be seen from Eq. (5); for example, in Fig. 6.10 the maximum load occurs at $\epsilon = 0.2$, corresponding to $e = 0.221$.

Two categories of applications. Most applications of plasticity theory fall into two categories, as follows.

1. ***Contained Plastic Deformation.*** When large deformations are prevented (e. g., by surrounding elastic material or by elastic redundant members of a statically indeterminate structure), the plastic deformation is said to be contained. The problem may be to design the structure so that the plastic deformation will be contained, i. e., so that "collapse" will not occur even though some members are loaded beyond the elastic limit. For such contained deformation it is reasonable to neglect work-hardening, and to use as the deformation variable the small-strain tensor of Sec. 4.2, whose increments at small strain are negligibly different from the increments of natural strain.

2. ***Uncontained Plastic Flow.*** In metal-forming processes (e. g., drawing, extrusion, rolling) the problem is to produce a desired large deformation by cold-working. The cold-working involves hardening, which may be desired in order to have a stronger finished product, but which interferes with the process

itself not only by necessitating greater working forces, but also because nonuniform hardening may leave residual stresses. It may be reasonable to neglect the elastic-strain increments in comparison to the plastic-strain increments at large strains and to take as the deformation variable either the rate-of-deformation tensor D_{ij} of Sec. 4.4 or the natural-strain increment, which is equal to $D_{ij}\,dt$. This avoids the use of the more complicated finite-strain tensors.

Fortunately, with metals, most practically important problems fall into one of the two categories above, which represent two opposite extremes of possible problems. Intermediate problems, where the elastic strains are still important in comparison to the plastic deformations even up to strains where the small-strain tensor is not a good approximation to the finite-strain tensor, seldom occur in metals but may occur in nonmetals, for example, in soil mechanics or in polymers. Stresses associated with deformations of such materials are also usually much more rate-dependent than is the case with metals, where the usual theories of plasticity neglect rate-dependence altogether. For these reasons, satisfactory theories and methods of analysis for applications to such intermediate problems have not yet been developed. See, for example, Green and Naghdi (1965) for one proposed formulation including finite elastic strains and possibly anisotropic hardening.†

Idealized stress-strain curves that are frequently used for applications are illustrated in Fig. 6.11. Most nonuniform distributions of stress and plastic deformation have been analyzed by an appropriate two- or three-dimensional combined-stress generalization of one of these idealized uniaxial curves. Note that the first two curves neglect hardening altogether; the name ***perfectly plastic*** is used for such nonhardening idealized materials. The last two curves approximate the hardening part of the curve by a straight line at a slope smaller than the elastic slope. This line represents ***linear hardening***.

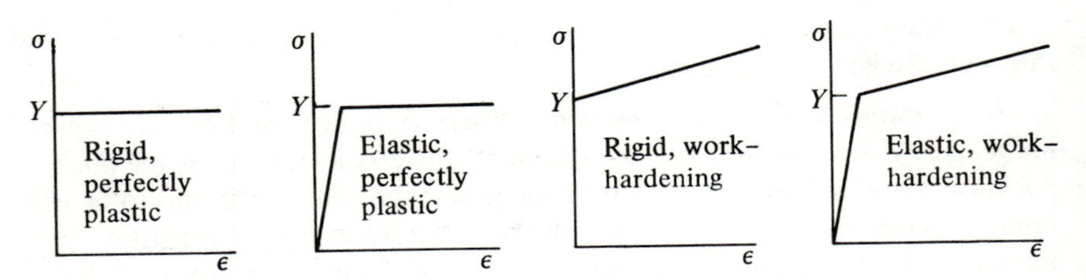

Fig. 6.11 Idealized stress-strain curves (piecewise linear).

Two of the curves neglect elastic strains altogether, the rigid, perfectly plastic and the rigid, linear-hardening idealized materials. We have already seen that in contained plastic deformation it is reasonable to neglect the hardening, while in uncontained plastic flow it is reasonable to neglect the elastic strains. It

†See also Lee (1966, 1968).

might seem that there would be no problem in which it is reasonable to neglect both as in the rigid, perfectly plastic idealized material. But, surprisingly enough, the rigid, perfectly plastic material has proved to be quite useful both in ***limit analysis*** of contained deformation and in such ***confined*** metal-forming processes as extrusion. Limit analysis, without analyzing the detailed contained deformation up to the limit load, seeks to find the limit load at which uncontained plastic flow would occur. It turns out that, if the geometry change of a structure below the limit load is negligible, the same limit load is predicted by rigid, plastic theory as by elastic, plastic theory. Since the rigid, plastic limit analysis does not actually calculate the geometry change (to see whether or not it is negligible), some engineering judgment or experimental verification may be needed to supplement rigid, plastic limit analysis. The rigid, perfectly plastic material is used in confined metal-forming analysis with surprising success, if a constant value for Y considerably higher than the actual measured initial yield stress is used. Usually, in fact, the value used for Y is the highest stress that would be required in a uniform compression to the desired reduced thickness. It appears that the flow pattern in such confined operations is more dependent on the boundary conditions than on the constitutive equation. It may be that the high temperatures developed by the rapid deformation and wall friction in extrusion processes, for example, cause a softening which offsets the local greater hardening to be expected from the local high strains in cold-working. Then, in some way not yet at all clear, the local severe deformations may induce metallurgical changes occurring a short time after the deformation and leaving the finished piece nonuniformly hardened even though the flow pattern and extrusion force during deformation agree well with some perfectly plastic theory "upper-bound solutions." See, for example, Thomsen, Yang and Kobayashi, (1965).

Bauschinger effect. Before proceeding to the formulation of the three-dimensional combined-stress equations, we note for uniaxial behavior one additional feature usually neglected in formulations intended for nonuniform stress and deformation analysis. Figure 6.12 shows uniaxial stress-strain curves with the possibility of either tension or compression. In metals, if true stress is plotted versus true strain, the compression curve for loading from the virgin state is the same as the tension curve: thus $OA'B'$ is the reflection of OAB in the origin. But if the direction of loading is reversed after loading to stress Y_1 at point B, the compression yield stress $-Y_R$ after the reversal is not $-Y_1$ or even $-Y_0$, but usually a much smaller magnitude, as indicated in Fig. 6.12(a). This lowering of the compression yield following a first loading in tension is called the ***Bauschinger effect***. It may even be so pronounced that the curve BCD deviates from the elastic unloading line BC while the stress is still positive. There is some evidence that the effect of the diminished reversed-yield stress tends to disappear when the reversed deformation is continued to large strains, the

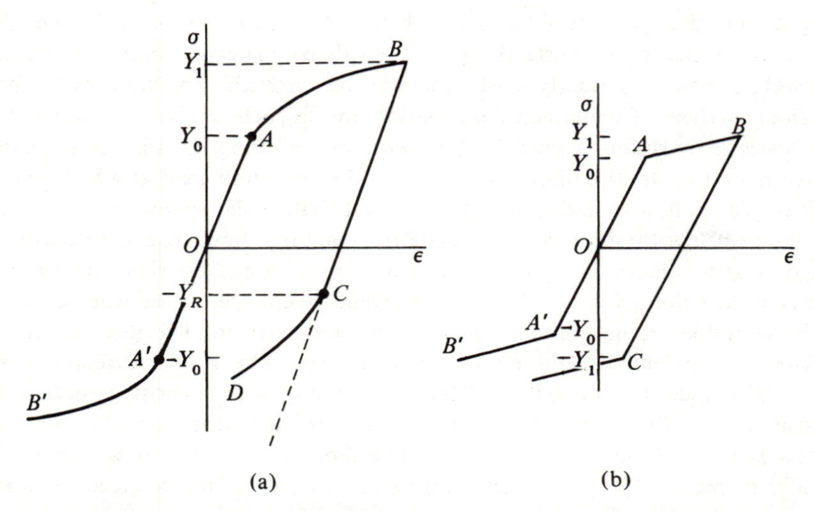

Fig. 6.12 (a) Illustrating Bauschinger effect; (b) idealized linear-hardening curve without Bauschinger effect.

curve tending eventually toward the curve which would have been obtained by the same amount of work all input in compression. For example, Templin and Sturm (1940) found that a small plastic tensile prestrain results in a decrease in compression yield stress, while a large plastic tensile prestrain results in an increase in compression yield. See, for example, Batdorf and Budiansky (1954) for a discussion of the relevance of this phenomenon to combined-stress theories.

If only uniaxial behavior is to be considered, much better approximations to the actual curves are possible than the piecewise linear idealized curves of Figs. 6.11 and 6.12(b). For example, a power law could be used, as in Eq. (7), or one of the curve-fitting formulas given by Ramberg and Osgood (1943). Another possibility is a piecewise linear approximation with more than two straight segments [see Phillips (1956)]. These improved approximate formulations can also be incorporated into the moment-curvature relationships of elementary beam-bending theories. But for continuum stress and deformation analysis they present great difficulty, and moreover their increased accuracy is likely to prove illusory in the light of the other shortcomings of the theory.

Part 2. Three-Dimensional Theory

For analysis of continuum stress and strain distributions, a constitutive theory for plasticity must specify the yield condition under combined stresses, since the uniaxial condition $|\sigma| = Y$ is inadequate when there is more than one stress component, and it must specify the postyield behavior, answering the following questions:

1. *Yield Condition:*
   What stress combinations permit inelastic response?
2. *Postyield Behavior:*
   (a) How are the plastic deformation increments or plastic rate-of-deformation components related to the stress components?
   (b) How does the yield condition change with work-hardening?

In Sec. 6.6, these questions will be discussed more fully in connection with plastic-potential theory and work-hardening. For the present, we consider only idealized materials with no work-hardening, assuming isotropy and assuming that plastic deformation does not produce any volume change. Neglecting elastic strains then it is reasonable to postulate a constitutive relation similar in form to the Newtonian viscosity equation $\mathbf{T}' = 2\mu\mathbf{D}'$. But such an assumption would give stresses linearly proportional to the deformation rates, the stresses doubling if the rates are doubled, while experience with metals indicates that in a tension test doubling the rate of extension induces a scarcely noticeable increase in stress.

The Levy-Mises perfectly plastic constitutive equations. These equations replace the constant multiplier 2μ of the Newtonian viscosity equations by $k/\sqrt{\mathrm{II}_D}$, where k is a constant and II_D is the second invariant of the rate-of-deformation tensor $\mathbf{D}$. They also assume incompressibility and neglect elastic strains. The rates of deformation D_{ij} are then interpreted as the rate of change of the plastic part of the natural strains, written $D_{ij} = d\epsilon_{ij}/dt$, since $\epsilon^e_{ij} = 0$. Thus

Levy-Mises Perfect Plasticity

Levy-Mises Equations:
$$\mathbf{T}' = \frac{k}{\sqrt{\mathrm{II}_D}}\mathbf{D} \quad \text{or} \quad T'_{ij} = \frac{k}{\sqrt{\mathrm{II}_D}} D_{ij} \tag{6.5.8}$$

implying

Plastic Incompressibility:
$$\operatorname{tr}\mathbf{D} = 0 \quad \text{or} \quad D_{kk} = 0 \tag{6.5.9}$$

(elastic strains neglected), and

Mises Yield Condition:
$$J'_2 = k^2 \quad \text{or} \quad \tfrac{1}{2}T'_{ij}T'_{ij} = k^2 \tag{6.5.10}$$

where $J'_2 \equiv \mathrm{II}_{T'}$

The component form in each of the equations is for rectangular Cartesians. The Mises yield condition, Eq. (10), may be seen to follow from Eq. (8) by squaring and adding the nine components. Thus

$$T'_{ij}T'_{ij} = \frac{k^2}{\mathrm{II}_D} D_{ij}D_{ij} \tag{6.5.11}$$

leads to Eq. (10), when plastic flow is occurring ($D_{ij} \neq 0$), since

$$J_2' \equiv \text{II}_{T'} = \tfrac{1}{2}T'_{ij}T'_{ij}, \quad \text{and} \quad \text{II}_D = \tfrac{1}{2}D_{ij}D_{ij} \qquad (\text{for } D_{kk} = 0). \qquad (6.5.12)$$

Since k is assumed to be constant, the yield condition does not change at all with plastic deformation in this perfectly plastic theory. The physical significance of the constant k may be understood by considering special states of stress. For a condition of pure shear, say $T_{xy} = T_{yx}$ with all other stress components equal to zero, Eq. (10) shows that $T_{xy}^2 = k^2$. Hence k ***is the yield stress in pure shear***. In a state of uniaxial stress $T_{xx} = Y$, Eq. (10) shows that $Y^2/3 = k^2$. Hence the Mises yield condition implies that the yield stress Y in tension is $\sqrt{3}$ times the yield stress in pure shear

$$k = \frac{Y}{\sqrt{3}}, \qquad (6.5.13)$$

a result closely approximated in many tests of polycrystalline metals [see, for example, Hill (1950), p. 22].

Since the Mises yield condition of Eq. (10) depends only on the stress deviator, the yield is independent of the hydrostatic or spherical part of the stress. Moreover, the Levy-Mises Eqs. (8) are unaffected by the presence of a hydrostatic stress. Despite some evidence to the contrary [see, for example, Hu (1958, 1960)], most experience with metals confirms that the plastic behavior is independent of the confining pressure, at confining pressures of the order of magnitude of the stresses producing the deformations [see, for example, Hill (1950), p. 16]. But one should not expect the deformations in metals to be independent of confining pressures several orders of magnitude greater than the ordinary tensile yield stress [see Bridgman (1952)], nor is it safe to generalize from the results for metals to other materials. In soil mechanics, for example, it is known that the deformation is highly dependent on the confining pressure, and moreover significant permanent volume changes occur in the compaction of soils.

Tresca yield condition. This condition, sometimes used instead of the Mises condition, predicts that yield will occur when the ***maximum shear stress*** reaches the value k. The Tresca yield condition is also independent of a superposed hydrostatic pressure. For uniaxial tension, the Tresca condition predicts that $k = Y/2$, while the Mises condition predicts $k = Y/\sqrt{3} = 0.577Y$, which is closer to most experimental determinations. As the plastic-potential theory of Sec. 6.6 shows, there are also theoretical reasons for preferring the Mises condition for use with the Levy-Mises equations. For design purposes, if no plastic yielding is to be permitted, the Tresca criterion is a conservative one provided it is set up so that it gives the correct results in tension, i. e., as

$$\text{Tresca yield condition:} \quad T_{\max} - T_{\min} = Y \qquad (6.5.14)$$

Then, for any other combined stress situation, the Tresca condition predicts yield at a lower stress level than does the Mises yield condition.

Levy-Mises perfect plasticity is a three-dimensional, isotropic, incompressible generalization of the rigid, perfectly plastic material illustrated in the uniaxial curve of Fig. 6.11. In terms of ***natural-strain increments*** $d\epsilon_{ij}$, the Eqs. (8) may be written in the form

$$\frac{d\epsilon_{xx}}{T'_{xx}} = \frac{d\epsilon_{yy}}{T'_{yy}} = \frac{d\epsilon_{zz}}{T'_{zz}} = \frac{d\epsilon_{yz}}{T'_{yz}} = \frac{d\epsilon_{zx}}{T'_{zx}} = \frac{d\epsilon_{xy}}{T'_{xy}} \tag{6.5.15}$$

with the understanding that if any denominator vanishes in Eq. (15), the numerator vanishes too and that fraction is omitted from the continued equation. This kind of equation originated with St.-Venant (1870) for plane strain; the general relationship of Eq. (15) was proposed by Levy (1871) and again independently by von Mises (1913).

Prandtl-Reuss elastic, plastic equations. Since Levy and von Mises used the total-strain increment $d\epsilon_{ij}$ and not the plastic-strain increment $d\epsilon^p_{ij}$, Eqs. (8) or (15) describe a fictitious rigid, plastic material (with infinite elastic modulus). An elastic, plastic generalization proposed by Prandtl (1924) for the plane problem, and by Reuss (1930) in general, assumed that

$$d\epsilon^p_{ij} = T'_{ij}d\lambda, \tag{6.5.16}$$

where $d\lambda$ is a scalar multiplier (not a constant). If the Mises yield condition of Eq. (10) is assumed in addition, then a procedure like that of Eq. (11) shows that

$$d\lambda = \frac{1}{k}\sqrt{\mathrm{II}_{d\epsilon^p}}, \qquad \text{where} \quad \mathrm{II}_{d\epsilon^p} = \tfrac{1}{2}d\epsilon^p_{ij}d\epsilon^p_{ij} \qquad (\text{for} \quad d\epsilon^p_{kk} = 0). \tag{6.5.17}$$

The complete natural-strain increment is then assumed to be

$$d\epsilon'_{ij} = (d\epsilon'_{ij})^e + d\epsilon^p_{ij} \quad \text{or} \quad d\epsilon_{ij} = d\epsilon^e_{ij} + d\epsilon^p_{ij}, \tag{6.5.18}$$

where the elastic natural-strain increment $d\epsilon^e_{ij}$ is given by (see Sec 6.2):

$$\left.\begin{aligned} (d\epsilon'_{ij})^e &= \frac{1}{2G}dT'_{ij} \qquad d\epsilon^e_{kk} = \frac{(1-2\nu)}{3E}dT_{kk} \quad \text{so that} \\ d\epsilon^e_{ij} &= \frac{1}{2G}dT'_{ij} + \frac{1-2\nu}{E}\delta_{ij}dT_{kk} \quad \text{or} \\ d\epsilon^e_{ij} &= \frac{1+\nu}{E}dT_{ij} - \frac{\nu}{E}\delta_{ij}dT_{kk}. \end{aligned}\right\} \tag{6.5.19}$$

Since these are ***natural-strain increments***, the reference configuration is implied to be the current configuration at the beginning of the increment. Because the elastic strain recoverable upon unloading in metals is small, $d\epsilon_{ij}^e/dt$ may alternatively be considered as the rate of change of small strain ϵ_{ij}^e, which may be interpreted as either the Eulerian small strain (defined in the deformed configuration) or the Lagrangian small strain defined relative to the configuration to which the material would return upon unloading (not the initial configuration before plastic deformation). Whether it is considered as Eulerian or Lagrangian, it measures the strain from the new unloaded configuration to the deformed configuration. (See the remarks on changing unloaded configuration in the discussion of postyield behavior following Fig. 6.8.) These elastic strains ϵ_{ij}^e need not satisfy the compatibility equations of Sec. 4.7, since there may be no single-valued continuous displacement field that would take the whole body from the deformed configuration to an unstressed configuration. Equations (19) are explicitly integrable to give a form of Hooke's law relating these small strains from the unloaded configuration to the stresses, which is the same as the form of Hooke's law given in Sec. 6.2 for strains from the initial configuration in elasticity.

But the plastic components of Eqs. (16) are not explicitly integrable. If Eqs. (16), (18), and (19) are divided by dt, the Reuss theory can also be written in terms of rates of deformation, with

$$D_{ij}^p = \frac{d\lambda}{dt} T'_{ij} \qquad \text{where} \quad \frac{d\lambda}{dt} = \frac{1}{k}\sqrt{\text{II}_{D^p}} \tag{6.5.20}$$

and D_{ij}^p is now only the plastic part of the rate of deformation. Despite the formal similarity of Eqs. (8) and (20) to the Newtonian viscosity equation $T'_{ij} = 2\mu D'_{ij}$, there is really no rate-dependence or viscosity implied by the plasticity equation. Doubling all the plastic rate-of-deformation components, for example, does not change the stress-deviator components, because the quantity $\sqrt{\text{II}_{D^p}}$ is doubled when all the D_{ij}^p are doubled. An idealized material combining some of the features of Levy-Mises plasticity and Newtonian viscosity is the ideal viscoplastic material.

Part 3. Viscoplasticity

A viscoplastic material differs from a fluid in that it can sustain a shear stress even when it is at rest, but when the stress intensity reaches a critical value (yield condition) the material flows with viscous stresses proportional to the excess of the stress intensity over the yield-stress intensity.

Consider first the case of a simple shear in the x-direction and suppose that no motion takes place until the stress intensity $|T_{xy}|$ reaches the critical value k, after which the magnitude $|D_{xy}|$ is proportional to the amount by which $|T_{xy}|$ exceeds k. Let the quantity F^1 be defined by

$$F^1 = 1 - \frac{k}{|T_{xy}|}. \tag{6.5.21}$$

Then for simple shear the viscoplastic equation is

$$2\eta D_{xy} = \begin{cases} 0 & \text{for} \quad F^1 < 0 \\ F^1 T_{xy} & \text{for} \quad F^1 \geq 0, \end{cases} \tag{6.5.22}$$

where η is a physical constant analogous to the viscosity coefficient μ of a fluid. This equation characterizes a so-called Bingham material introduced by Bingham (1922).

For three-dimensional flows, the following generalization, due to Hohenemser and Prager (1932) [see Prager (1961), Chap. 7], assumes incompressibility and generalizes Eq. (22) to

$$2\eta D_{ij} = \begin{cases} 0 & \text{for} \quad F < 0 \\ F T'_{ij} & \text{for} \quad F \geq 0, \end{cases} \tag{6.5.23}$$

where the generalized definition of F (without superscript) requires us to assume some scalar measure of the intensity of the state of stress. Hohenemser and Prager chose $\sqrt{J'_2}$ as a measure of the intensity of stress in the three-dimensional case, where $J'_2 \equiv \mathrm{II}_{T'}$ is the second invariant of the stress deviator, and replaced Eq. (21) by

$$F = 1 - \frac{k}{\sqrt{J'_2}}. \tag{6.5.24}$$

Equation (23) is the viscoplastic analogue of the Newtonian viscosity equation $2\mu D'_{ij} = T'_{ij}$. A more instuctive comparison of the two is obtained by solving Eq. (23) for T'_{ij}. We first square both sides of each component equatiòn and add, obtaining

$$4\eta^2 D_{ij} D_{ij} = F^2 T'_{ij} T'_{ij} \qquad \text{for} \quad F \geq 0.$$

Since $D_{ij} D_{ij} = 2\mathrm{II}_D$ (for deviatoric D_{ij}) and $T'_{ij} T'_{ij} = 2J'_2$, this yields the relation

$$4\eta^2 \mathrm{II}_D = F^2 J'_2, \tag{6.5.25}$$

which with Eq. (24) permits us to express F in terms of II_D as

$$F = \frac{2\eta\sqrt{\mathrm{II}_D}}{k + 2\eta\sqrt{\mathrm{II}_D}} \tag{6.5.26}$$

and, finally, allows us to solve Eq. (23) for T'_{ij}, obtaining

$$T'_{ij} = \left(2\eta + \frac{k}{\sqrt{\text{II}_D}}\right) D_{ij} \qquad (\text{for} \quad F \geq 0). \tag{6.5.27}$$

The coefficient of D_{ij} in this equation contains the viscosity term 2η plus an additional plasticity term, which vanishes if the yield stress k is zero, and which is not really a rate-dependent term because the product $kD_{ij}/\sqrt{\text{II}_D}$ is homogeneous of degree zero in the velocity components. Thus, for example, doubling all the velocity components would not change the contribution of this term to the stress, while the viscous contribution $2\eta D_{ij}$ would be doubled.

An elastic term $D^e_{ij} = d\epsilon_{ij}/dt$, given by Eq. (19), can be added to the viscoplastic term $D^p_{ij} = (F/2\eta)T'_{ij}$ of Eq. (23) as in Eq. (18). The physical behavior in pure shear D_{xy}, where

$$D^p_{xy}/\sqrt{\text{II}_{D_p}} = D^p_{xy}/|D^p_{xy}| = \text{sign of } D^p_{xy},$$

can be represented, at least for small displacement gradients where $D_{xy} \approx \dot{E}_{xy}$, by a model analogous to the three-parameter viscoelastic model of Fig. 6.7 (a), with the spring of the Kelvin-Voigt element replaced by a solid friction element acting as a rigid body for $|T_{xy}| < k$ and sliding at constant friction k when $|T_{xy}| \geq k$, as illustrated in Fig. 6.13.

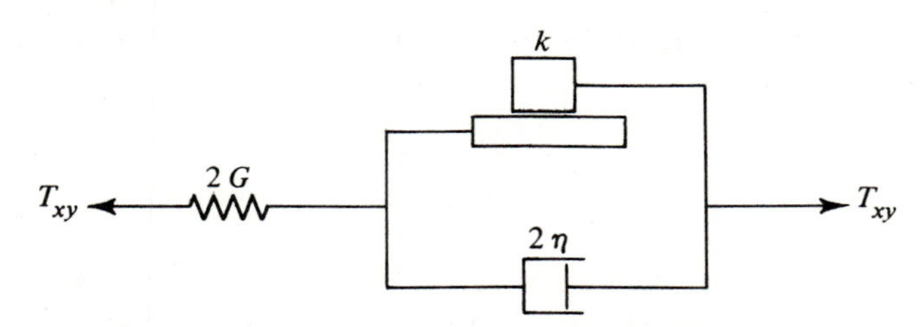

Fig. 6.13 Elastic, viscoplastic model in pure shear.

For pure shear, with $\dot{E}_{xy} = D_{xy}$, Eqs. (19) and (27) give

$$2GE^e_{xy} = T_{xy} = 2\eta\dot{E}^p_{xy} + (\text{sign of } \dot{E}^p_{xy})k$$

in agreement with the model. For generalization to three-dimensional small strain, the model does not give the correct result in a completely obvious way, since a friction block is required, contributing to T'_{ij} not [(sign of $\dot{E}^p_{ij}$) k] but instead contributing $k\dot{E}^p_{ij}/\sqrt{\text{II}_{\dot{E}^p}}$. We may think of the block as sliding over a nine-dimensional surface in a direction with direction cosines $\dot{E}^p_{ij}/\sqrt{\text{II}_{\dot{E}^p}}$. The model illustrates the fact that even without work-hardening the stress intensity during plastic deformation can be greater than the yield-stress intensity, because of the viscous contribution.

The simple model of Fig. 6.13 exhibits qualitatively the behavior of some actual materials, but it cannot be expected to fit accurately the quantitative response of most real materials except for a very limited range of strains and strain rates in situations where the rotation is small. In this respect the model shares the shortcomings of the three-parameter viscoelastic model in Fig. 6.7. More general viscoplastic equations have been proposed; see, for example, Perzyna (1963). But not much has yet been done to match them to the properties of real materials,

Rate effects are neglected in the work-hardening plasticity formulations discussed in Sec. 6.6.

EXERCISES

1. If the true-stress vs. true-strain curve in a uniaxial tension test is fitted to the formula $\sigma = (10^5)\epsilon^{1/4}$ in the plastic region, with $\sigma = 30(10^6)\epsilon$ in the elastic region (stresses in psi), determine (a) the true stress and true strain at yield, (b) the true stress and strain at the maximum load, (c) the nominal stress and strain at the ultimate strength, (d) the equation of the nominal stress vs. nominal strain in the plastic region, and (e) the equation relating the plastic strain ϵ_p to σ in the plastic region during the first loading.
2. If the slopes of the elastic and plastic portions of the elastic, linear work-hardening curve of Fig. 6.11 are E and B, respectively, determine (a) the equations of σ vs. ϵ in the plastic region, and (b) the equation of σ vs. the plastic strain ϵ_p.
3. Show that the Levy-Mises equations imply that the principal axes of the rate of deformation coincide with the principal axes of stress, and write Eqs. (6.5.8) and (10) in terms of the principal values D_j and σ_j. (*Note*: Not σ'_j.)
4. Interpret the Mises yield condition as a condition on
   (a) the octahedral shear stress of Ex. 10 of Sec. 3.3.
   (b) the elastic distortion energy of Ex. 11 (b) of Sec. 6.2.
5. (a) Combine Eqs. (6.5.16) through (6.5.19) to give the Prandtl-Reuss equations relating the total deviatoric strain increments $d\epsilon'_{ij}$ to the stress deviator T'_{ij}, dT'_{ij}, and $d\lambda$.
   (b) Now write out the six equations, without summation conventions, for the total strain increments $d\epsilon_{ij}$ in terms of stresses and stress-increments (not deviators) and $d\lambda$.
6. (a) Write the results of Ex. 5(a) in terms of deviatoric rates of deformation instead of strain deviator increments.
   (b) Show that in general the principal axes of **D** do not coincide with those of **T**. What is a sufficient condition on the stress history to ensure coincidence ?
   (c) Assuming that the principal axes do coincide, write the results of Ex. 5 in terms of principal values.
7. For plane-strain compression in the σ_1-direction with $\epsilon_3 \equiv 0$, and no restraint

in the σ_2-direction, so that $\sigma_2 \equiv 0$, and with no shears, determine the yield value of σ_1 in terms of the uniaxial yield stress Y and Poisson's ratio ν, if the material is governed by the Mises yield condition.

8. (a) Consider a state of combined stress produced by tension and torsion of a thin-walled circular cylinder. Assume that the stress on a cross-section is uniform and consists of an axial normal stress σ and a shear stress τ. If Y is the experimentally determined yield stress in uniaxial tension, derive the equations of the yield loci in the $\sigma\tau$-plane from the Mises and Tresca assumptions of Eq. (6.5.10) and Eq. (14). (Radial and circumferential normal stresses are assumed to be zero.)
   (b) Repeat (a) if the two theories both agree with experimentally determined yield stress k in pure shear, instead of with the measured tensile yield.
   (c) Sketch a plot of the two results of (a) on one set of axes, and the two results of (b) on another set of axes.
   (d) In many problems it is convenient to use some kind of generalized stress variables to simplify the representation. If in the tension-torsion of a thin-walled tube the stress variables plotted are σ/Y and $\sqrt{3}\tau/Y$, what is the yield locus obtained for the Mises yield condition of (a)?

9. If the loads in Ex. 8 (a) are controlled in such a way that $\sigma \equiv \tau$ during a test, what would be the ratios of the rates of deformation D_{33} in the axial direction, D_{11} in the radial direction, and D_{22} in the circumferential direction to the shear rate D_{32} if the material is governed by the Levy-Mises equations?

10. In plane plastic strain governed by the Levy-Mises equations, with $\epsilon_{zz} \equiv 0$ and elastic strains neglected, show that $T_{zz} = \frac{1}{2}(T_{xx} + T_{yy})$, while $T_{zx} = T_{zy} = 0$. Hence, write the Mises yield condition for plastic strain as a function of T_{xx}, T_{yy}, and T_{xy}, and show that it is identical to the Tresca yield condition for plane plastic strain.

11. (a) Show that in plane plastic strain with $T_{zz} = \frac{1}{2}(T_{xx} + T_{yy})$ as in Ex. 10, $T'_{xx} = -T'_{yy} = \frac{1}{2}(T_{xx} - T_{yy})$, whence $d\epsilon_{yy} = -d\epsilon_{xx}$ by the Levy-Mises equations.
    (b) Show that for plane plastic strain the Levy-Mises Eqs. (15) imply the following two equations in terms of velocities:

$$\frac{\partial v_x}{\partial x} + \frac{\partial v_y}{\partial y} = 0 \qquad \text{(incompressibility)}$$

$$2T_{xy}\left[\frac{\partial v_x}{\partial x} - \frac{\partial v_y}{\partial y}\right] = (T_{xx} - T_{yy})\left[\frac{\partial v_x}{\partial y} + \frac{\partial v_y}{\partial x}\right]$$

12. Carry out the details leading to Eq. (6.5.27), and write Eq. (27) for the case of pure shear T_{xy}.

13. Write the elastic, viscoplastic equations obtained by adding to Eq. (6.5.23) terms for the deviatoric elastic deformation rates.

14. What model replaces Fig. 6.13 to illustrate Prandtl-Reuss elastic, plastic behavior ?

15. If $k = 10{,}000$ psi, determine the value of η in Eq. (6.5.27) if it is observed

that, in a pure shear test, increasing D_{xy} from 10^{-2} sec^{-1} to 10^{2} sec^{-1} causes an increase in T_{xy} of: (a) 100%, (b) 10%.

References for collateral reading on plasticity

Drucker, D. C., *Introduction to the Mechanics of Deformable Solids.* New York: McGraw-Hill, 1967.

Ford, H., and J. M. Alexander, *Advanced Mechanics of Materials.* New York: Wiley, 1963.

Freudenthal, A. M., and H. Geiringer, "The Inelastic Continuum," in *Encyclopedia of Physics.* vol. 4, ed. S. Flügge. Berlin: Springer-Verlag, 1958.

Fung, Y. C., *Foundations of Solid Mechanics.* Englewood Cliffs, N. J.: Prentice-Hall, 1965.

Goodier, J. N., and P. G. Hodge, Jr., *Elasticity and Plasticity.* New York: Wiley, 1958.

Hill, R., *Plasticity.* London: Oxford University Press, 1950.

Hodge, P. G., Jr., *Plastic Analysis of Structures.* New York: McGraw-Hill, 1959.

Hodge, P. G., Jr., *Limit Analysis of Rotationally-Symmetric Plates and Shells.* Englewood Cliffs, N. J.: Prentice-Hall, 1963.

Hoffman, O., and G. Sachs, *Introduction to the Theory of Plasticity for Engineers.* New York: McGraw-Hill, 1953.

Johnson, W., and H. Kudo, *The Mechanics of Metal Extrusion.* Manchester, England: Manchester University Press, 1962.

Johnson, W., and P. B. Mellor, *Plasticity for Mechanical Engineers.* London: Van Nostrand, 1962.

Lee, E. H., and P. Symonds, eds., *Plasticity.* Oxford: Pergamon Press, 1960.

Lubahn, J. D., and R. P. Felgar, *Plasticity and Creep of Metals.* New York: Wiley, 1961.

Marin, J., *Mechanical Behavior of Engineering Materials.* Englewood Cliffs, N. J.: Prentice-Hall, 1962.

Mendelson, A., *Plasticity: Theory and Application.* New York: Macmillan, 1968.

Nadai, A., *Theory of Flow and Fracture of Solids*, 2nd ed. New York: McGraw-Hill, 1950.

Neal, B. G., *Plastic Methods of Structural Analysis.* London: Chapman and Hall, 1956.

Phillips, A., *Introduction to Plasticity.* New York: Ronald Press, 1956.

Polakowski, N. H., and E. J. Ripling, *Strength and Structure of Engineering Materials.* Englewood Cliffs, N. J.: Prentice-Hall, 1966.

Prager, W., *An Introduction to Plasticity*. Reading, Mass.: Addison-Wesley, 1959.

Prager, W., *Introduction to Mechanics of Continua*. Boston: Ginn & Co., 1961.

Prager, W., and P. G. Hodge, Jr., *Theory of Perfectly Plastic Solids*. New York: Wiley, 1951.

Thomas, T. Y., *Plastic Flow and Fracture in Solids*. New York: Academic Press, 1961.

Thomsen, E. G., C. T. Yang, and S. Kobayashi, *Mechanics of Plastic Deformation in Metal Processing*. New York: Macmillan, 1965.

6.6 Plasticity II. More advanced theories; yield conditions; plastic-potential theory; hardening assumptions; older total-strain theory (deformation theory)

In this section, we consider attempts to provide a more adequate theory than the classical, perfectly plastic theories. The theories presented here are for the most part incremental theories of plasticity (also called flow theories), but Part 4 gives a brief account of the older total-strain theory (also called deformation theory) of Hencky and Nadai.

As was pointed out in Sec. 6.5, any three-dimensional combined-stress theory of plasticity must answer three questions to determine the yield and postyield response:

1. *Yield Condition:*
   What stress combinations permit inelastic response?
2. *Postyield Behavior:*
   (a) How are the plastic deformation increments or plastic rate-of-deformation components related to the stress components?
   (b) How does the yield condition change with work-hardening?

For the Levy-Mises perfectly plastic idealized material, Questions 1 and 2 (a) above were answered by Eqs. (6.5.10) and (6.5.8), respectively, while perfect plasticity implied no work-hardening and therefore no change at all induced in the yield condition by the occurrence of plastic deformation. In this section, we consider some more general formulations that have been proposed in the attempt to give a more accurate representation of real materials, but throughout the section we neglect rate and temperature dependence. The first three parts of the section treat in order the three questions posed above.

Part 1. Yield Condition

Almost all rate-independent theories of continuum plasticity postulate;

> Assumption 1. *General Yield Condition*
>
> For a given state of a given material, there exists a function $f(\mathbf{T})$ of the stress such that the material is elastic for
>
> $$f(\mathbf{T}) < 0, \quad \text{or for} \quad f(\mathbf{T}) = 0 \quad \textbf{\textit{and}} \quad \frac{\partial f}{\partial T_{ij}} \dot{T}_{ij} < 0 \tag{6.6.1a}$$
>
> $$\text{and plastic for:} \quad f(\mathbf{T}) = 0 \quad \textbf{\textit{and}} \quad \frac{\partial f}{\partial T_{ij}} \dot{T}_{ij} \geq 0 \tag{6.6.1b}$$
>
> [$f(\mathbf{T}) \leq 0$ always].

For example, the Mises yield condition of Eq. (6.5.10) is

$$f(\mathbf{T}) \equiv J_2' - k^2 \equiv \tfrac{1}{2}T'_{ij}T'_{ij} - k^2 = 0, \tag{6.6.2}$$

where $J_2' \equiv \text{II}_{T'}$ is the second invariant of the stress deviator $\mathbf{T}'$. Besides the deviatoric stresses $T'_{ij} = T_{ij} - \frac{1}{3}T_{mm}\delta_{ij}$, the Mises yield function contains a single parameter k, which is assumed constant in a perfectly plastic theory, but which would increase during the plastic deformation of an isotropic-hardening Levy-Mises material (see Part 3). Then, even when $(\partial f/\partial T_{ij})\dot{T}_{ij}$ is positive during a finite time increment, f remains equal to zero in Eq. (2) because k^2 increases by the same amount as J_2'. In the general case of Eq. (1), there may be several parameters instead of the single parameter k, but during plastic deformation the parameters change in such a way that f remains equal to zero while plastic deformation continues. These parameters define the ***state of the material,*** which changes with deformation in a way that is in general difficult to specify. The parameters also depend on temperature and possibly on other factors not explicitly related to the deformation. Note that the general yield condition implies that whether the material at a point in space is elastic or plastic depends only on the stress at that point and not on the stress at neighboring points. Hence it does not contain any dependence on the stress gradient. Although dependence on the stress gradient is not inconceivable, no experience has yet indicated the need for it. Explicit dependence on the plastic strain components ϵ_{ij}^p is sometimes included as in Eq. (45a) on page 368. Here we are concerned only with a given state in which the prior plastic deformation enters only in determining the parameters defining the present state.

Yield surface in stress space. The yield condition of Eq. (1) can be interpreted geometrically in terms of a surface $f(\mathbf{T}) = 0$, i. e., a hypersurface in a nine-dimensional Euclidean space in which the stresses T_{ij} are interpreted as nine rectangular Cartesian coordinates. Since T_{ij} is symmetric, we may alternatively interpret the yield surface as a surface in the six-dimensional subspace in which the coordinates are

$$Q_1 = T_{11} \quad Q_2 = T_{22} \quad Q_3 = T_{33} \quad Q_4 = T_{23} \quad Q_5 = T_{31} \quad Q_6 = T_{12}, \tag{6.6.3}$$

but for some purposes it is more instructive to think of the nine-dimensional space. Special states of stress can be represented in terms of fewer dimensions. For uniaxial stress T_{xx}, the space is one-dimensional and the yield condition $|T_{xx}| = Y$ represents a "yield surface" consisting of only the two points $T_{xx} = Y$ and $T_{xx} = -Y$ in the one-dimensional subspace. These are the two points where the general yield surface $f(\mathbf{T}) = 0$ intersects the T_{xx}-axis in the nine-dimensional space. For biaxial stress, with $T_{xx} = Q_1$, $T_{yy} = Q_2$, and all other stress components zero, the Mises yield condition of Eq. (2) (with $k = Y/\sqrt{3}$) represents the two-dimensional yield "surface" or curve shown in Fig. 6.14, an ellipse defined by

$$Q_1^2 - Q_1Q_2 + Q_2^2 - Y^2 = 0. \tag{6.6.4}$$

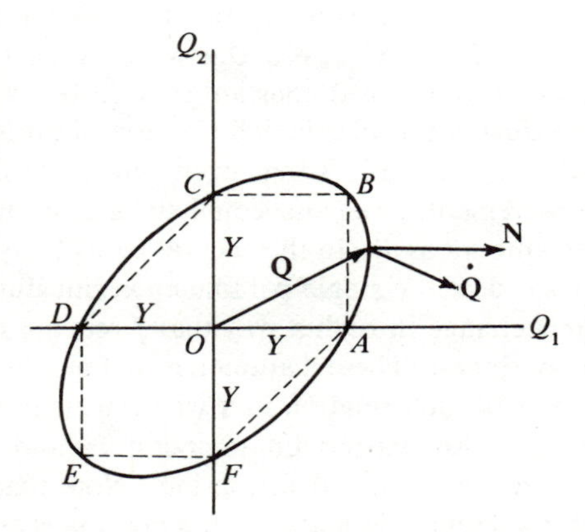

Fig. 6.14 Mises and Tresca yield curves for biaxial stress $Q_1 = T_{xx}$, $Q_2 = T_{yy}$.

Recall that by Exercise 6(c) of Sec. 3.3,

$$J_2' \equiv \mathrm{II}_{T'} = \tfrac{1}{6}[(T_{11} - T_{22})^2 + (T_{22} - T_{33})^2 + (T_{33} - T_{11})^2] + T_{23}^2 + T_{31}^2 + T_{12}^2 \tag{6.6.5}$$

whence the Mises condition $J_2' - \frac{1}{3}Y^2 = 0$ takes the form of Eq. (4) for the biaxial stress state. The hexagon *ABCDEF*, shown by dashed lines in Fig. 6.14, is the Tresca yield surface defined by the Tresca maximum shear stress yield condition of Eq. (6.5.14), namely

$$T_{\max} - T_{\min} - Y = 0. \tag{6.6.6}$$

Segments of the hexagon parallel to the Q_1- or Q_2- axis represent states of stress where $Q_3 = 0$ is either T_{max} or T_{min}.

The plastic loading condition $(\partial f/\partial T_{ij})\dot{T}_{ij} \geq 0$ of Eq. (1b) has the following interpretation in terms of the geometry of the yield surface in stress space, as illustrated for the biaxial stress state in Fig. 6.14. The vector $\mathbf{Q}$ in the stress space is the position vector whose components are the stresses; $\dot{\mathbf{Q}}$ is the time rate of change of $\mathbf{Q}$. The vector $\mathbf{N} = \nabla f$ is the outer normal with components $\partial f/\partial T_{ij}$ [or $\partial f/\partial Q_i$]. Hence $(\partial f/\partial T_{ij})\dot{T}_{ij} = \mathbf{N} \cdot \dot{\mathbf{Q}}$. When this is positive the angle between $\mathbf{N}$ and $\dot{\mathbf{Q}}$ is acute, and this means that $\dot{\mathbf{Q}}$ points toward the outside of the yield surface. When $\mathbf{N} \cdot \dot{\mathbf{Q}}$ is negative, $\dot{\mathbf{Q}}$ points toward the inside, and when $\mathbf{N} \cdot \dot{\mathbf{Q}} = 0$, $\dot{\mathbf{Q}}$ is tangent to the surface. Thus the yield condition requires for the possibility of plastic deformation not only that the stress point be on the yield surface at the instant in question, but also that it not be moving toward the elastic region inside the yield surface. At a corner on the Tresca yield surface, the criterion must be applied separately to the two surfaces intersecting at the corner; if either one of them gives $\mathbf{N} \cdot \dot{\mathbf{Q}}$ nonnegative, plastic deformation can occur according to the Tresca condition.

For the isotropic-hardening Levy-Mises theory, the yield surface maintains the same shape and the same center but enlarges with plastic deformation in such a way that the stress point never actually moves outside of the current surface. For more general theories, the surface may change its shape as well as its size, but we still suppose that the stress point never moves outside in the rate-independent plasticity theory considered in this section. In viscoplasticity, because of rate effects, the stress point may be outside of the limiting quasi-static or equilibrium yield surface.

We now consider additional assumptions, limiting the generality of the yield condition of Eq. (1), and some consequences following from the assumptions. Not all the additional assumptions are made in every plasticity theory, but all are included in the theories most widely used for metal plasticity. Experience teaches that to a first approximation the yielding of a metal is unaffected by a moderate hydrostatic pressure superposed on a combined-stress state. [See Hill (1950), p. 16 for references to original investigation.] Metal plasticity theories accordingly postulate:

ASSUMPTION 2. Yield is independent of the spherical part of the stress.

It follows that the function $f(\mathbf{T})$ in Eq. (1) must in fact be a function only of the stress deviator, say $f(\mathbf{T}')$. Note that both the Mises and the Tresca conditions satisfy this requirement. Because of the difficulties of loading in a state of pure hydrostatic tension, most experimental verifications have been limited to superposed hydrostatic compression, but the postulate is also assumed to apply to hydrostatic tension.

The yield function $f(\mathbf{T})$ in the general case (without Assumption 2) may be specified as a function of six independent variables, which may be the six independent stress components, as

$$f = f(T_{11}, T_{22}, T_{33}, T_{23}, T_{31}, T_{12}) \tag{6.6.7a}$$

or the three principal stresses σ_1, σ_2, σ_3 and the unit vectors $\hat{\mathbf{n}}_1$, $\hat{\mathbf{n}}_2$, $\hat{\mathbf{n}}_3$ defining a choice of the three principal axes, say as

$$f = f(\sigma_1, \sigma_2, \sigma_3, \hat{\mathbf{n}}_1, \hat{\mathbf{n}}_2, \hat{\mathbf{n}}_3). \tag{6.6.7b}$$

Naturally, the form of the function f in (7b) is different from the form in (7a). But, since the values of the T_{ij} in any other coordinate system are determined whenever the three principal stresses and the principal axes are given, we can say that the three σ_i and the three $\hat{\mathbf{n}}_i$ determine $\mathbf{T}$ completely and hence determine $f(\mathbf{T})$. In general, the form of the function f of (7a) will be different in one coordinate system than it is in another; that is, any parameters or constants appearing in the yield function may have different values in different coordinate systems. If the material is ***isotropic***, there can be no preferred directions, and the function should have the same form for all orientations of the rectangular Cartesian coordinate axes relative to the material. With regard to the form of (7b), isotropy implies that f cannot depend on the orientation of the principal-axis triad $\hat{\mathbf{n}}_1$, $\hat{\mathbf{n}}_2$, $\hat{\mathbf{n}}_3$ relative to the material but only on σ_1, σ_2, σ_3 and indeed it must be such a function that if $f(\sigma_1, \sigma_2, \sigma_3) = 0$ then so does $f(\sigma_2, \sigma_1, \sigma_3) = 0$, since the isotropic material is unable to distinguish which principal axis is labeled $\hat{\mathbf{n}}_1$ and which is $\hat{\mathbf{n}}_2$. This condition is met by requiring that ***in an isotropic material f is a symmetric function of*** $\sigma_1,\sigma_2,\sigma_3$.

ASSUMPTION 3. The material is isotropic.

As a consequence of Assumption 3, the yield function must be of the form

$$f = f(\sigma_1, \sigma_2, \sigma_3), \tag{6.6.8}$$

where f is a symmetric function of σ_1, σ_2, σ_3. Alternatively, since σ_1, σ_2, σ_3 are completely determined, according to Eq. (3.3.8), by the three scalar invariants I_T, II_T, III_T of the stress tensor, denoted J_1, J_2, J_3, respectively, in most of the plasticity literature, the yield condition equation must be expressible as

$$f = f(J_1, J_2, J_3). \tag{6.6.9}$$

Assumptions 2 and 3 together imply that f must be a symmetric function of σ_1', σ_2', σ_3' or of the invariants $J_2' = \mathrm{II}_{T'}$ and $J_3' = \mathrm{III}_{T'}$ of the stress deviator (since $J_1' \equiv \mathrm{I}_{T'} \equiv 0$). Thus

$$f = f(\sigma_1', \sigma_2', \sigma_3') \tag{6.6.10a}$$

or

$$f = f(J_2', J_3') \tag{6.6.10b}$$

in an isotropic material for which yield is independent of the hydrostatic part of the stress.

Evidently the Mises yield condition results from assuming the special case of Eq. (10) where f is independent of J_3'. The Mises version of Eqs. (8) or (10a) follows immediately as

$$\tfrac{1}{6}[(\sigma_1 - \sigma_2)^2 + (\sigma_2 - \sigma_3)^2 + (\sigma_3 - \sigma_1)^2] - k^2 = 0 \tag{6.6.11a}$$

or

$$\tfrac{1}{2}[\sigma_1'^2 + \sigma_2'^2 + \sigma_3'^2] - k^2 = 0. \tag{6.6.11b}$$

That the Tresca condition of Eq. (6) is also a special case of Eqs. (8) or (10) may be seen by writing it first as a symmetric function of σ_1, σ_2, σ_3. Since the maximum shear stress yield equality is satisfied by any one of the six linear equations

$$\tfrac{1}{2}(\sigma_1 - \sigma_2) = \pm k \qquad \tfrac{1}{2}(\sigma_2 - \sigma_3) = \pm k \qquad \tfrac{1}{2}(\sigma_3 - \sigma_1) = \pm k, \tag{6.6.12a}$$

where k is the yield stress in pure shear, the symmetric form of the yield condition is

$$[(\sigma_1 - \sigma_2)^2 - 4k^2][(\sigma_2 - \sigma_3)^2 - 4k^2][(\sigma_3 - \sigma_1)^2 - 4k^2] = 0. \tag{6.6.12b}$$

The Tresca yield surface in principal-stress space consists of six planes, and elastic states are represented by points where all three factors of Eq. (12b) are negative. In applications where it can be decided which one of the six plane faces of the yield surface in principal-stress space is operative, it is simpler to work with the appropriate linear Eq. (12a) than with the symmetric form. Equation (12b) can be transformed to an invariant form due to Reuss

$$\frac{27(J_2')^3}{64k^6}\left[\frac{4}{27} - \frac{(J_3')^2}{(J_2')^3}\right] - \left(1 - \frac{3J_2'}{4k^2}\right)^2 = 0, \tag{6.6.13}$$

but this form is too complicated for convenient application.

Note that $J_2'(-\mathbf{T}) = J_2'(\mathbf{T})$ while $J_3'(-\mathbf{T}) = -J_3'(\mathbf{T})$. Since J_3' occurs in Eq. (13) only as $(J_3')^2$, changing the signs of all stresses leaves the value of the yield function unchanged in both the Mises and Tresca formulations. This is usually postulated to apply to the general yield function.

Assumption 4. $f(-\mathbf{T}) = f(\mathbf{T})$ (6.6.14)

If Assumption 4 is supposed to apply to the changed yield condition after a plastic deformation, it rules out the possibility of representing the Bauschinger effect illustrated in Fig. 6.12 (a). Assumption 4 and the assumption of isotropy may both lead to significant departures from observed behavior when applied to materials after previous plastic deformation, but experience indicates that they work reasonably well for the initial yield in polycrystalline metals when the crystallites have a random orientation leading to an apparent isotropy in any sample large compared to crystallite dimensions. In soil mechanics, on the other hand, Assumption 4 is definitely incorrect, since soils are much weaker in tension than in compression.

For an isotropic material, the yield condition equation $f(\sigma_1, \sigma_2, \sigma_3) = 0$ can be represented as a surface in a three-dimensional, principal-stress space. See Hill (1950). Note that this kind of representation is not possible in general for anisotropic materials, although it can be used to record yield states produced in one set of experiments in which the principal-stress directions do not change. Representation in three dimensions can be visualized graphically, unlike the six- or nine-dimensional general case. Figure 6.15 shows part of the Mises and Tresca yield surfaces in the positive octant of $\sigma_1, \sigma_2, \sigma_3$-space. The lower portions near the plane *ABCDEF* are shown by dashed lines, since they are not in the positive octant. The surfaces are parallel to the

$$\text{Hydrostatic line } OH\text{:} \quad \sigma_1 = \sigma_2 = \sigma_3, \tag{6.6.15}$$

since they satisfy Assumption 2, requiring that if $(\sigma_1, \sigma_2, \sigma_3)$ is on the yield surface, then so is $(\sigma_1 + H, \sigma_2 + H, \sigma_3 + H)$ for any value of H. Hence the surfaces are right cylinders (or prisms) with elements perpendicular to the

$$\text{Deviatoric plane:} \quad \sigma_1 + \sigma_2 + \sigma_3 = 0 \tag{6.6.16}$$

(the plane *ABCDEF* in Fig. 6.15). The Mises yield surface $J_2' = Y^2/3$ is the right circular cylinder defined by Eq. (11a), while the Tresca yield surface is the regular hexagonal prism whose six plane faces are defined by Eqs. (12a), with k replaced by $Y/2$ in order that the Mises and Tresca conditions will agree in uniaxial tension or compression.

Uniaxial tension yield is represented by the marked points K, L, M in Fig. 6.15. Each coordinate plane intersects the yield surface in a curve like that shown in Fig. 6.14 for $\sigma_3 = 0$; the arcs KS and KTL in Fig. 6.15 are portions of two of these intersection curves with the Mises cylinder.

Since the hydrostatic part of the stress is assumed unimportant for the yield, the yield surface can be sufficiently represented by its intersection with the deviatoric plane of Eq. (16). Another possible two-dimensional representation plots the ***principal-stress differences***; for example, let

$$Q_1 = \sigma_1 - \sigma_3 \qquad Q_2 = \sigma_2 - \sigma_3 \tag{6.6.17}$$

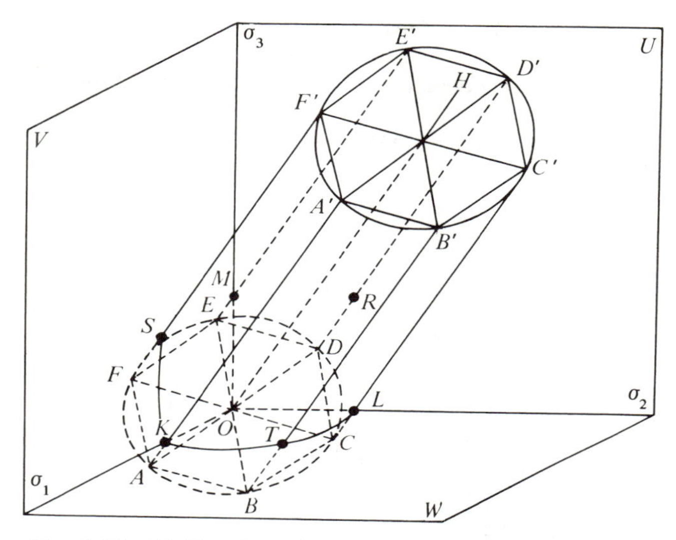

Fig. 6.15 Yield surfaces in principal-stress space, Mises circular cylinder and Tresca hexagonal prism, deviatoric plane *ABCDEF* and hydrostatic line *OH*.

to obtain the same curves as in Fig. 6.14 even when $\sigma_3 \neq 0$. The ***deviatoric plane*** plot is shown in Fig. 6.16. The three projected axes are labeled σ_1, σ_2, and σ_3. By the assumption of isotropy, f must be a symmetric function of σ_1, σ_2, σ_3, unchanged by the interchange of any two of the three arguments. Hence the locus in the deviatoric plane must be symmetric with respect to each of the three projected axes. By Assumption 4, it is also symmetric with respect to the origin. Hence it suffices to determine any 30-degree arc of the locus experimentally, e. g., *AT* in Fig. 6.16, in order to determine the whole locus, if the material satisfies Assumptions 3 and 4.

There are three projected axes shown in the two-dimensional deviatoric plane. If values of the principal deviatoric stresses are laid off as vectors parallel to the three-dimensional axes, their vector sum is a vector $\mathbf{OP}'$ lying in the deviatoric plane. Each of the three vectors projects as a vector in the deviatoric plane of length $\sqrt{\frac{2}{3}}$ times the actual vector length. For example, in Fig. 6.16

$$|OM'| = |\sigma_2'|\sqrt{\tfrac{2}{3}} \qquad |M'N'| = |\sigma_1'|\sqrt{\tfrac{2}{3}}$$

and

$$|N'P'| = |\sigma_3'|\sqrt{\tfrac{2}{3}}$$

and the vector sum

$$\mathbf{OP'} = \mathbf{OM'} + \mathbf{M'N'} + \mathbf{N'P'}$$

in the plane gives the same deviatoric point P' as the vector sum in space

$$\mathbf{OP'} = \mathbf{OM} + \mathbf{MN} + \mathbf{NP'}.$$

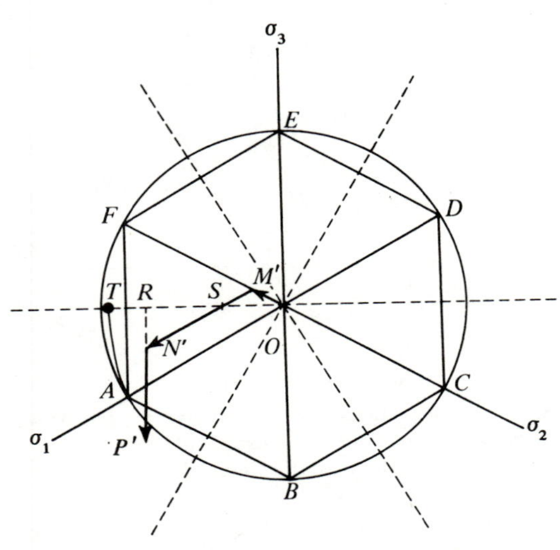

Fig. 6.16 Deviatoric plane, Mises circle and Tresca hexagon.

If the stresses are given, the corresponding point P' can be determined by laying off the three vectors in the appropriate directions in the deviatoric plane. The inverse problem of determining the stresses for a given point P' does not have a unique solution because of the undetermined hydrostatic stress. But the deviatoric σ_i' can be determined by the requirement that the three vectors add vectorially to $\mathbf{OP'}$, while the three algebraic values σ_i' add to zero. This can be accomplished graphically as follows. Draw the three bisectors of the angles between the projected axes (shown as dashed lines in Fig. 6.16). From P' drop a perpendicular $P'R$ to the next dashed line located clockwise from OP' ($P'R$ will be parallel to a projected axis). Divide $P'R$ at N' so that $|PN'| = 2|N'R|$ and draw $N'M'$ parallel to another axis to intersect at M' the third axis (σ_2-axis in the case illustrated). The algebraic sum will then be zero. (Note that in the case illustrated, $\sigma_2' = -|OM'|\sqrt{\frac{3}{2}}$, $\sigma_3' = -|N'P'|\sqrt{\frac{3}{2}}$, while $\sigma_1' = +|M'N'|\sqrt{\frac{3}{2}}$. Also $|M'N'| = |P'N'| + |OM'|$, because $|P'N'| = |N'S|$ and $|SM'| = |OM'|$.)

We have seen that when Assumptions 2, 3, and 4 are satisfied the yield condition can be represented by a two-dimensional plot in terms of the principal stresses. The yield condition is then completely independent of the orientation of the principal axes relative to the specimen. But when we come to the complete stress-strain relations it will be necessary to know the orientation of the axes relative to the specimen in order to compute the specimen deformation.

Other possible yield conditions. Other conditions that have been proposed besides those of Mises and Tresca include one of the form

$$f \equiv J_2'\left[1 - c\frac{(J_3')^2}{(J_2')^3}\right] - k^2 = 0. \tag{6.6.18}$$

By judicious choice of the parameters c and k, this gives a somewhat better fit not only to yield data, but also to some postyield deformation data. See, for example, Prager (1945) and Drucker (1949 and 1956). Equation (18) satisfies all of the assumptions listed above.

We shall see in Part 2 that the yield surface must be convex, i. e., so that through any point on the surface it is possible to pass a plane such that the entire interior of the surface lies to one side of the plane. The Mises and Tresca yield conditions satisfy this requirement either in principal stress space or in the nine-dimensional stress space. The convexity of the yield surface is a consequence of assuming that the material is ***stable*** in the sense of Drucker's definition given in Part 2.

Fung (1965), p. 143, lists several examples of yield conditions that include a Bauschinger effect but preserve isotropy, as well as a yield condition that reveals and preserves ***initial anisotropy***. Hill (1948) suggested the following generalization of the Mises quadratic yield condition for states of anisotropy possessing three mutually orthogonal planes of symmetry, when the reference Cartesian coordinate planes are chosen parallel to the three symmetry planes:

$$\begin{aligned} F(T_{22} - T_{33})^2 + G(T_{33} - T_{11})^2 + H(T_{11} - T_{22})^2 + 2LT_{23}^2 \\ + 2MT_{31}^2 + 2NT_{12}^2 - 1 = 0. \end{aligned} \tag{6.6.19}$$

When $F = G = H = 1/6k^2$ and $L = M = N = 1/2k^2$, Eq. (19) reduces to the Mises condition $J_2' - k^2 = 0$. This kind of anisotropy is representative of the behavior of sheet metal, where orthotropic symmetry is to be expected with symmetry axes perpendicular to the sheet, and parallel and perpendicular to the rolling direction used in manufacture of the sheet. See, for example, Hill (1950), Chap. 12.

Most of the discussion above is applicable to the initial yield of a statistically isotropic material. Formulations like Eq. (19) can include a known initial anisotropy, but the anisotropy and Bauschinger effects induced by an arbitrary plastic deformation are much more difficult to formulate satisfactorily. We will return to this question in Part 3 which considers the change in the yield condition brought about by plastic deformation. In Part 2, we treat plastic-potential theory, which shows how the next increment of deformation is related to the stress state at any given state of yield. It turns out that the yield function is also the plastic-potential function, and that the plastic strain-increment components are determined up to a scalar function multiple by the gradient of the yield function.

Part 2. Plastic-Potential Theory and Drucker's Definition of Stable Plastic Material

Generalized stress and deformation variables. [See, for example, Prager (1959), p. 13.] We may consider plastic continua of one, two, or three dimensions:

1-dim.: rods, beams, arches and rings;
2-dim.: membranes, plates and shells;
3-dim.: the three-dimensional solid.

We use three-dimensional language in the formulation of the mathematical theory, but the formulation applies to the one- or two-dimensional "engineering theories" by appropriate interpretation. For example, generalized "volume" means "length" in one dimension and "area" in two dimensions. And "energy density" means energy per unit generalized volume, e. g., per unit area in two dimensions.

In any case we define at each point

$$\text{Generalized stresses:} \quad Q_i \qquad i = 1, \ldots, n$$

and conjugate to them

$$\text{Generalized rates of deformation:} \quad \dot{q}_i \qquad i = 1, \ldots, n$$

such that, per unit generalized volume, we have

$$\left.\begin{aligned} &\text{Stress Power} \equiv D = Q_i \dot{q}_i \qquad (\text{sum} \quad i = 1 \text{ to } n) \\ &\text{or possibly} \quad D = C Q_i \dot{q}_i, \end{aligned}\right\} \tag{6.6.20}$$

where C is a common scalar factor, depending on how the conjugate generalized variables are defined. In elementary beam-bending theory, there is only one generalized stress, the bending moment, and the conjugate variable is the curvature rate. In elementary plate theory, there are several bending and twisting moments with conjugate curvature and twist rates. More refined theories include transverse shear forces and membrane forces and their conjugate variables. Here we consider primarily the three-dimensional formulation where the nine Q_i are usually the nine T_{ij} in some order and the $\dot{q}_i$ are the $D_{ij} = d\epsilon_{ij}/dt$, where $d\epsilon_{ij}$ is a natural-strain increment. We have already seen in Part 1, however, that because $\mathbf{T}$ is symmetric it may be convenient to list only six stress variables as in Eq. (3). When this is done ***the six conjugate deformation variables must be***

$$\dot{q}_1 = D_{11} \qquad \dot{q}_2 = D_{22} \qquad \dot{q}_3 = D_{33}$$
$$\dot{q}_4 = 2D_{23} \qquad \dot{q}_5 = 2D_{31} \qquad \dot{q}_6 = 2D_{12} \tag{6.6.21}$$

in order that summation of $Q_i\dot{q}_i$ for $i = 1$ to 6 will give the stress power D of Eq. (20). Thus each shear component $\dot{q}_i$ is twice a tensor component.

Plastic-potential function. This is the name given to a function $f(Q_1, \ldots, Q_n)$ of the stresses, such the plastic part $\dot{\mathbf{q}}^p$ of the rate of deformation has components proportional to the components of grad f,

$$\dot{q}_i^p = \dot{\lambda}\frac{\partial f}{\partial Q_i} \tag{6.6.22}$$

where $\dot{\lambda}$ is a scalar function (not a constant). Is there any reason to suppose that such a function exists? If it does exist, then by Eq. (22) it defines the stress versus plastic deformation rate relations except for the as yet unspecified multiplicative scalar function $\dot{\lambda}$. As a consequence of the assumption that the material is a ***stable plastic material*** according to Drucker's definition we shall see that

(1) a plastic potential function exists, and
(2) it is identical to the yield function, which must represent a convex surface in stress space.

Plastic-potential theory had already been used before Drucker's definition was introduced, and the plastic-potential function had often been assumed to be the same as the yield function because that assumption led to some useful variational theorems and limit analysis theorems. [See, for example, Hill (1950), Chap. 3.] Before considering Drucker's definition and its consequences (1) and (2) above, we note how the plastic-potential theory is applied.

Equation (22) may be written in terms of natural-strain increments dq_i^p and stresses Q_i as follows, eliminating the unspecified scalar function $d\lambda$:

$$\frac{dq_1^p}{\partial f/\partial Q_1} = \frac{dq_2^p}{\partial f/\partial Q_2} = \frac{dq_3^p}{\partial f/\partial Q_3} = \frac{dq_4^p}{\partial f/\partial Q_4} = \frac{dq_5^p}{\partial f/\partial Q_5} = \frac{dq_6^p}{\partial f/\partial Q_6} \tag{6.6.23}$$

with the understanding that if any denominator equals zero, then so does the numerator, and that fraction is omitted from Eq. (23). If $f = J_2' - k^2$ as in the Mises yield condition, then this leads to the Levy-Mises Eq. (6.5.15), since $\partial f/\partial Q_i = Q_i'$. [Note that if J_2' is considered a function of the six variables Q_i, then, for example, $\partial f/\partial Q_6 = 2Q_6$, whence $dq_6^p = d\lambda(2Q_6)$ or $2d\epsilon_{12}^p = d\lambda(2T_{12})$. On the other hand, if J_2' is considered a function of the nine variables T_{ij}, then, for example, T_{12}^2 should be replaced by $\frac{1}{2}(T_{12}^2 + T_{21}^2)$ in the expression for J_2', and the conjugate deformation variable to T_{12} will then be $d\epsilon_{12}$ instead of $2d\epsilon_{12}$.] This shows that the Mises yield condition is the appropriate one to use with the Levy-Mises equations.

A geometric interpretation of the plastic-potential theory of Eq. (22) in terms of the yield surface in stress space is instructive. If the stress is a bound vector $\mathbf{Q}$ with components Q_i drawn from the origin, and the plastic rate of deformation $\dot{\mathbf{q}}^p$ is a free vector in the same space [i. e., with components $\dot{q}_i^p$ parallel to the axes of Q_i], then Eq. (22) says that the plastic rate-of-deformation vector $\dot{\mathbf{q}}^p$ is parallel to the outer normal grad f at the stress point $\mathbf{Q}$ on the yield surface. Thus the plastic rate-of-deformation vector direction is determined by the given stress and the plastic-potential theory. Still to be determined is the magnitude of the rate of deformation, since the multiplier $\dot{\lambda}$ is not determined by the plastic-potential theory. For a perfectly plastic material, in fact, the rate-of-deformation magnitude is indeterminate in plasticity theory. This is true even in uniaxial stress; in the first two curves of Fig. 6.11, extension proceeds at constant stress Y at a rate undetermined by the stress-strain curve. This does not mean at an infinite rate; in a testing machine the rate would be governed by the rate at which energy was supplied to the test piece. In a work-hardening material the rate of deformation is still governed by the power input, but a work-hardening theory also furnishes a relationship between the rate of deformation and the rate of hardening (slope of the curve in the uniaxial stress case). Some proposals for this relationship in the case of isotropic hardening will be discussed in Part 3.

In the principal-stress space of Fig. 6.15, the outer normal to the yield surface is parallel to the deviatoric plane if hydrostatic stress does not influence yield [Assumption 2 of Part 1]; hence, plastic-potential theory implies that the plastic deformation is deviatoric. That is, there is no plastic volume change, a conclusion closely verified in metals deformed at hydrostatic pressures of the same order of magnitude as the deviatoric stresses. For the Mises theory, the outer normal to the yield surface is the continuation of the radius of the Mises circle in the deviatoric plane of Fig. 6.16, while for the Tresca yield surface the outer normal is not in general parallel to the deviatoric stress vector to the point on the surface. This shows that it is not appropriate to use the Levy-Mises equations with the Tresca yield condition as has sometimes been done. Note that with any yield condition like the Tresca condition the plastic-potential flow rule of Eq. (22) does not uniquely determine the direction of $\dot{\mathbf{q}}^p$ at a corner such as A in Fig. 6.16; the vector is assumed to fall between the outer normals to the faces AB and AF adjacent to the corner. Inversely, a rate-of-deformation vector perpendicular to one of the plane faces does not uniquely determine the deviatoric stress point on the hexagon side in Fig. 6.16. These instances of lack of uniqueness sometimes simplify the finding of an admissible field in limit analysis, but they make for an unsatisfactory fundamental theory.

Drucker's definition of a stable plastic material. In terms of the plastic work done by the incremental stress during an increment of plastic deformation, Drucker's definition of a stable plastic material is a generalization of the following stability considerations in a uniaxial stress test. Figure 6.17(a) shows a curve of true tensile stress Q versus natural strain q for a work-hardening material such that for any plastic-deformation increment the change dQ in the stress is positive. Figure 6.17(b) shows the same stresses plotted against plastic strain q^p [at each ordinate subtract the elastic strain $q^e = Q/E$ from the corresponding abscissa of Fig. 6.17(a)]. The rising curve denotes a stable work-hardening material, while in Fig. 6.17(c), from point Q_1 on, the curve falls so that further plastic strain occurs at smaller stresses, a clearly unstable situation. Note that this is a plot of true stress versus true strain; the falling curve represents a genuine material instability and not a geometric instability as in the curve of nominal stress versus conventional strain in Fig. 6.10. A perfectly plastic material exhibits neutral (or indifferent) stability, since the true stress neither increases nor decreases with increasing plastic deformation. How shall we generalize the concept of material stability to the case of combined stress where some stress components could increase while others decrease?

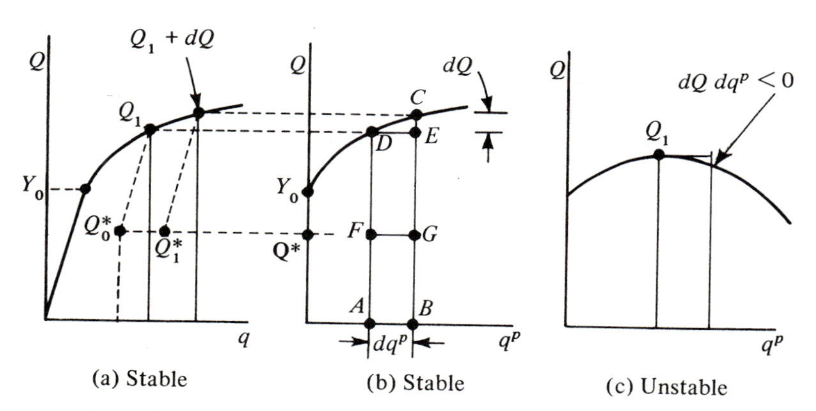

Fig. 6.17 Stable and unstable plastic materials (uniaxial stress).

Drucker (1951) formulated the material stability definition in terms of the work done by the stress increment on the plastic-deformation increment. For the increment dq^p shown in Fig. 6.17(b) the total plastic work is $\int Q\,dq^p$, equal to the area of the strip $ABCD$. But the major part of this area, namely $ABED$, is the work done by the stress Q_1 at which the increment began; the little area DEC is the work done by the stress increment dQ. In the stable case, this area is positive, having the same sign as $dQ\,dq^p > 0$, while in the unstable case $dQ\,dq^p < 0$ so that the total plastic work is less than would have been done by constant stress Q_1. For a perfectly plastic material, we would have the neutral condition $dQ\,dq^p = 0$. Note also that for the material in condition Q_0^* at the beginning of the increment, the current yield stress is Q_1; the material would be elastic for any stress Q^* in the interior of the current yield interval $-Q_1 \leq Q^* \leq Q_1$ (assuming no Bauschinger effect) and evidently

$$(Q - Q^*)dq^p > 0 \tag{6.6.24a}$$

for every Q^* in the interval

$$-Q_1 \leq Q^* \leq Q_1$$

during the increment in the stable case of Fig. 6.17(b). For $Q^* = Q_1$, this reduces to the condition already discussed. For stability,

$$dQ\,dq^p > 0. \tag{6.6.24b}$$

In the unstable case, however, it would be possible to choose an ***interior point*** Q^* sufficiently close to Q_1 so that $(Q - Q^*)dq^p < 0$ at some time during the increment. These observations motivate the following definition for combined stress.

Drucker (1951) considers a body in an initial state of equilibrium with stress state $\mathbf{Q}^*$. Suppose that to this body an external agency slowly applies a set of self-equilibrating forces and then slowly removes them. This external agency is to be understood as quite distinct from the agency causing the initially existing state of stress. The original configuration may or may not be restored after the cycle, but the stress is returned to the initial equilibrium value. Although the results are assumed to apply locally to a nonuniform stress state, it is easier to think of a whole body under a state of uniform stress being carried through the cycle of loading by an external agency.

> ***A stable work-hardening material*** is defined to be one such that:
> 1. the plastic work done by the external agency during the application of the additional stresses is positive, and
> 2. the net total work performed by the external agency during the cycle of adding and removing stresses is nonnegative.

This requirement leads to the following inequalities on the stress vector $\mathbf{Q}$ and rate-of-deformation vector $\dot{\mathbf{q}}^p$ (represented as vectors in nine-dimensional space).

$$\dot{\mathbf{Q}} \cdot \dot{\mathbf{q}}^p > 0 \quad \text{or} \quad \dot{Q}_i \dot{q}_i^p > 0 \tag{6.6.25a}$$

and

$$(\mathbf{Q} - \mathbf{Q}^*) \cdot \dot{\mathbf{q}}^p \geq 0 \quad \text{or} \quad (Q_i - Q_i^*)\dot{q}_i^p \geq 0 \tag{6.6.25b}$$

generalizing Eqs. (24).

The following demonstration that the definition given by Drucker implies the inequalities (25b) is adapted from Hodge's formulation in Goodier and Hodge (1958), p. 59. Figure 6.18 shows a schematic representation in stress space of the cycle of loading and unloading imposed by the external agency.

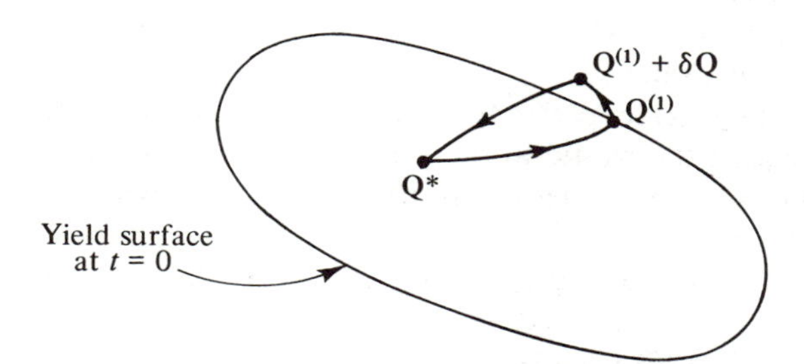

Fig. 6.18 Schematic representation of loading and unloading path in stress space.

We consider a case where at time $t = 0$ the stress is at $\mathbf{Q}^*$. Let the external agency change it to $\mathbf{Q}^{(1)}$ on the yield surface at time t_1, then to neighboring point $\mathbf{Q}^{(1)} + \delta\mathbf{Q}$ outside or on the initial yield surface at time $t_1 + \delta t$, and then follow an unloading path back to $\mathbf{Q}^*$ at t^*. The net work done by the external agency during the cycle is $\delta W_{\text{ext}} = \delta W_T - \delta W_0$, where δW_T is the total work during the cycle and δW_0 is the work that would be done during the same deformation by the initial stress $\mathbf{Q}^*$ if held constant. Let

$$\dot{\mathbf{q}} = \dot{\mathbf{q}}^e + \dot{\mathbf{q}}^p,$$

where $\dot{\mathbf{q}}^e$ is the elastic strain rate. Then

$$\begin{aligned}\delta W_T &= \int_0^{t_1} Q_i \dot{q}_i^e \, dt + \int_{t_1}^{t_1+\delta t} Q_i(\dot{q}_i^e + \dot{q}_i^p)\, dt + \int_{t_1+\delta t}^{t^*} Q_i \dot{q}_i^e \, dt \\ &= \oint Q_i \dot{q}_i^e \, dt + \int_{t_1}^{t_1+\delta t} Q_i \dot{q}_i^p \, dt \\ &= \int_{t_1}^{t_1+\delta t} Q_i \dot{q}_i^p \, dt\end{aligned}$$

since the net elastic work $\oint Q_i \dot{q}_i^e \, dt$ during the cycle is zero. Similarly

$$\delta W_0 = \int_{t_1}^{t\ +\delta t} Q_i^* \dot{q}_i^p \, dt,$$

whence

$$\delta W_{\text{ext}} = \delta W_T - \delta W_0 = \int_{t_1}^{t_1+\delta t} (Q_i - Q_i^*)\dot{q}_i^p \, dt. \tag{6.6.26}$$

The second part of Drucker's definition requires δW_{ext} to be nonnegative for any plastic deformation during an arbitrarily short δt following a loading to an arbitrary point $\mathbf{Q}^{(1)}$ on the yield surface. It follows that the integrand in Eq. (26) must be nonnegative, thus establishing Eq. (25b).

Equation (25a) follows from the first part of Drucker's definition; for an initial $\mathbf{Q}^* \equiv \mathbf{Q}^{(1)}$, $\mathbf{dQ} = \mathbf{Q} - \mathbf{Q}^*$, and the requirement of $(\mathbf{Q} - \mathbf{Q}^*)\cdot d\mathbf{q}^p > 0$ then implies $\dot{\mathbf{Q}}\cdot\dot{\mathbf{q}}^p > 0$.

For a neutrally stable, perfectly plastic material the Drucker definition is modified to require the plastic work of the external agency to be merely nonnegative during loading. Indeed it turns out to be zero when plastic-potential theory is used, since plastic-potential theory requires $\dot{\mathbf{q}}^p$ parallel to the outer normal, while $\dot{\mathbf{Q}}$ must be tangent to the yield surface during plastic deformation because $\mathbf{Q}$ can never move outside of the initial surface if perfect plasticity is assumed. Thus Eq. (25) is replaced by

$$\dot{\mathbf{Q}}\cdot\dot{\mathbf{q}}^p = 0 \quad \text{or} \quad \dot{Q}_i\dot{q}_i^p = 0 \tag{6.6.27a}$$

$$(\mathbf{Q} - \mathbf{Q}^*)\cdot\dot{\mathbf{q}}^p \geq 0 \quad \text{or} \quad (Q_i - Q_i^*)\dot{q}_i^p \geq 0 \tag{6.6.27b}$$

for perfectly plastic material.

Two important conclusions follow from Eqs. (25b) or (27b), namely (1) that in a stable plastic material the yield suraface must be convex, and (2) that the yield function is a plastic-potential function, as the following demonstration shows.

Figure 6.19 shows a schematic representation of the yield surface in stress space. If plastic deformation $\dot{\mathbf{q}}^p$ is occurring at stress $\mathbf{Q}$, draw the vector $\dot{\mathbf{q}}^p$ at the point $\mathbf{Q}$ on the yield surface, then Eqs. (25b) require that for every interior or boundary point $\mathbf{Q}^*$ the angle θ between the vectors $\mathbf{Q} - \mathbf{Q}^*$ and $\dot{\mathbf{q}}^p$ is not obtuse. Thus all interior points lie to one side of the hyperplane (dashed line in Fig. 6.19) normal to $\dot{\mathbf{q}}^p$, the side opposite to the one toward which $\dot{\mathbf{q}}^p$ points. Hence, through every point $\mathbf{Q}$ on the yield surface there is a plane such that all points interior to the surface lie on the same side of the plane. That is, the ***yield surface is convex.***

Now consider Fig. 6.19 again and assume that the convex yield surface is given but the vector $\dot{\mathbf{q}}^p$ is not. We ask what direction of the vector $\dot{\mathbf{q}}^p$ can occur under stress $\mathbf{Q}$ on the yield surface. By Eq. (25b), $\dot{\mathbf{q}}^p$ must make a nonobtuse angle with the vector $\mathbf{Q} - \mathbf{Q}^*$ from any interior or boundary point. At a smooth point on the surface the only direction for $\dot{\mathbf{q}}^p$ which satisfies this requirement for all $\mathbf{Q}^*$ is the direction of the outer normal. Hence $\dot{\mathbf{q}}^p$ is parallel to the gradient vector grad f, establishing the plastic-potential formulation of Eq. (22). At a corner on the yield surface, the outer normal is not well defined; then (25b) can be satisfied by any $\dot{\mathbf{q}}^p$ within or on the generalized cone formed by the limiting positions of the normals to the smooth surfaces intersecting at the corner. See the discussion in connection with Eq. (48) below, the generaliza-

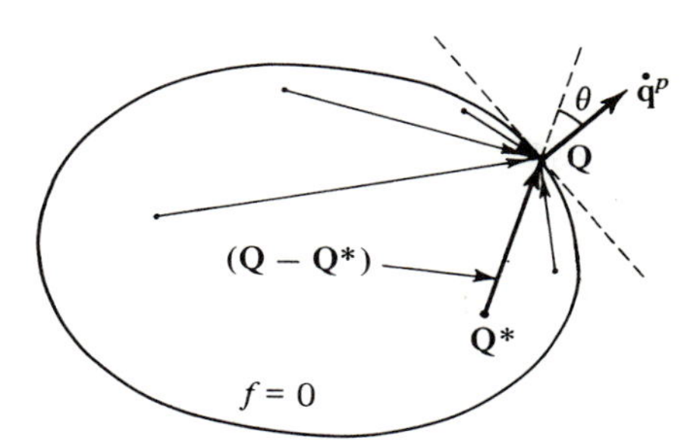

Fig. 6.19 Convex yield surface in stress space (schematic).

tion of the plastic potential Eq. (22) to the case of singular yield surfaces, given by Koiter (1953).

The ambiguity at a corner could be avoided by always using a smooth yield function: for example, the Mises yield condition. But, as we shall see in Part 3, there have been theoretical proposals that as the yield surface changes its size and shape during plastic deformation it will develop a corner at the active point. If this were true, the usefulness of plastic-potential theory would be greatly diminished. The question of just how the yield surface actually changes during plastic deformation is still unsettled, i. e., the theory of work-hardening behavior is yet to be established. The work-hardening theory must also determine the $\dot{\lambda}$ of the plastic potential Eq. (22).

Part 3. Work-Hardening Assumptions

The question of just how the yield function changes during plastic deformation is the most important open question in the mathematical theory of plasticity. The discussion here will be divided into four parts; the first three parts present three simple assumptions that have been made: (1) perfect plasticity, (2) isotropic hardening, and (3) kinematic hardening, while in (4) other possibilities, including slip theory, will be mentioned briefly. The implications of each of the three simple assumptions have been worked out at least for a few loading situations, but only the first two have been applied extensively. Consideration of slip theory is beyond the scope of this book, but some of its results will be mentioned.

(1) Perfect plasticity. The simplest possible assumption about how the yield surface changes during plastic deformation is that it does not change at all. This assumption, using the Mises yield condition, and often neglecting elastic strains, leads to the classical mathematical theory of perfectly plastic solids. (Some applications have used the Tresca yield condition with the perfectly plastic assumption.) For the general yield condition of Eq. (1), the form of the scalar multiplier $\dot{\lambda}$ of the plastic-potential theory Eq. (22) is determined by

$$\dot{q}_i^p \dot{q}_i^p = (\dot{\lambda})^2 \frac{\partial f}{\partial Q_i} \frac{\partial f}{\partial Q_i}, \quad \text{whence}$$

$$\dot{\lambda} = \left(\dot{q}_j^p \dot{q}_j^p \Big/ \frac{\partial f}{\partial Q_k} \frac{\partial f}{\partial Q_k}\right)^{1/2} = \sqrt{\mathrm{II}_{D^p}} \left(\frac{1}{2} \frac{\partial f}{\partial T_{km}} \frac{\partial f}{\partial T_{km}}\right)^{-1/2}. \tag{6.6.28}$$

When this is substituted back into Eq. (22) we see that the rates $\dot{q}_i^p$ are not completely determined by the stresses Q_i, because Eq. (22) continues to be satisfied if all the components $\dot{q}_m^p$ are multiplied by the same arbitrary constant. This typical perfectly plastic behavior has already been observed in the Levy-Mises theory of Sec. 6.5. The overall rate of deformation is limited by the rate at which energy can be supplied to the volume element, but the plasticity theory itself is rate-independent.

(2) Isotropic hardening. The next simplest hardening assumption—that the yield surface maintains its shape, while its size increase is controlled by a single parameter depending on the plastic deformation—is called ***isotropic hardening***. It remains only to specify the size-determining parameter and its dependence on the deformation. Two different schemes have been used for this, which are different in general but reduce to the same thing when used appropriately with the Mises yield condition. The first scheme discussed below is the assumption of a ***universal plastic stress-strain*** curve relating two scalar quantities—the ***effective stress*** $\bar{\sigma}$ (measuring the size of the yield surface) and the integral of the ***effective plastic-strain increment*** $\overline{d\epsilon^p}$. The second scheme proposes that $\bar{\sigma}$ is a single-valued function of $W_p = \int T_{ij} d\epsilon_{ij}^p$ the total plastic work. When used with the Mises yield condition, the appropriate ***effective stress*** $\bar{\sigma}$ is

$$\bar{\sigma} = \sqrt{3J_2'} = \sqrt{\tfrac{3}{2} T_{ij}' T_{ij}'} = \{\tfrac{1}{2}[(T_{11} - T_{22})^2 + (T_{22} - T_{33})^2 + (T_{33} - T_{11})^2] + 3(T_{23}^2 + T_{31}^2 + T_{12}^2)\}^{1/2}, \tag{6.6.29a}$$

defined so that in uniaxial stress T_{11}, we have $\bar{\sigma} = |T_{11}|$, making it more convenient than the equivalent often-used ***octahedral shear stress*** $\tau_o = \sqrt{2}\,\bar{\sigma}/3 = \sqrt{2J_2'/3}$. The appropriate effective-strain increment $\overline{d\epsilon^p}$ to use with $\bar{\sigma}$ is

$$\begin{aligned}\overline{d\epsilon^p} &= \sqrt{\tfrac{4}{3}\mathrm{II}_{d\epsilon^p}} = \sqrt{\tfrac{2}{3} d\epsilon_{ij}^p d\epsilon_{ij}^p} \\ &= \{\tfrac{2}{9}[(d\epsilon_{11}^p - d\epsilon_{22}^p)^2 + (d\epsilon_{22}^p - d\epsilon_{33}^p)^2 + (d\epsilon_{33}^p - d\epsilon_{11}^p)^2] \\ &\qquad + \tfrac{4}{3}[(d\epsilon_{23}^p)^2 + (d\epsilon_{31}^p)^2 + (d\epsilon_{12}^p)^2]\}^{1/2},\end{aligned} \tag{6.6.29b}$$

where the numerical factor has been chosen so that in ***uniaxial stress***, e. g., T_{11} (where $d\epsilon_{22}^p = d\epsilon_{33}^p = -\frac{1}{2} d\epsilon_{11}^p$), we have $\overline{d\epsilon^p} = d\epsilon_{11}^p$. Hence the assumption of a ***universal stress-strain curve***

$$\bar{\sigma} = H\left[\int \overline{d\epsilon^p}\right] \tag{6.6.30}$$

permits the determination of the form of the function H of the single independent variable $\int \overline{d\epsilon^p}$ in principle by a single tensile test where $\bar{\sigma} = T_{11}$ and

$$\int \overline{d\epsilon^p} = \int d\epsilon_{11}^p = \int \left(\frac{dL}{L} - \frac{dT_{11}}{E}\right) = \log_e \frac{L}{L_0} - \frac{T_{11} - Y_0}{E}. \tag{6.6.31}$$

The assumption that one universal stress-strain curve of the form of Eq. (30) governs all possible combined-stress loadings of a given material is obviously a very strong one. There is evidence that it works pretty well for combined loadings in which the stress-component ratios are held constant or nearly constant during loading, i. e., under conditions of ***radial loading*** or ***proportional loading*** where

$$\frac{T_{11}}{T_{11}^0} = \frac{T_{22}}{T_{22}^0} = \cdots = \frac{T_{12}}{T_{12}^0}, \tag{6.6.32}$$

(the T_{ij}^0 are the values of T_{ij} at the initial yield). See, for example, Lubahn and Felgar (1961), Chap. 8, for some examples of experimental data obtained under radial loading conditions and for references to the original investigations. Lubahn and Felgar also consider some other possible definitions of the effective stress, including maximum shear stress and a definition used by Drucker (1949),

$$\tau_{\text{eq}} = \tau_o\left[1 - c\left(\frac{J_3'^2}{J_2'^3}\right)^{1/2}\right].$$

When the loading deviates significantly from proportional loading, the universal stress-strain curve is less successful, especially when the deviation is so great that it amounts to a reversal of loading direction, where Bauschinger effects come into play as in Fig. 6.12. It is nevertheless remarkable that for a given material a single curve is able to correlate all radial loading paths so well regardless of the direction of the radial loading in stress space.

When the plastic-strain-increment component ratios are constant, as in ***radial straining*** or ***proportional straining***, say for small strains,

$$d\epsilon_{ij}^p = (d\alpha)(\epsilon_{ij}^p)^1 \quad \text{and} \quad \epsilon_{ij}^p = \alpha(\epsilon_{ij}^p)^1, \tag{6.6.33}$$

where the $(\epsilon_{ij}^p)^1$ are constants, then the effective plastic-strain increment $\overline{d\epsilon^p}$ can be integrated explicitly to give $\int \overline{d\epsilon^p}$ equal to

$$\bar{\epsilon}^p = \sqrt{\tfrac{4}{3}\mathrm{II}_{\epsilon^p}} = \sqrt{\tfrac{2}{3}\epsilon_{ij}^p\epsilon_{ij}^p} \qquad \text{for proportional straining (small strain } \epsilon_{ij}^p), \tag{6.6.34}$$

but when the straining is not proportional the $\bar{\epsilon}^p$ of Eq. (34) is not in general equal to the quantity $\int \overline{d\epsilon^p}$ needed for use in Eq. (30).

When the universal stress-strain curve of Eq. (30) is valid during a combined-stress plastic deformation, the multiplier $d\lambda$ of the plastic-potential equation [Eq. (22) with $f = J_2' - k^2$]

$$d\epsilon_{ij}^p = d\lambda T_{ij}'$$

can be determined as follows:

$$d\epsilon_{ij}^p d\epsilon_{ij}^p = (d\lambda)^2 T_{ij}' T_{ij}'$$

or

$$\tfrac{3}{2}(\overline{d\epsilon^p})^2 = (d\lambda)^2 \tfrac{2}{3}\bar{\sigma}^2$$

whence

$$d\lambda = \frac{3}{2}\frac{\overline{d\epsilon^p}}{\bar{\sigma}} = \frac{3}{2}\frac{d\bar{\sigma}}{\bar{\sigma}H'}, \qquad \text{where} \quad H' = d\bar{\sigma}/\overline{d\epsilon^p} \tag{6.6.35}$$

is the slope of the universal curve of Eq. (30) at the current value of $\bar{\sigma}$. Then the plastic-strain increment is given by

$$\left.\begin{aligned} d\epsilon_{ij}^p &= \frac{3}{2}\frac{\overline{d\epsilon^p}}{\bar{\sigma}} T_{ij}' \quad \text{or} \\ (\text{for } d\bar{\sigma} \geq 0) \quad d\epsilon_{ij}^p &= \frac{3}{2}\frac{d\bar{\sigma}}{\bar{\sigma}H'} T_{ij}' = \frac{3}{2\bar{\sigma}H'}\frac{\partial\bar{\sigma}}{\partial T_{pq}} dT_{pq} T_{ij}'. \end{aligned}\right\} \tag{6.6.36}$$

The last form exhibits explicitly the linear dependence of $d\epsilon_{ij}^p$ on the stress increments dT_{pq} during loading (linear in the sense that the coefficients do not contain the stress increments, although they do, of course, depend on the current stresses T_{pq}).

When used with the Mises yield condition, the alternative isotropic hardening assumption, that the ***size of the yield surface depends only on the total plastic work***, may be expressed as

$$\bar{\sigma} = F(W_p), \tag{6.6.37}$$

where

$$W_p = \int T_{ij} d\epsilon_{ij}^p. \tag{6.6.38}$$

This is equivalent to Eq. (30) when used with the Mises yield condition, since then W_p is a single-valued function of the abscissa $\int \overline{d\epsilon^p}$ of the universal curve, as the following argument shows. Let $\boldsymbol{\sigma}'$ denote the vector in nine-dimensional stress space with components $T'_{ij}\sqrt{\frac{3}{2}}$ and magnitude $\bar{\sigma}$, while $\boldsymbol{d\epsilon}^p$ denotes the vector with components $d\epsilon^p_{ij}\sqrt{\frac{2}{3}}$ and magnitude $\overline{d\epsilon^p}$. Then $dW_p = T_{ij}d\epsilon^p_{ij} = T'_{ij}d\epsilon^p_{ij} = \boldsymbol{\sigma}' \cdot \boldsymbol{d\epsilon}^p$. But the plastic-potential equation $d\epsilon^p_{ij} = d\lambda T'_{ij}$ then implies $\boldsymbol{d\epsilon}^p = (2d\lambda/3)\boldsymbol{\sigma}'$, i. e., $\boldsymbol{d\epsilon}^p$ is parallel to $\boldsymbol{\sigma}'$ so that $\boldsymbol{\sigma}' \cdot \boldsymbol{d\epsilon}^p = \bar{\sigma}\overline{d\epsilon^p}$. Then $dW_p = \bar{\sigma}\overline{d\epsilon^p}$ and W_p, equal to the area under the universal curve of $\bar{\sigma}$ versus $\int d\bar{\epsilon}^p$, is a single-valued function of $\int \overline{d\epsilon^p}$.

Under a generalization of the assumption of Eq. (37) the general yield condition $f(\mathbf{T}) = 0$ can be reformulated as

$$f(\mathbf{T}) - F(W_p) = 0 \tag{6.6.39}$$

by redefining f. Then the plastic-potential equation

$$d\epsilon^p_{ij} = d\lambda \frac{\partial f}{\partial T_{ij}} \tag{6.6.40}$$

leads to

$$dW_p = T_{ij}d\epsilon^p_{ij} = d\lambda\, T_{ij}\frac{\partial f}{\partial T_{ij}}$$

whence

$$d\lambda = \frac{dW_p}{T_{mn}\,\partial f/\partial T_{mn}}$$

and Eq. (40) becomes

$$d\epsilon^p_{ij} = \frac{dW_p}{T_{mn}\,\partial f/\partial T_{mn}}\frac{\partial f}{\partial T_{ij}}. \tag{6.6.41}$$

Equation (39) implies that in any loading increment, in order that Eq. (39) remain satisfied after the increment,

$$\frac{\partial f}{\partial T_{pq}}dT_{pq} = F'(W_p)dW_p, \tag{6.6.42}$$

whence Eq. (41) can be written as

$$d\epsilon^p_{ij} = \frac{\partial f}{\partial T_{ij}}\frac{\partial f}{\partial T_{pq}}dT_{pq} \Big/ F'(W_p)T_{mn}\,\frac{\partial f}{\partial T_{mn}}, \tag{6.6.43}$$

again exhibiting explicitly the linear dependence on dT_{pq} as in the last form of Eq. (36).

Since in isotropic hardening the yield surface maintains its shape, any radial loading from a yield state T^0_{ij} to αT^0_{ij} (for $\alpha \geq 1$) must, according to Eq. (39), require plastic work such that $f(\alpha T^0_{ij}) = \phi(\alpha) f(T^0_{ij})$, where $\phi(\alpha)$ is some function of α alone, not depending on what radial path (defined by T^0_{ij}) is used. But this implies that $f(\alpha T^0_{ij}) = \phi(\alpha) F(W^0_p)$ or $F(W_p) = \phi(\alpha) F(W^0_p)$, i. e., the amount of plastic work in loading from one yield surface to another successive one is independent of the particular radial loading path and depends only on α. In particular, if $f(T_{ij})$ is a homogeneous function of degree n in the stress components, i. e., such that

$$f(\alpha T_{ij}) \equiv \alpha^n f(T_{ij}),$$

we have $\phi(\alpha) = \alpha^n$ and $F(W_p) = \alpha^n F(W^0_p)$ for radial loading from the point T^0_{ij} where the accumulated work was W^0_p to $T_{ij} = \alpha T^0_{ij}$ where the accumulated work was W_p. This places some restrictions on the possible form of $F(W_p)$, which are difficult to make explicit. But assuming that they are satisfied so that f is homogeneous of degree n, Euler's theorem on homogeneous functions then requires that

$$T_{mn} \frac{\partial f}{\partial T_{mn}} \equiv n f(\mathbf{T})$$

whence Eq. (43) simplifies to

$$d\epsilon^p_{ij} = \frac{1}{n f(\mathbf{T}) F'(W_p)} \frac{\partial f}{\partial T_{ij}} \frac{\partial f}{\partial T_{pq}} dT_{pq}. \tag{6.6.44}$$

[Hill (1950) Eq. (27) is equivalent to this when his plastic-potential function g and yield function f are the same function.]

Explicit dependence of the yield function on the plastic deformation. Isotropic hardening has been illustrated by the two alternative simple assumptions: Eq. (30), $\bar{\sigma} = H[\int \overline{d\epsilon^p}]$, or Eq. (37), $\bar{\sigma} = F(W_p)$, or more generally Eq. (39), $f(\mathbf{T}) = F(W_p)$, with the additional stipulation that f is isotropic. In a third formulation Prager has formally written the yield equation in the form

$$f(T_{ij}, \epsilon^p_{ij}, K) = 0, \tag{6.6.45a}$$

where in addition to a work-hardening parameter K, which itself depends on the history of the deformation, the current plastic-deformation components appear explicitly in the yield function. The ***consistency equation***, Eq. (45a), must continue to hold identically during loading, whence the time-derivative

$$\dot{f} = 0 = \frac{\partial f}{\partial T_{ij}} \dot{T}_{ij} + \frac{\partial f}{\partial \epsilon^p_{ij}} \dot{\epsilon}^p_{ij} + \frac{\partial f}{\partial K} \frac{\partial K}{\partial \epsilon^p_{ij}} \dot{\epsilon}^p_{ij}. \tag{6.6.45b}$$

Substituting into this the plastic-potential equation, Eq. (22), $\dot{\epsilon}^p_{ij} = (\dot{\lambda}\partial f/\partial T_{ij})$, yields

$$\dot{\lambda} = -\frac{\partial f}{\partial T_{km}}\dot{T}_{km}\Big/\left[\frac{\partial f}{\partial \epsilon^p_{ij}} + \frac{\partial f}{\partial K}\frac{\partial K}{\partial \epsilon^p_{ij}}\right]\frac{\partial f}{\partial T_{ij}}$$

whence Eq. (22) takes the form

$$\dot{\epsilon}^p_{ij} = \hat{G}\frac{\partial f}{\partial T_{ij}}\frac{\partial f}{\partial T_{km}}\dot{T}_{km}, \tag{6.6.46}$$

where the scalar multiplier $\hat{G}$ is defined by

$$\hat{G} = -1\Big/\left[\frac{\partial f}{\partial \epsilon^p_{rs}} + \frac{\partial f}{\partial K}\frac{\partial K}{\partial \epsilon^p_{rs}}\right]\frac{\partial f}{\partial T_{rs}}.$$

Koiter (1953) generalized the plastic-potential theory by supposing the yield surface to be made up of several smooth surfaces, defined by several functions

$$f_1(T_{ij}), f_2(T_{ij}), \ldots, f_n(T_{ij}) \tag{6.6.47}$$

(for example, the Tresca yield surface composed of six surfaces). For elastic states all the functions are negative, while at yield at least one must equal zero. If more than one yield function is zero, the stress point is at some kind of a corner on the yield surface. If $f_h, \ldots, f_m$ are zero and all the others negative, the plastic-potential rule should be†

$$\dot{\epsilon}^p_{ij} = \dot{\lambda}_h\frac{\partial f_h}{\partial T_{ij}} + \cdots + \dot{\lambda}_m\frac{\partial f_m}{\partial T_{ij}}, \tag{6.6.48}$$

where $\dot{\lambda}_h, \ldots, \dot{\lambda}_m$ are arbitrary nonnegative factors (scalar functions, in general, not constants) of proportionality, subject usually to the incompressibility condition $\dot{\epsilon}^p_{kk} = 0$. Koiter also proposed that the individual yield surfaces harden independently, so that Eq. (46) is replaced by

$$\left.\begin{aligned} &\dot{\epsilon}^p_{ij} = \sum_{p=1}^{n} C_p\hat{G}_p\frac{\partial f_p}{\partial T_{ij}}\frac{\partial f_p}{\partial T_{km}}\dot{T}_{km}, \\ \text{where}\qquad &C_p = 0 \quad \text{if} \quad f_p < 0 \quad \text{or if} \quad \dot{T}_{km}(\partial f_p/\partial T_{km}) < 0 \\ &C_p = 1 \quad \text{if} \quad f_p = 0 \quad \text{and} \quad \dot{T}_{km}(\partial f_p/\partial T_{km}) \geq 0 \end{aligned}\right\} \tag{6.6.49}$$

and the $\hat{G}_p$ are n positive functions of stress, deformation, and deformation history, generalizing the $\hat{G}$ of Eqs. (46). No assumption of isotropy has been made

†In Eq. (48) the repeated indices h and m do not imply summation.

yet, and Eqs. (46) or (49) can be applied with nonisotropic hardening, e. g., with the kinematic-hardening assumption discussed below. Equations (46) or (49) established in general the linearity of the dependence of the $\dot{\epsilon}^p_{ij}$ on the $\dot{T}_{ij}$, as already observed in the particular isotropic-hardening cases of Eqs. (36) and (43). Some care is required in interpreting the meaning of the components ϵ^p_{ij} unless the displacement gradients are small enough to identify the ϵ_{ij} as small-strain components. For the general isotropic-hardening formulation

$$f(J_2', J_3') - K = 0,$$

Equation (46) takes the form

$$\dot{\epsilon}^p_{ij} = \hat{G}\left[\frac{\partial f}{\partial J_2'}T'_{ij} + \frac{\partial f}{\partial J_3'}M_{ij}\right]\frac{\partial f}{\partial T_{km}}\dot{T}_{km},$$

where

$$\left.\begin{aligned} M_{ij} &= \frac{\partial J_3'}{\partial T_{ij}} = T'_{ik}T'_{kj} - \frac{2}{3}J_2'\delta_{ij} \quad \text{or} \\ \mathbf{M} &= \mathbf{T}'\cdot\mathbf{T}' - \tfrac{2}{3}J_2'\mathbf{1}. \end{aligned}\right\} \tag{6.6.50}$$

($\mathbf{M}$ is the deviatoric part of the square of the deviatoric stress tensor $\mathbf{T}'$. In much of the literature, T'_{ij} is denoted by σ'_{ij} or s_{ij} while M_{ij} is denoted by t_{ij}.) The difficulty with this formulation, as with the more general Eq. (46), is the difficulty of specifying explicitly the dependence on ϵ^p_{ij}.

Attempts to construct a more complicated isotropic-hardening theory than that embodied in Eqs. (30) or (37) using the Mises yield condition do not appear worthwhile because of the inaccuracy inherent in the assumption of isotropic hardening itself. Instead what is needed is a nonisotropic hardening theory in which the yield surface in stress space can change its shape and/or its center as well as its size in order to represent anisotropic effects and Bauschinger effects caused by the deformation. The simplest proposal incorporating a Bauschinger effect is Prager's kinematic hardening.

(3) **Kinematic hardening.** Prager (1955, 1956) proposed a simple theory in which the yield surface does not change its size or its shape but merely translates in stress space in the direction of its normal (i. e., in the direction of the plastic-deformation-increment vector according to plastic-potential theory). Thus an initial yield surface $f(T_{ij}) = 0$ is changed into

$$f(T_{ij} - \alpha_{ij}) = 0, \tag{6.6.51}$$

where the α_{ij} are the nine coordinates of the new center of the yield surface and where f now has exactly the same dependence on $T_{ij} - \alpha_{ij}$ as the initial yield condition had on T_{ij}. All that remains is to specify how the α_{ij} depend on the deformation history. Almost all applications to date have been based on a ***linear-hardening assumption***

$$\dot{\alpha}_{ij} = c\dot{\epsilon}^p_{ij}, \tag{6.6.52}$$

where c is a constant [see, e. g., Shield and Ziegler (1958)]. Equations (22) and (45a) then give

$$\dot{\lambda} = \frac{\partial f}{\partial T_{ij}}\dot{T}_{ij} \Big/ c\frac{\partial f}{\partial T_{km}}\frac{\partial f}{\partial T_{km}}. \tag{6.6.53}$$

Some of the early applications of kinematic hardening to plane stress problems assumed that the center of the yield surface in a subspace of less than nine dimensions moved in the direction of the outer normal to the surface in the subspace. Hodge (1957) pointed out that the concept of kinematic hardening must be applied in the nine-dimensional stress space, which he called ***complete kinematic hardening*** in contrast to the ***direct hardening*** in the subspace. Shield and Ziegler (1958) studied the motions in various subspaces implied by complete kinematic hardening in nine-space. Since Prager's assumption that the center moved in the direction of the outer normal to the yield surface sometimes led to complications in applications, Ziegler (1959) proposed to substitute for Eq. (52) the equation

$$\dot{\alpha}_{ij} = \dot{\mu}(T_{ij} - \alpha_{ij}) \quad \text{for} \quad \dot{\mu} > 0, \tag{6.6.54}$$

which can be applied in the subspace as well as in nine-space. Naghdi (1960), Sec. 5, discusses this and other proposals and gives numerous references to original research. There is some evidence that for moderate plastic deformation the simple kinematic-hardening assumption comes closer than isotropic hardening to a representation of the changed yield surface needed for load reversals. See, for example, Jenkins (1965). But it is naturally too much to expect it to give an accurate account of the phenomenon. Hú, Markovitz, and Bartush (1966) in triaxial-stress experiments found that the hardening was neither isotropic nor kinematic. Hodge (1957) has discussed a possible combination of expansion and translation of the yield surface. What is really needed is a change of shape in addition, but how this would depend on the deformation history is an open question. Its answer may require a consideration of the individual crystallite deformations leading to preferred orientation development in the polycrystalline material as well as to residual stresses caused by differential deformation of neighboring crystallites.

(4) Other possibilities, slip theory. Batdorf and Budiansky (1949, 1954) proposed a mathematical theory of plasticity based on the concept of slip in crystal grains. Further developments and applications have been considered by Cicala (1950) and Lin (1958). For simple slip in a single crystal, without reversal, the plastic shear strain in a slip system is a function of the resolved shear stress in that slip system. Batdorf and Budiansky assumed that in a polycrystal the macroscopic shear strain in an element of volume will be a shear strain of the same orientation as that of the yielding grains, and depending somehow on the component of stress causing the deformation. But the relationship between the stress and strain was not taken from the single crystal measurements because of the effect of grain interactions. Instead, a power-series representation of the stress-strain relation with undetermined coefficients was assumed. After an averaging integration over all orientations, the coefficients were determined by fitting experimental polycrystalline tensile data. The theory, whose details are beyond the scope of this book, leads in the case of combined-stress deformation to the interpretation that the yield surface in stress space always contains all points contained in the yield surface corresponding to any previous stage of the loading and also any point within the cone of tangents from the current stress point back to any previous yield surface. This implies that for radial loading a corner should always develop on the yield surface at the loading point so that the plastic potential Eq. (22) does not determine uniquely the plastic rate-of-deformation vector direction. The concept of independent loading surfaces, in particular plane loading surfaces, developed by Koiter (1953) and Sanders (1954) incorporates this idea of a corner at the loading point, as do two papers by Kliushnikov (1959), without any derivation from slip theory. Kliushnikov gives an assumption which resolves the ambiguity about the plastic rate-of-deformation vector. These proposals have been discussed in the review paper by Naghdi (1960).

These theoretical predictions or assumptions of the existence of a corner have led to a number of experiments seeking to demonstrate the existence of a corner. This kind of experimentation is difficult, and the results so far are not conclusive despite considerable effort. The experimental technique as a rule makes it impossible to decide whether any corner is actually sharp or rounded. Most studies have been made for combined tension and torsion of a thin-walled tube. Results of Phillips and Gray (1961) suggest the existence of corners. Naghdi, Essenburg, and Koff (1958) and Ivey (1961) give results suggesting that a corner with a rounded vertex is developed at the torsion loading point in combined tension and torsion of a thin-walled tube. Careful measurements by Bertsch and Findley (1962) showed that combined tension and torsion produces rounded corners reminiscent of slip theory predictions, but that tension alone often leads to well-rounded curves with apparently no corner at all.

Budiansky (1959) and independently Kliushnikov (1959) have shown that a yield surface with a corner at the active point may imply a wider range of validity of the older ***total-strain theories*** of plasticity than had previously been thought possible. That is, the total-strain theories could still satisfy Drucker's definition of a stable plastic material for some loading paths departing from radial loading. The total-strain theories will be discussed briefly below. For an account of their use with a corner, see the papers by Budiansky or Kliushnikov.

Part 4. Total-Strain Theories (Deformation Theories)

In contrast to the ***incremental theories*** of plasticity discussed so far, in which the plastic-deformation increment $d\epsilon_{ij}^p$ (or the deformation rate $\dot{\epsilon}_{ij}^p$) is determined by the stress and stress increment (for a given state of the material), there is another type of theory called ***total-strain plasticity theory***. This gives the total strain ϵ_{ij} as a function of the current stress. (Total-strain theories are often called ***deformation theories*** in the literature, while the incremental theories are often called ***flow theories***.) A total-strain theory is evidently inappropriate to apply to a loading program where the stress path deviates so far from radial loading in stress space that it amounts to almost a reversal of the initial loading direction (with or without an intermediate elastic unloading). But for a single loading not deviating greatly from radial loading it may give almost the same results as an incremental theory, with a simpler mathematical boundary-value problem to solve for a nonuniform stress and deformation distribution. In the theory due to Hencky (1924), the Mises yield condition is assumed, and it is assumed that the small strain E_{ij} is the sum of an elastic part plus a plastic part given separately by the elastic and plastic stress-strain laws:

$$E_{ij} = E_{ij}^e + E_{ij}^p \tag{6.6.55a}$$

$$(E'_{ij})^e = \frac{1}{2G}T'_{ij} \qquad E_{kk}^e = \frac{1}{3K}T_{kk} = \frac{1-2\nu}{E}T_{kk} \tag{6.6.55b}$$

$$E_{ij}^p = \phi T'_{ij}, \tag{6.6.55c}$$

where ϕ is scalar function, positive during loading and zero during unloading. Nadai (1923) had used equations of this type in the special problem of torsion. Squaring and adding the components of Eq. (55c) gives

$$E_{ij}^p E_{ij}^p = \phi^2 T'_{ij} T'_{ij}$$

whence

$$\phi = \frac{3}{2}\frac{\bar{\epsilon}_p}{\bar{\sigma}}, \tag{6.6.56}$$

where

$$\bar{\epsilon}_p = \sqrt{\tfrac{4}{3}\mathrm{II}_{E^p}} \quad \text{and} \quad \bar{\sigma} = \sqrt{3J_2'} \tag{6.6.57}$$

reduce to $|E^p_{11}|$ and $|T_{11}|$ respectively in uniaxial stress T_{11}. This can be used with a universal stress-strain curve

$$\bar{\sigma} = H(\bar{\epsilon}^p) \tag{6.6.58}$$

replacing Eq. (30), the function H being determined in principle by a single tensile test.

In radial loading, the Reuss incremental plasticity equations may be integrated to give the form of the Hencky equations. For, if $T_{ij} = CT^0_{ij}$, where T^0_{ij} are the stresses at yield, the plastic-potential Eq.(22), with the Mises yield condition, becomes

$$dE^p_{ij} = CT'^0_{ij}d\lambda,$$

where C is an increasing scalar function, say $C = C(\lambda)$, during the deformation. Hence, for small strains, with $\bar{\lambda}$ as dummy integration variable,

$$E^p_{ij} = T'^0_{ij}\int C(\bar{\lambda})\,d\bar{\lambda} = T'_{ij}\left[\frac{1}{C(\lambda)}\int C(\bar{\lambda})\,d\bar{\lambda}\right].$$

[See Hill (1950), p. 47.] If we identify ϕ with $[1/C(\lambda)]\int C(\bar{\lambda})d\bar{\lambda}$ in the last equation, the equation takes the form of Eq. (55c). With the incremental theory, radial loading implies proportional straining as in Eq. (33), which has already been shown in Eq.(34) to lead to $\bar{\epsilon}^p = \int \overline{d\epsilon^p}$. When the loading is nonradial, the incremental and total-strain theories will in general lead to different results even when both use the Mises yield condition and a universal stress-strain curve based on the same tensile data. The radial loading condition is, however, a sufficient condition for integrability; the demonstration above does not prove that the boundary-value problem solutions for the two theories cannot agree closely even when the loading path deviates considerably from radial. Numerical solutions for torsion of a square bar, by Greenberg, Dorn, and Wetherell (1960) have shown close agreement in the stress distributions by the two theories even at points where the stress history was far from radial. Miller and Malvern (1967) have also shown in combined pure bending and torsion of a fully plastic bar examples where the two calculated stress distributions agree within the accuracy of the computation, even though the stress history is nonradial, provided the ratio of bending rate to twisting rate for the bar is held constant. When this ratio varied during the deformation, the examples calculated showed significant differences in the stress distributions calculated by the two theories (causing as much as 17% difference in the calculated twisting moment).

It should be apparent by this time that the theory of plasticity has not been definitively formulated. A great deal of research effort is still needed to build a work-hardening theory accurately accounting for most of the observed behavior. Most of the formulations for elastic, plastic behavior are not so constructed that they satisfy the principle of material frame-indifference to be discussed in Sec. 6.7. But the more serious shortcoming is that even in cases where there is no rotation the work-hardening assumptions do not account for all the observations.

EXERCISES

1. Show that the Tresca yield condition that the maximum shear stress equals $Y/2$ at yield implies the dashed yield curve of Fig. 6.14 for biaxial stress.

2. Show that the Mises yield condition represents a circular cylinder in principal stress space as was asserted in connection with Fig. 6.15. *Hint*: Show that the Mises yield surface has the same intersection with the deviatoric plane as has the sphere $\sigma_1^2 + \sigma_2^2 + \sigma_3^2 = 2k^2$.

3. What multiples of Y are the deviatoric principal stresses represented in Fig. 6.16 by points (a) A, B, C; (b) midpoints of arcs AB, BC, CD; (c) midpoints of straight segments AB, BC, CD?

4. Show that the Mises yield condition implies $\partial f/\partial T_{ij} = T'_{ij}$, whence Eq.(6.6.23) implies that the plastic strain increments are given by the Levy-Mises Eq. (6.5.15).

5. For the Mises yield surface in principal-stress space (see Fig.6.15), defined by Eq. (6.6.11a), find a set of normal direction numbers [numbers proportional to the components of the normal, and hence proportional to the principal plastic deformation rates $\dot{\epsilon}_1^p, \dot{\epsilon}_2^p, \dot{\epsilon}_3^p$, according to plastic-potential theory] for each of the following states of principal stress $\sigma_1, \sigma_2, \sigma_3$:
(a) Y, 0, 0; (b) 0, Y, Y; (c) $0.577Y$, $-0.577Y$, 0; and find the principal stresses if: (d) $\dot{\epsilon}_1^p = \dot{\epsilon}_3^p = -\frac{1}{2}\dot{\epsilon}_2^p$ while the mean stress is $-2Y/3$; (e) $\dot{\epsilon}_1^p = -0.8\,\dot{\epsilon}_3^p$, $\dot{\epsilon}_2^p = -0.2\,\dot{\epsilon}_3^p$ while mean stress is $0.2Y$.

6. For the Mises yield condition and plastic-potential flow law, determine numbers proportional to the plastic-strain increments for the following cases:
(a) uniaxial tension $\sigma_x = Y$, (b) balanced biaxial tension $\sigma_x = \sigma_y = Y$,
(c) biaxial stress $\sigma_x = -Y/\sqrt{3}$, $\sigma_y = +Y/\sqrt{3}$, (d) pure shear $\tau_{xy} = Y/\sqrt{3}$,
(e) combined tension-torsion of a thin-walled tube with axial stress $\sigma_x = 0.5Y$ and shear stress $\tau_{xy} = 0.5Y$.

7. Normals to the Tresca yield surface (hexagonal prism in principal-stress space, shown by its deviatoric plane section on p. 376) have direction numbers obtainable from the linear equation of the particular face in question. At the prism edge the normal is not uniquely defined. For example, at A the normal may be $[1, \lambda - 1, -\lambda]$ for any value of λ between 0 and 1. Using plastic-potential theory, complete the following table. (Stresses are multiples of Y.)

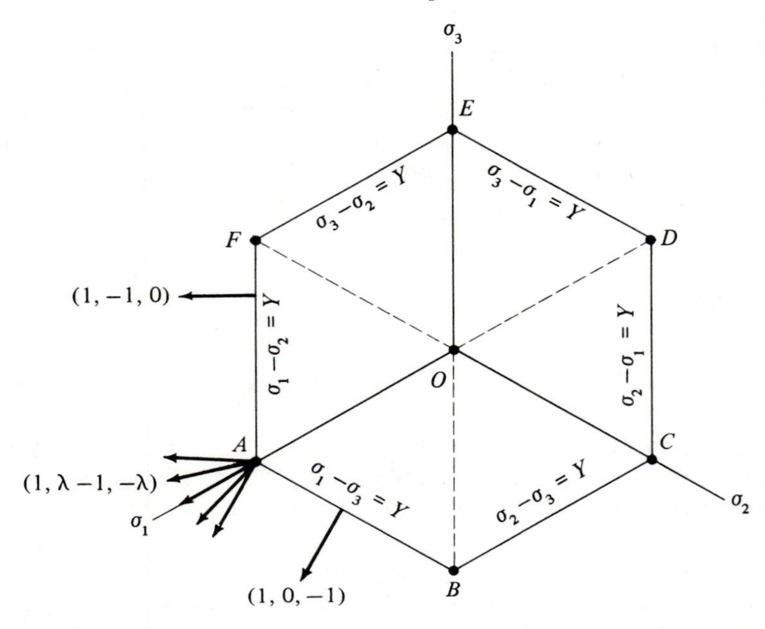

Principal Stresses	Mean Stress	Principal Deviator Stresses	Plane or Edge	Numbers proportional to $\dot{\epsilon}_1^p, \dot{\epsilon}_2^p, \dot{\epsilon}_3^p$
1, 0, 0	1/3	2/3, −1/3, −1/3	*A*	1, λ−1, −λ
0.5, −0.5, 0				
0, 0.2, −0.8				
, , −0.3	0.2			0, 1, −1
0.2, −0.8, 0.2				0.5, ,
	0.2			−0.8, −0.2, 1

8. Evaluate the generalized plastic strain increment $\overline{d\epsilon^p}$ of Eq.(6.6.29b) and also the plastic work increment dW_p [see Eq. (38) ff.] for the five loading cases of Ex. 6, if the testing-machine displacements are controlled so that $|d\epsilon_x^p|$ is 10^{-4} in./in. for each case except (d) where $d\epsilon_{xy}^p$ is 10^{-4}.

9. Assume that $\sigma = (10^5)(\epsilon^p)^{1/4}$ relates the true stress to the plastic part of the true strain in a tensile test. If Young's modulus E is 10^7 psi, express the total tensile strain (elastic plus plastic) in terms of σ. Using the universal stress-strain curve hypothesis for isotropic strain hardening, given in Eq. (6.6.30), express $\int \overline{d\epsilon^p}$ in terms of $\bar{\sigma}$, and also express $\overline{d\epsilon^p}$ in terms of $\bar{\sigma}$ and $d\bar{\sigma}$. Hence, calculate the change dY in the current yield stress Y produced by the hardening accompanying each of the increments $\overline{d\epsilon^p}$ of Ex. 8, assuming $Y = 25{,}000$ psi at the beginning of each increment.

10. In a biaxial tension and compression test, the two nonzero stresses are σ_x and

σ_y. Assume ***incompressible, perfectly plastic*** material obeying the ***Tresca yield condition*** and plastic-potential flow theory, with uniaxial tension yield at 40 kpsi. If the rate of deformation is controlled so that the rate of plastic work is 200 inch-pounds per cubic inch per second in Parts (a) and (b) below, find the plastic rate-of-deformation components for the two loading conditions:
(a) $\sigma_x = 40$ kpsi $\quad \sigma_y = 20$ kpsi,
(b) $\sigma_x = -10$ kpsi $\quad \sigma_y = 30$ kpsi.
Find the deviatoric stress components for the three plastic rate-of-deformation conditions below. If some of the deviatoric stress components are not completely determined, what is the range of possible values? [Note that the work rate of (a) and (b) does not apply to (c), (d) and (e).]
(c) $\dot{\epsilon}_x^p = 0.0015 \quad \dot{\epsilon}_y^p = -0.0025 \quad \dot{\epsilon}_z^p = 0.0010$,
(d) $\dot{\epsilon}_x^p = 0.01 \quad \dot{\epsilon}_y^p = 0 \quad \dot{\epsilon}_z^p = -0.01$
(e) $\dot{\epsilon}_x^p = -0.01 \quad \dot{\epsilon}_y^p = 0.01 \quad \dot{\epsilon}_z^p = 0$ in./in./sec.

11. Calculate $\overline{d\epsilon^p}/dt$ for each part of Ex. 10.

12. Verify Eq. (6.6.50).

13. Show that by use of Eq. (6.6.50), the equation preceding Eq. (50) can be obtained from Eq. (46), in the general isotropic-hardening case.

14. Show that if $f(\mathbf{T}) = J_2' - c(J_3'/J_2')^2$, where c is a constant, and the yield condition is $f(\mathbf{T}) - F(W_p) = 0$, as in Eq. (6.6.39), then the plastic-potential flow rule and the discussion leading up to Eq. (44) imply the following constitutive equation during loading:

$$\dot{\epsilon}_{ij}^p = \frac{\dot{f}}{2fF'(W_p)}\left\{\left[1 + 2c\,\frac{(J_3')^2}{(J_2')^3}\right] T'_{ij} - 2c\,\frac{J_3'}{(J_2')^2}\,M_{ij}\right\},$$

where $\dot{f} \equiv (\partial f/\partial T_{pq})\,\dot{T}_{pq}$.

15. Write out the equation of Ex. 14:
(a) for $\dot{\epsilon}_{11}^p$ and $\dot{\epsilon}_{22}^p$ in a uniaxial tension test where the only stress is T_{11},
(b) for $\dot{\epsilon}_{12}^p$ in a pure shear test where the only stresses are T_{12} and T_{21}.

16. For a hypothetical material whose uniaxial tension test is governed by $\sigma = Y_0 + B\epsilon^p$, where Y_0 and B are constants and ϵ^p is the plastic natural strain:
(a) Express W_p as a function of ϵ^p during the tension test, and invert this to give ϵ^p as a function of W_p, and hence finally express the new yield stress Y, after hardening, explicitly as a function of W_p.
(b) If the combined-stress yield condition is written as $\bar{\sigma} - F(W_p) = 0$, for isotropic hardening, what is the form of $F(W_p)$ for this material? If the same yield condition is written as $J_2' - F_1(W_p) = 0$, what is the $F_1(W_p)$?

17. For ***linear*** kinematic hardening ($d\alpha_{ij} = c\,d\epsilon_{ij}^p$) with the Mises yield condition

$$f \equiv \tfrac{1}{2}(T'_{ij} - \alpha_{ij})(T'_{ij} - \alpha_{ij}) - k^2 = 0$$

during plastic deformation:

(a) Show that the linear-hardening assumption and plastic-potential theory lead to $d\alpha_{ij} = (T'_{km} - \alpha_{km})\ dT'_{km}\ (T'_{ij} - \alpha_{ij})/2k^2$ for any plastic loading increments dT'_{km}.

(b) Show that for any ***radial loading path*** (defined by $T'_{ij} = sT'^{0}_{ij}$, where the T'^{0}_{ij} are the deviatoric stresses at first yield and s is a monotonically increasing parameter, equal to unity at first yield) $\alpha_{ij} = T'_{ij} - T'^{0}_{ij}$.

(c) Consider the following nonradial loading path. In tension-torsion of a thin-walled cylinder the axial tension T_{11} is first increased to $\frac{1}{2}k\sqrt{3}$ (still elastic) and then held constant while the shear stress is increased from zero with $T_{12} = T_{21} = 3s$, where s is a monotonically increasing parameter. Obtain a coupled set of ordinary differential equations for α_{11}, α_{22}, and α_{12} as functions of s for $s \geq s_0$, where s_0 is the value of s at first yield. It is not necessary to carry out the details of completing the integration of these equations, but show that;

(1) $\alpha_{22} = \alpha_{33}$ and $\alpha_{12} = \alpha_{21}$, so that there are just three independent equations for α_{11}, α_{22}, and α_{12}.

(2) The equation for α_{12} is uncoupled and integrable by the change of dependent variable to $u = 3s - \alpha_{12}$ followed by separation of variables.

(3) When α_{12} is known as a function of s, the other equations reduce to linear first-order equations, integrable by standard methods.

6.7 Theories of constitutive equations I: principle of equipresence; fundamental postulates of a purely mechanical theory; principle of material frame-indifference

When new constitutive equations are needed, especially when they represent materials whose behavior is nonlinear, the formulation of the new equations either on the basis of observed material in the large or from physical theories of the molecular behavior is a difficult and complicated task. The modern theories of constitutive equations can help considerably after the pertinent constitutive variables have been selected on the basis of experience and intuition. The theories can restrict the possible forms of the functional dependence on the assumed independent constitutive variables. This reduces the scope of the required experimental exploration of the material properties and may also reveal some of the limitations on the applicability of a formulation. For example, a proposed equation not satisfying the principle of material frame-indifference to be presented in this section may possibly still be applicable, at least approximately, to motions in which the rotations of material elements are always small.

We have already seen in the Clausius-Duhem inequality of thermodynamics one restriction on possible constitutive equations. Other restricitions are imposed by well-known requirements of dimensional homogeneity. In this section and in Sec. 6.8 we show some of the ways in which the modern theories of constitutive equations impose additional restrictions on the nature of mechanical response. Constitutive equations will in general depend also on the temperature θ (or alternatively on the entropy s), but in a purely mechanical theory we may either limit our considerations to isothermal processes (or alternatively isentropic processes), or else assume that the mechanical response is insensitive to the (small) changes in θ (or in s) occurring during an actual process. Coleman (1964) in his work on thermodynamics of materials with memory has applied the principle of material frame-indifference, in almost the same form we will give it, to materials whose independent constitutive variables include the temperature and the temperature gradient. When more than one independent constitutive variable is involved, the principle of equipresence may be invoked.

Principle of equipresence. Truesdell and Toupin (1960), generalizing earlier work of Truesdell (1951) have proposed the following principle of equipresence [reworded here in a manner similar to that of Coleman and Mizel (1964)].

> *Principle of Equipresence*
>
> An independent variable assumed to be present in one constitutive equation of a material should be assumed to be present in all constitutive equations of the same material, unless its presence contradicts an assumed symmetry of the material,or contradicts the principle of material frame-indifference or some other fundamental principle.

The principle of equipresence has not won universal acceptance. It runs counter to the customs of many physicists and engineering scientists who prefer to avoid complicating their equations with additional variables unless experience indicates that the variables are needed. Nevertheless, the principle seems to be worth adopting in the early stages of any formulation aiming at generality, for it reveals possible coupling effects between different kinds of phenomena that might otherwise be overlooked. In practice, one can still make simplifying assumptions including the omission of a variable from some of the equations, but one should at least be aware of the omission.

For simplicity, in this section and in Sec. 6.8 dealing with material-symmetry restrictions on constitutive equations, we consider only purely mechanical phenomena and have no occasion to invoke the principle of equipresence.

Fundamental postulates of a purely mechanical theory. Three fundamental postulates are assumed to be valid for any constitutive theory of purely mechanical phenomena in a continuous medium. [See Truesdell and Noll (1965), Secs. 19 and 26.] The three postulates are:

1. ***Principle of determinism for stress:*** The stress in a body is determined by the history of the motion of that body.
2. ***Principle of local action:*** In determining the stress at a given particle

X, the motion outside an arbitrary neighborhood of X may be disregarded.†

3. ***Principle of material frame-indifference:*** Constitutive equations must be invariant under changes of frame of reference. That is, two observers, even if in relative motion with respect to each other, observe the same stress in a given body. The principle of material frame-indifference is also called the principle of material objectivity. It will be given a mathematical form in connection with Eqs. (15) and (16) below after we have considered the problem of change of frame. It should be noted that the whole discussion is within the framework of the Galilean relativity of classical mechanics and does not consider the effects of Einsteinian relativity, which may become important for motions at speeds not small compared to the speed of light.

The first two principles seem obvious enough as postulates for a purely mechanical theory. The principle of determinism includes as special cases the classical history-independent Newtonian fluid and the classical elastic law whose only history dependence consists in possessing a natural state to which it will return upon unloading. In the sections on viscoelasticity and plasticity, we have considered some specific kinds of history dependence. But, of course, we have by no means exhausted the possibilities. The principle of local action applies to the formulation of constitutive equations defining the material response, excluding action-at-a-distance stress-strain relations. As a rule, the whole body must be considered in formulating a boundary-value problem to determine the stress and deformation distribution, which is governed by the general field equations of Chap. 5 in addition to the constitutive equations. The mathematical implications of the principle of material frame-indifference are not so obvious. The principal aim of this section is to spell out these implications and illustrate the use of the principle in restricting possible forms of constitutive equations. The restrictions on the form of constitutive equations will be illustrated by the example of finite elasticity near the end of this section. As stated above, verbally, the principle seems clear enough, and it is doubtful that anyone would question its validity for observers whose relative motion occurs at velocities small compared to the speed of light.

The principle of material frame-indifference is taken for granted in much of our physical thinking. For example, an elementary physics laboratory experiment demonstrating the formula for centripetal force makes use of a weight attached to a spring balance and spinning at steady speed on a horizontal disc. It is tacitly assumed that the ***response of the spring*** which measures the force is unaffected by the rigid rotation—the spring balance having previously been calibrated in a nonrotating condition. The postulate extends this assumption to

†In this section a particle will be de denoted by a capital letter not in boldface (without wavy underlining) to emphasize the distinction between the physical particle X and its place $\mathbf{X}$ in the reference configuration.

the most general mechanical constitutive equations, and plays an important role in some of them, while it is trivially satisfied in, for example, the classical Newtonian fluid, as we shall see at the end of this section. We begin with a digression intended to clarify the concept of a frame of reference.

Frame of reference. Unfortunately the term "frame of reference" is often used simply to mean a set of coordinate axes, which is not at all what it means in the statement of the principle of frame-indifference. As a two-dimensional example to illustrate the concept, consider two ships moving on a flat ocean, each carrying a clock. Each ship-clock combination may be taken as a frame of reference in which an event may be observed. An ***event*** is a pair $\{\mathbf{x}, t\}$ consisting of a point $\mathbf{x}$ in space and a time t. For example, from the two ships, say F and F^*, a common target may be observed as $\{\mathbf{x}, t\}$ in frame F and simultaneously as $\{\mathbf{x}^*, t^*\}$ in frame F^*. We will suppose the same units of measurement are used in both frames. Then the simultaneous readings of the two clocks will differ by a constant: $t^* = t - a$. The relationship between the points $\mathbf{x}^*$ and $\mathbf{x}$ (as seen in the two frames) can be specified by giving the relationship between position vectors $\mathbf{p}$ and $\mathbf{p}^*$ from origins in each of the frames to what is actually the same point. This relationship will be discussed below under the heading "Change of Reference Frame." A set of rectangular Cartesian coordinate axes may be imagined rigidly attached to one ship; but this may be done in infinitely many ways and ***the coordinate transformation equations*** (see Sec. 2.4) ***between these various coordinate systems, all in the same frame of reference, are time-independent.*** This, of course, does not mean that the coordinates of any target are time-independent, but merely that the coordinate change coefficients are time-independent. Curvilinear coordinates may also be used with coordinate surfaces and curves rigidly attached to the frame of reference and moving with it, and the coordinate transformation equations of Sec. I.4 between these various sets of curvilinear coordinates in the same reference frame will also be time-independent. Invariance under this kind of coordinate change is ensured merely by requiring the constitutive equations to be tensor equations, but this does not ensure invariance under the time-dependent change of reference frame. For example, Newton's Second Law, $\Sigma \mathbf{f} = k(d/dt)(m\mathbf{v})$, is not frame-invariant, even though it is a first-order tensor equation, unless certain "fictitious inertia forces" are supplied, including centrifugal force and coriolis force when the frame of reference is rotating relative to an inertial frame. Material response, however, is postulated to be frame-indifferent. In order to bring out the implications of the postulate, we need a mathematical representation of the change of frame.

Change of reference frame. To illustrate the ideas we first consider again the two-dimensional example of two ships observing the same target. We have made quite a point of the fact that a change of coordinate systems is not the same thing as a change of reference frame. We can, however, single out one rectangular Cartesian coordinate system in each frame, and the ***time-dependent*** transformation between these two chosen coordinate systems will characterize the change of frame when the time change $t^* = t - a$ is also made. For this purpose it is essential to use Cartesian axes, because their relative motion can be identified with the rigid-body relative motion of the two frames of reference; there seems to be no unambiguous way to do this with two curvilinear coordinate systems, one attached to each frame.

Figure 6.20 illustrates schematically the two ships F and F^*.

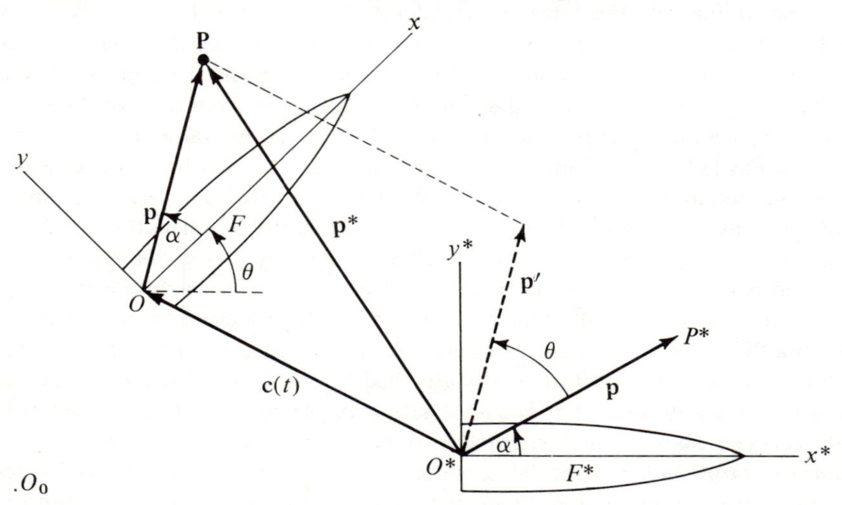

Fig. 6.20 Schematic illustration of two reference frames F and F^*.

Since the time change offers no difficulty, we will take $a = 0$ for this example and suppose that the two clocks read the same time. We may imagine that at $t = 0$ the two frames coincided, with O and O^* both at O_0 and both ships headed to the right, while at time t ship F has turned to the left through angle θ relative to F^*, which has maintained its original heading. An origin and a pair of Cartesian axes are shown fixed to each ship. The origin O has relative position vector $\mathbf{c}(t)$ from O^*. [In our later applications to constitutive equations we will often superpose a coordinate change in each frame onto the change of frame so that at the instant in question the two coordinate systems coincide, i.e., in the present example we would have $\mathbf{c}(t) = 0$ and $\theta(t) = 0$, but the time derivatives $\dot{\mathbf{c}}(t)$ and $\dot{\theta}(t)$ would not be zero.] A target P has position $\mathbf{p}$ in frame F and position $\mathbf{p}^*$ in frame F^*. We seek now an expression for $\mathbf{p}^*$ as seen in F^* in terms of $\mathbf{p}$ as seen in F. From our Olympian view of the picture we may be tempted to say simply $\mathbf{p}^* = \mathbf{c}(t) + \mathbf{p}$, but this is not what an observer in frame F^* must do. Since he cannot see $\mathbf{p}$, he must begin by drawing the vector $\mathbf{p}$ in his frame so that it appears relative to his frame in the same position as $\mathbf{p}$ appears in F (to fictitious point P^*). Perhaps he receives a radio message from F saying that $\mathbf{p}$ is drawn from O at angle α to the ship's heading. What must he do to this vector to get $\mathbf{p}^*$? He must evidently rotate it through angle θ (to position marked $\mathbf{p}'$) and add it to $\mathbf{c}(t)$. Hence

$$\mathbf{p}^* = \mathbf{c}(t) + (\mathbf{p} \text{ rotated through } \theta).$$

Observe that the rotation needed to bring $\mathbf{p}$ into position $\mathbf{p}'$ is precisely the rigid-body rotation of frame F relative to F^*. In three dimensions, if the orthogonal tensor $\mathbf{Q}(t)$ rotates frame F^* to the orientation of F so that the unit vectors of our two specially chosen Cartesian systems, one rigidly attached to each frame, are related by

$$\mathbf{i}_r = \mathbf{Q}(t)\cdot\mathbf{i}_r^*, \tag{6.7.1}$$

then at time t the position vectors $\mathbf{p}^*$ and $\mathbf{p}$ of the same target are related by

$$\mathbf{p}^* = \mathbf{c}(t) + \mathbf{Q}(t)\cdot\mathbf{p}. \tag{6.7.2a}$$

Since $\mathbf{Q}$ is an orthogonal tensor

$$\begin{gathered}\mathbf{Q}^{-1} = \mathbf{Q}^T, \quad \text{and} \quad \mathbf{Q}\cdot\mathbf{Q}^T = \mathbf{Q}^T\cdot\mathbf{Q} = 1 \\ \text{or} \quad Q_{kp}Q_{mp} = Q_{pk}Q_{pm} = \delta_{km}\end{gathered} \tag{6.7.2b}$$

where the Q_{ij} are Cartesian components of $\mathbf{Q}(t)$ with respect to any orthonormal basis $\mathbf{i}_j$ at time t. Hence the inverse relationship may be obtained by multiplying by $\mathbf{Q}^T$:

$$\mathbf{Q}^T\cdot[\mathbf{p}^* - \mathbf{c}] = \mathbf{Q}^T\cdot\mathbf{Q}\cdot\mathbf{p}$$

or

$$\mathbf{p} = \mathbf{Q}^T(t)\cdot[\mathbf{p}^* - \mathbf{c}(t)], \tag{6.7.2c}$$

where $\mathbf{Q}(\mathrm{t})$ is the orthogonal tensor rotating frame F^* into the orientation of frame F and $\mathbf{c}(t)$ is the vector from O^* to O. The change of reference frame given by Eqs. (2) is best thought of as a geometric operation, as illustrated schematically in the two-dimensional example of Fig. 6.20. Component representations of it are usually more confusing than helpful. If we write a Cartesian component form of the equations, what Cartesian coordinate system are we referring to ? The orthogonal tensor $\mathbf{Q}$ relates our two sets of Cartesian unit vectors $\mathbf{i}_r$ and $\mathbf{i}_r^*$ according to Eq. (1). If we write a Cartesian component form of Eq. (1), shall we write it with reference to the basis $\mathbf{i}_r$ or with reference to the basis $\mathbf{i}_r^*$? The answer is, of course, that at any one time we can write it with respect to whatever orthonormal basis we wish, not necessarily one of the two previously singled out. If we choose the basis $\mathbf{i}_r^*$, the component form of the equation for $\mathbf{i}_1$, for example, is

$$\begin{bmatrix}(\mathbf{i}_1)_{x_1^*}\\(\mathbf{i}_1)_{x_2^*}\\(\mathbf{i}_1)_{x_3^*}\end{bmatrix}=\begin{bmatrix}Q_{11}^* & Q_{12}^* & Q_{13}^*\\Q_{21}^* & Q_{22}^* & Q_{23}^*\\Q_{31}^* & Q_{32}^* & Q_{33}^*\end{bmatrix}\begin{bmatrix}1\\0\\0\end{bmatrix}, \tag{6.7.3a}$$

where the rather awkward notation in the first column has been used to imply that although the new vector is the base vector $\mathbf{i}_1$ of frame F, the components obtained for it are components relative to the basis $\mathbf{i}_r^*$ of F^*, so that

$$\mathbf{i}_1 = Q_{11}^*\,\mathbf{i}_1^* + Q_{21}^*\,\mathbf{i}_2^* + Q_{31}^*\,\mathbf{i}_3^*. \tag{6.7.3b}$$

Evidently, $Q_{21}^* = \cos(\mathbf{i}_2^*,\mathbf{i}_1)$ and similar considerations show that

$$Q_{mn}^* = \cos(\mathbf{i}_m^*,\mathbf{i}_n). \tag{6.7.4a}$$

If we evaluate the same equation relative to the basis $\mathbf{i}_r$ of F, the column matrix on the left would have elements (1,0,0) while the column matrix on the right would have $[(\mathbf{i}_1^*)_{x_1}, (\mathbf{i}_1^*)_{x_2}, (\mathbf{i}_1^*)_{x_3}]$ and the square matrix would have elements Q_{ij}. Alternatively, if we dot multiply $\mathbf{Q}^T$ into Eq. (1) we obtain

$$\mathbf{Q}^T\cdot\mathbf{i}_r = \mathbf{i}_r^*$$

whence a development similar to that of Eqs. (3) shows that

$$\mathbf{i}_1^* = Q_{11}^T\mathbf{i}_1 + Q_{21}^T\mathbf{i}_2 + Q_{31}^T\mathbf{i}_3,$$

etc. Hence, evidently $Q_{21}^T = \cos(\mathbf{i}_2,\mathbf{i}_1^*)$, etc., or

$$Q_{mn} = Q_{nm}^T = \cos(i_n,i_m^*). \tag{6.7.4b}$$

Equations (2a) may also be written in component form relative to either of our special bases. Component forms will be given below for the transformation equations for particle velocity and acceleration. For that purpose we will suppose that the two frames instantaneously coincide so that any given vector or tensor will have the same components relative to the two special coordinate systems. But the velocity vector relative to F^* will be a ***different vector*** from the velocity vector relative to F.

Event, vector, tensor, and deformation-gradient transformations under a change of frame. A scalar point function remains unchanged by a change of frame, but the change of frame induces transformations, for each time t, on vectors and tensors. The results are summarized in Eqs. (5) through (7) below.

The point transformation of Eq. (5) is the same as Eq. (2a) except that the point is denoted by $\mathbf{x}$ and $\mathbf{x}^*$ instead of $\mathbf{p}$ and $\mathbf{p}^*$. Equations (6) to (8) are derived below.

Change of Frame

1. Events $\{\mathbf{x}, t\}$:
$$\left.\begin{aligned} \mathbf{x}^* &= \mathbf{c}(t) + \mathbf{Q}(t)\cdot\mathbf{x} \\ t^* &= t - a \end{aligned}\right\} \tag{6.7.5}$$
2. Vectors† $\mathbf{v}$:
$$\mathbf{v}^* = \mathbf{Q}(t)\cdot\mathbf{v} \tag{6.7.6}$$
3. Second-order tensors $\mathbf{T}$ or $\mathbf{S}$ regarded as linear vector transformations:
$$\mathbf{u} = \mathbf{T}\cdot\mathbf{v} \quad \text{or} \quad \mathbf{u} = \mathbf{v}\cdot\mathbf{S}$$
$$\left.\begin{aligned} \mathbf{T}^* &= \mathbf{Q}(t)\cdot\mathbf{T}\cdot\mathbf{Q}(t)^T \\ \mathbf{S}^* &= \mathbf{Q}(t)\cdot\mathbf{S}\cdot\mathbf{Q}(t)^T \end{aligned}\right\} \tag{6.7.7}$$
4. Deformation gradient $\mathbf{F}(X, t) \equiv \boldsymbol{\chi}(X, t)\overleftarrow{\nabla}$:
$$\mathbf{F}^* = \mathbf{Q}(t)\cdot\mathbf{F} \tag{6.7.8}$$
(This two-point tensor transforms like a vector under change of frame at time t.)

The vector transformation follows immediately from Eq. (5) and the fact that any vector may be represented as a directed line from point $\mathbf{x}$ to point $\mathbf{y}$ by the ***point-difference relations***

$$\mathbf{v} = \mathbf{y} - \mathbf{x} \quad \text{and} \quad \mathbf{v}^* = \mathbf{y}^* - \mathbf{x}^*.$$

The second-order tensor transformation follows by use of the vector transformation and its inverse $\mathbf{v}^* = \mathbf{Q}^T(t)\cdot\mathbf{v}$:

$$\mathbf{u} = \mathbf{T}\cdot\mathbf{v} \quad \text{and} \quad \mathbf{u}^* = \mathbf{T}^*\cdot\mathbf{v}^*.$$

Thus

$$\begin{aligned} \mathbf{u}^* &= \mathbf{Q}(t)\cdot\mathbf{u} \\ &= \mathbf{Q}(t)\cdot(\mathbf{T}\cdot\mathbf{v}) \\ &= \mathbf{Q}(t)\cdot\mathbf{T}\cdot[\mathbf{Q}^T(t)\cdot\mathbf{v}^*] \\ &= [\mathbf{Q}(t)\cdot\mathbf{T}\cdot\mathbf{Q}^T(t)]\cdot\mathbf{v}^*. \end{aligned}$$

†Here $\mathbf{v}$ and $\mathbf{v}^*$ represent the ***same vector*** in the two different frames. In Eqs. (11) to (13) the velocity $\mathbf{v}^*$ is a ***different vector*** from $\mathbf{v}$. There we will denote by $(\mathbf{v})^*$ the vector $\mathbf{v}$ as seen in the starred frame.

Hence $\mathbf{T}^* = \mathbf{Q}(t)\cdot\mathbf{T}\cdot\mathbf{Q}^T(t)$ as stated in Eq. (7). If the tensor operates as

$$\mathbf{u} = \mathbf{v}\cdot\mathbf{S} \quad \text{and} \quad \mathbf{u}^* = \mathbf{v}^*\cdot\mathbf{S}^*,$$

then use of the vector transformation formulas

$$\mathbf{u}^* = \mathbf{u}\cdot\mathbf{Q}^T(t) \quad \text{and} \quad \mathbf{v} = \mathbf{v}^*\cdot\mathbf{Q}(t),$$

which are the same as

$$\mathbf{u}^* = \mathbf{Q}(t)\cdot\mathbf{u} \quad \text{and} \quad \mathbf{v} = \mathbf{Q}^T(t)\cdot\mathbf{v}^*,$$

leads similarly to

$$\mathbf{u}^* = (\mathbf{v}\cdot\mathbf{S})\cdot\mathbf{Q}^T(t) = \mathbf{v}^*\cdot[\mathbf{Q}(t)\cdot\mathbf{S}\cdot\mathbf{Q}^T(t)]$$

whence

$$\mathbf{S}^* = \mathbf{Q}(t)\cdot\mathbf{S}\cdot\mathbf{Q}^T(t).$$

The transformation formula of Eq. (8) for the deformation gradient $\mathbf{F}$ under a change of frame at time t, assuming that the two frames had the same orientation at time t_0 when the neighborhood of X was in the reference configuration (e.g., the unstrained state of an elastic body) so that $d\mathbf{X}^* = d\mathbf{X}$ at t_0, follows from

$$d\mathbf{x}^* = \mathbf{F}^*\cdot d\mathbf{X} \quad \text{and} \quad d\mathbf{x} = \mathbf{F}\cdot d\mathbf{X}.$$

By Eq. (6)

$$\begin{aligned} d\mathbf{x}^* &= \mathbf{Q}(t)\cdot d\mathbf{x} \\ &= \mathbf{Q}(t)\cdot[\mathbf{F}\cdot d\mathbf{X}] \\ &= [\mathbf{Q}(t)\cdot\mathbf{F}]\cdot d\mathbf{X}. \end{aligned}$$

Thus

$$[\mathbf{F}^* - \mathbf{Q}(t)\cdot\mathbf{F}]\cdot d\mathbf{X} = 0$$

for all $d\mathbf{X}$ in the (infinitesimal) reference neighborhood of X. Hence we obtain Eq. (8)

$$\mathbf{F}^* = \mathbf{Q}(t)\cdot\mathbf{F},$$

showing that the two-point-tensor deformation gradient transforms like a vector under change of frame at time t.

Frame-indifference. Functions and fields whose values are scalars, vectors, or tensors are called ***frame-indifferent*** or ***objective*** if both the dependent and the independent vector and tensor variables transform according to Eqs. (6) to (8), while the scalar variables are unchanged.

Motion, velocity, and acceleration. The motion of a medium is defined relative to some frame—as, for example, in Eq. (4.4.1) it is defined by

$$\mathbf{x} = \boldsymbol{\chi}(X, t), \tag{6.7.9}$$

where for additional clarity we have used different symbols to denote the function $\boldsymbol{\chi}$ (Greek chi) and its value $\mathbf{x}$, and where the argument X denotes a particle. According to Eqs. (5), the same motion is defined in another frame by

$$\mathbf{x}^* = \boldsymbol{\chi}^*(X, t^*) = \mathbf{c}(t) + \mathbf{Q}(t)\boldsymbol{\cdot}\boldsymbol{\chi}(X, t) \qquad t^* = t - a. \tag{6.7.10a}$$

Physically, $\boldsymbol{\chi}$ and $\boldsymbol{\chi}^*$ describe the same motion, but mathematically $\boldsymbol{\chi}^*$ is the motion obtained from $\boldsymbol{\chi}$ by the superposition of a time-dependent rigid transformation and a shift in the time scale. Two motions are called equivalent motions when they are related by Eq. (10a), which we will usually write for brevity as

$$\mathbf{x}^* = \mathbf{c} + \mathbf{Q}\boldsymbol{\cdot}\mathbf{x}, \tag{6.7.10b}$$

dropping the symbol $\boldsymbol{\chi}$.

The velocity relationship follows from material time differentiation of Eq. (10b)

$$\mathbf{v}^* = \frac{d\mathbf{x}^*}{dt} = \dot{\mathbf{c}} + \dot{\mathbf{Q}}\boldsymbol{\cdot}\mathbf{x} + \mathbf{Q}\boldsymbol{\cdot}\frac{d\mathbf{x}}{dt}.$$

But $d\mathbf{x}/dt = \mathbf{v}$, $\mathbf{Q}\boldsymbol{\cdot}\mathbf{v} = (\mathbf{v})^*$ and $\mathbf{x} = \mathbf{Q}^T\boldsymbol{\cdot}(\mathbf{x}^* - \mathbf{c})$. Hence

$$\left.\begin{aligned} \mathbf{v}^* &= \dot{\mathbf{c}} + \mathbf{A}\boldsymbol{\cdot}(\mathbf{x}^* - \mathbf{c}) + \mathbf{Q}\boldsymbol{\cdot}\mathbf{v} \\ \mathbf{v}^* - (\mathbf{v})^* &= \dot{\mathbf{c}} + \mathbf{A}\boldsymbol{\cdot}(\mathbf{x}^* - \mathbf{c}). \end{aligned}\right\} \tag{6.7.11a}$$

Thus† $\mathbf{A}$ is the skew tensor representing the angular velocity of the unstarred frame relative to the starred one:

$$\mathbf{A} = \dot{\mathbf{Q}}\boldsymbol{\cdot}\mathbf{Q}^T \quad \text{or} \quad A_{km} = \dot{Q}_{kp}Q_{mp}. \tag{6.7.12}$$

†Note that if the two frames instantaneously coincide at time t, we have $\mathbf{Q} = \mathbf{1}$, and the first Eq. (11a) takes the form commonly used in elementary kinematics, provided that angular velocity $\boldsymbol{\omega}$ of the unstarred frame relative to the "fixed" frame F^* is given by $\boldsymbol{\omega} = -A_{23}\mathbf{i}_1 - A_{31}\mathbf{i}_2 - A_{12}\mathbf{i}_3$.

The last Cartesian component form relates to an arbitrary orthonormal basis $[\dot{Q}_{kp} = (\dot{\mathbf{Q}})_{kp} = (d/dt)(Q_{kp})$ with respect to the same basis]. Note that $\mathbf{v}^*$ and $(\mathbf{v})^*$ are not the same thing; $\mathbf{v}^*$ is the velocity vector relative to the starred frame, while $(\mathbf{v})^*$ is the velocity vector relative to the unstarred frame as it appears to an observer in the starred frame.

To obtain the acceleration relationship we differentiate again the first form of Eq. (11a) to obtain

$$\mathbf{a}^* = \ddot{\mathbf{c}} + \dot{\mathbf{A}}\cdot(\mathbf{x}^* - \mathbf{c}) + \mathbf{A}\cdot(\mathbf{v}^* - \dot{\mathbf{c}}) + \dot{\mathbf{Q}}\cdot\mathbf{v} + \mathbf{Q}\cdot\mathbf{a}.$$

Now

$$\begin{aligned}\mathbf{Q}\cdot\mathbf{a} &= (\mathbf{a})^* \quad \text{and}\\ \dot{\mathbf{Q}}\cdot\mathbf{v} &= \dot{\mathbf{Q}}\cdot[\mathbf{Q}^T\cdot(\mathbf{v})^*] = \mathbf{A}\cdot(\mathbf{v})^*\\ &= \mathbf{A}\cdot[\mathbf{v}^* - \dot{\mathbf{c}} - \mathbf{A}\cdot(\mathbf{x}^* - \mathbf{c})],\end{aligned}$$

where the last step follows from Eq. (11a). Hence

$$\mathbf{a}^* - (\mathbf{a})^* = \ddot{\mathbf{c}} + 2\mathbf{A}\cdot(\mathbf{v}^* - \dot{\mathbf{c}}) + (\dot{\mathbf{A}} - \mathbf{A}^2)\cdot(\mathbf{x}^* - \mathbf{c}). \qquad (6.7.13a)$$

The following Cartesian component forms are obtained relative to the common axes if at time t the two origins and specially selected Cartesian axis systems in the two frames coincide:

$$v_k^* - v_k = \dot{c}_k + A_{km}p_m \qquad (6.7.11b)$$

$$a_k^* - a_k = \ddot{c}_k + 2A_{km}[v_m^* - \dot{c}_m] + (\dot{A}_{km} - A_{kr}A_{rm})p_m, \qquad (6.7.13b)$$

where the $p_m = x_m = x_m^*$ are the components of the position vector relative to these instantaneously coincident Cartesian axes. For most purposes the symbolic forms (11a) and (13a) are more useful.

Forces and stresses under a change of frame. The body and surface forces described in Sec. 3.1 are vector functions, which transform according to Eq. (6) under a change of frame if we assume that forces are frame-indifferent. But then the fundamental laws of dynamics are not form-invariant, since the acceleration **a** is not frame-indifferent, as Eq. (13) shows. The classical way of resolving the difficulty is to introduce "absolute space" and suppose that the dynamical equations hold only relative to this absolute space. {Noll has proposed a different interpretation [see Truesdell and Noll (1965), Sec. 18]. All forces are assumed frame-indifferent. Inertia is regarded as that mutual body force describing the interaction between the bodies considered and the masses in the rest of the universe. The term $\rho\, d\mathbf{v}/dt$ in the equation of motion, Eq.

(5.3.4), must be replaced by $\rho\mathbf{a}^\star$, where $\mathbf{a}^\star$ is the acceleration relative to the fixed stars. $\mathbf{a}$ is not frame-indifferent, but $\mathbf{a}^\star$ is; not only the position of the observed particle but also the positions of the "fixed stars" must be transformed according to Eq. (5) under a change of frame. Other external body forces, e.g., gravity exerted by external masses, require similar consideration. Because the dependence of these body forces on the configuration of the masses outside the body studied is not explicitly introduced, these forces appear not to be frame-indifferent and the equation of motion has the form of Eq. (5.3.4) only in a frame where $\mathbf{a} = \mathbf{a}^\star$.}

Since internal contact forces depend only on the configuration of the body considered and not on the configuration of the masses outside it, internal contact forces are formally invariant under change of frame and this implies that stresses are frame-indifferent, whence the stress tensor satisfies Eq. (7) under a change of frame, with Cartesian component form

$$T^*_{km}(t^*) = T_{km}(t) \tag{6.7.14}$$

if the special Cartesian axis systems are chosen as before so that they instantaneously coincide.

Equivalent processes. A ***dynamical process*** for a body means a pair $\{\boldsymbol{\chi}, \mathbf{T}\}$ of fields consisting of a motion $\mathbf{x} = \boldsymbol{\chi}(X, t)$ and a symmetric stress tensor $\mathbf{T}$ defined over the body. Note that for arbitrary, sufficiently smooth fields $\boldsymbol{\chi}$ and $\mathbf{T}$ and a given density distribution ρ, the equation of motion determines a body-force distribution which would make $\{\boldsymbol{\chi}, \mathbf{T}\}$ an admissible process. The required body force may be rather artificial, but the equation of motion itself does not in principle impose any limitation on the admissible processes $\{\boldsymbol{\chi}, \mathbf{T}\}$. Two dynamical processes $\{\boldsymbol{\chi}^*, \mathbf{T}^*\}$ and $\{\boldsymbol{\chi}, \mathbf{T}\}$ are said to be ***equivalent processes*** when $\boldsymbol{\chi}^*$ and $\boldsymbol{\chi}$ are related to each other by Eq. (6), while $\mathbf{T}^*$ and $\mathbf{T}$ are related by Eq. (5). They are really two different mathematical descriptions of the same physical process. We are now ready to make a formal mathematical statement of the third postulate assumed at the beginning of this section.

Principle of Material Frame-Indifference

Constitutive equations must be invariant under changes of reference frame. If a constitutive equation is satisfied for a dynamical process with a motion and a symmetric stress tensor given by

$$\mathbf{x} = \boldsymbol{\chi}(X, t) \qquad \mathbf{T} = \mathbf{T}(X, t), \tag{6.7.15}$$

then it must also be satisfied for any equivalent process $\{\boldsymbol{\chi}^*, \mathbf{T}^*\}$. This means that the constitutive equation must also be satisfied by the motion and stress tensor given by

$$\left.\begin{aligned} \mathbf{x}^* &= \boldsymbol{\chi}^*(X, t^*) = \mathbf{c}(t) + \mathbf{Q}(t)\cdot\boldsymbol{\chi}(X, t) \\ \mathbf{T}^* &= \mathbf{T}^*(X, t^*) = \mathbf{Q}(t)\cdot\mathbf{T}\cdot Q(t)^T \\ t^* &= t - a \end{aligned}\right\} \tag{6.7.16}$$

for any arbitrary point function of the time $\mathbf{c}(t)$, any arbitrary orthogonal tensor function of the time $\mathbf{Q}(t)$, and any arbitrary real number a.

(In this context a "point function" is a function whose value is a point. It can be considered as a "position-vector function.") Note that as defined here $\mathbf{Q}(t)$ could involve a reflection as well as a rotation. While the kind of rigid-body motion of two reference frames implied by the introductory illustrative example of Fig. 6.20, where the two ships started out coincident with each other, excludes such a reflection, there is no reason why we could not have begun with two ships that were mirror images of each other and proceeded from there by rigid relative motion. Whether invariance should be required also for improper orthogonal tensors has been questioned. We will make use only of the invariance with respect to proper $\mathbf{Q}(t)$ in applying Eqs. (6.7.22) and (6.7.28).

Simple materials. According to the principles of determinism and local action postulated at the beginning of this section, the stress at a particle X depends only on the history of the motion of an arbitrarily small neighborhood of X. If we make a special choice of the arbitrary change of frame such that the moving origin moves with the particle X and that

$$\mathbf{Q}(\tau) \equiv \mathbf{1}, \qquad a = 0, \quad \text{and} \quad \mathbf{c}(\tau) \equiv -\boldsymbol{\chi}(X, \tau)$$

for all past times $\tau \leq t$, where t is the present time, then

$$\begin{gathered} \boldsymbol{\chi}^*(Z, \tau) = \boldsymbol{\chi}(Z, \tau) - \boldsymbol{\chi}(X, \tau) \\ \mathbf{T}^*(\tau) = \mathbf{T}(\tau) \end{gathered} \tag{6.7.17}$$

by Eq. (16), since $\tau^* = \tau$. Thus the stress at X and t depends only on the history for $\tau \leq t$ of the relative motion (relative to X) of the set of all particles Z in an arbitrarily small neighborhood of X. For a sufficiently small neighborhood, assuming differentiability, the relative motion can be approximated to any desired accuracy by

$$\boldsymbol{\chi}(Z, \tau) - \boldsymbol{\chi}(X, \tau) \simeq \mathbf{F}(X, \tau)\cdot d\mathbf{X}, \tag{6.7.18}$$

where $\mathbf{F}(X, \tau)$ is the deformation gradient at X at time τ.

$$\mathbf{F}(X, \tau) \equiv \boldsymbol{\chi}(X, \tau)\overset{\leftarrow}{\nabla} \tag{6.7.19}$$

(denoted $\mathbf{x}\overset{\leftarrow}{\nabla}$ in Secs. 4.5 and 4.6, where the reference configuration was the natural state of an elastic body), and $d\mathbf{X}$ is the vector from X to Z. This suggests that, since the relative motion history of an infinitesimal neighborhood of X is completely determined by the history of the deformation gradient at X, then the stress $\mathbf{T}(X, t)$ must be determined by the history of $\mathbf{F}(X, \tau)$ for $\tau \leq t$. This motivates the definition of a ***simple material*** as one where the stress at each particle is a functional of the deformation gradient at the particle relative to some reference configuration of the neighborhood of the particle. Thus the stress at X depends on the history up to t of the single two-point-tensor-valued function $\mathbf{F}$ of the single argument τ (for fixed X). Coleman (1964) also includes the history of the temperature and the temperature gradient, and he adds the heat flux, internal energy, and entropy to the dependent variables. Truesdell and Noll (1965), for more general materials in Secs. 26 and 27, and for simple materials in Sec. 28ff, examine the restrictions imposed on the forms of the constitutive functionals by considering separately three special changes of frame, which taken successively can represent an arbitrary change of frame:

1. Translation with X:

$$\mathbf{Q}(\tau) \equiv \mathbf{1}, \qquad a = 0, \qquad \mathbf{c}(\tau) = -\boldsymbol{\chi}(X, \tau) \tag{6.7.20}$$

2. Time shift such that the present time t becomes the reference time:

$$\mathbf{Q}(\tau) \equiv \mathbf{1}, \qquad \mathbf{c}(\tau) \equiv \mathbf{0}, \quad \text{and} \quad a = t \tag{6.7.21}$$

3. Time-dependent rotation of frame:

$$\mathbf{c}(\tau) \equiv 0, \qquad a = 0, \qquad \mathbf{Q}(\tau) \text{ arbitrary orthogonal.} \tag{6.7.22}$$

We have already seen how the first special change of frame motivates the definition of a simple material. The second one places some restrictions on the form of the functional equation to embody the fact that the stress depends only on past history and not on future history. The requirement of form-invariance under the third special class of frame changes (rotations) leads to important reductions in the constitutive functional equation forms for simple materials. Some of the results will be presented in Eqs. (59) to (62). But first the ideas will be illustrated below for finite elasticity where constitutive functions can be used instead of functionals.

Application to finite elasticity. For an elastic material the principles of determinism and local action imply that the stress at a particle X is determined by the present local configuration of an arbitrary small neighborhood of X, relative to its initial configuration. That is, the stress $\mathbf{T}(X, t)$ at X is determined by the set

$$\{\boldsymbol{\chi}(Z, t) - \boldsymbol{\chi}(X, t)\} \tag{6.7.23}$$

of relative motions $\boldsymbol{\chi}(Z, t) - \boldsymbol{\chi}(X, t)$ of all particles Z in an arbitrarily small neighborhood of X. Now for a sufficiently small neighborhood, assuming differentiability,

$$\boldsymbol{\chi}(Z, t) - \boldsymbol{\chi}(X, t) \approx \mathbf{F}(X, t) \cdot d\mathbf{X}, \tag{6.7.24}$$

where

$$\mathbf{F}(X, t) = \mathbf{x}\overleftarrow{\nabla}_X = \mathbf{x}(X, t)\overleftarrow{\nabla} \tag{6.7.25}$$

is the deformation gradient at X at time t and $d\mathbf{X}$ is the vector from particle X to an arbitrary particle Z in the neighborhood considered. Hence, to any desired degree of accuracy, the whole configuration of a sufficiently small neighborhood of X is determined by the value of $\mathbf{F}(X, t)$, and we may therefore say that the stress $\mathbf{T}(X, t)$, which was assumed to be determined by the local configuration is completely determined by $\mathbf{F}(X, t)$. That is, the most general constitutive equation of an elastic material is for given X, of the form

$$\mathbf{T} = \mathbf{g}(\mathbf{F}), \tag{6.7.26}$$

where the ***response function*** $\mathbf{g}$ is a tensor-valued function (in general nonlinear) of the single tensor argument $\mathbf{F}$. If the material is not homogeneous, the function $\mathbf{g}$ for a different particle may have a different dependence on $\mathbf{F}$, but that will not affect the following argument. For brevity of notation we omit the explicit dependence on X and t in our consideration of the restrictions imposed on $\mathbf{g}$ by the principle of material frame-indifference under arbitrary rotations such that

$$t^* = t \qquad \boldsymbol{\chi}^*(X, t) = \mathbf{Q}(t) \cdot \boldsymbol{\chi}(x, t). \tag{6.7.27}$$

Recall that by Eq. (8), for a change of frame at time $t = t^*$ [or $\tau = \tau^*$ in Eq. (16)], the deformation gradient $\mathbf{F}$, a two-point tensor, behaves like a vector. Then Eq. (16) requires that when the independent variable $\mathbf{F}$ is replaced by $\mathbf{Q} \cdot \mathbf{F}$ in Eq. (26), the dependent variable $\mathbf{T}$ must be replaced by $\mathbf{Q} \cdot \mathbf{T} \cdot \mathbf{Q}^T$ to ensure frame-indifference. We are thus led to the following identity, Eq. (28), that must be satisfied by admissible (frame-indifferent) response functions $\mathbf{g}$ ($\mathbf{F}$).

$$\boxed{\mathbf{g}(\mathbf{Q} \cdot \mathbf{F}) = \mathbf{Q} \cdot \mathbf{g}(\mathbf{F}) \cdot \mathbf{Q}^T \qquad \text{for arbitrary orthogonal } \mathbf{Q}.} \tag{6.7.28}$$

This identity, for a special choice of $\mathbf{Q}$, yields further important information about suitable constitutive equations. Note that

$$\mathbf{F} = \mathbf{R} \cdot \mathbf{U} \tag{6.7.29a}$$

by the polar decomposition of Eq. (4.6.1a), where $\mathbf{U}$ is the right stretch tensor and $\mathbf{R}$ is the rotation tensor. From Eq. (29a),

$$\mathbf{R}^T \cdot \mathbf{F} = (\mathbf{R}^T \cdot \mathbf{R}) \cdot \mathbf{U} = \mathbf{U} \tag{6.7.29b}$$

since $\mathbf{R}^T \cdot \mathbf{R} = \mathbf{1}$. Now make the special choice

$$\mathbf{Q}(t) = \mathbf{R}(t)^T. \tag{6.7.30}$$

Then Eq. (28) becomes

$$\mathbf{g}(\mathbf{R}^T \cdot \mathbf{F}) = \mathbf{R}^T \cdot \mathbf{g}(\mathbf{F}) \cdot \mathbf{R}.$$

Multiplying from the left by $\mathbf{R}$ and from the right by $\mathbf{R}^T$ transforms this to

$$\mathbf{R} \cdot \mathbf{g}(\mathbf{R}^T \cdot \mathbf{F}) \cdot \mathbf{R}^T = \mathbf{1} \cdot \mathbf{g}(\mathbf{F}) \cdot \mathbf{1}$$

or, by Eq. (29b),

$$\mathbf{R} \cdot \mathbf{g}(\mathbf{U}) \cdot \mathbf{R}^T = \mathbf{g}(\mathbf{F}).$$

Hence, Eq. (26) takes the reduced form

$$\mathbf{T} = \mathbf{R} \cdot \mathbf{g}(\mathbf{U}) \cdot \mathbf{R}^T \tag{6.7.31}$$

$$\text{or} \quad \mathbf{T} = \mathbf{R} \cdot \mathbf{f}(\mathbf{C}) \cdot \mathbf{R}^T \tag{6.7.32}$$

$$\text{or} \quad \mathbf{T} = \mathbf{R} \cdot \mathbf{G}(\mathbf{E}) \cdot \mathbf{R}^T, \tag{6.7.33}$$

where

$$\mathbf{f}(\mathbf{C}) = \mathbf{g}(\mathbf{C}^{1/2}) \equiv \mathbf{g}(\mathbf{U}) \tag{6.7.34}$$

and

$$\mathbf{G}(\mathbf{E}) = \mathbf{f}(\mathbf{1} + 2\mathbf{E}). \tag{6.7.35}$$

Thus frame-indifference requires according to Eq. (31) that the dependence on $\mathbf{F}$ must take the form of an arbitrary function of $\mathbf{U}$ with the additional explicit dependence on $\mathbf{R}$ as shown. Many other reduced forms are possible. Equation (32) is a reduced form, following immediately from Eq. (31) by use of $\mathbf{U} = \mathbf{C}^{1/2}$, according to Eq. (4.6.3), while Eq. (33) is another reduced form, following immediately from Eq. (32), since $\mathbf{C} = \mathbf{1} + 2\mathbf{E}$ by Eq. (4.5.5a).

Additional reduced forms may be obtained if the Piola-Kirchhoff stress tensors of Sec. 5.3 are used in the constitutive equation instead of $\mathbf{T}$. By Eq. (5.3.24),

$$\frac{\rho}{\rho_0} \mathbf{F} \cdot \mathbf{T}^0 = \mathbf{T}, \tag{6.7.36}$$

where $\mathbf{T}^0$ is the first Piola-Kirchhoff tensor. [The first Piola-Kirchhoff tensor, $\mathbf{T}_R$, used in Truesdell and Noll (1965), Sec. 43, is the transpose of $\mathbf{T}^0$.] Hence, by Eq. (31),

$$\rho\mathbf{F}\cdot\mathbf{T}^0 = \rho_0\mathbf{R}\cdot\mathbf{g}(\mathbf{U})\cdot\mathbf{R}^T.$$

Multiply from the left by $\mathbf{R}^T$ to obtain

$$\rho\mathbf{R}^T\cdot\mathbf{F}\cdot\mathbf{T}^0 = \rho_0\mathbf{g}(\mathbf{U})\cdot\mathbf{R}^T$$

or, by Eq. (29b),

$$\rho\mathbf{U}\cdot\mathbf{T}^0 = \rho_0\mathbf{g}(\mathbf{U})\cdot\mathbf{R}^T$$

whence

$$\boxed{\mathbf{T}^0 = \mathbf{h}^0(\mathbf{U})\cdot\mathbf{R}^T,} \tag{6.7.37}$$

where

$$\mathbf{h}^0(\mathbf{U}) = \frac{\rho_0}{\rho}\mathbf{U}^{-1}\cdot\mathbf{g}(\mathbf{U})$$

is an arbitrary function of $\mathbf{U}$. Note that

$$\mathbf{T}_R = (\mathbf{T}^0)^T = \mathbf{R}\cdot[\mathbf{h}^0(\mathbf{U})]^T,$$

since for any two tensors $\mathbf{M}$ and $\mathbf{N}$, $(\mathbf{M}\cdot\mathbf{N})^T = \mathbf{N}^T\cdot\mathbf{M}^T$. Hence

$$\boxed{\mathbf{T}_R = \mathbf{R}\cdot\mathbf{h}(\mathbf{U}),} \tag{6.7.38}$$

where $\mathbf{h}(\mathbf{U})$ is an arbitrary function of $\mathbf{U}$. The symmetric second Piola-Kirchhoff tensor $\tilde{\mathbf{T}}$ is related to $\mathbf{T}^0$ by Eq. (5.3.25):

$$\mathbf{T}^0 = \tilde{\mathbf{T}}\cdot\mathbf{F}^T. \tag{6.7.39}$$

Hence

$$\mathbf{T}_R = (\mathbf{T}^0)^T = \mathbf{F}\cdot\tilde{\mathbf{T}}, \tag{6.7.40}$$

since $\tilde{\mathbf{T}}$ is symmetric ($\tilde{\mathbf{T}}^T = \tilde{\mathbf{T}}$). Equations (38) and (40) yield

$$\mathbf{F}\cdot\tilde{\mathbf{T}} = \mathbf{R}\cdot\mathbf{h}(\mathbf{U})$$

whence

$$\mathbf{R}^T \cdot \mathbf{F} \cdot \tilde{\mathbf{T}} = \mathbf{h}(\mathbf{U})$$

or, by Eq. (29b),

$$\mathbf{U} \cdot \tilde{\mathbf{T}} = \mathbf{h}(\mathbf{U})$$

whence

$$\tilde{\mathbf{T}} = \mathbf{U}^{-1} \cdot \mathbf{h}(\mathbf{U}) = \tilde{\mathbf{h}}(\mathbf{U}) = \tilde{\mathbf{f}}(\mathbf{C}),$$

where

$$\tilde{\mathbf{h}}(\mathbf{U}) = \mathbf{U}^{-1} \cdot \mathbf{h}(\mathbf{U}) \quad \text{and} \quad \tilde{\mathbf{f}}(\mathbf{C}) = \tilde{\mathbf{h}}(\mathbf{C}^{1/2})$$

giving the alternative reduced forms, with $\tilde{\mathbf{G}}(\mathbf{E}) = \tilde{\mathbf{f}}(\mathbf{1} + 2\mathbf{E})$,

$$\tilde{\mathbf{T}} = \tilde{\mathbf{h}}(\mathbf{U}) \tag{6.7.41}$$

$$\text{or} \quad \tilde{\mathbf{T}} = \tilde{\mathbf{f}}(\mathbf{C}) \tag{6.7.42}$$

$$\text{or} \quad \tilde{\mathbf{T}} = \tilde{\mathbf{G}}(\mathbf{E}). \tag{6.7.43}$$

The functions $\tilde{\mathbf{h}}$, $\tilde{\mathbf{f}}$, and $\tilde{\mathbf{G}}$ remain arbitrary. The principle of material frame-indifference has led to the conclusion that in terms of the second Piola-Kirchhoff tensor the response function of finite elasticity depends only on one of the strain tensors and not on the rotation. When the Cauchy stress $\mathbf{T}$ is used, on the other hand, the stress-strain relation with the strain $\mathbf{U}$, $\mathbf{C}$, or $\mathbf{E}$ must also contain the rotation $\mathbf{R}$ as in Eqs. (31) to (33). For the special case of isotropic elastic solids, explicit dependence of $\mathbf{T}$ on the rotation can be avoided; see Eqs. (6.8.35) and (6.8.36). Most treatments of finite elasticity from the Lagrangian point of view in the past have used one of these three strain tensors. Material frame-indifference is then ensured by any one of the stress-strain relations (31) to (33), or (41) to (43) for arbitrary choice of the response function appearing in the equation. It may sometimes turn out to be more convenient to take $\mathbf{F}$ itself as the independent variable as in Eq. (26) and choose the function $\mathbf{g}(\mathbf{F})$ so that it satisfies Eq. (28) identically, in order to satisfy frame-indifference.

We can also, as an alternative to Eq. (37), get an identity analogous to Eq. (28) that must be satisfied by the response function if $\mathbf{T}^0$ is taken as a function of $\mathbf{F}$. From Eqs. (26) and (36),

$$\frac{\rho}{\rho_0} \mathbf{F} \cdot \mathbf{T}^0 = \mathbf{g}(\mathbf{F}).$$

If we write

$$\mathbf{T}^0 = \mathbf{h}^0(\mathbf{F}), \tag{6.7.44}$$

then, since $\rho_0/\rho = \det \mathbf{F}$,

$$\mathbf{h}_0(\mathbf{F}) = (\det \mathbf{F})\mathbf{F}^{-1}\cdot\mathbf{g}(\mathbf{F}) \tag{6.7.45}$$

whence

$$\mathbf{h}^0(\mathbf{Q}\cdot\mathbf{F}) = (\det (\mathbf{Q}\cdot\mathbf{F}))(\mathbf{Q}\cdot\mathbf{F})^{-1}\mathbf{g}(\mathbf{Q}\cdot\mathbf{F})$$

or

$$\mathbf{Q}\cdot\mathbf{F}\cdot\mathbf{h}^0(\mathbf{Q}\cdot\mathbf{F}) = (\det \mathbf{F})\mathbf{Q}\cdot\mathbf{g}(\mathbf{F})\cdot\mathbf{Q}^T$$

by Eq. (28). Successive multiplication by $\mathbf{Q}^T$ and $\mathbf{F}^{-1}$ then yields

$$\mathbf{h}^0(\mathbf{Q}\cdot\mathbf{F}) = (\det \mathbf{F})\mathbf{F}^{-1}\cdot\mathbf{g}(\mathbf{F})\cdot\mathbf{Q}^T$$

or, by Eq. (45), we obtain

$$\mathbf{h}^0(\mathbf{Q}\cdot\mathbf{F}) = \mathbf{h}^0(\mathbf{F})\cdot\mathbf{Q}^T \tag{6.7.46}$$

as the identity the function $\mathbf{h}^0(\mathbf{F})$ must satisfy for arbitrary orthogonal $\mathbf{Q}$. Note that this differs from Eq. (28) in lacking the factor $\mathbf{Q}$ premultiplied into the function on the right side. For the transposed tensor $\mathbf{T}_R = (\mathbf{T}_0)^T$ assumed given by

$$\mathbf{T}_R = \mathbf{h}(\mathbf{F}) \tag{6.7.47}$$

the comparable identity is

$$\mathbf{h}(\mathbf{Q}\cdot\mathbf{F}) = \mathbf{Q}\cdot\mathbf{h}(\mathbf{F}) \tag{6.7.48}$$

for arbitrary orthogonal $\mathbf{Q}$. Similarly if the equation

$$\tilde{\mathbf{T}} = \tilde{\mathbf{h}}(\mathbf{F}) \tag{6.7.49}$$

is assumed, Eqs. (40) and (47) yield

$$\mathbf{F}\cdot\tilde{\mathbf{h}}(\mathbf{F}) = \mathbf{h}(\mathbf{F})$$

or

$$\tilde{\mathbf{h}}(\mathbf{F}) = \mathbf{F}^{-1}\cdot\mathbf{h}(\mathbf{F}) \tag{6.7.50}$$

whence

$$\tilde{\mathbf{h}}(\mathbf{Q}\cdot\mathbf{F}) = (\mathbf{Q}\cdot\mathbf{F})^{-1}\cdot\mathbf{h}(\mathbf{Q}\cdot\mathbf{F})$$

or

$$\mathbf{Q}\cdot\mathbf{F}\cdot\tilde{\mathbf{h}}(\mathbf{Q}\cdot\mathbf{F}) = \mathbf{h}(\mathbf{Q}\cdot\mathbf{F}) = \mathbf{Q}\cdot\mathbf{h}(\mathbf{F})$$

by Eq. (48). Hence

$$\tilde{\mathbf{h}}(\mathbf{Q}\cdot\mathbf{F}) = \mathbf{F}^{-1}\cdot\mathbf{h}(\mathbf{F})$$

or, by Eq. (50),

$$\tilde{\mathbf{h}}(\mathbf{Q}\cdot\mathbf{F}) = \tilde{\mathbf{h}}(\mathbf{F}) \tag{6.7.51}$$

is the identity to be satisfied by $\tilde{\mathbf{h}}$ for arbitrary orthogonal $\mathbf{Q}$.

Additional forms are given by Truesdell and Noll (1965) and in references cited by them, including one for the "convected stress" $\bar{\mathbf{T}} \equiv \mathbf{F}^T\cdot\mathbf{T}\cdot\mathbf{F}$. They also give reduced forms for the functional constitutive equations of history-dependent simple materials.

Small rotation case. When there is no rotation, the rotation tensor $\mathbf{R}$ is the unit tensor $\mathbf{1}$ (see Sec. 4.6). For rotation sufficiently small that all the rectangular Cartesian components $R_{km} - \delta_{km}$ of $\mathbf{R} - \mathbf{1}$ are small compared to unity, $\mathbf{R}$ may be replaced by $\mathbf{1}$ in the reduced forms of Eqs. (31) to (33), giving

$$\mathbf{T} = \mathbf{g}(\mathbf{U}) \qquad \mathbf{T} = \mathbf{f}(\mathbf{C}) \qquad \mathbf{T} = \mathbf{G}(\mathbf{E}) \tag{6.7.52}$$

to sufficient accuracy. Thus the problem of frame-indifference is not important when the rotations are required to be small, as they are in classical small-displacement elasticity, for example. Applications with finite strain but small rotations can also be treated by one of the Eqs. (52).

Hyperelastic or Green-elastic solid. The foregoing discussion of elasticity made no assumption of the existence of a strain-energy function. When the strain-energy function or elastic potential function is assumed to exist, the reduction is even simpler. In Eq. (6.2.17b) the hyperelastic formulation of Truesdell and Noll (1965) was given,

$$T_{km} = \rho x_{k,J}\frac{\partial\sigma(\mathbf{F})}{\partial x_{m,J}} = \rho x_{m,J}\frac{\partial\sigma(\mathbf{F})}{\partial x_{k,J}} \tag{6.7.53}$$

where σ is the strain energy per unit mass. If the scalar σ is frame-indifferent, then for any orthogonal $\mathbf{Q}$

$$\sigma(\mathbf{Q}\cdot\mathbf{F}) = \sigma(\mathbf{F}).$$

In particular, choosing for $\mathbf{Q}$ the inverse $\mathbf{R}^{-1} = \mathbf{R}^T$, we have, since $\mathbf{F} = \mathbf{R}\cdot\mathbf{U}$ whence $\mathbf{R}^{-1}\cdot\mathbf{F} = \mathbf{1}\cdot\mathbf{U} = \mathbf{U}$, the result that

$$\sigma(\mathbf{U}) = \sigma(\mathbf{F}),$$

that is σ depends on $\mathbf{F}$ only through its dependence on $\mathbf{U}$. Since $\mathbf{U} = \mathbf{C}^{1/2}$, we may define

$$\bar{\sigma}(\mathbf{C}) = \sigma(\mathbf{C}^{1/2})$$

and alternatively satisfy frame-indifference by taking the strain energy per unit mass to be a function $\bar{\sigma}(C)$. Now

$$\frac{\partial \bar{\sigma}}{\partial x_{k,J}} = \frac{\partial \bar{\sigma}}{\partial C_{IJ}} x_{k,I} + \frac{\partial \bar{\sigma}}{\partial C_{JI}} x_{k,I}.$$

Hence Eq. (53) gives the Cauchy stress T_{km} as

$$T_{km} = \rho x_{k,I} x_{m,J} \left[\frac{\partial \bar{\sigma}}{\partial C_{IJ}} + \frac{\partial \bar{\sigma}}{\partial C_{JI}} \right]. \tag{6.7.54}$$

In terms of the Piola-Kirchhoff stresses $\mathbf{T}^0$ and $\tilde{\mathbf{T}}$, we have

$$T^0_{Pm} = \rho_0 x_{m,J} \left[\frac{\partial \bar{\sigma}}{\partial C_{PJ}} + \frac{\partial \bar{\sigma}}{\partial C_{JP}} \right] \tag{6.7.55a}$$

and

$$\tilde{T}_{PQ} = \rho_0 \left[\frac{\partial \bar{\sigma}}{\partial C_{PQ}} + \frac{\partial \bar{\sigma}}{\partial C_{QP}} \right] \tag{6.7.55b}$$

or, if $\bar{\sigma}$ is symmetrized in C_{IJ} and C_{JI} (by replacing each by half their sum),

$$\tilde{T}_{PQ} = 2\rho_0 \frac{\partial \bar{\sigma}}{\partial C_{PQ}} \tag{6.7.55c}$$

as given in Eq. (6.2.17c). Equations (55a) and (55b) follow from Eq. (54), since

$$T^0_{Pm} = \frac{\rho_0}{\rho} X_{P,k} T_{km} \text{ and } \tilde{T}_{PQ} = X_{Q,m} T^0_{Pm}$$

by Eqs. (5.3.22) and (5.3.23), and since

$$X_{P,k} x_{k,I} = \delta_{PI} \quad \text{and} \quad X_{Q,m} x_{m,J} = \delta_{QJ}.$$

Reduced form for the functional of a simple material. We let a superscript t on a function symbol denote the history of the function. For example, let

$$\mathbf{F}^t \equiv \mathbf{F}^t(s) = \mathbf{F}(t - s) \quad \text{for all } s \geq 0. \tag{6.7.56}$$

Then the functional definition of a simple material is

$$\mathbf{T}(t) = \mathop{\mathscr{G}}_{s=0}^{\infty}\{\mathbf{F}^t(s)\}. \tag{6.7.57}$$

We often omit the limits $s = 0$ and ∞ and the argument s of the histories for simplicity in the discussion that follows.

The principle of material frame-indifference requires the response functional $\mathscr{G}$ to satisfy

$$\mathscr{G}\{\mathbf{Q}^t \cdot \mathbf{F}^t\} = \mathbf{Q}(t) \cdot \mathscr{G}\{\mathbf{F}^t\} \cdot \mathbf{Q}(t)^T \tag{6.7.58}$$

identically in the histories $\mathbf{F}^t$ of any nonsingular tensor $\mathbf{F}$ and $\mathbf{Q}^t$ of any orthogonal tensor $\mathbf{Q}$. This leads to a reduced form for $\mathscr{G}$ in exactly the same way as in the derivation leading up to Eq. (31), with $\mathbf{F}^t = \mathbf{R}^t \cdot \mathbf{U}^t$ and a particular choice $\mathbf{Q}^t = (\mathbf{R}^t)^T$. The result is analogous to Eq. (31):

$$\mathbf{T}(t) = \mathbf{R}(t) \cdot \mathscr{G}\{\mathbf{U}^t(s)\} \cdot \mathbf{R}(t)^T \tag{6.7.59}$$

[see Truesdell and Noll (1965), Eq. (29.3)], or equivalently, since $\mathbf{U}^t = (\mathbf{C}^{1/2})^t$,

$$\mathbf{T}(t) = \mathbf{R}(t) \cdot \mathscr{H}\{\mathbf{C}^t(s)\} \cdot \mathbf{R}(t)^T \tag{6.7.60}$$

where $\mathscr{H}\{\mathbf{C}^t\} \equiv \mathscr{G}\{(\mathbf{C}^{1/2})^t\}$.

This important result shows that only the present value of the rotation, $\mathbf{R}(t)$, enters instead of the rotation history, while the stress, initially assumed to depend on the history $\mathbf{F}^t$ of the deformation gradient $\mathbf{F}$, can only depend on the history of one of the strain measures, such as the right stretch $\mathbf{U}$, the deformation tensor $\mathbf{C}$, or the strain $\mathbf{E} = \frac{1}{2}(\mathbf{C} - \mathbf{1})$. Hence it is possible, at least in principle, to reduce the number of tests needed to determine the response functional. If we can determine the form of the response functional giving the stress $\mathbf{T}$ for all homogeneous pure stretch histories (with $\mathbf{R}^t \equiv \mathbf{1}$), then we can calculate $\mathbf{T}$ for any history by including the present rotation as in Eq. (59) or Eq. (60).

Many other reduced forms are possible. Since $\mathbf{U} \cdot \mathbf{U}^{-1} = \mathbf{U}^{-1} \cdot \mathbf{U} = \mathbf{1}$, Eq. (60) may be written as

$$\mathbf{T} = \mathbf{R} \cdot \mathbf{U} \cdot \mathbf{U}^{-1} \cdot \mathscr{H}\{\mathbf{C}^t\} \cdot \mathbf{U}^{-1} \cdot \mathbf{U} \cdot \mathbf{R}^T.$$

Since $\mathbf{R} \cdot \mathbf{U} = \mathbf{F}$ and $\mathbf{U} \cdot \mathbf{R}^T = \mathbf{F}^T$ ($\mathbf{U}$ is symmetric) and $\mathbf{U}^{-1} = \mathbf{C}^{-1/2}$, if we define $\mathscr{L}$ by

$$\mathscr{L}\{\mathbf{C}^t(s)\} \equiv \mathbf{C}^{-1/2}(t) \cdot \mathscr{H}\{\mathbf{C}^t(s)\} \cdot \mathbf{C}^{-1/2}(t) \tag{6.7.61a}$$

we obtain the reduced form

$$\mathbf{T}(t) = \mathbf{F}(t) \cdot \mathscr{L}\{\mathbf{C}^t(s)\} \cdot \mathbf{F}^T(t) \tag{6.7.61b}$$

[see Truesdell and Noll (1965), Eq. (29.11)]. Reduced forms analogous to those given for elasticity in Eqs. (37) to (43) are also possible in terms of the Piola-Kirchhoff stress tensors. We give here only the analogue to Eq. (42). If we replace $\mathbf{C}^t(s)$ by $\mathbf{C}(t - s)$ [see Eq. (56)], we can write the analogue as

$$\tilde{\mathbf{T}}(t) = \mathscr{F}\{\mathbf{C}(t - s)\}. \tag{6.7.62}$$

In Sec. 6.8, the component form of this equation, Eq. (6.8.17a) will be taken as the starting point for a reference to Rivlin's further reduction of the equation of a simple material with memory by imposing material symmetry conditions. We now return to a consideration of more elementary constitutive equations, and show first that the rate-of-deformation tensor is frame-indifferent, and then that this makes the constitutive equations of Newtonian viscosity frame-indifferent.

Rate of deformation is frame-indifferent. We have seen in Eq. (11a) that particle velocity is not frame-indifferent. But we can show that the rate of deformation $\mathbf{D} = \frac{1}{2}(\mathbf{v}\overleftarrow{\nabla} + \overrightarrow{\nabla}\mathbf{v})$ is frame-indifferent, as follows. Equation (6) shows that at time t we have $d\mathbf{x} = \mathbf{Q}(t)^T \boldsymbol{\cdot} d\mathbf{x}^*$, but the formula for the total differential of a function $\mathbf{x}$ in terms of the spatial variables $\mathbf{x}^*$ shows that $d\mathbf{x} = (\mathbf{x}\overleftarrow{\nabla}_{\mathbf{x}^*}) \boldsymbol{\cdot} d\mathbf{x}^*$. Hence

$$\mathbf{x}\overleftarrow{\nabla}_{\mathbf{x}^*} = \mathbf{Q}^T \quad \text{and} \quad \overrightarrow{\nabla}_{\mathbf{x}^*}\mathbf{x} = \mathbf{Q}. \tag{6.7.63}$$

Furthermore, at time t, evaluating the spatial gradients of the ***same vector*** $\mathbf{v}$ in the two special Cartesian coordinate systems, one attached to each frame, shows that

$$\mathbf{v}\overleftarrow{\nabla}_{\mathbf{x}^*} = \mathbf{v}\overleftarrow{\nabla}_{\mathbf{x}} \boldsymbol{\cdot} \mathbf{Q}(t)^T \quad \text{and} \quad \overrightarrow{\nabla}_{\mathbf{x}^*}\mathbf{v} = \mathbf{Q}(t) \boldsymbol{\cdot} \overrightarrow{\nabla}_{\mathbf{x}}\mathbf{v}, \tag{6.7.64}$$

where, for example, $\mathbf{v}\overleftarrow{\nabla}_{\mathbf{x}^*}$ (for fixed t) can be represented by the component matrix $[\partial v_i/\partial x_k^*]$ while $\mathbf{v}\overleftarrow{\nabla}_{\mathbf{x}}$ is represented by $[\partial v_i/\partial x_j]$. (See Ex. 7.)

Now consider $\mathbf{v}^*$, which is a different vector from $\mathbf{v}$. From the equation before Eq. (11a),

$$\mathbf{v}^* = \dot{\mathbf{c}} + \dot{\mathbf{Q}} \boldsymbol{\cdot} \mathbf{x} + \mathbf{Q} \boldsymbol{\cdot} \mathbf{v}. \tag{6.7.65}$$

Then, since $\dot{\mathbf{c}}$, $\dot{\mathbf{Q}}$, and $\mathbf{Q}$ are independent of $\mathbf{x}^*$, we can take the gradient of Eq. (65) with respect to the spatial variable $\mathbf{x}^*$ to obtain

$$\mathbf{v}^*\overleftarrow{\nabla}_{\mathbf{x}^*} = \dot{\mathbf{Q}} \boldsymbol{\cdot} (\mathbf{x}\overleftarrow{\nabla}_{\mathbf{x}^*}) + \mathbf{Q} \boldsymbol{\cdot} (\mathbf{v}\overleftarrow{\nabla}_{\mathbf{x}^*}).$$

Substituting the first Eq. (63) and the first Eq. (64) in this, we obtain

$$\mathbf{v}^*\overleftarrow{\nabla}_{\mathbf{x}^*} = \dot{\mathbf{Q}} \boldsymbol{\cdot} \mathbf{Q}^T + \mathbf{Q} \boldsymbol{\cdot} (\mathbf{v}\overleftarrow{\nabla}_{\mathbf{x}}) \boldsymbol{\cdot} \mathbf{Q}^T. \tag{6.7.66}$$

We now rewrite Eq. (65) as

$$\mathbf{v}^* = \dot{\mathbf{c}} + \mathbf{x}\cdot\dot{\mathbf{Q}}^T + \mathbf{v}\cdot\mathbf{Q}^T$$

and calculate the other gradient with respect to $\mathbf{x}^*$, namely

$$\vec{\nabla}_{\mathbf{x}^*}\mathbf{v}^* = (\vec{\nabla}_{\mathbf{x}^*}\mathbf{x})\cdot\dot{\mathbf{Q}}^T + (\vec{\nabla}_{\mathbf{x}^*}\mathbf{v})\cdot\mathbf{Q}^T.$$

Substituting the second Eq. (63) and the second Eq. (64) in this, we obtain

$$\vec{\nabla}_{\mathbf{x}^*}\mathbf{v}^* = \mathbf{Q}\cdot\dot{\mathbf{Q}}^T + \mathbf{Q}\cdot(\vec{\nabla}_{\mathbf{x}}\mathbf{v})\cdot\mathbf{Q}^T. \tag{6.7.67}$$

Eqs. (66) and (67) show that the velocity gradients are not frame-indifferent because of the extra term $\dot{\mathbf{Q}}\cdot\mathbf{Q}^T$ or $\mathbf{Q}\cdot\dot{\mathbf{Q}}^T$ on the right sides. But if we add Eqs. (66) and (67), since

$$\dot{\mathbf{Q}}\cdot\mathbf{Q}^T + \mathbf{Q}\cdot\dot{\mathbf{Q}}^T = \frac{d}{dt}(\mathbf{Q}\cdot\mathbf{Q}^T) = 0,$$

we obtain, after dividing by two,

$$\mathbf{D}^* = \mathbf{Q}\cdot\mathbf{D}\cdot\mathbf{Q}^T, \tag{6.7.68a}$$

which shows that the rate-of-deformation tensor is frame-indifferent.

The spin tensor $\mathbf{W}$ is of course not frame-indifferent. In fact, half the difference of Eqs. (66) and (67) gives

$$\mathbf{W}^* = \tfrac{1}{2}(\dot{\mathbf{Q}}\cdot\mathbf{Q}^T - \mathbf{Q}\cdot\dot{\mathbf{Q}}^T) + \mathbf{Q}\cdot\mathbf{W}\cdot\mathbf{Q}^T$$

or

$$\mathbf{W}^* = \mathbf{A} + \mathbf{Q}\cdot\mathbf{W}\cdot\mathbf{Q}^T \tag{6.7.68b}$$

where $\mathbf{A} = \dot{\mathbf{Q}}\cdot\mathbf{Q}^T = -\mathbf{Q}\cdot\dot{\mathbf{Q}}^T$ is the angular velocity tensor of the unstarred frame relative to the starred frame, as in Eq. (12).

Newtonian viscosity is frame-indifferent. This is true because its linear constitutive equation, Eq. (6.3.14),

$$\mathbf{T}' = 2\mu\mathbf{D}'$$

automatically requires

$$(\mathbf{T}')^* = \mathbf{Q}\cdot\mathbf{T}'\cdot\mathbf{Q}^T$$

whenever the independent tensor variable $\mathbf{D}'$ is replaced by

$$(\mathbf{D}')^* = \mathbf{Q}\cdot\mathbf{D}'\cdot\mathbf{Q}^T.$$

Here the primes denote deviators. The scalar equation for the mean pressure is unaffected by the change of frame. Frame-indifference does place restrictions on possible nonlinear viscous constitutive equations. In fact frame-indifference implies that the function $\mathbf{f}(\mathbf{D})$ in Eq. (6.3.8) must be an isotropic tensor function. See, for example, Truesdell and Toupin (1960), Sec. 299.

Co-rotational and convected stress rates, stress flux. Some particular constitutive equations that have been proposed include the material time derivative $\dot{\mathbf{T}}$ of the stress $\mathbf{T}$ among the constitutive variables, for example the Maxwell model and the standard linear solid of linear viscoelasticity (see Sec. 6.4) and the Prandtl-Reuss elastic plastic material (Sec. 6.5). Such equations do not satisfy the principle of material frame-indifference for arbitrary motions. One part of the difficulty is that the material derivative $\dot{\mathbf{T}}$ does not transform according to Eq. (7) under an orthogonal change of the spatial reference frame even though $\mathbf{T}$ does. That is, $\dot{\mathbf{T}}$ is not frame-indifferent, even though $\mathbf{T}$ is. We shall demonstrate that $\dot{\mathbf{T}}$ is not frame-indifferent, and in the demonstration we shall discover a group of terms which together are frame-indifferent. This group is called the ***co-rotational stress rate*** and denoted by $\mathring{\mathbf{T}}$ with a superposed small circle instead of a dot. It is equal to the material derivative of the stress as it would appear to an observer in a frame of reference attached to the particle and rotating with it at an angular velocity equal to the instantaneous value of the angular velocity $\mathbf{w}$ of the material. We define:

$$\text{Co-rotational stress rate} \qquad \mathring{\mathbf{T}} = \dot{\mathbf{T}} - \mathbf{W}\cdot\mathbf{T} + \mathbf{T}\cdot\mathbf{W}, \tag{6.7.69}$$

where $\mathbf{W}$ is the spin tensor of Sec. 4.4.

To lead to the definition of Eq. (69) we begin by differentiating the equation

$$\mathbf{T}^* = \mathbf{Q}\cdot\mathbf{T}\cdot\mathbf{Q}^T \tag{6.7.70}$$

given by Eq. (7), obtaining

$$\dot{\mathbf{T}}^* = \mathbf{Q}\cdot\dot{\mathbf{T}}\cdot\mathbf{Q}^T + \dot{\mathbf{Q}}\cdot\mathbf{T}\cdot\mathbf{Q}^T + \mathbf{Q}\cdot\mathbf{T}\cdot\dot{\mathbf{Q}}^T. \tag{6.7.71}$$

Because of the last two terms in Eq. (71), the tensor $\dot{\mathbf{T}}$ is not frame-indifferent. We now eliminate $\dot{\mathbf{Q}}$ and $\dot{\mathbf{Q}}^T$ from Eq. (71) as follows. We substitute $\mathbf{A} = \dot{\mathbf{Q}}\cdot\mathbf{Q}^T$ in Eq. (68b) and then multiply Eq. (68b) by $\mathbf{Q}$ from the right, to obtain

$$\mathbf{W}^*\cdot\mathbf{Q} = \dot{\mathbf{Q}}\cdot\mathbf{Q}^T\cdot\mathbf{Q} + \mathbf{Q}\cdot\mathbf{W}\cdot\mathbf{Q}^T\cdot\mathbf{Q},$$

whence, since $\mathbf{Q}^T\cdot\mathbf{Q} = \mathbf{1}$, we obtain

$$\begin{aligned} \dot{\mathbf{Q}} &= \mathbf{W}^*\cdot\mathbf{Q} - \mathbf{Q}\cdot\mathbf{W} \quad \text{with transpose} \\ \dot{\mathbf{Q}}^T &= -\mathbf{Q}^T\cdot\mathbf{W}^* + \mathbf{W}\cdot\mathbf{Q}^T. \end{aligned} \tag{6.7.72}$$

(Recall that $\mathbf{W}$ and $\mathbf{W}^*$ are skew, so that, for example $\mathbf{W}^T = -\mathbf{W}$.) Substituting Eqs. (72) into Eq. (71) and collecting terms, we obtain by using Eq. (70) the result

$$(\dot{\mathbf{T}}^* - \mathbf{W}^*\cdot\mathbf{T}^* + \mathbf{T}^*\cdot\mathbf{W}^*) = \mathbf{Q}\cdot(\dot{\mathbf{T}} - \mathbf{W}\cdot\mathbf{T} + \mathbf{T}\cdot\mathbf{W})\cdot\mathbf{Q}^T, \qquad (6.7.73a)$$

which takes the following form after simplification by use of Eq. (69):

$$\overset{\circ}{\mathbf{T}}{}^* = \mathbf{Q}\cdot\overset{\circ}{\mathbf{T}}\cdot\mathbf{Q}^T. \qquad (6.7.73b)$$

This shows that the co-rotational stress rate $\overset{\circ}{\mathbf{T}}$ is frame-indifferent. When the material is not rotating with respect to the observer's frame of reference, we have $\mathbf{W} = \mathbf{0}$, and $\overset{\circ}{\mathbf{T}}$ reduces to the usual material derivative $\dot{\mathbf{T}}$. A proposed constitutive equation involving $\dot{\mathbf{T}}$ may sometimes be made frame-indifferent by merely replacing $\dot{\mathbf{T}}$ by $\overset{\circ}{\mathbf{T}}$, provided that all the other independent and dependent variables are frame-indifferent, but the reader is warned that the material derivatives of strain $\mathbf{E}$ and deformation $\mathbf{C}$ are not frame-indifferent. The replacement of $\dot{\mathbf{T}}$ by $\overset{\circ}{\mathbf{T}}$ may be accomplished either by actually using the symbol $\overset{\circ}{\mathbf{T}}$ or alternatively by substituting the expression $\dot{\mathbf{T}} - \mathbf{W}\cdot\mathbf{T} + \mathbf{T}\cdot\mathbf{W}$ in place of $\dot{\mathbf{T}}$, thus modifying the form of the constitutive dependence on $\dot{\mathbf{T}}$.

A frame-indifferent stress rate such as $\overset{\circ}{\mathbf{T}}$ is sometimes called a ***stress flux***. There are many different kinds of stress fluxes. Any one is as good as another, though they are not equal to each other. Besides the co-rotational stress rate $\overset{\circ}{\mathbf{T}}$ another frequently used stress flux is the ***convected stress rate*** $\overset{\triangle}{\mathbf{T}}$, which is obtained by adding the frame-indifferent quantity $\mathbf{D}\cdot\mathbf{T} + \mathbf{T}\cdot\mathbf{D}$ to $\overset{\circ}{\mathbf{T}}$. Thus, since $\mathbf{D} - \mathbf{W} = \mathbf{L}^T$ and $\mathbf{D} + \mathbf{W} = \mathbf{L}$, where $\mathbf{L}$ is the velocity-gradient tensor,

$$\overset{\triangle}{\mathbf{T}} = \overset{\circ}{\mathbf{T}} + \mathbf{D}\cdot\mathbf{T} + \mathbf{T}\cdot\mathbf{D} \qquad (6.7.74)$$

yields

$$\text{Convected stress rate} \qquad \overset{\triangle}{\mathbf{T}} = \dot{\mathbf{T}} + \mathbf{L}^T\cdot\mathbf{T} + \mathbf{T}\cdot\mathbf{L}. \qquad (6.7.75)$$

Higher-order co-rotational and convected stress rates may also appear in constitutive equations of the "rate type." See Truesdell and Noll (1965), Sec.36.

Rivlin-Ericksen tensors. That the material derivatives $d\mathbf{E}/dt$ and $d\mathbf{C}/dt$ of the strain tensor and the deformation tensor are not frame-indifferent may be seen from Eqs. (4.5.13). For example,

$$\frac{d\mathbf{C}}{dt} = 2\mathbf{F}^T\cdot\mathbf{D}\cdot\mathbf{F}. \qquad (6.7.76)$$

The rate-of-deformation tensor $\mathbf{D}$ is frame-indifferent, but the deformation gradients $\mathbf{F}$ and $\mathbf{F}^T$ are not. Hence $d\mathbf{C}/dt$ is not frame-indifferent. Higher-order material time derivatives of $\mathbf{T}$, $\mathbf{E}$, and $\mathbf{C}$ also fail to be frame-indifferent. A set of higher-order frame-indifferent tensors related to the derivatives of **C** was introduced by Rivlin and Ericksen (1955). We denote these tensors, known as the ***Rivlin-Ericksen tensors*** by $\mathbf{A}^{(n)}$ for $n = 1,2,\ldots$. These tensors arise naturally when Eq. (4.4.11) is generalized by taking higher-order material time derivatives of $(ds)^2$. The Rivlin-Ericksen tensors can be defined by

$$\frac{d^n}{dt^n}[(ds)^2] = d\mathbf{x}\cdot\mathbf{A}^{(n)}\cdot d\mathbf{x}. \qquad (6.7.77)$$

By comparison with Eq. (4.4.11) we see that

$$\mathbf{A}^{(1)} = 2\mathbf{D}. \qquad (6.7.78)$$

Material differentiation of Eq. (77) and use of Eq. (4.4.10b) lead to the recursion formula

$$\mathbf{A}^{(n+1)} = \frac{d\mathbf{A}^{(n)}}{dt} + \mathbf{A}^{(n)}\cdot\mathbf{L} + \mathbf{L}^T\cdot\mathbf{A}^{(n)}. \qquad (6.7.79)$$

Since $(ds)^2 = d\mathbf{X}\cdot\mathbf{C}\cdot d\mathbf{X}$ by Eq. (4.5.4a), the higher-order material derivatives $(ds)^2$ can also be expressed as

$$\frac{d^n}{dt^n}[(ds)^2] = d\mathbf{X}\cdot\frac{d^n\mathbf{C}}{dt^n}\cdot d\mathbf{X}, \qquad (6.7.80)$$

while, by use of Eq. (4.5.2a), we can write Eq. (77) as

$$\frac{d^n}{dt^n}[(ds)^2] = (d\mathbf{X}\cdot\mathbf{F}^T)\cdot\mathbf{A}^{(n)}\cdot(\mathbf{F}\cdot d\mathbf{X}) = d\mathbf{X}\cdot(\mathbf{F}^T\cdot\mathbf{A}^{(n)}\cdot\mathbf{F})\cdot d\mathbf{X}.$$

Comparison of this with Eq. (80) then shows that

$$\frac{d^n\mathbf{C}}{dt^n} = \mathbf{F}^T\cdot\mathbf{A}^{(n)}\cdot\mathbf{F}. \qquad (6.7.81)$$

For further discussion of the Rivlin-Ericksen tensors see, for example, Truesdell and Toupin (1960), Eringen (1967), and Eringen (1962). Truesdell and Noll (1965) define the $\mathbf{A}^{(n)}$ in a different (but equivalent) way in terms of the rates of the ***relative*** deformation tensor $\mathbf{C}_t$ as

$$\mathbf{A}^{(n)}(t) = \frac{d^n}{d\tau^n}[\mathbf{C}_t(\tau)]\bigg|_{\tau=t}. \qquad (6.7.82)$$

They also define a different set of frame-indifferent higher-order rates, which they call the ***stretching tensors*** $\mathbf{D}_n$, defined in terms of the relative right stretch $\mathbf{U}_t$ by

$$\mathbf{D}_n(t) = \frac{d^n}{d\tau^n}[\mathbf{U}_t(\tau)]\bigg|_{\tau=t} \cdot \tag{6.7.83}$$

In particular

$$\mathbf{D}_1 = \mathbf{D} = \tfrac{1}{2}\mathbf{A}^{(1)}. \tag{6.7.84}$$

These definitions in terms of the relative tensors $\mathbf{C}_t$ and $\mathbf{U}_t$ are especially suitable, since the Rivlin-Ericksen tensors $\mathbf{A}^{(n)}$ and the stretching tensors $\mathbf{D}_n$, like the rate of deformation $\mathbf{D}$, are usually applied in the spatial description of the motion.

Summary. In this section we began by mentioning briefly the principle of equipresence which requires that any constitutive variable present in one constitutive equation of a material should be assumed present in all constitutive equations of that material, unless it can be shown that its presence is prohibited by some other principle. For the major part of this section, however, we have limited ourselves to purely mechanical phenomena in simple materials for which there is only one independent constitutive variable, the deformation gradient, and one dependent variable, the stress. We stated three fundamental postulates for any constitutive theory of purely mechanical phenomena. The principle of determinism for stress and the principle of local action together imply that the stress at a particle is determined by the history of the motion of an arbitrarily small neighborhood of the particle. This motivated the definition of a simple material where the stress at a particle depends only on the history of the deformation gradient at the particle. The third postulate was the principle of material frame-indifference which was presented in detail and used to obtain reduced forms for the functions appearing in the constitutive equations of elasticity and for the functionals appearing in the constitutive equations of a simple material with memory.

We have seen that the classical theories of linear viscosity and of small-rotation elasticity satisfy the requirements of frame-indifference. In linear viscoelasticity also, difficulty is only encountered when the rotation tensor is too large to be approximated satisfactorily by the unit tensor. But we have seen that the material derivatives $\dot{\mathbf{T}}$, $\dot{\mathbf{E}}$, and $\dot{\mathbf{C}}$ are not frame-indifferent, so that constitutive equations containing them do not in general satisfy the principle of material frame-indifference for arbitrary motions. Finally we introduced the co-rotational stress rate $\mathring{\mathbf{T}}$, another stress rate called the convected stress rate $\hat{\mathbf{T}}$, and the Rivlin-Ericksen tensors $\mathbf{A}^{(n)}$ and stretchings $\mathbf{D}_n$(related to the strain rates), which are frame-indifferent.

In Sec. 6.8 we give a brief account of some of the further reductions the modern theories of constitutive equations make possible in the forms of the constitutive equations of materials with isotropic material properties or with material symmetries less than isotropy.

EXERCISES

1. Carry out details of the development leading to Eq. (6.7.4b).
2. Show that $\mathbf{Q}\cdot\dot{\mathbf{Q}}^T = -\dot{\mathbf{Q}}\cdot\mathbf{Q}^T$, so that Eq. (6.7.12) can alternatively be written as $\mathbf{A} = -\mathbf{Q}\cdot\dot{\mathbf{Q}}^T$.
3. Use Eq. (6.7.63) to derive Eq. (6.7.64).
4. (a) If the ***convected stress tensor*** $\mathbf{T}_C$ (sometimes denoted $\bar{\mathbf{T}}$) is defined by $\mathbf{T}_C = \mathbf{F}^T\cdot\mathbf{T}\cdot\mathbf{F}$, show that the principle of material frame-indifference when applied to Eq. (31) leads to the reduced form $\mathbf{T}_C = \boldsymbol{\phi}(\mathbf{U})$.

   (b) Show that the result of (a) can be written $\mathbf{T}_C = \boldsymbol{\psi}(\mathbf{C})$.
5. Make an alternative derivation of Eq. (6.7.41) by beginning with $\tilde{\mathbf{T}} = \tilde{\mathbf{h}}\,(\mathbf{F})$ as in Eq. (49) and showing that the identity of Eq.(51) then leads to the form of Eq. (41) upon use of Eq. (29b).
6. (a) Show that the equation preceding Eq. (6.7.54) is correct.
   (b) Show how Eqs. (55) follow from Eq. (54).
7. Verify Eqs. (6.7.63) and (64) for Cartesian component matrices. Note that
   $\mathbf{x}=\mathbf{Q}^T\cdot(\mathbf{x}^* - \mathbf{c}) = (\mathbf{x}^* - \mathbf{c})\cdot\mathbf{Q}$ or $x_k = Q_{mk}(x_m^* - c_m) = (x_m^* - c_m)Q_{mk}$.
8. Write component forms of Eqs. (6.7.69) and (75).
9. Show that Eqs. (6.7.73) follow from substituting Eqs. (72) into Eq. (71).
10. Derive Eq. (6.7.79) from Eq. (77).

6.8 Theories of constitutive equations II: material symmetry restrictions on constitutive equations of simple materials; isotropy

In the theory of purely mechanical phenomena a ***simple material*** was defined as one in which the stress $\mathbf{T}$ at a particle is a functional of the deformation gradient $\mathbf{F}$ at the particle relative to some reference configuration of the neighborhood of the particle. [See Sec. 6.7, following Eq. (6.7.19).] In this definition no restrictions are placed on the nature of the functional giving the dependence of $\mathbf{T}$ on the history of $\mathbf{F}$. Special cases include the linear hereditary integrals of Sec. 6.4 and also special nonlinear materials in which the history $\mathbf{F}(t - s)$ of $\mathbf{F}$ at times $\tau = t - s$ can be adequately represented by a power series in τ. When the history can be represented by such a power series, the history is determined by $\mathbf{F}$ and its time derivatives (of all orders) with respect to τ evaluated at $\tau = t$, and therefore the stress at time t, which we assumed to depend only on the history of $\mathbf{F}$ up to t, is determined by $\mathbf{F}$ and its derivatives at time t. When the stress depends on only a finite number of these derivatives (truncated power series case), the material is called a ***material of the differential type.*** The constitutive equation is then assumed to give $\mathbf{T}$ as a function of the values at time t of $\mathbf{F}$ and its time derivatives. For this special case the stress is given by an ordinary function of several tensor arguments (the arguments are $\mathbf{F}$ and the derivatives up to the order required), instead of by a functional of the history of the one tensor $\mathbf{F}$ at all previous times.

The basic physical property of a simple material is that its response to deformations homogeneous in a neighborhood of a particle X determines its response to every deformation at X. This follows from the general definition that $\mathbf{T}$ depends only on the history of $\mathbf{F}$ at X, since that history is not affected by whether or not the deformation is homogeneous in the neighborhood of X. A motion $\chi_\kappa(X, t)$ gives rise to homogeneous deformation with respect to a configuration κ if the motion relative to this configuration is described by an equation of the form

$$\mathbf{x} = \chi_\kappa(X,t) = \mathbf{F}(t)\cdot(\mathbf{X} - \mathbf{X}_0) + \mathbf{x}_0(t), \tag{6.8.1}$$

where $\mathbf{x}_0(t)$ is a place in the deformed configuration, and $\mathbf{X}_0$ is a fixed place in the reference configuration κ. Experimenters investigating material properties for possible use in constitutive equations almost always attempt to devise a test specimen such that both the material properties and the deformation are homogeneous throughout the test section of the specimen, for example the gage length of a tensile specimen. This can lead to a constitutive equation applicable to a nonhomogeneous deformation of the material only if the material is simple. When the material is nonsimple, the stress may depend on higher-order gradients, and when the constitutive equations include strain gradients, for example, as independent variables, other types of stress variables are also needed, such as couple stress. See, for example, the discussion of multipolar media in Jaunzemis (1967). Here we consider only nonpolar simple materials and purely mechanical phenomena. Even for this class of materials the full development of modern constitutive theories and their application is beyond the scope of this book. We will only indicate briefly some of the important results concerning material symmetry, omitting most of the proofs, and will cite the original research papers or more recent advanced treatises where further details may be found.

We wish to determine the restrictions placed on the constitutive equations of simple materials by the assumption of isotropy, or of some material symmetry less than isotropy. If possible, also, we wish to obtain explicit reduced forms of the constitutive equations meeting these restrictions. Modern investigations into these problems divide mainly into two groups, those of R. S. Rivlin and his associates, to be discussed in Part 1 of this section, and those of W. Noll and his associates to be discussed in Part 2. Other modern investigators have followed one or the other of these two approaches.† The program initiated by Noll (1958) and presented at length in Truesdell and Noll (1965) has the merit of

†See also Oldroyd (1950).

great generality and elegance. The material symmetry group or isotropy group of the material is defined in terms of changes of reference configurations without introducing coordinate systems. The methods used in the pioneering work of Rivlin and his associates still offer powerful tools to researchers working on particular new constitutive equations. Unfortunately no comprehensive account of them is available. A brief summary is given in Rivlin (1966).

Part 1. Material Symmetry Formulation of Rivlin and Others

The work of Rivlin and his associates, beginning with important papers by Rivlin and Ericksen (1955) and Rivlin (1955) on stress-deformation relations for isotropic materials, deals with the problems of material symmetry by considering orthogonal transformations of the material coordinate systems. The earlier investigations in this series also made continuity assumptions permitting the approximation of functions by polynomials [see, for example, Rivlin (1955) and Pipkin and Rivlin (1959)] or for history-dependent materials continuity assumptions justifying the use of integral representations for functionals [Green and Rivlin (1957); Green, Rivlin, and Spencer (1959); and Rivlin (1960)]. Pipkin and Wineman (1963) and Wineman and Pipkin (1964) have proved that the main results obtained for polynomial constitutive equations remain valid even if the functions involved in the constitutive equations cannot be approximated by polynomials. Thus the assumption of polynomial constitutive equations in order to determine the effects of material symmetry is merely a matter of mathematical convenience. Wineman and Pipkin (1964) also generalize the earlier results of Rivlin and his associates on materials with memory by removing the continuity restrictions. In Noll's theory, on the other hand, the functionals were assumed to be quite general from the outset and included both dependence on the past history of the independent tensor variables and also on their present values, as in Eqs. (6.7.61). Before giving additional references to more general results we consider a few examples to illustrate the method used by Rivlin and his associates.

Hyperelasticity. If as in Sec. 6.2 we write $W = \rho_0\bar{\sigma}$ for the elastic potential or strain-energy function, the constitutive equation of a hyperelastic or Green-elastic material, which we obtained in the reduced form of Eq. (6.7.55a) satisfying the principle of material frame-indifference, is

$$T^0_{Pm} = x_{m,Q}\left(\frac{\partial W}{\partial C_{PQ}} + \frac{\partial W}{\partial C_{QP}}\right). \tag{6.8.2}$$

This constitutive equation is completely determined by the single scalar function $W(C_{PQ})$ of the components of the deformation tensor $\mathbf{C}$. The material symmetry at a particle may be represented by the group $\mathscr{S}$ of orthogonal transformations of the material coordinates which leaves the function W unaltered in form. That is the group of transformations (see Sec. 2.4)

$$\bar{X}_R = A_R^S X_S \qquad A_P^S A_Q^S = \delta_{PQ} \qquad A_R^M A_R^N = \delta_{MN} \tag{6.8.3}$$

for which

$$W(\bar{C}_{PQ}) = W(C_{PQ}) \quad \text{when} \quad \bar{C}_{PQ} = A_P^M A_Q^N C_{MN} \tag{6.8.4}$$

where $A_J^I = \cos(\bar{X}_J, X_I)$. In discussing material symmetry for the linearized theory of elasticity in Sec. 6.2, we characterized the symmetry by the coordinate transformations such that the elastic constants of the generalized Hooke's law were the same in both coordinate systems. For the nonlinear theory here, we require the functional dependence of W on $\bar{C}_{PQ}$ to be the same as that of W on C_{PQ} whenever the $\bar{X}_R$ are related to the X_S by a transformation of the group $\mathscr{S}$. Equation (4) is an implicit statement of the material-symmetry restrictions. We can determine an explicit form for the possible dependence of W on $\mathbf{C}$ under these symmetry restrictions.

Integrity basis. Any polynomial scalar invariant function of any number of tensors under the full orthogonal group of coordinate transformations or under any subgroup of the orthogonal group can be expressed as a polynomial in a ***finite*** number of polynomial scalar invariants, none of which is expressible as a polynomial in the remaining ones. Such a set of polynomial scalar invariants is called an ***irreducible integrity basis*** for the tensors under the transformation group. ***Any scalar invariant, whether polynomial or not, may be expressed as a single-valued function of the elements of this irreducible integrity basis.***

For the hyperelastic solid of Eqs. (2) and (4), we seek an irreducible integrity basis for the single tensor $\mathbf{C}$ under the material symmetry group $\mathscr{S}$ of transformations of the material coordinates. For the ***isotropic case*** the symmetry group $\mathscr{S}$ is the full orthogonal group, and an irreducible integrity basis consists of the three invariants [see Eq. (3.3.9)]:

$$\mathrm{I}_C = \operatorname{tr} \mathbf{C} \qquad \mathrm{II}_C = \tfrac{1}{2}(\mathbf{C}{:}\mathbf{C} - \mathrm{I}_C^2) \qquad \mathrm{III}_C = \det \mathbf{C}. \tag{6.8.5}$$

For the isotropic hyperelastic material, then, W can depend on $\mathbf{C}$ only through these three invariants and, for a given particle X,

$$W = W(\mathbf{I}_C, \mathbf{II}_C, \mathbf{III}_C). \tag{6.8.6}$$

By substituting Eq. (6) into Eq. (2) we obtain the equation for the first Piola-Kirchhoff stress T_{Pm}^{0} in the form

$$T^0_{Pm} = x_{m,Q} \sum_{J=\mathrm{I}}^{\mathrm{III}} \frac{\partial W}{\partial J_C}\left[\frac{\partial J_C}{\partial C_{PQ}} + \frac{\partial J_C}{\partial C_{QP}}\right]. \tag{6.8.7}$$

It can be shown that this leads to the following expression for the Cauchy stress.

$$T_{km} = 2\frac{\rho}{\rho_0}\left\{\left[\frac{\partial W}{\partial \mathrm{I}_C} + \mathrm{I}_C\frac{\partial W}{\partial \mathrm{II}_C}\right]B_{mk} + \frac{\partial W}{\partial \mathrm{II}_C}B_{ms}B_{sk} + \frac{\partial W}{\partial \mathrm{III}_C}\mathrm{III}_C\delta_{km}\right\}, \tag{6.8.8}$$

where $B_{ij} = x_{i,K}x_{j,K}$ is the Finger deformation tensor component or left Cauchy-Green deformation tensor component of Sec. 4.6. See, for example, Rivlin (1966), Eq. (18.4), where I_2 is the negative of our II_C and where W/ρ_0 is taken as the free energy in an isothermal deformation.

For aelotropic hyperelastic material the symmetry group $\mathscr{S}$ is less than the full orthogonal group. The appropriate integrity basis for the group $\mathscr{S}$ then replaces that of Eq. (5). Fortunately an irreducible integrity basis is known for most of the kinds of symmetry of interest. Smith and Rivlin (1958) obtained an integrity basis for the symmetry characterizing each of the 32 crystallographic point groups, and Smith (1962) showed that it was irreducible. The crystal classes have a finite group of symmetries, while isotropy and transverse isotropy are characterized by continuous groups. A material has ***transverse isotropy*** if it has a single axis of rotational symmetry and if moreover every plane containing this axis is a plane of reflection symmetry. Transverse isotropy has been treated by Ericksen and Rivlin (1954). For this case, if the X_3-axis is the axis of rotational symmetry, an irreducible integrity basis consists of

$$\mathrm{I}_C,\ \mathrm{II}_C,\ \mathrm{III}_C,\ C_{33},\quad \text{and} \quad C_{31}^2 + C_{23}^2 \tag{6.8.9}$$

according to Rivlin (1966). Integrity bases for the crystal classes are also summarized in Rivlin (1966). See also Green and Adkins (1960). When the integrity basis is known, the elastic-potential function W of the hyperelastic material can be taken as a function of the basis elements, say $J_1, J_2, \ldots, J_B$,

$$W = W(J_1, J_2, \ldots, J_B) \tag{6.8.10a}$$

and Eq. (2) then yields

$$T^0_{Pm} = x_{m,Q}\sum_{\beta=1}^{B}\frac{\partial W}{\partial J_\beta}\left[\frac{\partial J_\beta}{\partial C_{PQ}} + \frac{\partial J_\beta}{\partial C_{QP}}\right]. \tag{6.8.10b}$$

Tensor-valued constitutive functions. The stress may depend on one or more tensor or vector arguments. An example is the material of the differential type mentioned in the first paragraph of this section, where $\mathbf{T}$ is a function of $\mathbf{F}$ and of a finite number (say N) of the time derivatives of $\mathbf{F}$. Application of the principle of material frame-indifference of Sec. 6.7 [see, for example, Rivlin

(1966)] shows that a reduced form of the equation for a material of the differential type may be obtained giving the second Piola-Kirchhoff stress $\tilde{T}_{AB}$ as a function of the N material derivatives $C_{PQ}^{(n)}$ of the deformation tensor components C_{PQ}.

$$\tilde{T}_{AB} = G_{AB}(C_{PQ}^{(n)}) \qquad n = 0, 1, 2, \ldots, N \tag{6.8.11}$$

where G_{AB} is a tensor-valued function (in general nonlinear) of the N tensors $\mathbf{C}^{(n)}$, and where

$$C_{PQ}^{(n)} = \left.\frac{d^n C_{PQ}(\tau)}{d\tau^n}\right|_{\tau=t}.$$

If the orthogonal transformation of Eq. (3) is in the symmetry group $\mathscr{S}$ of the material, then the components G_{AB} of the tensor-valued response function must satisfy

$$G_{AB}(\bar{C}_{PQ}^{(n)}) = A_A^I\, A_B^J\, G_{IJ}(C_{PQ}^{(n)}) \tag{6.8.12}$$

identically in the tensors $\mathbf{C}^{(n)}$. Recall that $\bar{C}_{PQ}^{(n)} = A_P^R\, A_Q^S\, C_{RS}^{(n)}$. If G_{AB} satisfies Eq. (12) for each transformation of the group $\mathscr{S}$, then G_{AB} is said to be ***form-invariant*** under the group $\mathscr{S}$. We wish to find the most general form of the function G_{AB} satisfying this property of form invariance. If G_{AB} is a polynomial in the components of the argument tensors the reduction can be performed in the following way. [See Rivlin (1966), Pipkin and Wineman (1963), or Pipkin and Rivlin(1959).] Let $\bar{\psi}_{AB}$ and ψ_{AB} be components of an arbitrary symmetric second-order tensor $\boldsymbol{\psi}$ in the two coordinate systems $\bar{X}_R$ and X_S related by the symmetry transformation. Then

$$\bar{\psi}_{AB}G_{AB}(\bar{C}_{PQ}^{(n)}) = \psi_{AB}G_{AB}(C_{PQ}^{(n)}) = f \quad \text{(say)} \tag{6.8.13a}$$

is a scalar invariant polynomial function under the transformation group $\mathscr{S}$. Moreover the invariant scalar function f is linear in the components of $\boldsymbol{\psi}$, though not in those of the $\mathbf{C}^{(n)}$. Because f is linear in the components of $\boldsymbol{\psi}$, we have

$$\tilde{T}_{AB} = G_{AB}(\mathbf{C}^{(n)}) = \frac{1}{2}\left(\frac{\partial f}{\partial \psi_{AB}} + \frac{\partial f}{\partial \psi_{BA}}\right) \tag{6.8.13b}$$

and this equation is independent of the components of $\boldsymbol{\psi}$.

Now suppose that we know the elements of a finite integrity basis for polynomials in the components of the $N + 2$ tensors $\boldsymbol{\psi}$ and the $\mathbf{C}^{(n)}$ for $n = 0,1, \ldots, N$. Let $J_1, J_2, \ldots, J_A$ be those elements of the irreducible integrity basis which are functions only of the components of the $\mathbf{C}^{(n)}$ (independent of the components of $\boldsymbol{\psi}$). Then $J_1, J_2, \ldots, J_A$ are an irreducible integrity basis for the

$N + 1$ tensors $\mathbf{C}^{(n)}$. Let $K_1, K_2, \ldots, K_B$ be the remainder of the elements of an irreducible integrity basis for the $N + 2$ tensors ($\boldsymbol{\psi}$ and the $\mathbf{C}^{(n)}$); these are the elements which depend on the components of $\boldsymbol{\psi}$ as well as on the components of the $\mathbf{C}^{(n)}$, and, since f is linear in the components of ψ, the elements $K_1, K_2, \ldots, K_B$ will be linear in the components of $\boldsymbol{\psi}$. Hence f is expressible in the form

$$f = \sum_{\beta=1}^{B} K_\beta f^{(\beta)}(J_1, \ldots, J_A) \tag{6.8.14a}$$

where the B scalar functions $f^{(\beta)}$ depend only on the J_α and not on the K_β. Eq. (13b) then takes the form

$$\tilde{T}_{AB} = \frac{1}{2}\sum_{\beta=1}^{B} \left(\frac{\partial K_\beta}{\partial \psi_{AB}} + \frac{\partial K_\beta}{\partial \psi_{BA}}\right) f^{(\beta)}(J_1, \ldots, J_A). \tag{6.8.14b}$$

In order to proceed further we need to know the elements J_α and K_β of the irreducible integrity basis. These are known for the isotropic case where $\mathscr{S}$ is the full orthogonal group, and also for the cases of orthotropy and transverse isotropy. We do not reproduce them here, but refer the reader to the original papers. For isotropy [according to Rivlin (1966)], the original derivations of Rivlin and Ericksen (1955), Rivlin (1955), Spencer and Rivlin (1959a, 1959b, 1960), and Spencer (1961) have been more directly obtained by Rivlin and Smith [in press (1969)]. The cases of orthotropy and transverse isotropy were treated by Adkins (1960a, 1960b). A simple example, the case of isotropic elasticity, where G_{AB} is replaced by $\tilde{f}_{AB}(\mathbf{C})$, a tensor-valued function of a single symmetric tensor argument, will be treated briefly below. The foregoing discussion, beginning with Eq. (11), was presented for a material of the differential type, since the $N + 1$ argument tensors were the derivatives $\mathbf{C}^{(n)}$. But the invariance discussion applies equally well to any number of symmetric second-order tensor arguments, which need not be derivatives of the same function. Wineman and Pipkin (1964) consider very general tensor-valued functions of any order, which are functions of N tensors of various orders. They derive quite similar results. The only problem is in knowing the required integrity bases. Wineman and Pipkin give references to papers where the integrity bases are found for the following cases.

Type of Symmetry	*Independent Variables and Source*
Crystal Classes	1 vector, 1 second-order tensor Smith, Smith, and Rivlin (1963)
Crystal Classes	M vectors Smith and Rivlin (1964)
Orthotropy	N second-order tensors Adkins (1960a)

Transverse isotropy	N second-order tensors Adkins (1960a)
Transverse isotropy	M vectors, N second-order tensors Adkins (1960b)
Isotropy	N second-order tensors Spencer and Rivlin (1959, 1960), Spencer (1961)
Isotropy	M vectors, N second-order tensors Spencer and Rivlin (1962)

As an example of the use of Eqs. (11) to (14), consider the case of isotropic finite elasticity without assuming the existence of a strain-energy function. Then by Eq. (6.7.42)

$$\tilde{T}_{AB} = \tilde{f}_{AB}(\mathbf{C}) \quad \text{and} \quad \bar{\psi}_{AB}\tilde{f}_{AB}(\bar{C}_{PQ}) = \psi_{AB}\tilde{f}_{AB}(C_{PQ}) = \tilde{f} \qquad (6.8.15a)$$

replace Eqs. (11) and (13). An irreducible integrity basis for the representation of any polynomial scalar invariant of two symmetric tensors $\boldsymbol{\psi}$ and $\mathbf{C}$ consists of

$$\left.\begin{array}{l} \operatorname{tr}\boldsymbol{\psi}, \quad \operatorname{tr}(\boldsymbol{\psi}^2), \quad \operatorname{tr}(\boldsymbol{\psi}^3), \quad \operatorname{tr}\mathbf{C}, \quad \operatorname{tr}(\mathbf{C}^2), \quad \operatorname{tr}(\mathbf{C}^3) \\ \operatorname{tr}(\boldsymbol{\psi}\cdot\mathbf{C}), \quad \operatorname{tr}(\boldsymbol{\psi}\cdot\mathbf{C}^2), \quad \operatorname{tr}(\boldsymbol{\psi}^2\cdot\mathbf{C}), \quad \operatorname{tr}(\boldsymbol{\psi}^2\cdot\mathbf{C}^2). \end{array}\right\} \qquad (6.8.15b)$$

See, for example, Rivlin (1966) or Eringen (1967). Since $\tilde{f}$ is linear in $\boldsymbol{\psi}$, only those elements which either do not depend on $\boldsymbol{\psi}$ at all or else are linear in the components of $\boldsymbol{\psi}$ are needed to represent $\tilde{f}$. Those linear in the components of $\boldsymbol{\psi}$ are

$$\begin{aligned} K_1 &= \operatorname{tr}\boldsymbol{\psi} = \psi_{IJ}\delta_{IJ}, \qquad K_2 = \operatorname{tr}(\boldsymbol{\psi}\cdot\mathbf{C}) = \psi_{IM}C_{MI}, \\ K_3 &= \operatorname{tr}(\boldsymbol{\psi}\cdot\mathbf{C}^2) = \psi_{IM}C_{MN}C_{NI} \end{aligned} \qquad (6.8.15c)$$

Hence, the formula $\frac{1}{2}[\partial K_\beta/\partial\psi_{AB} + \partial K_\beta/\partial\psi_{BA}]$ yields, for $\beta = 1, 2,$ and 3, respectively, the components of the tensors $\mathbf{1}$, $\mathbf{C}$, and $\mathbf{C}^2$, whence Eq. (14b) becomes

$$\tilde{\mathbf{T}} = \tilde{f}^{(1)}\mathbf{1} + \tilde{f}^{(2)}\mathbf{C} + \tilde{f}^{(3)}\mathbf{C}^2 \qquad (6.8.16)$$

where $\tilde{f}^{(1)}$, $\tilde{f}^{(2)}$, and $\tilde{f}^{(3)}$ are scalar functions of $\operatorname{tr}\mathbf{C}$, $\operatorname{tr}(\mathbf{C}^2)$, and $\operatorname{tr}(\mathbf{C}^3)$. Alternatively the $\tilde{f}^{(\beta)}$ may be taken as functions of I_C, II_C, and III_C or of the three principal values C_1, C_2, C_3. Equation (16) is a representation of the isotropic tensor function $\tilde{\mathbf{f}}$. In the representation the tensorial form is completely determined, and it remains only to determine the three scalar functions $\tilde{f}^{(\beta)}$. This representation is due to Rivlin and Ericksen (1955). Note that the tensorial form of Eq. (16) is the same as that of Eq. (8), which gives the dependence of $\mathbf{T}$ on $\mathbf{B}$

in terms of the tensors **1**, **B**, and $\mathbf{B}^2$ for the case of hyperelasticity. But in the case of hyperelasticity the scalar coefficients are derived from a single scalar invariant function W, while in Eq. (16) there are three undetermined scalar functions $f^{(\beta)}$. Representations like Eq. (16) will be mentioned again at the close of Part 2 of this section. See Eqs. (38) to (40).

Wineman and Pipkin also proved that the consequences of material symmetry obtained for polynomial constitutive equations remain valid even when the functions involved in the constitutive equations cannot be approximated by polynomials. When the G_{AB} are polynomials, the f of Eqs. (13) and (14) is a polynomial, and the B scalar functions $f^{(\beta)}$ are polynomials in the J_α. When G_{AB} is not a polynomial, the result is changed only in that the $f^{(\beta)}$ are single-valued scalar functions of the J_α, but not in general polynomials.

An alternative discussion of materials of the differential type is given in Truesdell and Noll (1965), Sec. 35.

Materials with memory. Starting from the reduced form of Eq. (6.7.62), with $\tau = t - s$,

$$\tilde{T}_{AB} = \mathop{\mathscr{F}_{AB}}_{\tau=-\infty}^{t}\{\mathbf{C}(\tau)\} \tag{6.8.17a}$$

which was obtained by requiring the simple material of Eq. (6.7.57) to satisfy the principle of material frame-indifference, Rivlin (1966) discusses material symmetry restrictions on the functional $\mathscr{F}_{AB}$ in a manner analogous to that of the tensor function G_{AB} of Eq. (6.8.11). Introducing again the arbitrary symmetric tensors $\boldsymbol{\psi}$, and for simplicity omitting the limits on the functionals, we obtain

$$\bar{\psi}_{AB}\mathscr{F}_{AB}\{\bar{C}_{PQ}(\tau)\} = \psi_{AB}\mathscr{F}_{AB}\{C_{PQ}(\tau)\} = \mathscr{F} \quad \text{(say)} \tag{6.8.17b}$$

where $\mathscr{F}$ is a scalar invariant functional depending on the present values of $\boldsymbol{\psi}$ and on the history of **C** from $\tau = -\infty$ to $\tau = t$. According to Rivlin (1966) it can be shown that any such functional may be expressed as a linear functional in the multilinear elements of an irreducible integrity basis for the $N + 1$ tensors $\boldsymbol{\psi}(t)$, $\mathbf{C}(\tau_1)$, $\mathbf{C}(\tau_2)$, ..., $\mathbf{C}(\tau_N)$, where $N + 1$ is the maximum number of tensors entering a table of typical invariants for the irreducible integrity basis under $\mathscr{S}$ of an arbitrary number of symmetric second-order tensors, and as a functional (in general nonlinear) of the multilinear elements of an irreducible integrity basis for $\mathbf{C}(\tau_1)$, $\mathbf{C}(\tau_2)$, ..., $\mathbf{C}(\tau_{N+1})$, where $\tau_1, \tau_2, \ldots, \tau_{N+1}$ are $N + 1$ instants in the time τ in $-\infty < \tau < t$. This was shown by Green and Rivlin (1957) subject to certain continuity restrictions and by Wineman and Pipkin (1964) without the restrictions. Results for the isotropic case are given, for example, in Rivlin (1966). In Part 2 of this section we cite another approach to isotropic materials with memory deriving from the work of Noll (1958).

Part 2. Noll's Formulation

Noll's treatment of the concept of material symmetry is made not in terms of coordinate transformations, but in terms of changes in the reference configurations of the neighborhood of a particle X in the physical body. Consider a simple material defined by the response functional $\mathscr{G}\{\mathbf{F}^t(s)\}$ of Eq. (6.7.57) with respect to a local reference configuration κ of the neighborhood of the particle in question.

$$\mathbf{T} = \mathop{\mathscr{G}}_{s=0}^{\infty}\{\mathbf{F}^t(s)\}. \tag{6.8.18}$$

For a different choice of the reference configuration the response functional is in general different. Let κ and $\hat{\kappa}$ be two reference configurations related by the mapping $\hat{\mathbf{X}} = \boldsymbol{\lambda}(\mathbf{X})$, where $\mathbf{X}$ and $\hat{\mathbf{X}}$ are the positions of the particle X in the configurations κ and $\hat{\kappa}$, respectively. Then, if $\mathbf{P} = \nabla\boldsymbol{\lambda}$, the deformation gradients $\mathbf{F} = \mathbf{x}\overleftarrow{\nabla}_{\mathbf{x}}$ and $\hat{\mathbf{F}} = \mathbf{x}\overleftarrow{\nabla}_{\hat{\mathbf{x}}}$ are related by

$$\mathbf{F} = \hat{\mathbf{F}}\boldsymbol{\cdot}\mathbf{P} \qquad \hat{\mathbf{F}} = \mathbf{F}\boldsymbol{\cdot}\mathbf{P}^{-1} \tag{6.8.19a}$$

[see Truesdell and Noll (1965), Sec. 28 or Truesdell (1965)].

Then Eq. (18) becomes

$$\mathbf{T} = \mathscr{G}\{\mathbf{F}^t\} = \mathscr{G}\{\hat{\mathbf{F}}^t\boldsymbol{\cdot}\mathbf{P}\} \tag{6.8.19b}$$

where for simplicity the limits $s = 0$ and $s = \infty$ and the time-lapse argument s of the histories $\mathbf{F}^t(s) \equiv \mathbf{F}(t - s)$ have been omitted. The gradient $\mathbf{P}$ of the transformation between the two reference configurations is constant in time. Hence $\mathscr{G}\{\hat{\mathbf{F}}^t\boldsymbol{\cdot}\mathbf{P}\}$ is a functional of the history of the deformation gradient $\hat{\mathbf{F}}^t$ with respect to the configuration $\hat{\kappa}$, which we may as well denote by $\hat{\mathscr{G}}\{\mathbf{F}^t\}$

$$\hat{\mathscr{G}}\{\mathbf{F}^t\} = \mathscr{G}\{\hat{\mathbf{F}}^t\boldsymbol{\cdot}\mathbf{P}\}. \tag{6.8.20}$$

Thus, if the material is a simple material in its response to deformation from any one configuration κ, it is a simple material in its response from any other configuration $\hat{\kappa}$, although the form of the response functional is in general different. The functional form may also be different for a different particle. In the present discussion we do not explicitly indicate the dependence on the particle. The whole discussion is concerned with the response functional at a particle.

Equation (20) says that the response functionals $\mathscr{G}$ and $\hat{\mathscr{G}}$ describe the response of the ***same material*** to the deformation-gradient history from reference configurations κ and $\hat{\kappa}$, respectively.

For simple materials a particle is said to be ***materially isomorphic*** to itself in a nontrivial manner [see Truesdell and Noll (1965), Secs. 27 and 30] if there exist two local reference configurations κ and $\hat{\kappa}$ of the same particle X such that:

(a) κ and $\hat{\kappa}$ have the ***same uniform density***

$$\rho_\kappa = \rho_{\hat{\kappa}} \tag{6.8.21}$$

(b) the response functional $\mathscr{G}$ with respect to κ coincides with the response functional with respect to $\hat{\kappa}$.

The last requirement implies that $\hat{\mathscr{G}}$ is the same functional of the argument $\hat{\mathbf{F}}^t = \mathbf{F}^t \cdot \mathbf{P}^{-1}$ as $\mathscr{G}$ is of $\mathbf{F}^t$,

$$\mathscr{G}\{\mathbf{F}^t \cdot \mathbf{P}^{-1}\} = \mathscr{G}\{\mathbf{F}^t\}. \tag{6.8.22}$$

In Eq. (22) $\mathbf{P}$ is a ***unimodular tensor*** (i.e., det $\mathbf{P} = \pm 1$), since $\hat{\kappa}$ and κ have the same density. Conversely any ***unimodular tensor*** $\mathbf{H}$ such that, for all $\mathbf{F}^t$,

$$\mathscr{G}\{\mathbf{F}^t \cdot \mathbf{H}\} = \mathscr{G}\{\mathbf{F}^t\} \tag{6.8.23}$$

gives rise to a material isomorphism of the particle onto itself. (Notice that $\mathscr{G}$ appears on both sides of Eqs. (22) and (23), not $\hat{\mathscr{G}}$.)

Isotropy group. The set of all unimodular tensors $\mathbf{H}$ for which Eq. (23) holds forms a group [see, for example, Truesdell (1965), Eq. (6.3)]. This group is called the ***isotropy group*** or ***material symmetry group*** $\mathscr{g}_\kappa$ of the material at the particle whose place in the reference configuration κ is $\mathbf{X}$. [A ***group*** is a set of abstract elements $\mathbf{A}, \mathbf{B}, \mathbf{C}, \ldots$ with a defined product $\mathbf{A} \cdot \mathbf{B}$ such that: (a) $\mathbf{A} \cdot \mathbf{B}$ is in the set for every $\mathbf{A}, \mathbf{B}$, (b) the set contains a unit $\mathbf{1}$ such that $\mathbf{1} \cdot \mathbf{A} = \mathbf{A} = \mathbf{A} \cdot \mathbf{1}$ for every $\mathbf{A}$, (c) $\mathbf{A} \cdot (\mathbf{B} \cdot \mathbf{C}) = (\mathbf{A} \cdot \mathbf{B}) \cdot \mathbf{C}$, and (d) every element has an inverse $\mathbf{A}^{-1}$ in the set, such that $\mathbf{A} \cdot \mathbf{A}^{-1} = \mathbf{A}^{-1} \cdot \mathbf{A} = \mathbf{1}$.] From its definition $\mathscr{g}_\kappa$ is a subgroup of the ***unimodular group*** $\mathscr{u}$. The isotropy group $\mathscr{g}_\kappa$ is the group of all static density-preserving deformations of the reference configuration κ such that the material at the particle X is indistinguishable in its response from the same material after it has been deformed into a new configuration. If $\mathbf{H}$ is a member of the isotropy group, then $\mathbf{H} = \nabla \boldsymbol{\lambda}$ where $\boldsymbol{\lambda}$ is the motion which carries one configuration into another in which the material is indistinguishable, and $\mathbf{H}$ is unimodular. The unimodular tensors $\mathbf{H}$ need not be orthogonal, but they may be, and the definition of an isotropic simple material is given in terms of the orthogonal group.

If an orthogonal tensor $\mathbf{Q}$ is in the isotropy group $\mathscr{g}_\kappa$, then so is its inverse $\mathbf{Q}^T = \mathbf{Q}^{-1}$, since $\mathscr{g}_\kappa$ is a group. Hence, by Eq. (23), which holds for any nonsingular $\mathbf{F}^t$ (including $\mathbf{Q} \cdot \mathbf{F}^t$ for any $\mathbf{F}^t$)

$$\mathscr{G}\{\mathbf{Q} \cdot \mathbf{F}^t \cdot \mathbf{Q}^T\} = \mathscr{G}\{\mathbf{Q} \cdot \mathbf{F}^t\} \tag{6.8.24}$$

where $\mathbf{Q}$ is time-independent (not a history). But if in Eq. (6.7.58) expressing the principle of material frame-indifference

is a solid, corresponding to the triclinic system. All the classical crystallographic groups (if extended to include $-\mathbf{1}$) correspond to solids. See Truesdell and Noll (1965), Sec. 33, for a listing of the transformation groups for the crystal classes and for transverse isotropy.

Simple liquid crystals. The simple fluid and the simple solid do not exhaust the possible types of simple materials. For the simple fluid $\mathscr{g} = \mathscr{u}$, while for the simple solid $\mathscr{g} \subset \mathscr{o} \subset \mathscr{u}$. It is possible that there could be a material such that some, but not all, density-preserving deformations belong to the isotropy group as well as some rotations. Coleman (1965) proposed the name simple liquid crystal for a simple material which is neither a simple fluid nor a simple solid.

When the simple material is isotropic, however, it is either a simple fluid ($\mathscr{g} = \mathscr{u}$) ***or an isotropic simple solid*** ($\mathscr{g}_{\kappa} = \mathscr{o}$ for any undistorted $\boldsymbol{\kappa}$). In the remainder of this discussion we consider only isotropic simple materials.

For an isotropic simple material, Eq. (25) becomes an identity to be satisfied by all orthogonal tensors $\mathbf{Q}$ instead of an equation to be solved to determine those orthogonal $\mathbf{Q}$ belonging to the isotropy group, and likewise Eq. (23) is an identity for all orthogonal $\mathbf{H}$. Truesdell and Noll [(1965) Eqs. (31.9) and (31.10)] show that the constitutive equation for any isotropic simple material can be written in the forms

$$\mathbf{T}(t) = \mathscr{H}\{\mathbf{U}_t^t(s);\ \mathbf{V}(t)\} \tag{6.8.30}$$

or

$$\mathbf{T}(t) = \mathbf{f}[\mathbf{B}(t)] + \mathscr{F}\{\mathbf{C}_t^t(s) - \mathbf{1};\ \mathbf{B}(t)\} \tag{6.8.31}$$

such that

$$\mathbf{T}(t) = \mathbf{f}[\mathbf{B}(t)] \quad \text{if} \quad \mathbf{C}_t^t(s) \equiv \mathbf{1},$$

where $\mathbf{U}_t^t(s)$ and $\mathbf{C}_t^t(s)$ denote the histories of the relative right stretch tensor $\mathbf{U}_t$ and the relative deformation tensor $\mathbf{C}_t$ [relative to the instantaneous configuration at time t; see Eq. (4.6.22)], while $\mathbf{V}(t)$ and $\mathbf{B}(t)$ are instantaneous values of the left stretch $\mathbf{V}$ and of $\mathbf{B}$ (see Sec. 4.6). In Eq. (31), $\mathbf{f}$ is an ordinary function, not a functional, while the functional $\mathscr{F}$ vanishes if $\mathbf{C}_t^t(s) \equiv \mathbf{1}$ for all s in the history prior to t.

For a simple fluid it can be shown that the dependence on $\mathbf{B}(t)$ reduces to dependence only on the density ρ at time t and that $\mathbf{f}(\mathbf{B})$ is hydrostatic:

$$\text{Simple fluid} \quad \mathbf{T}(t) = -p(\rho)\mathbf{1} + \mathscr{F}\{\mathbf{C}_t^t(s) - \mathbf{1}, \rho\}, \tag{6.8.32}$$

where $p(\rho)$ is a scalar function of ρ. Dependence on temperature is excluded by the restriction of a simple material to purely mechanical phenomena.

For simple solids in general no especially simple form of the constitutive equations has been found. Truesdell and Noll (1965), Secs. 38 to 40, consider a class of materials called ***materials with fading memory***. These considerations are beyond the scope of this book. We conclude this section with a brief reference to isotropic elastic solids, for which simple forms have been found.

Representation theorems for isotropic elastic solids. For an elastic solid, the functionals are replaced by functions as in Eqs. (6.7.26) to (6.7.43). An elastic solid is isotropic if and only if its response function $\mathbf{g}(\mathbf{F})$ relative to some undistorted state satisfies the identity

$$\mathbf{Q}\cdot\mathbf{g}(\mathbf{F})\cdot\mathbf{Q}^T = \mathbf{g}(\mathbf{Q}\cdot\mathbf{F}\cdot\mathbf{Q}^T) \tag{6.8.33}$$

which replaces Eq. (25) or the identity

$$\mathbf{g}(\mathbf{F}) = \mathbf{g}(\mathbf{F}\cdot\mathbf{Q}) \tag{6.8.34}$$

which replaces Eq. (23), for all orthogonal tensors $\mathbf{Q}$. Then, for example, with $\mathbf{Q} = \mathbf{R}$

$$\mathbf{R}\cdot\mathbf{g}(\mathbf{U})\cdot\mathbf{R}^T = \mathbf{g}(\mathbf{R}\cdot\mathbf{U}\cdot\mathbf{R}^T)$$

whence, since $\mathbf{R}\cdot\mathbf{U}\cdot\mathbf{R}^T = \mathbf{V}$, Eq. (6.7.31), namely $\mathbf{T} = \mathbf{R}\cdot\mathbf{g}(\mathbf{U})\cdot\mathbf{R}^T$, takes the form

$$\mathbf{T} = \mathbf{g}(\mathbf{V}). \tag{6.8.35}$$

Similarly Eq. (6.7.32), $\mathbf{T} = \mathbf{R}\cdot\mathbf{f}(\mathbf{C})\cdot\mathbf{R}^T$ takes the form

$$\mathbf{T} = \mathbf{f}(\mathbf{B}) \tag{6.8.36}$$

since $\mathbf{B} = \mathbf{R}\cdot\mathbf{C}\cdot\mathbf{R}^T$. The simple forms of Eqs. (35) and (36) (not explicitly containing the rotation $\mathbf{R}$) are correct only for elastic solids that are isotropic. We already had in Eq. (6.7.42) for the Piola-Kirchhoff stress tensor $\tilde{\mathbf{T}}$ (even without isotropy) the simple form

$$\tilde{\mathbf{T}} = \tilde{\mathbf{f}}(\mathbf{C}). \tag{6.8.37}$$

Any tensor-valued function satisfying the identity of Eq. (33) for all orthogonal $\mathbf{Q}$ is called an ***isotropic tensor function***. Fortunately there are representation theorems available for isotropic tensor functions of one tensor variable. See Truesdell and Noll (1965), Sec. 12. In three dimensions, for example, the function $\mathbf{f}(\mathbf{B})$ of Eq. (36) has a representation of the form

$$\mathbf{T} = \mathbf{f}(\mathbf{B}) = f_0\mathbf{1} + f_1\mathbf{B} + f_2\mathbf{B}^2, \tag{6.8.38}$$

where f_0, f_1, and f_2 are scalar invariant functions of the three principal invariants of $\mathbf{B}$, namely I_B, II_B, III_B, or equivalently of the principal values of $\mathbf{B}$.

Thus the tensorial form of $\mathbf{f(B)}$ is completely known, and it remains only to determine the three scalar functions f_0, f_1, and f_2. This theorem is due to Rivlin and Ericksen (1955). See Eq. (16) in Part 1 of this section. Similar representations are available for $\mathbf{g(V)}$ and $\tilde{\mathbf{f}}(\mathbf{C})$ and indeed for any isotropic tensor function of one tensor independent variable. Of course Eq. (38) does not imply that the stress-strain equation can only be a quadratic in the components B_{km}, since the general nonlinear scalar functions f_0, f_1, and f_2 are not limited to quadratic polynomials, and in any case the argument III_B is already cubic in the components.

When the argument tensor has an inverse, another representation is possible. For example, in place of Eq. (38) we may write

$$\mathbf{T} = \mathbf{f(B)} = F_0\mathbf{1} + F_1\mathbf{B} + F_{-1}\mathbf{B}^{-1}, \tag{6.8.39}$$

where again F_0, F_1, and F_{-1} are scalar invariant functions of I_B, II_B, III_B or alternatively of the three principal values of $\mathbf{B}$. Similar representations are available for Eqs. (35) and (37) because $\mathbf{V}$ and $\mathbf{C}$ also possess inverses. Truesdell and Noll (1965), Sec. 47, give formulas expressing the scalar functions of each of these representations of $\mathbf{f(B)}$ in terms of those of the other and also formulas relating the scalar functions $\tilde{F}_0$, $\tilde{F}_1$, and $\tilde{F}_{-1}$ for

$$\tilde{\mathbf{T}} = \tilde{\mathbf{f}}(\mathbf{C}) = \tilde{F}_0\mathbf{1} + \tilde{F}_1\mathbf{C} + \tilde{F}_{-1}\mathbf{C}^{-1}, \tag{6.8.40}$$

to the scalar functions of Eq. (39). Equation (40) is a representation of Eq. (37) of the same type as Eq. (39) is of Eq. (36).

Since the principal axes of $\mathbf{B}$ and $\mathbf{B}^{-1}$ coincide (in the deformed configuration; see Sec. 4.6), Eq. (39) shows that the principal axes of the Cauchy stress $\mathbf{T}$ coincide with those of $\mathbf{B}$ at the deformed position $\mathbf{x}$ of the particle, for any possible choice of nonlinear isotropic elastic constitutive equations. (The principal axes of $\mathbf{B} = \mathbf{V}^2$ are the same as those of $\mathbf{V}$.) Similarly Eq. (40) shows that the principal axes of the second Piola-Kirchhoff stress tensor $\tilde{\mathbf{T}}$ coincide with those of $\mathbf{C}$ at $\mathbf{X}$ in the reference configuration. (The principal axes of $\mathbf{C}$ are also those of $\mathbf{E}$ and $\mathbf{U}$.)

This concludes our brief sketch of the two principal approaches to the representation of isotropy and other types of material symmetry in modern theories of constitutive equations. The application of the reduced equations obtained to nonlinear problems of continuum mechanics is beyond the scope of this book. See the references cited in this section (listed in the general bibliography at the end of the book), especially the comprehensive treatise by Truesdell and Noll (1965). See also the four-volume collection of reprints of original papers, *Continuum Mechanics*, edited by Truesdell (1965, 1966).

We turn now to examples of the application of the linear constitutive equations formulated in the first three sections of this chapter. In Chap. 7 we consider fluid mechanics, and in Chap. 8 linearized elasticity.

EXERCISES

1. Write a component form of Eq. (6.8.1), and show that it leads to deformation gradient components that are independent of the material coordinates.
2. Derive Eq. (6.8.8) from Eq. (7).
3. Write out Eq. (6.8.10b) for the case that the J_β are the five elements of Eq. (9).
4. Carry out the details leading from Eq. (6.8.15) to Eq. (6.8.16).
5. Verify the details leading to Eq. (6.8.35) and (6.8.36).
6. Write component forms of: (a) Eq. (6.8.38), (b) Eq. (6.8.39).
7. Write an expansion similar to Eq. (6.8.38) for the equation $\tilde{\mathbf{T}} = \tilde{\mathbf{G}}(\mathbf{E})$ of Eq. (6.7.43). If each of the scalar functions $\tilde{G}_0$, $\tilde{G}_1$, and $\tilde{G}_2$ appearing in your result is assumed to be a polynomial in its arguments I_E, II_E, III_E, and the component equations of $\tilde{\mathbf{T}} = \tilde{\mathbf{G}}(\mathbf{E})$ are assumed to be quadratic polynomials in the components of $\mathbf{E}$, write the forms assumed by $\tilde{G}_0$, $\tilde{G}_1$, and $\tilde{G}_2$. How many constants appear? Note that this is a severe approximation as compared with Eq. (38), for example, where the general tensorial form includes only terms up to the square of $\mathbf{B}$, but where the scalar coefficients are not limited in degree.

CHAPTER 7

Fluid Mechanics

7.1 Field equations of Newtonian fluid; Navier-Stokes equations; example: parallel plane flow of incompressible fluid between flat plates

As a first example of the formulation of a complete particular theory in continuum mechanics by the combination of the general principles of Chap. 5, valid for all continua, with a particular constitutive equation from Chap. 6, we consider the case of the Newtonian fluid defined by Eqs. (6.3.2) and (6.3.10). We shall also consider some of the special theories obtained by making one or more additional simplifying assumptions, for example, Stokes condition, Eq. (6.3.17b); incompressibility; or inviscid (frictionless) fluid. A brief bibliography on fluid mechanics is given at the end of the chapter. The general problem is usually formulated as a boundary-value problem or an initial-and-boundary-value problem for a system of partial differential equations.

A steady-flow boundary-value problem, for example, might seek to determine the steady-flow configuration through a container for given steady inlet and outlet conditions. This means determining the velocities, the pressure, and the density as functions of the spatial coordinates inside the region considered. An airplane flying level at constant velocity through uniform stationary air presents a nonsteady flow pattern to a stationary observer. But to an observer on the airplane, the flow can be considered as a steady flow of fluid over a stationary airplane. The boundary-value problem is to determine this steady relative-flow configuration. We may also wish to calculate the total force (lift and drag) on the airplane, and we may be interested in similar total-force calculations also in the contained flow problems. These types of problems are both examples of steady-flow boundary-value problems to determine time-independent solutions of a set of partial differential equations, satisfying certain boundary conditions at the stationary boundary surfaces and prescribed inlet and outlet conditions (uniform relative velocity at infinity in the case of the airplane).

Transient problems include the starting-up problem in a pipe or channel, for example, and also such problems as the propagation through air or water of sound waves from an explosion. The starting-up problem is much more difficult to solve than the steady-flow problem, and exact solutions to this type of initial-and-boundary-value problem are rare. The wave-propagation problem often permits a number of simplifying assumptions based on the assumed smallness of the motion from the initial state. The type of mathematical analysis for the wave-propagation problem is quite different from that for a steady flow. As a rule, boundary conditions must be specified all the way around a finite volume of fluid in order to properly set a steady-flow problem. With the transient wave propagation, on the other hand, at any given time the solution at a certain point may be determined by the initial conditions in only a small region of the fluid and the boundary conditions on only a small part of the boundary. This limited dependence on the boundary and initial conditions is typical of ***hyperbolic systems*** of partial differential equations, while the steady-flow problem, requiring boundary conditions all around, is typical of ***elliptic systems*** of partial differental equations. The mathematical classification of partial differential equation systems in general into the categories called ***hyperbolic systems***, ***elliptic systems***, and ***parabolic systems*** is beyond the scope of this section [see, for example, Courant and Hilbert, vol. 2 (1962)], but the reader is warned that the physical boundary conditions and the differential equations employed in fluid mechanics fall sometimes into one category and sometimes into another. An example will be considered in Sec. 7.2; see Eq. (7.2.42). There is no such thing as a "typical" mathematical formulation; the best we could hope for would be several different typical formulations. In this chapter, we present a few sample formulations without pretending to cover all the possibilities. For other formulations and for advanced analytical and numerical methods of solving the problems, the reader should consult the fluid-mechanics literature. We give very few actual solutions in this book, but limit ourselves to problems where analytical complexity does not obscure the ideas.

The field equations of a Newtonian fluid obeying Fourier's law of heat conduction and a caloric equation of state are summarized below in the spatial description, along with their rectangular Cartesian component forms. The numbers in parentheses at the left indicate the number of independent component equations.

General Theorems

Continuity equation (Sec. 5.2):

[1 eq.]

$$\left.\begin{aligned} \frac{d\rho}{dt} + \rho\nabla\cdot\mathbf{v} &= 0 \\ \frac{d\rho}{dt} + \rho\frac{\partial v_k}{\partial x_k} &= 0 \end{aligned}\right\} \tag{7.1.1}$$

Equations of motion (Sec. 5.3):

[3 eqs.] $$\nabla \cdot \mathbf{T} + \rho \mathbf{b} = \rho \frac{d\mathbf{v}}{dt} \qquad \frac{\partial T_{rm}}{\partial x_r} + \rho b_m = \rho \frac{dv_m}{dt} \tag{7.1.2}$$

Energy equation (Sec. 5.4):

[1 eq.] $$\mathbf{T}:\mathbf{D} + \rho r = \nabla \cdot \mathbf{q} + \rho \frac{du}{dt}$$

$$T_{rm} D_{rm} + \rho r = \frac{\partial q_k}{\partial x_k} + \rho \frac{du}{dt} \tag{7.1.3}$$

CONSTITUTIVE EQUATIONS

Navier-Poisson law of Newtonian fluid (Sec. 6.3):

[6 eqs.] $$\mathbf{T} = -p\mathbf{1} + \lambda\,(\mathrm{tr}\,\mathbf{D})\,\mathbf{1} + 2\mu\mathbf{D}$$

$$T_{rm} = -p\delta_{rm} + \lambda D_{kk}\,\delta_{rm} + 2\mu D_{rm} \tag{7.1.4}$$

Kinetic equation of state (Sec. 6.3):

[1 eq.] $$F\,(p, \rho, \theta) = 0 \tag{7.1.5}$$

Fourier's law of heat conduction [due to Fourier (1822)]:

[3 eqs.] $$\mathbf{q} = -k\nabla\theta \qquad q_m = -k\frac{\partial \theta}{\partial x_m} \tag{7.1.6}$$

Caloric equation of state (Sec.6.3):

[1 eq.] $$u = u\,(\theta, \rho) \tag{7.1.7}$$

This is a total of 16 equations, although the actual forms of the kinetic and caloric equations of state to be assumed as part of the constitutive equations have not been specified. In fact an explicit form for Eq. (7) is seldom known. If we suppose that in these equations the rate of deformation **D** is everywhere expressed in terms of the velocity by

$$\mathbf{D} = \frac{1}{2}(\mathbf{L} + \mathbf{L}^T) \qquad D_{rm} = \frac{1}{2}\left(\frac{\partial v_r}{\partial x_m} + \frac{\partial v_m}{\partial x_r}\right) \tag{7.1.8}$$

and that the body forces and heat source strengths are given, then the variable unknowns are:

6	stress components	T_{rm}
3	velocity components	v_m

3	heat-flux components	q_m
1	thermodynamic pressure	p
1	density	ρ
1	internal energy	u
1	temperature	θ
16		

for a total of 16 unknowns. The number can be reduced immediately to seven equations in seven unknowns by eliminating the T_{rm} and q_m through use of Eqs. (4) and (6), as follows. From Eq. (8)

$$2\frac{\partial D_{rm}}{\partial x_r} = \frac{\partial}{\partial x_r}\left(\frac{\partial v_r}{\partial x_m} + \frac{\partial v_m}{\partial x_r}\right) = \frac{\partial}{\partial x_m}\left(\frac{\partial v_r}{\partial x_r}\right) + \frac{\partial^2 v_m}{\partial x_r \partial x_r}$$
$$= \frac{\partial}{\partial x_m}(\nabla \cdot \mathbf{v}) + \nabla^2 v_m. \tag{7.1.9}$$

Hence, by Eq. (4),

$$\frac{\partial T_{rm}}{\partial x_r} = -\frac{\partial p}{\partial x_r}\delta_{rm} + \lambda \frac{\partial}{\partial x_r}(\nabla \cdot \mathbf{v})\,\delta_{rm} + 2\mu \frac{\partial D_{rm}}{\partial x_r}$$
$$= -\frac{\partial p}{\partial x_m} + \lambda \frac{\partial}{\partial x_m}(\nabla \cdot \mathbf{v}) + \mu \frac{\partial}{\partial x_m}(\nabla \cdot \mathbf{v}) + \mu \nabla^2 v_m. \tag{7.1.10}$$

We substitute this into the equations of motion, Eqs. (2), to obtain the generalized Navier-Stokes equations of motion for fluids with bulk viscosity

$$-\frac{\partial p}{\partial x_m} + (\lambda + \mu)\frac{\partial}{\partial x_m}(\nabla \cdot \mathbf{v}) + \mu \nabla^2 v_m + \rho b_m = \rho \frac{dv_m}{dt}, \tag{7.1.11}$$

which are rewritten in symbolic vector form as the second Eq. (14) below. We also rewrite the energy equation as follows. Besides eliminating $\mathbf{q}$ by means of Eq. (6), we use Eq. (6.3.18) to write

$$\mathbf{T} : \mathbf{D} = -p\nabla \cdot \mathbf{v} + 2W_D, \tag{7.1.12}$$

where $2W_D$ is the dissipation power of Eq. (6.3.19) and can be written as

$$2W_D = (\lambda + \tfrac{2}{3}\mu)\,\mathrm{I}_D^2 + 4\mu \mathrm{II}_{D'}, \tag{7.1.13}$$

where we suppose that the invariants I_D and $\mathrm{II}_{D'}$ are expressed in terms of the velocities. The remaining equations are then as follows:

Continuity equation:	$\frac{d\rho}{dt} + \rho \nabla \cdot \mathbf{v} = 0$	(7.1.14a)

Generalized Navier-Stokes equation for fluids with bulk viscosity:

$$-\nabla p + (\lambda + \mu)\nabla(\nabla\cdot\mathbf{v}) + \mu\nabla^2\mathbf{v} + \rho\mathbf{b} = \rho\frac{d\mathbf{v}}{dt} \tag{7.1.14b}$$

Energy equation:

$$\rho\frac{du}{dt} = \nabla\cdot(k\nabla\theta) - p\nabla\cdot\mathbf{v} + \rho r + 2W_D \tag{7.1.14c}$$

Equations of state:

$$F(p, \rho, \theta) = 0 \tag{7.1.14d}$$

$$u = u(\theta, \rho). \tag{7.1.14e}$$

Equations (14) are a vector equation and four scalar equations for the seven unknowns v_m, p, ρ, u, θ.

Boundary conditions imposed on $\mathbf{v}$ in viscous-flow theory usually require that the fluid adhere to any rigid wall. At a stationary wall then, $\mathbf{v} = 0$ in viscous-flow theory, while for ideal frictionless fluid only the normal component is required to vanish at a rigid wall. Boundary conditions are also required for the temperature. These may take the form of holding the wall at a constant temperature, but may require a consideration of heat transfer at the boundary. See, for example, Rohsenow and Choi (1961) for further discussion of thermal boundary conditions. In most of the examples to be discussed in this book the thermal variables will be ignored.

If the flow is barotropic, so that the kinetic equation of state is independent of the temperature, say

$$F(p, \rho) = 0 \tag{7.1.15}$$

as in Eq. (6.3.4), a purely mechanical boundary-value problem can be formulated for the five unknowns v_m, p, ρ by using only the Navier-Stokes equations of motion, the continuity equation, and the barotropic equation of state. We will see several examples of this type in which further simplifications appear in special cases.

The nonlinearity of the field equations makes their integration difficult without special assumptions. The nonlinearity appears in the inertial acceleration terms

$$\rho\frac{d\mathbf{v}_m}{dt} = \rho\left[\frac{\partial v_m}{\partial t} + v_k\frac{\partial v_m}{\partial x_k}\right] \tag{7.1.16}$$

of the Navier-Stokes equations, in the energy equation, and in the equations of state. The nonlinearity does not disappear even if an ideal frictionless fluid is assumed; indeed the only nonlinear term directly involving the viscosity is the term $2W_D$ in the energy equation.

If the term $\rho\,(d\mathbf{v}/dt)$ is moved to the left side of the Navier-Stokes equation, the various terms may be interpreted as forces (per unit volume) acting on a volume element. The first term $-\nabla p$ is the pressure-gradient force per unit volume, acting in a direction opposite to the vector ∇p. The next two terms give the viscous forces on the volume element. The prescribed body force per unit volume is $\rho\mathbf{b}$, equal to $-\rho g\mathbf{i}_3$ if the body force is that of a uniform gravitational field in the negative x_3-direction. The last term is the inertia force $-\rho(d\mathbf{v}/dt)$ per unit volume. In many problems one or more of these forces may be negligible in comparison to the others. Omission of various terms then leads to the different kinds of special simplified theories.

Before we begin looking at some of these special cases, it may be well to remind ourselves that even Eqs. (14) are not general enough to cope with many common flow situations in real engineering problems. The most noteworthy shortcoming of the equations is that they may be used to find the desired velocity field only in ***laminar-flow*** situations. When ***turbulent flow*** occurs, as is quite common in practice, the mean flow is ***not*** adequately described by Eqs. (14).

The Navier-Stokes equation of motion in its usual form is the equation replacing the second Eq. (14) under either of the assumptions: (a) incompressible fluid, or (b) Stokes condition $\lambda = -\frac{2}{3}\mu$ (implying no bulk viscosity). We write the two equations below, after dividing by ρ:

Navier-Stokes Equations

Incompressible fluid:

$$-\frac{1}{\rho}\nabla p + \nu\nabla^2\mathbf{v} + \mathbf{b} = \frac{d\mathbf{v}}{dt} \tag{7.1.17a}$$

Compressible fluid with no bulk viscosity:

$$-\frac{1}{\rho}\nabla p + \frac{\nu}{3}\nabla(\nabla\cdot\mathbf{v}) + \nu\nabla^2\mathbf{v} + \mathbf{b} = \frac{d\mathbf{v}}{dt} \tag{7.1.17b}$$

with $\lambda = -\frac{2}{3}\mu$,

where

$$\nu = \frac{\mu}{\rho} \tag{7.1.18}$$

is the ***kinematic viscosity***. The equations are due to Navier (1822) and Stokes (1845).

The connection between viscosity and vorticity is suggested by the alternative form of the equation of motion for an incompressible fluid, given in Eq. (22) below. The ***vorticity vector*** is usually† defined as twice the angular-velocity vector $\mathbf{w}$,

†Some authors, however, define the vorticity vector as equal to $\mathbf{w}$ instead of $2\,\mathbf{w}$.

$$\mathbf{w} = \frac{1}{2} \nabla \times \mathbf{v}, \tag{7.1.19}$$

which was introduced in Sec. 4.4, where it was related to the spin tensor $\mathbf{W}$ by

$$\mathbf{w} = -W_{23}\mathbf{i}_1 - W_{31}\mathbf{i}_2 - W_{12}\mathbf{i}_3 \quad \text{or} \quad w_i = -\tfrac{1}{2}e_{ijk}W_{jk}. \tag{7.1.20}$$

When the vector identity

$$\nabla^2\mathbf{v} \equiv \nabla(\nabla\cdot\mathbf{v}) - \nabla \times (\nabla \times \mathbf{v}) \tag{7.1.21}$$

of Eq. (2.5.27j) is substituted into Eqs. (17a) [or into Eq. (14b) with $\nabla\cdot\mathbf{v} = 0$ for incompressibility], the result is

$$\text{Incompressible flow:} \quad -\frac{1}{\rho}\nabla p - 2\nu\nabla \times \mathbf{w} + \mathbf{b} = \frac{d\mathbf{v}}{dt}. \tag{7.1.22}$$

When the flow is irrotational ($\mathbf{w} = 0$), this equation reduces to the equation of motion for a perfect fluid (inviscid fluid with $\nu = 0$):

$$\text{Perfect fluid:} \quad -\frac{1}{\rho}\nabla p + \mathbf{b} = \frac{d\mathbf{v}}{dt}, \tag{7.1.23}$$

since the second term in Eq. (22) vanishes either when the vorticity vector $2\mathbf{w}$ vanishes or when the kinematic viscosity ν vanishes. This means that any irrotational motion of an incompressible perfect fluid satisfies the full Navier-Stokes equations, but it should not be concluded from this that such solutions are useful directly for viscous fluid flow, because in general such solutions do not satisfy the boundary condition $\mathbf{v} = 0$ at a fixed boundary, required for a viscous fluid. This may help to make plausible, however, the use of an irrotational perfect-fluid solution in the free-stream portion of a flow, with a thin ***boundary layer*** of viscous rotational flow between the free-stream flow and the wall, as discussed in Sec. 7.5.

Steady laminar flow between parallel plates (incompressible viscous fluid). As an example of an exact solution† of the Navier-Stokes equations, for incompressible fluid, consider the steady plane horizontal flow between two parallel plates of infinite extent parallel to the xy-plane. Let the xz-plane be parallel to the plane of flow, and let the only body force be gravity in the negative z-direction. Then all y-components and y-derivatives vanish, and the continuity equation and the two remaining components of the Navier-Stokes equation, Eq. (17a), take the following forms for steady incompressible flow, where v_x and v_z are velocity components:

†A comprehensive review of solutions of the Navier-Stokes equations has been given by Berker (1963).

Steady Incompressible Plane Flow

Continuity:

$$\frac{\partial v_x}{\partial x} + \frac{\partial v_z}{\partial z} = 0 \tag{7.1.24}$$

Navier-Stokes:

$$\left.\begin{aligned} -\frac{1}{\rho}\frac{\partial p}{\partial x} + \nu\left(\frac{\partial^2 v_x}{\partial x^2} + \frac{\partial^2 v_x}{\partial z^2}\right) &= v_x\frac{\partial v_x}{\partial x} + v_z\frac{\partial v_x}{\partial z} \\ -\frac{1}{\rho}\frac{\partial p}{\partial z} + \nu\left(\frac{\partial^2 v_z}{\partial x^2} + \frac{\partial^2 v_z}{\partial z^2}\right) - g &= v_x\frac{\partial v_z}{\partial x} + v_z\frac{\partial v_z}{\partial z} \end{aligned}\right\} \tag{7.1.25}$$

For the laminar flow between parallel plates, as illustrated in Fig. 7.1, we further assume parallel flow in the x-direction with $v_z = 0$ everywhere. Then Eq. (24) implies

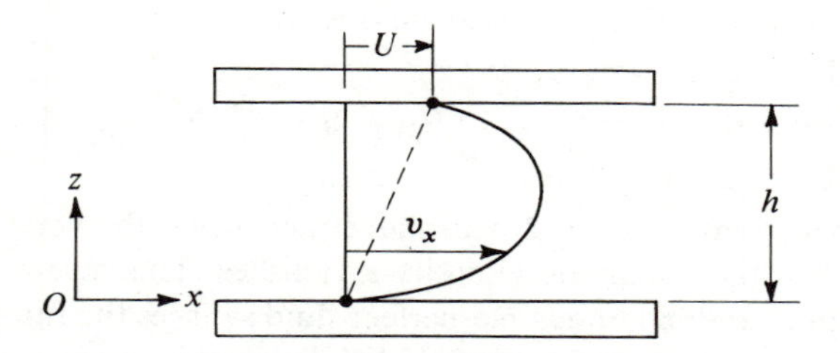

Fig. 7.1 Laminar flow between parallel plates.

$$\frac{\partial v_x}{\partial x} = 0 \tag{7.1.26}$$

and hence v_x is independent of x. We recall that in an incompressible Newtonian fluid p equals the mean normal pressure $\bar{p}$ [see Eq. (6.3.13)]. But its value is not usually determinate from the equations and velocity boundary conditions. With $v_z = 0$ and v_x independent of x, the Navier-Stokes equations of Eqs. (25) take the especially simple form:

$$-\frac{1}{\rho}\frac{\partial p}{\partial x} + \nu\frac{\partial^2 v_x}{\partial z^2} = 0 \tag{7.1.27}$$

$$-\frac{1}{\rho}\frac{\partial p}{\partial z} - g = 0. \tag{7.1.28}$$

Note that the acceleration terms on the right are all zero under these assumptions, so that inertia plays no role in this special flow, and that all of the nonlinear terms are zero. Equation (28) now shows that

$$p = -\gamma z + f_1(x), \tag{7.1.29}$$

where $\gamma = \rho g$ is the specific weight and $f_1(x)$ is an arbitrary function of x, equal to the (generally unknown) pressure distribution on the lower plate. Equation (27) now takes the form

$$\left.\begin{aligned} \mu\frac{\partial^2 v_x}{\partial z^2} &= f_1'(x), \quad \text{whence} \\ \mu v_x &= \frac{z^2}{2}f_1'(x) + zf_2(x) + f_3(x), \end{aligned}\right\} \tag{7.1.30}$$

where $f_2(x)$ and $f_3(x)$ are arbitrary functions of x. If the lower plate is stationary, then the boundary condition for a viscous fluid requires $v_x \equiv 0$ at $z = 0$, whence $f_3(x) \equiv 0$. Then the second Eq. (30) yields

$$\mu\frac{\partial v_x}{\partial x} = \frac{z^2}{2}f_1''(x) + zf_2'(x).$$

But we saw in Eq. (20) that $\partial v_x/\partial x \equiv 0$. Hence $zf_1''(x) = -2f_2'(x)$ for all x and z, which implies that

$$f_1''(x) = 0 \quad \text{and} \quad f_2'(x) = 0$$

whence

$$f_1(x) = C_1 x + C_3 \qquad f_2(x) = C_2$$

where C_1, C_2, and C_3 are constants. The second Eq. (30) becomes

$$\mu v_x = \frac{z^2}{2}C_1 + zC_2. \tag{7.1.31}$$

Since

$$C_1 = \frac{\partial p}{\partial x} \quad \text{and} \quad C_2 = \mu\left[\frac{\partial v_x}{\partial z}\right]_{z=0} \tag{7.1.32}$$

it follows that the pressure gradient is independent of x and so is the wall shear stress. If the upper plate is moving to the right at speed U, then the boundary condition at $z = h$ is $v_x = U$, whence Eq. (31) gives $hC_2 = \mu U - (h^2/2)C_1$, so that Eq. (31) takes the form

$$\mu v_x = \frac{z^2}{2}C_1 + \frac{z}{h}\left(\mu U - \frac{h^2}{2}C_1\right), \tag{7.1.33}$$

where the constant $C_1 = \partial p/\partial x$ remains undetermined. If the pressure at two points separated by horizontal distance Δx can be measured, then $C_1 = \Delta p/\Delta x$. (Note that in this special case the constitutive equations imply that the normal stress across any surface is equal to $-p$, so that an instrument measuring normal stress can determine p.)

Two special cases are of interest. In ***plane Couette flow*** the horizontal pressure gradient is zero. With $C_1 = 0$, the solution is

$$\text{Plane Couette flow:} \quad v_x = \frac{Uz}{h}, \tag{7.1.34}$$

giving a linear distribution of velocity as in Fig. 7.2(a). In ***plane Poiseuille flow***, the speed of the upper plate is zero. With $U = 0$, the parabolic velocity distribution of Eq. (33) is symmetric about $z = h/2$, and is given by

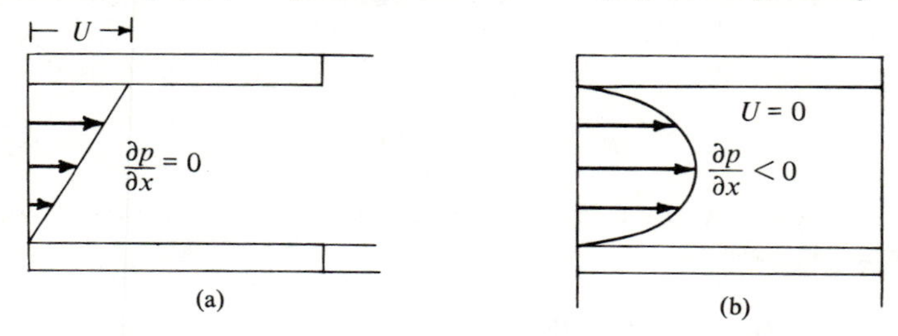

Fig. 7.2 Velocity profiles for two special cases of plane parallel flow of incompressible Navier-Stokes fluid: (a) plane Couette flow; (b) plane Poiseuille flow.

$$\left.\begin{aligned} &\text{Plane Poiseuille flow:} \quad \mu v_x = \frac{C_1}{2}(z^2 - hz) \quad \text{or} \\ &\mu v_x = -\frac{1}{2}\frac{\partial p}{\partial x}\left[\left(\frac{h}{2}\right)^2 - \left(z - \frac{h}{2}\right)^2\right], \end{aligned}\right\} \tag{7.1.35}$$

in which $C_1 = \partial p/\partial x$ is the constant pressure gradient in the x-direction, a negative quantity for flow in the positive direction. The Couette flow is approximated in a Couette viscometer, where the flow is actually between two coaxial circular cylinders, whose radii are both large compared to the gap between the two cylinders. This type of flow was first studied by Couette (1890).

Hagen-Poiseuille flow*† *in a horizontal cylinder of radius a results from a comparable solution of the equations in cylindrical coordinates. The result is

$$\mu v_z = -\frac{1}{4}\frac{\partial p}{\partial z}(a^2 - r^2) \tag{7.1.36}$$

if the z-axis is along the axis of the horizontal tube. Solution details are omitted.

†This type of flow was observed by Hagen (1839) and Poiseuille (1840).

These special flows of a viscous fluid were easily obtained because the geometric constraints and the condition of incompressibility removed all nonlinear terms, so that the inertia forces played no role. We consider now another kind of simplification, the ***perfect fluid*** idealization, where viscous forces play no role.

EXERCISES

1. Show that the vector identity of Eq. (7.1.21) leads to Eq. (22) for incompressible flow and to Eq. (23) when the flow is also irrotational.
2. For the case of plane flow parallel to the xy-plane, write out component forms of the Navier-Stokes equation for compressible flow [Eq. (7.1.17b)], and also write out the equation of continuity. How many unknowns appear? What additional equations are needed to formulate the problem?
3. Determine the viscous friction drag (per unit area) on the upper and lower plates of Fig. 7.1 for the flow of Eq. (7.1.33), and verify that the total horizontal force (including the pressure forces on the ends) is zero for the rectangular fluid region $0<x<L$, $0<z<h$, $0<y<a$.
4. Obtain an expression for the dissipation power $2W_D$ per unit volume for the parallel-flow example of Eq. (7.1.33).
5. Verify Eqs. (7.1.32).
6. Obtain the solution of Eqs. (7.1.24) and (25), modified to have body-force components $b_x = g \sin \theta$ and $b_z = -g \cos \theta$ for the case of parallel flow inclined at angle θ below the horizontal, with a free surface at $z = h$.
7. Consider ***nonsteady*** plane flow of incompressible fluid parallel to the x-axis. For fluid of infinite depth over the boundary plane $z = 0$, assume that $\partial p/\partial x = 0$, and that the fluid was initially at rest, but that at $t = 0$ the boundary plane $z = 0$ is set in motion with constant velocity of magnitude U in the x-direction. Obtain a partial differential equation for v_x as a function of z.
8. Use the formulas of Appendix Sec. II.4 to derive cylindrical component forms of the continuity equation and the components of Eq. (7.1.22) for the following special case. Assume incompressible, radially symmetric, steady flow parallel to the horizontal axis (z-axis) of a circular cylinder. Show that the equations can be combined to give for v_z the equation
$$-\frac{\partial p}{\partial z} + \mu \frac{1}{r} \frac{\partial}{\partial r}\left(r \frac{\partial v_z}{\partial r}\right) = 0$$
where $\partial p/\partial z$ is independent of z.
9. If the cylinder axis (z-axis) is not horizontal in Ex. 8, show that the final equation can be written in the same form, except that p is replaced by $p+\rho g h$ where h is the height above an arbitrary horizontal plane.
10. Neglecting the variation of p over the cross-section, integrate the result of Ex. 8 to obtain the result of Eq. (7.1.36) for flow through a stationary horizontal tube.

11. Use the formulas of appendix Sec. II.4 to derive cylindrical component forms of: (a) the continuity equation, (b) Eq. (7.1.23), (c) Eq. (22).

7.2 Perfect fluid: Euler equation; Kelvin's theorem; Bernoulli equation; irrotational flow: velocity potential; acoustic waves; gas dynamics

While no real fluids are actually perfect fluids (nonviscous), the free-stream flow of a fluid outside of a thin boundary layer near the solid boundaries of the flow is often little affected by the viscosity. When the boundary layer is thin, practical solutions can be obtained for the free-stream flow pattern by assuming it to be a perfect fluid and applying the perfect-fluid boundary conditions at the solid boundaries instead of at the interface between the boundary layer and the free stream. The equation to be solved, subject to these conditions, is the ***Euler equation***

$$-\frac{1}{\rho}\nabla p + \mathbf{b} = \frac{d\mathbf{v}}{dt}, \tag{7.2.1}$$

obtained by putting the viscosity coefficients λ and μ equal to zero in the Navier-Stokes equation (7.1.14b).

The boundary condition at a stationary solid boundary is that the normal component of velocity be zero. The tangential component is not affected by the boundary; this differs from the viscous-flow boundary condition, which required both components to vanish. Examples of the use of the perfect-fluid boundary conditions are given in Sec. 7.3 for incompressible flow. In the present discussion, we consider some properties of perfect-fluid flows without actually solving any boundary-value problems.

Conservation of circulation in a barotropic perfect fluid under conservative body forces (Kelvin's theorem). The line integral

$$\Gamma = \oint \mathbf{v}\cdot\delta\mathbf{r} \equiv \oint \mathbf{v}\cdot\hat{\mathbf{t}}\, ds \tag{7.2.2}$$

around a closed curve in space is called the ***velocity circulation*** around that contour. Here $\delta\mathbf{r} = \hat{\mathbf{t}}ds$ is the differential of the position vector and $\hat{\mathbf{t}}$ is a unit vector tangent to the curve. By Stokes theorem, Eq. (5.1.11),

$$\oint_C \mathbf{v}\cdot\hat{\mathbf{t}}\, ds = \int_S (\nabla\times\mathbf{v})\cdot\hat{\mathbf{n}}\, dS$$

for every oriented surface S within the fluid bounded by a curve C. Hence, ***if***

and only if the flow is irrotational, the circulation Γ ***is zero around every closed curve bounding a surface within the fluid.*** This result is based on the assumed continuity of the integrands.

Consider the rate of change $d\Gamma/dt$ for a "material curve" of the fluid:

$$\frac{d\Gamma}{dt} = \frac{d}{dt}\oint \mathbf{v}\cdot\delta\mathbf{r} = \oint \frac{d\mathbf{v}}{dt}\cdot\delta\mathbf{r} + \oint \mathbf{v}\cdot\frac{d(\delta\mathbf{r})}{dt}.$$

Now

$$\oint \mathbf{v}\cdot\frac{d(\delta\mathbf{r})}{dt} = \oint \mathbf{v}\cdot\delta\frac{d\mathbf{r}}{dt} = \oint \mathbf{v}\cdot\delta\mathbf{v} = \oint \delta\left(\frac{1}{2}v^2\right) = 0,$$

since the integral of the total differential $\delta(\frac{1}{2}v^2)$ around the closed contour is zero. Hence

$$\frac{d\Gamma}{dt} = \oint \mathbf{a}\cdot\delta\mathbf{r} \equiv \oint \mathbf{a}\cdot\hat{\mathbf{t}}\, ds, \tag{7.2.3}$$

where $\mathbf{a} = d\mathbf{v}/dt$ is the acceleration. If the acceleration $\mathbf{a}$ is the gradient of a scalar potential, then $\nabla\times\mathbf{a}$ vanishes and Stokes theorem implies that $d\Gamma/dt$ is zero for every closed material contour. If barotropic flow is assumed, so that $\rho = \rho(p)$, we can introduce a ***pressure function***

$$P(p) = \int_{p_0}^{p} \frac{dp}{\rho(p)} \tag{7.2.4a}$$

such that

$$\nabla P = \frac{1}{\rho}\nabla p, \tag{7.2.4b}$$

and, if the body-force field is conservative, a scalar body-force potential Ω exists such that

$$\mathbf{b} = -\nabla\Omega. \tag{7.2.5}$$

The perfect-fluid equation of motion, Eq. (1), can then be written as

$$\mathbf{a} = -\nabla(\Omega + P), \tag{7.2.6}$$

showing that under these assumptions $\mathbf{a}$ is the gradient of a potential. Hence ***in barotropic flow under conservative body forces, the velocity circulation around any closed material contour is independent of time.*** This is Kelvin's theorem.†

†W. Thomson, Lord Kelvin (1869).

A consequence of Kelvin's theorem is that if the flow of such a fluid is once irrotational it remains irrotational, as can be seen by another use of Stokes theorem. This explains why most flow analyses for barotropic flow of a perfect fluid assume irrotationality.

Another useful form of the equation of motion can be obtained by rewriting the convective part $\mathbf{v}\cdot\nabla\mathbf{v}$ of the acceleration $\mathbf{a} = \partial\mathbf{v}/\partial t + \mathbf{v}\cdot\nabla\mathbf{v}$ as follows. In rectangular Cartesians the convective part is $v_j v_{k,j}$. We add and subtract $v_j v_{j,k}$ to obtain

$$\begin{aligned} v_j v_{k,j} &= v_j(v_{k,j} - v_{j,k}) + v_j v_{j,k} \\ &= 2v_j W_{kj} + (\tfrac{1}{2}v_j v_j)_{,k}. \end{aligned}$$

Hence

$$\mathbf{v}\cdot\nabla\mathbf{v} = 2\mathbf{W}\cdot\mathbf{v} + \nabla(\tfrac{1}{2}v^2) = 2\mathbf{w}\times\mathbf{v} + \nabla(\tfrac{1}{2}v^2), \tag{7.2.7}$$

where

$$\mathbf{w} = -W_{23}\mathbf{i}_1 - W_{31}\mathbf{i}_2 - W_{12}\mathbf{i}_3 = \frac{1}{2}\nabla\times\mathbf{v} \tag{7.2.8}$$

is the angular-velocity vector and $\mathbf{W}$ is the spin tensor of Sec. 4.4. Hence†

$$\mathbf{a} = \frac{\partial\mathbf{v}}{\partial t} + 2\mathbf{w}\times\mathbf{v} + \nabla\left(\frac{1}{2}v^2\right) \tag{7.2.9}$$

and the equation of motion, Eq. (6), for a barotropic flow of a perfect fluid under conservative body forces can be written in the form

$$-\nabla(\Omega + P + \tfrac{1}{2}v^2) = \frac{\partial\mathbf{v}}{\partial t} + 2\mathbf{w}\times\mathbf{v} \tag{7.2.10}$$

[see Lamb (1945), Sec. 146]. This equation simplifies even further in the case of steady flow or in the case of irrotational flow.

Bernoulli equation for steady flow. In steady flow the term $\partial\mathbf{v}/\partial t$ is zero in Eq. (10). The other term on the right can be eliminated by forming the scalar product of $\mathbf{v}$ with Eq. (10), since $\mathbf{v}\cdot(\mathbf{w}\times\mathbf{v}) = 0$, because $\mathbf{w}\times\mathbf{v}$ is perpendicular to $\mathbf{v}$. Hence

$$\mathbf{v}\cdot\nabla(\Omega + P + \tfrac{1}{2}v^2) = 0,$$

showing that the gradient of $\Omega + P + \frac{1}{2}v^2$ is perpendicular to $\mathbf{v}$ and hence to

†The result of Eq. (9) is due to Lagrange (1783, 1788).

the streamline through the point where the gradient is taken. Hence the ***Bernoulli equation***†

$$\Omega + P + \tfrac{1}{2}v^2 = \text{const.} \tag{7.2.11}$$

holds along any streamline in the steady barotropic flow of a perfect fluid under conservative body forces. In general the constant on the right side of the Bernoulli equation varies from one streamline to another, but ***in the special case of irrotational steady flow the same constant applies throughout the flow field***, since the right side of Eq. (10) is then identically zero. The terms in this form of the Bernoulli equation have the same dimensions as energy per unit mass (specific energy); the last and first terms are, respectively the specific kinetic energy and body-force potential energy. To interpret P we note that with the barotropic flow and perfect-fluid assumptions, which disregard all heat conduction and frictional dissipation, we may consider the special case where $r = 0$ and $\mathbf{q} = 0$ in the energy equation, Eq. (7.1.3), and $W_D = 0$ in Eq. (7.1.12). Then the energy equation becomes

$$-p\nabla \cdot \mathbf{v} = \rho \frac{du}{dt},$$

which transforms by use of the continuity equation to

$$\frac{p}{\rho^2}\frac{d\rho}{dt} = \frac{du}{dt}.$$

An integration by parts then yields

$$P(p) \equiv \int_{p_0}^{p} \frac{dp}{\rho} = \left[u + \frac{p}{\rho}\right]_{p_0}^{p}. \tag{7.2.12}$$

Since the specific internal energy is only defined up to an arbitrary additive constant anyway, when $r = 0$ and $\mathbf{q} = 0$ we may interpret $P(p)$ as equal to the ***specific enthalpy*** $u + (p/\rho)$; see Sec. 5.7. Despite these remarks about the Bernoulli equation having the dimensions of specific energy it should be remembered that the Bernoulli equation is not an energy equation. It was derived from the Euler equation, a momentum equation, and not from the energy equation. The energy equation is another independent equation that can provide additional information provided that the energy terms can all be accounted for.

†This equation is called the Bernoulli equation because it expresses explicitly some relationships observed by D. Bernoulli (1738), although he did not actually give the equation. For a discussion of various "Bernoullian theorems" see Truesdell and Toupin (1960), Secs. 120 to 126.

The Bernoulli equation may be given another often-used form when the only body force is a uniform gravity field with

$$\Omega = gh,$$

where h is the ***gravity head***, the elevation above an arbitrary datum, and g is the free-fall acceleration due to gravity. Dividing Eq. (11) by g we obtain:

$$h + \int \frac{dp}{\rho g} + \frac{v^2}{2g} = \text{const. or } h + h_p + h_v = \text{const.} \tag{7.2.13}$$

along any streamline, where h_p denotes the ***pressure head*** and h_v the ***velocity head***, defined by

$$h_p = \frac{1}{g} P(p) \equiv \frac{1}{g} \int_{p_0}^{p} \frac{dp}{\rho} \qquad h_v = \frac{v^2}{2g}, \tag{7.2.14}$$

and h is the gravity head or potential head. Each term in this form of the Bernoulli equation now has dimensions of length. The Bernoulli equation (13) then states that the ***total head*** is constant along any streamline in the barotropic flow of a perfect fluid when the only body force is a uniform gravitational field. The total head varies in general from one streamline to another, but in the special case of irrotational steady flow the same constant applies throughout the flow field.

Irrotational flow or potential flow. We saw in connection with Kelvin's theorem, Eq. (6), that if the barotropic flow of a perfect fluid is at any time irrotational it will remain irrotational. This motivates the study of irrotational flows, which are also known as ***potential flows*** because $\nabla \times \mathbf{v} = 0$ is a necessary and sufficient condition for the existence of a scalar ***velocity potential*** ϕ, such that†

$$\mathbf{v} = \nabla \phi. \tag{7.2.15}$$

Introduction of the velocity potential replaces the three unknown velocity components by the single unknown scalar function ϕ and thus greatly simplifies many problems. The only other unknowns in barotropic flow are the pressure p and density ρ. Before making the further simplifying assumption of incompressibility we consider first two special cases of compressible flow: (1) acoustic waves of small amplitude, and (2) steady flow.

Acoustic waves of small amplitude. As our only example of nonsteady

†Many authors introduce a minus sign, defining ϕ differently, so that $\mathbf{v} = -\nabla \phi$. According to Lamb (1945), the velocity potential is due to Lagrange (1781). A more rigorous demonstration was given by Cauchy (1827).

flow, we consider acoustic waves involving only small barotropic fluctuations of pressure and density about an equilibrium state p_0, ρ_0. We also assume that the velocity fluctuations are "small" in a sense to be specified later in order to linearize the equations. (Since we take the equilibrium velocity $\mathbf{v}_0 = 0$, we obviously cannot require fluctuations small compared to v_0. It turns out that what we must require is that v be small in comparison to the sound speed in the fluid.) Consider first rectilinear motion in the x-direction, with $v_y = v_z = 0$ and all quantities assumed independent of y and z. The Euler equation (neglecting body forces) and the continuity equation are then

$$-\frac{1}{\rho}\frac{\partial p}{\partial x} = \frac{\partial v_x}{\partial t} + v_x\frac{\partial v_x}{\partial x} \qquad \frac{\partial \rho}{\partial t} + v_x\frac{\partial \rho}{\partial x} + \rho\frac{\partial v_x}{\partial x} = 0, \tag{7.2.16}$$

to which must be added the barotropic equation of state

$$\rho = \rho(p).$$

We linearize these equations as follows. We approximate the barotropic equation of state by

$$p - p_0 = \left(\frac{dp}{d\rho}\right)_0 (\rho - \rho_0) \tag{7.2.17}$$

where $(dp/d\rho)_0$ is the derivative $dp/d\rho$ evaluated at the initial state. Then, introducing u', p', ρ' for the small disturbances in v_x, p, ρ, we have

$$v_x = u' \qquad p = p_0 + p' \qquad \rho = \rho_0 + \rho' \tag{7.2.18}$$
$$\frac{\partial p}{\partial x} = \left(\frac{dp}{d\rho}\right)_0 \frac{\partial \rho}{\partial x} = \left(\frac{dp}{d\rho}\right)_0 \frac{\partial \rho'}{\partial x}.$$

The nonlinear equations take the form

$$\frac{\partial u'}{\partial t} + u'\frac{\partial u'}{\partial x} + \frac{1}{1 + \rho'/\rho_0}\frac{1}{\rho_0}\left(\frac{dp}{d\rho}\right)_0 \frac{\partial \rho'}{\partial x} = 0$$
$$\frac{\partial \rho'}{\partial t} + u'\frac{\partial \rho'}{\partial x} + \rho_0\left(1 + \frac{\rho'}{\rho_0}\right)\frac{\partial u'}{\partial x} = 0.$$

Under the stated assumption that $\rho' \ll \rho_0$, we are clearly justified in neglecting the terms ρ'/ρ_0 in comparison to unity. Linearization can then be accomplished provided

$$\left|u'\frac{\partial u'}{\partial x}\right| \ll \frac{\partial u'}{\partial t} \quad \text{and} \quad \left|u'\frac{\partial \rho'}{\partial x}\right| \ll \frac{\partial \rho'}{\partial t}. \tag{7.2.19}$$

The physical meaning of this assumption will become clear only after we have obtained the solution. We proceed to make the assumption (19) in order to get a linear system, and inquire afterward into its validity. Since $(dp/d\rho)_0$ is a positive quantity, we denote it by c_0^2,

$$c_0^2 = \left(\frac{dp}{d\rho}\right)_0. \tag{7.2.20}$$

Then the linearized system is

$$\rho_0 \frac{\partial u'}{\partial t} + c_0^2 \frac{\partial \rho'}{\partial x} = 0 \qquad \rho_0 \frac{\partial u'}{\partial x} + \frac{\partial \rho'}{\partial t} = 0 \tag{7.2.21}$$

along with

$$p' = c_0^2 \rho'. \tag{7.2.22}$$

We can eliminate ρ' from Eqs. (21) by differentiating the first equation with respect to t and the second with respect to x and combining them to obtain the first Eq. (23a). Similar elimination of u', and use of (22), give the other two Eqs. (23a):

$$\frac{\partial^2 u'}{\partial t^2} = c_0^2 \frac{\partial^2 u'}{\partial x^2} \qquad \frac{\partial^2 \rho'}{\partial t^2} = c_0^2 \frac{\partial^2 \rho'}{\partial x^2} \qquad \frac{\partial^2 p'}{\partial t^2} = c_0^2 \frac{\partial^2 p'}{\partial x^2}. \tag{7.2.23a}$$

Thus each of the unknown functions satisfies the wave equation

$$\frac{\partial^2 \psi}{\partial t^2} = c_0^2 \frac{\partial^2 \psi}{\partial x^2}, \tag{7.2.23b}$$

whose general solution is known to be

$$\psi = f(x - c_0 t) + g(x + c_0 t), \tag{7.2.24}$$

where f and g are arbitrary twice-differentiable functions of the single variables

$$\xi = x - c_0 t \quad \text{and} \quad \eta = x + c_0 t,$$

respectively.

The solution

$$u' = f_1(x - c_0 t), \tag{7.2.25}$$

for example, gives the particle velocity u' in a wave traveling in the positive x-direction at speed c_0, since at any time $t + \Delta t$ the same particle velocity is found at $x + c_0 \Delta t$ as occurred at (x, t). A solution $\psi = g(x + c_0 t)$ similarly represents a wave traveling in the negative x-direction at speed c_0. Hence c_0 is the ***acoustic speed*** or ***sound speed*** for small disturbances about the equilibrium state. Let us examine the first assumption (19) in the light of the solution (25). We find

$$\frac{\partial u'}{\partial x} = f_1'(x - c_0 t) \qquad \frac{\partial u'}{\partial t} = -c_0 f_1'(x - c_0 t),$$

where $f_1'(x - c_0 t)$ denotes the derivative with respect to the argument $\xi = x - c_0 t$. Thus

$$\left|\frac{u'(\partial u'/\partial x)}{\partial u'/\partial t}\right| = \left|\frac{u'}{c_0}\right|. \tag{7.2.26}$$

Hence the requirement that $u'(\partial u'/\partial x)$ be small compared to $\partial u'/\partial t$ is satisfied by this solution, provided that u' is small compared to c_0. The same condition also applies to a wave traveling in the opposite direction with $u' = g_1(x + c_0 t)$. A similar consideration with, for example, $\rho' = f_2(x - c_0 t)$ shows that the other assumption (19) is satisfied also for $|u'| \ll c_0$.

Three-dimensional wave equation. Guided by our experience in the one-dimensional example above, we formulate the three-dimensional small-amplitude acoustics problem as follows. Neglecting body forces, and neglecting $\mathbf{v}\cdot\nabla\mathbf{v}$ in comparison to $\partial\mathbf{v}/\partial t$ and $\mathbf{v}\cdot\nabla\rho$ in comparison to $\partial\rho/\partial t$, as in (19) above, and writing ρ_0 for ρ in the coefficients, we linearize the Euler equation and the continuity equation, and obtain

$$\rho_0\frac{\partial \mathbf{v}}{\partial t} + \nabla p = 0 \qquad \frac{\partial \rho}{\partial t} + \rho_0 \nabla\cdot\mathbf{v} = 0. \tag{7.2.27}$$

Also

$$\nabla p = \left(\frac{dp}{d\rho}\right)_0 \nabla\rho = c_0^2\,\nabla\rho$$

and

$$\mathbf{v} = \nabla\phi$$

for irrotational motion. The first Eq. (27) then takes the form

$$\nabla\left[\rho_0\frac{\partial\phi}{\partial t} + c_0^2\rho\right] = 0,$$

whence $\rho_0(\partial\phi/\partial t) + c_0^2\rho$ is independent of position and therefore at most a function of time only. But adding an arbitrary function of time to ϕ or $\partial\phi/\partial t$ does not affect the velocities; hence we lose no generality in taking

$$\rho_0\frac{\partial\phi}{\partial t} + c_0^2\rho = 0. \tag{7.2.28a}$$

With the continuity equation, the second Eq. (27), in the form

$$\frac{\partial\rho}{\partial t} + \rho_0\nabla^2\phi = 0, \tag{7.2.28b}$$

we have a system of two linear equations for ϕ and ρ. We can eliminate ρ by differentiating the first equation with respect to t. We obtain†

$$\frac{\partial^2\phi}{\partial t^2} = c_0^2\nabla^2\phi, \tag{7.2.29}$$

which is the three-dimensional wave equation; p and ρ satisfy the same equation. We will not consider here any examples of solutions of the three-dimensional wave equation, but remark that these are hyperbolic differential equations for which boundary-and-initial conditions are not prescribed all the way around the four-dimensional solution region of x-y-z-t-space. We turn our attention now to steady motion.

Steady irrotational flow of a barotropic perfect fluid. In steady flow, the Euler equation and the continuity equation become

$$-\frac{1}{\rho}\nabla p + \mathbf{b} = \mathbf{v}\cdot\nabla\mathbf{v} \qquad \mathbf{v}\cdot\nabla\rho + \rho\nabla\cdot\mathbf{v} = 0. \tag{7.2.30}$$

For most problems of interest in gas dynamics the body force is negligible in comparison to the other forces. We neglect $\mathbf{b}$ and eliminate ρ from the equation of continuity by using the barotropic equation of state and the Euler equation as follows.

$$\nabla p = \frac{dp}{d\rho}\nabla\rho = c^2\nabla\rho, \tag{7.2.31}$$

where

$$c = \left(\frac{dp}{d\rho}\right)^{1/2} \tag{7.2.32}$$

is the ***local sound velocity***. Hence

$$\frac{1}{\rho}\nabla\rho = \frac{1}{\rho c^2}\nabla p = -\frac{1}{c^2}\mathbf{v}\cdot\nabla\mathbf{v}$$

†See, for example, Lamb (1945), Sec. 287.

by the first Eq. (30). Hence the second Eq. (30) becomes

$$-\mathbf{v}\cdot\left(\frac{1}{c^2}\mathbf{v}\cdot\nabla\mathbf{v}\right) + \nabla\cdot\mathbf{v} = 0$$

or

$$c^2\nabla\cdot\mathbf{v} - \mathbf{v}\cdot(\mathbf{v}\cdot\nabla\mathbf{v}) = 0, \tag{7.2.33}$$

the ***gas dynamical equation***.† In this equation we suppose that c is a function of the magnitude v; its dependence on v is in principle determined by the Bernoulli equation, Eq. (11),

$$P(p) + \tfrac{1}{2}v^2 = \text{const.} \tag{7.2.34}$$

(neglecting body forces), since this gives p as a function of v, while ρ is a function of p and hence of v, and $c^2 = dp/d\rho$. For example, in adiabatic isentropic flows of an ideal gas, governed by the isentropic equation

$$\frac{p}{\rho^\gamma} = \text{const.}, \tag{7.2.35}$$

it can be shown that

$$c^2 = c_0^2 - \frac{\gamma - 1}{2}v^2, \tag{7.2.36}$$

where c_0 is the sonic velocity at the reference condition p_0, ρ_0 used in defining $P(p) = \int_{p_0}^{p} dp/\rho$, provided $v = 0$ in the reference state (stagnation condition). In terms of the potential function ϕ, Eq. (33) takes the following rectangular Cartesian form,‡ where subscripts denote partial derivatives,

$$\begin{aligned}(c^2 - \phi_x^2)\phi_{xx} + (c^2 - \phi_y^2)\phi_{yy} + (c^2 - \phi_z^2)\phi_{zz}\\ - 2(\phi_x\phi_y\phi_{xy} + \phi_y\phi_z\phi_{yz} + \phi_z\phi_x\phi_{zx}) = 0\end{aligned} \tag{7.2.37a}$$

or, in indicial notation,

$$(c^2\delta_{km} - \phi_{,k}\phi_{,m})\phi_{,km} = 0. \tag{7.2.37b}$$

If c is supposed known as a function of $v^2 = \phi_x^2 + \phi_y^2 + \phi_z^2$ through the Bernoulli equation and the barotropic equation of state, then Eq. (37) is a nonlinear differential equation for the unknown velocity-potential function ϕ. Because of the nonlinearity, solutions are difficult to obtain.

†See, for example, Prager (1961), p. 103.
‡See, for example, Shapiro (1953), Sec. 9.8.

If all the velocity components ϕ_x, ϕ_y, ϕ_z are very small in comparison to c, the first line of Eq. (37) may be replaced by $c^2\nabla^2\phi$. If the mixed second derivatives in the second line are not an order of magnitude greater than ϕ_{xx}, ϕ_{yy}, and ϕ_{zz}, then the whole second line may be neglected, and the equation reduces to the linear equation

$$\nabla^2\phi = 0, \tag{7.2.38}$$

which is the form taken by the continuity equation in irrotational flow of an incompressible fluid. Thus at sufficiently low velocities (compared to the local sound velocity), the steady potential flow of a compressible fluid is approximately described by the governing equation for potential flow of an incompressible fluid. At higher flow speeds, a linearization can sometimes still be performed, but a different equation from Eq. (38) is obtained, as in Eq. (45) below. First, however, we examine the nonlinear equation in the special case of two-dimensional flow.

For flow with ϕ independent of z, Eq. (37) takes the form

$$\begin{gathered} A\phi_{xx} + B\phi_{xy} + C\phi_{yy} = 0, \quad \text{where} \\ A = c^2 - \phi_x^2 \qquad B = -2\phi_x\phi_y \qquad C = c^2 - \phi_y^2 . \end{gathered} \tag{7.2.39}$$

The character of this quasi-linear, partial differential equation is determined by the sign of the discriminant $B^2 - 4AC$, which can be put in the form†

$$B^2 - 4AC = 4c^4(M^2 - 1), \tag{7.2.40}$$

where

$$M = \frac{v}{c} \tag{7.2.41}$$

is the local ***Mach number***, the ratio of the flow speed $v = |\nabla\phi|$ to the local sound speed c. Hence the equation is

$$\left.\begin{array}{lll} \text{elliptic} & \text{for } M < 1 \text{ (subsonic)} & B^2 - 4AC < 0, \\ \text{hyperbolic} & \text{for } M > 1 \text{ (supersonic)} & B^2 - 4AC > 0, \\ \text{parabolic} & \text{for } M = 1 \text{ (transonic)} & B^2 - 4AC = 0. \end{array}\right\} \tag{7.2.42}$$

The full significance of the classification of quasi-linear, second-order, partial differential equations as elliptic, hyperbolic, or parabolic cannot be developed here. See, for example, Courant and Hilbert (1962), vol. 2. But the solution methods differ markedly from one classification to another. For example, bound-

†See, for example, Pai (1959), p. 193.

ary conditions are generally imposed all the way around the xy-region of flow when the equation is elliptic, and the solution must have no discontinuities in the second derivatives, except possibly at singular points where the differential equation does not hold. In the hyperbolic case, on the other hand, boundary conditions are not usually imposed all around, and certain kinds of discontinuities in the second derivatives are admissible across certain curves, called ***characteristic curves***, or ***characteristics***, in such a way that the differential equation continues to hold there.

Since the classification of the differential equation changes within the solution region wherever the speed reaches the sound speed, this creates great difficulties in solution when part of the flow is supersonic and part of it is subsonic. For completely subsonic flows, the solution of boundary-value problems with the nonlinear elliptic equation is also difficult, although it may be possible to accomplish by numerical finite-difference methods. Fortunately, a linearization may give good results in many subsonic flows. Completely supersonic flows are easier to solve by Eq. (39), with the method of characteristics for hyperbolic equations. But in supersonic flow shock waves may occur; these are surfaces (or curves in two-dimensional analysis) across which the velocity, pressure, and density are discontinuous and on which the differential equation does not hold. Physically, these surfaces are actually thin layers through which the parameters vary so rapidly that the continuum analysis is invalid; they are represented in continuum analysis as actual surfaces separating regions of continuum flow. The dissipative process involved in flow through a shock wave introduces vorticity into a flow which was irrotational and puts in question the use of the gas dynamics equation in terms of a velocity potential. It is found in many cases, however, that solution is still possible in terms of the velocity potential. The solutions are easier to obtain than for subsonic flow, but usually require a numerical-graphical technique unless linearization is possible. See, for example, Shapiro (1953), Chap. 14.

Example of linearization of steady compressible potential flow. In analyzing the relative flow over an object such as an airplane or a missile, we consider the object to be stationary, while the fluid velocity at infinity is $\mathbf{v} = v_1\mathbf{i}$, a uniform flow in the x-direction at pressure p_1, density ρ_1, and sound speed c_1. We assume that in the flow over the object the perturbations in $\mathbf{v}$, p, and ρ are small in comparison to v_1, p_1, ρ_1. Let ϕ' be the potential of the perturbation velocity; then

$$\mathbf{v} = v_1\mathbf{i} + \nabla\phi' = \nabla(v_1 x + \phi'). \tag{7.2.43}$$

We assume that

$$|\nabla\phi'| \ll v_1 \quad \text{and} \quad |\nabla\phi'| \ll c, \tag{7.2.44}$$

where c is the local sound speed, although v_1 may be of the same order of magnitude as c. Since $c^2 - c_1^2$ is small in comparison to c_1^2, we replace c^2 by c_1^2 in the linearization. Neglecting squares and products of the quantities assumed small, and assuming again that the mixed derivatives ϕ'_{xy}, etc., do not become an order of magnitude larger than ϕ'_{xx}, etc., we obtain from Eq. (37a) the linearized equation†

$$(c_1^2 - v_1^2)\phi'_{xx} + c_1^2(\phi'_{yy} + \phi'_{zz}) = 0$$

or

$$(1 - M_1^2)\phi'_{xx} + \phi'_{yy} + \phi'_{zz} = 0, \tag{7.2.45}$$

where

$$M_1 = \frac{v_1}{c_1} \tag{7.2.46}$$

is now the ***free-stream Mach number*** (at infinity), which is a constant so that the equation does not change its character in the solution region.

A careful examination of the assumptions involved in the linearization for the case of two-dimensional flow of an isentropic adiabatic gas [see Eqs. (35) and (36)] shows that we must require

$$\frac{M_1^2}{1 - M_1^2}\frac{v'_x}{v_1} \ll 1 \quad \text{and} \quad M_1^2\frac{v'_y}{v_1} \ll 1 \tag{7.2.47}$$

[see Shapiro (1953), Chap. 10]. This implies that for flow over thin bodies the linearized equation can be used for subsonic flows up to about $M_1 = 0.8$. Equation (45) is definitely not valid in transonic flow regions where $M_1 \cong 1$. When the flow is entirely supersonic and satisfies (47), the linearized equation can be used again, if strong shock waves do not occur, when M_1 is not too large. For supersonic flow over a slender body with thickness-to-length ratio δ, Shapiro (1953), Chap. 14, suggests the limitation

$$M_1^2\,\delta \ll 1. \tag{7.2.48}$$

Notice, however, that at the stagnation point on the body, where the flow divides, the relative velocity is zero, and hence the perturbation component $v'_x = -v_1$ is not small compared to v_1. This means that near the stagnation point the linearized equation is not valid. The region of invalidity is small for pointed slender bodies with their long dimension parallel to the flow, and may sometimes

†See, for example, Pai (1959), p. 97.

be ignored. Alternatively, a power-series solution of the nonlinear equation in a small neighborhood of the stagnation point may be used to match to the solution of the linearized equation outside of the small neighborhood.

In Sec. 7.3, we consider the two-dimensional flow of an incompressible fluid, recalling that at sufficiently small velocities compared to sonic speed, as was remarked after Eq. (38), the steady potential flow of a compressible fluid is approximately described by the governing equation of an incompressible fluid.

EXERCISES

1. Write out explicitly the demonstration that Kelvin's theorem implies the consequence that if the flow of a barotropic perfect fluid in a simply connected region is once irrotational it remains irrotational.
2. Show that in a perfect fluid the rate of change of the circulation Γ around a curve C is given by

$$\frac{d\Gamma}{dt} = -\int_S \left[\nabla\left(\frac{1}{\rho}\right) \times \nabla p\right] \cdot \hat{\mathbf{n}}\, dS$$

   where $\hat{\mathbf{n}}$ is the normal to an oriented surface S bounded by C and lying within the fluid. See Eq. (7.2.3) and Eqs. (2.5.27).
3. Carry out the details of deriving Eq. (7.2.12).
4. Use the Bernoulli equation to determine the flow velocity of liquid issuing into the atmosphere through a hole at depth h below the free surface of an open tank.
5. Carry out the details leading from Eq. (7.2.21) to Eq. (23).
6. Verify that Eq. (7.2.24) gives the solution of the preceding wave equation for arbitrary choices of the sufficiently differentiable functions f and g.
7. Express the acoustic wave speed c_0 as a function of the equilibrium p_0 and ρ_0 for the adiabatic isentropic perfect gas of Eq. (7.2.35).
8. Carry out the details leading to Eq. (7.2.29).
9. Show that $\phi = f(\hat{\mathbf{n}} \cdot \mathbf{x} - c_0 t)$ and $\phi = g(\hat{\mathbf{n}} \cdot \mathbf{x} + c_0 t)$ are solutions of the three-dimensional wave equation, Eq. (7.2.29), where f and g are arbitrary twice differentiable functions of their arguments, $\hat{\mathbf{n}}$ is an arbitrary unit vector, and $\mathbf{x}$ is the variable position vector. Interpret these solutions as ***plane waves*** such that at any time t the value of ϕ is constant over a plane with normal $\hat{\mathbf{n}}$ and such that the plane on which ϕ has any constant value is propagating in the direction of its normal (or the opposite direction) at speed c_0.
10. What kind of functions F and G furnish product solutions $\psi = F(x)G(t)$ for the one-dimensional wave equation?

11. Evaluate $P(p)$ for the adiabatic isentropic fluid of Eq. (7.2.35), and show that this leads to Eq. (36).

12. Show that Eqs. (7.2.37) follow from Eq. (33) when the velocity potential is introduced.

13. Express the local Mach number M as a function of v, γ, and c_0 for the adiabatic isentropic gas of Eq. (7.2.36).

14. If the steady ***plane flow*** of a compressible fluid is given by

$$v_x = \frac{\rho_0}{\rho}\frac{\partial \psi}{\partial y} \qquad v_y = -\frac{\rho_0}{\rho}\frac{\partial \psi}{\partial x}$$

where $\psi(x,y)$ is a scalar function called the stream function, ρ is the density, and ρ_0 is the constant density in some reference state:
(a) Show that the continuity equation is satisfied identically.
(b) If the flow is irrotational, derive from the irrotationality condition a partial differential equation for ψ. (The equation also contains ρ and its partial derivatives.) How does this equation simplify for incompressible flow?

15. Carry out the linearization details leading to Eqs. (7.2.45) and (46).

16. Show that the function

$$\phi'(x,y) = \frac{v_1}{\beta} h \sin\left(\frac{2\pi x}{L}\right) e^{-2\pi\beta y/L},$$

where $\beta = \sqrt{1 - M_1^2}$, h, and L are constants, represents a plane flow satisfying the linearized gas-dynamics equation for irrotational barotropic flow, Eq. (7.2.45). If $h \ll L$, show that the total flow of Eq. (43) then satisfies the boundary condition for flow in the semi-infinite region above the boundary curve $y = h \cos(2\pi x/L)$, if small quantities are appropriately neglected.

7.3 Potential flow of incompressible perfect fluid

Although no fluids are actually incompressible, in many important situations the compressibility of the fluid may safely be negleccted. Liquids may be considered incompressible for flow studies under usual conditions, and we saw in connection with Eq. (7.2.38) that a gas may be treated approximately as incompressible for flow studies at speeds very small compared to the sonic speed. Under the assumption of incompressibility, the perfect-fluid field equations simplify greatly, especially when the flow is irrotational, as it will be assumed to be in most of this section.

In the Bernoulli equation, Eq. (7.2.11), we can write

$$P(p) \equiv \int_{p_0}^{p} \frac{1}{\rho} dp = \frac{p}{\rho} + \text{const.}$$

for incompressible flow (ρ = const.), whence the Bernoulli equation becomes

$$\Omega + \frac{p}{\rho} + \frac{1}{2}v^2 = \text{const.} \tag{7.3.1}$$

along any streamline. For irrotational flow the same constant applies throughout the field. When the body-force potential is that of a uniform gravitational field, $\Omega = gh$, it is often convenient to separate the pressure into the sum of two parts, the ***static pressure***

$$p_s = -\rho g h + \text{const.}$$

and the ***dynamic pressure***† p_d. Since

$$\Omega + \frac{p_s}{\rho} = \text{const.}$$

in this case, the Bernoulli equation may be written as

$$p_d - (p_d)_0 = \frac{\rho}{2}(v_0^2 - v^2), \tag{7.3.2}$$

which is usually applied to conditions where $(p_d)_0 = 0$ in the "undisturbed region," where $v = v_0$. The subscript d is often dropped from p_d when the context makes it clear that p is the dynamic pressure. With this equation, the pressure can be determined throughout the flow field when the velocity has been determined. The velocity solution is obtained from the ***continuity equation for incompressible potential flow***

$$\nabla^2\phi = 0, \tag{7.3.3}$$

where ϕ is the velocity potential, such that $\mathbf{v} = \nabla\phi$. Equation (3) is solved subject to the ***boundary condition***

$$\frac{\partial\phi}{\partial n} = v_n^B \tag{7.3.4}$$

where v_n^B is the normal component of the velocity of the rigid boundary, equal to zero if the boundary is stationary. The boundary-value problem of Eqs. (3) and (4) is called a ***Neumann problem*** in potential theory, the solution of Laplace's equation for ϕ with given normal derivative on the boundary. Before considering examples of solutions of such boundary-value problems, we establish some general properties of such flows.

†Some authors use the term "dynamic pressure" for the expression $\frac{1}{2}\rho v^2$ instead of for that part of the pressure not associated with hydrostatic effects.

Green's identities† are useful in establishing many of the properties of potential flow. In the following discussion of Green's identities ϕ and ψ are assumed to be single-valued functions. The velocity potential ϕ is single-valued in any simply connected region. In multiply connected regions, it may not be single-valued and a revised version of the identities may be needed: see, for example, Milne-Thomson (1960), Sec. 2.62. By Eq. (2.5.27d)

$$\nabla\cdot(\mathbf{a}\phi) = \phi(\nabla\cdot\mathbf{a}) + \mathbf{a}\cdot(\nabla\phi) \tag{7.3.5a}$$

for any (sufficiently differentiable) vector $\mathbf{a}$ and scalar ϕ. In particular, if $\mathbf{a} = \nabla\psi$, this yields

$$\nabla\cdot[\phi\nabla\psi] = \phi\nabla^2\psi + \nabla\psi\cdot\nabla\phi, \tag{7.3.5b}$$

whence by the divergence theorem of Sec. 5.1 (note $\hat{\mathbf{n}}\cdot\nabla\psi = \partial\psi/\partial n$)

$$\int_S \phi\frac{\partial\psi}{\partial n}dS = \int_V \nabla\cdot[\phi\nabla\psi]\,dV = \int_V [\phi\nabla^2\psi + \nabla\psi\cdot\nabla\phi]\,dV.$$

This establishes ***Green's First Identity***

$$\int_V \nabla\psi\cdot\nabla\phi\,dV = \int_S \phi\frac{\partial\psi}{\partial n}\,dS - \int_V \phi\nabla^2\psi\,dV, \tag{7.3.6}$$

where $\hat{\mathbf{n}}$ is the ***outer normal direction***. The roles of ϕ and ψ can be interchanged in Eq. (6); if we subtract the result from Eq. (6), we get ***Green's Second Identity***:

$$\int_S \left(\phi\frac{\partial\psi}{\partial n} - \psi\frac{\partial\phi}{\partial n}\right)dS = \int_V (\phi\nabla^2\psi - \psi\nabla^2\phi)\,dV. \tag{7.3.7}$$

If both ϕ and ψ are harmonic, $\nabla^2\phi = 0$ and $\nabla^2\psi = 0$, and the right side of Eq. (7) vanishes. In particular if $\psi = 1/r$, where r is the distance from a point P in the interior to a variable point on the surface S, then ***Green's Third Identity***

$$\phi_P = -\frac{1}{4\pi}\int_S\left[\phi\frac{\partial}{\partial n}\left(\frac{1}{r}\right) - \frac{1}{r}\frac{\partial\phi}{\partial n}\right]dS$$
$$\text{for}\quad \nabla^2\phi = 0 \quad \text{inside } S, \tag{7.3.8}$$

where n is the ***outward normal*** direction, may be established as follows. We apply Eq. (7) to the region between S and a sphere of radius R about P, entirely inside S. In this region $\nabla^2\psi = 0$, for $\psi = 1/r$, and the volume integral of Eq. (7) vanishes. The surface integral over S plus the sphere gives

†Due to George Green (1828).

$$\int_S \left[\phi \frac{\partial}{\partial n}\left(\frac{1}{r}\right) - \frac{1}{r}\frac{\partial \phi}{\partial n}\right] dS - \int_{\text{sphere}} \left[\phi \frac{\partial}{\partial R}\left(\frac{1}{R}\right) - \frac{1}{R}\frac{\partial \phi}{\partial R}\right] dS = 0.$$

The minus sign appears before the second integral because $\partial/\partial R = -\partial/\partial n$ on the sphere, since the outer normal for the region between the two surfaces points toward the inside of the sphere. Since the first integral is independent of R, so is the second; we evaluate the second by taking its limit as $R \to 0$.

$$\int_{\text{sphere}} \left[\phi \frac{\partial}{\partial R}\left(\frac{1}{R}\right) - \frac{1}{R}\frac{\partial \phi}{\partial R}\right] dS = \lim_{R\to 0} 4\pi R^2 \left[\phi\left(-\frac{1}{R^2}\right) - \frac{1}{R}\frac{\partial \phi}{\partial R}\right]_{\text{Av.}} = -4\pi\phi_P$$

where the subscript Av. denotes the mean value on the sphere. When this limit is substituted for the second integral, Eq. (8) results.

The mean-value theorem for a harmonic function ϕ states that the mean value of ϕ over any spherical surface S of radius r, inside which $\nabla^2\phi = 0$, is equal to the value ϕ_P at the center:

$$\phi_P = \frac{1}{4\pi r^2}\int_S \phi \, dS. \tag{7.3.9}$$

This theorem is due to Gauss (1839). The theorem is an immediate consequence of Green's Third Identity, since if we take the S of Eq. (8) to be a sphere with center P, then r is constant on S, so that

$$\int_S \frac{1}{r}\frac{\partial \phi}{\partial n} dS = \frac{1}{r}\int_S \frac{\partial \phi}{\partial n} dS,$$

and, by the divergence theorem,

$$\int_S \frac{\partial \phi}{\partial n} dS = 0 \quad \text{for} \quad \nabla^2\phi = 0 \quad \text{inside } S. \tag{7.3.10}$$

The first term of Eq. (8) then yields Eq. (9), proving the mean-value theorem. If ϕ is the velocity potential, Eq. (10) expresses the fact that in an incompressible flow the net outward volume flux through any closed surface (not necessarily a sphere) within the flow must be zero. If this condition is not satisfied, there must be a fluid source or sink inside S, a singularity where $\nabla^2\phi = 0$ does not hold.

An immediate corollary of the mean-value theorem is that ***a harmonic function*** ϕ ***cannot have a maximum or a minimum in the interior.*** Another corollary is that ***in incompressible potential flow the maximum speed must occur on the boundary.*** For if an interior point P were a point of maximum v, we could choose Cartesian axes with the x-axis parallel to $\mathbf{v}_P$, and conclude that $\partial\phi/\partial x$ attained a maximum at P. But this violates the first corollary, since $\partial\phi/\partial x$ is harmonic, because $\nabla^2(\partial\phi/\partial x) = \partial(\nabla^2\phi)/\partial x = 0$. From Eq. (2) it follows then that ***the minimum dynamic pressure must occur on the boundary.***

The total kinetic energy K of the flow inside a surface S is given by

$$K = \frac{1}{2}\rho\int_S \phi\frac{\partial\phi}{\partial n}dS, \tag{7.3.11}$$

where n is the outward normal. This follows from Green's First Identity, Eq. (6), with $\phi = \psi$, since $K = (\rho/2)\int_V \nabla\phi\cdot\nabla\phi\, dV$, and $\nabla^2\phi = 0$. If $\partial\phi/\partial n = 0$ everywhere on S, it follows from Eq. (11) that $K = 0$ and hence that $\mathbf{v} = 0$ everywhere in V. Hence there is no nontrivial single-valued potential ϕ in a region bounded entirely by a surface on which $\partial\phi/\partial n = 0$ identically. A uniqueness theorem follows from this result.

Uniqueness theorem. Any two single-valued velocity potentials ϕ_1 and ϕ_2 differ at most by a constant (not affecting the flow) in a region bounded by S, if they satisfy the same boundary condition on S, i.e., if $\partial\phi_1/\partial n = \partial\phi_2/\partial n$ on S. For if this were not true, then $\phi = \phi_1 - \phi_2$ would be a nontrivial potential with $\partial\phi/\partial n = 0$ on S, violating the conclusion just established. (Note that, because $\nabla^2\phi = 0$ is a linear equation, the difference of any two solutions will also be a solution of the equation.) A similar uniqueness theorem can be established for flow in a simply connected infinite region exterior to an immersed finite solid, if the velocity at infinity is given and uniform. See, for example, Milne-Thomson (1960). Note, however, the possibility of nonuniqueness in a multiply connected region where the potential may not be single-valued. This will be illustrated later in this section by the two-dimensional flow in the multiply connected region exterior to an infinitely long circular cylinder.

Stream function in two-dimensional flow. The continuity equation for incompressible fluid in plane flow parallel to the xy-plane is

$$\frac{\partial v_x}{\partial x} + \frac{\partial v_y}{\partial y} = 0. \tag{7.3.12}$$

Hence, whether or not the flow is irrotational, there always exists a function $\psi(x,y)$, called the ***stream function,*** such that

$$v_x = \frac{\partial\psi}{\partial y} \quad \text{and} \quad v_y = -\frac{\partial\psi}{\partial x}, \tag{7.3.13}$$

satisfy Eq. (12) identically.† If we substitute Eq. (13) into the differential equation of a streamline

†According to Lamb (1945), the stream function was introduced in this way by Lagrange (1781), while the kinematical interpretation of Eq. (16) is due to Rankine (1864).

$$\frac{dx}{v_x} = \frac{dy}{v_y} \tag{7.3.14}$$

and clear of fractions, we obtain

$$\frac{\partial \psi}{\partial x} dx + \frac{\partial \psi}{\partial y} dy = 0$$

or

$$\psi(x,y) = \text{const.} \tag{7.3.15}$$

Thus ***the stream function is constant along any streamline.*** Consider any curve C joining P_0 to P in the xy-plane. The volume flux Q through a cylindrical surface with elements perpendicular to the xy-plane, intersecting the xy-plane in C, is (per unit length perpendicular to the plane)

$$Q = \int_{P_0 \atop C}^{P} v_n \, ds = \int_{P_0 \atop C}^{P} (v_x n_x + v_y n_y) \, ds = \int_{P_0 \atop C}^{P} (v_x \, dy - v_y \, dx)$$

if the positive normal is from left to right across the curve when we face along the curve from P_0 toward P. Then, using Eqs. (13), we find

$$Q = \int_{P_0}^{P} d\psi = \psi(P) - \psi(P_0). \tag{7.3.16}$$

The total flux from left to right across the arc is thus equal to the increase in the stream function ψ along the arc. [Authors who define the velocity potential ϕ by $\mathbf{v} = -\nabla\phi$ usually also define ψ as the negative of our ψ, so that the minus sign appears on the other Eq. (13); then Eq. (16) would give the flux from right to left.]

Complex-function formulation of plane potential flow. Up to this point in our discussion of the stream function, we have not assumed irrotationality. When the plane flow is irrotational, the methods of complex function theory can be used. These methods furnish some useful solutions by elementary methods. More advanced methods, including conformal mapping methods, have provided solutions for many practically important problems, for example the flows over a large class of airfoils that can be mapped conformally onto the exterior of a circular cylinder. See, for example, Milne-Thomson (1960). If the plane flow is irrotational, then

$$v_x = \frac{\partial \phi}{\partial x} \qquad v_y = \frac{\partial \phi}{\partial y}, \tag{7.3.17}$$

where ϕ is the velocity potential. Hence the two functions ϕ and ψ satisfy the two equations

$$\frac{\partial \phi}{\partial x} = \frac{\partial \psi}{\partial y} \qquad \frac{\partial \phi}{\partial y} = -\frac{\partial \psi}{\partial x}. \tag{7.3.18}$$

These two equations are the Cauchy-Riemann conditions.

From complex-function theory, it is known that a single-valued complex function of the complex variable $z = x + iy$, say $f(z) = \phi(x, y) + i\psi(x, y)$, whose real part $\phi(x, y)$ and imaginary part $\psi(x, y)$ have continuous first partial derivatives in a region is ***a holomorphic function of*** z if and only if the Cauchy-Riemann equations, Eqs. (18), are satisfied. A holomorphic function of a complex variable is one for which the derivative $f'(z)$ exists. Holomorphic functions are sometimes called ***analytic functions***, but the term analytic is more often applied to a wider class of functions, including multiple-valued functions, while holomorphic applies only to a single-valued function or to one single-valued branch of a multiple-valued function. The fact that $f'(z)$ exists for a holomorphic function means that if

$$f(z) = \phi(x, y) + i\psi(x, y),$$

then

$$\frac{df}{dz} = \lim_{\Delta z \to 0} \frac{\Delta\phi + i\Delta\psi}{\Delta x + i\Delta y},$$

and this limit is independent of the path along which $\Delta z \to 0$. It is easy to see that the Cauchy-Riemann equations are a necessary condition. For, assuming that a unique limit is obtained, we can take the limit in two special ways and equate the results, as follows

1. Put $\Delta y = 0$ first, and let $\Delta x \to 0$. Then

$$\frac{df}{dz} = \lim_{\Delta x \to 0} \frac{\Delta\phi}{\Delta x} + i\frac{\Delta\psi}{\Delta x} = \frac{\partial \phi}{\partial x} + i\frac{\partial \psi}{\partial x}. \tag{7.3.19a}$$

2. Put $\Delta x = 0$ first, and let $\Delta y \to 0$. Then

$$\frac{df}{dz} = \lim_{\Delta y \to 0} \frac{\Delta\phi}{i\Delta y} + i\frac{\Delta\psi}{i\Delta y} = \frac{1}{i}\frac{\partial \phi}{\partial y} + \frac{\partial \psi}{\partial y} = \frac{\partial \psi}{\partial y} - i\frac{\partial \phi}{\partial y}, \tag{7.3.19b}$$

since

$$\frac{1}{i} = -i.$$

In order that the real and imaginary parts of the two results be equal it is necessary that Eqs. (18) be satisfied. The quality of being holomorphic is thus a very restrictive one. It also has a number of remarkable consequences. A holomorphic function of z has continuous derivatives $f^{(n)}(z) = d^n f/dz^n$ of all orders with respect to z. A function $f(z)$ is holomorphic in a region if and only if it can be represented by a convergent power series $f(z) = a_0 + a_1(z - z_0) + a_2(z - z_0)^2 + \cdots$ in some neighborhood of every point z_0 in the interior of the region. Moreover the line integral

$$\oint_C f(z)\,dz \equiv \oint_C (\phi + i\psi)(dx + i\,dy)$$
$$\equiv \oint_C [\phi\,dx - \psi\,dy] + i\oint_C [\psi\,dx + \phi\,dy] = 0, \qquad (7.3.20)$$

i.e., the line integral vanishes around every closed curve in the interior of a simply connected region if and only if $f(z)$ is holomorphic in the region, since the Cauchy-Riemann equations are the necessary and sufficient conditions that the two real line integrals be independent of the path between any two points and therefore vanish for a closed curve (see Sec. 2.5). By eliminating ψ and ϕ, respectively, from Eqs. (18), we see that the real and imaginary parts each satisfy the two-dimensional Laplace equation:

$$\frac{\partial^2\phi}{\partial x^2} + \frac{\partial^2\phi}{\partial y^2} = 0 \qquad \frac{\partial^2\psi}{\partial x^2} + \frac{\partial^2\psi}{\partial y^2} = 0. \qquad (7.3.21)$$

We already knew that the velocity potential ϕ satisfied the equation, and now see that the stream function also satisfies the same equation when the flow is irrotational. Of course ψ satisfies a different boundary condition. Equation (13) implies that the tangential derivative of ψ must vanish on any stationary solid boundary. Hence ψ is constant on any simply connected smooth stationary boundary, and satisfies a Dirichlet problem of potential theory.

The ***complex potential function*** $f(z)$ is the holomorphic function whose real and imaginary parts are the velocity potential ϕ and the stream function ψ, respectively. The ***complex velocity*** is related to the derivative $f'(z)$, but is defined in different ways by different writers. We take the ***complex velocity*** to be the derivative†

$$f'(z) = v_x - iv_y \qquad (7.3.22)$$

The inverse method of solution for flow problems consists of examining various holomorphic complex potential functions and choosing one that is useful. Any streamline in such an inviscid flow field may be replaced by a stationary

† Writers who define ϕ so that $\mathbf{v} = -\nabla\phi$ would obtain $f'(z) = -v_x + iv_y$; they also define ψ as the negative of ours, so that Eq. (16) gives the flux from right to left. Some writers define the complex velocity as $v_x + iv_y$, and hence as the complex conjugate of $f'(z)$ [or of $-f'(z)$, if they have taken $\mathbf{v} = -\nabla\phi$]. Also some writers use the symbol w for the complex velocity, while some use it for the complex potential function.

solid boundary to obtain a boundary-value problem solution. Direct methods for finding the $f(z)$ for a given boundary exist but require more effort than the inverse methods.

All of the usual elementary single-valued functions of z are holomorphic in some region. Let us examine the two-dimensional flows represented by some of them. The linear function

$$f(z) = (a + ib)z \tag{7.3.23a}$$

is holomorphic in the entire finite plane, and represents the uniform flow with

$$f'(z) \equiv a + ib = v_x - iv_y \tag{7.3.23b}$$

or, for a and b real,

$$v_x = a \qquad v_y = -b. \tag{7.3.23c}$$

Every positive integral power function z^n is also holomorphic in the entire finite plane, and hence so is every polynomial in z. Any power series in z is holomorphic within its circle of convergence. Several singular potential functions $f(z)$ furnish useful flow patterns away from their points of singularity.

The logarithmic potentials,

$$\text{Source line:} \quad f(z) = m \ln(z - z_0) \tag{7.3.24a}$$

$$\text{Sink line:} \quad f(z) = -m \ln(z - z_0) \tag{7.3.24b}$$

$$\text{Vortex:} \quad f(z) = -i\kappa \ln(z - z_0), \tag{7.3.24c}$$

for positive m and real κ, are multiple-valued functions. We introduce polar coordinates centered at z_0

$$x - x_0 = r \cos\theta \qquad y - y_0 = r \sin\theta \quad \text{or} \quad z - z_0 = re^{i\theta} \tag{7.3.25}$$

by the Euler relation, for real θ,

$$e^{i\theta} = \cos\theta + i \sin\theta. \tag{7.3.26}$$

Then, since $\ln(e^{i\theta}) = i\theta$, we have

$$\ln(z - z_0) = \ln r + i\theta. \tag{7.3.27}$$

Evidently z_0 is a singular point, since $\ln r \to -\infty$ as $r \to 0$. Moreover the logarithm is multiple-valued because at any point other than z_0, the angle θ may have an infinite number of values, differing by multiples of 2π. The complex velocity corresponding to this potential is, however, single-valued.

$$\frac{d}{dz}[\ln (z - z_0)] = \frac{1}{z - z_0} = \frac{1}{r} e^{-i\theta} = \frac{1}{r} \cos \theta - \frac{i}{r} \sin \theta. \qquad (7.3.28)$$

Hence, if we restrict attention to one single-valued branch of $\ln (z - z_0)$ by restricting θ to satisfy $0 < \theta < 2\pi$, $f(z)$ will be single-valued and holomorphic in the finite cut plane, with the points on the branch cut $\theta = 0$ excluded from the region. The flow field may even include the points on the branch cut, except for $r = 0$, since the complex velocity is unambiguous there. The flow field for each case of Eq. (24) may be represented by drawing representative streamlines and equipotential lines, as in Fig. 7.3.

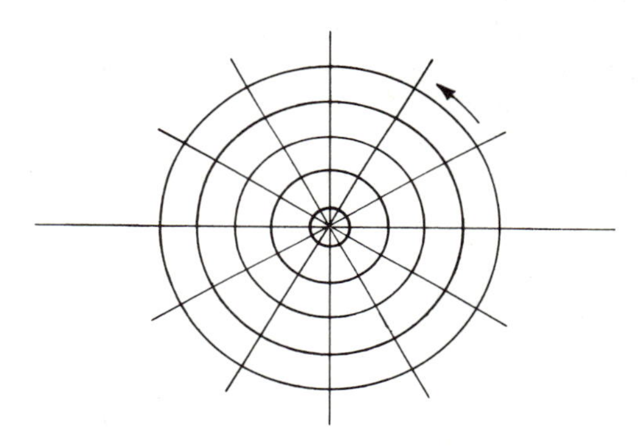

Fig. 7.3 Flow net for source, sink or vortex (arrow indicates sense of vortex for positive κ).

For the ***source line***, the streamlines are the radial lines $\theta = $ const., while the equipotential lines are the circles $r = $ const. By Eq. (16), the outward flux (per unit thickness perpendicular to the plane) across each circle is $2\pi m$, the amount by which the stream function $\psi = m\theta$ increases on the circle. This flux is the same for every circle, and evidently the singularity represents the idealized situation where fluid is being fed into the flow along a source line perpendicular to the plane of the flow at the rate of $2\pi m$ volume units per unit length of the source line, which we call a ***source strength*** of $2\pi m$. (Some writers call m the source strength and then multiply the "strength" by 2π to get the volume flux per unit length.) Equation (24b) evidently represents a ***sink line*** of strength $2\pi m$. Notice that the minus sign would move to the source with the convention $\mathbf{v} = -\nabla\phi$.

The ***vortex*** of Eq. (24c) has the complex potential

$$f(z) = \kappa\theta - i\kappa \ln r \qquad \text{with complex velocity}$$

$$v_x - iv_y = f'(z) = -\frac{i\kappa}{r} e^{-i\theta} = -\frac{\kappa}{r} \sin \theta - i\frac{\kappa}{r} \cos \theta \qquad (7.3.29)$$

in the polar representation of Eq. (25). Hence the streamlines are the circles, and the flow is a circulating flow with counterclockwise velocity circulation

$$\Gamma = 2\pi\kappa \tag{7.3.30}$$

around each circle, a vortex of strength $2\pi\kappa$. Since any streamline may be replaced by a solid boundary in inviscid flow, this can represent the circulating flow around an infinitely long circular cylinder. Note that for any two different values of κ we obtain two different flows exterior to the cylinder with the same zero velocity at infinity and the same normal velocity on the cylinder wall, namely $v_n = 0$. But this does not contradict the uniqueness theorem, since the theorem as stated applies only to single-valued potentials.

A ***doublet*** is defined as the singularity obtained by taking a source and sink of the same strength $2\pi m$ separated by distance $2a$, and letting $a \to 0$ and $m \to \infty$ in such a way that $D = 2ma$ remains constant. Consider a source at $z = ae^{i\alpha}$ and sink at $z = -ae^{i\alpha}$. The complex potential of the combination is

$$\begin{aligned} f(z) &= m \ln (z - ae^{i\alpha}) - m \ln (z + ae^{i\alpha}) \\ &= m \ln \left[z\left(1 - \frac{ae^{i\alpha}}{z}\right)\right] - m \ln \left[z\left(1 + \frac{ae^{i\alpha}}{z}\right)\right] \\ &= m \ln \left(1 - \frac{ae^{i\alpha}}{z}\right) - m \ln \left(1 + \frac{ae^{i\alpha}}{z}\right). \end{aligned}$$

Using the infinite series expansion for the two logarithms,

$$\ln (1 + t) = t - \tfrac{1}{2}t^2 + \tfrac{1}{3}t^3 - \cdots,$$

and keeping only the first term in each, since we expect to let $a \to 0$, we obtain, with $D = 2ma$,

$$f(z) = -\frac{De^{i\alpha}}{z}$$

for a ***doublet at the origin*** or

$$f(z) = -\frac{De^{i\alpha}}{(z - z_0)} \tag{7.3.31}$$

for a ***doublet at*** z_0, with axis inclined at angle α to the x-direction. Figure 7.4 shows some of the streamlines for a doublet. They consist of circles tangent to the doublet axis at z_0, of diameter D/ψ for constant D and various values of ψ.

Naturally these singular points cannot actually occur within a real flow. But their flow patterns are sometimes useful as approximations valid sufficiently far from the singular point. Or, as in the case of the circulating flow exterior to a

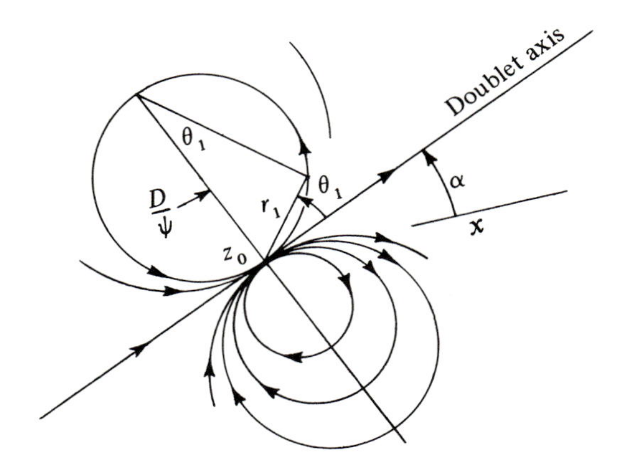

Fig. 7.4 Streamlines for doublet at z_0.

cylinder in Eq. (29), the singularity is not in the flow field at all. They are usually combined with a nonsingular complex potential. Then any streamline of the combined flow represents a possible boundary, and the solution is admissible if the singularity is not within the flow. We consider one further example, combining a uniform flow and a doublet at the origin with $\alpha = \pi$.

Symmetrical flow past a circular cylinder. For real v_1, the complex potential

$$f(z) = v_1 \left(z + \frac{a^2}{z}\right) = v_1 \left(re^{i\theta} + \frac{a^2}{r}e^{-i\theta}\right) \tag{7.3.32}$$

has real and imaginary parts

$$\phi = v_1 \left(r + \frac{a^2}{r}\right) \cos\theta \qquad \psi = v_1 \left(r - \frac{a^2}{r}\right) \sin\theta. \tag{7.3.33}$$

The streamline $\psi = 0$ is given by $\sin\theta = 0$ or $r = a$. We choose the part $r = a$ as a solid boundary to obtain a flow exterior to a cylinder of radius a; the only singularity is then inside the cylinder. The complex velocity is

$$\begin{aligned} v_x - iv_y = f'(z) &= v_1 \left(1 - \frac{a^2}{z^2}\right) \\ &= v_1 \left(1 - \frac{a^2}{r^2}\cos 2\theta\right) + iv_1\frac{a^2}{r^2}\sin 2\theta. \end{aligned} \tag{7.3.34}$$

The flow is symmetrical about the ***dividing streamline*** $\psi = 0$, and reduces to the uniform flow $\mathbf{v} = v_1\mathbf{i}$ at infinity, as illustrated in Fig. 7.5. The velocity is zero at the front and rear stagnation points F and R and maximum at the top

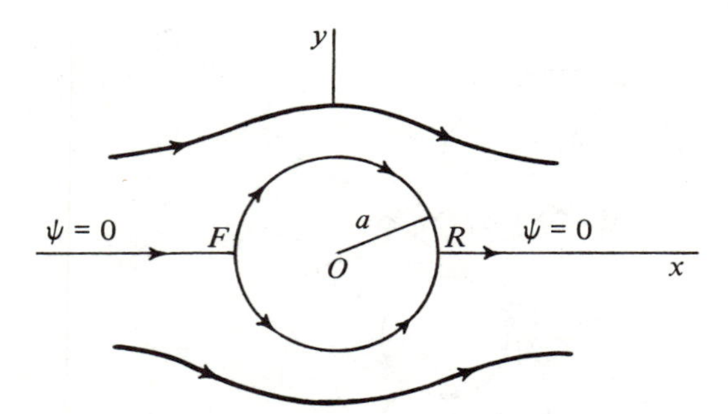

Fig. 7.5 Symmetrical flow past a circular cylinder.

and bottom, where the speed is $2v_1$. Actual flows past circular cylinders are observed to differ from this idealized flow significantly on the downstream side of the cylinder. Because the cylinder is a badly streamlined body, the dynamic pressure [given by the Bernoulli Eq. (2) if the potential flow is actually correct] rises rapidly on the cylinder surface downstream from the point of maximum speed. The viscous boundary layer near the cylinder, whose existence has been ignored in the analysis, then tends to thicken, and separation of the flow occurs from the cylinder, leaving a turbulent wake downstream from the cylinder. [See, for example, Yuan (1967), p. 343.]

For well-streamlined bodies, the turbulent wake is much reduced in significance, and the potential-flow solution gives good predictions of the pressure distribution on the bodies. The circular-cylinder solution is useful not so much for actual flows over circular cylinders as for calculating the flow over streamlined cylinders which can be mapped onto a circle by means of a conformal mapping. [See, for example, Milne-Thomson (1960).] This technique is useful for a wide class of two-dimensional airfoils; the nonuniqueness represented by the arbitrariness of the circulating flow about the cylinder is usually resolved by adding just enough circulation to cause the rear stagnation point on the cylinder to be the image of the airfoil trailing edge under the mapping, for normal flight conditions.

Direct methods of boundary-value problem solution will not be presented here. Any of the methods of potential theory may be used. For finite flow fields, numerical analysis by finite-difference equations can be used. Methods of the calculus of variations are also sometimes employed.

Experimental analysis is also widely used, employing reduced-scale models of the flow channels, and of aircraft or ships in wind tunnels or towing tanks. This type of analysis has the advantage of using real fluids instead of the idealized, inviscid incompressible fluid treated in this section. Conditions required for dynamic similarity will be discussed in Sec. 7.4.

EXERCISES

1. If the velocity potential ϕ is constant over the bounday of any simply connected region occupied by incompressible fluid in irrotational flow, prove that ϕ is constant throughout the region.

2. (a) Determine the condition on the constants A, B, C, D that ensures that $v_x = Ax + By$, $v_y = Cx + Dy$ can be an incompressible plane flow.
   (b) Determine $\psi(x, y)$ for this flow, and show that the streamlines are conic sections. If the flow is also irrotational, show that the streamlines are rectangular hyperbolas.

3. Show that if $f(z)$ is the complex potential, then $f'(z) = v_x - iv_y$ as stated in Eq. (7.3.22).

4. Sketch the streamlines and equipotential lines for the complex potential $f(z) = Az^2$, where A is a real constant.

5. Write the velocity components in terms of rectangular coordinates x and y for the vortex of Eq. (7.3.29), and show that the flow is irrotational. Why does this not contradict the fact that the circulation is not zero around circles r=constant?

6. Show that the circles in Fig. 7.4 are in fact streamlines for the doublet flow as was stated after Eq. (7.3.31).

7. Write the complex potential and complex velocity for flow past a circular cylinder with circulation by superposing the vortex of Eq. (7.3.29) (with $z_0 = 0$) onto the flow of Eq. (32).

8. Express the dynamic pressure as a function of θ on the cylinder of Ex. 7, and calculate the resultant force exerted on the cylinder by the fluid.

9. The ***Theorem of Blasius*** states that if a fixed cylinder (in general noncircular) is placed in a plane potential flow of incompressible fluid, then the force components (X, Y) of the total pressure force on unit length of the cylinder and the moment M of the pressure forces are given by

$$X - iY = \frac{1}{2}i\rho \oint \left(\frac{df}{dz}\right)^2 dz \qquad M = \text{real part of } [-\frac{1}{2}\rho \oint z\left(\frac{df}{dz}\right)^2 dz]$$

   where f is the complex potential. Derive the first result, for $X - iY$. *Hint*: Show that the force on boundary element $dx + i\,dy$ is $dX - i\,dY = -ip\,d\bar{z}$. Notice that $v^2 = f'(z)\overline{f'(z)} = (df/dz)(d\bar{f}/d\bar{z})$, and that $d\bar{f} = df$ on the boundary streamline, since ψ = constant on a streamline.

10. Derive the second part of the Theorem of Blasius of Ex. 9, to obtain the expression for M.

11. If there are no singularities of df/dz between the boundary curve of Ex. 9 and a large circle outside of the boundary curve, show that the integrals appearing

in the Theorem of Blasius can be evaluated around the large circle instead of around the boundary. *Hint:* Use Eq. (20), and choose the path C to consist of the cylinder boundary, the large circle traversed in the negative sense, and a straight-line cut joining the cylinder boundary to the large circle. The cut is traversed once in each direction to make a continuous traverse around a simply connected region.

12. Use the result of Ex. 9 to find the resultant force in Ex. 8.

7.4 Similarity of flow fields in experimental model analysis; characteristic numbers; dimensional analysis

Solutions to boundary-value problems for viscous fluids are difficult to obtain. In the limiting case of very small viscosity, boundary-layer theory may furnish useful results, and the other limiting case of the slow flow of fluids of very large viscosity can also be treated analytically; see Sec. 7.5 for a discussion of these two limiting cases. In the intermediate range, it is much more difficult to solve the nonlinear equations. For this reason, most studies of viscous fluids are experimental investigations, using scale models in wind tunnels or towing tanks, for example, or studying flow through models of rivers and over dams. Experimental studies are especially helpful when turbulent flow conditions occur, since these are difficult to analyze. The most general formulation of Sec. 7.1 cannot be solved directly for the mean velocity and pressure in turbulent flow. The discussion of similitude given below is based on the Navier-Stokes equation, which does not apply directly to the mean velocity and pressure in turbulent flow. But the dimensionless characteristic numbers introduced here are also used in turbulent flow studies, usually along with additional characteristic numbers. Experimental analysis is usually performed on a reduced-scale model geometrically similar to the prototype, and possibly with a different fluid. The question arises under what conditions the two flows exhibit dynamic similarity.

Dynamic similarity means that two systems with geometrically similar boundaries have geometrically similar flow patterns at corresponding instants of time, and hence that all the different forces (pressure-gradient force, viscous force, inertia force, and body force) acting on a fluid element in the model flow at a certain time must be the same multiple of the corresponding force acting on the corresponding fluid element in the prototype flow at the corresponding time. Complete dynamic similarity is rarely achieved, but when one or more of the forces is negligible, it may be possible to achieve similarity with the important ones. When thermal phenomena are important, the energy equation and the equations of state must be considered in studying the similarity. We will consider later the use of these equations in the case of a perfect gas. But our first similarity study will consider just the Navier-Stokes equation and the continuity equation, which contain only mechanical variables.

Dimensionless variables are introduced to facilitate the comparison of model and prototype flows. Since the two flows are to be geometrically similar, a single ***characteristic length*** L establishes the scale; this could be the diameter of a pipe, the length of a ship, or the wingspan of an airplane. We also introduce a ***characteristic time*** T for any periodic or transient process; this could be the period of a sinusoidal wave, the revolution time of a propeller, or a time required for a starting-up process to reach a specified fraction of its steady-state value. The dimensionless independent variables $\mathbf{x}'$ and t' are then defined as

$$\mathbf{x}' = \frac{\mathbf{x}}{L} \qquad t' = \frac{t}{T}. \tag{7.4.1}$$

A ***characteristic speed*** V (e.g., the free-stream speed at infinity in the flow over an airfoil) is selected, and ***characteristic values*** p_0, ρ_0, θ_0 of the pressure, density, and temperature are chosen at a reference point and time so that they obey the equation of state. Then the dimensionless dependent variables are

$$\mathbf{v}' = \frac{\mathbf{v}}{V} \qquad p' = \frac{p}{p_0} \qquad \rho' = \frac{\rho}{\rho_0} \qquad \theta' = \frac{\theta}{\theta_0}. \tag{7.4.2}$$

The Navier-Stokes equation [Eq. (7.1.17)] is

$$-\nabla p + \mu[\nabla^2\mathbf{v} + \frac{1}{3}\nabla(\nabla\cdot\mathbf{v})] - \rho g\nabla h = \rho\frac{\partial\mathbf{v}}{\partial t} + \rho\mathbf{v}\cdot\nabla\mathbf{v} \tag{7.4.3}$$

if the only body force is the gravitational force $\mathbf{b} = -\rho g\nabla h$, where $h = Lh'$ is the elevation. When the dimensionless variables (1) and (2) are introduced, the equation takes the form

$$\begin{aligned} -\left(\frac{p_0}{L}\right)\nabla'p' + \frac{\mu V}{L^2}[\nabla'^2\mathbf{v}' + \frac{1}{3}\nabla'(\nabla'\cdot\mathbf{v}')] - (\rho_0 g)\rho'\nabla'h' \\ -\left(\frac{\rho_0 V}{T}\right)\rho'\frac{\partial\mathbf{v}'}{\partial t'} - \left(\frac{\rho_0 V^2}{L}\right)\rho'\mathbf{v}'\cdot\nabla'\mathbf{v}' = 0, \end{aligned} \tag{7.4.4}$$

where ∇' denotes the gradient with respect to the dimensionless $\mathbf{x}'$. The dimensional coefficients in parentheses depend only on the characteristic values selected, and the viscosity μ, which, in this discussion, is assumed constant in time and independent of position in any one flow, although it may differ between prototype flow and model flow. The gravitational constant appears, but it is not adjustable. The dimensional coefficients are summarized in Table 7-1. Some idea of the relative importance of the different kinds of forces can be obtained by comparing the magnitudes of these coefficients, but only if the dimensionless quantities they multiply are all of the same order of magnitude. Engineering judgment should be used in the selection of the characteristic values L, T, V, etc., in order to try to make the dimensionless quantities in the equation all of the same order, say of order unity, in the region of interest. See, for example, Kline (1965), Chap. 4.

TABLE 7-1

DIMENSIONAL COEFFICIENTS IN NAVIER-STOKES EQUATION†

Pressure-force term	p_0/L
Viscous-force terms	$\mu V/L^2$
Gravity-force term	$\rho_0 g$
Steady-flow inertia-force term	$\rho_0 V^2/L$
Additional inertia-force term in nonsteady flow	$\rho_0 V/T$

†Each term has dimensions of force per unit volume.

The Navier-Stokes equation may be made dimensionless by dividing through by any one of the terms in Table 7-1. The usual procedure is to divide by the steady-flow inertia-force coefficient $\rho_0 V^2/L$. The result is shown in Eqs. (5) and (6) along with the dimensionless continuity equation, which is obtained similarly.

Dimensionless Navier-Stokes and Continuity Equations

$$-\left(\frac{p_0}{\rho_0 V^2}\right)\nabla' p' + \left(\frac{\mu}{\rho_0 V L}\right)\left[\nabla'^2 \mathbf{v}' + \frac{1}{3}\nabla'(\nabla'\cdot\mathbf{v}')\right] - \left(\frac{gL}{V^2}\right)\rho'\nabla' h' = \left(\frac{L}{TV}\right)\frac{\rho'\partial\mathbf{v}'}{\partial t'} + \rho'\mathbf{v}\cdot\nabla\mathbf{v}' \tag{7.4.5}$$

$$\left(\frac{L}{TV}\right)\frac{\partial\rho'}{\partial t'} + \nabla'\cdot(\rho'\mathbf{v}') = 0. \tag{7.4.6}$$

In order that the model flow be governed by equations identical to the prototype-flow equations it is necessary that the dimensionless coefficients in parentheses have the same values in the two systems. These coefficients (or their reciprocals) are called ***characteristic numbers*** of the flow; they are ratios of the dimensional coefficients shown in Table 7-1. It is not necessary that the viscosity μ, the reference values p_0, ρ_0, and the characteristic speed have the same magnitudes in the two systems (although the reference quantities should be defined at corresponding points and times); what is required is that the four dimensionless groups formed from these quantities have the same values in the two systems. A different fluid could be used in the model system, and the ambient pressure could be varied to meet these requirements, in addition to adjusting the length and time scales. But it is not in general feasible to satisfy all four of these requirements. Fortunately, in many engineering problems one or more of them may be disregarded with negligible effect on the flow similarity. In steady-flow problems, L/TV drops out, and only three requirements remain, but the time scale drops out too and we must in general still neglect one of the other requirements. The characteristic numbers appearing in the dimensionless Navier-Stokes equation and the continuity equation are tabulated in Table 7-2 with their customary names and interpretation as the ratio of two kinds of force coefficients from Table 7-1. (Note that Reynolds number and the Froude number are actually the reciprocals of the coefficients.)

TABLE 7-2

CHARACTERISTIC NUMBERS IN NAVIER-STOKES AND CONTINUITY EQUATIONS

$E = p_0/\rho_0 V^2$	Euler number	pressure/steady-flow inertia
$Re = \rho_0 VL/\mu$	Reynolds number	steady-flow inertia/viscous
$F = V^2/gL$	Froude number	steady-flow inertia/gravity
$S = L/TV$	Strouhal number	transient/steady-flow inertia

As an illustration of the difficulty in matching all these numbers suppose that we wish to use the same incompressible liquid in both flows. Then ρ_0, g, and μ must be the same in both flows, and, for a scale model with length $1/n$ times that of the prototype, equality of the Reynolds numbers requires a model characteristic speed $V_m = nV_p$, where V_p is the prototype characteristic speed, while equality of the Froude numbers requires $V_m = V_p/\sqrt{n}$. Evidently, both requirements cannot be satisfied if the same incompressible fluid is used in both flows. With the limited range of density and viscosity combinations available in liquids it is not usually possible to match both numbers even by choosing a different liquid.

The gravity term may be eliminated completely for confined flows without a free surface, or for flows in which the free surface is approximately all at the same level, by writing $p = p_s + p_d$, the sum of the static and dynamic pressures, since then $-\nabla p_s = \rho g \nabla h$ exactly cancels the gravity term. We may then omit the gravity term and interpret p as the dynamic pressure in the Navier-Stokes equation. In this case, for steady flow the two important characteristic numbers are the Reynolds number and the Euler number, while the Froude number is unimportant. The Froude number is important, however, in gravity waves (surface waves on liquids) and in ship model analysis, where the drag is significantly affected by the bow wave.

Surface-tension or capillarity effects occurring in small-scale flows have been neglected throughout our analysis. When they are significant, other characteristic numbers not considered here will appear.

The ***dimensionless boundary conditions*** must also be the same for the two flows. This requirement is met by the geometric similarity requirement, if the boundaries are rigid bodies, either stationary or having a motion prescribed in advance. But when the boundary motion is affected by the fluid forces developed, additional characteristic numbers, containing parameters such as the mass of the rigid body, will appear. When transitions from laminar to turbulent flow are studied, and for friction losses in pipes, characteristic numbers involving the average height of the roughness projections on the boundary walls are often used.

The energy equation for a perfect gas obeying Fourier's law of heat conduction, with no internal heat sources, may be modified as follows. For a perfect gas, the internal energy u is a function only of the temperature θ, and $du = c_v d\theta$, as we saw in Sec. 5.6, where the right-hand side must be converted to mechanical units by multiplication by J, the mechanical equivalent of heat. It can be shown that the gas constant R of Eq. (5.6.3a), $pv = RT$, is given by $R = c_p - c_v$. Hence, including the constant J to make explicit the conversion to mechanical units, we write

$$u = Jc_v\theta \tag{7.4.7}$$

$$p = J(c_p - c_v)\rho\theta \tag{7.4.8}$$

where c_v and c_p are the specific heats at constant volume and at constant pressure, respectively, and $\rho = 1/v$. The heat-conduction law, Eq. (7.1.6), is

$$\mathbf{q} = -k\nabla\theta. \tag{7.4.9}$$

The energy equation, Eq. (7.1.14c), then takes the form, in mechanical energy units,

$$\rho Jc_v\frac{d\theta}{dt} = Jk\nabla^2\theta - p\nabla\cdot\mathbf{v} + 2W_D \tag{7.4.10}$$

if there is no internal heat supply and if, for simplicity, c_v and k are assumed to be constants, which limits the treatment to moderate temperature changes. With the Stokes condition $\lambda = -2\mu/3$, the dissipation $2W_D$ is given by Eq. (7.1.13) as

$$2W_D = 4\mu\mathrm{II}_{\tilde{D}} = 2\mu\tilde{D}_{ij}\tilde{D}_{ij},$$

where $\tilde{D}_{ij}$ is used to denote the deviatoric part of the rate of deformation instead of D'_{ij} to avoid confusion with the dimensionless variables, which are denoted by primes. Then

$$2W_D = 2\mu\left(\frac{V^2}{L^2}\right)\tilde{D}'_{ij}\tilde{D}'_{ij},$$

where the $\tilde{D}'_{ij}$ are the dimensionless deviatoric parts of the rate of deformation, and the energy equation becomes

$$\rho_0\rho' Jc_v\left[\left(\frac{\theta_0}{T}\right)\frac{\partial\theta'}{\partial t'} + \left(\frac{V\theta_0}{L}\right)\mathbf{v}'\cdot\nabla'\theta'\right] = \left(\frac{k\theta_0}{L^2}\right)\nabla'^2\theta'$$
$$- \left(\frac{p_0 p' V}{L}\right)\nabla'\cdot\mathbf{v}' + 2\mu\left(\frac{V^2}{L^2}\right)\tilde{D}'_{ij}\tilde{D}'_{ij}, \tag{7.4.11}$$

from which the dimensionless energy equation is obtained by dividing by $J\rho_0\theta_0 c_v V/L$:

$$\left(\frac{L}{TV}\right)\rho'\frac{\partial\theta'}{\partial t'} + \rho'\mathbf{v}'\cdot\nabla'\theta' = \left(\frac{k}{\rho_0 c_v VL}\right)\nabla'^2\theta' - \left(\frac{p_0}{J\rho_0\theta_0 c_v}\right)p'\nabla'\cdot\mathbf{v}'$$
$$+ \left(\frac{2\mu V}{\rho_0 Jc_v\theta_0 L}\right)\tilde{D}'_{ij}\tilde{D}'_{ij}. \tag{7.4.12}$$

The dimensionless coefficient $p_0/J\rho_0\theta_0 c_v$ can be transformed by substituting $p_0 = J\rho_0\theta_0(c_p - c_v)$ according to Eq. (8). Thus

$$\frac{p_0}{J\rho_0\theta_0 c_v} = \frac{c_p}{c_v} - 1 = \gamma - 1, \tag{7.4.13}$$

where

$$\gamma = \frac{c_p}{c_v}. \tag{7.4.14}$$

A necessary condition for similarity is thus that the ratio γ of the two specific heats be the same in both flows. The other new dimensionless coefficients may be transformed as follows:

$$\frac{k}{\rho_0 c_v VL} = \left(\frac{c_p}{c_v}\right)\left(\frac{k}{\rho_0 c_p VL}\right) = \frac{\gamma}{Pe}, \tag{7.4.15}$$

where Pe is the ***Péclet number***

$$Pe = \frac{\rho_0 c_p VL}{k}, \tag{7.4.16}$$

which may be written as the product of the Reynolds number and the ***Prandtl number***

$$Pr = \frac{c_p\mu}{k}, \tag{7.4.17}$$

which is a measure of the relative importance of viscosity and heat conduction. In most gases Pr is near unity, which indicates that whenever one effect is important the other is too, according to Pai (1959). Also

$$\frac{2\mu V}{\rho_0 J c_v \theta_0 L} = 2\left(\frac{\mu}{\rho_0 V L}\right)\left(\frac{V^2}{J c_v \theta_0}\right).$$

The factor $\mu/\rho_0 VL$ is the reciprocal of the Reynolds number. The factor $V^2/Jc_v\theta_0$ can be transformed by noting that for the perfect gas $p/\rho^\gamma = \text{const.}$ in isentropic flow, so that the sonic speed c is given by

$$c^2 = \frac{dp}{d\rho} = \frac{\gamma p}{\rho}. \tag{7.4.18}$$

Hence, using Eqs. (8) and (18), we write

$$\frac{V^2}{J c_v \theta_0} = \frac{V^2}{c_0^2}\frac{\gamma p_0}{\rho_0}\frac{1}{J c_v \theta_0} = \gamma \frac{V^2}{c_0^2}(\gamma - 1). \tag{7.4.19}$$

If the reference velocity V is chosen at the same reference point as p_0 and ρ_0, then

$$M_0 = \frac{V}{c_0} \tag{7.4.20}$$

is the reference Mach number, and the similarity of the two energy equations is assured by the equality of γ and M_0 for the two flows, in addition to the characteristic numbers of Table 7-2. The Euler number $E = p_0/\rho_0 V^2$ can also be transformed as follows for the perfect-gas case:

$$E = \frac{p_0}{\rho_0 V^2} = \frac{c_0^2}{V^2} = \frac{1}{M_0^2}. \tag{7.4.21}$$

Hence equality of the Euler numbers for the two perfect-gas flows is assured by equality of the ratios γ of specific heats and of the reference Mach numbers.

It follows that, in the steady flow of a perfect gas with negligible gravity effects, similarity of the energy equation and the momentum equation (Navier-Stokes) is assured whenever the specific heat ratio γ, the Reynolds number Re, the Mach number M, and the Prandtl number Pr have the same values in the model flow as in the prototype flow. Fortunately the Prandtl number is nearly the same for all gases and is not strongly temperature-dependent, according to Shapiro (1953). Kinetic theory leads, according to Patterson (1956), to the semi-empirical formula

$$Pr = \frac{4\gamma}{9\gamma - 5} \tag{7.4.22}$$

due to Eucken (1913). Thus the Prandtl number depends only on the specific heat ratio γ, and similarity of the energy equation and the momentum equation in the steady flow of a perfect gas with negligible gravity effects is assured by the equality of the three characteristic numbers: γ, Re, and M.

Thermal boundary conditions must also be considered for similarity. An insulated boundary or a constant-temperature boundary offers no difficulty in principle. But in fact a thermal boundary layer may form, so that the boundary temperature seen by the free stream outside the boundary layer is not the same as that at the wall. For this case, and for other heat-transfer conditions, consult the literature on heat transfer, where other characteristic numbers also appear.

The method of establishing the characteristic numbers by working from the governing differential equations, which has been illustrated by the foregoing examples, is the most useful method when the governing equations are known. Besides establishing conditions for dynamic similarity when model and prototype are geometrically similar, this approach can be used to extract important information when a solution of the governing boundary-value problem is difficult to obtain. If some engineering judgment can be used to select the characteristic values in such a way that the normalized dimensionless quantities appearing in the equations are all of order unity, as suggested in the discussion of Table 7-1, then the magnitudes of the characteristic numbers may indicate terms in the equations that can be omitted in an approximate analysis. For further details and many practical suggestions on approximation methods, see, for example, Kline (1965).

Elementary fluid mechanics textbooks often present two other methods of establishing the characteristic numbers: (1) dimensional analysis using the pi theorem, and (2) the method of similitude using force ratios. These methods give less information than the method based on the governing equations, but they can sometimes be used when the governing equations are not known.

Dimensional analysis, the pi theorem. The first step, and the most difficult one in dimensional analysis of a problem, is to write down a list of the parameters that are important in the problem. In the absence of knowledge of the governing equation this may sometimes be accomplished by intuition based on experience. We may have, for example, one dependent parameter Q_1 depending on $m - 1$ independent parameters Q_j

$$Q_1 = f_1(Q_2, \ldots, Q_m) \tag{7.4.23a}$$

where the actual form of the function f_1 may not be known. This may alternatively be written in implicit form as

$$f_2(Q_1, \ldots, Q_m) = 0 \tag{7.4.23b}$$

where f_2 is another unspecified function. The Buckingham pi theorem then says that there is an equivalent relation in terms of n nondimensional parameters $\pi_1, \ldots, \pi_n$, the characteristic numbers of the problem, also called pi's, formed from the Q's

$$f_3(\pi_1, \ldots, \pi_n) = 0 \quad \text{or alternatively} \quad \pi_1 = f_4(\pi_2, \ldots, \pi_n). \tag{7.4.24}$$

Here

$$n = m - k \tag{7.4.25}$$

where k is the largest number of parameters contained in the original list that will ***not*** combine to form a nondimensional product of the form $Q_1^{a_1} Q_2^{a_2} \ldots, Q_k^{a_k}$ with nonzero exponents a_j. If r denotes the minimum number of independent dimensions required to construct the dimensions of all the original parameters, then

$$k \leq r. \tag{7.4.26}$$

Most commonly it turns out that $k = r$, but there are exceptions. In purely mechanical problems we have $r = 3$, and the number $k \leq 3$ can usually be determined by trial as in the example below, which also illustrates the formation of the pi's. We omit the proof of the theorem, which is attributed to Buckingham (1914). See, for example, Bridgman (1931) or Sedov (1959). A recent careful statement and rigorous proof have been given by Brand (1957).

Example. Consider the problem of determining the drag force F on a stationary sphere of diameter d in a steady flow of incompressible viscous fluid. We assume

$$F = f_1(\rho, \mu, d, V) \quad \text{or} \quad f_2(F, \rho, \mu, d, V) = 0 \tag{7.4.27}$$

where ρ and μ are the fluid density and viscosity and V is the uniform free-stream velocity far from the sphere. Since only mechanical quantities enter, we have $r = 3$. Furthermore there is no nondimensional product of the form $\rho^a V^b d^c$. Hence $k = 3$, and $n = 5 - 3 = 2$. We seek to determine two pi's as follows

$$\pi_1 = \rho^{a_1} V^{b_1} d^{c_1} F \qquad \pi_2 = \rho^{a_2} V^{b_2} d^{c_2} \mu. \tag{7.4.28}$$

If π_1 and π_2 have dimensions $[M^0 L^0 T^0]$ in mass M, length L, and time T, a little elementary algebra gives

$$a_1 = -1, \qquad b_1 = -2, \qquad c_1 = -2$$

and

$$a_2 = b_2 = c_2 = -1.$$

Hence the relationship may be represented as $f_3(\pi_1, \pi_2) = 0$ or $\pi_1 = f_4(\pi_2)$ as follows

$$f_3\left(\frac{F}{\rho V^2 d^2}, \frac{\mu}{\rho V d}\right) = 0 \quad \text{or} \quad \frac{F}{\rho V^2 d^2} = f_4\left(\frac{\mu}{\rho V d}\right). \tag{7.4.29}$$

The advantage of the reduction in the number of independent parameters from four to one can hardly be overstated, when it comes to correlating data in order to determine the functional dependence empirically. In this example $\pi_2 = \mu/\rho V d$ is recognized as the reciprocal of the Reynolds number, while $\pi_1 = F/\rho V^2 d^2$ is a "force coefficient." We could, of course, equally well write $F/\rho V^2 d^2 = f_5(\rho V d/\mu)$. We have happened upon the two most commonly used characteristic numbers for this problem. But the choice is not unique. We could equally well have begun by seeking parameters π_1^* and π_2^* as follows

$$\pi_1^* = \mu^{a_1} d^{b_1} V^{c_1} F \qquad \pi_2^* = \mu^{a_2} d^{b_2} V^{c_2} \rho,$$

which would have led to $\pi_1^* = F/\mu V d$ and $\pi_2^* = \rho V d/\mu$, or

$$\frac{F}{\mu V d} = f_6\left(\frac{\rho V d}{\mu}\right). \tag{7.4.30}$$

Thus $\pi_2^* = 1/\pi_2$ is the Reynolds number, but π_1^* is an apparently quite different kind of force coefficient. Note that in fact $\pi_1^* = \pi_1/\pi_2$.

The method of similitude using force ratios. This method will not be presented at length here. See, for example, Vennard (1954) for further details. The method requires a formulation of the representative forces on a fluid element, for example pressure force F_P, gravity force F_G, viscous force F_V, and inertia force F_I. If all forces are included, dynamic similarity requires the force polygons for corresponding elements in model and prototype to be geometrically similar. For example, the ratio F_I/F_V must have the same value at corresponding points in model and prototype flows. If we multiply the first four dimensional coefficients of Table 7-1 by L^3 to give dimensions of force we obtain the expressions usually listed for F_P, F_V, F_G, and F_I in steady flow. The requirement that the force ratios have the same values for model and prototype will be satisfied if the Euler, Reynolds, and Froude numbers each have the same value in both systems. As we have seen, this complete dynamic similarity is difficult to achieve. We may have to settle for equality only of the one characteristic number deemed most important. For enclosed steady flow of an incompressible fluid, or even a compressible fluid at low Mach numbers, the most important characteristic number is the Reynolds number. When free-surface effects are important, however, as in surface-wave formation studies or in studies of the bow-wave drag on a surface vessel, the most important characteristic number is the Froude number.

An additional characteristic number called the ***Weber number,*** involving surface tension, may also become important in free-surface systems. This does not enter the governing equations, but may affect the boundary conditions at a free surface or at an interface between immiscible fluids. We will obtain the Weber number by a direct use of the force ratio F_I/F_T, where F_T is the surface tension force. The surface tension coefficient σ is the force per unit length acting across a line or curve in the surface. Then

$$F_T = \sigma L \quad \text{and} \quad F_I = \rho_0 V^2 L^2, \tag{7.4.31}$$

(F_I is obtained by multiplying the steady-flow inertial coefficient of Table 7-1 by L^3.) Hence

$$\text{Weber number} \quad W = \frac{F_I}{F_T} = \frac{\rho_0 V^2 L}{\sigma}. \tag{7.4.32}$$

Equality of Weber numbers in prototype and model flows would be required if we wished to obtain similarity of surface-tension effects. For example, this might be the most important effect in the formation of droplets from a nozzle, in the generation of very small surface waves, or in capillarity.

In this section we have emphasized the use of the governing equation, which is the most powerful approach to similarity when complete governing equations are known. It is sometimes advantageous to apply comparable methods to ***distorted models*** instead of geometrically similar models, for example by using different characteristic lengths for the different coordinates. Kline (1965) illustrates this procedure in a study of the two-dimensional boundary layer by introducing dimensionless coordinates $x' = x/L$ and $y' = y/\delta$, where L is the length of the body and δ is a length to be chosen appropriately. In Sec. 7.5 we will make our initial formulation of the boundary-layer equations by using $x' = x/L$ and $y' = y/L$, but we will subsequently make the change of variable $X = x'$, $Y = y'\sqrt{Re}$ which, for large values of the Reynolds number Re, greatly distorts the boundary-layer region by stretching it in the vertical direction. Besides distorting the region, this change of variables will illustrate another procedure, which Kline calls ***absorption of parameters***, and which will lead to a boundary-layer equation not containing Re explicitly.

Internal and external similitude. The type of similarity we have discussed in this section is sometimes called ***external similitude***; see, for example, Kline (1965). External similitude relates performance in one system to performance in a similar system (model and prototype). Sometimes a relationship between values of the dependent parameter at different points within a single system can be obtained by suitable changes of variable, leading to what Kline calls ***internal similitude.*** This will be illustrated in Sec. 7.5 by the similar profiles in Fig. 7.8 in connection with the boundary layer in the flow over a semi-infinite flat plate.

We turn now to a discussion of the two limiting cases where analysis is possible, as mentioned in the first paragraph of this section, the case of very low viscosity and the case of very high viscosity, which we are now able to characterize more precisely by the use of the Reynolds number $Re = \rho_0 VL/\mu$ formed with appropriate characteristic numbers. As compared with inertia effects, high viscosity is characterized by small Re and vice versa.

EXERCISES

1. Show that Eqs. (7.4.5) and (6) are obtained from Eq. (3) and the continuity equation by introducing the dimensionless variables as stated.
2. Carry out the details of deriving Eq. (7.4.12), the dimensionless energy equation, under the assumptions stated in the text.
3. Make a dimensional analysis of steady laminar incompressible flow through a circular pipe under the assumption that the pressure drop Δp depends on pipe length L and diameter d and on fluid viscosity μ, velocity V, and density ρ. Show that the results can be put in the form
$$\Delta p/\tfrac{1}{2}\rho V^2 = f(\rho Vd/\mu,\ L/d).$$
4. Intuition suggests that the pressure drop per unit length should be independent of the actual length in Ex. 3. Hence if the length is measured in units of diameters, show that the pressure drop per unit length is a function only of the Reynolds number, and that this implies that the pressure drop Δp in length L (L/d in units of diameters) is given by
$$\frac{\Delta p/(L/d)}{\tfrac{1}{2}\rho V^2} = f_2\left(\frac{\rho Vd}{\mu}\right).$$
5. Show that the known solution for Poiseuille flow, Eq. (7.1.36) can be put into the form of the result of Ex. 4. *Hint*: Define V as the averge of v_z over the cross-sectional area A, and integrate Eq. (7.1.36) times dA over the area A.
6. Assume that the propagation speed c of surface waves on a shallow liquid depends only on the depth h of liquid, the density ρ, and the gravitational constant g. Show by dimensional analysis that $c/\sqrt{gh} = \text{constant}$.
7. Assume that the thrust force T of a propeller is given by: $T = f(d, \rho, \mu, N, V)$, where the function f is unknown, d is propeller diameter, N is the number of revolutions per unit time, V is the advance velocity, and ρ and μ are fluid density and viscosity. Use dimensional analysis to determine an equivalent form $\pi_1 = f_1(\pi_2, \ldots, \pi_n)$ where π_1 involves T.
8. Air flows through a pipe three inches in diameter at an average velocity of 100 ft/sec. For dynamically similar flows at the same temperature and pressure what should be the average velocity of air flow through a 10-in. diameter pipe?

9. A scale model 10 feet long is used in a towing tank to produce surface waves dynamically similar to those produced by a ship 250 feet long. If the ship's speed is 10 knots, what should be the model's speed?

10. In studying the action of bubbles rising in a liquid it is judged that the important parameters are the bubble's diameter d, density ρ_1, and velocity V relative to the liquid; the liquid's density ρ and viscosity μ; and the surface tension coefficient σ. Determine a set of convenient dimensionless parameters.

11. The height h of the capillary rise in a small tube is assumed to depend on the tube diameter d, the specific weight γ of the liquid, and on the surface tension coefficient σ. By dimensional analysis find a representation of the form $\pi_1 = f(\pi_2)$.

12. Consider the problem of heat conduction into a solid block: $-A \leq x \leq A$, $-B \leq y \leq B$, $-C \leq z \leq C$. The block is initially at temperature θ_i. Its bounday surfaces are raised at $t = 0$ to θ_a, and the surfaces are subsequently held at constant temperature. The heat-conduction equation is obtained from Eq. (7.4.10) by setting $\mathbf{v} \equiv 0$. The result is $(1/\alpha)\partial\theta/\partial t = \nabla^2\theta$, where $\alpha = k/\rho c_v$ is the thermal diffusivity. Normalize the variables to obtain a distorted model in which all dimensionless coordinates run from -1 to $+1$, take the zero level of temperature to be θ_a, and normalize the variable temperature by taking θ_i (measured from θ_a as zero) as characteristic temperature. Arbitrarily define the characteristic time T as the (unknown) time required for the temperature at the origin to change by $1/e$ times the amount it would change in infinite time. Write the equation and boundary conditions in nondimensional form by dividing through by the coefficient of $\partial\theta'/\partial x'$ to obtain a set of of nondimensional characteristic numbers for the problem.

13. In the approach to similitude by use of force ratios, an elastic force $F_E = E_v L^2$ is often introduced, where E_v is the bulk modulus

$$E_v = -\frac{dp}{dv/v} = +\frac{dp}{d\rho/\rho}.$$

(Complete prescription requires a statement of the process, e.g., isentropic or isothermal.) The Cauchy number is then defined by the force ratio F_I/F_E.
(a) Express the Cauchy number in terms of E_v and other characteristic values.
(b) Show that in a barotropic flow the Cauchy number is equal to the square of the Mach number.

14. Carry out the details leading to Eq. (7.4.19).

15. For an inviscid compressible fluid, assume that the dynamic force F depends only on ρ, V, L, the bulk modulus E_v (see Ex. 13), and the specific heats c_p and c_v. Use dimensional analysis to show that $F/\rho V^2 L^2 = f(\rho V^2/E_v, \gamma) = f(M, \gamma)$, where M is the Mach number and $\gamma = c_p/c_v$.

16. For a viscous compressible fluid, add μ and the thermal conductivity coefficient k (of the Fourier heat-conduction law) to the list of independent variables given in Ex. 15, and conclude by dimensional analysis that $F/\rho V^2 L^2$ is a function of Mach number, Reynolds number, Prandtl number, and the specific heat ratio.

7.5 Limiting cases: creeping-flow equation and boundary-layer equation for plane flow of incompressible viscous fluid

For two extreme limiting cases, the equations of laminar viscous flow simplify considerably. These two limiting cases are the creeping flow of fluids of very high viscosity (very small Reynolds number $Re = \rho VL/\mu$), and the opposite limit of very low viscosity (large Reynolds number). In the creeping-flow case, a linear equation system results, as $Re \to 0$. In the low-viscosity case, the effects of viscosity are pronounced only in a small region called the ***boundary layer***, where the velocity gradients are so large that viscous forces are important even for small values of μ, while the main flow can be considered frictionless. The boundary-layer equations are still nonlinear, and solution is still difficult, but not quite so bad as with the full Navier-Stokes equations; in some cases the integration can be accomplished. In this section we formulate each of the limiting cases for two-dimensional flow of incompressible fluid. We also present one example of solution of the boundary-layer equations for flow over a thin flat plate parallel to the uniform free-stream flow. The creeping-flow equation will be derived from the vorticity-transport equation, which is derived below.

Vorticity-transport equation. For plane flow parallel to the xy-plane, the Navier-Stokes Eqs. (7.1.17) for incompressible fluid under gravitational body force $\mathbf{b} = -\nabla h$ become

$$\left.\begin{aligned} -\frac{1}{\rho}\frac{\partial p}{\partial x} - g\frac{\partial h}{\partial x} + \nu\left(\frac{\partial^2 v_x}{\partial x^2} + \frac{\partial^2 v_x}{\partial y^2}\right) &= \frac{\partial v_x}{\partial t} + v_x\frac{\partial v_x}{\partial x} + v_y\frac{\partial v_x}{\partial y} \\ -\frac{1}{\rho}\frac{\partial p}{\partial y} - g\frac{\partial h}{\partial y} + \nu\left(\frac{\partial^2 v_y}{\partial x^2} + \frac{\partial^2 v_y}{\partial y^2}\right) &= \frac{\partial v_y}{\partial t} + v_x\frac{\partial v_y}{\partial x} + v_y\frac{\partial v_y}{\partial y}. \end{aligned}\right\} \tag{7.5.1}$$

The pressure and body-force terms can be eliminated by cross-differentiation and subtraction. Noting that the angular velocity vector $\mathbf{w}$ is

$$\mathbf{w} = \omega\mathbf{k}, \qquad \text{where} \quad \omega = \frac{1}{2}\left(\frac{\partial v_y}{\partial x} - \frac{\partial v_x}{\partial y}\right) \tag{7.5.2}$$

we obtain

$$2\nu\nabla^2\omega = 2\left[\frac{\partial \omega}{\partial t} + v_x\frac{\partial \omega}{\partial x} + v_y\frac{\partial \omega}{\partial y}\right] + \left(\frac{\partial v_x}{\partial x} + \frac{\partial v_y}{\partial y}\right)\left(\frac{\partial v_y}{\partial x} - \frac{\partial v_x}{\partial y}\right).$$

The last term vanishes by virtue of the continuity equation

$$\frac{\partial v_x}{\partial x} + \frac{\partial v_y}{\partial y} = 0 \tag{7.5.3}$$

for incompressible fluid, and we obtain the ***vorticity-transport equation***

$$\nu\nabla^2\omega = \frac{\partial\omega}{\partial t} + v_x\frac{\partial\omega}{\partial x} + v_y\frac{\partial\omega}{\partial y} \quad \text{or} \quad \nu\nabla^2\omega = \frac{d\omega}{dt}. \tag{7.5.4}$$

See, for example, Schlichting (1968), Eq. (4.6). If we introduce the stream function ψ, as in Sec. 7.3, such that

$$v_x = \frac{\partial\psi}{\partial y} \qquad v_y = -\frac{\partial\psi}{\partial x}, \tag{7.5.5a}$$

the continuity equation is satisfied identically, while

$$\omega = -\tfrac{1}{2}\nabla^2\psi, \tag{7.5.5b}$$

so that the vorticity-transport equation (4) furnishes the following nonlinear equation for the single unknown ψ:

$$\nabla^4\psi = \frac{1}{\nu}\left[\frac{\partial(\nabla^2\psi)}{\partial t} + \frac{\partial\psi}{\partial y}\frac{\partial(\nabla^2\psi)}{\partial x} - \frac{\partial\psi}{\partial x}\frac{\partial(\nabla^2\psi)}{\partial y}\right]. \tag{7.5.6}$$

Creeping-flow equation. When the kinematic viscosity ν is very large, the terms on the right in Eq. (6) are neglected and the remaining equation is

$$\nabla^4\psi = 0 \quad \text{or} \quad \frac{\partial^4\psi}{\partial x^4} + 2\frac{\partial^4\psi}{\partial x^2\partial y^2} + \frac{\partial^4\psi}{\partial y^4} = 0 \tag{7.5.7}$$

for creeping flow of a highly viscous fluid. This is a linear equation, known as the ***biharmonic equation,*** which can be solved readily in many cases. It is important in lubrication theory. We note that the terms retained are those of highest order; hence it is possible to satisfy the viscous-flow boundary conditions. It is left as an exercise for the student to make Eq. (6) nondimensional as in Sec. 7.4, and verify that the nondimensional limiting condition required is that the Reynolds number $Re \to 0$ provided that the Strouhal number $S = L/TU$ remains finite. Pai (1956), Chap. 7, gives examples of solutions of a comparable linearization of Eqs. (1).

We turn now to the other limiting case of very small viscosity and introduce the boundary-layer concept.

Boundary-layer concept.† Despite the fact that the perfect-fluid equations do not in general possess solutions satisfying the viscous-flow boundary condition of no relative tangential velocity at a solid boundary, experience indicates that with fluids of low viscosity, for example air or water, the flow patterns

†The boundary-layer concept and the resulting simplification of the Navier-Stokes equations, leading to Eqs. (16) to (18) are due to Prandtl (1904).

predicted by perfect-fluid solutions in many cases agree closely with experimentally observed flow patterns. For example, in the two-dimensional flow over a well-streamlined airfoil at speeds well below the sonic speed, the streamline pattern observed agrees closely with that calculated from potential theory when the multiple-valued potential solution is made to have just enough circulation to cause the rear stagnation point to be at the trailing edge. But this potential solution predicts that the airfoil boundary is a streamline on which the velocity attains its maximum value, contrary to the known boundary condition that the velocity is zero on the boundary. It also predicts no shear traction between fluid and boundary and therefore no skin-friction drag, contrary to experience. Closer observation reveals that there is actually a thin boundary layer of fluid near the airfoil and a wake extending downstream from the airfoil in which the potential flow is not valid, as illustrated schematically in Fig. 7.6 with the boundary-layer thickness shown greatly exaggerated. The actual thickness for a streamlined slender body is of the order of one per cent of the length, increasing somewhat from the leading edge toward the trailing edge. Experiment indicates that the fluid outside the boundary layer and wake behaves like a perfect fluid with very little vorticity and may be considered irrotational. It may therefore be treated by potential-flow theory. Only when the fluid enters the boundary layer does it acquire significant vorticity because of the retarding effect of the boundary. In the wake, the vorticity gradually decays because of viscosity and diffusion. Note that the outer edge of the boundary layer is not a streamline, since fluid crosses it.

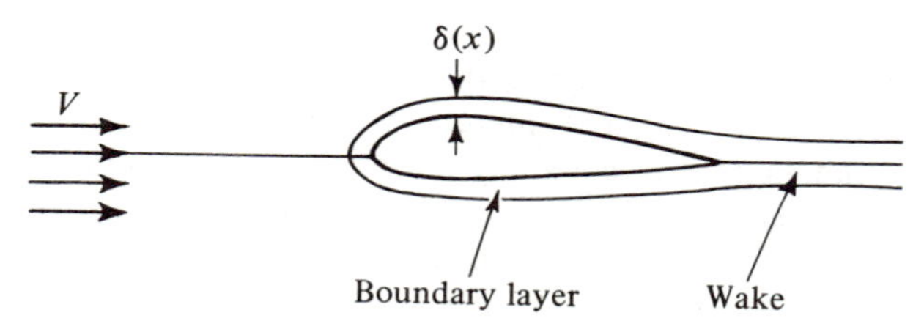

Fig. 7.6 Boundary layer and wake on an airfoil (schematic).

When the boundary layer is so thin, the potential-flow solution outside the boundary layer can be determined with negligible error as though its boundary streamline were actually the airfoil. Then the solution inside the boundary layer is determined by using the boundary-layer equations, with boundary conditions requiring $\mathbf{v} = 0$ on a stationary boundary and requiring that $\mathbf{v}$ approach the external-flow velocity at the outer edge of the boundary layer. In practice this means, for example, if the flow in the xy-plane over a horizontal surface $y = 0$ has the potential-flow component $v_x(x, y, t)$, we let $U(x, t) = v_x(x, 0, t)$ and wish to impose the condition that the boundary-layer x-component of velocity approach $U(x, t)$ as $y \to \delta(x)$, where $\delta(x)$ is the boundary-layer thickness. It should, of course, approach $v_x(x, \delta(x), t)$ as $y \to \delta(x)$, but the thickness $\delta(x)$ is not known *a priori*. Experimentally, it is observed that the potential-flow velocity is never exactly reached. The boundary condition actually imposed on the boundary-layer solution is then that the boundary layer $v_x \to U(x, t)$ as $y \to \infty$. ***Note that it is not the free-stream velocity at infinity which is approached as*** $y \to \infty$, ***however, but the potential-flow velocity*** $U(x, t)$ ***at the wall*** $y = 0$. Later in this section we will apply this procedure to the flow over a flat plate at zero angle of incidence; for this special case the potential flow is uniform and $U(x, t) \equiv V$, where V is the free-stream velocity at infinity, but that is an exceptional case.

In experimental measurements of boundary-layer thickness, $\delta(x)$ is often defined as the distance at which the measured speed differs by one per cent from the calculated potential-flow speed at the wall.

Boundary-layer equations for plane flow of incompressible fluid. We begin with the nondimensional Navier-Stokes and continuity equations, Eqs. (7.4.5) and (7.4.6) modified as follows. We omit the body-force terms; this means that the pressure will be interpreted as the dynamic pressure only, which we assume is not affected by the body force. We also put $\nabla \cdot \mathbf{v}' = 0$ and $\rho \equiv \rho_0$ for incompressibility and choose the pressure scale p_0 and the time scale T, so that $p_0 = \rho_0 V^2$ and $T = L/V$. Hence

$$\rho' \equiv 1 \qquad \frac{p_0}{\rho_0 V^2} = 1 \qquad \frac{L}{TV} = 1. \tag{7.5.8}$$

We choose the length scale L so that the dimensionless derivative $\partial v'_x/\partial x'$ does not exceed unity in the region of interest. We choose a local coordinate system with x' measured along the wall and y' normal to it, and assume that the wall curvature is small enough that these can be treated as rectangular Cartesian coordinates. [The analysis is strictly valid only for a flat wall. But analysis by curvilinear coordinates leads to the same final boundary-layer equations when the boundary-layer thickness is small compared to the radius of curvature R and when no rapid variations of curvature occur, so that dR/dx is of order not greater than unity. See, for example, Pai (1956) or Schlichting (1960).] Since $\partial v'_x/\partial x'$ is of order unity, the continuity equation implies that $\partial v'_y/\partial y'$ is also of order unity. Since $v'_y = 0$ at $y' = 0$, it follows that the maximum value of v'_y in the boundary layer is of the order of the dimensionless thickness $\delta(x)/L$ of the boundary layer. We assume that $\partial v'_y/\partial x'$ and $\partial^2 v'_y/\partial x'^2$ are also of order δ/L. The component equations are written below with the estimated order of magnitude of each factor involving a velocity printed under it. We assume that some of the viscous terms are of the same order of magnitude as the inertia terms in the boundary layer, which requires $1/Re$ to be of order $(\delta/L)^2$.

$$\underset{1}{\frac{\partial v'_x}{\partial t'}} + \underset{1}{v'_x}\underset{1}{\frac{\partial v'_x}{\partial x'}} + \underset{\frac{\delta}{L}}{v'_y}\underset{\frac{L}{\delta}}{\frac{\partial v'_x}{\partial y'}} = -\frac{\partial p'}{\partial x'} + \underset{(\frac{\delta}{L})^2}{\frac{1}{Re}}\left(\underset{1}{\frac{\partial^2 v'_x}{\partial x'^2}} + \underset{(\frac{L}{\delta})^2}{\frac{\partial^2 v'_x}{\partial y'^2}}\right) \tag{7.5.9}$$

$$\underset{\frac{\delta}{L}}{\frac{\partial v'_y}{\partial t'}} + \underset{1}{v'_x}\underset{\frac{\delta}{L}}{\frac{\partial v'_y}{\partial x'}} + \underset{\frac{\delta}{L}}{v'_y}\underset{1}{\frac{\partial v'_y}{\partial y'}} = -\frac{\partial p'}{\partial y'} + \underset{(\frac{\delta}{L})^2}{\frac{1}{Re}}\left(\underset{\frac{\delta}{L}}{\frac{\partial^2 v'_y}{\partial x'^2}} + \underset{\frac{L}{\delta}}{\frac{\partial^2 v'_y}{\partial y'^2}}\right) \tag{7.5.10}$$

$$\underset{1}{\frac{\partial v'_x}{\partial x'}} + \underset{1}{\frac{\partial v'_y}{\partial y'}} = 0 \tag{7.5.11}$$

with boundary conditions desired to be

$$\left.\begin{aligned} v'_x = v'_y = 0 \quad &\text{at} \quad y' = 0 \\ v'_x \approx U'(x', t) \quad &\text{at} \quad y' \approx \frac{\delta(x')}{L} \end{aligned}\right\} \tag{7.5.12}$$

As we have remarked previously, the last boundary condition cannot be applied explicitly because $\delta(x')$ is not known.

The order estimations shown in Eqs. (9) to (11) involve some assumptions. Mathematically it is possible for $\partial v'_y/\partial x'$ to be much greater than δ/L even though v'_y is of order δ/L. We are ruling out high-frequency periodic variations with respect to x'. We have also assumed the nonsteady acceleration terms are of the same order of magnitude as the convective terms, ruling out sudden accelerations such as those in large pressure waves. Since v'_x increases from zero at $y' = 0$ to $U' = U/V$ at $y' = \delta/L$, and U' is of order unity (although U' may be somewhat greater than unity), we have $\partial v'_x/\partial y'$ of order L/δ and $\partial^2 v'_x/\partial y'^2$ of order $(L/\delta)^2$, while $\partial v'_y/\partial y'$ and $\partial^2 v'_y/\partial y'^2$ are of order unity and L/δ, respectively. Our assumption that at least some of the viscous terms are of the same order of magnitude as the inertia terms in the boundary layer can then be satisfied by assuming that $1/Re$ is of order $(\delta/L)^2$. If this order estimate on Re is written as an equality it furnishes an estimate of the boundary-layer thickness for a given Reynolds number, namely

$$\frac{\delta}{L} \approx \frac{1}{\sqrt{Re}} = \sqrt{\frac{\mu}{\rho V L}}. \tag{7.5.13}$$

Equation (10) implies that $|\partial p'/\partial y'|$ is of order δ/L. The pressure change across the boundary-layer thickness is then of order $(\delta/L)^2$, which is negligible. We will assume that p varies only with x in the boundary layer. The variation of p with x can be determined from the potential-flow conditions at the outer edge of the boundary layer, where the viscous terms in Eq. (9) are negligible and where $\partial v'_x/\partial y'$ is also small, and $v_x \approx U(x, t)$. Equation (9) in dimensional form then gives at the edge of the boundary layer

$$\frac{\partial U}{\partial t} + U\frac{\partial U}{\partial x} = -\frac{1}{\rho}\frac{\partial p}{\partial x} \tag{7.5.14}$$

from which $\partial p/\partial x$ and hence p can be determined from the known potential-flow solution. For steady flow $\partial U/\partial t = 0$, and this takes a particularly simple form.

$$\text{Steady flow:} \quad p + \tfrac{1}{2}\rho U(x, t)^2 = \text{const.} \tag{7.5.15}$$

Since we are interested in boundary layers only when Re is very large so that δ/L is very small, we omit terms of order δ/L or smaller in Eqs. (9) to (11). For the steady-flow case we then have $\partial p'/\partial y' = 0$ and the following

Steady-Flow Nondimensional Laminar
Boundary-Layer Equations
(incompressible fluid)

$$v'_x\frac{\partial v'_x}{\partial x'} + v'_y\frac{\partial v'_x}{\partial y'} = -\frac{dp'}{dx'} + \frac{1}{Re}\frac{\partial^2 v'_x}{\partial y'^2} \tag{7.5.16}$$

$$\frac{\partial v'_x}{\partial x'} + \frac{\partial v'_y}{\partial y'} = 0 \tag{7.5.17}$$

with boundary conditions

$$v'_x = v'_y = 0 \quad \text{at} \quad y' = 0, \qquad v'_x \to U'(x') \quad \text{as} \quad y' \to \infty. \tag{7.5.18}$$

The last boundary condition replaces the condition $v'_x = U'(x')$ at $y' = \delta(x')/L$ by $v'_x \to U'(x')$ as $y' \to \infty$ because, as we have previously remarked, $\delta(x')$ is unknown, and moreover experiment indicates that the potential-flow value is never exactly attained. Solutions of Eqs. (16) to (18) would not represent a genuine boundary-layer behavior, however, unless v'_x is very close to $U'(x')$ when y' reaches the values of about $1/\sqrt{Re}$.

Equation (16) is still nonlinear, but the problem has been simplified by eliminating one unknown, namely p', which is assumed to be known from the potential solution or possibly from experimental measurements. The number of equations has also been reduced by one. But there remain some other difficulties besides the nonlinearity and the difficulty of applying the boundary condition at $y = \delta$ when $\delta(x)$ is unknown. It turns out that Eq. (16) is parabolic. This means that additional boundary conditions or initial conditions are needed at the inlet section of the boundary layer—for example, a specification of the velocity profile $v'_x(x'_0, y')$ between $y' = 0$ and $y' = \delta/L$ at the inlet section $x' = x'_0$. This profile is unknown, unless a local solution of the full Navier-Stokes equations can be found—for example, by power series valid in the neighborhood of the leading edge.† For approximate solutions, the inlet profile is sometimes assumed to be uniform, $v'_x(x'_0, y') = \text{const.}$ If the inlet velocity and the thickness $\delta(x')/L$ were known, Eqs. (16) to (18) would in principle determine the solution in a boundary layer extending indefinitely in the x'-direction. But in practical problems, the flow often separates from the body well before the rear stagnation point is reached, so that the thin-layer assumption is violated. When this happens, the solution still may be approximately valid almost to the separation point. But the difficulty is that the separation point is unknown. A related problem is the onset of turbulence in the boundary layer; in the turbulent case our equations, based on the Navier-Stokes equations, do not describe the mean velocities. These questions are beyond the scope of this book, but the difficulties mentioned should make it clear why so many of the available analyses are approximate. We conclude this section with one example of solution of the boundary-layer equations.

Boundary layer along a thin flat plate at zero angle of incidence. Figure 7.7 shows schematically the boundary layer along a plate of length L placed parallel to the uniform flow at infinity (zero incidence). The plate is assumed

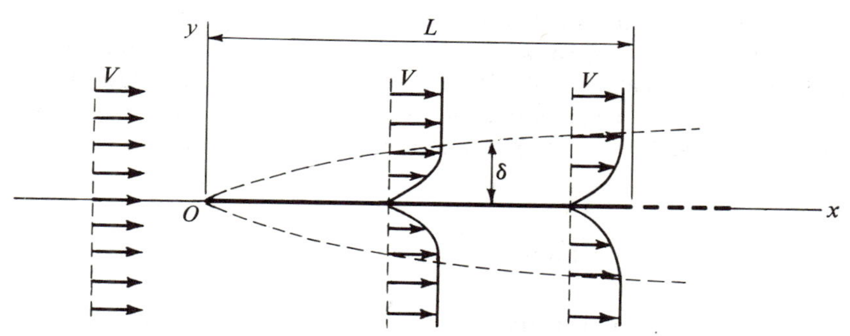

Fig. 7.7 Boundary layer along a flat plate at zero incidence (schematic).

to be infinitely wide in the direction perpendicular to the figure, or at any rate the flow is assumed two-dimensional parallel to the xy-plane at sections sufficiently far from the ends. The plate is very thin relative to L and will be idealized as of zero thickness; the potential-flow solution is then trivial, namely $\mathbf{v} \equiv V\mathbf{i}$ everywhere; in particular, the quantity $U(x, t)$ appearing in the boundary conditions for the boundary-layer flow will be $U(x, t) \equiv V$ in this case, or in the dimensionless form of Eq. (18), $U'(x') \equiv 1$. For this case, Eq. (15)

†According to Schlichting (1968), attempts to use the full Navier-Stokes equations near the leading edge of a flat plate have been made by Carrier and Lin (1948) and by Boley and Friedman (1959).

implies $\partial p'/\partial x' = 0$, so that the pressure drops out of the equations. For large Reynolds numbers, the boundary layer will be very thin. One might hope that the boundary layer is little influenced by what goes on in the very thin wake downstream from the plate. If this is the case, the boundary-layer solution for a semi-infinite plate extending in the downstream direction should closely approximate that for the real plate in $0 < x < L$, although it would not, of course, represent the flow in the wake of the real plate. Experience indicates that the substitute problem with the semi-infinite plate does give a good representation of the boundary layer for a finite plate at zero incidence. The semi-infinite problem permits a simplification, reducing the nonlinear partial differential equation system to a single nonlinear ordinary differential equation by a suitable change of variables so that a similarity solution is obtained. This is an example of what Kline (1956) calls ***internal similitude***. The solution for the semi-infinite plate is the only one presented here; it is widely used for finite plates at zero incidence. The similarity solution has the advantage of not requiring a knowledge of the inlet velocity profile, but it cannot give an exact account of the flow at the leading edge.

Before attempting the solution, we modify the boundary-layer equations to obtain a form not explicitly containing the Reynolds nmber, $Re = \rho VL/\mu$, since when $L \to \infty$, this implies that $Re \to \infty$ also, causing the highest-order term in Eq. (16) to disappear, and making it impossible to satisfy the required boundary conditions at the wall. If we make the change of variables $X = x'$, $Y = y'(Re)^a$ and at the same time denote the new velocity components by u, v, where $u = v'_x$ and $v = v'_y(Re)^a$, the continuity equation retains its form while Eq. (16) becomes

$$u\frac{\partial u}{\partial X} + v\frac{\partial u}{\partial Y} = (Re)^{2a-1}\frac{\partial^2 u}{\partial Y^2}. \tag{7.5.19}$$

If this equation is not to contain Re explicitly and is still to retain the highest-order term as $Re \to \infty$, we must choose $a = \frac{1}{2}$. We then obtain Eqs. (20) to (22) in which the parameter Re does not occur explicitly. This is an example of what Kline (1965) calls ***absorption of parameters.***

Boundary-Layer Equations for Flat Plate

$$X = x' \qquad Y = y'\sqrt{Re} \qquad u = v'_x \qquad v = v'_y\sqrt{Re}$$

$$u\frac{\partial u}{\partial X} + v\frac{\partial u}{\partial Y} = \frac{\partial^2 u}{\partial Y^2} \tag{7.5.20}$$

$$\frac{\partial u}{\partial X} + \frac{\partial v}{\partial Y} = 0 \tag{7.5.21}$$

with boundary conditions

$$u = v = 0 \quad \text{at} \quad Y = 0, \qquad u \to 1 \quad \text{as} \quad Y \to \infty. \tag{7.5.22}$$

If we introduce the dimensionless stream function ψ in these variables, such that

$$u = \frac{\partial \psi}{\partial Y} \qquad v = -\frac{\partial \psi}{\partial X}, \tag{7.5.23}$$

the continuity equation (21) is identically satisfied, while Eq. (20) furnishes a single third-order equation for ψ.

$$\frac{\partial \psi}{\partial Y}\frac{\partial^2 \psi}{\partial X \partial Y} - \frac{\partial \psi}{\partial X}\frac{\partial^2 \psi}{\partial Y^2} = \frac{\partial^3 \psi}{\partial y^3}. \tag{7.5.24}$$

Since the semi-infinite plate has no characteristic length, it is reasonable to seek a solution in which the shapes of the profile plots of ψ versus Y are the same at all abscissas, i. e., such that the profile plots are geometrically similar.† This turns out to be impossible in the variables X, Y, but one more change of variable makes it possible. We seek to determine the function $\phi(X)$, so that after a change of variable,

$$\bar{X} = \phi(X) \qquad \bar{Y} = Y, \tag{7.5.25}$$

the transformed Equation (24) will possess a solution $\bar{\psi}(\bar{X}, \bar{Y})$ exhibiting similar profiles such that

$$\bar{\psi}(\bar{X}, \bar{Y}) = \bar{X}\bar{\psi}\left(1, \frac{\bar{Y}}{\bar{X}}\right), \tag{7.5.26}$$

as shown in Fig. 7.8. This means that the value of $\bar{\psi}(\bar{X}, \bar{Y})$ at point $P(\bar{X}, \bar{Y})$ on the profile plot of $\bar{\psi}$ at the abscissa $\bar{X}$ is the same multiple of $\bar{\psi}$ at $P_1(1, \bar{Y}/\bar{X})$ as the ordinate $\bar{Y}$ at P is of the ordinate $\bar{Y}/\bar{X}$ at P_1. The quantity $\bar{\psi}/\bar{X} = \bar{\psi}(1, \bar{Y}/\bar{X})$ would then have the same value everywhere on the extended

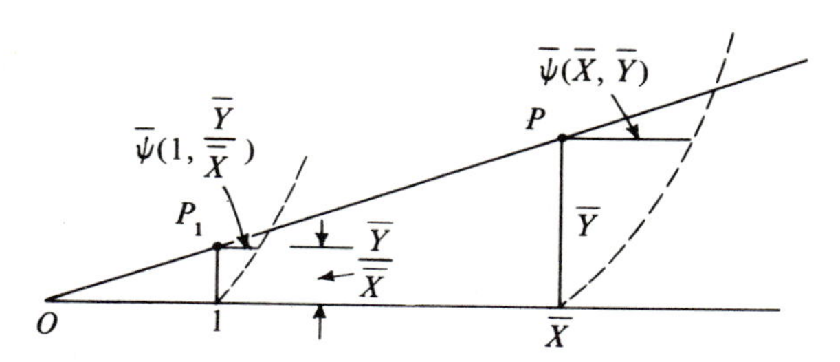

Fig. 7.8 Similar profiles for $\bar{\psi}(\bar{X}, \bar{Y}) = \bar{X}\bar{\psi}(1, \bar{Y}/\bar{X})$

†See, for example, Pai (1956), p. 162 for a somewhat more general discussion of similar solutions of boundary-layer equations.

line OP in Fig. 7.8, and would be a function only of the single variable $\eta = \bar{Y}/\bar{X}$, the slope of OP. In terms of the unbarred variables X, Y, ψ, we seek to determine a function $\phi(X)$ such that the quantity $\psi(X)/\phi(X)$ is a function of the single variable η, say

$$\frac{\psi[\phi(X), Y]}{\phi(X)} = f(\eta) \qquad \text{where} \quad \eta = \frac{Y}{\phi(X)}. \tag{7.5.27a}$$

Then Eq. (24) would become an ordinary differential equation for $f(\eta)$. Noting that then

$$\frac{\partial \psi}{\partial X} = \phi'(X)f(\eta) - \frac{y\phi'(X)f'(\eta)}{\phi}$$

$$\frac{\partial \psi}{\partial Y} = f'(\eta) \qquad \frac{\partial^2 \psi}{\partial Y^2} = \frac{f''(\eta)}{\phi}$$

$$\frac{\partial^3 \psi}{\partial Y^3} = \frac{f'''(\eta)}{\phi^2} \qquad \frac{\partial^2 \psi}{\partial X \partial Y} = -\frac{Y\phi'(X)f''(\eta)}{\phi},$$

we obtain from Eq. (24) the equation

$$f'''(\eta) + \phi(X)\phi'(X)f''(\eta)f(\eta) = 0. \tag{7.5.27b}$$

This will be an ordinary differential equation for $f(\eta)$ provided $\phi(X)\phi'(X)$ is a constant. Hence, if we take

$$\phi(X) = \sqrt{X} \qquad \phi(X)\phi'(X) = \frac{1}{2}, \tag{7.5.28}$$

then we obtain

$$\frac{\psi(X)}{\sqrt{X}} = f(\eta) \qquad \text{where} \quad \eta = \frac{Y}{\sqrt{X}}, \tag{7.5.29}$$

and we have to solve the nonlinear ordinary differential equation

$$2f'''(\eta) + f(\eta)f''(\eta) = 0 \qquad 0 \leq \eta < \infty \tag{7.5.30}$$

with end conditions

$$f(0) = f'(0) = 0, \qquad f'(\eta) \to 1 \quad \text{as} \quad \eta \to \infty. \tag{7.5.31}$$

The solution of this equation is still not easy, but it can be accomplished. In 1908, H. Blasius obtained a power-series solution of the form

$$f(\eta) = \sum_{k=0}^{\infty} \frac{A_k}{k!} \eta^k, \tag{7.5.32}$$

where the coefficients A_k were to be determined. The end conditions at $\eta = 0$ require $A_0 = A_1 = 0$. If we let

$$\alpha = A_2 = f''(0) \tag{7.5.33}$$

(which is unknown), the other coefficients can be determined successively in terms of α by substituting into the differential equation, Eq. (30), and requiring the collected coefficients of each power of η to vanish. It turns out that only terms of the form $k = 2 + 3n$ have nonzero coefficients for $n = 0, 1, 2, \ldots$. Hence the series takes the form

$$f(\eta) = \sum_{n=0}^{\infty}\left(-\frac{1}{2}\right)^n \frac{C_n\alpha^{n+1}}{(3n+2)!}\eta^{3n+2} \tag{7.5.34}$$

where the first six coefficients are

$$\left.\begin{array}{lll} C_0 = 1 & C_1 = 1 & C_2 = 11 \\ C_3 = 375 & C_4 = 27{,}897 & C_5 = 3{,}817{,}137 \end{array}\right\} \tag{7.5.35}$$

and the parameter α is still to be determined. The power series cannot be used to determine α from the boundary condition at infinity, but it furnishes some information on the relationship of α to the behavior as η increases. Equation (34) may be written in the form

$$f(\eta) = \alpha^{1/3}F(\alpha^{1/3}\eta) \tag{7.5.36}$$

whence, assuming the relationship of Eq. (36) to remain valid for large η, we have

$$\lim_{\eta\to\infty} f'(\eta) = \alpha^{2/3}\lim_{\eta\to\infty} F'(\alpha^{1/3}\eta).$$

But $\lim F'(\alpha^{1/3}\eta)$ as $\eta \to \infty$ is the same as $\lim F'(\eta)$, since in both cases the argument of F' increases without bound. Hence, since $f'(\eta) \to 1$ by Eq. (31),

$$1 = \alpha^{2/3}\lim_{\eta\to\infty} F'(\eta) \quad \text{or} \quad \alpha = [\lim_{\eta\to\infty} F'(\eta)]^{-3/2}, \tag{7.5.37}$$

where $F(\eta)$ is the solution obtained for $f(\eta)$ by assuming $\alpha = 1$. $F(\eta)$ can be found by numerical integration, using starting values for F and some of its derivatives obtained from the series solution with $\alpha = 1$ and continuing the numerical integration until F' is constant to a sufficient approximation. Very accurate values have been obtained by Howarth (1938). With $\alpha = 0.33206$, he calculated and tabulated values for f, f', f'' to $\eta = 8.8$. From plotted velocity profiles, with this solution it is found that the boundary-layer thickness $\delta(x)$ where v_x is within one per cent of V is approximately

$$\delta(x) = 5.2\sqrt{\frac{\mu x}{\rho V}} \tag{7.5.38}$$

[see, for example, Pai (1956)]. The shear traction on one side of the plate is

$$\tau_{yx} = \mu\left(\frac{\partial v_x}{\partial y}\right)_{y=0} = \mu V f''(0)\sqrt{\frac{\rho V}{\mu x}}. \tag{7.5.39}$$

Although the solution has been obtained for a semi-infinite plate with $Re \to \infty$, its results are often successfully applied to finite plates. For example, the drag D_1 on one side of a plate of length L parallel to the flow (per unit width perpendicular to the flow plane) is

$$D_1 = \int_0^L \tau_{yx}\, dx = 2\alpha\sqrt{\mu\rho L V^3}. \tag{7.5.40}$$

The total drag $D = 2D_1$ (per unit width) on two sides is thus given, with $\alpha = 0.332$, by

$$D = 1.328\sqrt{\mu\rho L V^3}$$

Additional examples of boundary-layer solutions are given in the fluid mechanics literature, including boundary layers in compressible flow. See, for example, Schlichting (1968). In heat-transfer studies a thermal boundary layer is also often important near a heated or cooled wall. For the most part these solutions are approximate in character, since complete exact solutions of the nonlinear boundary-layer equations are not easy to obtain. We will refer briefly to some of the approximations in connection with the momentum integral equations to be presented in Eqs. (47) to (49) below, but first we consider some alternative definitions of boundary-layer thickness.

Displacement thickness and momentum thickness. We have seen that the definition of boundary-layer thickness δ is somewhat arbitrary. We would like to interpret δ as the distance to the point where v_x reaches the value $U(x, t)$ predicted for the slip flow at the boundary by the potential theory. Unfortunately this value is not usually actually reached. Experimenters frequently define δ as the distance to the point where v_x reaches 99 per cent of U. We introduce here two other arbitrary definitions frequently used for boundary-layer thickness, especially in connection with the von Kármán momentum integral equations of the boundary layer, Eqs. (47) to (49) below, namely the ***displacement thickness*** δ^* and the ***momentum thickness*** θ. (Do not confuse this θ with the temperature, which does not enter the present discussion.)

The velocity retardation in the boundary layer causes the mass flux parallel to the boundary to be less than it would have been in the absence of a boundary layer. Because of this the streamlines outside the boundary layer are displaced a certain distance from the wall. The ***displacement thickness*** δ^* is defined as the thickness required for a layer of fluid moving at the rate U to give a mass flux equal to the defect caused by the boundary layer's presence. Here $U(x, t)$ is the boundary tangential velocity of the potential flow that would exist in the absence of viscosity. Assuming that δ is the distance to the point where v_x actually reaches the magnitude U, we can calculate δ^* as follows for incompressible flow:

$$\rho\delta^* U = \rho\int_0^\delta (U - v_x)\,dy \quad \text{or} \quad \delta^* = \int_0^\delta \left(1 - \frac{v_x}{U}\right)dy. \tag{7.5.41}$$

If δ were actually equal to the distance to the point where $v_x = U$, and if the potential flow were uniform for this distance, then the δ^* given by Eq. (41) would actually represent the thickness of a layer whose mass flux (at speed U) was equal to the mass-flux defect caused by the presence of the boundary layer. But since neither of these conditions is exactly fulfilled, and since the integral is usually evaluated with a plausible assumed profile for $v_x(y)$ instead of the actual profile, the result is an approximation.

The ***momentum thickness*** is defined by

$$\rho\theta U^2 = \rho\int_0^\delta (Uv_x - v_x^2)\,dy \quad \text{or} \quad \theta = \int_0^\delta \frac{v_x}{U}\left(1 - \frac{v_x}{U}\right)dy. \tag{7.5.42}$$

This defines θ by equating the momentum $\rho\theta U^2$ in a layer of thickness θ moving at the potential-flow boundary speed U to an integral which will appear in the momentum integral equation, Eq. (47) below, and will lead to a convenient simplification.

Momentum integral equations of boundary-layer flow. The boundary-layer equations discussed in this section are still nonlinear and difficult to solve. Von Kármán (1921) introduced an approximation method that gives a momentum relation, which can be obtained by integrating the boundary-layer equation through its thickness and by using some plausible approximation to the velocity profile. The method yields approximations to such desired information as the boundary-layer thickness and the skin-friction drag coefficient, in terms of the characteristic Reynolds number for a given problem, as will be illustrated by an example in the discussion of Eqs. (50) to (57) below.

The dimensional form of the boundary-layer equation replacing Eq. (16) for nonsteady incompressible flow is

$$\frac{\partial v_x}{\partial t} + v_x\frac{\partial v_x}{\partial x} + v_y\frac{\partial v_x}{\partial y} = -\frac{1}{\rho}\frac{dp}{dx} + \frac{\mu}{\rho}\frac{\partial^2 v_x}{\partial y^2}. \tag{7.5.43}$$

We integrate this equation from $y = 0$ to $y = \delta$, as follows:

$$\frac{\partial}{\partial t}\int_0^\delta v_x\,dy + \int_0^\delta v_x\frac{\partial v_x}{\partial x}\,dy + \int_0^\delta v_y\frac{\partial v_x}{\partial y}\,dy = -\frac{1}{\rho}\int_0^\delta \frac{dp}{dx}\,dy + \frac{\mu}{\rho}\int_0^\delta \frac{\partial^2 v_x}{\partial y^2}\,dy. \tag{7.5.44a}$$

The third term on the left can be integrated by parts:

$$\int_0^\delta v_y\frac{\partial v_x}{\partial y}\,dy = [v_y v_x]_0^\delta - \int_0^\delta v_x\frac{\partial v_y}{\partial y}\,dy = U\int_0^\delta \frac{\partial v_y}{\partial y}\,dy - \int_0^\delta v_x\frac{\partial v_y}{\partial y}\,dy \tag{7.5.44b}$$

under the assumption that $v_x = U$ at $y = \delta$, and $v_x = 0$ at $y = 0$. By the continuity equation, $\partial v_y/\partial y = -\,\partial v_x/\partial x$. Hence Eq. (44b) may be written as

$$\int_0^\delta v_y\frac{\partial v_x}{\partial y}\,dy = -U\int_0^\delta \frac{\partial v_x}{\partial x}\,dy + \int_0^\delta v_x\frac{\partial v_x}{\partial x}\,dy.$$

When we substitute this into Eq. (44a) we obtain

$$\frac{\partial}{\partial t}\int_0^\delta v_x\,dy + \int_0^\delta \frac{\partial(v_x^2)}{\partial x}\,dy - U\int_0^\delta \frac{\partial v_x}{\partial x}\,dy = -\frac{1}{\rho}\frac{dp}{dx}\delta - \frac{\tau_0}{\rho} \tag{7.5.45}$$

where

$$\tau_0 = \mu\left(\frac{\partial v_x}{\partial y}\right)_{y=0}$$

is the wall shear stress, and where we have assumed that $\partial v_x/\partial y = 0$ at $y = \delta$, and that p (and hence dp/dx) is independent of y in the boundary layer in accordance with the discussion of Eq. (10). The second and third terms of Eq. (45) can be written in the following way by using the Leibnitz rule for the derivative of an integral

$$\int_0^\delta \frac{\partial(v_x^2)}{\partial x}\,dy = \frac{d}{dx}\int_0^\delta v_x^2\,dy - U^2\frac{d\delta}{dx}$$

$$U\int_0^\delta \frac{\partial v_x}{\partial x}dy = U\frac{d}{dx}\int_0^\delta v_x\,dy - U^2\frac{d\delta}{dx}.$$

Hence Eq. (45) takes the form

Momentum Integral Equation of the Boundary Layer

$$\frac{\partial}{\partial t}\int_0^\delta v_x dy + \frac{d}{dx}\int_0^\delta v_x^2 dy - U\frac{d}{dx}\int_0^\delta v_x dy = -\frac{\delta}{\rho}\frac{dp}{dx} - \frac{\tau_0}{\rho}. \tag{7.5.46}$$

Equation (46) is one form of the von Kármán integral relation; it is also called the momentum integral equation of the boundary layer. Our order-of-magnitude arguments leading to the Prandtl boundary-layer equations and thence to the von Kármán momentum integral equations neglected any contributions due to turbulent fluctuations. The momentum integral equation is, however, often used for turbulent flow as well as for laminar flow.

The momentum integral relation can be written in terms of the displacement thickness δ^* and the momentum thickness θ defined in Eqs. (41) and (42) as follows. With use of Eq. (14), the momentum integral equation [Eq. (46)] takes the form

$$\frac{\partial}{\partial t}\int_0^\delta v_x\,dy - \frac{\partial U}{\partial t}\delta + \frac{d}{dx}\int_0^\delta v_x^2\,dy - U\frac{d}{dx}\int_0^\delta v_x\,dy - U\frac{dU}{dx}\delta = -\frac{\tau_0}{\rho}$$

which can be written as

$$\frac{\partial}{\partial t}\int_0^\delta (U - v_x)\,dy + \frac{d}{dx}\int_0^\delta v_x(U - v_x)\,dy + \frac{dU}{dx}\int_0^\delta (U - v_x)\,dy = \frac{\tau_0}{\rho}. \qquad (7.5.47)$$

Use of the definitions of Eqs. (41) and (42) then gives Eq. (47) the form:

Momentum Integral Equations in Terms of δ^* and θ

$$\frac{\partial}{\partial t}(U\delta^*) + \frac{d}{dx}(U^2\theta) + U\frac{dU}{dx}\delta^* = \frac{\tau_0}{\rho}, \quad \text{or} \qquad (7.5.48)$$

$$\text{for steady flow} \quad \frac{d\theta}{dx} + (2\theta + \delta^*)\frac{1}{U}\frac{dU}{dx} = \frac{\tau_0}{\rho U^2}. \qquad (7.5.49)$$

These forms are convenient for applications, as the following simple example indicates.

Example: frictional drag coefficient for a flat plate at zero incidence. For the flat plate at zero incidence, the potential flow is uniform: $U \equiv V$, and the steady-flow momentum integral equation, Eq. (49), takes the simple form

$$\frac{d\theta}{dx} = \frac{\tau_0}{\rho V^2}\,. \qquad (7.5.50)$$

The momentum thickness, defined by Eq. (42), can then be written as

$$\frac{\theta}{\delta} = \int_0^1 \left(1 - \frac{v_x}{V}\right)\frac{v_x}{V}\,d\eta = A, \qquad (7.5.51)$$

where

$$\eta = \frac{y}{\delta}\,.$$

We assume that in the boundary layer v_x is a function only of η. Then A is a constant. If we substitute Eq. (51) into Eq. (50), we obtain

$$A \frac{d\delta}{dx} = \frac{\tau_0}{\rho V^2}.$$

Now

$$\tau_0 = \left(\frac{\partial v_x}{\partial y}\right)_{y=0} = \mu \frac{V}{\delta} \left[\frac{\partial}{\partial \eta}\left(\frac{v_x}{V}\right)\right]_{\eta=0} = B\mu \frac{V}{\delta}, \tag{7.5.52}$$

where

$$B = \left[\frac{\partial(v_x/V)}{\partial \eta}\right]_{\eta=0}. \tag{7.5.53}$$

Hence

$$A \frac{d\delta}{dx} = B \frac{\mu}{\rho V \delta},$$

which integrates to give

$$\delta = x\sqrt{2 \frac{B}{A} \frac{\mu}{\rho V x}} \tag{7.5.54}$$

if we assume $\delta = 0$ at $x = 0$. Hence

$$\tau_0 = \mu V \sqrt{\frac{AB}{2}} \sqrt{\frac{\rho V}{\mu x}}. \tag{7.5.55}$$

The skin-friction drag coefficient C_f is defined by

$$C_f = \frac{D_1}{\frac{1}{2}\rho V^2 L} \tag{7.5.56}$$

where D_1 is the total drag on one side of a plate of unit width and length L. Use of Eq. (55) then gives

$$C_f = 2\sqrt{2AB} \sqrt{\frac{\mu}{\rho V L}}. \tag{7.5.57}$$

Equations (54) and (57) give an estimate for $\delta(x)$ and for C_f in terms of the Reynolds numbers $Re = \rho v L/\mu$ and $(Re)_x = \rho v x/\mu$, if A and B can be determined from Eqs. (51) and (52) by using some plausible assumption for the velocity profile $v_x(\eta)$. It will be left as an exercise for the reader to fit a cubic of the form

$$\frac{v_x}{V} = a\eta + c\eta^3 \tag{7.5.58a}$$

that satisfies the boundary conditions

$$\left.\begin{aligned} v_x &= 0 \quad \text{at } \eta = 0, \text{ and} \\ v_x &= V \quad \text{and} \quad \frac{\partial v_x}{\partial \eta} = 0 \quad \text{at } \eta = 1 \quad (y = \delta), \end{aligned}\right\} \tag{7.5.58b}$$

and to show that this leads to the estimates

$$\delta = 4.64\sqrt{\frac{\mu x}{\rho V}} \quad \text{and} \quad C_f = 1.292\sqrt{\frac{\mu}{\rho V L}}, \tag{7.5.59}$$

which may be compared with the values

$$\delta = 5.2\sqrt{\frac{\mu x}{\rho V}} \quad \text{and} \quad C_f = 1.328\sqrt{\frac{\mu}{\rho V L}} \tag{7.5.60}$$

based on the Blasius solution of Eqs. (38) and (40).

Von Kármán-Pohlhausen method. In the example just considered the potential flow was uniform, whence we had $\partial p/\partial x = 0$ along the boundary. At von Kármán's suggestion, Pohlhausen (1921) approximated the boundary layer in steady flow by a fourth-degree polynomial of the form

$$\frac{v_x}{U(x)} = a\eta + b\eta^2 + c\eta^3 + d\eta^4 \tag{7.5.61}$$

where

$$\eta = \frac{y}{\delta}.$$

The four polynomial coefficients were determined by requiring

$$\begin{aligned} &\text{at } y = 0: \quad v_x = 0 \quad \text{and} \quad -\frac{\partial^2 v_x}{\partial y^2} = \frac{1}{\mu}\frac{dp}{dx} = -U\frac{dU}{dx} \\ &\text{at } y = \delta: \quad v_x = U(x) \quad \text{and} \quad \frac{\partial v_x}{\partial y} = \frac{\partial^2 v_x}{\partial y^2} = 0. \end{aligned} \tag{7.5.62}$$

The second condition at $y = 0$, for steady flow, is obtained from the boundary-layer equation, Eq. (43), by use of Eq. (14) and the conditions $v_x = v_y = 0$ at $y = 0$. The condition $\partial^2 v_x/\partial y^2 = 0$ at $y = \delta$ ensures a smooth join between the assumed boundary-layer profile and the potential-flow solution at $y = \delta$. These boundary conditions are all satisfied by

$$a = 2 + \frac{\lambda}{6}, \quad b = -\frac{\lambda}{2}, \quad c = -2 + \frac{\lambda}{2}, \quad d = 1 - \frac{\lambda}{6}, \tag{7.5.63}$$

where

$$\lambda = \frac{\rho\delta^2}{\mu}\frac{dU}{dx} = -\frac{dp}{dx}\frac{\delta^2}{\mu U}. \tag{7.5.64}$$

The approximation profile may be written as

$$\frac{v_x}{U(x)} = F(\eta) + \lambda G(\eta), \tag{7.5.65}$$

where

$$F(\eta) = 2\eta - 2\eta^3 + \eta^4 \quad \text{and} \quad G(\eta) = \frac{\eta}{6}(1 - \eta)^3. \tag{7.5.66}$$

In practice the adjustable parameter λ, called the ***Pohlhausen parameter***, is often used to vary the shape of the velocity profile, and finally chosen to match experimental data. It is found that λ must fall in the range $-12 \leq \lambda \leq 12$, since for $\lambda = -12$ the boundary layer separates from the wall, as evidenced by $\partial v_x/\partial y = 0$ at $y = 0$, while $\lambda > 12$ implies $v_x > U$. For details of application, see, for example, Yuan (1967), Pai (1956), or Schlichting (1968). The last two references also give additional types of approximations and applications of boundary-layer theory.

This concludes our brief survey of fluid mechanics. In Sec. 7.1 we derived the Navier-Stokes equations for laminar flow of a viscous fluid and saw an example of their solution when inertia terms vanished because of the geometry. Secs. 7.2 and 7.3 treated nonviscous fluids; Sec. 7.2 formulated a linearized wave-propagation theory as our only example of nonsteady flow, and formulated the gas-dynamic equation for steady flow, while Sec. 7.3 was mainly concerned with irrotational motion (potential flow) of incompressible fluid. Section 7.4 treated dynamic similarity, essential for planning and interpretation of experimental studies, and introduced some of the dimensionless characteristic numbers of a flow. And, finally, in Sec. 7.5 we discussed briefly the two extreme limiting cases of very low and very high viscosity. The aim has been to offer some conception of the extent and variety of the subject of fluid mechanics as well as to illustrate how a complete theory is put together by adding the constitutive equations to the general field equations valid for all continua. The survey is far from complete; notable omissions are any treatment of turbulent flow, heat production by dissipation, heat transfer, hypersonic flow, or magnetofluid mechanics. For these subjects and also for details on the subjects introduced here, consult the specialized literature. A short bibliography is appended to this chapter; additional references are cited in the books listed and in the recent periodical literature. As for the other aim of illustrating how a complete theory is put together, the conclusion must be that even within the domain of fluid mechanics there is no such thing as a ***typical*** formulation. We have seen instead a variety of formulations, depending on the idealizations assumed. If there is no formulation typical even of fluid mechanics, then we can expect even greater variety when other constitutive equations are assumed. There is no unified theory of boundary-value problem formulation in continuum mechanics. There are occasional resemblances between the boundary-value problems formulated in linear elasticity and those of fluid mechanics, but these resemblances are more accidental than fundamental. Laplace's equation, for example, turns up in the study of elastic torsion of a prismatic bar and in connection with the Papkovich-Neuber potentials of three-dimensional elasticity (Sec. 8.4) as well as in incompressible potential flow (and also in heat conduction and in electrostatics). The biharmonic equation turns up in two-dimensional elasticity, in connection with the Galerkin vector (Sec. 8.4), and (with an inhomogeneous term) in the bending of elastic thin plates as well as in the creeping flow treated at the beginning of this section. These coincidences have led to the possibility of useful analogies to deduce a solution in one field from experiments in another field where measurements are easier. But the real unifying thread in continuum theory is the general field equations and some general principles on constitutive relations (see Secs. 6.7 and 6.8) rather than any similarities in the way the complete theories are put together. At the level of problem formulation and solution, it is necessary to turn to specialized expositions for each field.

EXERCISES

1. Carry out the details leading to Eq. (7.5.4), the vorticity transport equation for incompressible plane flow, and show that introduction of the stream function then leads to Eq. (6).
2. Introduce dimensionless variables in Eq. (7.5.6), and verify that the nondimensional limiting condition required to obtain Eq. (7) is that the Reynolds number approach zero while the Strouhal number is bounded.
3. (a) If the nonlinear terms in a three-dimensional version of Eqs. (7.5.1) are neglected in the creeping flow, and h is linear in x, y, and z, show that the pressure satisfies Laplace's equation $\nabla^2 p = 0$.
   (b) If the inertia terms are neglected completely, and the pressure is written as $p = p_d + p_s = p_d + (\text{constant} - \rho g h)$, show that the three-dimensional version of Eqs. (1) may be written as $\nabla p_d = \mu \nabla^2 \mathbf{v}$.
4. Carry out the details of the change of variable leading to Eqs. (7.5.20) to (22).
5. Show that introduction of the dimensionless stream function ψ leads to Eq. (7.5.24). What are the boundary conditions on ψ?
6. Carry out the details of showing that Eq. (7.5.27a) leads to Eq. (27b).
7. Show that Eq. (7.5.28) leads from Eqs. (27) to Eqs. (29) and (30), and that the end conditions of Eqs. (31) are obtained.

8. For the power series of Eq. (7.5.32), show that the end conditions require $A_0 = A_1 = 0$, and that the differential equation implies $A_3 = A_4 = 0$ and $A_5 = -\frac{1}{2}\alpha^2$. Show that these results for A_3, A_4, and A_5 are consistent with Eqs. (34) and (35).
9. Show that Eq. (7.5.34) can be written in the form of Eq. (36). Exhibit the form of the power series in u for $F(u)$.
10. (a) Show that the shear stress is given by Eq. (7.5.39).
    (b) Show that this leads to the value for D given in the equation following Eq. (40) when $\alpha = 0.332$.
11. (a) Verify the details of the derivation leading from Equation (43) to the momentum integral equation, Eq. (7.5.46).
    (b) Show that introduction of δ^* and θ then leads to Eqs. (48) and (49).
12. Verify the calculations of δ and C_f in Eqs. (7.5.54) and (57).
13. Based on the momentum integral equation, calculate $\delta(x)$ and C_f according to Eqs. (7.5.54) and (57) for the flat plate at zero incidence with the following assumed profiles for v_x/U:
    (a) $v_x/U = \eta$,
    (b) Eq. (7.5.58a) with the boundary conditions (58b),
    (c) $v_x/U = 2\eta - 2\eta^3 + \eta^4$.
14. Verify the results of Eqs. (7.5.63) through (66) for the von Kármán-Pohlhausen method.

Brief bibliography on fluid mechanics

General Fluid Mechanics

Flügge, S. (ed.), *Encyclopedia of Physics,* vols. 8(1), 8(2), and 9. Berlin: Springer, 1959, 1963, and 1960.

Landau, L. D. and E. M. Lifshitz, *Fluid Mechanics.* Oxford: Pergamon Press, (Reading, Mass.: Addison-Wesley) 1959.

Loitsyanskii, L. G., *Mechanics of Liquids and Gases.* Oxford: Pergamon Press, 1966.

Daily, J. W. and D. R. F. Harleman, *Fluid Dynamics.* Reading, Mass.: Addison-Wesley, 1966.

Yuan, S. W., *Foundations of Fluid Mechanics.* Englewood Cliffs, N. J.: Prentice-Hall, Inc., 1967.

Compressible Fluids

Courant, R. and K. O. Friedrichs, *Supersonic Flow and Shock Waves.* New York: Interscience, 1948.

Ferri, A., *Elements of Aerodynamics of Supersonic Flows.* New York: Macmillan, 1949.

Howarth, L. (ed.), *Modern Developments in Fluid Dynamics, High Speed Flow,* 2 vols. London: Oxford University Press, 1953.

Pai, S.-I., *Introduction to the Theory of Compressible Flow.* Princeton, N. J.: Van Nostrand, 1959.

Shapiro, Ascher H., *The Dynamics and Thermodynamics of Compressible Fluid Flow,* 2 vols. New York: Ronald Press, 1953–1954.

HYDRODYNAMICS (Incompressible Frictionless Fluids)

Lamb. H., *Hydrodynamics,* 6th ed. New York: Dover, 1945.

Milne-Thomson, L. M., *Theoretical Hydrodynamics,* 4th ed. New York: Macmillan, 1960.

Robertson, J. M., *Hydrodynamics in Theory and Application.* Englewood Cliffs, N. J.: Prentice-Hall, 1965.

VISCOUS FLUIDS AND BOUNDARY-LAYER THEORY

Pai. S.-I., *Viscous Flow Theory,* vol. 1, *Laminar Flow,* vol. 2, *Turbulent Flow.* Princeton, N. J. : Van Nostrand, 1956–1957.

Schlichting, H., *Boundary Layer Theory,* 4th ed. New York: McGraw-Hill, 1960; 6th ed. New York: McGraw-Hill, 1968

TRANSPORT PHENOMENA, INCLUDING HEAT TRANSFER

Bird, R. B., W. E. Stewart, and E. N. Lightfoot, *Transport Phenomena.* New York: Wiley, 1961.

Eckert, E. R. G. and R. M. Drake, *Heat and Mass Transfer,* 2nd ed. New York: McGraw-Hill, 1959.

Kutateladze, S. S., *Fundamentals of Heat Transfer.* New York: Academic Press, 1963.

Rohsenow, W. M. and H. Choi, *Heat, Mass and Momentum Transfer.* Englewood Cliffs, N. J.: Prentice-Hall, 1961.

TURBULENT FLOW

Hinze, J., *Turbulence; An Introduction to its Mechanism and Theory.* New York: McGraw-Hill, 1959.

Townsend, A. A., *The Structure of Turbulent Shear Flow.* Cambridge: University Press, 1956.

MAGNETOFLUID-MECHANICS

Cambel, A. B., *Plasma Physics and Magnetofluid-Mechanics.* New York: McGraw-Hill, 1963.

Jeffrey, A., *Magnetohydrodynamics.* Edinburgh: Oliver and Boyd (New York: Interscience), 1966.

Pai, S.-I, *Magnetogasdynamics and Plasma Physics.* Vienna: Springer (Englewood Cliffs, N. J.: Prentice-Hall), 1962.

STABILITY OF FLOW

Chandrasekhar, S., *Hydrodynamic and Hydromagnetic Stability.* Oxford: Clarendon Press, 1961.

Lin, C. C., *The Theory of Hydrodynamic Stability.* Cambridge: University Press, 1955.

CHAPTER 8

Linearized Theory of Elasticity

8.1 Field equations

This chapter presents a formulation of the field equations of linearized isotropic elasticity theory and a few examples of problem formulation and solution. The formulation uses the Lagrangian referential description in terms of material coordinates X_J of a particle in the natural state. A completely linear theory results, based on the assumptions that the displacement-gradient components are small compared to unity, that the generalized Hooke's law of Sec. 6.2 is the constitutive equation, and that the equations of motion or equilibrium can be considered to be satisfied in the undeformed reference configuration. This contrasts with the Navier-Stokes equations of fluid mechanics, where the convective acceleration terms of the spatial description are nonlinear.

The smallness of the displacement-gradient components permits the use of the small-strain tensor in place of the finite strain of Sec. 4.5, so that the geometric strain-displacement equations are linearized. This leaves for discussion only the assumption that the equations of motion are "considered satisfied in the reference configuration." This is not a defensible assumption on strictly logical grounds. An examination of the nonlinear equations of motion in the reference state will reveal some of the implications of the assumption. In terms of Cartesian components, the first Piola-Kirchhoff stress tensor T^0_{Ji} is related to the Cauchy stress T_{ri} at the point $\mathbf{x} = \mathbf{X} + \mathbf{u}$ by

$$T^0_{Ji} = \frac{\rho_0}{\rho} \frac{\partial X_J}{\partial x_r} T_{ri}; \tag{8.1.1}$$

see Eq. (5.3.22). The equations of motion in the reference state are given by Eqs. (5.3.27) as

$$\frac{\partial T^0_{Ji}}{\partial X_J} + \rho_0 b_{0i} = \rho_0 \frac{d^2 x_i}{dt^2}. \tag{8.1.2}$$

The right-hand side of the equation of motion is equal to $\rho_0 \partial^2 u_i/\partial t^2$ (partial derivative with the material coordinates held constant); hence Eq. (2) is a linear partial differential equation. The difficulty comes when we substitute for T^0_{Ji} by Eq. (1) in order to get equations for the Cauchy stress T_{ri} considered as a function of the X_J and t; this leads to a nonlinear equation of motion because of the nonlinearity of Eq. (1). To make the linearization process clear, we write out three of the components of Eq. (1). Noting that $X_1 = x_1 - u_1$, etc., we obtain

$$T^0_{11} = \frac{\rho_0}{\rho}\left[\left(1 - \frac{\partial u_1}{\partial x_1}\right) T_{11} - \frac{\partial u_1}{\partial x_2} T_{21} - \frac{\partial u_1}{\partial x_3} T_{31}\right]$$

$$T^0_{21} = \frac{\rho_0}{\rho}\left[- \frac{\partial u_2}{\partial x_1} T_{11} + \left(1 - \frac{\partial u_2}{\partial x_2}\right) T_{21} - \frac{\partial u_2}{\partial x_3} T_{31}\right]$$

$$T^0_{12} = \frac{\rho_0}{\rho}\left[\left(1 - \frac{\partial u_1}{\partial x_1}\right) T_{12} - \frac{\partial u_1}{\partial x_2} T_{22} - \frac{\partial u_1}{\partial x_3} T_{32}\right],$$

where

$$\frac{\rho_0}{\rho} = 1 + \frac{\partial u_1}{\partial X_1} + \frac{\partial u_2}{\partial X_2} + \frac{\partial u_3}{\partial X_3} + \text{higher-order terms,}$$

If we neglect both $\partial u_i/\partial X_J$ and $\partial u_i/\partial x_j$ in comparison to unity, these equations are simplified somewhat, but complete linearization requires also neglect of terms like $(\partial u_1/\partial x_2)T_{21}$ in comparison to T_{11}. This is reasonable only if T_{21} is not an order of magnitude larger than T_{11}, etc. With this assumption $T^0_{11} = T_{11}$, $T^0_{12} = T_{12} = T_{21} = T^0_{21}$, and to this degree of accuracy we may write the equation of motion as in Eq. (3) below, where the subscript zero has been dropped from ρ_0 and b_0 for simplicity, although these are to be understood as evaluated as functions of $\mathbf{X}$, and ρ is the initial density. [Hooke's law is Eq. (6.2.4), while Eq. (5) comes from Sec. 4.2.]

Field Equations of Linearized Isotropic Isothermal Elasticity
(in material coordinates)

3 Eqs. of Motion	$\dfrac{\partial T_{ji}}{\partial X_j} + \rho b_i = \rho \dfrac{\partial^2 u_i}{\partial t^2}$	(8.1.3)
6 Hooke's Law Eqs.	$T_{ij} = \lambda E_{kk}\delta_{ij} + 2\mu E_{ij}$	(8.1.4)
6 Geometric Eqs.	$E_{ij} = \dfrac{1}{2}\left(\dfrac{\partial u_i}{\partial X_j} + \dfrac{\partial u_j}{\partial X_i}\right)$	(8.1.5)

15 eqs. for 6 stresses, 6 strains, 3 displacements

Boundary conditions for the field equations may be:

1. ***Displacement boundary conditions***, with the three components u_i prescribed on the boundary.
2. ***Traction boundary conditions***, with the three traction components $t_i = T_{ji} n_j$ prescribed at a boundary point where the boundary unit normal is $\hat{\mathbf{n}}$.
3. ***Mixed boundary conditions*** include cases where
   (a) Displacement boundary conditions are prescribed on a part of the bounding surface, while traction boundary conditions are prescribed on the remainder, or
   (b) at each point of the boundary we choose local rectangular Cartesian axes $\bar{X}_i$ (usually with one axis along the normal) and then prescribe: (Note $\bar{t}_i = \bar{T}_{ji}\bar{n}_j$.)
   (1) $\bar{u}_1$ or $\bar{t}_1$, but not both,
   (2) $\bar{u}_2$ or $\bar{t}_2$, but not both, and
   (3) $\bar{u}_3$ or $\bar{t}_3$, but not both.

The last case 3(b) includes the others as special cases. Traction boundary conditions are sometimes called stress boundary conditions, but this is misleading. It is not physically possible to impose boundary conditions from outside on all the stress tensor components at a boundary point. Only the traction components are accessible. For example, at a bounding plane $X_3 =$ const., T_{31}, T_{32}, T_{33} may be prescribed, but not T_{11}, T_{22}, or T_{12}.

Superposition of solutions is possible because the equations are linear. This means that if fields $\mathbf{T}^{(1)}$, $\mathbf{E}^{(1)}$, $\mathbf{u}^{(1)}$ are a solution of Eqs. (3) through (5) for given body forces $\mathbf{b}^{(1)}$, while $\mathbf{T}^{(2)}$, $\mathbf{E}^{(2)}$, $\mathbf{u}^{(2)}$ are a solution for given $\mathbf{b}^{(2)}$, then $A\mathbf{T}^{(1)} + B\mathbf{T}^{(2)}$, $A\mathbf{E}^{(1)} + B\mathbf{E}^{(2)}$, $A\mathbf{u}^{(1)} + B\mathbf{u}^{(2)}$ are solutions corresponding to given $\mathbf{b} = A\mathbf{b}^{(1)} + B\mathbf{b}^{(2)}$, where A and B are arbitrary dimensionless constants (small enough to stay within the small-displacement assumptions). The boundary conditions are superposed in the same way. In the case of a mixed boundary-value problem, we use the same mixture for both solutions, e.g., at a point if we choose to prescribe $\bar{u}_1^{(1)}$ and $\bar{u}_1^{(2)}$, then the superposed solution $\bar{u}_1$ must equal the prescribed $A\bar{u}_1^{(1)} + B\bar{u}_1^{(2)}$ at that point.

Navier's displacement equations of motion. Solving a boundary-value problem involving 15 equations for 15 unknowns is a formidable task. There are several way of formulating the problem in terms of fewer unknowns and fewer equations. The most straightforward method is to substitute Eqs. (5) into (4) to obtain the stress in terms of displacement gradients, and then substitute the result into Eqs. (3) to obtain three second-order partial differential equations for the three displacement components. The result is

$$(\lambda + \mu)\frac{\partial^2 u_k}{\partial X_i \partial X_k} + \mu \frac{\partial^2 u_i}{\partial X_k \partial X_k} + \rho b_i = \rho \frac{\partial^2 u_i}{\partial t^2}. \tag{8.1.6}$$

This is especially convenient with displacement boundary conditions. By means of Eqs. (4) and (5) the traction boundary condition for $t_i = T_{ji} n_j$ may be given the form

$$\lambda \frac{\partial u_k}{\partial X_k} n_i + \mu\left(\frac{\partial u_i}{\partial X_j} + \frac{\partial u_j}{\partial X_i}\right) n_j = \text{Prescribed function} \tag{8.1.7}$$

in terms of the displacements. This may be awkward to apply, but can often be used successfully. In vector notation, Eqs. (6) and (7) take the form

Navier Equation†

$$(\lambda + \mu)\nabla(\nabla \cdot \mathbf{u}) + \mu \nabla^2 \mathbf{u} + \rho \mathbf{b} = \rho \frac{\partial^2 \mathbf{u}}{\partial t^2} \tag{8.1.8}$$

Traction Boundary Condition

$$\lambda(\nabla \cdot \mathbf{u})\hat{\mathbf{n}} + \mu(\mathbf{u}\overleftarrow{\nabla} + \overrightarrow{\nabla}\mathbf{u}) \cdot \hat{\mathbf{n}} = \text{Prescribed function} \tag{8.1.9}$$

The vector forms are independent of the coordinate system; they may be evaluated in curvilinear coordinates by the methods of App. I or II. By the vector identity of Eq. (2.5.27j) we have

$$\nabla^2 \mathbf{u} = \nabla(\nabla \cdot \mathbf{u}) - \nabla \times (\nabla \times \mathbf{u}).$$

If we introduce

$$e = \nabla \cdot \mathbf{u} \quad \text{and} \quad \boldsymbol{\omega} = \tfrac{1}{2} \nabla \times \mathbf{u}$$

as in Sec. 4.2, we may rewrite the vector form of the Navier equation as

$$(\lambda + 2\mu)\nabla e - 2\mu \nabla \times \boldsymbol{\omega} + \rho \mathbf{b} = \rho \frac{\partial^2 \mathbf{u}}{\partial t^2}, \tag{8.1.10}$$

a form more convenient for evaluation in orthogonal curvilinear coordinates. (See Sec. II.4 for cylindrical and spherical coordinate forms.)

In elastostatics the acceleration $\partial^2 \mathbf{u}/\partial t^2$ is zero in Eq. (3) or Eq. (8), which are then equations of equilibrium in terms of stresses or displacements, respectively. Since Eqs. (3) then do not involve displacements, an alternative formulation of the elastostatic problem in terms of stresses alone is possible; see Eqs. (25) and (26) below. First, however, we consider the question of the uniqueness of solution for an elastostatic boundary-value problem.

†Equations of this form were given by Navier in a memoir of 1821, published in 1827, but they contained only one elastic constant because they were deduced from an inadequate molecular model. The two-constant version was given by Cauchy in 1822.

Uniqueness of the elastostatic stress and strain field solutions can be proved under the assumption that the equilibrium equations are satisfied in the (undeformed) reference configuration. This rules out buckling situations; it is well known that thin columns or plates under axial compression sometimes buckle elastically, exhibiting a behavior where more than one elastic state can exist (e.g., an unbuckled state or a buckled state) under the same external loads. Such possibilities will not appear in the linearized theory presented above. The study of such physically possible nonuniqueness is called the theory of ***elastic stability***; it requires examination of the equilibrium equations in the deformed state and leads to nonlinear equations. What we prove here mathematically is that the linearized problem has a unique solution for $\mathbf{T}$ and $\mathbf{u}$. The proof depends on the assumption that there exists a strain-energy function W, which is a positive-definite, homogeneous quadratic function of the strains, with coefficients appropriately symmetrized so that

$$T_{ij} = \frac{\partial W}{\partial E_{ij}} \tag{8.1.11}$$

as in Eq. (6.2.18) whence it follows that

$$W = \tfrac{1}{2} T_{ij} E_{ij}. \tag{8.1.12a}$$

For the isotropic Hooke's law, we saw in Sec. 6.2 that there always does exist such a function, given by Eq. (6.2.40) as

$$W = \tfrac{1}{2}\lambda e^2 + G E_{ij} E_{ij}, \tag{8.1.12b}$$

which is positive-definite provided the elastic modulus E and Poisson's ratio satisfy $E > 0$ and $-1 < \nu < \frac{1}{2}$. The following uniqueness proof† applies to the more general Hooke's law of Eq. (11) under the ***assumption*** that W is a positive-definite homogeneous quadratic form in the E_{ij}. Since Eq. (11) is also linear, the superposition of solutions discussed above for the system of Eqs. (3), (4), and (5) applies also to the system of Eqs. (3), (11), and (5). If there did exist two solutions $\mathbf{T}^{(1)}$, $\mathbf{E}^{(1)}$, $\mathbf{u}^{(1)}$ and $\mathbf{T}^{(2)}$, $\mathbf{E}^{(2)}$, $\mathbf{u}^{(2)}$ corresponding to the same boundary conditions and the same body forces, then $\mathbf{T} = \mathbf{T}^{(2)} - \mathbf{T}^{(1)}$, $\mathbf{E} = \mathbf{E}^{(2)} - \mathbf{E}^{(1)}$, $\mathbf{u} = \mathbf{u}^{(2)} - \mathbf{u}^{(1)}$ would also be a solution of the equations for $\mathbf{b} = 0$ satisfying boundary conditions such that $\mathbf{u} \cdot \mathbf{t} = 0$ at every boundary point, since at each point and for each k, either $\bar{u}_k = 0$ or $\bar{t}_k = 0$. Hence, integrating over the boundary S, we have

$$\int_S \mathbf{u} \cdot \mathbf{t} \, dS = 0. \tag{8.1.13}$$

† The basic ideas of this uniqueness proof are due to Kirchhoff (1859).

We substitute $\mathbf{t} = \mathbf{n} \cdot \mathbf{T} = \mathbf{T} \cdot \mathbf{n}$ (for symmetric $\mathbf{T}$) to obtain

$$\int_S \mathbf{u} \cdot \mathbf{T} \cdot \mathbf{n} \, dS = 0$$

whence by the divergence theorem

$$\int_V \operatorname{div} (\mathbf{u} \cdot \mathbf{T}) \, dV = 0. \tag{8.1.14}$$

In rectangular Cartesian components

$$\operatorname{div} (\mathbf{u} \cdot \mathbf{T}) = \frac{\partial}{\partial X_j}(u_i T_{ij}) = \frac{\partial u_i}{\partial X_j} T_{ij} + u_i \frac{\partial T_{ij}}{\partial X_j}.$$

Now $\partial T_{ij}/\partial X_j = 0$ by the equilibrium equation with $\mathbf{b} = 0$, and $(\partial u_i/\partial X_j) T_{ij} = T_{ij} E_{ij} = 2W$ by Eq. (11) since $\partial u_i/\partial X_j = E_{ij} + \Omega_{ij}$ and $T_{ij}\Omega_{ij} = 0$. Hence Eq. (14) takes the form

$$\int_V 2W \, dV = 0. \tag{8.1.15}$$

But under the assumption that W is positive-definite and continuous the integral can only vanish if $W = 0$ everywhere, and this is only possible if all the $E_{ij} = 0$ everywhere, so that

$$E_{ij}^{(2)} = E_{ij}^{(1)} \quad \text{and therefore} \quad T_{ij}^{(2)} = T_{ij}^{(1)} \tag{8.1.16}$$

everywhere, by Hooke's law. Hence there cannot be two different stress and strain fields corresponding to the same externally imposed body forces and boundary conditions and satisfying the linearized elastostatic Eqs. (3), (11), (5).

Stress equations of elastostatics. When the displacements do not explicitly appear as dependent field variables the compatibility equations must be satisfied to ensure the existence of a displacement field. With $\partial f/\partial X_k$ denoted by $f_{,k}$, etc., the compatibility Eqs. (4.7.14) may be written as

$$E_{ij,km} + E_{km,ij} - E_{ik,jm} - E_{jm,ik} = 0, \tag{8.1.17}$$

representing only six linearly independent equations (and in fact only three functionally independent conditions, as was pointed out in Sec. 4.7). Hooke's law, Eq. (6.2.8), is

$$E_{ij} = -\frac{\nu}{E} T_{pp}\delta_{ij} + \frac{1+\nu}{E} T_{ij}. \tag{8.1.18}$$

We substitute this in Eq. (17) to obtain

$$T_{ij,km} + T_{km,ij} - T_{ik,jm} - T_{jm,ik}$$
$$= \frac{\nu}{1+\nu}[\delta_{ij} T_{pp,km} + \delta_{km} T_{pp,ij} - \delta_{ik} T_{pp,jm} - \delta_{jm} T_{pp,ik}]. \quad (8.1.19)$$

Since only six of these 81 equations are linearly independent, they are equivalent to the six linearly independent equations obtained by putting $k = m$ and summing to obtain

$$T_{ij,kk} + T_{kk,ij} - T_{ik,jk} - T_{jk,ik}$$
$$= \frac{\nu}{1+\nu}[\delta_{ij} T_{pp,kk} + \delta_{kk} T_{pp,ij} - \delta_{ik} T_{pp,jk} - \delta_{jk} T_{pp,ik}] \quad (8.1.20)$$

which is a set of nine equations, but only six ***distinct*** equations because of the symmetry in the free indices i and j. These equations can be simplified by using the equations of equilibrium as follows:

$$-T_{ik,jk} \equiv -(T_{ki,k})_{,j} = (\rho b_i)_{,j} \quad \text{and} \quad -T_{jk,ik} \equiv -(T_{kj,k})_{,i} = (\rho b_j)_{,i}. \quad (8.1.21)$$

Hence, recalling that for any function F we have $F_{,kk} = \nabla^2 F$, we obtain from Eqs. (20)

$$\nabla^2 T_{ij} + \frac{1}{1+\nu} T_{pp,ij} - \frac{\nu}{1+\nu}\delta_{ij}\nabla^2 T_{pp} = -(\rho b_i)_{,j} - (\rho b_j)_{,i}. \quad (8.1.22)$$

Eqs. (22) can be further simplified by getting an expression for $\nabla^2 T_{pp}$ in terms of the body forces as follows. We put $k = i$ and $m = j$ in Eq. (19). The resulting summations give (after we replace T_{ii} by T_{pp} and collect terms)

$$T_{ij,ij} = \frac{1-\nu}{1+\nu}\nabla^2 T_{pp}. \quad (8.1.23)$$

But from the equations of equilibrium we have

$$T_{ij,ij} = -(\rho b_i)_{,i}.$$

Hence Eq. (23) gives

$$\nabla^2 T_{pp} = -\frac{1+\nu}{1-\nu}\nabla\cdot(\rho \mathbf{b}), \quad (8.1.24)$$

and the compatibility equations, Eqs. (22), take the simplified form of Eqs. (25) below. Since Eqs. (25) still represent only three functionally independent conditions, we must add the three equilibrium equations to obtain a complete formulation.

Stress Equations of Elastostatics

Beltrami-Michell Compatibility Equations†

$$\nabla^2 T_{ij} + \frac{1}{1+\nu} T_{pp,ij} = -\frac{\nu}{1-\nu}\delta_{ij}\nabla\cdot(\rho\mathbf{b}) - (\rho b_i)_{,j} - (\rho b_j)_{,i} \tag{8.1.25}$$

Equilibrium Equations

$$T_{ji,j} + \rho b_i = 0 \tag{8.1.26}$$

Equations (25) and (26) are nine distinct equations for only six unknown stresses, but Eqs. (25) represent only three independent conditions. It is not convenient to solve nine equations for only six unknowns, and very few solutions of the full three-dimensional boundary-value problem have been attempted. A more convenient formulation in terms of ***stress functions*** is possible in certain special cases. We have already met the example of the torsion stress function in exercises of Sec. 3.2 and Sec. 4.7. Timoshenko and Goodier (1951) use a stress function for axially symmetric problems (see their Chap. 13). See also Love (1944) for definitions of stress functions in three dimensions. Except for the torsion problem, these problems are more often formulated in terms of displacement potentials, such as the Galerkin vector or the Neuber-Papkovich potentials, to yield solutions of the Navier equations (see Sec. 8.4). The most widely used stress function, the Airy stress function of two-dimensional elasticity, will be discussed in Secs. 8.2 and 8.3.

EXERCISES

1. For the ***elastostatic case*** make an independent derivation of the Navier equation and traction boundary condition, assuming again small displacements and small displacement gradients, but using this time the ***Eulerian formulation*** in ***spatial coordinates.***
2. Discuss the extension of the results of Ex. 1 to the dynamic case.
3. (a) Show that when Poisson's ratio has the value $\nu = \frac{1}{2}$, the Hooke's law of Eq. (8.1.18) may be written as $T_{ij} = (2E/3)E_{ij} + \frac{1}{3} T_{pp}\delta_{ij}$, and that when this is

†Equations (25) were obtained by Michell in 1899. Beltrami (1892) had earlier given the equations for the case of no body forces.

substituted into the equilibrium equations we get the three equations $Eu_{j,ii} + 3\rho b_j + T_{pp,j} = 0$, which together with the equation $u_{i,i} = 0$ (also a consequence of $\nu = \frac{1}{2}$) form a set of four equations for the three displacement components u_i and the first stress invariant T_{pp} in the case of equilibrium of an incompressible elastic material.

(b) What values must be assigned to the Lamé elastic constant λ and to the bulk modulus K when $\nu = \frac{1}{2}$, and how is μ related to E? Why can you not obtain the result of Part (a) by setting $\nabla \cdot \mathbf{u} = 0$ in the equilibrium version of Eq. (8.1.8)?

(c) Write the traction components $t_i = T_{ji} n_j$ in terms of the displacements and T_{pp} for the case that $\nu = \frac{1}{2}$.

4. Carry out the details of the derivations: (a) from Eq. (8.1.20) to Eq. (22), (b) from Eq. (22) to Eq. (25).

5. (a) Write the form assumed by the Beltrami-Michell compatibility equations, Eqs. (8.1.25), when the body force per unit volume is derivable from a harmonic potential function F, so that $\rho\mathbf{b} = -\nabla F$ and $\nabla^2 F = 0$.

   (b) Show from Eq. (8.1.24) that in the case described in Part (a) the first stress invariant T_{pp} is a harmonic function, and then show from the results of Part (a) that the stress components are biharmonic functions, that is, functions satisfying $\nabla^4 T_{ij} = 0$.

6. Write out, without indicial notation, two typical Beltrami-Michell equations: for T_{xx} and for T_{xy}.

7. Obtain the solution to Ex. 7 of Sec. 4.7 by using the Beltrami-Michell equations.

8. A prismatic bar of length L and uniform density ρ hangs in equilibrium in a vertical position subject only to its own weight (acting in the negative z-direction) and a uniformly distributed $T_{zz} = \sigma$ at its top end $z = L$, and constrained so that the displacement and rotation of the element at (0, 0, L) are zero. Assuming that all stresses except T_{zz} vanish, integrate Eqs. (8.1.3) to (5) to determine the stress, strain, and displacement distributions.

8.2 Plane elasticity in rectangular coordinates

In ***plane deformation***, the assumptions $u_z = 0$ and u_x and u_y independent of z lead to only three independent strain components ϵ_x, ϵ_y, and $\epsilon_{xy} = \frac{1}{2}\gamma_{xy}$, which are independent of z, a state of ***plane strain*** parallel to the xy-plane. The isotropic Hooke's law, Eq. (6.2.4), reduces to

$$\left.\begin{aligned} \sigma_x &= \lambda e + 2G\epsilon_x \\ \sigma_y &= \lambda e + 2G\epsilon_y \\ \tau_{xy} &= 2G\epsilon_{xy} \end{aligned}\right\} \tag{8.2.1}$$

with, in addition,

$$\sigma_z = \lambda e = \nu(\sigma_x + \sigma_y), \tag{8.2.2}$$

where

$$e = \epsilon_x + \epsilon_y, \qquad G = \frac{E}{2(1+\nu)}, \quad \text{and} \quad \lambda = \frac{E\nu}{(1+\nu)(1-2\nu)} \tag{8.2.3}$$

or, inversely,

$$\left.\begin{aligned} \epsilon_x &= \frac{1+\nu}{E}[(1-\nu)\sigma_x - \nu\sigma_y] \\ \epsilon_y &= \frac{1+\nu}{E}[-\nu\sigma_x + (1-\nu)\sigma_y] \\ \epsilon_{xy} &= \frac{\tau_{xy}}{2G}. \end{aligned}\right\} \tag{8.2.4}$$

To these must be added two equations of motion

$$\left.\begin{aligned} \frac{\partial\sigma_x}{\partial x} + \frac{\partial\tau_{xy}}{\partial y} + \rho b_x &= \rho\frac{\partial^2 u_x}{\partial t^2} \\ \frac{\partial\tau_{xy}}{\partial x} + \frac{\partial\sigma_y}{\partial y} + \rho b_y &= \rho\frac{\partial^2 u_y}{\partial t^2} \end{aligned}\right\} \tag{8.2.5}$$

and one compatibility equation

$$\frac{\partial^2\epsilon_x}{\partial y^2} + \frac{\partial^2\epsilon_y}{\partial x^2} = 2\frac{\partial^2\epsilon_{xy}}{\partial x\,\partial y}, \tag{8.2.6}$$

if for small displacements we ignore the difference between the material coordinates X, Y and the spatial coordinates x, y. The third equation of motion is satisfied for $u_z \equiv 0$ and σ_z independent of z by Eq. (2).

In elastostatics, the acceleration terms drop out of Eqs. (5), and the Eqs. (1), (5), (6) are then six equations for the six independent unknowns

$$\sigma_x,\ \sigma_y,\ \tau_{xy},\ \epsilon_x,\ \epsilon_y,\ \epsilon_{xy}$$

as functions of x and y, which may presumably be solved when suitable boundary conditions and body forces are given. We can state the compatibility equations in terms of stresses by substituting Eqs. (4) into Eq. (6) and eliminating the shear stress by using the equation

$$2\frac{\partial^2\tau_{xy}}{\partial x\,\partial y} = -\frac{\partial^2\sigma_x}{\partial x^2} - \frac{\partial^2\sigma_y}{\partial y^2} - \frac{\partial(\rho b_x)}{\partial x} - \frac{\partial(\rho b_y)}{\partial y}$$

obtained by differentiating the first equilibrium equation with respect to x and the second with respect to y and adding the two. The resulting compatibility equation for stresses in plane strain is

$$\nabla_1^2 (\sigma_x + \sigma_y) = -\frac{1}{1-\nu}\left[\frac{\partial (\rho b_x)}{\partial x} + \frac{\partial (\rho b_y)}{\partial y}\right]$$

(plane strain) (8.2.7)

where ∇_1^2 is the two-dimensional Laplace operator $(\partial^2/\partial x^2) + (\partial^2/\partial y^2)$. An example of suitable boundary conditions would be to prescribe at each point of the boundary curve in the xy-plane either the normal traction component or the normal displacement and either the tangential traction component or the tangential displacement component. If tractions are prescribed all the way around, they must form an equilibrium system of forces with the given body forces. Other boundary conditions are possible, e.g., the normal component of traction might be proportional to the normal displacement at a spring support, but we will not consider here any of these more general support conditions. Prescribed displacements should be continuous, though normal traction components may exhibit finite jump discontinuities.

Plane deformation is approximately realized in a long cylindrical or prismatic body if on the lateral surfaces (parallel to the z-axis) the boundary conditions involve only displacements or tractions parallel to the xy-plane, and if these imposed boundary conditions are uniform along the length, i. e., they are independent of z. On the ends of the cylinder or prism, the appropriate boundary conditions are $u_z = 0$ and $\tau_{zx} = \tau_{zy} = 0$. But Eq. (2) shows that the normal traction on the ends, $t_z = \sigma_z n_z = \pm\sigma_z$, is not zero, except in special cases where $\sigma_x + \sigma_y = 0$. If the plane-strain solution gives $\sigma_x + \sigma_y < 0$, the required compressive loading on the ends can be imagined provided as the reaction of a frictionless rigid wall. Alternatively, the actual strain can be sought as the superposition of a plane-strain solution, and another solution, since the boundary-value problem is linear. For example, if the lateral boundary conditions are all traction boundary conditions, while the ends are free, we could find the plane-strain solution for these lateral boundary conditions. Then, if in this plane-strain solution $\sigma_x + \sigma_y = f(x, y)$, requiring $\sigma_z = \nu f(x, y)$ on the ends, we seek to find a second solution with $\sigma_z = -\nu f(x, y)$, $\tau_{zx} = \tau_{zy} = 0$ on the ends and with traction-free lateral boundaries. This second solution is usually difficult. However, for a cylinder or prism that is long in comparison to its cross-sectional dimensions, a good approximation is furnished (except near the ends) by a superposed, uniform uniaxial stress equal to the average value of the required σ_z over the cross-section, plus a bending solution corresponding to a linear distribution of σ_z. This is an example of Saint-Venant's principle.

Saint-Venant's Principle

In elastostatics, if the boundary tractions on a part S_1 of the boundary S are replaced by a statically equivalent traction distribution, the effects on the stress distribution in the body are negligible at points whose distance from S_1 is large compared to the maximum distance between points of S_1.

This principle is of great importance in applied elasticity, where it is frequently invoked to justify solutions in long slender structural members, where the end traction boundary conditions are satisfied only in an average sense, so that the correct stress resultants act on the ends. In such solutions, the actual stress distribution near the ends may differ considerably from the calculated stress distribution. There are two reasons for using such approximate solutions in applied elasticity. In the first place it is not usually known in detail just how the loads applied at the end of a structural member will be distributed. Also, simple solutions, obtained by inverse methods, can sometimes be applied, if not all the boundary conditions are exactly satisfied, while the exact solutions may require elaborate calculations.

B. de Saint-Venant enunciated his principle in 1855. It has been widely accepted on empirical grounds, and a precisely stated version of it was proved in 1954 by Sternberg. See the discussion in Fung (1965), Chap. 10.

Particular solution for body forces. The linearity can also be used to construct the solution in two parts: $\sigma_{ij} = \sigma_{ij}^H + \sigma_{ij}^P$, $\epsilon_{ij} = \epsilon_{ij}^H + \epsilon_{ij}^P$. Here the ***particular solution*** σ_{ij}^P, ϵ_{ij}^P satisfies the given equations with given body-force distributions but not the boundary conditions, while the distribution σ_{ij}^H, ϵ_{ij}^H satisfies the ***homogeneous differential equations*** (with no body force) and suitably modified boundary conditions, so that the sum of the two solutions satisfies the original boundary conditions. (The ***homogeneous solution*** is sometimes called the ***complementary solution***.)

When the ***body force is simply the weight***, say $b_x = 0$, $b_y = -g$, then a possible particular solution is

$$\sigma_y^P = \rho g y - C, \qquad \sigma_x^P = \tau_{xy}^P = 0, \tag{8.2.8}$$

where $-C$ is the value of σ_y^P at $y = 0$, ρ is the density, and g is the acceleration of gravity.

When the body force is inertia force due to rotation about the z-axis with constant angular velocity ω,

$$b_x = \omega^2 x \qquad b_y = \omega^2 y, \tag{8.2.9}$$

a suitable particular solution is

$$\sigma_x^P = -\frac{\rho\omega^2}{2}\left(x^2 + \frac{\nu}{1-\nu}y^2\right) \qquad \sigma_y^P = -\frac{\rho\omega^2}{2}\left(y^2 + \frac{\nu}{1-\nu}x^2\right) \qquad \tau_{xy}^P = 0. \tag{8.2.10}$$

These two stress distributions satisfy the equilibrium differential equations for two common body-force distributions; and the plane-strain state corresponding to each of these two body-force distributions according to Hooke's law satisfies the compatibility equations.

Since a particular solution may usually be found for the body forces, we will in the following suppose either that this has already been done and the boundary conditions appropriately modified, or else that the body forces are negligible, and that we have still to solve only the homogeneous equations with no body forces. We will drop the superscript H, but it should be remembered that we are dealing with the homogeneous solution only.

Airy stress function. For plane strain with no body forces, the equilibrium equations are identically satisfied if the stresses are related to a scalar function $\phi(x, y)$, called Airy's stress function, by the equations†

$$\sigma_x = \frac{\partial^2\phi}{\partial y^2}, \qquad \sigma_y = \frac{\partial^2\phi}{\partial x^2}, \qquad \tau_{xy} = -\frac{\partial^2\phi}{\partial x\,\partial y}. \tag{8.2.11}$$

The compatibility equation then becomes the biharmonic equation

$$\nabla_1^2(\nabla_1^2)\phi = 0 \tag{8.2.12a}$$

or

$$\nabla_1^4\phi = 0, \tag{8.2.12b}$$

which written out in rectangular coordinates is

$$\frac{\partial^4\phi}{\partial x^4} + 2\frac{\partial^4\phi}{\partial x^2\,\partial y^2} + \frac{\partial^4\phi}{\partial y^4} = 0. \tag{8.2.12c}$$

If a function ϕ can be found satisfying the biharmonic equation in the interior of the region and such that the stresses computed from ϕ lead to boundary tractions or displacements satisfying the prescribed boundary conditions, then that function ϕ furnishes the solution to the plane-strain boundary-value prob-

†The stress function of Eqs. (11) was introduced by G. B. Airy in 1862.

lem. The Airy stress function is especially well-suited for use when all boundary conditions are for tractions.

Boundary conditions for Airy stress function when boundary tractions are prescribed. Consider the simply connected region shown in Fig. 8.1 representing a cross-section of a body long in the z-direction and assumed to

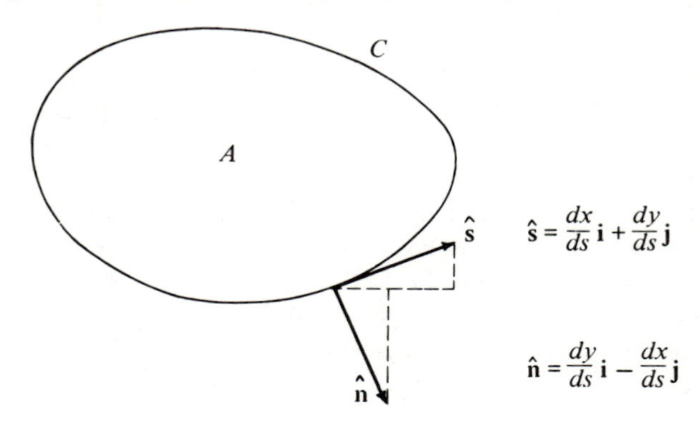

Fig. 8.1 Region for plane strain solution.

be in a state of plane strain with boundary tractions $\mathbf{t}$ given at each point of the lateral boundary. (There may be a finite number of corners or jump discontinuities in $\hat{\mathbf{n}}$ and $\hat{\mathbf{s}}$ as functions of arc length s, and there may also be finite jump discontinuities in $\mathbf{t}$, but the functions $\hat{\mathbf{n}}$, $\hat{\mathbf{s}}$, and $\mathbf{t}$ are assumed to be piecewise smooth.)

Since

$$t_i = \sigma_{ji} n_j,$$

we have

$$t_x = \sigma_x n_x + \tau_{xy} n_y$$
$$t_y = \tau_{xy} n_x + \sigma_y n_y$$

or

$$\left.\begin{aligned} t_x &= \frac{\partial^2 \phi}{\partial y^2}\frac{dy}{ds} + \frac{\partial^2 \phi}{\partial x\,\partial y}\frac{dx}{ds} = \frac{d}{ds}\left(\frac{\partial \phi}{\partial y}\right) \\ t_y &= -\frac{\partial^2 \phi}{\partial x\,\partial y}\frac{dy}{ds} - \frac{\partial^2 \phi}{\partial x^2}\frac{dx}{ds} = -\frac{d}{ds}\left(\frac{\partial \phi}{\partial x}\right). \end{aligned}\right\} \tag{8.2.13}$$

Hence, integrating along the boundary, we obtain

$$\frac{\partial \phi}{\partial x} = -\int_C t_y\, ds + C_1 \qquad \frac{\partial \phi}{\partial y} = \int_C t_x\, ds + C_2. \tag{8.2.14}$$

From the results of Eqs. (14), we can calculate

$$\frac{d\phi}{ds} = \frac{\partial\phi}{\partial x}\frac{dx}{ds} + \frac{\partial\phi}{\partial y}\frac{dy}{ds}$$

$$\frac{d\phi}{dn} = \frac{\partial\phi}{\partial x}\frac{dy}{ds} - \frac{\partial\phi}{\partial y}\frac{dx}{ds} \tag{8.2.15}$$

and integrate the first of these two to obtain

$$\phi = \int_C \frac{d\phi}{ds}\,ds + C_3. \tag{8.2.16}$$

We now know ϕ and $d\phi/dn$ at each point of the boundary, and these are precisely the mathematical boundary conditions needed to formulate a boundary-value problem for the solution of the biharmonic equation in the interior. The boundary conditions involve three arbitrary constants, whose choice does not affect the stress distribution, but which may be selected in such a way as to simplify the calculations. For example, if the stress distribution is known to be symmetric with respect to the y-axis, and the origin of s is on the y-axis, we would choose $C_1 = 0$ so that ϕ could be an even function of x with $\partial\phi/\partial x = 0$ on the y-axis. Symmetry with respect to the x-axis will dictate the choice of C_2 so that $\partial\phi/\partial y$ can vanish on the x-axis. When there is no symmetry the choice is completely arbitrary, and C_3 is completely arbitrary in any case.

Plane stress. If the only nonzero stresses in a body are σ_x, σ_y, τ_{xy} (plus $\tau_{yx} = \tau_{xy}$), the isotropic Hooke's law in the form of Eqs. (6.2.9) becomes

$$\left.\begin{aligned} \epsilon_x &= \frac{1}{E}(\sigma_x - \nu\sigma_y) \qquad \epsilon_y = \frac{1}{E}(\sigma_y - \nu\sigma_x) \\ \epsilon_{xy} &\equiv \frac{1}{2}\gamma_{xy} = \frac{\tau_{xy}}{2G} \end{aligned}\right\} \tag{8.2.17}$$

plus

$$\epsilon_z = -\frac{\nu}{E}(\sigma_x + \sigma_y) \tag{8.2.18}$$

or, inversely,

$$\left.\begin{aligned} \sigma_x &= \frac{E}{1-\nu^2}(\epsilon_x + \nu\epsilon_y) \qquad \sigma_y = \frac{E}{1-\nu^2}(\epsilon_y + \nu\epsilon_x) \\ \tau_{xy} &= 2G\,\epsilon_{xy}. \end{aligned}\right\} \tag{8.2.19}$$

The equilibrium equations are the same as for plane strain, but the first compatibility equation, in terms of stress, becomes

$$\nabla_1^2(\sigma_x + \sigma_y) = -(1 + \nu)\left[\frac{\partial (\rho b_x)}{\partial x} + \frac{\partial (\rho b_y)}{\partial y}\right]$$
(plane stress). (8.2.20)

The mathematical formulation of plane stress problems (ignoring all but the first of the compatibility equations) parallels that of plane strain. In fact ***all of the plane stress equations reduce to the plane strain equations if in the plane stress equations we replace E by $E/(1 - \nu^2)$ and ν by $\nu/(1 - \nu)$***. This leaves $G = E/2\,(1 + \nu)$ unchanged. Inversely, replace λ in the plane strain equations by $2\lambda G/(\lambda + 2G)$ to get the plane stress equations. The stresses are related to the stress function ϕ in exactly the same way as in plane strain and the stress function (in the absence of body forces) satisfies the same biharmonic differential equation. It is only when strains and displacements are considered and in the particular solution for body forces that the difference appears; it may appear in the boundary conditions of the homogeneous problem, as a consequence of displacement boundary conditions or because of the modifications in the boundary conditions introduced by the particular solution.

Equation (20) is the first compatibility equation in terms of the plane stresses, but the other compatibility equations, Eqs. (4.7.5), cannot be satisfied by assuming that the problem is truly two-dimensional, i.e., assuming σ_x, σ_y, τ_{xy} independent of z, because that would require the strains also to be independent of z, by Hooke's law, and therefore Eqs. (4.7.5) would require in addition to Eq. (20)

$$\frac{\partial^2 \epsilon_z}{\partial y^2} = 0 \qquad \frac{\partial^2 \epsilon_z}{\partial x^2} = 0 \qquad \frac{\partial^2 \epsilon_z}{\partial x\, \partial y} = 0, \tag{8.2.21}$$

which, by Eq. (18), can only be satisfied if $\sigma_x + \sigma_y$ is a linear function of x and y. Except for a few rather trivial cases, the compatibility conditions cannot be satisfied by plane stresses independent of z. For example, the particular solution proposed for inertia body forces [Eqs. (10) modified for plane stress as in Ex. 2(b) at the end of this section] does not meet this requirement.

It can be shown† that when the z-dependence is symmetric with respect to the xy-plane (e.g., the middle plane of a thin plate), and there are no body forces, the compatibility conditions can all be satisfied by a stress function $\phi(x, y, z)$ of the form

$$\phi\,(x, y, z) = \phi_0\,(x, y) - \frac{1}{2}\frac{\nu z^2}{1 + \nu}\nabla_1^2\phi_0, \tag{8.2.22}$$

if $\phi_0\,(x, y)$ is biharmonic, i.e., if

$$\nabla_1^4\phi_0 = 0. \tag{8.2.23}$$

†See, for example, Timoshenko and Goodier (1951), Sec. 84.

The dependence of ϕ on z appears only as z^2 in the last term of Eq. (22), and, if the plate is sufficiently thin, this term may be negligible compared with the first one. A rigorous demonstration that it is in fact negligible is difficult when the solution is unknown, since the function multiplying z^2 may become large or make a large contribution to the stresses calculated from ϕ. Nevertheless, the literature is filled with plane stress solutions assumed independent of z, and many of these solutions, for thin plates under edge loads parallel to the plane of the plate, have proved to be extremely useful. It should always be remembered, however, that such solutions are approximate and should be subjected to experimental verifications when the accuracy of the approximation cannot be otherwise determined.

Generalized plane stress. The approximate two-dimensional plane stress distributions may more rigorously be regarded as average stresses, averaged through the thickness. This average stress distribution is sometimes called ***generalized plane stress.*** Each component is here a stress resultant (obtained by integrating the stress through the thickness), divided by the thickness, e.g.,

$$\bar{\sigma}_x = \frac{1}{h}\int_{-h/2}^{h/2} \sigma_x \, dz$$

in a thin plate of uniform thickness h. Although σ_x may be a function of x, y and z, the average $\bar{\sigma}_x$ is a function of x and y only. We assume all loads and also the solution to be symmetric with respect to the middle plane, and assume that τ_{zx} and τ_{zy} vanish on the surfaces $z = \pm h/2$, while σ_z vanishes everywhere. Then if we integrate all the equations of three-dimensional elasticity through the thickness and divide by h, we find that the plane stress equations are in fact rigorously satisfied by the averaged quantities $\bar{\sigma}_{ij}$, $\bar{\epsilon}_{ij}$, and $\bar{u}_i$. The boundary conditions must also apply to averages, e.g., $\bar{\sigma}_{ji}n_i = \bar{t}_i$. Now, dropping off the bars, we may simply interpret all the stresses, strains, and displacements in the previous plane stress formulation as averages. It is to be expected that in practical problems of thin plates under edge loads parallel to the plate, we know at best only the thickness average of the boundary tractions in any case. Photoelastic measurements made by polarized light transmitted through a transparent thin-plate model also respond to some kind of an average through the thickness, but photostress coatings and strain gages mounted on the surface measure surface strains rather than average strains, and this fact may lead to disagreement between the calculated average strains and the measured surface strains, especially in moderately thick plates.

Strain-energy function. In an isotropic material, the strain-energy density function (or elastic-potential function) W of Sec. 6.2. has the form

$$W = \tfrac{1}{2}\lambda e^2 + G\epsilon_{ij}\epsilon_{ij} \tag{8.2.24}$$

giving the stored strain energy W per unit volume. In ***plane strain,*** this reduces to

$$W = \tfrac{1}{2}\lambda e^2 + G\,(\epsilon_x^2 + \epsilon_y^2 + 2\epsilon_{xy}^2), \tag{8.2.25}$$

where

$$e = \epsilon_x + \epsilon_y.$$

The total strain energy in a slice of unit thickness is obtained by integrating W over the area of a cross-section of the body perpendicular to the z-axis.

By using Hooke's law, Eq. (24) can be put in the form of Eq. (6.2.21):

$$W = \tfrac{1}{2}\,\sigma_{ij}\,\epsilon_{ij}. \tag{8.2.26}$$

In plane stress or plane strain, this becomes

$$W = \tfrac{1}{2}(\sigma_x\epsilon_x + \sigma_y\epsilon_y + 2\tau_{xy}\epsilon_{xy}). \tag{8.2.27}$$

By using Hooke's law again we can convert Eq. (27) into an expression for W in terms of stresses. However, two different results are obtained, since the plane stress Hooke's law differs from the plane strain Hooke's law.

Plane Strain:

$$W = \frac{1-\nu^2}{2E}\,(\sigma_x^2 + \sigma_y^2) - \frac{\nu(1+\nu)}{E}\,\sigma_x\sigma_y + \frac{1}{2G}\,\tau_{xy}^2 \tag{8.2.28}$$

Plane Stress:

$$W = \frac{1}{2E}(\sigma_x^2 + \sigma_y^2) - \frac{\nu}{E}\,\sigma_x\sigma_y + \frac{1}{2G}\tau_{xy}^2 \tag{8.2.29}$$

Observe again that the plane strain form can be obtained by substituting $E/(1-\nu^2)$ for E in the plane stress form and $\nu/(1-\nu)$ for ν. Finally we can write the W in terms of strains for the plane stress case, either by using the plane stress Hooke's law in Eq. (27) or by replacing λ in Eq. (24) by $2\lambda G/(\lambda + 2G)$. The result is

Plane Stress:

$$W = G\left[\frac{1}{1-\nu}(\epsilon_x^2 + \epsilon_y^2 + 2\nu\,\epsilon_x\epsilon_y) + 2\epsilon_{xy}^2\right]. \tag{8.2.30}$$

Navier's displacement equations of motion may be used in two-dimensional problems, instead of the Airy function formulation. For ***plane strain,*** the equations and traction boundary conditions have exactly the same form as in the three-dimensional case, but now we have $u_z = 0$ while u_x and u_y are independent of z. Since $\lambda = 2\nu G/(1 - 2\nu)$ and $\mu = G$, the two equations for plane strain take the following form, which can also be obtained by substituting Eqs. (1) and (2) into the equations of motion, Eqs. (5).

Plane Strain Navier Equations

$$\left.\begin{aligned} G\nabla_1^2 u_x + \frac{G}{1-2\nu}\frac{\partial e}{\partial x} + \rho b_x &= \rho\frac{\partial^2 u_x}{\partial t^2} \\ G\nabla_1^2 u_y + \frac{G}{1-2\nu}\frac{\partial e}{\partial y} + \rho b_y &= \rho\frac{\partial^2 u_y}{\partial t^2} \end{aligned}\right\} \tag{8.2.31}$$

where

$$e = \frac{\partial u_x}{\partial x} + \frac{\partial u_y}{\partial y}$$

Traction Boundary Conditions in Terms of u_x, u_y (Plane Strain)

$$\left.\begin{aligned} t_x &= \frac{2G}{1-2\nu}\left[(1-\nu)\frac{\partial u_x}{\partial x} + \nu\frac{\partial u_y}{\partial y}\right]n_x + G\left(\frac{\partial u_x}{\partial y} + \frac{\partial u_y}{\partial x}\right)n_y \\ t_y &= G\left[\frac{\partial u_x}{\partial y} + \frac{\partial u_y}{\partial x}\right]n_x + \frac{2G}{1-2\nu}\left[\nu\frac{\partial u_x}{\partial x} + (1-\nu)\frac{\partial u_y}{\partial y}\right]n_y \end{aligned}\right\} \tag{8.2.32}$$

$$t_z = \frac{2\nu G}{1-2\nu}\left(\frac{\partial u_x}{\partial x} + \frac{\partial u_y}{\partial y}\right)n_z \tag{8.2.33}$$

For elastostatic problems, the accelerations on the right side of Eqs. (31) are zero. Equation (33) gives the normal traction distribution required on the ends of a cylinder to maintain it in a state of plane strain when the lateral-surface boundary conditions are independent of z.

For ***plane stress*** the Navier equations have a slightly different form, which can be obtained by substituting the plane stress Hooke's law, Eqs. (19), into the equations of motion.

Plane Stress Navier Equations

$$\left.\begin{aligned} G\nabla_1^2 u_x + G\frac{1+\nu}{1-\nu}\frac{\partial}{\partial x}\left(\frac{\partial u_x}{\partial x} + \frac{\partial u_y}{\partial y}\right) + \rho b_x &= \rho\frac{\partial^2 u_x}{\partial t^2} \\ G\nabla_1^2 u_y + G\frac{1+\nu}{1-\nu}\frac{\partial}{\partial y}\left(\frac{\partial u_x}{\partial x} + \frac{\partial u_y}{\partial y}\right) + \rho b_y &= \rho\frac{\partial^2 u_y}{\partial t^2} \end{aligned}\right\} \tag{8.2.34}$$

Traction Boundary Conditions in Terms of u_x, u_y (Plane Stress)

$$\left.\begin{aligned} t_x &= \frac{2G}{1-\nu}\left(\frac{\partial u_x}{\partial x} + \nu\frac{\partial u_y}{\partial y}\right)n_x + G\left(\frac{\partial u_x}{\partial y} + \frac{\partial u_y}{\partial x}\right)n_y \\ t_y &= G\left(\frac{\partial u_x}{\partial y} + \frac{\partial u_y}{\partial x}\right)n_x + \frac{2G}{1-\nu}\left(\nu\frac{\partial u_x}{\partial x} + \frac{\partial u_y}{\partial y}\right)n_y \end{aligned}\right\} \tag{8.2.35}$$

The reader can check that (as suggested in the discussion following Eq. 20) replacing ν by $\nu/(1-\nu)$ in the plane stress Eqs. (34) and (35) reduces them to the corresponding plane strain Eqs. (31) and (32).

Solutions. Practical solutions of the two-dimensional boundary-value problems in simply connected regions can be accomplished by numerical integration of the finite-difference approximations to the differential equations. The elegant and powerful methods of complex-function theory are also useful. [See Muskhelishvili (1953), Sokolnikoff (1956), Green and Zerna (1954), Sneddon and Berry (1958), and Milne-Thomson (1960).] These methods have found important applications to the solutions for stress concentrations around holes in an infinite region which can be mapped conformally onto the exterior of a circle; see Savin (1961). Variational methods have also been developed; see Sokolnikoff (1956) and Fung (1965). A few special solutions have been found by inverse methods, assuming a simple solution (e.g., a polynomial) for the Airy stress function, which satisfies the biharmonic equation, and trying to find a region for which the function satisfies reasonable boundary conditions; see Timoshenko and Goodier (1951), Wang (1953), and Biezeno and Grammel (1955). An example of these special solutions will be given below, in which Saint-Venant's principle is invoked to justify failure to satisfy all the boundary conditions.

Polynomial solutions. Inverse method. Any polynomial of degree three or less in x and y satisfies the biharmonic equation, Eq. (12), identically and is therefore a possible stress function. Since the stresses are given by second derivatives, only terms of second degree or higher are significant. A second-degree polynomial for ϕ gives a uniform state of stress by Eqs. (11), while a third-degree polynomial gives a stress field linear in x and y. Polynomials of higher degree may be used also, but then it is necessary to choose the coefficients so that the biharmonic equation is satisfied. A systematic way of doing this was proposed by Neou (1957). If

$$\phi = \sum_{m=0}^{\infty}\sum_{n=0}^{\infty} C_{mn}x^m y^n, \tag{8.2.36a}$$

then, by Eqs. (11),

$$
\left.\begin{aligned}
\sigma_x &= \sum_{m=0}^{\infty} \sum_{n=2}^{\infty} n(n-1)\, C_{mn}\, x^m y^{n-2} \\
\sigma_y &= \sum_{m=2}^{\infty} \sum_{n=0}^{\infty} m(m-1)\, C_{mn}\, x^{m-2} y^n \\
\tau_{xy} &= -\sum_{m=1}^{\infty} \sum_{n=1}^{\infty} mn\, C_{mn}\, x^{m-1} y^{n-1}.
\end{aligned}\right\} \qquad (8.2.36b)
$$

The compatibility Eq. (12), after substitution according to Eq. (36a) and regrouping, takes the form

$$
\sum_{m=2}^{\infty} \sum_{n=2}^{\infty} [(m+2)(m+1)m(m-1)\, C_{m+2,n-2} + 2m(m-1)n(n-1)\, C_{mn} \\
+ (n+2)(n+1)n(n-1) C_{m-2,n+2}]\, x^{m-2} y^{n-2} = 0. \qquad (8.2.37)
$$

Since the equation must be identically satisfied by the proposed solution, the coefficient in brackets must vanish. This furnishes relationships among the coefficients. If an infinite power series were used, as indicated in Eqs. (36) and (37), it would be necessary to establish its convergence after the coefficients had been determined. And it would be of interest to establish how general a boundary condition on a rectangular region could be represented by the resulting power series, if it turned out to be convergent. In practice, one usually stops with a polynomial of fairly low degree and is happy to discover that there are some special boundary conditions of practical interest that can be satisfied in this way.

The recursion relation

$$
(m+2)(m+1)m(m-1)C_{m+2,n-2} + 2m(m-1)n(n-1)C_{mn} \\
+ (n+2)(n+1)n(n-1)C_{m-2,n+2} = 0 \qquad (8.2.38)
$$

obtained from Eq. (37) establishes relationships among groups of three alternate coefficients in the diagonals running from lower left to upper right in the array (39) of coefficients C_{mn}.

$$
\begin{bmatrix}
0 & 0 & C_{02} & C_{03} & \boxed{C_{04}} & C_{05} & \underset{\sim}{\underline{C_{06}}} & \cdots \\
0 & C_{11} & C_{12} & C_{13} & C_{14} & \enclose{circle}{C_{15}} & \cdots & \\
C_{20} & C_{21} & \boxed{C_{22}} & C_{23} & \underset{\sim}{\underline{C_{24}}} & \cdots & & \\
C_{30} & C_{31} & C_{32} & \enclose{circle}{C_{33}} & \cdots & & & \\
\boxed{C_{40}} & C_{41} & \underset{\sim}{\underline{C_{42}}} & \cdots & & & & \\
C_{50} & \enclose{circle}{C_{51}} & \cdots & & & & & \\
\underline{C_{60}} & \cdots & & & & & &
\end{bmatrix} \qquad (8.2.39)
$$

For example, in the diagonal from C_{40} to C_{04} containing coefficients of fourth-degree terms, we have one relation (for $m = n = 2$)

$$(4)(3)(2)(1)C_{40} + (2)(2)(1)(2)(1)C_{22} + (4)(3)(2)(1)C_{04} = 0, \qquad (8.2.40)$$

while in the diagonal from C_{60} to C_{06} there are three such equations, each relating three elements in the array, which are circled, underlined, or identified by a wavy underline. There will be $m - 3$ such relationships for the subsequent diagonals beginning with C_{m0}, leaving four disposable coefficients on the diagonal containing coefficients of terms of degree m, which may be chosen in an attempt to match boundary conditions.

For example, consider the homogeneous fourth-degree polynomial

$$\phi_4 = C_{40}x^4 + C_{31}x^3y + C_{22}x^2y^2 + C_{13}xy^3 + C_{04}y^4 \qquad (8.2.41)$$

with

$$3C_{40} + C_{22} + 3C_{04} = 0 \qquad (8.2.42)$$

by Eq. (40). The stresses are

$$\left.\begin{aligned} \sigma_x &= 2C_{22}x^2 + 6C_{13}xy + 12C_{04}y^2 \\ \sigma_y &= 12C_{40}x^2 + 6C_{31}xy + 2C_{22}y^2 \\ \tau_{xy} &= -\,3C_{31}x^2 - 4C_{22}xy - 3C_{13}y^2. \end{aligned}\right\} \qquad (8.2.43)$$

These can be used for the end-loaded cantilever of Fig. 8.2 with rectangular cross-section of thickness b in the z-direction. If all coefficients except C_{13} are

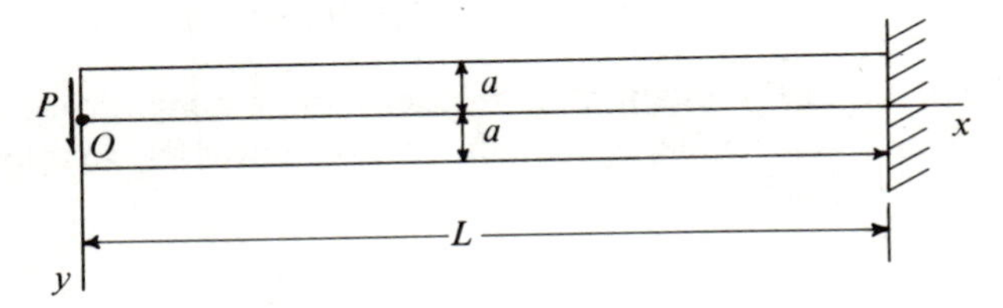

Fig. 8.2 End-loaded cantilever.

taken to be zero in Eq. (41), the compatibiiity condition (42) is satisfied and the stress distribution is

$$\sigma_x = 6C_{13}xy \qquad \sigma_y = 0 \qquad \tau_{xy} = -\,3C_{13}y^2.$$

But this gives not only a parabolic shear traction on the loaded end $x = 0$, but also uniform shear tractions $\tau_{yx} = -\,3C_{13}a^2$ on top and bottom. These can be removed by superposing uniform shear stress $\tau_{xy} = +3C_{13}a^2$ corresponding to $\phi_2 = -3C_{13}a^2xy$. Note that with $C_{20} = C_{02} = 0$ and $C_{11} = -3C_{13}a^2$, ϕ_2 does not give any additional contribution to σ_x or σ_y. The solution is then

$$\sigma_x = 6C_{13}xy \qquad \sigma_y = 0 \qquad \tau_{xy} = 3C_{13}(a^2 - y^2).$$

The constant C_{13} is determined by requiring

$$P = b\int_{-a}^{a} -\tau_{xy}\,dy = -3bC_{13}\int_{-a}^{a}(a^2 - y^2)\,dy,$$

whence

$$C_{13} = \frac{-P}{4a^3b}$$

and the solution is

$$\left.\begin{aligned} \phi &= +\frac{3P}{4ab}xy - \frac{P}{4a^3b}xy^3 \\ \sigma_x = -\frac{3P}{2a^3b}xy \qquad \tau_{xy} &= -\frac{3P}{4a^3b}(a^2 - y^2) \qquad \sigma_y = 0. \end{aligned}\right\} \tag{8.2.44}$$

The second moment of area of the cross-section is $I = 2a^3b/3$. Hence the plane stress solution agrees with the elementary beam-theory solution

$$\sigma_x = -\frac{P}{I}xy \qquad \tau_{xy} = -\frac{P}{2I}(a^2 - y^2) \qquad \sigma_y = 0. \tag{8.2.45}$$

The negative sign appears because of the sign convention for shear stresses. A positive τ_{xy} would act in the negative y-direction on the negative side of the loaded end. The plane stress solution satisfies the loaded-end boundary condition exactly only if the loads are actually applied in the form of a parabolic distribution of shear traction. Since this is not likely to be the way the load is applied, the boundary condition satisfied is only statically equivalent to the actual loading. According to Saint-Venant's principle, replacing the actual loads by a statically equivalent traction distribution on the end does not significantly affect the stress distribution in a long beam, except near the end. Note also that the validity of the solution near the support depends on the reactions at the built-in end being distributed in the manner indicated by the solution, which is not likely to be the case.

Some other biharmonic functions that may be useful in inverse-method solutions may be obtained as follows. If $f(x, y)$ is a plane harmonic function (satisfying $\nabla_1^2 f = 0$), then the following are biharmonic functions ϕ (satisfying $\nabla_1^4 \phi = 0$:

$$f(x, y), \qquad (x^2 + y^2)f(x, y),$$
$$xf(x, y), \qquad yf(x, y). \tag{8.2.46}$$

For example, since the functions in (47) are harmonic for arbitrary α

$$e^{\alpha x}\sin \alpha y, \qquad e^{\alpha x}\cos \alpha y, \qquad e^{\alpha y}\sin \alpha x, \qquad e^{\alpha y}\cos \alpha x, \tag{8.2.47}$$

they and any products of them by x, y, or $x^2 + y^2$ are biharmonic, as is the linear combination in (48) for arbitrary choices of the constants C_i,

$$\sin \alpha x\,[C_1 \cosh \alpha y + C_2 \sinh \alpha y + C_3\, y \cosh \alpha y + C_4\, y \sinh \alpha y], \tag{8.2.48}$$

or a similar function with $\cos \alpha x$ instead of $\sin \alpha x$. The roles of x and y can also be interchanged.

Trigonometric series solutions may be obtained by suitable combinations of the functions (48) for an infinite number of choices of α. For example, in the rectangular domain of Fig. 8.2, if we take $\alpha = m\pi/L$ for $m = 1, 2, \ldots,$ an infinite series of the functions (48) will reduce to trigonometric sine series in x on the top and bottom boundaries where y is constant. The coefficients can be determined by matching the coefficients of the calculated series for the stresses σ_x and τ_{xy} with the Fourier sine or cosine series expansions of the prescribed traction components on the top and the bottom. Polynomial terms may be added to ϕ to provide constant terms for the Fourier cosine series. Since traction distributions with finite jump discontinuities can be expanded in Fourier series, this offers the possibility of obtaining solutions which will be continuous in the interior even when the tractions are discontinuous on the boundary. Fourier series for discontinuous functions are notorious for slow convergence, so that for discontinuous tractions many terms may be needed, and it is necessary to check the convergence carefully. With the functions (48) there are just enough constants available, for each m, to match the four constants appearing in the four sine or cosine Fourier expansions of the prescribed σ_x and τ_{xy} on the top and bottom. This leaves none available to match the end conditions. For certain cases of symmetrical loading on the top and bottom, the solution gives on each end self-equilibrating tractions that are statically equivalent to zero traction there. For these cases, the solution is valid for prescribed traction-free ends at distances from the ends large compared to the end depth, according to Saint-Venant's principle. Examples and calculation details may be found, for example, in Timoshenko and Goodier (1951), Sec. 23. This procedure is not valid, however, unless L is much greater than the depth of the rectangle, although its use has sometimes been attempted when this is not the case.

For rectangles where the two dimensions are of the same order of magnitude, a solution can be obtained by using a double series, formed with functions involving trigonometric terms in both x and y. The procedure here is much more complicated, leading to the simultaneous solution of an infinite number of linear algebraic equations for an infinite number of unknowns. If the procedure converges, this actually means the solution of a large finite number of equations. This double series method, which permits satisfaction of boundary conditions all the way around, has been used extensively by Pickett. See, for example, Pickett (1944).

Trigonometric solutions of plane elastostatic Navier equations for no body forces. Pickett (1944) also listed for the case of no body forces the two following displacement solution forms that can be combined for a sequence of values of the parameter α to obtain Fourier series representations of prescribed boundary conditions. Either term may be chosen from each bracket, but if, for example, the upper term in the first bracket and the lower term in the second bracket are chosen in the expression for u_x, then the same choice should be made in the expression for u_y, etc. The first form is

$$u_x = \frac{1}{\alpha}\begin{Bmatrix}\sin \alpha x\\ \cos \alpha x\end{Bmatrix}\begin{Bmatrix}\sinh \alpha y\\ \cosh \alpha y\end{Bmatrix} \qquad u_y = \frac{1}{\alpha}\begin{Bmatrix}-\cos \alpha x\\ \sin \alpha x\end{Bmatrix}\begin{Bmatrix}\cosh \alpha y\\ \sinh \alpha y\end{Bmatrix}. \tag{8.2.49}$$

The second form is

$$\begin{aligned} u_x &= y\begin{Bmatrix}\sin \alpha x\\ \cos \alpha x\end{Bmatrix}\begin{Bmatrix}\sinh \alpha y\\ \cosh \alpha y\end{Bmatrix} \\ u_y &= \frac{1}{\alpha}\begin{Bmatrix}-\cos \alpha x\\ \sin \alpha x\end{Bmatrix}\begin{Bmatrix}[\alpha y \cosh \alpha y - K \sinh \alpha y]\\ [\alpha y \sinh ay - K \cosh \alpha y]\end{Bmatrix}\end{aligned} \tag{8.2.50}$$

where

$$\begin{aligned} K &= 3 - 4\nu \text{ for plane strain} \\ K &= \frac{3-\nu}{1+\nu} \quad \text{for plane stress.}\end{aligned} \tag{8.2.51}$$

Two more forms, one similar to Eq. (49) and one similar to Eq. (50) may be obtained by interchanging the variables x and y and also at the same time interchanging u_x and u_y.

In order to attempt to match arbitrarily prescribed boundary conditions variable with x on the top and bottom of a rectangular region, it is necessary in general to use a series containing all four of the combinations given in Eqs. (49) and (50) for a sequence of values α_m of the parameter α, as in Eq. (48) and the discussion following Eq. (48). This would leave no arbitrary constants available to match end conditions, unless the four additional combinations obtained by interchanging x and y and also interchanging u_x and u_y in Eqs. (49) and (50) were included. By suitable choices for problems symmetric in x

or y, or antisymmetric, the problem can be simplified somewhat, but it remains a complicated task to sort out all the coefficients.

Pickett (1944) gave comparable three-dimensional forms in Cartesian coordinates, which we omit. He also gave, for the special case of radial symmetry, cylindrical coordinate forms involving Bessel functions in r and trigonometric or hyperbolic functions in z. These forms for radial symmetry will be given at the end of Sec. 8.3, which is concerned with cylindrical coordinate formulation and especially with plane problems in polar coordinates.

EXERCISES

1. Verify the derivation of: (a) Eq. (8.2.7); (b) Eq. (8.2.20).
2. (a) Verify that the stresses of the particular solutions of Equations (8.2.8) and (8.2.10) satisfy equilibrium and compatibility for ***plane strain*** under the stated body forces.
   (b) How should Eq. (10) be modified to give a particular solution that satisfies equilibrium and the ***plane stress*** compatibility equation, Eq. (20)? Show that the modified solution is not, however, consistent with Eqs. (21).
3. State boundary conditions in terms of α for the traction components t_x and t_y of the homogeneous problem solution that must be added to the particular solution for gravity body forces:

$$\sigma_y^P = \rho g y, \qquad \sigma_x^P = \tau_{xy}^P = 0.$$

   The complete problem is for a circular cylinder in plane strain, loaded only by its weight and the concentrated upward support reaction at $x = 0$, $y = -R$ as shown in the figure.

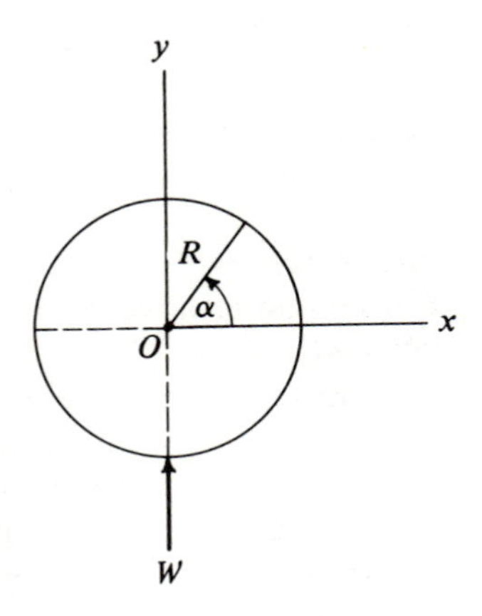

4. Transform the stresses of the particular solution for rotational inertia forces, Eqs. (8.2.10), to $\bar{x}\bar{y}$-axes such that the $\bar{x}$-axis makes angle θ with the x-axis. Your results will be the physical components σ_r, σ_θ, $\tau_{r\theta}$ of stress in cylindrical coordinates. What are the boundary conditions to be imposed on the cylindrical coordinate components of stress for the homogeneous solution that must be added to this particular solution if the circular boundary is traction-free in the complete solution?
5. Show that Eqs. (8.2.4) are the inverse of Eqs. (1) in plane strain.
6. Show that Eqs. (8.2.19) are the inverse of Eqs. (17) in plane stress.
7. If ϕ_0 is given by the first Eq. (8.2.44) for the end-loaded cantilever of Fig. 8.2, compare the magnitudes of the stresses computed from Eq. (22) with those of Eqs. (44). How are the boundary conditions changed if Eq. (22) is taken as the solution instead of simply the ϕ_0 of the first Eq. (44)? That is, what additional tractions will have to be supplied to make the modified solution correct?
8. Verify that Eq. (8.2.27) leads to Eq. (28) for the strain energy density in plane strain, as a function of the stresses.
9. Verify that Eq. (8.2.27) leads to Eqs. (29) and (30) for the strain energy density in plane stress. Verify also that the plane stress forms reduce to the plane strain forms upon substitution of $E/(1 - \nu^2)$ in place of E and $\nu/(1 - \nu)$ in place of ν in the plane stress forms.
10. Show that the three-dimensional forms of the Navier equations and traction boundary conditions lead to Eqs. (8.2.31) through (33) in the case of plane strain.
11. Show that substituting the plane stress Hooke's law into the equations of motion and the traction boundary conditions leads to the forms of Eqs. (8.2.34) and (35) for the Navier equations and traction boundary conditions in plane stress.
12. Verify that replacing ν by $\nu/(1 - \nu)$ in the plane stress Navier equations and traction boundary conditions of Eqs. (8.2.34) and (35) reduces these equations to the plane strain forms of Eqs. (8.2.31) and (32).
13. Check that the compatibility equation leads to the recursion relation of Eq. (8.2.38) when the power series of Eq. (8.2.36a) is assumed as the solution.
14. Any polynomial of degree less than four satisfies the biharmonic equation and is therefore a possible stress function ϕ for some plane region, in the absence of body forces. What kind of stress distribution is represented by $\phi = C_{20}x^2 + C_{11}xy + C_{02}y^2$? What traction boundary conditions does it satisfy on the boundaries of the rectangular region of Fig. 8.2?
15. (a) Write the strain distribution for the cantilever of Fig. 8.2, if the stresses are given by Eqs. (8.2.45).
    (b) If this is a plate in plane strain (long in the z-direction) instead of a beam in plane stress, write the strain distribution.
16. (a) Determine the displacements corresponding to the strain distribution of Ex. 15(a), if the usual fixed-end support conditions are assumed at $x = L$, $y = 0$, namely zero displacement there and zero slope for the deformed centerline. Sketch the shape of the deformed end $x = L$, according to this solution.

(b) If the end condition is changed to permit nonzero slope of the centerline at the end but to require the vertical element through $x = L$, $y = 0$ to remain vertical, how does the displacement solution differ? This additional deflection, not predicted by the simplest elementary beam theory is usually called the effect of shear deformation on the beam deflection.

17. Show that if $f(x, y)$ is a plane harmonic function then the following are possible stress functions: (a) $f(x, y)$; (b) $xf(x, y)$; (c) $yf(x, y)$; (d) $(x^2 + y^2)f(x, y)$.

18. Verify that the function $(y \cosh \alpha y \sin \alpha x)$ in the third term of Eq. (8.2.48) is biharmonic.

19. Consider the plane stress problem illustrated in the figure below. The circular plate of radius R is loaded by prescribed boundary tractions $\mathbf{t} = \sigma\mathbf{i}$ on arc DAB $(-\pi/4 < \alpha < \pi/4)$ and $\mathbf{t} = -\sigma\mathbf{i}$ on the opposite arc $B'D'$, where σ is a constant. Arcs BB' and DD' are traction-free, and body force is negligible. In terms of the angle α obtain expressions for the Airy stress function ϕ and its normal derivative on the two arcs AB and BB'. [See Eqs. (8.2.15) and (16).]

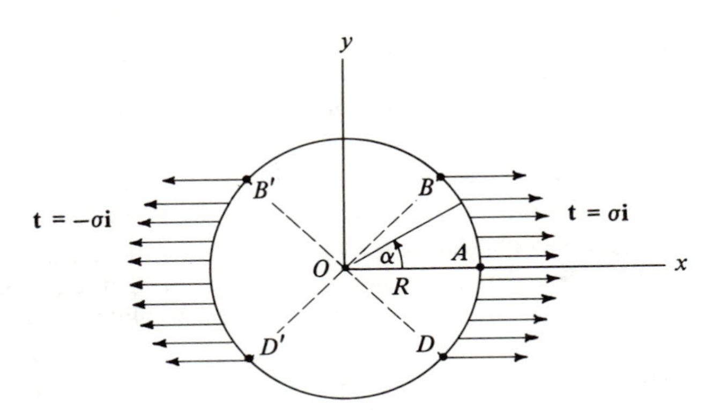

20. By the method of Neou, Eqs. (8.2.36) to (38), determine a polynomial solution for the stresses in a beam of rectangular cross-section. The beam, of length $2L$, is loaded by uniform downward pressure p on its top edge $y = -a$. (Take the origin at the center and the y-axis downward.) The exact distribution of end tractions is not prescribed, but they must be statically equivalent to supporting half the load at each end, and must imply no resultant axial force or bending moment on the ends.

21. (a) Verify that each term of the series

$$u_x = \sum_{m=1}^{\infty} \frac{A_m}{\alpha_m} \sin \alpha_m x \sinh \alpha_m y$$

$$u_y = \sum_{m=1}^{\infty} \frac{-A_m}{\alpha_m} \cos \alpha_m x \cosh \alpha_m y$$

is a solution of the plane stress Navier equation for equilibrium with no body forces as asserted in Eq. (8.2.49).

(b) Show that the parameters α_m can be selected so that both u_x and τ_{xy} vanish on the sides $x = 0$ and $x = L$ of a square. With this choice of α_1 consider only the first term for $m = 1$ and sketch the traction distributions on the four sides $x = 0$, $x = L$, $y = 0$, and $y = L$ of the square, and verify that the tractions provided by the first term satisfy the requirement that the total force in the x-direction and the total force in the y-direction each vanish. *Remark:* Since only one set of arbitrary constants A_m is available, this series solution is not complete enough to match arbitrarily prescribed boundary conditions even on one side.

22. (a) In the solution of Ex. 20 of Sec. 3.2 it was shown that when the yz-plane is a plane of symmetry of stress, then σ_x and σ_y are even functions of x while τ_{xy} is an odd function of x. Show that this implies that in plane stress or plane strain the Airy stress function $\phi(x, y)$ must be of the form $\phi(x, y) = f(x, y) + Cx$, where $f(x, y)$ is even in x and C is a constant. Show also that there is no loss of generality in choosing $C = 0$, so that $\phi(x, y)$ is even in x.
(b) Make a comparable statement for the case that the xz-plane is a plane of symmetry of stress.
(c) If a plane stress state is antisymmetric with respect to the yz-plane (so that σ_x and σ_y are odd in x while τ_{xy} is even), what conclusions can you draw about the Airy function? Make a comparable statement about a stress state antisymmetric with respect to the xz-plane.

8.3 Cylindrical coordinate components; plane elasticity in polar coordinates

Appendix II shows how to evaluate the displacement-gradient tensor $\mathbf{u}\overleftarrow{\nabla}$ in cylindrical coordinates, where

$$\mathbf{u} = u_r\hat{\mathbf{e}}_r + u_\theta\hat{\mathbf{e}}_\theta + u_z\hat{\mathbf{e}}_z$$

by differentiating the variable unit vectors $\hat{\mathbf{e}}_r$, $\hat{\mathbf{e}}_\theta$ as well as the coefficients of the three unit vectors. It is not necessary to master the derivations of that appendix before proceeding with the following development, although we will quote some of the results from the appendix. An independent derivation will be given below for the special case of plane polar coordinates; the derivations in this section are less rigorous than those in the appendix, but may give more insight into the physical origin of the extra terms appearing in the formulas. Equations (II.4.C8) and (II.4.C9) give the components of $\mathbf{v}\overleftarrow{\nabla}$ and $\mathbf{D}$. By replacing $\mathbf{v}$ by $\mathbf{u}$ and $\mathbf{D}$ by $\mathbf{E}$ we obtain the results given in Eq. (8.3.1) below. The coordinates obtained are the ***physical components*** (see Sec. I.5) of the tensors. The physical components associated with the r-, θ-, and z-directions are equal to the components in a local rectangular Cartesian system whose directions coincide with the r-, θ-, and z-directions. But since these directions change from point to point the physical components in cylindrical coordinates differ from the rectangular Cartesian components in any single set of reference axes.

Small-strain components in cylindrical coordinates (physical components) are

$$\left.\begin{aligned} E_{rr} &= \frac{\partial u_r}{\partial r} \qquad E_{\theta\theta} = \frac{1}{r}\frac{\partial u_\theta}{\partial \theta} + \frac{u_r}{r} \qquad E_{zz} = \frac{\partial u_z}{\partial z} \\ E_{r\theta} &= \frac{1}{2}\left[\frac{1}{r}\frac{\partial u_r}{\partial \theta} + \frac{\partial u_\theta}{\partial r} - \frac{u_\theta}{r}\right] \qquad E_{\theta z} = \frac{1}{2}\left(\frac{\partial u_\theta}{\partial z} + \frac{1}{r}\frac{\partial u_z}{\partial \theta}\right) \\ E_{zr} &= \frac{1}{2}\left(\frac{\partial u_z}{\partial r} + \frac{\partial u_r}{\partial z}\right), \end{aligned}\right\} \tag{8.3.1}$$

where (r, θ, z) should be interpreted as material coordinates, although the distinction between material and spatial coordinates is usually ignored in small-displacement elasticity. The rotation components are given in Eq. (II.4.C14). Recalling that arc length in the θ-direction is given by $ds = r\,d\theta$, we see that these formulas are analogous to those of rectangular Cartesians except for the extra terms u_r/r in $E_{\theta\theta}$ and $-u_\theta/2r$ in $E_{r\theta}$. The following geometrical analysis for the plane-motion case may give some insight into the extra terms.

Plane-strain components will be denoted for

$$\left.\begin{aligned} &\mathbf{u} = u\mathbf{e}_r + v\mathbf{e}_\theta \\ \text{by}\quad &\epsilon_r = \frac{\partial u}{\partial r} \qquad \epsilon_\theta = \frac{1}{r}\frac{\partial v}{\partial \theta} + \frac{u}{r} \\ &\epsilon_{r\theta} = \frac{1}{2}\left(\frac{1}{r}\frac{\partial u}{\partial \theta} + \frac{\partial v}{\partial r} - \frac{v}{r}\right). \end{aligned}\right\} \tag{8.3.2}$$

In Fig. 8.3, the material elements $AB = dr$ and $AC = r\,d\theta$ move to positions $A'B'$ and $A'C'$. The strains are

$$\epsilon_r = \frac{A'B' - AB}{AB}$$

$$\epsilon_\theta = \frac{A'C' - AC}{AC}$$

$$\epsilon_{r\theta} = \tfrac{1}{2}(\theta_2 + \theta_1) = \tfrac{1}{2}(\theta_2 + \theta_4 - \theta_3).$$

If u and v are the radial- and transverse-displacement components, then

$$A'H = dr + \frac{\partial u}{\partial r}dr, \qquad FC' = \frac{\partial u}{\partial \theta}d\theta, \qquad B'H = \frac{\partial v}{\partial r}dr \quad \text{and}$$

$$A'F = DE - DA' + EF = (r + u)\,d\theta - v + \left(v + \frac{\partial v}{\partial \theta}d\theta\right).$$

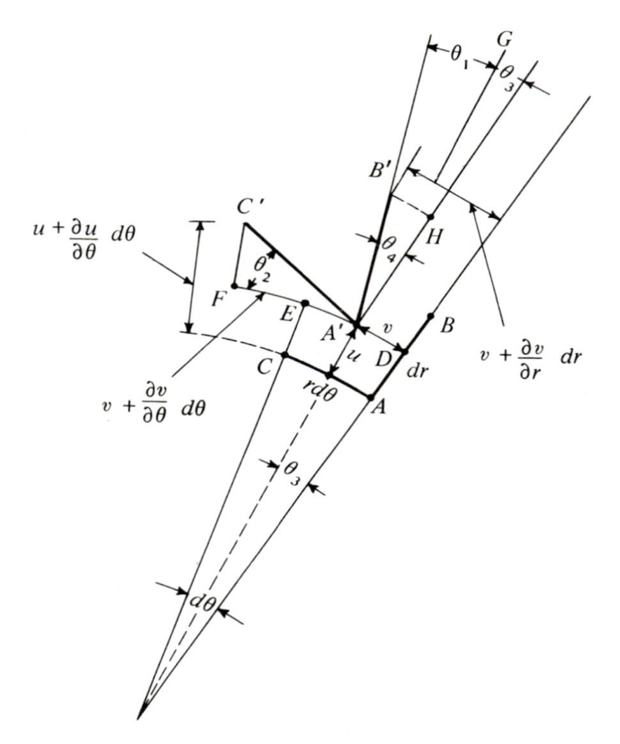

Fig. 8.3 Small strain in polar coordinates.

From the expression for ϵ_θ we have

$$(A'C')^2 = (1 + \epsilon_\theta)^2(AC)^2 = (1 + \epsilon_\theta)^2(r\,d\theta)^2.$$

Also in the right triangle $A'FC'$, we have

$$(A'C')^2 = (A'F)^2 + (FC')^2.$$

If we equate these two expressions for $(A'C')^2$, substitute for $A'F$ and FC', and expand, neglecting products and powers of the small quantities ϵ_θ, u, and $\partial v/\partial\theta$, we get

$$(1 + 2\epsilon_\theta)(r\,d\theta)^2 = \left(r^2 + 2ru + 2r\frac{\partial v}{\partial \theta}\right)(d\theta)^2$$

whence

$$\epsilon_\theta = \frac{1}{r}\frac{\partial v}{\partial \theta} + \frac{u}{r}$$

as in Eq. (2). The extra term u/r in ϵ_θ thus represents the circumferential strain associated with purely radial displacements. In the absence of transverse displacements, a uniform radial displacement u would increase the circumference of a circle from $2\pi r$ to $2\pi(r + u)$ so that its change in length per unit initial length is u/r. The radial strain is $\epsilon_r = \partial u/\partial r$ with no extra term needed.

For the shear-strain calculation we replace, for small angle changes,

$$\theta_2 \approx \tan \theta_2 = \frac{FC'}{A'F} = \frac{(\partial u/\partial \theta)\, d\theta}{(r + u)\, d\theta + (\partial v/\partial \theta)\, d\theta}.$$

Hence $\theta_2 = (1/r)(\partial u/\partial \theta)$, neglecting in the denominator terms containing products $u\, d\theta$ and $(\partial v/\partial \theta)\, d\theta$. A similar derivation yields $\theta_4 = \partial v/\partial r$ and $\theta_3 = v/r$. Hence we obtain

$$\epsilon_{r\theta} = \frac{1}{2}\left(\frac{1}{r}\frac{\partial u}{\partial \theta} + \frac{\partial v}{\partial r} - \frac{v}{r}\right).$$

The extra term $-v/r$ comes from θ_3. If there were no shear strain, $C'A'B'$ would be a right angle, which could be made to coincide with $FA'G$ by a rigid rotation; then we would have $\theta_2 = 0$, $\theta_1 = 0$, and $\theta_4 = \theta_3$ so that $\partial u/\partial \theta = 0$ and the term $-v/r$ is needed to correct for the nonzero $\partial v/\partial r$.

Equations of motion in cylindrical coordinates are given in Eqs. (II.4.C11). For small displacements, as in Sec. 8.1 for Cartesians, we may interpret the r, θ, z as material coordinates on the left-hand side of these equations, while the acceleration components on the right are then simply partial derivatives:

$$a_r = \frac{\partial^2 u_r}{\partial t^2} \qquad a_\theta = \frac{\partial^2 u_\theta}{\partial t^2} \qquad a_z = \frac{\partial^2 u_z}{\partial t^2}. \tag{8.3.3}$$

The Navier displacement equations of motion are given in Eqs. (II.4.C12) to (II.4.C14). Some examples of solutions of the elastostatic Navier equations in cylindrical coordinates are given in Eqs. (41) to (44) at the end of this section for the special case of radial symmetry. First, however, we consider plane elasticity in polar coordinates.

Equilibrium equations in plane stress then take the following form, if we denote physical components of stress T_{rr} by σ_r, $T_{\theta\theta}$ by σ_θ, $T_{r\theta}$ by $\tau_{r\theta}$, and $\rho\mathbf{b}$ by $\mathbf{f}$.

$$\left.\begin{aligned} \frac{\partial \sigma_r}{\partial r} + \frac{1}{r}\frac{\partial \tau_{\theta r}}{\partial \theta} + \frac{1}{r}(\sigma_r - \sigma_\theta) + f_r = 0 \\ \frac{\partial \tau_{r\theta}}{\partial r} + \frac{1}{r}\frac{\partial \sigma_\theta}{\partial \theta} + \frac{1}{r}(\tau_{r\theta} + \tau_{\theta r}) + f_\theta = 0. \end{aligned}\right\} \tag{8.3.4}$$

Extra terms $(\sigma_r - \sigma_\theta)/r$ and $(\tau_{r\theta} + \tau_{\theta r})/r$ appear in these equations as compared with the analogous rectangular Cartesians. For symmetric stress $(\tau_{r\theta} + \tau_{\theta r})/r$ may be written as $2\tau_{r\theta}/r$. The physical significance of the extra terms may be seen by considering the equilibrium of a cylindrical coordinate volume element. Figure 8.4 shows the two-dimensional stresses on the faces of a polar-

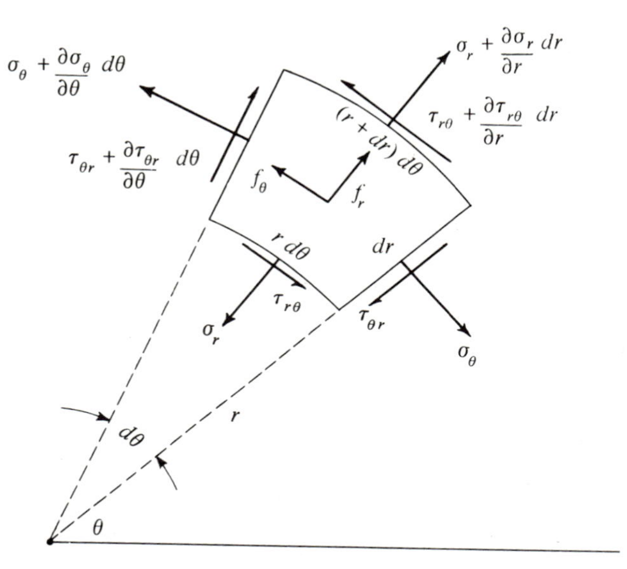

Fig. 8.4 Equilibrium in polar coordinates.

coordinate element of volume of unit thickness in the z-direction. Summation of forces parallel to the radial direction through the center of the element yields

$$\left(\sigma_r + \frac{\partial \sigma_r}{\partial r} dr\right)(r + dr)\, d\theta - \sigma_r\,(r\, d\theta) - \left[\sigma_\theta + \frac{\partial \sigma_\theta}{\partial \theta} d\theta + \sigma_\theta\right] dr \sin \frac{d\theta}{2}$$
$$+ \left[\tau_{\theta r} + \frac{\partial \tau_{\theta r}}{\partial \theta} d\theta - \tau_{\theta r}\right] dr \cos \frac{d\theta}{2} + f_r\, r\, dr\, d\theta = 0.$$

We approximate $\sin (d\theta/2)$ by $d\theta/2$, and $\cos (d\theta/2)$ by unity and divide through by $r\, dr\, d\theta$ to obtain

$$\frac{1}{r}\sigma_r + \frac{\partial \sigma_r}{\partial r}\left(1 + \frac{dr}{r}\right) - \frac{\sigma_\theta}{r} - \frac{\partial \sigma_\theta}{\partial \theta}\frac{d\theta}{r} + \frac{1}{r}\frac{\partial \tau_{\theta r}}{\partial \theta} + f_r = 0$$

whence we get the first Eq. (4) by dropping the terms in dr/r and $d\theta/r$ in the limit. The second equation follows similarly from summation in the θ-direction. This derivation is not rigorous, since we are not justified in summing all the radial stresses σ_r as though they were parallel to the radial direction through the center of the element. In fact, this seems inconsistent with our recognition of the lack of parallelism between the traction components on the two sides corresponding to θ and $\theta + d\theta$. Nevertheless it does yield the same result as the rigorous derivation of App. II and furnishes some insight into the extra terms provided automatically by the tensor formulas. We see that two effects are present. In the first equation, the term σ_r/r comes from the fact that the outer face length $(r + dr)\,d\theta$ is not the same as the inner one $r\,d\theta$, while the term $-\sigma_\theta/r$ is the radial contribution of the stresses on the transverse faces, which are not exactly perpendicular to the radial direction through the center. The extra term in the second equation is $(\tau_{r\theta} + \tau_{\theta r})/r$, the first part coming from the two unequal faces and the second from lack of parallelism of the $\tau_{\theta r}$ stresses on the two transverse faces. The equality $\tau_{r\theta} = \tau_{\theta r}$ can be seen by taking moment equilibrium about the center of the element, assuming all four normal forces and the body force act through the center, and neglecting higher-order terms. The tensor derivations establish the same result rigorously, since a tensor which is symmetric in one coordinate system is symmetric in all coordinate systems and we have previously seen that, in the absence of couple stresses, the stress tensor is symmetric in Cartesian coordinates. Then $(\tau_{r\theta} + \tau_{\theta r})/r = 2\tau_{r\theta}/r$.

As usual, in the equilibrium partial differential equations the derivatives are with respect to spatial coordinates, since equilibrium occurs in the deformed configuration. The small-strain definitions used material coordinates, but in small-displacement elasticity we will ignore the difference and denote both by r, θ.

Airy stress function in polar coordinates. In polar coordinates, the physical stress components are given in terms of the Airy stress function $\phi(r, \theta)$ for the case of no body forces, by

$$\left.\begin{aligned} \sigma_r &= \frac{1}{r}\frac{\partial \phi}{\partial r} + \frac{1}{r^2}\frac{\partial^2 \phi}{\partial \theta^2} \qquad \sigma_\theta = \frac{\partial^2 \phi}{\partial r^2} \\ \tau_{r\theta} &= \frac{1}{r^2}\frac{\partial \phi}{\partial \theta} - \frac{1}{r}\frac{\partial^2 \phi}{\partial r\,\partial \theta} = -\frac{\partial}{\partial r}\left[\frac{1}{r}\left(\frac{\partial \phi}{\partial \theta}\right)\right]. \end{aligned}\right\} \tag{8.3.5}$$

It is easy to verify by direct substitution that for any sufficiently differentiable function $\phi(r, \theta)$, stresses defined by Eqs. (5) do satisfy identically the polar-coordinate equilibrium Eqs. (4) for no body forces. We might be clever enough to guess the form of Eqs. (5) needed to satisfy the equilibrium equations, but it is not immediately obvious. A tensor derivation† leads automati-

†The general tensor relation between the contravariant stress components and the Airy stress functions is $\sigma^{\alpha\beta} = \epsilon^{\gamma\alpha}\epsilon^{\rho\beta}\,\phi|_{\rho\gamma}$, where the Greek indices range from 1 to 2 and $\epsilon^{\gamma\alpha}$ denotes the value on the surface $x^3 = 0$ of the component $\epsilon^{\gamma\alpha 3}$ of the contravariant three-index tensor defined by Eq. (I.6.22), and where $\phi|_{\rho\gamma}$ denotes the covariant second derivative with respect to x^ρ and x^γ.

cally to the correct form, and the tensor approach would furnish the proper form in any system of curvilinear coordinates. The tensor approach also makes clear that there is not one Airy function in Cartesian coordinates and another in polar coordinates, but that it is the same scalar function, merely evaluated in the two different systems.

Since it is the same function, we already know what differential equation it must satisfy, the biharmonic equation. It merely remains to write the biharmonic equation in polar coordinates. The form of the Laplace operator in polar coordinates is well-known [see Eq. (II.4.C4] to be

$$\nabla_1^2 = \frac{\partial^2}{\partial r^2} + \frac{1}{r}\frac{\partial}{\partial r} + \frac{1}{r^2}\frac{\partial^2}{\partial \theta^2}. \tag{8.3.6}$$

Hence the biharmonic equation becomes

$$\left(\frac{\partial^2}{\partial r^2} + \frac{1}{r}\frac{\partial}{\partial r} + \frac{1}{r^2}\frac{\partial^2}{\partial \theta^2}\right)\left(\frac{\partial^2 \phi}{\partial r^2} + \frac{1}{r}\frac{\partial \phi}{\partial r} + \frac{1}{r^2}\frac{\partial^2 \phi}{\partial \theta^2}\right) = 0. \tag{8.3.7}$$

Hooke's law in polar coordinates. In orthogonal curvilinear coordinates, the physical components of a tensor at a point are merely the Cartesian components in a local rectangular Cartesian system at the point with its axes tangent to the coordinate curves. Any algebraic relationship between physical components is therefore identical in form to the Cartesian component relationship. For example, in plane polar coordinates, the plane-strain Hooke's law is

$$\left.\begin{aligned} \sigma_r &= \lambda e + 2G\epsilon_r \qquad \sigma_\theta = \lambda e + 2G\epsilon_\theta \\ \tau_{r\theta} &= 2G\epsilon_{r\theta} \qquad \text{where} \quad e = \epsilon_r + \epsilon_\theta \\ \sigma_z &= \nu(\sigma_r + \sigma_\theta). \end{aligned}\right\} \tag{8.3.8}$$

This and its inverse form, and also the plane stress forms, may be written down immediately from the corresponding Cartesian formulas of Sec. 8.2 by replacing x by r and y by θ. Extra terms enter only when differentiation appears, as in the definitions of strains in terms of displacement and in the equations of motion.

Strain-energy function. The expressions for the strain-energy density are also algebraic and can be written down immediately by replacing x and y by r and θ in the equations of Sec. 8.2.

Solutions. In circular regions or in annular regions bounded by two concentric circles, the plane elastostatic problem can be solved by using the known general solution of the biharmonic equation in polar coordinates for such regions. This solution will be presented below; it is an infinite series of functions with undetermined constants, which can be determined in a straightforward manner by appropriate boundary conditions on the circular boundaries. Some of the terms in this general solution also furnish interesting special results in noncircular regions. Before considering the general solution for the Airy stress function in the annular ring, we obtain one simple solution in the ring in the case of radial symmetry, by integrating directly the Navier displacement equation.

Lamé solution† for cylindrical tube under internal and external pressure. Consider a tube long in the z-direction, loaded by internal pressure p_i and external pressure p_o, with negligible body forces, and assume plane deformation with radial symmetry, independent of z and θ in the plane region $a \leq r \leq b$. The Navier equations, Eqs. (II.4.C12) then become, with $u_\theta = u_z = 0, u_r = u$, simply $(\lambda + 2G)\, \partial e/\partial r = 0$ where $e = (1/r)(d/dr)(ru)$, whence the one Navier equation is satisfied if

$$\frac{d}{dr}\left[\frac{1}{r}\frac{d}{dr}(ru)\right] = 0. \tag{8.3.9}$$

Two integrations with respect to r then give

$$u = Ar + \frac{B}{r}, \tag{8.3.10}$$

where A and B are arbitrary constants. The arbitrary constants can be determined from the boundary conditions as follows. By Eqs. (2),

$$\epsilon_r = \frac{\partial u_r}{\partial r} = A - \frac{B}{r^2} \qquad \epsilon_\theta = \frac{u_r}{r} = A + \frac{B}{r^2} \qquad \epsilon_{r\theta} = 0 \tag{8.3.11}$$

whence Hooke's law, Eqs. (8), gives

$$\left.\begin{aligned} \sigma_r &= 2A\lambda + 2GA - \frac{2GB}{r^2} \\ \sigma_\theta &= 2A\lambda + 2GA + \frac{2GB}{r^2} \\ \tau_{r\theta} &= 0 \qquad \sigma_z = 4A\nu(\lambda + G). \end{aligned}\right\} \tag{8.3.12}$$

With the first Eq. (12) we apply the boundary conditions at $r = a$ and $r = b$.

$$\left.\begin{aligned} r = a \qquad -p_i &= 2A(\lambda + G) - \frac{2GB}{a^2} \\ r = b \qquad -p_o &= 2A(\lambda + G) - \frac{2GB}{b^2} \end{aligned}\right\} \tag{8.3.13}$$

whence

†Due to G. Lamé (1852).

$$\left.\begin{aligned} 2GB &= \frac{(p_i - p_o)\,a^2 b^2}{b^2 - a^2} \\ 2(\lambda + G)A &= \frac{p_i a^2 - p_o b^2}{b^2 - a^2} \end{aligned}\right\} \tag{8.3.14}$$

so that the stress distribution is

$$\left.\begin{aligned} \sigma_r &= \frac{p_i a^2 - p_o b^2}{b^2 - a^2} - \frac{p_i - p_o}{b^2 - a^2}\frac{a^2 b^2}{r^2} \\ \sigma_\theta &= \frac{p_i a^2 - p_o b^2}{b^2 - a^2} + \frac{p_i - p_o}{b^2 - a^2}\frac{a^2 b^2}{r^2} \\ \sigma_z &= \frac{2\nu(p_i a^2 - p_o b^2)}{b^2 - a^2} \qquad \tau_{r\theta} = 0. \end{aligned}\right\} \tag{8.3.15}$$

The general solution of the biharmonic equation in polar coordinates in an annular ring or a solid circle was found by Michell (1899). Timpe (1905) indicated a physical meaning for the multivalued displacements corresponding to some of the terms. The mathematical procedure of finding the solution and proving that it is in fact the general solution is beyond the scope of this book. It can be verified that each term of the series does satisfy the biharmonic equation; hence, if the series converges uniformly in a region, it is a solution of the equation in the region. Convergence depends, of course, on the coefficients and these in turn depend on the boundary conditions. The general solution is

$$\left.\begin{aligned} \phi = {} & [a_0 \ln r + b_0 r^2 + c_0 r^2 \ln r] + d_0 r^2 \theta + a_0' \theta \\ & + \frac{a_1}{2} r\theta \sin\theta + (b_1 r^3 + a_1' r^{-1} + b_1' r \ln r)\cos\theta \\ & - \frac{c_1}{2} r\theta \cos\theta + (d_1 r^3 + c_1' r^{-1} + d_1' r \ln r)\sin\theta \\ & + \sum_{n=2}^{\infty} [(a_n r^n + b_n r^{n+2} + a_n' r^{-n} + b_n' r^{-n+2})\cos n\theta \\ & + (c_n r^n + d_n r^{n+2} + c_n' r^{-n} + d_n' r^{-n+2})\sin n\theta]. \end{aligned}\right\} \tag{8.3.16}$$

Stresses can be calculated by Eqs. (5). The stresses will be multivalued in a complete ring unless

$$d_0 = 0. \tag{8.3.17}$$

The displacements will be multivalued unless

$$b_1' = -\frac{a_1(1-\nu)}{4} \qquad d_1' = -\frac{c_1(1-\nu)}{4} \qquad \text{and} \quad c_0 = 0. \tag{8.3.18}$$

(Proofs of these statements are omitted.) There are some important cases in which the displacement is not single-valued, namely in elastic dislocation theory, but, for the time being, we require that both stresses and displacements be single-valued.

The stress solution depends upon Poisson's ratio ν, because of the first two conditions of Eqs. (18), even though the biharmonic equation does not contain any elastic constants. An exception to this statement occurs if $a_1 = 0$ and $c_1 = 0$, which also implies $b_1' = 0$ and $d_1' = 0$; this will be the case when the resultant force on each boundary is zero. [See the discussion below leading to Eqs. (21).] When the resultant force on each boundary is zero the stress distribution for a traction boundary-value problem in plane elasticity is independent of the elastic constants. If, however, the resultant force on one boundary must be balanced by the resultant on the other boundary, the stress distribution will depend on ν. The elastic constants also enter through the boundary conditions, if some of these are conditions on boundary displacements. The displacement and strain distributions in general depend on the elastic constants. When the stress distribution is independent of the elastic constants, an experimentally determined stress distribution on a model of one kind of material is valid for a body of the same size and shape made of any elastic material. This is the basis for the success of photoelastic analysis of plane elastic problems, since the results obtained on the transparent plastic models are applicable to bodies made of structural metals.

Boundary conditions. When traction boundary conditions are prescribed on the two boundary circles of the ring-shaped region $a \leq r \leq b$, then σ_r and $\tau_{r\theta}$ will be known as functions of θ (for $-\pi \leq \theta \leq \pi$) on the boundaries $r = a$ and $r = b$. (Note that σ_θ is not a traction component.) These known boundary values may be expanded in Fourier series:

$$\left.\begin{aligned}
\sigma_r(a, \theta) &= A_0 + \sum_{n=1}^{\infty} (A_n \cos n\theta + B_n \sin n\theta) \\
\sigma_r(b, \theta) &= A_0' + \sum_{n=1}^{\infty} (A_n' \cos n\theta + B_n' \sin n\theta) \\
\tau_{r\theta}(a, \theta) &= C_0 + \sum_{n=1}^{\infty} (C_n \cos n\theta + D_n \sin n\theta) \\
\tau_{r\theta}(b, \theta) &= C_0' + \sum_{n=1}^{\infty} (C_n' \cos n\theta + D_n' \sin n\theta),
\end{aligned}\right\} \tag{8.3.19}$$

where the coefficients are calculated in terms of the boundary values. For example,

$$\left.\begin{aligned} A_0 &= \frac{1}{2\pi}\int_{-\pi}^{\pi} \sigma_r(a, \theta)\, d\theta \\ A_n &= \frac{1}{\pi}\int_{-\pi}^{\pi} \sigma_r(a, \theta) \cos n\theta\, d\theta \\ B_n &= \frac{1}{\pi}\int_{-\pi}^{\pi} \sigma_r(a, \theta) \sin n\theta\, d\theta. \end{aligned}\right\} \tag{8.3.20}$$

The requirement that the tractions on the boundaries must be in overall equilibrium imposes some restrictions on the Fourier coefficients. For example,

$$0 = \int_{\substack{-\pi \\ S}}^{\pi} t_x\, r\, d\theta = \int_{-\pi}^{\pi} [\sigma_r(b, \theta) \cos \theta - \tau_{r\theta}(b, \theta) \sin \theta]\, b\, d\theta$$

$$- \int_{-\pi}^{\pi} [\sigma_r(a, \theta) \cos \theta - \tau_{r\theta}(a, \theta) \sin \theta]\, a\, d\theta.$$

When the expressions of Eqs. (19) are substituted into this equation, the first Eq. (21) results after the integrals are performed. The second Eq. (21) follows similarly from requiring the y-component of the resultant traction to be zero. Only the coefficients of $\sin \theta$ and $\cos \theta$ enter, since

$$\int_{-\pi}^{\pi} \cos n\theta \cos \theta = 0 \quad \text{for} \quad n \neq 1$$

and similar orthogonality conditions apply to the other terms. The force equilibrium furnishes finally the two Eqs. (21):

$$\left.\begin{aligned} a(A_1 - D_1) &= b(A_1' - D_1') \\ a(B_1 + C_1) &= b(B_1' + C_1'). \end{aligned}\right\} \tag{8.3.21}$$

Moment equilibrium

$$\int_{-\pi}^{\pi} b\tau_{r\theta}(b, \theta) b\, d\theta - \int_{-\pi}^{\pi} a\tau_{r\theta}(a, \theta) a\, d\theta = 0$$

furnishes the condition

$$b^2 C_0' = a^2 C_0. \tag{8.3.22}$$

We now calculate the stresses in the interior by using Eqs. (5), with the ϕ given by Eq. (16), taking account of Eqs. (17) and (18). To determine the constants in the series for ϕ, we put $r = a$ or $r = b$ in the results for σ_r and $\tau_{r\theta}$, and require coefficients of $\cos n\theta$ and $\sin n\theta$ to be equal to the corresponding coefficients in the boundary series. For $n \geq 2$, this yields eight conditions for each n (four for $\cos n\theta$ and four for $\sin n\theta$) and determines the eight constants $a_n, \ldots, d_n'$.

Constants a_0 and b_0 are determined in terms of A_0 and A_0' by the boundary conditions on σ_r, while a_0' is determined in terms of C_0 by the boundary condition on $\tau_{r\theta}$ at $r = a$. This leaves only the eight coefficients with subscript one in Eq. (16). Equating coefficients of $\sin\theta$ and $\cos\theta$ with the boundary values yields two sets of four equations, but these do not add up to eight independent equations. For example, from the $\cos\theta$ terms in $\sigma_r(a, \theta)$ and the $\sin\theta$ terms in $\tau_{r\theta}(a, \theta)$ we get the two equations

$$(a_1 + b_1')a^{-1} = 2b_1 a - 2a_1' a^{-3} = A_1$$
$$b_1' a^{-1} + 2b_1 a - 2a_1' a^{-3} = D_1,$$

which imply that $a_1 = a(A_1 - D_1)$. Similar consideration of three other pairs of equations leads to three other results. In all we have four results:

$$\begin{aligned} a_1 &= a(A_1 - D_1) \qquad & c_1 &= a(B_1 + C_1) \\ a_1 &= b(A_1' - D_1') \qquad & c_1 &= b(B_1' + C_1'). \end{aligned} \tag{8.3.23}$$

The two solutions for a_1 are equivalent by Eqs. (21), and the same is true of c_1. When the results of Eq. (23) are substituted back into the equations obtained from the boundary conditions, the two sets of four equations become identical, and we have only four independent equations for the six coefficients $b_1, a_1', b_1', d_1, c_1, d_1'$, but the first two Eqs. (18) supply the deficiency, if the displacements are required to be single-valued. This completes the determination of all coefficients in terms of arbitrarily prescribed equilibrium traction boundary conditions.

When the region is a solid circle, the boundary conditions on the inner circle are replaced by finiteness conditions on the stresses at the origin $r = 0$; all coefficients of negative powers of r in the series representations of $\sigma_r, \sigma_\theta, \tau_{r\theta}$ must vanish. This furnishes enough conditions to make up for the missing inside boundary.

When single-valued displacements are not required, the first two Eqs. (18) no longer apply and we are left with two arbitrary constants. The boundary loading does not completely determine the stress distribution in this case; there may be residual stresses even when there are no external loads. Examples of this are elastic dislocations; see, for example, Timoshenko and Goodier (1951), Sec. 39, for this case and for interpretations of some of the individual terms.

The solution is also sometimes used to represent approximately the solution in a partial ring, e.g., a curved beam initially occupying a sector of the annulus. The solution is not sufficiently general to satisfy distributed traction boundary conditions on the ends of the beam, but it can yield a distribution with prescribed resultant end forces and couples. In such a simply connected partial ring the multivalued difficulties do not arise; hence, Eqs. (17) and (18) do not apply and there are four more constants at our disposal than we had in the complete ring with single-valued displacements.

It is left as an exercise for the reader to verify that the stress function

$$\phi = a_0 \ln r + b_0 r^2, \tag{8.3.24}$$

using only the first two terms of Eq. (16), furnishes the same stress distribution as the Lamé solution of Eq. (15), provided that

$$2b_0 = \frac{p_i a^2 - p_o b^2}{b^2 - a^2} \quad \text{and} \quad a_0 = -\frac{(p_i - p_o)a^2 b^2}{b^2 - a^2}. \tag{8.3.25}$$

Stress concentration on the boundary of a circular hole. Another interesting special case of Eq. (16) is obtained by a suitable combination of the radially symmetric Lamé solution with the term containing $\cos 2\theta$ in Eq. (16). With different notation for the four arbitrary constants the term containing $\cos 2\theta$ is

$$\phi = (Ar^2 + Br^4 + Cr^{-2} + D) \cos 2\theta \tag{8.3.26}$$

with stresses, according to Eqs. (5),

$$\left.\begin{aligned} \sigma_r &= -(2A + 6Cr^{-4} + 4Dr^{-2}) \cos 2\theta \\ \sigma_\theta &= (2A + 12Br^2 + 6Cr^{-4}) \cos 2\theta \\ \tau_{r\theta} &= (2A + 6Br^2 - 6Cr^{-4} - 2Dr^{-2}) \sin 2\theta. \end{aligned}\right\} \tag{8.3.27}$$

If the outer boundary $r = b$ is allowed to become infinite, Eqs. (27) are a stress solution in the infinite region exterior to the hole $r = a$. If we choose $B = 0$ to keep the stresses finite as r increases, the stress approaches

$$\left.\begin{aligned} \sigma_r &= -2A \cos 2\theta \\ \sigma_\theta &= 2A \cos 2\theta \\ \tau_{r\theta} &= 2A \sin 2\theta \end{aligned}\right\} \quad \text{as} \quad r \to \infty. \tag{8.3.28}$$

These stresses at infinity are reminiscent of a uniaxial stress $\sigma_x = S$, $\sigma_y = \tau_{xy} = 0$, which gives by the plane stress transformation, Eqs. (3.5.5) and (3.5.6) (with $\sigma_r = \sigma_x'$, etc.),

$$\left.\begin{aligned} \sigma_r &= \tfrac{1}{2} S(1 + \cos 2\theta) \\ \sigma_\theta &= \tfrac{1}{2} S(1 - \cos 2\theta) \\ \tau_{r\theta} &= -\tfrac{1}{2} S \sin 2\theta. \end{aligned}\right\} \tag{8.3.29}$$

Hence if we take $2A = -\frac{1}{2}S$, and superpose a plane hydrostatic tension at infinity $\sigma_r = \sigma_\theta = \frac{1}{2}S$ on Eqs. (28), the stress at infinity would be uniaxial tension S in the x-direction. The solution with $\sigma_r = \sigma_\theta = \frac{1}{2}S$ at infinity can be obtained from the Lamé solution of Eqs. (15) by taking $-p_o = \frac{1}{2}S$ and $p_i = 0$ and then letting $b \to \infty$ (after first dividing each numerator and denominator by b^2). The result is

$$\sigma_r = \frac{1}{2}S\left(1 - \frac{a^2}{r^2}\right) \qquad \sigma_\theta = \frac{1}{2}S\left(1 + \frac{a^2}{r^2}\right) \qquad \tau_{r\theta} = 0. \tag{8.3.30}$$

When this is superposed on the stresses (27), with $2A = -\frac{1}{2}S$, the stress at infinity is uniaxial tension $\sigma_x = S$. For a traction-free hole, the boundary conditions at $r = a$ then require, by Eqs. (27), with $B = 0$ and $2A = -\frac{1}{2}S$.

$$\left.\begin{aligned} \sigma_r &= 0 \qquad & \tfrac{1}{2}S - 6Ca^{-4} - 4Da^{-2} &= 0 \\ \tau_{r\theta} &= 0 \qquad & -\tfrac{1}{2}S - 6Ca^{-4} - 2Da^{-2} &= 0 \end{aligned}\right\} \tag{8.3.31}$$

whence

$$C = -\tfrac{1}{4}Sa^4 \qquad D = \tfrac{1}{2}Sa^2. \tag{8.3.32}$$

The solution presented to give Eqs. (15) was for plane strain, but the same solution for σ_r, σ_θ, and $\tau_{r\theta}$ by Eqs. (24) and (25) can apply also to plane stress. The superposed solutions (27) and (30) can thus be used to study the stress concentration around a circular hole in a large plate under uniform tension at infinity. The superposed solution, due to G. Kirsch (1898), is given in Eqs. (33).

Stresses in Plate with Circular Hole $r = a$ for Uniaxial $\sigma_x = S$ at Infinity

$$\begin{aligned} \sigma_r &= \frac{S}{2}\left(1 - \frac{a^2}{r^2}\right) + \frac{S}{2}\left(1 + \frac{3a^4}{r^4} - \frac{4a^2}{r^2}\right)\cos 2\theta \\ \sigma_\theta &= \frac{S}{2}\left(1 + \frac{a^2}{r^2}\right) - \frac{S}{2}\left(1 + \frac{3a^4}{r^4}\right)\cos 2\theta \\ \tau_{r\theta} &= -\frac{S}{2}\left(1 - \frac{3a^4}{r^4} + \frac{2a^2}{r^2}\right)\sin 2\theta. \end{aligned} \tag{8.3.33}$$

The maximum stress occurring is $\sigma_\theta(a, \pi/2) = 3S$, three times the tension at infinity. Thus the ***stress-concentration factor*** is three. Some other loading conditions at infinity can be treated by superposing another solution with uniaxial stress $\sigma_y = \pm S$ at infinity. With $\sigma_y = +S$ this represents all-around radial tension S at infinity, leading to maximum stress $\sigma_\theta = 2S$ on the edge of the hole. With $\sigma_y = -S$, the superposed solution at infinity represents a pure shear on a rectangular element at 45° to the x- and y-axes; for this case, the maximum $\sigma_\theta = 4S$ on the edge of the hole. See, for example, Timoshenko and Goodier (1950), Sec. 32.

Flamant solution for concentrated force on a straight boundary. By taking only the single term with coefficient $\frac{1}{2}a_1$ in Eq. (16) and choosing $a_1 = -2p/\pi$ we obtain the stress function and stresses

$$\left.\begin{aligned} \phi &= -\frac{p}{\pi} r\theta \sin\theta \\ \sigma_r &= -\frac{2p}{\pi}\frac{\cos\theta}{r}, \qquad \tau_{r\theta} = \sigma_\theta = 0. \end{aligned}\right\} \tag{8.3.34}$$

By use of the transformations of Sec. 3.5 it can be shown that in Cartesian coordinates this is the stress distribution:

$$\sigma_x = -\frac{2p}{\pi}\frac{\cos^3\theta}{r}, \qquad \sigma_y = -\frac{2p}{\pi}\frac{\sin^2\theta\cos\theta}{r}, \qquad \tau_{xy} = -\frac{2p}{\pi}\frac{\sin\theta\cos^2\theta}{r} \tag{8.3.35}$$

discussed in Exercises 3 to 5 of Sec. 3.2, where this stress distribution was shown to satisfy the traction boundary conditions for a concentrated normal force p (per unit length in the z-direction) acting in the x-direction on the edge of a semi-infinite plate as shown in Fig. 8.5. The solution was originally obtained by Flamant in 1892 from the three-dimensional solution of Boussinesq

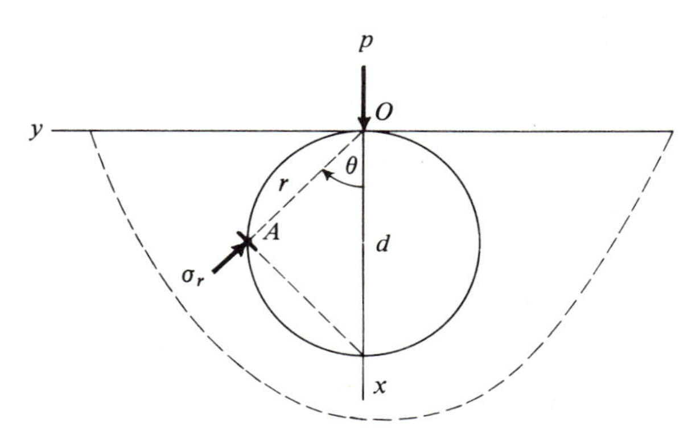

Fig. 8.5 Concentrated normal force on edge of semi-infinite plate.

to be presented at the end of Sec. 8.4. Figure 8.5 shows the traction component σ_r acting on a surface element perpendicular to the radius OA at A (not a surface element of the circle). Note that the magnitude of σ_r is the same for all points A on any one circle tangent to the y-axis at O. For this reason the stress distribution is sometimes called a ***simple radial distribution***.

Superposition with the Flamant solution. The Flamant solution for the concentrated normal force on the edge of a semi-infinite plate can be used to give by superposition the stress function for the distributed normal pressure $q(s)$ on the edge of the region as shown in Figure 8.6. Consider the load $q(s)ds$ at the point $y = s$ on the boundary, and denote the stress function for this increment of the distributed load (considered as a concentrated force) by $d\phi$. Then, by Eq. (34),

$$d\phi = -\frac{q(s)\,ds}{\pi} r\theta \sin\theta = -\frac{1}{\pi} q(s)\,(y - s) \arctan\frac{y - s}{x}\,ds$$

Hence

Superposition with Flamant Solution

$$\phi = \int_c^d -\frac{1}{\pi} q(s)\,(y - s) \arctan\frac{y - s}{x}\,ds$$

$$\frac{\partial\phi}{\partial x} = \frac{1}{\pi}\int_c^d q(s)\,\frac{(y - s)^2}{x^2 + (y - s)^2}\,ds$$

$$\frac{\partial\phi}{\partial y} = -\frac{1}{\pi}\int_c^d q(s)\left[\arctan\frac{y - s}{x} + \frac{(y - s)x}{x^2 + (y - s)^2}\right]ds$$

$$\sigma_x = \frac{\partial^2\phi}{\partial y^2} = -\frac{1}{\pi}\int_c^d q(s)\left[\frac{2x}{x^2 + (y - s)^2} - \frac{2(y - s)^2 x}{[x^2 + (y - s)^2]^2}\right]ds$$

$$= -\frac{1}{\pi}\int_c^d q(s)\,\frac{2x^3}{[x^2 + (y - s)^2]^2}\,ds \tag{8.3.36}$$

$$\sigma_y = \frac{\partial^2\phi}{\partial x^2} = -\frac{1}{\pi}\int_c^d q(s)\,\frac{2x(y - s)^2}{[x^2 + (y - s)^2]^2}\,ds$$

$$\tau_{xy} = -\frac{\partial^2\phi}{\partial x\,\partial y} = \frac{1}{\pi}\int_c^d q(s)\left[\frac{-(y - s)}{x^2 + (y - s)^2} + \frac{y - s}{x^2 + (y - s)^2}\right.$$

$$\left. - \frac{2x^2(y - s)}{[x^2 + (y - s)^2]^2}\right]ds = -\frac{1}{\pi}\int_c^d q(s)\,\frac{2x^2(y - s)}{[x^2 + (y - s)^2]^2}\,ds$$

It can be shown that the stresses σ_x and τ_{xy} approach the given boundary values for normal approach to the boundary. That is,

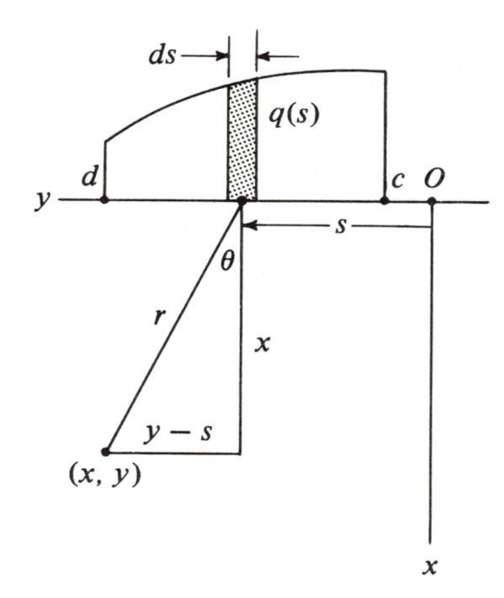

Fig. 8.6 Superposition with the Flamant solution.

$$\text{for fixed } y = b, \text{ as } x \to 0 \quad \begin{cases} \sigma_x \to -\, q(b) & \text{if } c < b < d, \\ \sigma_x \to 0 & \text{if } y < c \text{ or } y > d, \text{ and} \\ \tau_{xy} \to 0 & \text{for all } y. \end{cases} \tag{8.3.37}$$

But the limiting process must be performed carefully, since the integrals become improper when x is zero, because the denominators vanish at $s = y$. It is not possible just to put x equal to zero, because this would apparently cause all the stresses to be zero on the boundary, while σ_x is prescribed to be nonzero. Consider, for example, the limit for σ_x at $y = b$ for the case that $c < b < d$:

$$\sigma_x(x, b) = -\frac{1}{\pi}\int_c^d q(s)\frac{2x^3}{[x^2 + (b - s)^2]^2}\,ds.$$

Let

$$R^4 = [x^2 + (b - s)^2]^2.$$

Then

$$\sigma_x(x, b) = -\frac{1}{\pi}\int_c^{b-\delta} q(s)\frac{2x^3}{R^4}\,ds - \frac{1}{\pi}\int_{b-\delta}^{b+\delta} q(s)\frac{2x^3}{R^4}\,ds - \frac{1}{\pi}\int_{b+\delta}^{d} q(s)\frac{2x^3}{R^4}\,ds. \tag{8.3.38}$$

The first and last terms in Eq. (38) approach zero with x, because the denominator is always positive for s outside the range from $b - \delta$ to $b + \delta$, where δ is a small positive number. We need only consider the middle term, which we denote by I. Since the factor $2x^3/R^4$ is nonnegative on the interval, the mean value theorem of the integral calculus implies that for some value s^* in the interval from $b - \delta$ to $b + \delta$

$$I = -\frac{1}{\pi} q(s^*) \int_{b-\delta}^{b+\delta} \frac{2x^3}{R^4} \, ds.$$

Let

$$s - b = x \tan \alpha \quad \text{and} \quad \delta = x \tan \alpha_1.$$

Then

$$ds = + \, x \sec^2 \alpha \, d\alpha \quad \text{and} \quad x = R \cos \alpha \quad \text{(see Fig. 8.7).}$$

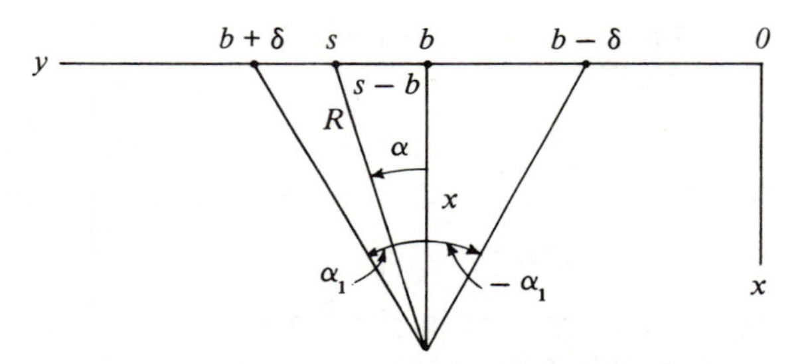

Fig. 8.7 Variables for the limiting process of normal approach to the boundary.

After this change of variables in the integral, we have

$$I = -\frac{1}{\pi} q(s^*) \int_{-\alpha_1}^{\alpha_1} 2 \cos^2 \alpha \, d\alpha = -\frac{2}{\pi} q(s^*)(\alpha_1 + \tfrac{1}{2} \sin 2\alpha_1).$$

Hence, as x approaches zero and α_1 approaches $\pi/2$, I approaches $-q(s^*)$. The value of s^*, to be sure, may depend upon x and may vary during the limiting process, but it is always constrained to lie between $b - \delta$ and $b + \delta$. Hence, by choosing δ sufficiently small, the limiting value can be made as near to $-q(b)$ as we please, while the first and last integrals in Eq. (38) still approach zero. This implies that $\sigma_x(x, b)$ approaches $-q(b)$ as x approaches zero and demonstrates that σ_x approaches the correct boundary value when the boundary is approached along the normal, as was asserted in Eq. (37). The last Eq. (37) can be demonstrated similarly.

For arbitrary prescription of the boundary loading the integrals in Eqs. (36) may have to be evaluated numerically. But in certain cases explicit evaluation is possible; for example the following results may be obtained.

For Uniform Pressure q on $-c<y<c$

$$\phi = \frac{q}{2\pi}\left\{[(y-c)^2 + x^2]\arctan\frac{y-c}{x} - [(y+c)^2 + x^2]\arctan\frac{y+c}{x} + 2cx\right\}$$

$$\sigma_x = \frac{q}{\pi}\left[\arctan\frac{y-c}{x} - \arctan\frac{y+c}{x} + \frac{x(y-c)}{x^2+(y-c)^2} - \frac{x(y+c)}{x^2+(y+c)^2}\right]$$

$$\sigma_y = \frac{q}{\pi}\left[\arctan\frac{y-c}{x} - \arctan\frac{y+c}{x} - \frac{x(y-c)}{x^2+(y-c)^2} + \frac{x(y+c)}{x^2+(y+c)^2}\right]$$

$$\tau_{xy} = \frac{q}{\pi}\left[\frac{(y-c)^2}{x^2+(y-c)^2} - \frac{(y+c)^2}{x^2+(y+c)^2}\right].$$

(8.3.39)

Inclined concentrated force on boundary. Boussinesq showed in 1892 that the Flamant solution of Eqs. (34) is also valid for an inclined concentrated force on the boundary as shown in Fig. 8.8, provided that the angle θ is measured from the line of action of the force (the line OB in Fig. 8.8). Then the Cartesian components of stress with respect to the xy-axes shown in Fig. 8.8 are

$$\sigma_x = -\frac{2p}{\pi}\frac{\cos\theta\cos^2(\theta+\alpha)}{r}$$

$$\sigma_y = -\frac{2p}{\pi}\frac{\cos\theta\sin^2(\theta+\alpha)}{r} \qquad (8.3.40)$$

$$\tau_{xy} = -\frac{2p}{\pi}\frac{\cos\theta\sin(\theta+\alpha)\cos(\theta+\alpha)}{r}.$$

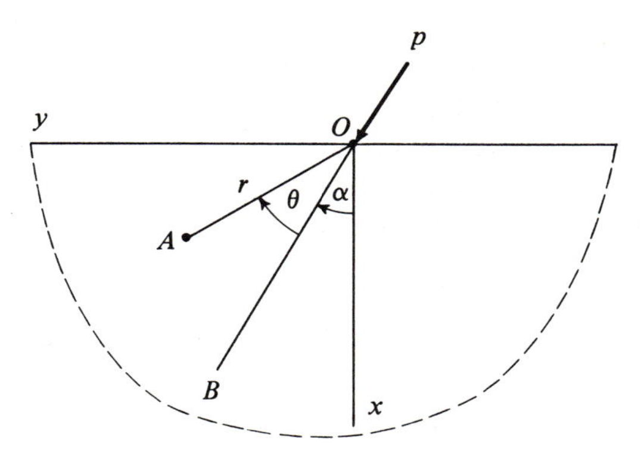

Fig. 8.8 Inclined concentrated force on a semi-infinite plate.

This solution may also lead to superposition solutions for distributed tractions, generalizing the distribution of Fig. 8.6. These are best handled by superposing two separate solutions, normal traction with $\alpha = 0$ in Eqs. (40), and tangential tractions with $\alpha = \pi/2$.

Application of Flamant solution to finite body. The foregoing discussion, beginning with Eqs. (34), is exclusively concerned with a semi-infinite plate (or half-space in the case of plane strain) where the necessary reactions to hold the body in equilibrium may be supposed to act at infinity. The same solutions may also be used as part of the solution for a finite body with a straight boundary loaded in the manner indicated in the foregoing discussion. The straight boundary is then traction-free except under the given concentrated load (or under the given distributed load). But an additional solution must be superposed to cancel the unwanted tractions furnished by the Flamant solution on the remainder of the boundary (the part of the boundary of the finite body which falls in the interior of the semi-infinite region to which the Flamant solution properly applies) and to satisfy the prescribed boundary conditions on this boundary. This additional solution must provide zero traction on the straight boundary, and the boundary conditions on the remaining boundary must be such that the body as a whole is in equilibrium. It may be necessary to resort to numerical methods such as finite difference approximation of the partial differential equations to solve the additional problem. but this will be easier than solving the whole problem by finite differences, since the boundary conditions for the additional problem will be continuous, while the concentrated-force boundary conditions impose on the solutions a singularity not readily treatable by finite differences. And even the finite jump discontinuities at the ends of distributed tractions such as those in Eq. (39) or in Fig. 8.6 are not as easy to handle as continuous distributions.

Radially symmetric displacement solutions in cylindrical coordinates. Pickett (1944) has listed the following pairs of solutions of the Navier equations for no body forces in the case of radial symmetry. In each solution either term in the bracket may be selected, but if the upper term is chosen for u_r, then the upper term should also be chosen for u_z, etc.

$$u_r = \frac{1}{\alpha} J_1(\alpha r) \begin{Bmatrix} \sinh \alpha z \\ \cosh \alpha z \end{Bmatrix}, \qquad u_z = -\frac{1}{\alpha} J_0(\alpha r) \begin{Bmatrix} \cosh \alpha z \\ \sinh \alpha z \end{Bmatrix} \tag{8.3.41}$$

$$u_r = z J_1(\alpha r) \begin{Bmatrix} \sinh \alpha z \\ \cosh \alpha z \end{Bmatrix}, \tag{8.3.42a}$$

$$u_z = -\frac{1}{\alpha}\, J_0(\alpha r) \begin{Bmatrix} \left[\alpha z \cosh \alpha z - \dfrac{\lambda + 3G}{\lambda + G} \sinh \alpha z\right] \\ \left[\alpha z \sinh \alpha z - \dfrac{\lambda + 3G}{\lambda + G} \cosh \alpha z\right] \end{Bmatrix} \tag{8.3.42b}$$

$$u_r = -\frac{1}{\alpha} I_1(\alpha r) \begin{Bmatrix} \sin \alpha z \\ \cos \alpha z \end{Bmatrix}, \qquad u_z = \frac{1}{\alpha} I_0(\alpha r) \begin{Bmatrix} -\cos \alpha z \\ \sin \alpha z \end{Bmatrix} \tag{8.3.43}$$

$$\left. \begin{aligned} u_r &= r I_0(\alpha r) \begin{Bmatrix} \sin \alpha z \\ \cos \alpha z \end{Bmatrix}, \\ u_z &= \frac{1}{\alpha}\left[\alpha r\, I_1(\alpha r) + 2\frac{\lambda + 2G}{\lambda + G} I_0(\alpha r)\right] \begin{Bmatrix} \cos \alpha z \\ -\sin \alpha z \end{Bmatrix} \end{aligned} \right\} \tag{8.3.44}$$

In these equations J_0 and J_1 are Bessel functions of the first kind of order zero and one, respectively, while $I_0(\alpha r) = J_0(i\alpha r)$ and $I_1(\alpha r) = -iJ_1(i\alpha r)$ are the modified Bessel functions.

In principle the solutions of Eqs. (41) to (44) can be combined for a sequence of values α_m to give series solutions inside a circular cylinder. Such solutions are, however, very complicated. According to Lur'e (1964), aside from a few trivial solutions, not a single solution is known that satisfies completely and rigorously all the boundary conditions on the lateral surface and the ends of a finite cylinder. Examples where the end conditions are not satisfied in detail, but only to the extent that the end tractions of the solution are statically equivalent to the prescribed end tractions, have been treated. For a long cylinder, St.-Venant's principle then implies that the solution is a good approximation at distances from the ends large compared to the diameter. See, for example, Lur'e (1964), Sec. 7.5. Lur'e also discusses in his Sec. 7.6 the case of an infinitely long cylinder with certain kinds of lateral surface tractions. For the infinite cylinder the series are replaced by Fourier integrals with the integration variable α.

In a semi-infinite cylinder ($z>0$), the hyperbolic functions occurring in the particular solutions are replaced by negative exponentials $e^{-\alpha z}$ in order to have finite displacements at infinity. And for a hollow cylinder additional forms are possible using Bessel functions of the second kind, Y_0, Y_1, and the modified Bessel functions of the second kind, K_0 and K_1. Both of these ideas are illustrated in a paper by Iyengar and Yogananda (1966) on the problem of end loading of a semi-infinite hollow cylinder.

The paper by Iyengar and Yogananda formulates the problem in terms of Love's strain function, a special case of the Galerkin vector to be discussed briefly in Sec. 8.4. Lur'e (1964), Chap. 7, approaches the problems of a symmetrically loaded circular cylinder by using the Papkovich-Neuber potentials, which will also be introduced in Sec. 8.4 in our discussion of the solution for displacements in three-dimensional elasticity.

EXERCISES

1. Carry out the details of the derivation based on Fig. 8.3 leading to the expression for $\epsilon_{r\theta}$.
2. Carry out the details of the derivation based on Fig. 8.4 for the second equilibrium equation given in Eqs. (8.3.4).
3. Verify that Eqs. (8.3.5) satisfy the plane polar coordinate equilibrium equations for no body forces.
4. Carry out the details of obtaining the Lamé solution for a tube loaded by internal and external pressure, Eqs. (8.3.9) to (15).
5. Show that Eq. (8.3.24) gives the same stress solution for a hollow cylinder loaded by internal pressure p_i at $r = a$ and outside pressure p_o at $r = b$ as is given by Eqs. (15), provided that a_0 and b_0 are as given in Eq. (25). For the case $p_o = 0$, where does the maximum shear stress attain its greatest value?
6. Derive Eqs. (8.3.21) from the force balance requirement for the general solution in a circular ring.
7. Derive Eq. (8.3.22) from the moment balance requirement for the general solution in a circular ring.
8. Verify that the term containing $\cos 2\theta$ in Eq. (8.3.16) is a solution of the biharmonic equation.
9. Consider only the term containing $\cos 2\theta$ for $n = 2$ in the general solution for a ring. Obtain expressions for the stresses represented by this term, thus verifying Eqs. (8.3.27) except for the change in notation of the constants.
10. For a solid circle, what coefficients of the solution of Ex. 9 must be zero in order to ensure finiteness of the stresses at the origin? Then what are the traction components on the boundary $r = b$ for this term? How does σ_r vary with r?
11. If a circular ring $a \le r \le b$ is loaded by tractions

$$\sigma_r = \sigma_a \cos 2\theta, \qquad \tau_{r\theta} = \tau_a \sin 2\theta \quad \text{on } r = a$$

$$\sigma_r = \sigma_b \cos 2\theta, \qquad \tau_{r\theta} = \tau_b \sin 2\theta \quad \text{on } r = b$$

where σ_a, σ_b, τ_a, and τ_b are constants, write the four simultaneous linear algebraic equations that determine a_2, a_2', b_2 and b_2' (see Ex. 9) in terms of σ_a, σ_b, τ_a, and τ_b.

12. Show that the stresses given by the term $d_0 r^2 \theta$ in Eq. (8.3.16) are multivalued.

13. What plane strain problem in the ring $a \leq r \leq b$ is represented by the term $a_0' \theta$ in Eq. (8.3.16) ? Show that the stresses are single-valued even though the stress function is not, and express a_0' in terms of a prescribed boundary traction at $r = b$.

14. If the displacements are constrained to vanish at $r = a$ in Ex. 13, show that the radial displacement u is identically zero, and determine the distribution of the θ-displacement v.

15. Show that the Lamé solution of Eq. (8.3.15) leads to the result of Eq. (30) for a plane hydrostatic tension of $\frac{1}{2} S$ at infinity and that when the stress fields of Eqs. (27) and (30) are superposed (with $2A = -\frac{1}{2} S$) the superposed field represents uniaxial tension at infinity.

16. Sketch the distribution of σ_θ on the line $\theta = \pi/2$ according to the stress-concentration solution of Eq. (8.3.33). At what distance from the center is σ_θ reduced to 1.1 S?

17. Verify the stress-concentration factors stated in the paragraph following Eq.(33) for the following cases of loading at infinity: (a) uniaxial tension S, (b) all-around radial tension S, (c) pure shear S.

18. Verify that the stress function of Eqs. (8.3.34) gives the stresses of Eqs. (34) for the Flamant solution. Show that σ_r is constant on the circle of diameter d shown in Fig. 8.5. Verify that the boundary ($x = 0$) is traction-free everywhere except at the origin.

19. Verify the calculation of σ_x in Eq. (8.3.36).

20. Show that τ_{xy} approaches zero as x approaches zero, for any fixed value of y, as asserted in Eq. (8.3.37).

21. Verify the calculation of ϕ for uniform pressure q in Eq. (8.3.39).

22. Show that the transformations of Sec. 3.5 lead from the stresses of Eqs. (8.3.34) to the Cartesian components of Eqs. (40) for the case of an inclined concentrated load on the boundary.

23. Set up an expression according to the first Eq. (36) for the stress function ϕ for a distributed pressure $q(s)$ in the form of an isosceles triangle of height q_0 on $-c \leq s \leq c$. It is not necessary to evaluate the integrals.

24. Verify that Eqs. (8.3.41), with the lower term selected in each bracket, is a solution of the elastostatic Navier equations for no body forces and radial symmetry. Calculate the corresponding strain and stress distributions.

25. Calculate the strain distribution corresponding to the displacements of Eq.(8.3.44), with the upper term selected in each bracket.

8.4 Three-dimensional elasticity; solution for displacements; vector and scalar potentials; wave equations; Galerkin vector; Papkovich-Neuber potentials; examples, including Boussinesq problem

Direct solution of Navier equations. Three-dimensional elasticity problems are most compactly formulated to solve the ***Navier equations*** for the displacement field. The rectangular Cartesian form of these equations, given in vector form in Eq. (8.1.8), is

$$(\lambda + G)\frac{\partial e}{\partial X_k} + G\nabla^2 u_k + \rho b_k = \rho \frac{\partial^2 u_k}{\partial t^2} \qquad \text{where} \quad e = \frac{\partial u_k}{\partial X_k} \tag{8.4.1}$$

is the dilatation. For elastostatic problems, with $\partial^2 u_k/\partial t^2 \equiv 0$, the solution of such a set of three equations in three dimensions by finite-difference or finite-element methods is beginning to be a possibility with the development of digital computer systems with very large memory storage capacity. Few three-dimensional solutions have yet been published, however. Two-dimensional elastostatic problems are well within the capability of computer systems now widely available, and several studies of such solutions for the displacement equations have appeared.

Dynamic problems involving two space dimensions and time, requiring computer capacity comparable to the three-dimensional elastostatic problems, are also just beginning to be solved by numerical methods. As larger computer systems become available and as experience accumulates to indicate the accuracy and practicality of numerical analysis of such large problems, we may expect numerical methods to become increasingly important.

One advantage of solving directly for the displacements, when numerical methods are used, is that strains (and hence stresses by Hooke's law) can then be obtained in terms of the first partial derivatives of the displacement field. This contrasts with the Airy stress function of Secs. 8.2 and 8.3 and with the displacement potentials to be presented below, where stress calculation involves evaluation of second derivatives, or even third derivatives in the case of the Galerkin vector. Since numerical evaluation of derivatives is an inherently inaccurate process, it is desirable to keep the numerical differentiation to as low an order as possible. When an analytical solution can be obtained for the potentials, however, no loss of accuracy is introduced by differentiation.

Scalar and vector potentials (Helmholtz representation). It is well-known that any continuously differentiable vector field $\mathbf{u}$ can be represented as the sum of an ***irrotational vector field*** which is the gradient of a ***scalar potential*** ϕ plus a ***solenoidal*** (***equivoluminal***) ***vector field*** which is the curl of a ***vector potential*** $\boldsymbol{\psi}$. Thus

$$\mathbf{u} = \nabla\phi + \nabla \times \boldsymbol{\psi} \tag{8.4.2a}$$

or, in rectangular Cartesian components,

$$u_1 = \phi_{,1} + \psi_{3,2} - \psi_{2,3} \qquad u_2 = \phi_{,2} + \psi_{1,3} - \psi_{3,1} \qquad u_3 = \phi_{,3} + \psi_{2,1} - \psi_{1,2}$$

or

$$u_k = \frac{\partial\phi}{\partial X_k} + e_{krs}\frac{\partial\psi_s}{\partial X_r}. \tag{8.4.2b}$$

This representation, known as the Helmholtz representation, is especially useful in wave-propagation studies; in an unbounded elastic medium, the dilatational and equivoluminal waves propagate independently. Note that there are now four unknown potential functions to solve for: ϕ and the three components of $\boldsymbol{\psi}$. Since it turns out that Eq. (2) places no restrictions on the divergence of $\boldsymbol{\psi}$, for definiteness we impose the requirement

$$\nabla \cdot \boldsymbol{\psi} \equiv 0 \quad \text{or} \quad \psi_{k,k} \equiv 0, \tag{8.4.3}$$

which will simplify the representation of $\nabla \times \mathbf{u}$ in terms of $\boldsymbol{\psi}$. The proof that the Helmholtz representation is always possible is omitted here [see, for example, Fung (1965), p. 184 or Aris (1962), p. 70]. The reader can easily demonstrate from Eqs. (2) that when $\mathbf{u}$ is the small-displacement field, the vector potential $\boldsymbol{\psi}$ does not contribute to the volume change, and that the dilatation e is given by

$$e = \nabla \cdot \mathbf{u} = \nabla^2\phi. \tag{8.4.4}$$

That the rotation vector $\boldsymbol{\omega}$ of Eq. (4.2.4), is given by

$$\boldsymbol{\omega} = \tfrac{1}{2}\nabla \times \mathbf{u} = -\tfrac{1}{2}\nabla^2\boldsymbol{\psi} \tag{8.4.5a}$$

or, in rectangular Cartesian components,

$$\omega_k = -\tfrac{1}{2}\nabla^2\psi_k, \tag{8.4.5b}$$

when the condition $\nabla \cdot \boldsymbol{\psi} = 0$ is imposed, can be seen as follows. Since $\nabla\phi$ is irrotational, we have from Eq. (2),

$$2\boldsymbol{\omega} = \nabla \times (\nabla \times \boldsymbol{\psi}) = \begin{vmatrix} \mathbf{i}_1 & \mathbf{i}_2 & \mathbf{i}_3 \\ \partial_1 & \partial_2 & \partial_3 \\ \psi_{3,2} - \psi_{2,3} & \psi_{1,3} - \psi_{3,1} & \psi_{2,1} - \psi_{1,2} \end{vmatrix}. \tag{8.4.6}$$

Thus, for example,

$$2\omega_1 = \psi_{2,12} - \psi_{1,22} - \psi_{1,33} + \psi_{3,13}.$$

From the identity of Eq. (3), we have $\psi_{k,k1} = 0$ or $\psi_{1,11} + \psi_{2,21} + \psi_{3,31} = 0$ whence by interchanging the order of differentiation we obtain

$$\psi_{2,12} + \psi_{3,13} = -\psi_{1,11}. \tag{8.4.7}$$

Thus

$$2\omega_1 = -\psi_{1,11} - \psi_{1,22} - \psi_{1,33} = -\nabla^2\psi_1.$$

The other components of Eq. (5) follow similarly from Eq. (6).

Equations of motion in terms of the potentials are obtained by substituting the component form of Eq. (2b) into the Navier Eqs. (1). We omit the body-force term. (If necessary, a particular solution can be added to account for a simple body-force distribution, e.g., for a uniform gravitational field.) The resulting equation of motion after an interchange of order of differentiation is

$$(\lambda + 2G)(\nabla^2\phi)_{,k} + Ge_{krs}(\nabla^2\psi_s)_{,r} = \rho\left(\frac{\partial^2\phi}{\partial t^2}\right)_{,k} + \rho e_{krs}\left(\frac{\partial^2\psi_s}{\partial t^2}\right)_{,r}. \tag{8.4.8}$$

In the most general case, all of the potentials are coupled together by Eq. (8). Certain special cases where they uncouple are of interest.

Wave equations. Equation (8) will be satisfied if the following wave equations, obtained by requiring certain groups of terms to vanish separately, are satisfied, although this is not the most general solution of Eq. (8).

$$\nabla^2\phi = \frac{1}{c_1^2}\frac{\partial^2\phi}{\partial t^2} \qquad \nabla^2\psi_k = \frac{1}{c_2^2}\frac{\partial^2\psi_k}{\partial t^2}, \tag{8.4.9}$$

where

$$c_1 = \left[\frac{\lambda + 2G}{\rho}\right]^{1/2} \qquad c_2 = \left[\frac{G}{\rho}\right]^{1/2}. \tag{8.4.10}$$

This indicates that a possible dynamic solution is one where two types of waves propagate separately, a ***dilatational wave*** governed by the scalar potential ϕ and propagating at speed c_1, and an ***equivoluminal wave*** (also called a ***shear wave***, a ***distortional wave*** or a ***rotational wave***) governed by the three vector potential components ψ_k propagating at speed c_2. If we apply the ∇^2 operator again to both sides of Eqs. (9) and substitute $e = \nabla^2\phi$ and $2\omega_k = -\nabla^2\psi_k$, we obtain

$$\nabla^2 e = \frac{1}{c_1^2}\frac{\partial^2 e}{\partial t^2} \qquad \nabla^2 \omega_k = \frac{1}{c_2^2}\frac{\partial^2 \omega_k}{\partial t^2} \tag{8.4.11}$$

as wave equations for the dilatation e and the rotation components. It should be remembered that these are only special cases of the general Eq. (8). In a specific problem, it may not be possible to satisfy boundary conditions with the solutions uncoupled in this manner.

Let us return to the general case of Eq. (8) and differentiate it with respect to X_m to obtain

$$(\lambda + 2G)(\nabla^2 \phi)_{,km} + G e_{krs}(\nabla^2 \psi_s)_{,rm} = \rho\left(\frac{\partial^2 \phi}{\partial t^2}\right)_{,km} + \rho e_{krs}\left(\frac{\partial^2 \psi_s}{\partial t^2}\right)_{,rm} = 0. \tag{8.4.12}$$

A contraction by placing $m = k$ and summing yields an equation for ϕ. The terms involving ψ_s sum to zero, since, for example, $e_{krs}(\nabla^2 \psi_s)_{,rk} = 0$, because the first factor is antisymmetric in kr, while the second is symmetric. We thus obtain

$$\nabla^2\left[(\lambda + 2G)\nabla^2 \phi - \rho\frac{\partial^2 \phi}{\partial t^2}\right] = 0. \tag{8.4.13}$$

We obtain equations for the ψ_k by multiplying Eq. (12) by e_{kmp}, summing on m and k, and using the fact that $\psi_{k,k} = 0$. The terms involving ϕ sum to zero because e_{kmp} is antisymmetric in km. For $p = 1$, the terms in ψ_s then expand to

$$G\nabla^2(\psi_{2,12} - \psi_{1,22} - \psi_{1,33} + \psi_{3,13}) = \rho\frac{\partial^2}{\partial t^2}(\psi_{2,12} - \psi_{1,22} - \psi_{1,33} + \psi_{3,13}).$$

By using Eq. (7), we can write this as

$$\nabla^2\left[G\nabla^2 \psi_1 - \rho\frac{\partial^2 \psi_1}{\partial t^2}\right] = 0,$$

and similar equations may be obtained for $p = 2$ and $p = 3$. Thus, changing the free index from p to k, we have the three equations

$$\nabla^2\left[G\nabla^2 \psi_k - \rho\frac{\partial^2 \psi_k}{\partial t^2}\right] = 0. \tag{8.4.14}$$

If we define functions Φ, Ψ_1, Ψ_2, Ψ_3 by

$$\nabla^2 \phi + \frac{1}{c_1^2}\frac{\partial^2 \phi}{\partial t^2} = \Phi \tag{8.4.15}$$

$$\nabla^2 \psi_k + \frac{1}{c_2^2}\frac{\partial^2 \psi_k}{\partial t^2} = \Psi_k, \tag{8.4.16}$$

then Φ and Ψ_k must be harmonic functions, since by Eqs. (13) and (14),

$$\nabla^2 \Phi = 0 \qquad \nabla^2 \Psi_k = 0. \tag{8.4.17}$$

The Ψ_k and Φ are not independent, since they are connected through Eq. (8) by

$$\frac{c_1^2}{c_2^2}\frac{\partial \Phi}{\partial X_k} + e_{krs}\frac{\partial \Psi_s}{\partial X_r} = 0 \qquad \text{or} \qquad \frac{c_1^2}{c_2^2}\nabla\Phi + \nabla \times \Psi = 0. \tag{8.4.18}$$

Solutions of Eq. (8) can be obtained by solving the system of Eqs. (15) through (18). The special case of the separated wave equations, Eqs. (9) and (10), corresponds to choosing $\Phi \equiv 0$ and the $\Psi_k \equiv 0$.

Elastostatic equations can be obtained by putting the time derivatives equal to zero in Eqs. (15) and (16), which then combine with Eqs. (17) to give

$$\nabla^4 \phi = 0 \quad \text{and} \quad \nabla^4 \psi_k = 0, \tag{8.4.19}$$

showing that the potentials are biharmonic functions in the elastostatic case.

In certain special cases, the elastostatic solutions may be found by taking Φ and the Ψ_k as constants, so that ϕ and the ψ_k satisfy Poisson's equations:

$$\nabla^2 \phi = \text{const.} \qquad \nabla^2 \psi_k = \text{const.} \tag{8.4.20}$$

If the constants are zero, the elastostatic scalar and vector potentials are then harmonic. This may lead to some useful solutions, but there is, of course, no guarantee that every elastostatic solution can be represented in terms of four harmonic potential functions by Eqs. (2). We shall see later, however, that the Papkovich-Neuber potentials [see Eqs. (26) ff. below] provide a representation for every elastostatic displacement field in terms of four harmonic functions. A special case of Eqs. (20) is ***Lamé's strain potential***, which can be obtained by choosing all the $\psi_k = 0$ and choosing ϕ as a harmonic function. Fung (1965), Sec. 8.3, gives explicit formulas for stress in rectangular and cylindrical coordinates in terms of Lamé's strain potential (with a factor of $2G$ introduced into the definition, so that $2Gu_k = \partial\phi/\partial X_k$).

We saw in connection with Eqs. (19) that the elastostatic field can be represented by four biharmonic potential functions. The four are not independent, since they must satisfy Eq. (18) identically. The Galerkin vector provides an explicit representation in terms of just three independent functions, which are biharmonic in the absence of body forces.

Galerkin vector. In papers published in 1930, B. Galerkin introduced three potential functions from which displacements are obtained in terms of second derivatives of the potentials. P. F. Papkovich pointed out in 1932 that the three Galerkin functions are components of a vector, now known as the Galerkin vector. The presentation here is similar to that of Papkovich; see also Westergaard (1952), Sec. 66. We attempt to choose the components of a vector field **f** and the constant c in such a way that the equation

$$2G\mathbf{u} = c\nabla^2\mathbf{f} - \nabla(\nabla\cdot\mathbf{f}) \tag{8.4.21}$$

is an elastostatic solution of the Navier Eq. (1), which for the elastostatic case can, by the use of Eqs. (6.2.7), be brought into the form

$$G\left[\nabla^2\mathbf{u} + \frac{1}{1-2\nu}\nabla(\nabla\cdot\mathbf{u})\right] + \rho\mathbf{b} = 0. \tag{8.4.22}$$

Substitution of Eq. (21) into Eq. (22) yields

$$c\nabla^4\mathbf{f} - \nabla^2[\nabla(\nabla\cdot\mathbf{f})] + \frac{c}{1-2\nu}\nabla[\nabla\cdot\nabla^2\mathbf{f}] - \frac{1}{1-2\nu}\nabla\{\nabla\cdot[\nabla(\nabla\cdot\mathbf{f})]\} + 2\rho\mathbf{b} = 0.$$

Now the last three operators in this equation are identically equal, i.e.,

$$\nabla^2[\nabla(\nabla\cdot\mathbf{f})] \equiv \nabla[\nabla\cdot\nabla^2\mathbf{f}] \equiv \nabla\{\nabla\cdot[\nabla(\nabla\cdot\mathbf{f})]\},$$

as can be verified by component expansions. Hence, if we choose $c = 2(1-\nu)$, the equation for **f** simplifies greatly. Then any **f** satisfying

$$\nabla^4\mathbf{f} = -\frac{\rho\mathbf{b}}{1-\nu} \tag{8.4.23}$$

provides an equilibrium solution of the Navier equations. The solution, Eq. (21), takes the form

$$2G\mathbf{u} = 2(1-\nu)\nabla^2\mathbf{f} - \nabla(\nabla\cdot\mathbf{f}). \tag{8.4.24}$$

Typical rectangular Cartesian component forms resulting are

$$2Gu_x = 2(1-\nu)\nabla^2 f_x - \frac{\partial}{\partial X}(\nabla\cdot\mathbf{f}) \qquad 2Ge = (1-2\nu)\nabla^2(\nabla\cdot\mathbf{f})$$

$$T_{xx} = 2(1-\nu)\frac{\partial}{\partial X}(\nabla^2 f_x) + \left(\nu\nabla^2 - \frac{\partial^2}{\partial X^2}\right)(\nabla\cdot\mathbf{f}) \qquad T_{kk} = (1+\nu)\nabla^2(\nabla\cdot\mathbf{f})$$

$$T_{xy} = (1-\nu)\left[\left(\frac{\partial}{\partial Y}\nabla^2 f_x\right) + \left(\frac{\partial}{\partial X}\nabla^2 f_y\right)\right] - \frac{\partial^2}{\partial X\,\partial Y}(\nabla\cdot\mathbf{f}). \tag{8.4.25}$$

Other Cartesian components can be written by cyclic permutation of the indices. Note that the stresses involve third derivatives.

Because of the vector character of the Galerkin potential, evaluation in curvilinear coordinates is a straightforward, though possibly complicated, process using the results of App. II. The Papkovich-Neuber potentials, to be discussed shortly, can also be represented in vector form, and have the advantage of involving only first derivatives in the computation of displacements, and also of being in terms of harmonic functions instead of biharmonic functions. Direct solutions for the potentials in either case are complicated by the fact that boundary conditions are in general not easily formulated in terms of the potential functions. However, a number of interesting special solutions have been obtained by inverse methods. And in certain special cases where a solution is possible in terms of a single function, e.g., with two components of the Galerkin vector identically zero, it may be fairly easy to establish boundary conditions for a direct solution for the remaining component. An example of this is ***Love's strain function***, which may be interpreted as a Galerkin vector with $f_1 = f_2 = 0$. Explicit formulas for the components of stress and displacement in this case may be found, for example, in Fung (1965) for both rectangular Cartesian coordinates and cylindrical coordinates. When f_3 is a function only of the cylindrical r and z, this reduces to the function introduced by Love in 1906 to treat solids of revolution under axisymmetric loading.

Several useful special solutions can be obtained by means of Love's strain function. See, for example, Fung (1965) or Westergaard (1952) for a solution to Kelvin's problem of a concentrated force in the interior of an infinite region and for solutions to the Boussinesq and Cerruti problems of concentrated force loading on the boundary of a semi-infinite region. These last two solutions do not follow directly from the Love strain function except when Poisson's ratio has the special value of one-half. The general case can, however, be solved by use of what Westergaard calls the " twinned gradient " to account for the perturbation of an elasticity solution by a change in Poisson's ratio. These problems can also be solved by other methods; some examples using the Papkovich-Neuber potentials will be presented at the end of this section.

The Papkovich-Neuber functions can be related to the Galerkin vector components; see, for example, Fung (1965), Sec. 8.12. But we will follow a different procedure, beginning with the Helmholtz representation of Eq. (2).

Papkovich-Neuber potentials. When the Helmholtz representation of Eq. (2) is substituted into the elastostatic Navier Eq. (22), the resulting third-order differential equation is, after collecting terms,

$$\nabla^2[\alpha \nabla \phi + \nabla \times \boldsymbol{\psi}] = -\frac{\rho \mathbf{b}}{G}, \tag{8.4.26}$$

where

$$\alpha = 1 + \frac{1}{1-2\nu} = \frac{2(1-\nu)}{1-2\nu}. \tag{8.4.27}$$

The ***Papkovich-Neuber potentials*** are a scalar potential function ϕ_0 and a vector potential $\boldsymbol{\phi}$ (not to be confused with the scalar potential ϕ of the Helmholtz representation, which will disappear from the discussion). The vector $\boldsymbol{\phi}$ is defined by

$$\boldsymbol{\phi} = \alpha\nabla\phi + \nabla \times \boldsymbol{\psi}. \tag{8.4.28}$$

Then Eq. (26) becomes

$$\nabla^2\boldsymbol{\phi} = -\frac{\rho\mathbf{b}}{G}. \tag{8.4.29}$$

From Eq. (28),

$$\nabla\cdot\boldsymbol{\phi} = \alpha\nabla^2\phi.$$

If this last equation is considered as an equation for the scalar ϕ, we can verify that

$$2\alpha\phi = \phi_0 + \mathbf{r}\cdot\boldsymbol{\phi}, \tag{8.4.30}$$

where $\mathbf{r}$ is the position vector, is a solution for the scalar ϕ in terms of the vector $\boldsymbol{\phi}$ and another scalar function ϕ_0, provided that ϕ_0 satisfies

$$\nabla^2\phi_0 = \frac{\rho\mathbf{b}\cdot\mathbf{r}}{G}. \tag{8.4.31}$$

The verification is as follows: from the proposed solution, Eq. (30), we get

$$\begin{aligned} 2\alpha\nabla^2\phi &= \nabla^2\phi_0 + 2\nabla\cdot\boldsymbol{\phi} + \mathbf{r}\cdot\nabla^2\boldsymbol{\phi} \\ &= \nabla^2\phi_0 + 2\nabla\cdot\boldsymbol{\phi} - \frac{\rho\mathbf{b}\cdot\mathbf{r}}{G} \end{aligned}$$

by Eq. (29). Hence we see that ϕ satisfies the equation $\nabla\cdot\boldsymbol{\phi} = \alpha\nabla^2\phi$, provided that Eq. (31) is satisfied, as was asserted above. Equations (29) and (31) are Poisson's equations for the Papkovich-Neuber potentials $\boldsymbol{\phi}$ and ϕ_0. When the body forces are zero, these reduce to Laplace's equations, so that, in the absence of body forces, $\boldsymbol{\phi}$ and ϕ_0 are harmonic functions. From Eq. (28),

$$\begin{aligned} \nabla \times \boldsymbol{\psi} &= \boldsymbol{\phi} - \alpha\nabla\phi \\ &= \boldsymbol{\phi} - \tfrac{1}{2}\nabla(\phi_0 + \mathbf{r}\cdot\boldsymbol{\phi}) \qquad \text{by Eq. (30).} \end{aligned}$$

When we substitute this and Eq. (30) into the Helmholtz relation, Eq. (2), we obtain, after collecting terms and using Eq. (27) to eliminate α, the displacement solution.

Displacement Solution in Terms of
Papkovich-Neuber Potentials ϕ_0 and $\boldsymbol{\phi}$

$$\mathbf{u} = \boldsymbol{\phi} - \frac{1}{4(1-\nu)} \nabla(\phi_0 + \mathbf{r}\cdot\boldsymbol{\phi})$$

with rectangular Cartesian components

$$u_k = \phi_k - \frac{1}{4(1-\nu)} \frac{\partial}{\partial X_k}(\phi_0 + X_1\phi_1 + X_2\phi_2 + X_3\phi_3), \tag{8.4.32}$$

where

$$\nabla^2\phi_0 = \frac{\rho\mathbf{b}\cdot\mathbf{r}}{G} \quad \text{and} \quad \nabla^2\boldsymbol{\phi} = -\frac{\rho\mathbf{b}}{G}. \tag{8.4.33}$$

Typical rectangular Cartesian stress components are [Recall that $\alpha = 2(1-\nu)/(1-2\nu)$.]

$$\begin{aligned} T_{11} &= \frac{\lambda}{\alpha}\nabla\cdot\boldsymbol{\phi} + 2G\left[\frac{\partial\phi_1}{\partial X_1} - \frac{1}{4(1-\nu)}\frac{\partial^2}{\partial X_1^2}(\phi_0 + X_1\phi_1 + X_2\phi_2 + X_3\phi_3)\right] \\ T_{23} &= G\left[\frac{\partial\phi_2}{\partial X_3} + \frac{\partial\phi_3}{\partial X_2} - \frac{1}{2(1-\nu)}\frac{\partial^2}{\partial X_2 \partial X_3}(\phi_0 + X_1\phi_1 + X_2\phi_2 + X_3\phi_3)\right]. \end{aligned} \tag{8.4.34}$$

Note that $e = \nabla\cdot\mathbf{u} = (\nabla\cdot\boldsymbol{\phi})/\alpha$.

The Papkovich-Neuber potentials ϕ_0 and $\boldsymbol{\phi}$ satisfy Poisson's equations in the interior of the region, so that $\nabla^2\phi_0$ and $\nabla^2\boldsymbol{\phi}$ are known throughout the region when the body forces are prescribed. Hence the values of ϕ_0 and $\boldsymbol{\phi}$ could be found by a Green's formula, Eq. (7.3.8) (with an added term equal to the volume integral of $-\nabla^2\phi_0/r_1$ or $-\nabla^2\boldsymbol{\phi}/r_1$, since the functions are not harmonic), if the boundary values of the potentials were known. Unfortunately, the boundary values of ϕ_0 and $\boldsymbol{\phi}$ are not usually known; instead the potentials satisfy complicated boundary conditions involving derivatives of the potential functions. For displacement boundary conditions, for example, Eqs. (32) provide relations among the values of the first derivatives of the potential functions on the boundary. Traction boundary conditions, involving second derivatives, may be obtained by substituting Eq. (32) into the boundary condition of Eq. (8.1.9). But the traction boundary conditions are too complicated in the general case to be of any use in formulating a problem for direct solution. Some special cases can be handled.

Before treating any examples, we introduce some remarks on the solution.

Papkovich showed in 1932 that every sufficiently regular solution of the elastostatic Navier Eq. (22) admits a representation of the form of Eq. (32) (in terms of harmonic functions for the case of no body forces). A more direct proof was given by Mindlin in 1936. The solution was independently discovered by Neuber in 1934. Some elements of the solution had already been given by Boussinesq in 1885 but not the general case. For the case of no body forces, the solution is in terms of four scalar harmonic functions, ϕ_0 and the three Cartesian components ϕ_1, ϕ_2, ϕ_3 of $\boldsymbol{\phi}$. Several authors, beginning with Papkovich and Neuber, have claimed that the number of required independent harmonic functions can be reduced to three—for example, by arbitrarily choosing one of the four to be identically zero. Some of the early proofs of this were not correct, but in 1956 Eubanks and Sternberg published a proof that ***in any convex region containing the origin :*** (1) any one of the Cartesian components ϕ_1, ϕ_2, or ϕ_3 may be set equal to zero without loss of completeness, or (2) if 4ν is not an integer, ϕ_0 may be set equal to zero without loss of completeness. In these cases, the solution for no body force can be represented in terms of just three independent harmonic functions.

Green's formula. By Eq. (7.3.7), two sufficiently differentiable functions f and g of position in a volume V bounded by a surface S with ***outer normal*** $\hat{\mathbf{n}}$ satisfy ***Green's Second Identity:***

$$\int_S \left[f\frac{\partial g}{\partial n} - g\frac{\partial f}{\partial n} \right] dS = \int_V [f\nabla^2 g - g\nabla^2 f]\, dV. \tag{8.4.35}$$

We now obtain a modified version of Green's Third Identity, Eq. (7.3.8) by a derivation similar to that given in Sec. 7.3. The modification consists in not requiring f to be harmonic.

Let $P(X, Y, Z)$ be an ***arbitrary field point***, considered a fixed point for the integrations, which are carried out with respect to a variable ***source point*** $Q(\xi, \eta, \zeta)$. In rectangular coordinates, for example, we would have $dV = d\xi\, d\eta\, d\zeta$. We denote the distance PQ by r_1. The reader can verify that, if we define $g = 1/r_1$, then $\nabla^2 g = 0$ for $r_1 \neq 0$, by evaluating the Laplacian in rectangular coordinates with respect to the variable point Q. ($1/r_1$ is also harmonic with respect to the variable point P, if Q is fixed.) We apply Eq. (35) with this choice of g to the region $V - V'$ between the surface S and a small sphere of radius ϵ, centered at P, whose volume and surface we denote by V' and S'. See Fig. 8.9.

The result is, since $\nabla^2 g = 0$ in the region between S' and S,

$$\int_S \left[f\frac{\partial}{\partial n}\left(\frac{1}{r_1}\right) - \frac{1}{r_1}\frac{\partial f}{\partial n} \right] dS - \int_{S'} \left[f\frac{\partial}{\partial r_1}\left(\frac{1}{r_1}\right) - \frac{1}{r_1}\frac{\partial f}{\partial r_1} \right] dS = -\int_{V-V'} \frac{1}{r_1}\nabla^2 f\, dV. \tag{8.4.36}$$

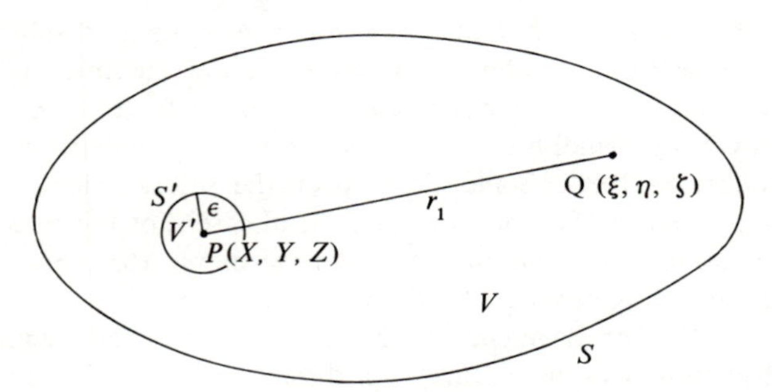

Fig. 8.9 Source point Q and field point P for Green's formula.

The minus sign appears before the second integal because $\partial/\partial r_1 = -\partial/\partial n$ on the sphere S', since the outer normal from the region $V - V'$ points toward P, the origin of r_1. We seek the limiting form of Eq. (36) as the radius ϵ of the sphere tends to zero. Just as in Sec. 7.3,

$$\lim_{\epsilon\to 0}\int_{S'}\left[f\frac{\partial}{\partial r_1}\left(\frac{1}{r_1}\right)-\frac{1}{r_1}\frac{\partial f}{\partial r_1}\right]dS=\lim_{\epsilon\to 0}4\pi\epsilon^2\left[-\frac{1}{\epsilon^2}f-\frac{1}{\epsilon}\frac{\partial f}{\partial r_1}\right]_{\text{Av.}}=-4\pi f_P,$$

where the subscript "Av." denotes the mean value on the sphere. The limit of the volume integral over $V - V'$ will be the improper volume integral over V:

$$\lim_{\epsilon\to 0}\int_{V-V'}\frac{1}{r_1}\nabla^2 f\,dV=\int_V\frac{1}{r_1}\nabla^2 f\,dV.$$

The existence of this improper integral follows from the fact that the integral of $\nabla^2 f/r_1$ over V' is bounded for any ϵ and tends to zero as the radius ϵ tends to zero, ***assuming that*** $\nabla^2 f$ ***is bounded,*** since if $|\nabla^2 f|\leq M$ in V, then

$$\left|\int_{V'}\frac{1}{r_1}\nabla^2 f\,dV\right|\leq M\int_{V'}\frac{1}{r_1}\,dV=2\pi M\epsilon^2.$$

(The last integral over the sphere can be evaluated explicitly, for example, by using spherical coordinates r_1, θ, ϕ.) When these limits are substituted into Eq. (36), the result is ***Green's formula:***

$$4\pi f_P=\int_S\left[\frac{1}{r_1}\frac{\partial f}{\partial n}-f\frac{\partial}{\partial n}\left(\frac{1}{r_1}\right)\right]dS-\int_V\frac{1}{r_1}\nabla^2 f\,dV. \tag{8.4.37}$$

The explicit application of this formula requires a knowledge of f and $\partial f/\partial n$ on S and of $\nabla^2 f$ inside V. We consider first the case of an infinite solid, where the boundary disappears.

Infinite solid loaded by body forces. Consider the problem of a body occupying volume V and loaded over a small part V_1 of V by a continuous body-force distribution $\rho\mathbf{b}$ per unit volume, with zero body force in the remainder of V. We take the origin inside V_1 and try to find a solution for the Papkovich-Neuber potentials $\boldsymbol{\phi}$ and ϕ_0 in an infinite region, such that on any surface S outside a large sphere of radius R^*, centered at the origin, $\boldsymbol{\phi}$, $\partial\boldsymbol{\phi}/\partial n$, ϕ_0, and $\partial\phi_0/\partial n$ all tend to zero as R^* becomes infinite. This is consistent with assuming that displacements and tractions tend to zero at infinity, although the total force on the infinite boundary need not tend to zero. Under these assumptions, Eqs. (33) and Green's formula, Eq. (37), yield

$$\phi_0 = -\frac{1}{4\pi G}\int_{V_1}\frac{\mathbf{r}'\cdot\rho\mathbf{b}}{r_1}\,dV \qquad \boldsymbol{\phi} = +\frac{1}{4\pi G}\int_{V_1}\frac{\rho\mathbf{b}}{r_1}\,dV, \tag{8.4.38}$$

where $\mathbf{r}'$ is the position vector of the variable source point Q in V_1 and r_1 is the distance from Q to the field point where ϕ_0 and $\boldsymbol{\phi}$ are evaluated.

Kelvin's problem of concentrated force in the interior of an infinite solid may be solved by specializing Eqs. (38) as follows. Assume that as the volume V_1 shrinks down to a point at the origin, the body-force intensity $\rho\mathbf{b}$ becomes infinite in such a way that

$$\lim_{V_1\to 0}\int_{V_1}\rho\mathbf{b}\,dV = P\hat{\mathbf{e}}_P, \tag{8.4.40}$$

where $\hat{\mathbf{e}}_P$ is a unit vector and P is the finite magnitude of the concentrated force at the origin. Although our derivation of Eq. (37), and hence Eqs. (38), with the integrand $\nabla^2 f$ assumed to be bounded, does not rigorously include this case, we assume that the concentrated force can be handled as the limiting case of Eqs. (38), subject to Eq. (40). Thus as $V_1 \to 0$, it follows that $Q \to 0$, $\mathbf{r}' \to \mathbf{0}$, and $r_1 \to R$, where R is the magnitude of the position vector to the field point where the functions ϕ_0 and $\boldsymbol{\phi}$ are being evaluated. In the limit then, as $V_1 \to 0$, we obtain for Kelvin's problem

$$\boldsymbol{\phi} = \frac{P}{4\pi GR}\hat{\mathbf{e}}_P \qquad \phi_0 = 0. \tag{8.4.41a}$$

If P acts along the z-axis, in ***cylindrical coordinates*** r, θ, z, $\hat{\mathbf{e}}_P = \hat{\mathbf{e}}_z$, and

$$\boldsymbol{\phi} = \frac{P}{4\pi GR}\hat{\mathbf{e}}_z \quad \phi_0 = 0, \qquad \text{where} \quad R^2 = r^2 + z^2. \tag{8.4.41b}$$

The displacements are given by Eq. (32),

$$\mathbf{u} = \boldsymbol{\phi} - \frac{1}{4(1-\nu)}\nabla(\phi_0 + \mathbf{r}\cdot\boldsymbol{\phi}).$$

Hence we have in cylindrical coordinates

$$\mathbf{u} = \frac{P}{4\pi GR}\hat{\mathbf{e}}_z - \frac{1}{4(1-\nu)}\nabla\left(\frac{zP}{4G\pi R}\right) = \frac{P}{4\pi G}\left[\frac{1}{R}\hat{\mathbf{e}}_z - \frac{1}{4(1-\nu)}\nabla\left(\frac{z}{R}\right)\right].$$

Now, with radial symmetry,

$$\nabla\left(\frac{z}{R}\right) = \left[\hat{\mathbf{e}}_r\frac{\partial}{\partial r} + \hat{\mathbf{e}}_z\frac{\partial}{\partial z}\right]\frac{z}{R} \qquad \text{by Eq. (II.4.C2).}$$

Hence

$$\nabla\left(\frac{z}{R}\right) = -\frac{rz}{R^3}\hat{\mathbf{e}}_r + \left(\frac{1}{R} - \frac{z^2}{R^3}\right)\hat{\mathbf{e}}_z,$$

and the displacement is given in cylindrical coordinates by

$$\mathbf{u} = \frac{P}{16\pi G(1-\nu)}\left[\frac{rz}{R^3}\hat{\mathbf{e}}_r + \left(\frac{3-4\nu}{R} + \frac{z^2}{R^3}\right)\hat{e}_z\right] \tag{8.4.42}$$

where

$$R^2 = r^2 + z^2.$$

This result agrees with that obtained in Fung (1965), Sec. 8.8, by means of the Galerkin vector. Fung also gives explicit formulas for the stresses. (Note that his concentrated force magnitude is $2P$.)

We now solve the problem of a distributed normal load on the plane boundary of a half-space by determining the Green's function.

Green's function for the half-space. In potential theory the Green's function $G(P, Q)$ for a region is a function of two points P and Q, symmetric in P and Q, and representable as

$$G(P, Q) = \frac{1}{r_1} + g(P, Q) \qquad \nabla^2 g = 0, \tag{8.4.43}$$

where r_1 is the distance from P to Q and $g(P, Q)$ is a harmonic function in the interior of the region (as a function of either P or Q with the other point held fixed) and such that $G(P, Q) \to 0$ as either P or Q approaches a point on the boundary along a path for which the final approach to the boundary is along a normal, with the other point held fixed. Note that $G(P, Q)$ is itself harmonic, except at $r_1 = 0$. If the harmonic function $g(P, Q)$ is substituted for g in Green's Second Identity, Eq. (35), the result is

$$0 = \int_S \left[g\frac{\partial f}{\partial n} - f\frac{\partial g}{\partial n}\right] dS - \int_V g\,\nabla^2 f\,dV.$$

We add this to Eq. (37) to eliminate $\partial f/\partial n$, since $(1/r_1) + g = G$, which vanishes on S. We obtain then the ***formula for f in terms of Green's function:***

$$4\pi f_P = -\int_S f\frac{\partial G}{\partial n}\,dS - \int_V G\,\nabla^2 f\,dV, \tag{8.4.44}$$

where the integrations are again with respect to the variable source point Q. This formula provides the solution at an interior point P for a function f satisfying Poisson's equation ($\nabla^2 f =$ known function) in V and given boundary values on S, provided that the Green's function for the region is known. Unfortunately the determination of the Green's function for a region is usually not easy. For the half-space, however, it is readily found, as follows.

Let the plane boundary of the half-space be the XY-plane, with the body occupying the region $Z > 0$. Let $Q_2(\xi, \eta, -\zeta)$ be the image point of $Q(\xi, \eta, \zeta)$ in the XY-plane, and let r_2 denote the distance PQ_2. For any P in the interior, $r_2 \neq 0$, since Q_2 is not in the interior. And $1/r_2$ is a harmonic function of Q in the interior of the body, which takes on boundary values on S equal to those of $1/r_1$. Hence we may take $g(P, Q) = -1/r_2$ to obtain the ***Green's function for the half-space:***

$$G(P, Q) = \frac{1}{r_1} - \frac{1}{r_2}.$$

On S

$$\frac{\partial G}{\partial n} = -\left[\frac{\partial G}{\partial \zeta}\right]_{\zeta=0} = -\left[\frac{Z-\zeta}{r_1^3} + \frac{Z+\zeta}{r_2^3}\right]_{\zeta=0} = -\frac{2Z}{r_0^3}$$

$$= +2\frac{\partial}{\partial Z}\left(\frac{1}{r_0}\right),$$

where

$$r_1^2 = (X-\xi)^2 + (Y-\eta)^2 + (Z-\zeta)^2$$
$$r_2^2 = (X-\xi)^2 + (Y-\eta)^2 + (Z+\zeta)^2$$

and we let

$$r_0 = [r_1]_{\zeta=0} = [r_2]_{\zeta=0}. \tag{8.4.45}$$

We thus obtain the following special ***Green's formula for the half-space:***

$$f(P) = -\frac{1}{2\pi}\frac{\partial}{\partial Z}\int_S \frac{f}{r_0}\,dS - \frac{1}{4\pi}\int_V G\,\nabla^2 f\,dV, \tag{8.4.46}$$

where r_0, r_1, r_2, and G are given by (45).

Problem of normal traction on the plane boundary of the half-space. Given on part S_1 of the boundary S ($Z = 0$) of the half-space a distributed normal pressure of intensity q (in general a variable function of position on S_1), the traction boundary conditions are

$$T_{zz} = -q \qquad T_{zx} = T_{zy} = 0 \quad \text{on } S_1 \tag{8.4.47}$$

with zero tractions on the remainder of the boundary $Z = 0$. For zero body force, we seek to determine the Papkovich-Neuber potentials ϕ_0 and $\boldsymbol{\phi}$. We will try to solve the problem for ϕ_0 and ϕ_3, assuming that $\phi_1 \equiv 0$ and $\phi_2 \equiv 0$. If we succeed in this, then we will have, by Eqs. (32),

$$\mathbf{u} = \phi_3 \hat{\mathbf{e}}_z - \frac{1}{4(1-\nu)} \nabla(\phi_0 + z\phi_3) \quad \text{and} \quad \nabla^2 \phi_0 = 0 \qquad \nabla^2 \phi_3 = 0 \tag{8.4.48}$$

for no body force. In terms of this solution, we will have the following.

Typical Cartesian components when $\phi_2 = \phi_1 = 0$:

$$\left.\begin{aligned}
u_x &= -\frac{1}{4(1-\nu)} \frac{\partial}{\partial X} [\phi_0 + Z\phi_3] \\
u_z &= \phi_3 - \frac{1}{4(1-\nu)} \frac{\partial}{\partial Z} [\phi_0 + Z\phi_3] \\
E_{xx} &= -\frac{1}{4(1-\nu)} \frac{\partial^2}{\partial X^2} [\phi_0 + Z\phi_3] \qquad e = \frac{1-2\nu}{2(1-\nu)} \frac{\partial \phi_3}{\partial Z} \\
E_{zz} &= \frac{\partial \phi_3}{\partial Z} - \frac{1}{4(1-\nu)} \left[\frac{\partial^2 \phi_0}{\partial Z^2} + Z \frac{\partial^2 \phi_3}{\partial Z^2} + 2 \frac{\partial \phi_3}{\partial Z} \right] \\
E_{xy} &= -\frac{1}{4(1-\nu)} \left[\frac{\partial^2 \phi_0}{\partial X \partial Y} + Z \frac{\partial^2 \phi_3}{\partial X \partial Y} \right] \\
E_{xz} &= \frac{1}{2} \frac{\partial \phi_3}{\partial X} - \frac{1}{4(1-\nu)} \left[\frac{\partial^2 \phi_0}{\partial X \partial Y} + Z \frac{\partial^2 \phi_3}{\partial X \partial Z} + \frac{\partial \phi_3}{\partial X} \right] \\
T_{zx} &= G \left[\frac{\partial \phi_3}{\partial X} - \frac{1}{2(1-\nu)} \left(\frac{\partial^2 \phi_0}{\partial Z \partial X} + Z \frac{\partial^2 \phi_3}{\partial Z \partial X} + \frac{\partial \phi_3}{\partial X} \right) \right] \\
T_{zz} &= G \left[\frac{\partial \phi_3}{\partial Z} - \frac{1}{2(1-\nu)} \left(\frac{\partial^2 \phi_0}{\partial Z^2} + Z \frac{\partial^2 \phi_3}{\partial Z^2} \right) \right].
\end{aligned}\right\} \tag{8.4.49}$$

The boundary conditions on $Z = 0$, Eqs. (47), then take the form

$$\left.\begin{aligned}
\left[\frac{\partial \phi_3}{\partial Z} - \frac{1}{2(1-\nu)} \frac{\partial^2 \phi_0}{\partial Z^2} \right]_{Z=0} &= \frac{-q}{G} \\
\left[(1-2\nu) \frac{\partial \phi_3}{\partial X} - \frac{\partial^2 \phi_0}{\partial Z \partial X} \right]_{Z=0} &= 0 \\
\left[(1-2\nu) \frac{\partial \phi_3}{\partial Y} - \frac{\partial^2 \phi_0}{\partial Z \partial Y} \right]_{Z=0} &= 0.
\end{aligned}\right\} \tag{8.4.50}$$

We still do not know the boundary values of ϕ_0 and ϕ_3; therefore we cannot apply the special Green's formula for the half-space, Eq. (46), directly to calculate ϕ_0 and ϕ_3. But we recognize from examining the boundary conditions of Eq. (50) that if we introduce three new functions f_1, f_2, f_3, equal, respectively, to the contents of the three brackets in Eqs. (50), then each of these new functions will be harmonic in the interior. And we know the boundary values for f_1, f_2, and f_3 by Eqs. (50). Hence we can use Eq. (46) to determine each of these new functions. The result is

$$\frac{\partial}{\partial Z}\left[\phi_3 - \frac{1}{2(1-\nu)}\frac{\partial\phi_0}{\partial Z}\right] = \frac{1}{2\pi G}\frac{\partial}{\partial Z}\int_{S_1}\frac{q(\xi,\eta)}{r_0}\,dS$$

$$\frac{\partial}{\partial X}\left[(1-2\nu)\phi_3 - \frac{\partial\phi_0}{\partial Z}\right] = 0 \qquad \frac{\partial}{\partial Y}\left[(1-2\nu)\phi_3 - \frac{\partial\phi_0}{\partial Z}\right] = 0. \tag{8.4.51}$$

We integrate the first Eq. (51) with respect to Z; the arbitrary function of X and Y, which should be added, must vanish in order that the displacements vanish at infinity. The two second equations, when integrated with respect to X and Y, respectively (with arbitrary additive functions set equal to zero for displacements vanishing at infinity), both yield the same result. We obtain

$$2(1-\nu)\phi_3 - \frac{\partial\phi_0}{\partial Z} = \frac{1-\nu}{\pi G}\int_{S_1}\frac{q}{r_0}\,dS \qquad (1-2\nu)\phi_3 - \frac{\partial\phi_0}{\partial Z} = 0.$$

Simultaneous solution of these for ϕ_3 and $\partial\phi_0/\partial Z$ yields the following result.

Solution for distributed normal load:

$$\phi_3 = \frac{1-\nu}{\pi G}\int_{S_1}\frac{q}{r_0}\,dS \qquad \frac{\partial\phi_0}{\partial Z} = \frac{(1-\nu)(1-2\nu)}{\pi G}\int_{S_1}\frac{q}{r_0}\,dS. \tag{8.4.52}$$

Calculation of the stresses requires only $\partial\phi_0/\partial Z$, not ϕ_0. But for displacements we need ϕ_0 too, which requires another integration with respect to Z of the second Eq. (52). This will be illustrated below for the specific example of the Boussinesq problem.

Boussinesq problem of concentrated normal force on boundary of half-space. This problem can be solved by taking the limit of Eqs. (52) as S_1 shrinks down to the origin O, while q becomes infinite in such a way that

$$\lim_{S_1\to O}\int_{S_1} q\,dS = P,$$

where P is a finite concentrated load at O in the positive Z-direction (downward); see Fig. 8.10. As $S_1 \to 0$, $r_0 \to R$, and we obtain the following from Eqs. (52).

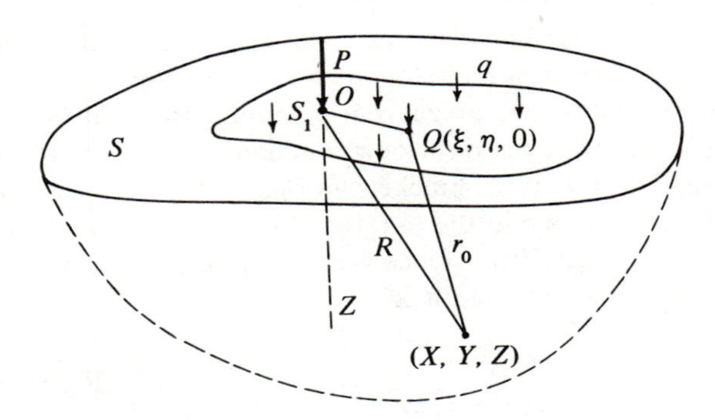

Fig. 8.10 Boussinesq problem.

Boussinesq problem solution:

$$\phi_3(X, Y, Z) = \frac{(1-\nu)P}{\pi G R}$$

$$\frac{\partial}{\partial Z}[\phi_0(X, Y, Z)] = \frac{(1-\nu)(1-2\nu)P}{\pi G R} \tag{8.4.53a}$$

whence

$$\phi_0(X, Y, Z) = \frac{(1-\nu)(1-2\nu)P}{\pi G}\log_e(R+Z) \tag{8.4.53b}$$

since

$$\int \frac{dz}{\sqrt{r^2+z^2}} = \log_e[\sqrt{r^2+z^2}+z]$$

if the arbitrary additive function of X and Y is chosen to be zero so that the displacements vanish at infinity. An arbitrary linear function of X and Y could be added here and an arbitrary constant added to ϕ_3 to adjust the rigid-body displacement of the whole body in such a way that the displacement components would vanish at any specified finite point except the origin, where the solution is singular because of the concentrated load there. This would not change the stress distribution.

In cylindrical coordinates, the physical components of displacement and stress follow from Eq. (32), the strain definitions, and Hooke's law. The results are as follows.

Boussinesq problem solution in cylindrical coordinates (r, θ, z)**:**

$$
\left.\begin{aligned}
u_r &= \frac{Pr}{4\pi GR}\left[\frac{z}{R^2} - \frac{1-2\nu}{R+z}\right] \qquad u_\theta = 0.\\
u_z &= \frac{P}{4\pi GR}\left[2(1-\nu) + \frac{z^2}{R^2}\right] \qquad R^2 = r^2 + z^2\\
T_{rr} &= \frac{P}{2\pi}\left[\frac{1-2\nu}{R(R+z)} - \frac{3r^2 z}{R^5}\right]\\
T_{\theta\theta} &= \frac{P(1-2\nu)}{2\pi}\left[\frac{z}{R^3} - \frac{1}{R(R+z)}\right]\\
T_{zz} &= -\frac{P}{2\pi}\frac{3z^3}{R^5} \qquad T_{rz} = -\frac{P}{2\pi}\frac{3rz^2}{R^5} \qquad T_{r\theta} = T_{\theta z} = 0.
\end{aligned}\right\} \tag{8.4.54}
$$

With this example we conclude our introduction to the use of displacement potentials in three-dimensional elasticity. A number of other special solutions can be found by similar methods. The examples we have treated by Papkovich-Neuber potentials can also be treated by other methods, and are discussed in most standard works on elasticity. Direct solution for the potentials in more complicated problems is also sometimes possible by expanding them in series of eigenfunctions appropriate to the problem and seeking to determine the coefficients from the boundary conditions. But these methods are beyond the scope of this book.

In this section we have introduced the representation of the displacement field by potentials. The vector and scalar potentials of the Helmholtz representation are especially useful in dynamic problems, where they lead to separate wave equations for the dilatational wave and the shear wave. For elastostatic problems the Galerkin vector and the Papkovich-Neuber potentials are more convenient. We have seen examples of the use of the Papkovich-Neuber potentials for body-force loading of an infinite solid and for normal loads on the plane boundary of a half-space.

This section on three-dimensional solutions completes our brief introduction to the linearized theory of elasticity. Section 8.1 formulated the theory, and Secs. 8.2 and 8.3 presented plane problems. Although some of our methods of solution were similar to those of perfect-fluid theory, it is evident that there are more differences than similarities, and this is even more true when other materials are considered, e.g., viscoelastic or plastic solids. No unified method exists for all materials. For further solution methods, consult specialized works in the separate theories. A brief bibliography of books on elasticity is given at the end of this chapter.

EXERCISES

1. Show that when **u** is the small-displacement field, the vector potential $\boldsymbol{\psi}$ of Eqs. (8.4.2) does not contribute to the volume change, and show that the dilatation e is given by Eq. (4).
2. Verify that the special cases of the wave equations, Eqs. (8.4.9) and (10), satisfy Eq. (8), and that they lead to Eqs. (11).
3. Verify the derivation of the first Eq. (8.4.14) for $k = 1$.
4. Verify Eq. (8.4.18)
5. For the case that ϕ and $\boldsymbol{\psi}$ are functions only of X_1 and t (independent of X_2 and X_3), write out the expressions for the displacements according to Eq. (8.4.2b) Show that in this case differentiation of Eqs. (9) leads to wave equations for the components u_1, u_2, and u_3. Interpret this case physically.
6. Show that the ratio c_2/c_1 of the two wave speeds given by Eqs. (8.4.10) depends only on Poisson's ratio.
7. Show that the choice $c = 2(1 - \nu)$ leads to Eqs. (8.4.23) and (24), and verify that these equations lead to the expression given for T_{xx} in Eq. (25).
8. Determine the body forces, stresses, and displacements defined by the Galerkin vector $\mathbf{f} = A(X^2 + Y^2 + Z^2)\mathbf{i} + Bz^4\mathbf{k}$, where A and B are constants and $\mathbf{i}$ and $\mathbf{k}$ are unit vectors in the X-direction and Z-direction, respectively.
9. Carry out the details of showing that Eq. (8.4.32) is a solution of the Navier equation whenever ϕ_0 and $\boldsymbol{\phi}$ satisfy Eqs. (33).
10. Carry out the details of the solution of Kelvin's problem, Eqs. (8.4.41) and (42).
11. Carry out the details of deriving Green's formula for the half-space, Eq.(8.4.46).
12. Consider the problem of uniform normal pressure $q =$ Const. on a circular portion of the boundary of a half-space, with center at the origin and radius a. Set up the integral of Eqs. (8.4.52) for a typical interior point with cylindrical coordinates (r, θ, z). Evaluate the stress T_{zz} for points on the z-axis ($r = 0$).
13. Carry out the details of obtaining the Boussinesq problem solution of Eqs.(8.4.53).
14. Show that the Boussinesq solution of Eqs. (8.4.53) implies the formulas for the displacements u_r and u_z given in Eqs. (54).
15. Show that the displacement formulas of the Boussinesq solution, Eqs. (8.4.54), lead to the stresses given in Eqs. (54).

References for collateral reading on elasticity

BOOKS SLANTED TOWARD ENGINEERS

Biezeno, C.B., and R. Grammel, *Engineering Dynamics*, Vol. I, *Theory of Elasticity*, (translated from the German by M. L. Meyer). London: Blackie and Son, Ltd., 1955.

Boresi, A. P., *Elasticity in Engineering Mechanics*. Englewood Cliffs, N.J.: Prentice-Hall, Inc., 1965.

Chou, P. C., and N. J. Pagano, *Elasticity; Tensor, Dyadic and Engineering Approaches*. Princeton, N. J.: D. Van Nostrand Co., 1967.

Fung, Y. C., *Foundations of Solid Mechanics*. Englewood Cliffs, N. J.: Prentice-Hall, Inc., 1965.

Jaeger, J. C., *Elasticity, Fracture, and Flow*. London: Methuen & Co., Ltd.; New York: Barnes & Noble, 2nd ed., 1962.

Nadeau, G., *Introduction to Elasticity*. New York: Holt, Rinehart & Winston, 1964.

Neuber, H., *Kerbspannungslehre*, 2nd German ed. Berlin: Springer-Verlag, 1958, 1st ed; translated into English as *Theory of Notch Stresses*, Ann Arbor, Michigan: J. W. Edwards Co., 1946.

Southwell, R., *An Introduction to the Theory of Elasticity for Engineers and Physicists*. London: Oxford University Press, 1941.

Timoshenko, S., and J. N. Goodier, *Theory of Elasticity*, 2nd ed. New York: McGraw-Hill Book Company, 1951.

Wang, C. T., *Applied Elasticity*. New York: McGraw-Hill Book Company, 1953.

Westergaard, H. M., *Theory of Elasticity and Plasticity*. New York: Dover Publications, Inc., 1952.

MORE MATHEMATICAL TREATMENTS

Green, A. E., and W. Zerna, *Theoretical Elasticity*. London: Oxford University Press, 1954.

Landau, L. D., and E. M. Lifshitz, *Theory of Elasticity*. Oxford: Pergamon Press, Ltd., 1959.

Love, A. E. H., *Mathematical Theory of Elasticity*, 4th ed. New York: Dover Publications, Inc., 1944.

Lur'e, A. I., *Three-Dimensional Problems of the Theory of Elasticity*. New York: Interscience Publishers, Inc., 1964.

Milne-Thomson, L. M., *Plane Elastic Systems*. Berlin: Springer-Verlag, 1960.

Milne-Thomson, L. M., *Antiplane Elastic Systems*. Berlin: Springer-Verlag, 1962.

Muskhelishvili, N. I., *Some Basic Problems of the Mathematical Theory of Elasticity*. Groningen, The Netherlands: P. Noordhoff Ltd., 2nd English ed., 1963.

Pearson, C. E., *Theoretical Elasticity*. Cambridge, Mass.: Harvard University Press, 1959.

Savin, G. N., *Stress Concentrations Around Holes*. Oxford: Pergamon Press, Ltd., 1959.

Sneddon, I. N., and D. S. Berry, "The Classical Theory of Elasticity," pp. 1-126 in *Encyclopedia of Physics*, v. 6, ed. S. Flügge. Berlin: Springer-Verlag, 1958.

Sokolnikoff, I. S., *Mathematical Theory of Elasticity*, 2nd ed. New York: McGraw-Hill Book Company, 1956.

FINITE DEFORMATIONS

Green, A. E., and W. Zerna, *Theoretical Elasticity*. London: Oxford University Press, 1954.

Green, A. E., and Adkins, J. E., *Large Elastic Deformations and Non-Linear Continuum Mechanics*. London: Oxford University Press, 1960.

Murnaghan, F. D., *Finite Deformation of an Elastic Solid*. New York: John Wiley & Sons, Inc., 1951.

Novozhilov, V. V., *Foundations of the Nonlinear Theory of Elasticity*. Baltimore: Graylock Press, 1953.

Truesdell, C., ed., *Continuum Mechanics* IV, *Problems of Non-linear Elasticity*. New York: Gordon & Breach, 1965.

Truesdell, C., and W. Noll, "The Non-Linear Field Theories of Mechanics," in *Encyclopedia of Physics*, v. III/3, ed. S. Flügge. Berlin: Springer-Verlag, 1965.

THERMAL STRESS

Boley, B. A., and J. H. Weiner, *Theory of Thermal Stresses*. New York: John Wiley & Sons, Inc., 1960.

Gatewood, B. E., *Thermal Stresses*. New York: McGraw-Hill Book Company, 1957.

Parkus, H., *Thermoelasticity*. Waltham, Mass: Blaisdell Publishing Co., 1968.

ELASTIC STABILITY

Bolotin, V. V., *The Dynamic Stability of Elastic Systems*. San Francisco: Holden-Day, Inc., 1964.

Bolotin, V. V., *Nonconservative Problems of the Theory of Elastic Stability*. New York: The Macmillan Co., 1963.

Timoshenko, S., and J. M. Gere, *Theory of Elastic Stability*, 2nd ed. New York: McGraw-Hill Book Company, 1961.

Ziegler, H., *Principles of Structural Stability*. Waltham, Mass.: Blaisdell Publishing Co., 1968.

APPENDIX I

Tensors

I.1 Introduction; vector space axioms; linear independence; basis; contravariant components of a vector; Euclidean vector space; dual base vectors; covariant components of a vector

In Sec. 2.4, we defined a second-order tensor as a linear vector function associating with each argument vector in a certain set of vectors (the domain of definition of the function) another vector, the value of the function. This definition was quite independent of any coordinate system or any basis for the representation of the vectors and the tensor. Throughout the book we have, however, almost exclusively employed a rectangular Cartesian coordinate system and used the orthogonal unit base vectors $\mathbf{i}_k$ of the coordinate system as the basis for representation of vectors and tensors. This basis is independent of position and provides an especially simple formulation.

As is shown in Sec. I. 4, the natural basis of a curvilinear coordinate system is not independent of position, and this makes the formulation considerably more complicated. Why then do we use curvilinear coordinates at all in continuum mechanics? One reason is that the geometric shape of a solid body or of a fluid container may be such that the boundary-value problem to be solved is simpler in curvilinear coordinates. Another important reason is that the constitutive equations defining the material behavior may be easier to formulate relative to a coordinate net of material particles than to a coordinate net of geometric points in space. When the material undergoes deformation, what started out as a rectangular Cartesian net of material particles will be deformed into a (not necessarily orthogonal) curvilinear net, which may be useful as the reference system for the next increment of deformation.

If we have to deal only with orthogonal curvilinear coordinates, the physical components of a tensor relative to the orthogonal system can be defined without actually introducing the machinery of general curvilinear tensor components.† This appendix, however, presents general curvilinear tensor components in Euclidean space. Because so much of the modern literature on continuum mechanics is published in the language of general tensor components, it is almost essential to learn the language.

Before undertaking the study of tensors in curvilinear coordinates, we shall consider the concepts of vector space, base vectors, change of basis, tensors as linear vector operators, and the definitions of contravariant and covariant components relative to a given basis. All of this will be applicable to curvilinear tensor components when we choose the basis to be the natural basis of the system of curvilinear coordinates.

A ***vector space*** E over the set of real numbers is a set of elements $\mathbf{u}$, $\mathbf{v}$, etc. obeying the following two sets of rules.

ADDITION AXIOMS

To every pair $\mathbf{u}$, $\mathbf{v}$ there corresponds an element $\mathbf{u} + \mathbf{v}$ having the properties:

$$\mathbf{u} + \mathbf{v} = \mathbf{v} + \mathbf{u} \qquad \text{commutative} \tag{I.1.1a}$$

$$\mathbf{u} + (\mathbf{v} + \mathbf{w}) = (\mathbf{u} + \mathbf{v}) + \mathbf{w} \qquad \text{associative} \tag{I.1.1b}$$

There exists a zero vector denoted $\mathbf{0}$ such that $\mathbf{u} + \mathbf{0} = \mathbf{u}$ for every $\mathbf{u}$. (I.1.1c)

For each $\mathbf{u}$ there exists another, denoted $-\mathbf{u}$, such that $\mathbf{u} + (-\mathbf{u}) = \mathbf{0}$. (I.1.1d)

SCALAR MULTIPLE AXIOMS

To every combination of an element, $\mathbf{u}$, and a real number, a, there corresponds an element $a\mathbf{u}$ having the properties:

$$1\mathbf{u} = \mathbf{u} \tag{I.1.2a}$$

$$a(b\mathbf{u}) = (ab)\mathbf{u} \qquad \text{associative} \tag{I.1.2b}$$

$$\left.\begin{aligned} (a + b)\mathbf{u} &= a\mathbf{u} + b\mathbf{u} \qquad &\text{(I.1.2c)} \\ a(\mathbf{u} + \mathbf{v}) &= a\mathbf{u} + a\mathbf{v}. \qquad &\text{(I.1.2d)} \end{aligned}\right\} \text{ distributive}$$

A set of ordinary vectors in three dimensions is an example of a vector space, but there are many other sets of elements for which the operation "+" can be defined, as can multiplication by a real number, so that the above axioms are obeyed. Another example might be the set of all positive integral powers x^n of a real number x. Since second-order tensors satisfy these axioms, (see Sec. 2.4), the set of all second-order tensors is a vector space. Throughout this section the word "vector" will mean "element of a vector space," unless otherwise specified.

†See App. II, where the simpler formulation is given.

Linearly independent vectors. A set of m vectors $\mathbf{v}_1, \mathbf{v}_2, \ldots, \mathbf{v}_m$ is said to form a ***linearly independent system of order*** m if there is no linear combination of the m vectors which equals $\mathbf{0}$,

$$a^1\mathbf{v}_1 + a^2\mathbf{v}_2 + \cdots + a^m\mathbf{v}_m = \mathbf{0}, \tag{I.1.3}$$

except the trivial one with all coefficients a^i equal to zero. If there does exist such a linear combination with at least one $a^i \neq 0$, then the m vectors are ***linearly dependent***.

n-dimensional vector space E_n. If in a vector space there exists at least one set of n linearly independent vectors but no set of $n + 1$ linearly independent vectors, the vector space is said to be ***n-dimensional***, denoted E_n. For example, in two dimensions, if we have two noncollinear vectors $\mathbf{v}_1$ and $\mathbf{v}_2$, any other vector $\mathbf{v}_3$ can be represented as a linear combination $\mathbf{v}_3 = c_1\mathbf{v}_1 + c_2\mathbf{v}_2$ of them, so that with $c_3 = -1$, we have $c_1\mathbf{v}_1 + c_2\mathbf{v}_2 + c_3\mathbf{v}_3 = 0$, showing that three vectors cannot be linearly independent in a two-dimensional space.

We will be concerned mainly with three-dimensional space, sometimes with two-dimensional, and occasionally with spaces of more than three dimensions. For example, in plasticity the nine components of the stress tensor at a point are sometimes represented as the nine components of a vector in nine-dimensional space. We will therefore allude briefly here to the base vectors of an n-dimensional space, although in what follows we will be mainly concerned only with $n = 3$ or $n = 2$.

Basis. In an n-dimensional vector space any set of n linearly independent vectors $\mathbf{b}_1, \mathbf{b}_2, \ldots, \mathbf{b}_n$ is called a basis. Any vector $\mathbf{v}$ in the space can be expressed as a unique linear combination of the n ***base vectors*** of the basis:

$$\mathbf{v} = v^k\mathbf{b}_k \equiv v^1\mathbf{b}_1 + v^2\mathbf{b}_2 + \cdots + v^{(n)}\mathbf{b}_{(n)}. \tag{I.1.4}$$

The summation convention used here is that when a letter index is diagonally repeated (once as a superscript and once as a subscript) in a term, it denotes the sum of all the terms obtained by giving the letter index the values $1, 2, \ldots, (n)$. Note that the superscript k is not an exponent. When we occasionally need to raise a superscripted variable to a power, we enclose it in parentheses. For example, $(x^3)^2$ denotes x^3 squared. The result of Eq. (4) that an arbitrary vector $\mathbf{v}$ can be expressed in terms of the $\mathbf{b}_r$ can be seen as follows. We need consider only the case that $\mathbf{v}$ is not one of the $\mathbf{b}_r$. Then, since the space is n-dimensional, the $n + 1$ vectors $\mathbf{v}, \mathbf{b}_1, \mathbf{b}_2, \ldots, \mathbf{b}_n$ must be linearly dependent, so that

$$c\mathbf{v} + a^1\mathbf{b}_1 + a^2\mathbf{b}_2 + \cdots + a^n\mathbf{b}_n = 0$$

with $c \neq 0$ (otherwise the n base vectors would be linearly dependent). Hence we obtain Eq. (4) with $v^k = -a^k/c$.

The coefficients v^k are called the ***contravariant components*** of the vector $\mathbf{v}$ with respect to the basis. (The significance of the adjective "contravariant" will appear in Sec. I.2.) Note that the base vectors $\mathbf{b}_k$ need not be unit vectors, and they need not be orthogonal. Hence the v^k are ***not the orthogonal projections*** of $\mathbf{v}$ onto the $\mathbf{b}_k$. Many authors use the notation $\mathbf{e}_k$ for these general base vectors, but in this book $\mathbf{e}_k$ will always denote unit vectors tangential to curvilinear, coordinate curves, while $\mathbf{i}_k$ will denote rectangular Cartesian unit vectors. For most of our purposes we use only Euclidean vector spaces.

A Euclidean vector space of n dimensions is a vector space over the real numbers for which there is defined a ***scalar product***, a rule of composition having the properties (5) below, which associates a real number $\mathbf{u}\cdot\mathbf{v}$ with every pair of elements $\mathbf{u}$, $\mathbf{v}$. A ***proper Euclidean vector space*** is one such that $\mathbf{u}\cdot\mathbf{u}$ is positive for $\mathbf{u} \neq \mathbf{0}$.

SCALAR-PRODUCT AXIOMS

$\mathbf{u}\cdot\mathbf{v} = \mathbf{v}\cdot\mathbf{u}$	commutative	(I.1.5a)
$(a\mathbf{u})\cdot\mathbf{v} = \mathbf{u}\cdot(a\mathbf{v}) = a(\mathbf{u}\cdot\mathbf{v})$	associative with respect to multiplication by real a	(I.1.5b)
$\mathbf{u}\cdot(\mathbf{v} + \mathbf{w}) = \mathbf{u}\cdot\mathbf{v} + \mathbf{u}\cdot\mathbf{w}$	distributive	(I.1.5c)

If $\mathbf{u}\cdot\mathbf{v} = 0$ for arbitrary $\mathbf{u}$, then $\mathbf{v} = \mathbf{0}$. (I.1.5d)

An example of a Euclidean vector space is the set of all ordinary vectors, i.e., all directed line segments joining two points in the three-dimensional space of elementary geometry, which is a Euclidean point space.

In a proper Euclidean vector space, the ***magnitude*** $|\mathbf{v}|$ of $\mathbf{v}$ is defined by

$$|\mathbf{v}| = \sqrt{\mathbf{v}\cdot\mathbf{v}}.$$

Scalar product in terms of contravariant components. If the Euclidean vector space is referred to a basis, then

$$\mathbf{u} = u^r\mathbf{b}_r \quad \text{and} \quad \mathbf{v} = v^s\mathbf{b}_s$$

and by the scalar-product axioms 5(b) and 5(c)

$$\mathbf{u}\cdot\mathbf{v} = u^r v^s \mathbf{b}_r\cdot\mathbf{b}_s.$$

Let

$$g_{rs} = \mathbf{b}_r\cdot\mathbf{b}_s \qquad (\text{note } g_{rs} = g_{sr}). \tag{I.1.6}$$

In the next section, we will see that the g_{rs} are the covariant components (relative to the given basis) of the unit tensor **1**. Then

$$\mathbf{u} \cdot \mathbf{v} = g_{rs} u^r v^s \tag{I.1.7}$$

gives the scalar product as a symmetric bilinear form in the contravariant components. That the form is ***nondegenerate***, i.e., that its determinant is nonzero,

$$\det |g_{rs}| \neq 0, \tag{I.1.8}$$

is a consequence of the scalar-product axiom (5d), as follows. If $\mathbf{u} \cdot \mathbf{v} = 0$ for arbitrary $\mathbf{u}$, this means by Eq. (7) that

$$g_{rs} u^r v^s = 0$$

for arbitrary choices of the u^r. Successively choose first $u^1 = 1$ with all other $u^r = 0$, second $u^2 = 1$ with all other $u^r = 0$, etc. to obtain n equations

$$g_{rs} v^s = 0,$$

which, by Axiom (5d), have only the trivial solution with all the v^s zero. Hence the determinant must not be zero, since the vanishing of the determinant is a necessary and sufficient condition for the existence of a nontrivial solution.

Orthogonality. If $\mathbf{u} \cdot \mathbf{v} = 0$, the two vectors $\mathbf{u}$ and $\mathbf{v}$ are said to be orthogonal.

The selection of a basis is rather arbitrary and may be done in infinitely many ways. If we choose a new basis, then each of the n new base vectors $\bar{\mathbf{b}}_j$ must be expressible as a linear combination of the original base vectors $\mathbf{b}_k$, since every vector is so expressible. Before considering the change-of-basis formulas in Sec. I.2, we define a ***dual basis*** (or ***reciprocal basis***) associated with each given basis in a Euclidean vector space.

Dual (or ***reciprocal***) ***base vectors*** $\mathbf{b}^q$ $(q = 1, 2, \ldots, n)$ are defined for each given set $\mathbf{b}_p$ $(p = 1, 2, \ldots, n)$ of base vectors in Euclidean vector space as the set of vectors satisfying the conditions

$$\mathbf{b}_p \cdot \mathbf{b}^q = \delta_p^q \qquad \begin{matrix} p = 1, \ldots, n \\ q = 1, \ldots, n \end{matrix} \tag{I.1.9}$$

where the Kronecker delta δ_s^r is defined by

$$\delta_s^r = \begin{cases} 1 & \text{if} \quad r = s \\ 0 & \text{if} \quad r \neq s. \end{cases} \tag{I.1.10}$$

Equations (9) may be regarded as a set of n^2 linear equations for the n^2 unknown components of the base vectors $\mathbf{b}^q$, where the coefficients are the known components of the given base vectors $\mathbf{b}_p$. For the important case of ordinary vectors with $n = 3$ we can express the dual base vectors in terms of the original basis by using the cross product as follows.

For $n = 3$, the equations $\mathbf{b}_2 \cdot \mathbf{b}^1 = 0$ and $\mathbf{b}_3 \cdot \mathbf{b}^1 = 0$, express the fact that $\mathbf{b}^1$ is perpendicular to both $\mathbf{b}_2$ and $\mathbf{b}_3$. Hence

$$E\mathbf{b}^1 = \mathbf{b}_2 \times \mathbf{b}_3,$$

where E is a numerical factor to be determined. Then by Eq. (9), with $p = q = 1$,

$$E\mathbf{b}_1 \cdot \mathbf{b}^1 = E = \mathbf{b}_1 \cdot (\mathbf{b}_2 \times \mathbf{b}_3). \tag{I.1.11}$$

Thus E is equal to the mixed triple product of the three base vectors (equal to the volume of the parallelepiped formed on the three vectors as edges, if the system of base vectors is right-handed). Similar treatment of each of the new dual base vectors yields finally

$$\mathbf{b}^1 = \frac{1}{E}\mathbf{b}_2 \times \mathbf{b}_3 \qquad \mathbf{b}^2 = \frac{1}{E}\mathbf{b}_3 \times \mathbf{b}_1 \qquad \mathbf{b}^3 = \frac{1}{E}\mathbf{b}_1 \times \mathbf{b}_2 \tag{I.1.12a}$$

and, inversely,

$$\mathbf{b}_1 = E\mathbf{b}^2 \times \mathbf{b}^3 \qquad \mathbf{b}_2 = E\mathbf{b}^3 \times \mathbf{b}^1 \qquad \mathbf{b}_3 = E\mathbf{b}^1 \times \mathbf{b}^2. \tag{I.1.12b}$$

If the given basis is ***orthonormal*** (i.e., composed of mutually orthogonal unit vectors), then $E = 1$ (for a right-handed system), and the dual basis is identical to the given basis. When the base vectors of the given basis are mutually orthogonal but not orthonormal, Eq. (9) implies that $|\mathbf{b}^k| = 1/|\mathbf{b}_k|$; the magnitude of each of the dual base vectors is then the reciprocal of the corresponding base vector in the given basis, but this reciprocal relation is not satisfied when the given basis is not orthogonal.

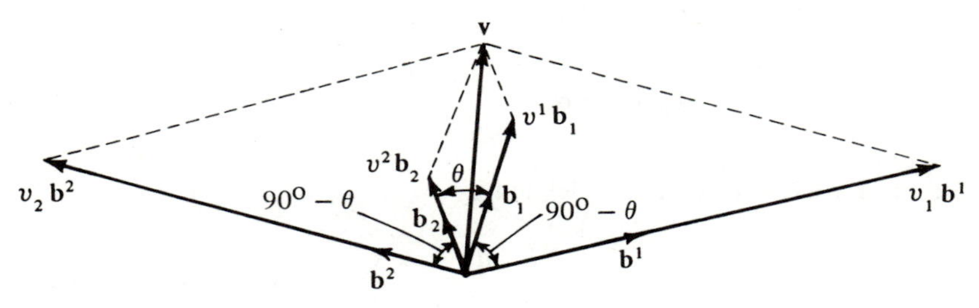

Fig. I.1 Dual base vectors in two dimensions.

Figure I.1 shows the relationship of the dual base vectors $\mathbf{b}^1$ and $\mathbf{b}^2$ to the given base vectors $\mathbf{b}_1$ and $\mathbf{b}_2$ in a plane vector space ($n = 2$) for a case where the given basis is not orthogonal, since the angle θ between $\mathbf{b}_1$ and $\mathbf{b}_2$ is acute. Equation (9) requires $\mathbf{b}^1$ to be perpendicular to $\mathbf{b}_2$. Thus $\mathbf{b}^1$ must be at angle $90° \pm \theta$ to $\mathbf{b}_1$. But Eq. (9) also requires $\mathbf{b}_1 \cdot \mathbf{b}^1 = 1$, so that the angle between $\mathbf{b}^1$ and $\mathbf{b}_1$ must be acute; therefore we must select $90° - \theta$ as shown. For this coplanar case, the angle between $\mathbf{b}^2$ and $\mathbf{b}_2$ must be the same, namely $90° - \theta$. Note that Eq. (9) implies $|\mathbf{b}_1|\,|\mathbf{b}^1| \cos(90° - \theta) = 1$ whence $|\mathbf{b}^1| = 1/|\mathbf{b}_1| \sin\theta$, emphasizing that the reciprocal relation between the magnitudes of $\mathbf{b}_1$ and $\mathbf{b}^1$ holds only for the orthogonal case where $\theta = 90°$.

The figure also shows, for a vector $\mathbf{v}$, its vector components $v^1\mathbf{b}_1$ and $v^2\mathbf{b}_2$ parallel to the given basis and its components $v_1\mathbf{b}^1$ and $v_2\mathbf{b}^2$ parallel to the dual basis. In the representation $\mathbf{v} = v_1\mathbf{b}^1 + v_2\mathbf{b}^2$ the coefficients v_j are called the ***covariant components*** of $\mathbf{v}$. Covariant components are always written with subscripts, while contravariant components are written with superscripts. We thus have two different representations for the same vector $\mathbf{v}$, namely

$$\mathbf{v} = v^k\mathbf{b}_k = v_j\mathbf{b}^j \tag{I.1.13}$$

with summation on diagonally repeated indices.

Some authors call the set of components v_j a ***covariant vector*** and the set v^k a ***contravariant vector***, but from our point of view ***there is just one vector***, namely $\mathbf{v}$, which, for a given basis, may be represented in terms of contravariant components with respect to the given base vectors, or, alternatively, in terms of covariant components.

The reason for the names "covariant" and "contravariant" will appear in Sec. I.2, where we shall see that in a change of basis from the given basis $\mathbf{b}_k$ to a new basis, say $\bar{\mathbf{b}}_j$, the covariant components v_k are transformed by the ***same*** transformation coefficients a_j^k as are used to transform the given base vectors $\mathbf{b}_k$, while the transformation coefficients for the contravariant components v^j are the coefficients of the backward change from the $\bar{\mathbf{b}}_j$ to the $\mathbf{b}_k$.

EXERCISES

1. If a vector space consists of all polynomials in x of degree five or less, with real coefficients, determine a basis for the space.
2. Derive the first Eq. (I.1.12b).
3. In ordinary three-dimensional space, determine the dual base vectors $\mathbf{b}^r$ in terms of the rectangular Cartesian $\mathbf{i}$, $\mathbf{j}$, $\mathbf{k}$, if $\mathbf{b}_1 = \mathbf{j} + \mathbf{k}$, $\mathbf{b}_2 = \mathbf{i} + \mathbf{k}$, and $\mathbf{b}_3 = \mathbf{i} + \mathbf{j}$.
4. (a) Evaluate the components g_{rs} defined by Eq. (I.1.6) for the basis $\mathbf{b}_j$ given in Ex. 3.

(b) If $\mathbf{u} = 2\mathbf{b}_1 + 3\mathbf{b}_2 - \mathbf{b}_3$ and $\mathbf{v} = \mathbf{b}_1 - \mathbf{b}_2 + \mathbf{b}_3$, evaluate $\mathbf{u}\cdot\mathbf{v}$ by means of Eq. (7).

(c) Determine the covariant components of the vectors $\mathbf{u}$ and $\mathbf{v}$ of (b).

5. Sketch the vector resolutions of Fig. I.1, and compute v_1 and v_2, for the case that $v^1 = 3$, $v^2 = 2$, and $\theta = 30°$.

Suggested reference books

Brillouin, L., *Tensors in Mechanics and Elasticity*. New York: Academic Press, 1964 (translation of *Les Tenseurs en Mécanique et en Elasticité*. Paris: Masson et Cie, 1938.)

Block, H. D., *Introduction to Tensor Analysis*. New York: Charles E. Merrill, 1962 (paperback).

Hawkins, G. A., *Multilinear Analysis for Students in Engineering and Science*. New York: John Wiley & Sons, Inc., 1963.

Lichnerowicz, A., *Elements of Tensor Calculus*. London: Methuen & Co. Ltd. (New York: John Wiley & Sons, Inc.), 1962.

I.2 Change of basis; unit-tensor components

We have already noted that the selection of a basis for a vector space is rather arbitrary. If we have any two bases, say

$$\bar{\mathbf{b}}_j, \quad j = 1, \ldots, n \quad \text{and} \quad \mathbf{b}_k, \quad k = 1, \ldots, n,$$

then each vector in one basis must be a linear combination of the vectors of the other basis

$$\begin{aligned} \bar{\mathbf{b}}_j &= a_j^p \mathbf{b}_p & \mathbf{b}_k &= b_k^q \bar{\mathbf{b}}_q \\ \det |a_j^p| &\neq 0 & \det |b_k^q| &\neq 0 \end{aligned} \tag{I.2.1}$$

(summed on p and q, respectively), where the a_j^p are the coefficients for the forward change of basis, and b_k^q the coefficients for the backward change of basis from $\bar{\mathbf{b}}_q$ to $\mathbf{b}_k$. The determinants must be different from zero in order that each set of base vectors be a linearly independent set.

Contravariant component transformations under the change of basis may be derived as follows. We require that the representations of the vector in the two systems have exactly the same form. Then

$$\bar{v}^q \bar{\mathbf{b}}_q = \mathbf{v} = v^k \mathbf{b}_k = v^k (b^q_k \bar{\mathbf{b}}_q)$$
$$\therefore (\bar{v}^q - v^k b^q_k)\bar{\mathbf{b}}_q = \mathbf{0}.$$

Since the base vectors $\bar{\mathbf{b}}_q$ are linearly independent, the coefficients of $\bar{\mathbf{b}}_q$ in the last equation must all be zero. Hence

$$\bar{v}^q = b^q_k v^k \quad \text{and, similarly,} \quad v^p = a^p_j \bar{v}^j, \tag{I.2.2a}$$

showing that the forward change from components v^k to $\bar{v}^q$ used the coefficients b^q_k of the backward change from base vectors $\bar{\mathbf{b}}_q$ to the original $\mathbf{b}_k$, while the backward change of components uses the coefficients of the forward change of basis. This is why these components are called ***contravariant***. If b^q_k is the element in the qth row and kth column of a square matrix B, while a^p_j is an element of A, Eqs. (2a) take the ***matrix form***:

$$\bar{v} = Bv \quad \text{and} \quad v = A\bar{v} \qquad B = A^{-1}, \tag{I.2.2b}$$

where v and $\bar{v}$ denote column matrices of contravariant components in both cases. When the change of basis is merely a rotation of an orthonormal basis, as in the change of rectangular Cartesian axes in Sec. 2.4, we have $A^{-1} = A^T$, so that $B = A^T$ as in Sec. 2.4, but in general $B = A^{-1} \neq A^T$.

Before deriving the transformation formula for covariant components, we note an important result, which some authors take as the definition of covariant and contravariant components:

$$\text{Contravariant Components:} \quad v^q = \mathbf{v} \cdot \mathbf{b}^q \tag{I.2.3a}$$

$$\text{Covariant Components:} \quad v_p = \mathbf{v} \cdot \mathbf{b}_p. \tag{I.2.3b}$$

Equation (3b) follows from expressing $\mathbf{v}$ in terms of its covariant components to evaluate the scalar product

$$\mathbf{v} \cdot \mathbf{b}_p = (v_q \mathbf{b}^q) \cdot \mathbf{b}_p = v_q (\mathbf{b}^q \cdot \mathbf{b}_p) = v_q \delta^q_p = v_p.$$

Equation (3a) is derived similarly.

The ***covariant component transformation*** may now be derived as follows. By Eq. (3b) $\bar{v}_j = \mathbf{v} \cdot \bar{b}_j$, which, using $\bar{\mathbf{b}}_j = a^p_j \mathbf{b}_p$ from Eq. (1), we write as

$$\bar{v}_j = \mathbf{v} \cdot (a^p_j \mathbf{b}_p) = a^p_j (\mathbf{v} \cdot \mathbf{b}_p) = a^p_j v_p.$$

A similar derivation yields the inverse transformation, so that we have for the ***covariant component transformation***

$$\bar{v}_j = a^p_j v_p \quad \text{and} \quad v_k = b^q_k \bar{v}_q. \tag{I.2.4a}$$

We see that the coefficients of the covariant component transformation under the change of basis are the same as the coefficients in Eq. (1) for the change of basis. If the covariant vector components are written as row matrices, the matrix form of Eq. (4) is

$$\bar{v} = vA \quad \text{and} \quad v = \bar{v}B \qquad B = A^{-1}. \tag{I.2.4b}$$

Covariant components may be expressed in terms of contravariant components, and vice versa by using Eq. (3). Thus

$$\left.\begin{aligned} v_i &= \mathbf{v}\cdot\mathbf{b}_i = (v^j\mathbf{b}_j)\cdot\mathbf{b}_i = (\mathbf{b}_i\cdot\mathbf{b}_j)v^j \quad \text{and} \\ v^i &= \mathbf{v}\cdot\mathbf{b}^i = (v_j\mathbf{b}^j)\cdot\mathbf{b}^i = (\mathbf{b}^i\cdot\mathbf{b}^j)v_j. \end{aligned}\right\} \tag{I.2.5}$$

Define the fundamental-tensor components g_{ij}, g^{ij}, $g^i{}_j$ and $g_j{}^i$ as follows:

$$\left.\begin{aligned} g_{ij} &= \mathbf{b}_i\cdot\mathbf{b}_j \quad \text{and} \quad g^{ij} = \mathbf{b}^i\cdot\mathbf{b}^j \\ g^i{}_j &= \mathbf{b}^i\cdot\mathbf{b}_j = \delta^i_j = \mathbf{b}_j\cdot\mathbf{b}^i = g_j{}^i. \end{aligned}\right\} \tag{I.2.6}$$

Note that $g_{ij} = g_{ji}, g^{ij} = g^{ji}, g^i{}_j = g_j{}^i$. Then Eqs. (5) yield

$$\left.\begin{aligned} v^i &= g^{ij}v_j \quad \text{and} \quad & v_i &= g_{ij}v^j \\ v_j &= g_j{}^i v_i & v^i &= g^i{}_j v^j \end{aligned}\right\} \tag{I.2.7}$$

Evidently the various sets of quantities $g_{ij}, g^{ij}, g^i{}_j, g_j{}^i$ have the property that when they are used as the coefficients of a linear transformation operating on the covariant or contravariant components of a vector, they yield as a result of the operation the components ***of the same vector*** (covariant or contravariant components, depending on which set is used). These quantities are therefore the components of the ***unit tensor*** **1** such that

$$\mathbf{v} = \mathbf{1}\cdot\mathbf{v} = \mathbf{v}\cdot\mathbf{1}. \tag{I.2.8}$$

The unit tensor is also called the ***fundamental tensor*** or the ***metric tensor*** of the space. The g_{ij} are its covariant components, g^{ij} its contravariant components, and $g^i{}_j$ and $g_j{}^i$ are two kinds of mixed components. Since $g^i{}_j = g_j{}^i$, it is customary to write their common value as $g^i_j = \delta^i_j$, without indicating whether it is the superscript or the subscript which is the first suffix.

The process illustrated in the first two Eqs. (7) is often called ***raising or lowering indices***; it involves a linear transformation whose coefficients are a suitable set of components of the unit tensor to convert from covariant components to contravariant components of the same vector or vice versa. The process of raising or lowering indices can also be performed on the base vectors themselves, but this produces a different vector, the dual base vector. Thus

$$v_j \mathbf{b}^j = \mathbf{v} = v^i \mathbf{b}_i = (g^{ij} v_j)\mathbf{b}_i = v_j(g^{ij}\mathbf{b}_i) = v_j(g^{ji}\mathbf{b}_i)$$

or

$$v_j(\mathbf{b}^j - g^{ji}\mathbf{b}_i) = 0$$

for arbitrary v_j. Hence

$$\mathbf{b}^j = g^{ji}\mathbf{b}_i \quad \text{and similarly} \quad \mathbf{b}_j = g_{ji}\mathbf{b}^i. \tag{I.2.9}$$

Since the first two Eqs. (7) are inverse transformations to each other, the matrices $[g^{ij}]$ and $[g_{ij}]$ must be inverse.

$$[g^{ij}] = [g_{ij}]^{-1}. \tag{I.2.10a}$$

Thus, if D^{rs} denotes the cofactor (signed minor subdeterminant) of g_{rs} in the determinant $|g_{rs}|$, then (note $D^{ij} = D^{ji}$, since $[g_{ij}]$ is symmetric)

$$g^{ij} = \frac{D^{ij}}{g}, \tag{I.2.10b}$$

where

$$g = \det |g_{ij}| \tag{I.2.11}$$

is the determinant of the matrix $[g_{ij}]$. The determinant of the inverse matrix is the reciprocal of g:

$$\det |g^{ij}| = \frac{1}{g}. \tag{I.2.12}$$

The determinant g is nonzero by Eq. (I.1.8). Hence, for a given basis, the dual basis is uniquely defined, and the relationships of Eqs. (7) are a one-to-one transformation with a unique inverse.

Tensors as linear vector functions. In Secs. 2.1 and 2.4, we defined a second-order (or second-rank) tensor as a linear vector function associating with each argument vector another vector, e.g.,

$$\mathbf{u} = \mathbf{T} \cdot \mathbf{v}. \tag{I.2.13}$$

For any given basis $\mathbf{b}_1, \mathbf{b}_2, \ldots, \mathbf{b}_n$ either of the two vectors may be represented by either covariant or contravariant components v_j or v^j and u_i or u^i. There are thus four possible sets of n^2 coefficients for the four different linear transformations

$$u_i = T_{ij}v^j \qquad u^i = T^{ij}v_j$$
$$u_i = T_i^{\ j}v_j \qquad u^i = T^i_{\ .j}v^j \tag{I.2.14}$$

involving, respectively, the

covariant components T_{ij}
contravariant components T^{ij}
or mixed components $T^i_{.j}$ or $T_i^{\ j}$.

Since in general $T^i_{.j} \neq T_j^{\ i}$, it is necessary to observe carefully the order of the indices. For this reason it is common practice to put a dot above or below the first index in handwritten text. The dot is frequently omitted in printed text where the order is clear.

Second-order tensor component transformations under a change of basis can be derived as follows. Since there are four kinds of components (covariant, contravariant, and two kinds of mixed components), four kinds of transformations are needed. Suppose that a tensor **T** exists that associates with each arbitrary vector **v** in the space another vector **u**. For any two given bases $\bar{\mathbf{b}}_j$ and $\mathbf{b}_k$ related by Eq. (1), the covariant components of **u** are related by Eq. (4)

$$\begin{aligned} \bar{u}_i &= a_i^p u_p \\ &= a_i^p T_{pq} v^q && \text{by Eq. (14)} \\ &= a_i^p T_{pq} a_j^q \bar{v}^j && \text{by Eq. (2).} \end{aligned}$$

Thus

$$\bar{T}_{ij}\bar{v}^j = a_i^p T_{pq} a_j^q \bar{v}^j \qquad \text{by Eq. (11), or}$$
$$(\bar{T}_{ij} - a_i^p a_j^q T_{pq})\bar{v}^j = 0$$

for arbitrary $\bar{v}^j$. Hence

$$\bar{T}_{ij} = a_i^p a_j^q T_{pq}$$

relates the covariant components T_{ij} to the covariant components $\bar{T}_{pq}$. Comparing this with the first Eq. (4a) we see that it is formed in the same way, with one coefficient a_n^m of the forward change of basis for each covariant subscript m. The free indices on the transformation coefficients for the covariant components must be subscripts, designated by the same letters as the free indices on the left-hand side of the equation. Similar derivations yield the other Eqs. (15) below.

Second-order tensor transformation formulas under basis change.

$$\text{Covariant:} \qquad \bar{T}_{ij} = a_i^p a_j^q T_{pq} \qquad \text{(I.2.15a)}$$

$$\text{Contravariant:} \qquad \bar{T}^{ij} = b_p^i b_q^j T^{pq} \qquad \text{(I.2.15b)}$$

$$\text{Mixed:} \quad \begin{cases} \bar{T}_{i}^{\cdot j} = a_i^p b_q^j T_{p}^{\cdot q} & \text{(I.2.15c)} \\ \bar{T}_{\cdot j}^{i} = b_p^i a_j^q T_{\cdot q}^{p} & \text{(I.2.15d)} \end{cases}$$

The coefficients b_n^m are the coefficients of the backward change of basis from $\bar{\mathbf{b}}_j$ to $\mathbf{b}_k$ of Eq. (1), while the a_n^m are from the forward change. These tensor transformation formulas are quite easy to remember, once the vector component transformation formulas of Eqs. (2) and (4) have been learned. There must be one forward-change coefficient a_n^m for each covariant index and one backward-change coefficient b_n^m for each contravariant index. In the last formula, for example, the free indices on the right are the contravariant superscript i and the covariant subscript j, the same as the indices on the left. The other two indices p and q are each diagonally repeated, once on a coefficient and once on $T_{\cdot q}^{p}$, implying summation over the n^2 possible combinations of pq for $p = 1, 2, \ldots, n$ and $q = 1, 2, \ldots, n$. For three-dimensional space, $n = 3$, there are nine equations for any one type of transformation, e.g., nine Eqs. (15a), and nine terms in the sum on the right-hand side of each equation.

Matrix forms of the equations. Since each second-order tensor has four component matrices, it is necessary to indicate which one is used in any case. Note that as written, Eqs. (15) are not matrix equations. The matrix forms of Eqs. (15) are:

$$\text{Covariant:} \qquad [\bar{T}_{ij}] = [a_i^p]^T [T_{pq}][a_j^q] \qquad \text{(I.2.16a)}$$

$$\text{Contravariant:} \qquad [\bar{T}^{ij}] = [b_p^i][T^{pq}][b_q^j]^T \qquad \text{(I.2.16b)}$$

$$\text{Mixed:} \quad \begin{cases} [\bar{T}_{i}^{\cdot j}] = [a_i^p]^T [T_{p}^{\cdot q}][b_q^j]^T & \text{(I.2.16c)} \\ [\bar{T}_{\cdot j}^{i}] = [b_p^i][T_{\cdot q}^{p}][a_j^q]. & \text{(I.2.16d)} \end{cases}$$

where the bracket with superscript T indicates the transpose of the matrix shown. Note that in each of the four cases premultiplication by an A matrix requires use of the transpose A^T, while it is postmultiplication by a B matrix that requires use of B^T. For the transformation matrix elements (e.g., a_i^p and b_q^j), the superscript identifies the row while the subscript identifies the column, while in the matrices of tensor components the first suffix identifies the row, while the second identifies the column. Then matrix multiplication is accomplished by the indicial scheme shown only when the transposed matrices are used as indicated. The indicial forms of Eqs. (15) are easier to remember, keeping in mind the rules of covariance and contravariance, but there may be some advantage to the matrix forms for actual numerical calculations, or even for indicated numerical calculations using letters to represent the numerical values.

Remark on second-order tensor components. Equations (3) and (14) can be used to get an operational representation of the second-order tensor components. Given

$$\mathbf{u} = \mathbf{T}\cdot\mathbf{v}, \tag{I.2.17}$$

use $u_i = \mathbf{b}_i \cdot \mathbf{u}$ by Eq. (3) and $\mathbf{v} = \mathbf{b}_k v^k$ to obtain

$$\begin{aligned} u_i &= \mathbf{b}_i \cdot [\mathbf{T}\cdot\mathbf{b}_k v^k] \\ &= (\mathbf{b}_i \cdot \mathbf{T}\cdot\mathbf{b}_k) v^k. \end{aligned}$$

Comparison with Eq. (14) yields the first Eq. (18).

$$\left.\begin{aligned} T_{ik} &= \mathbf{b}_i \cdot \mathbf{T}\cdot\mathbf{b}_k \\ T^{ik} &= \mathbf{b}^i \cdot \mathbf{T}\cdot\mathbf{b}^k \\ T_i^{\cdot k} &= \mathbf{b}_i \cdot \mathbf{T}\cdot\mathbf{b}^k \\ T^i_{\cdot k} &= \mathbf{b}^i \cdot \mathbf{T}\cdot\mathbf{b}_k \end{aligned}\right\} \tag{I.2.18}$$

These operational representations [analogous to Eqs. (3) for vector components] may be taken as definitions of the components with respect to a given basis for a second-order tensor operating in the manner of Eqs. (14) and (17).

Transpose of a tensor. If the linear vector function S operates by postmultiplication instead of premultiplication, i.e., if

$$\mathbf{u} = \mathbf{v}\cdot\mathbf{S} \tag{I.2.19}$$

with component forms

$$\left.\begin{aligned} u_i &= v^j S_{ji} \qquad & u^i &= v_j S^{ji} \\ u_i &= v_j S^j_{\cdot i} \qquad & u^i &= v^j S_j^{\cdot i}, \end{aligned}\right\} \tag{I.2.20}$$

the operational representations still have the same form as Eqs. (18), namely

$$\left.\begin{aligned} S_{ji} &= \mathbf{b}_j \cdot \mathbf{S}\cdot\mathbf{b}_i \qquad & S^{ji} &= \mathbf{b}^j \cdot \mathbf{S}\cdot\mathbf{b}^i \\ S^j_{\cdot i} &= \mathbf{b}^j \cdot \mathbf{S}\cdot\mathbf{b}_i \qquad & S_j^{\cdot i} &= \mathbf{b}_j \cdot \mathbf{S}\cdot\mathbf{b}^i. \end{aligned}\right\} \tag{I.2.21}$$

Comparison of Eqs. (14) and (20) shows that if the two operations $\mathbf{v}\cdot\mathbf{S}$ and $\mathbf{T}\cdot\mathbf{v}$ always produce the same $\mathbf{u}$ for every $\mathbf{v}$, that is if

$$\mathbf{T} \cdot \mathbf{v} = \mathbf{v} \cdot \mathbf{S} \tag{I.2.22}$$

for every $\mathbf{v}$, then the matrices of covariant or contravariant components of $\mathbf{T}$ are the transposes of those of $\mathbf{S}$

$$[T_{ij}] = [S_{ij}]^T \qquad [T^{ij}] = [S^{ij}]^T. \tag{I.2.23}$$

In this case, we say that the tensor $\mathbf{T}$ is the ***transpose*** of $\mathbf{S}$ and vice versa and denote this by writing

$$\mathbf{T} = \mathbf{S}^T \qquad \mathbf{S} = \mathbf{T}^T. \tag{I.2.24}$$

But in general when $\mathbf{S} = \mathbf{T}^T$, we have $S_j^{\cdot i} = T^i_{\cdot j} \neq T_i^{\cdot j}$, so that the matrix with element $S_j^{\cdot i}$ in jth row and ith column is ***not*** the transpose of the matrix with $T_i^{\cdot j}$ in the ith row and jth column.

A ***symmetric*** tensor of second order is one that is operationally identical to its transpose, so that

$$\text{For Symmetric } \mathbf{T}: \quad \mathbf{T} \cdot \mathbf{v} = \mathbf{v} \cdot \mathbf{T} \tag{I.2.25}$$

for all $\mathbf{v}$. It follows that

$$\text{For Symmetric } \mathbf{T}: \quad \begin{Bmatrix} T_{ij} = T_{ji} \\ T^{ij} = T^{ji} \\ T^i_{\cdot j} = T_j^{\cdot i} \end{Bmatrix}, \tag{I.2.26}$$

so that, for a symmetric tensor, the matrix of covariant or contravariant components is symmetric, while the mixed component matrices are not in general symmetric, because the third Eq. (26) relates elements of the two different matrices of mixed components instead of symmetrically placed elements of the same matrix.

A ***skew tensor*** of second order (sometimes called ***skew-symmetric***) is one such that, for all $\mathbf{v}$,

$$\left.\begin{aligned} &\text{For Skew } \mathbf{T}: \quad \mathbf{T} \cdot \mathbf{v} = -\mathbf{v} \cdot \mathbf{T} \\ &\qquad T_{ij} = -T_{ji} \qquad T^{ij} = -T^{ij} \qquad T^i_{\cdot j} = -T_j^{\cdot i}. \end{aligned}\right\} \tag{I.2.27}$$

Higher-order tensors may be introduced by an operation analogous to the definition of a second-order tensor as a linear vector operator. A fourth-order tensor is a linear function whose argument (input) is a second-order tensor and whose value (output) is also a second-order tensor. We use the following square-bracket notation for a fourth-order tensor

$$\mathbf{T} = \mathbf{C}[\mathbf{E}], \tag{I.2.28a}$$

where **C** is the fourth-order tensor relating the argument **E** to the value **T**. (In printed text it is also possible to distinguish **C** by using boldface sans serif type instead of the boldface italic used for second-order tensors.) We will find it convenient to use most of the time a component form of Eq. (28), such as

$$\left.\begin{aligned} T^{mn} &= C^{mnrs} E_{rs} \\ T_{mn} &= C_{mnrs} E^{rs} \\ T_{mn} &= C_{mn}^{\cdot\cdot rs} E_{rs} \\ T^{m}_{.n} &= C^{m.rs}_{.n} E_{rs}, \text{ etc.} \end{aligned}\right\} \tag{I.2.28b}$$

There are, of course, many more kinds of mixed tensor components possible than are shown in Eq. (28b). From the known transformation equations for second-order tensor components, given in Eqs. (15), the transformation formulas for fourth-order tensor components may be derived; for example,

$$\bar{C}^{p.rs}_{.q} = b^p_i a^j_q b^r_k b^s_m C^{i.km}_{.j} \tag{I.2.29}$$

with one backward-transformation coefficient b^n_i for each contravariant index and one forward-transformation coefficient a^n_i for each covariant index.

Raising and lowering indices may be performed on higher-order tensor components in the same way as for second-order tensors, using appropriate components of the unit tensor **1**. For example,

$$C^{mnrs} = g^{nk} C^{m.rs}_{.k} \quad \text{and} \quad C^{m.rs}_{.k} = g_{kn} C^{mnrs}. \tag{I.2.30}$$

All of the many different sets of fourth-order components obtainable by raising and lowering indices are just different representations of the same fourth-order tensor.

Tensors of other orders can be defined similarly. For example, if the argument input is a vector, while the output is a second-order tensor, we have a third-order tensor. One component form for a third-order tensor is

$$S^{ij} = T^{ij}_{..k} v^k. \tag{I.2.31}$$

We will not use a symbolic notation analogous to Eq. (28a) except for the fourth-order case.

The linear and homogeneous character of the transformations for change of basis implies that ***if all components of a tensor with respect to one basis vanish, then the components with respect to any other basis must vanish.*** This in turn implies that if, for example, $A_{rs} = B_{rs}$ for one basis $\mathbf{b}_m$, then, for any other basis $\bar{\mathbf{b}}_n$, $\bar{A}_{rs} = \bar{B}_{rs}$, since this is equivalent to $A_{rs} - B_{rs} = 0$ implying $\bar{A}_{rs} - \bar{B}_{rs} = 0$. Thus if a tensor equation holds for one basis, it holds for any basis of the vector space. Actually these are just different ways of saying that $\mathbf{A} = \mathbf{B}$.

Contraction. We now define another tensor process called ***contraction***. In the mixed tensor components $A^{mr}_{\cdot\cdot sn}$, make one superscript and one subscript the same—for example, $A^{mr}_{\cdot\cdot sm}$. This implies summation, and if $B^{r}_{\cdot s}$ is used to denote the result,

$$B^{r}_{\cdot s} = A^{mr}_{\cdot\cdot sm} = A^{1r}_{\cdot\cdot s1} + A^{2r}_{\cdot\cdot s2} + A^{3r}_{\cdot\cdot s3}, \tag{I.2.32}$$

it can be shown that the $B^{r}_{\cdot s}$ are tensor components of the type indicated by the indices. Contraction always produces a tensor of order two less than that of the tensor to which it is applied. In particular, if $T^{m}_{\cdot n}$ are second-order tensor components, then $T^{m}_{\cdot m}$ is a scalar invariant, since

$$\bar{T}^{m}_{\cdot m} = b^{m}_{r} a^{s}_{m} T^{r}_{\cdot s} = \delta^{s}_{r} T^{r}_{\cdot s} = T^{r}_{\cdot r} \tag{I.2.33}$$

since $a^{s}_{m} b^{m}_{r} = \delta^{s}_{r}$, because the transformation matrices are inverse, $B = A^{-1}$. A ***scalar invariant function*** of the tensor components is a scalar-valued function that has the same form as a function of the tensor components in the two systems, e.g., $T^{m}_{\cdot m}$ and $\bar{T}^{n}_{\cdot n}$ and also the same numerical value.

The ***outer product*** of two sets of tensor components (not necessarily of the same order or type) is a set of tensor components that are obtained simply by writing the components of the two tensors beside each other with no repeated indices. For example, if A^{i} and $B_{j}^{\cdot k}$ are tensor components, and

$$C^{i\cdot k}_{\cdot j} = A^{i} B_{j}^{\cdot k}, \tag{I.2.34}$$

then the $C^{i\cdot k}_{\cdot j}$ are tensor components of the order and type indicated by the indices. An ***inner product*** is obtained from an outer product by contraction with a superscript from one factor and a subscript from the other. For example, from the outer product $C^{i\cdot k}_{\cdot j}$ above, we obtain the inner product

$$D^{k} = A^{i} B_{i}^{\cdot k}, \tag{I.2.35}$$

giving first-order contravariant tensor components. From the outer product $A_{ij} B^{m}_{\cdot n}$ we can form two inner products $A_{mj} B^{m}_{\cdot n}$ and $A_{im} B^{m}_{\cdot n}$, each of which gives second-order covariant tensor components, but the two products in general represent two different second-order tensors, namely $\mathbf{A}^{T}\boldsymbol{\cdot}\mathbf{B}$ and $\mathbf{A}\boldsymbol{\cdot}\mathbf{B}$, respectively.

The operational product $\mathbf{T}\boldsymbol{\cdot}\mathbf{S}$ of two second-order tensors was defined in Eq. (2.4.23) as the second-order tensor $\mathbf{P}$ such that

$$\mathbf{P}\boldsymbol{\cdot}\mathbf{v} \equiv (\mathbf{T}\boldsymbol{\cdot}\mathbf{S})\boldsymbol{\cdot}\mathbf{v} = \mathbf{T}\boldsymbol{\cdot}(\mathbf{S}\boldsymbol{\cdot}\mathbf{v})$$

for all vectors $\mathbf{v}$. Its covariant components P_{ij} are given by the inner products

$$P_{ij} = T_{i}^{\cdot k} S_{kj} = T_{ik} S^{k}_{\cdot j}. \tag{I.2.36}$$

The scalar products **T : S** and **T••S**. We define **T : S** as the scalar produced by a double contraction of the outer product:

$$\mathbf{T}:\mathbf{S} = T^{ij}S_{ij} = T_{ij}S^{ij} = T^{i}_{.j}S_{i}^{.j} = T_{i}^{.j}S^{i}_{.j}. \qquad \text{(I.2.37)}$$

Note that in all the possible ways of writing Eq. (37) the two first suffixes are the same, while the two second indices are the same. Another kind of scalar product can be defined putting the two inside indices equal and the two outside indices equal.

$$\mathbf{T}\cdot\cdot\mathbf{S} = T^{ij}S_{ji} = T_{ij}S^{ji} = T^{i}_{.j}S^{j}_{.i} = T_{i}^{.j}S_{j}^{.i}. \qquad \text{(I.2.38)}$$

In general $\mathbf{T}\cdot\cdot\mathbf{S} \neq \mathbf{T}:\mathbf{S}$, but if either one of the two tensors is symmetric, then $\mathbf{T}\cdot\cdot\mathbf{S} = \mathbf{T}:\mathbf{S}$. Notations for these two scalar products vary with different authors.

Quotient rule. Tensor tests. To prove that quantities are the components of a tensor, it is sometimes more convenient to use an indirect method than to verify that they transform according to the appropriate law.

Tensor tests. Quotient rules. We have defined a second-order tensor operationally as a linear vector function. Hence, if a set of second-order coefficients is known to relate the components of two vectors **u** and **v** with respect to the same basis by any of the Eqs. (14) or (20), then the coefficients are second-order tensor components with respect to that basis. These are examples of ***quotient rules*** as tests for tensor components. Equations (28b) and (31) are additional examples. To determine whether a physical entity with the appropriate number of components is a tensor, we may

1. show that the components transform appropriately under change of basis,
2. use a quotient rule, or
3. show that contraction produces tensor components of the indicated type. For example, Eq. (32) establishes that the $A^{mr}_{..sn}$ are mixed tensor components if it is known that the $B^{r}_{.s}$ are tensor components.

A frequently used example of the last type reduces the tensor components to a scalar by contraction as in Tests 1, 2, 3, and 4 below.

TEST 1. $[\bar{u}^m\bar{v}_m = u^s v_s]$ If an entity is defined by n quantities u^m with respect to any basis of the vector space and if $u^m v_m$ is a scalar invariant when formed with the covariant components of an arbitrary vector **v**, then the u^m are contravariant components of a vector.

Proof: By the assumed invariance, for any two bases, we have

$$\bar{u}^m\bar{v}_m = u^s v_s.$$

Substitute $v_s = b_s^m \bar{v}_m$ by Eq. (4a) to obtain

$$(\bar{u}^m - b_s^m u^s)\bar{v}_m = 0.$$

Since this equation holds for arbitrary $\bar{v}_s$ (arbitrary vector **v**), it follows that

$$\bar{u}^m = b_s^m u^s$$

and hence that the components u^r transform as contravariant vector components, according to Eq. (2a).

The other tests will be stated without giving the proofs, which are analogous to the proof of Test 1.

TEST 2. $[\bar{u}_m \bar{v}^m = u_s v^s]$ If an entity is defined by n quantities u_m with respect to any basis of the vector space and if $u_m v^m$ is a scalar invariant when formed with the contravariant components of an arbitrary vector **v**, then the u_m are the covariant components of a vector.

TEST 3. $[\bar{M}_{rs} \bar{T}^{rs} = M_{pq} T^{pq}]$ If an entity is defined by n^2 quantities M_{rs} with respect to any basis and if $M_{rs} T^{rs}$ is a scalar invariant when formed with the contravariant components of an arbitrary second-order tensor **T**, then the M_{rs} are covariant components of a second-order tensor.

TEST 4(a). $[\bar{M}_{rs} \bar{u}^r \bar{v}^s = M_{pq} u^p v^q]$ If an entity is defined by n^2 quantities M_{rs} with respect to any basis of the vector space and if $M_{rs} u^r v^s$ is a scalar invariant when formed with the contravariant components of any two arbitrary vectors **u** and **v**, then the M_{rs} are covariant components of a second-order tensor.

TEST 4(b). $[\bar{S}_{rs} \bar{v}^r \bar{v}^s = S_{pq} v^p v^q$ and $S_{rs} = S_{sr}]$ If an entity is defined by a ***symmetric*** set of n^2 quantities S_{rs} with respect to any basis of the vector space and if $S_{rs} v^r v^s$ is a scalar invariant when formed with the contravariant components of any one arbitrary vector **v**, then the S_{rs} are covariant components of a symmetric second-order tensor.

Remark: The symmetry is essential in the hypothesis of Test 4(b), since without it we are led only to

$$\bar{S}_{rs} + \bar{S}_{sr} = a_r^m a_s^n (S_{mn} + S_{nm}).$$

The reader should be able to write down analogous tests for contravariant components, analogous to Tests 3 and 4 for covariant components. Tests similar to Tests 3 and 4(a) can also be used for mixed components. And higher-order forms also exist.

Before applying these results to tensors referred to the natural base vectors of a curvilinear coordinate system we consider dyadic notation briefly in Sec. I.3.

EXERCISES

1. Derive the second Eq. (I.2.2a), giving the v^p in terms of the $\bar{v}^j$.
2. Derive the second Eq. (4a), giving the v_k in terms of the $\bar{v}_q$.
3. Verify that when the equation $\mathbf{b}_j = g_{ji}\,\mathbf{b}^i$ is applied to the results of Exercises 3 and 4(a) of Sec. I.1 the expressions given for the $\mathbf{b}_j$ in Exercise 3 of Sec. I.1 are recovered.
4. Begin with $\bar{u}_i = a_i^p u_p$. Then substitute $u_p = T_{p}^{\cdot q} v_q$ and proceed to derive Eq. (I.2.15c) for the transformation of mixed components.
5. Derive the last Eq. (I.2.18), giving the operational representation of $T^i_{.k}$.
6. Given the matrices $[T^{ij}]$ and $[g_{ij}]$ below, calculate the matrices $[T^i_{.j}]$ and $[T_i^{\cdot j}]$, and show that they are equal but not symmetric, even though $[T^{ij}]$ is symmetric.

$$[T^{ij}] = \begin{bmatrix} 1 & 2 & 3 \\ 2 & 4 & 5 \\ 3 & 5 & 6 \end{bmatrix} \qquad [g_{ij}] = \begin{bmatrix} 2 & 1 & 0 \\ 1 & 3 & 0 \\ 0 & 0 & 4 \end{bmatrix}$$

7. Prove that if the outer product $P_r^{\cdot s} = u_r v^s$ is formed with covariant components of a vector $\mathbf{u}$ and contravariant components of a vector $\mathbf{v}$, then the components $P_r^{\cdot s}$ transform under a change of basis as mixed tensor components of the type indicated by the index positions.
8. Prove that the two scalars $T^{ij}S_{ij}$ and $T_{ij}S^{ij}$ in Eq. (I.2.37) are equal.
9. Prove that if a symmetric set of quantities S_{pq} and $\bar{S}_{rs}$ satisfy Test 4(b), then they transform as covariant components of a second-order tensor.
10. Write (without proof) tensor tests comparable to Test 4(a) for contravariant components M^{pq} and mixed components $M^p_{.q}$.

I.3 Dyads and dyadics; dyadics as second-order tensors; determinant expansions; vector (cross) products

The ***open product*** or ***tensor product*** **ab** of two vectors is called a ***dyad***. A linear combination of such dyads is called a ***dyadic***, e.g., $T^{11}\mathbf{a}_1\mathbf{b}_1 + T^{12}\mathbf{a}_1\mathbf{b}_2 + T^{22}\mathbf{a}_2\mathbf{b}_2$, where the T^{rs} are numerical coefficients. We consider here only the case of real coefficients. [For complex dyadics see, for example, Drew (1961).] Higher-order open products are called ***polyads*** (e.g., triads and tetrads, open products of three vectors or four vectors) and linear combinations of polyads are called ***polyadics***. (All polyads in a polyadic must be of the same order.) We consider for the most part only dyads and dyadics.

Without assigning any physical significance yet to a polyad, we assume that all the usual multiplicative rules of elementary algebra hold for polyads, except that ***open multiplication is not commutative.*** For example, for any real numbers m and n we assume the ***associative laws***

$$\left.\begin{aligned} m(\mathbf{ab}) &= (m\mathbf{a})\mathbf{b} = \mathbf{a}(m\mathbf{b}) = m\mathbf{ab} \\ (\mathbf{ab})\mathbf{c} &= \mathbf{a}(\mathbf{bc}) = \mathbf{abc} \\ (m\mathbf{a})(n\mathbf{b}) &= (mn)\mathbf{ab}, \end{aligned}\right\} \tag{I.3.1}$$

and distributive laws

$$\begin{aligned} \mathbf{a}(\mathbf{b}+\mathbf{c}) &= \mathbf{ab}+\mathbf{ac} \\ (\mathbf{a}+\mathbf{b})\mathbf{c} &= \mathbf{ac}+\mathbf{bc} \\ m(\mathbf{ab}+\mathbf{cd}) &= m\mathbf{ab}+m\mathbf{cd}, \end{aligned} \tag{I.3.2}$$

and the unitary law

$$1(\mathbf{ab}) = \mathbf{ab}, \tag{I.3.3}$$

but in general

$$\mathbf{ab} \neq \mathbf{ba}. \tag{I.3.4}$$

The ***conjugate dyad***, denoted $(\mathbf{ab})_c$, to a given dyad **ab** is obtained by reversing the order of the vectors:

$$(\mathbf{ab})_c = \mathbf{ba}. \tag{I.3.5}$$

[Milne-Thomson (1962) denotes **ab** by **a**; **b**.]

Contraction, single-dot product, double-dot (scalar) product. If in a polyad we replace one of the open products by a scalar product of the two vectors, we obtain a polyad of order two less than that of the original polyad. This process is called ***contraction.*** For example, with the tetrad **abcd,** contraction can be done in three ways with adjacent vectors to yield ***three different dyadics***:

$$\begin{aligned} \mathbf{a}\cdot\mathbf{bcd} &= (\mathbf{a}\cdot\mathbf{b})\mathbf{cd} \\ \mathbf{ab}\cdot\mathbf{cd} &= (\mathbf{b}\cdot\mathbf{c})\mathbf{ad} \\ \mathbf{abc}\cdot\mathbf{d} &= (\mathbf{c}\cdot\mathbf{d})\mathbf{ab}, \end{aligned} \tag{I.3.6}$$

where the dot product in parentheses is a scalar factor multiplying the dyad. The second of these contractions defines the ***single-dot*** product of the dyad **ab** by the dyad **cd**. Note that

$$\mathbf{ab}\cdot\mathbf{cd} \neq \mathbf{cd}\cdot\mathbf{ab} \tag{I.3.7}$$

so that (unlike the dot product of two vectors) ***the single-dot product of two dyads is not commutative.***

The ***scalar product*** of two dyads (or ***double-dot product***) denoted $\mathbf{ab} : \mathbf{cd}$ is defined as the scalar obtained by multiplying together the two scalar products $\mathbf{a} \cdot \mathbf{c}$ and $\mathbf{b} \cdot \mathbf{d}$. Note that the first vector of the first dyad multiplies the first vector of the second dyad, and the second vector of the first dyad multiplies the second vector of the second dyad. ***The scalar product is commutative.***

$$\mathbf{ab} : \mathbf{cd} = (\mathbf{a} \cdot \mathbf{c})(\mathbf{b} \cdot \mathbf{d}) = (\mathbf{c} \cdot \mathbf{a})(\mathbf{d} \cdot \mathbf{b}) = \mathbf{cd} : \mathbf{ab}. \tag{I.3.8}$$

Some authors denote the scalar product of Eq. (8) by $\mathbf{ab} \cdot \cdot \mathbf{cd}$ instead of $\mathbf{ab} : \mathbf{cd}$. Also, some authors define the scalar product differently, multiplying the two outside vectors together and the two inside vectors together. For example, Milne-Thomson (1962) writes $(\mathbf{a} ; \mathbf{b}) \cdot \cdot (\mathbf{c} ; \mathbf{d}) = (\mathbf{a} \cdot \mathbf{d})(\mathbf{b} \cdot \mathbf{c})$. In this book, the double-dot notation with the two dots on the same level denotes the product obtained by multiplying the two outside vectors together and the two inside vectors together.

$$\mathbf{ab} \cdot \cdot \mathbf{cd} = (\mathbf{a} \cdot \mathbf{d})(\mathbf{b} \cdot \mathbf{c}). \tag{I.3.9}$$

Dyadics as second-order tensors. Since the single-dot product of a dyad $\mathbf{ab}$ by a vector $\mathbf{v}$ yields another vector, and since

$$(\mathbf{ab}) \cdot (m\mathbf{u} + n\mathbf{v}) = m(\mathbf{ab}) \cdot \mathbf{u} + n(\mathbf{ab}) \cdot \mathbf{v} \tag{I.3.10}$$

for arbitrary vectors $\mathbf{u}, \mathbf{v}$ and arbitrary real numbers m and n it follows that the dyad operates as a linear vector function and is therefore a second-order tensor when operating as $(\mathbf{ab}) \cdot$. If we denote it by $\mathbf{T}$, then $(\mathbf{ab}) \cdot \mathbf{v} = \mathbf{T} \cdot \mathbf{v}$. The operation $\mathbf{v} \cdot (\mathbf{ab})$ also produces a vector, and $\cdot (\mathbf{ab})$ operates as another tensor, say $\mathbf{S}$, such that $\mathbf{v} \cdot (\mathbf{ab}) = \mathbf{v} \cdot \mathbf{S}$. Since $\mathbf{S}$ is the transpose of the tensor $\mathbf{T}$ (i.e., $\mathbf{S} = \mathbf{T}^T$ because $\mathbf{v} \cdot \mathbf{S} = \mathbf{T} \cdot \mathbf{v}$ for all $\mathbf{v}$) and

$$\mathbf{v} \cdot (\mathbf{ab}) = (\mathbf{ab})_c \cdot \mathbf{v}, \tag{I.3.11}$$

we see that the conjugate dyad operates as the transpose of the tensor represented by the original dyad.

We have seen that every dyad operates as a second-order tensor. Is the converse true? Can every second-order tensor be represented as a dyad, i.e., as the open product of two vectors? The answer is no. But every second-order tensor can be represented as a dyadic, a linear combination of the n^2 dyads formed from n linearly independent base vectors of the n-dimensional vector space on which the tensor is defined. For example, in three dimensions, with base vectors $\mathbf{b}_1$, $\mathbf{b}_2$, $\mathbf{b}_3$, we may write any second-order tensor $\mathbf{T}$ as

$$\left.\begin{aligned}\mathbf{T} = T^{11}\mathbf{b}_1\mathbf{b}_1 + T^{12}\mathbf{b}_1\mathbf{b}_2 + T^{13}\mathbf{b}_1\mathbf{b}_3 \\ + T^{21}\mathbf{b}_2\mathbf{b}_1 + T^{22}\mathbf{b}_2\mathbf{b}_2 + T^{23}\mathbf{b}_2\mathbf{b}_3 \\ + T^{31}\mathbf{b}_3\mathbf{b}_1 + T^{32}\mathbf{b}_3\mathbf{b}_2 + T^{33}\mathbf{b}_3\mathbf{b}_3\end{aligned}\right\} \tag{I.3.12}$$

or

$$\mathbf{T} = T^{rs}\mathbf{b}_r\mathbf{b}_s,$$

where the T^{rs} are the contravariant components of the tensor with respect to the basis. In Euclidean vector space, by introducing the dual basis $\mathbf{b}^k$, we obtain additional representations for $\mathbf{T}$.

$$\mathbf{T} = T^{rs}\mathbf{b}_r\mathbf{b}_s = T_{rs}\mathbf{b}^r\mathbf{b}^s = T^{r}_{.s}\mathbf{b}_r\mathbf{b}^s = T_{r}^{.s}\mathbf{b}^r\mathbf{b}_s. \tag{I.3.13}$$

The convention of upper and lower indices does not guarantee a unique tensor for a given set of components, since, for example, in general

$$T^{r}_{.s}\mathbf{b}_r\mathbf{b}^s \neq T^{r}_{.s}\mathbf{b}^s\mathbf{b}_r. \tag{I.3.14}$$

For definiteness, ***we adopt the convention that the first index on the tensor component goes with the first vector of the dyad*** as on the left side of the inequality (14).

It can be shown that the coefficients of Eq. (13) transform according to the tensor-component transformation formulas, Eqs. (I.2.15), under a change of basis given by Eqs. (I.2.1), if we require, for example,

$$\bar{T}_{ij}\bar{\mathbf{b}}^i\bar{\mathbf{b}}^j = T_{rs}\mathbf{b}^r\mathbf{b}^s \tag{I.3.15}$$

for all possible choices of the matrix $[T_{rs}]$. Furthermore, if $\mathbf{T}$ is given by Eq. (12),

$$\mathbf{b}^i\cdot\mathbf{T}\cdot\mathbf{b}^j = T^{rs}(\mathbf{b}^i\cdot\mathbf{b}_r)(\mathbf{b}_s\cdot\mathbf{b}^j) = T^{rs}\delta^i_r\delta^j_s = T^{ij} \tag{I.3.16}$$

in agreement with Eqs. (I.2.18), etc.

Determinant expansions. The value of any third-order determinant $|a^m_n|$ with element a^m_n in the mth row and nth column is known to be given by the sum $\sum_{ijk} \pm a^i_1 a^j_2 a^k_3$ where ijk is a permutation of 1 2 3 and either the plus or minus sign is used according to whether the permutation is even or odd. This may be given a convenient form by using the three index symbols $e^{ijk} \equiv e_{ijk}$, where the e_{ijk} are defined, as in Eq. (2.3.22), by

$$\det|a^m_n| = e_{ijk}a^i_1a^j_2a^k_3.$$

A similar development may be made with other fixed subscripts, say pqr instead of 1 2 3, and it turns out that if pqr is an even permutation of 1 2 3 we get the value of the determinant, while if it is odd we get its negative (this amounts to interchanging two columns in the determinant). Hence, in general,

$$\left.\begin{aligned} e_{pqr} \det |a_n^m| &= e_{ijk} a_p^i a_q^j a_r^k \quad \text{and, similarly,} \\ e^{pqr} \det |a_n^m| &= e^{ijk} a_i^p a_j^q a_k^r. \end{aligned}\right\} \tag{I.3.17}$$

This determinant expansion will be used in evaluating cross products.

Cross products in three-dimensional Euclidean vector space. An orthonormal basis can always be chosen in a Euclidean vector space. For example, in three dimensions, given any basis $\mathbf{b}_1$, $\mathbf{b}_2$, $\mathbf{b}_3$ we can construct an orthonormal basis $\mathbf{i}_1$, $\mathbf{i}_2$, $\mathbf{i}_3$ by the ***Gram-Schmidt orthogonalization procedure***, as follows.

Let

$$\left.\begin{aligned} \mathbf{a}_1 &= \mathbf{b}_1 & \mathbf{i}_1 &= \frac{\mathbf{a}_1}{\sqrt{\mathbf{a}_1 \cdot \mathbf{a}_1}} \\ \mathbf{a}_2 &= \mathbf{b}_2 - (\mathbf{b}_2 \cdot \mathbf{i}_1)\mathbf{i}_1 & \mathbf{i}_2 &= \frac{\mathbf{a}_2}{\sqrt{\mathbf{a}_2 \cdot \mathbf{a}_2}} \\ \mathbf{a}_3 &= \mathbf{b}_3 - (\mathbf{b}_3 \cdot \mathbf{i}_1)\mathbf{i}_1 - (\mathbf{b}_3 \cdot \mathbf{i}_2)\mathbf{i}_2 & \mathbf{i}_3 &= \frac{\mathbf{a}_3}{\sqrt{\mathbf{a}_3 \cdot \mathbf{a}_3}}. \end{aligned}\right\} \tag{I.3.18}$$

For the orthonormal basis, the dual basis is identical with the given basis

$$\mathbf{i}^1 = \mathbf{i}_1 \qquad \mathbf{i}^2 = \mathbf{i}_2 \qquad \mathbf{i}^3 = \mathbf{i}_3.$$

In an abstract vector space, the concept of right-handedness does not exist. If we ***define*** the cross product of two of the vectors of the orthonormal basis as

$$\mathbf{i}_r \times \mathbf{i}_s = e_{rst}\,\mathbf{i}^t, \tag{I.3.19}$$

using the three-index symbol introduced in Eq. (2.3.22), this will give the same results we obtained in Eq. (2.3.15) for the cross products of Cartesian unit vectors, provided the coordinate system is right-handed. We now obtain expressions for the cross products of the base vectors $\mathbf{b}_r$ of an arbitrary basis in a three-dimensional Euclidean vector space, assuming that we have an orthonormal basis available for use and that the cross product is distributive and obeys the usual axiom that

$$a\mathbf{u} \times b\mathbf{v} = ab(\mathbf{u} \times \mathbf{v})$$

for arbitrary vectors $\mathbf{u}$, $\mathbf{v}$ and arbitrary real numbers a, b. Let e_r^k denote the backward-transformation coefficients from the orthonormal basis $\mathbf{i}_k$ to the basis $\mathbf{b}_r$.

$$\mathbf{b}_r = e_r^k \mathbf{i}_k. \tag{I.3.20}$$

Then

$$\left.\begin{aligned} \mathbf{b}_r \times \mathbf{b}_s &= (e_r^k \mathbf{i}_k) \times (e_s^j \mathbf{i}_j) \\ &= e_r^k e_s^j \mathbf{i}_k \times \mathbf{i}_j \\ &= e_r^k e_s^j e_{kjm} \mathbf{i}^m. \end{aligned}\right\} \tag{I.3.21}$$

Let c_t denote the covariant components of $\mathbf{b}_r \times \mathbf{b}_s$. Then, by Eq. (I.1.13),

$$\mathbf{b}_r \times \mathbf{b}_s = c_t \mathbf{b}^t \tag{I.3.22}$$

whence, by Eq. (I.2.3b),

$$\begin{aligned} \mathbf{b}_r \times \mathbf{b}_s \cdot \mathbf{b}_t = c_t \quad &\text{and, by Eqs. (20) and (21),} \\ c_t = \mathbf{b}_r \times \mathbf{b}_s \cdot \mathbf{b}_t &= e_r^k e_s^j e_{kjm} \mathbf{i}^m \cdot e_t^p \mathbf{i}_p \\ &= e_r^k e_s^j e_t^p e_{kjm} \delta_p^m \\ &= e_r^k e_s^j e_t^m e_{kjm} \\ &= e_{rst} \det |e_q^p|, \end{aligned} \tag{I.3.23}$$

where the last step follows from Eq. (17). Hence Eq. (22) takes the form

$$\mathbf{b}_r \times \mathbf{b}_s = e_{rst} \det |e_q^p| \mathbf{b}^t. \tag{I.3.24}$$

Now, by Eq. (I.2.6),

$$\begin{aligned} g_{rs} = \mathbf{b}_r \cdot \mathbf{b}_s &= (e_r^k \mathbf{i}_k) \cdot (e_s^j \mathbf{i}_j) \\ &= e_r^k e_s^j \delta_{kj} = \sum_k e_r^k e_s^k. \end{aligned}$$

Hence the matrix equation

$$[g_{rs}] = [e_r^k]^T [e_s^k] \tag{I.3.25a}$$

holds, whence

$$g \equiv \det |g_{rs}| = (\det |e_s^k|)^2, \tag{I.3.25b}$$

and Eq. (24) yields the following form for the cross product of two base vectors, which no longer contains any reference to the orthonormal basis:

$$\mathbf{b}_r \times \mathbf{b}_s = \sigma \sqrt{g}\, e_{rst} \mathbf{b}^t \qquad \text{where} \quad \sigma = \pm 1. \tag{I.3.26}$$

In ordinary geometric space, we choose the orthonormal basis of Eq. (19) to be right-handed, and then choose $\sigma = +1$ if the basis $\mathbf{b}_n$ is right-handed; otherwise, $\sigma = -1$ (thus $\sigma =$ sign of det $|e^p_q|$). In an abstract space, lacking the concept of right-handedness, we always choose $\sigma = +1$.

It follows that for any two arbitrary vectors

$$\mathbf{u} = u^r \mathbf{b}_r \qquad \mathbf{v} = v^s \mathbf{b}_s \qquad \mathbf{u} \times \mathbf{v} = \sigma\sqrt{g}\, e_{rst} u^r v^s \mathbf{b}^t,$$

which can be expressed formally as a determinant

$$\mathbf{u} \times \mathbf{v} = \sigma\sqrt{g}\, e_{rst} u^r v^s \mathbf{b}^t \equiv \sigma\sqrt{g} \begin{vmatrix} \mathbf{b}^1 & \mathbf{b}^2 & \mathbf{b}^3 \\ u^1 & u^2 & u^3 \\ v^1 & v^2 & v^3 \end{vmatrix} \tag{I.3.27a}$$

or, alternatively, by a similar derivation,

$$\mathbf{u} \times \mathbf{v} = \frac{\sigma}{\sqrt{g}} e^{rst} u_r v_s \mathbf{b}_t \equiv \frac{\sigma}{\sqrt{g}} \begin{vmatrix} \mathbf{b}_1 & \mathbf{b}_2 & \mathbf{b}_3 \\ u_1 & u_2 & u_3 \\ v_1 & v_2 & v_3 \end{vmatrix} \tag{I.3.27b}$$

where $\sigma = -1$ if the basis is left-handed. In all other cases $\sigma = +1$. How do the covariant components of the cross product of Eq. (27a) transform under an arbitrary basis change? Let

$$\mathbf{w} = \mathbf{u} \times \mathbf{v} \tag{I.3.28}$$

$$w_t = \sigma\sqrt{g}\, e_{rst} u^r v^s \tag{I.3.29a}$$

$$\bar{w}_n = \bar{\sigma}\sqrt{\bar{g}}\, e_{kmn} \bar{u}^k \bar{v}^m. \tag{I.3.29b}$$

Transform the barred expressions on the right-hand side, using Eq. (I.2.2a) and note that, since the matrix equation

$$[\bar{g}_{mn}] = [a^r_m]^T [g_{rs}][a^s_n]$$

holds, we have

$$\bar{g} = \det|\bar{g}_{mn}| = (\det A)^2 g$$

$$\sqrt{\bar{g}} = |\det A|\sqrt{g}\,. \tag{I.3.30}$$

Hence, multiplying Eq. (29b) by b^n_t we obtain

$$\begin{aligned} b^n_t \bar{w}_n &= \bar{\sigma}|\det A|\sqrt{g}\, e_{kmn} b^k_r b^m_s b^n_t u^r u^s \\ &= \bar{\sigma}|\det A|\sqrt{g}\, e_{rst}\, (\det B) u^r u^s \\ &= \bar{\sigma}|\det A|(\det B) w_t \sigma. \end{aligned}$$

Since A and B are inverse matrices [see Eq. (I.2.4b)], $|\det A|$ $(\det B) = \pm 1$, the plus sign applying if $\det B$ is positive. Hence, the backward-transformation formula for covariant components of the cross product is $w_t = \kappa b^n_t \bar{w}_n$, where κ is the sign of $\sigma\bar{\sigma} \det B$. In general

$$\left.\begin{array}{l}\bar{w}_n = \kappa a^t_n w_t \qquad w_t = \kappa b^n_t \bar{w}_n \\ \bar{w}^n = \kappa b^n_t w^t \qquad w^t = \kappa a^t_n \bar{w}^n. \\ \text{In abstract space:} \quad \kappa = \dfrac{\det A}{|\det A|} = \dfrac{\det B}{|\det B|}. \\ \text{In ordinary space:} \quad \kappa = +1,\end{array}\right\} \tag{I.3.31}$$

since in abstract space $\sigma = \bar{\sigma} = +1$, while in ordinary space $\sigma\bar{\sigma}$ = sign of $\det A$ = sign of $\det B$, according to the conventions following Eq. (26). With these conventions in ordinary space, the cross product components transform as ordinary vector components (components of a polar vector, see Sec. 2.4), while with σ always equal to $+1$, the transformation is that of an ***axial vector***. (Some authors always choose $\sigma = +1$, even in ordinary space, and then the cross product transforms as an axial vector in ordinary space.) Axial vector component transformation formulas in rectangular Cartesian components were given in Eqs. (2.4.7).

We now consider in Sec. I.4 the tensor components obtained when the basis is the natural basis of a system of curvilinear coordinates.

EXERCISES

1. (a) Show that Eq. (I.3.15) implies the covariant tensor component transformation formula for the coefficients T_{pq}.
   (b) State and prove a similar result for mixed components $T^p_{\cdot q}$.
2. Carry out the demonstration comparable to Eq. (I.3.16) for dyadics whose coefficients are: (a) $T^r_{\cdot s}$, (b) T_{rs}, (c) $T_s^{\cdot r}$.
3. Verify the determinant expansion in the final Eq. (I.3.17) for the cases that p, q, r are: (a) 1, 2, 3; and (b) 2, 1, 3.
4. Use the Gram-Schmidt orthogonalization procedure of Eqs. (I.3.18) to construct an orthonormal basis for the vector space of polynomials of degree up to five in x on the interval from $x = 0$ to $x = 1$, with real coefficients, if the scalar product of two elements is defined as the integral from zero to one of the product of the two polynomials. (See Ex. 1 of Sec. I.1.)
5. Verify that the expansion of the determinant in Eq. (I.3.27a) gives the same results as $\mathbf{u} \times \mathbf{v} = \sigma\sqrt{g}\, e_{rst}\, u^r v^s \mathbf{b}^t$.

Some references on dyadics

Block, H. D., *Introduction to Tensor Analysis.* Columbus, Ohio: Charles E. Merrill Books (paperback), 1962.

Drew, T. B., *Handbook of Vector and Polyadic Analysis.* New York: Reinhold Publishing Corp, 1961.

Gibbs, J. W., and E. B. Wilson, *Vector Analysis.* New Haven: Yale University Press, 1913.

Milne-Thomson, L. M., *Antiplane Elastic Systems.* New York: Academic Press (Berlin: Springer-Verlag), 1962.

Weatherburn, C. E., *Advanced Vector Analysis.* London: G. Bell and Sons, 1928.

I.4 Curvilinear coordinates; contravariant and covariant components relative to the natural basis; the metric tensor

The concepts of stress, strain, and deformation and some general principles applicable to all continuous media were developed in Chaps. 3 through 5, and constitutive equations in Chap. 6, using rectangular Cartesian coordinates almost exclusively and Cartesian tensor components, which are much simpler than general curvilinear components. For applications in orthogonal curvilinear coordinates, it is often sufficient to work with physical components, avoiding altogether the complications of general tensor components as in the simpler formulation of App. II. Here we consider the general formulation, which is helpful for reading much of the literature of continuum mechanics.

In the three-dimensional Euclidean space of elementary geometry, we can define a system of curvilinear coordinates by specifying three uniquely invertible functions of a system of rectangular Cartesian coordinates. For example, the equations

$$r = \sqrt{x^2 + y^2} \qquad \theta = \tan^{-1}\frac{y}{x} \qquad z = z$$

or, inversely,

$$x = r\cos\theta \qquad y = r\sin\theta \qquad z = z$$

define cylindrical coordinates r, θ, z, as in Sec. 2.5. For a suitably restricted range of θ (e.g., $0 \leq \theta < 2\pi$) these equations define a unique one-to-one relationship between the Cartesian coordinates (x, y, z) and the cylindrical coordinates (r, θ, z) except on the singular line $x = y = 0$, where θ is not determined.

The index notation used for general curvilinear coordinates employs superscripts instead of subscripts. For example, let

$$x^1 = r \qquad x^2 = \theta \qquad x^3 = z$$

for cylindrical coordinates x^1, x^2, x^3. These superscripts do not denote exponents; when we occasionally need to use an exponent on a superscripted variable, we enclose the base in parentheses, e.g., $(x^3)^2$ denotes x^3 squared. We distinguish rectangular Cartesian components by using z instead of x for the superscripted variable. Thus

$$z^1 = x \qquad z^2 = y \qquad z^3 = z.$$

Of course, the Cartesian coordinates z^m are contained within the general class of curvilinear coordinates x^m, but we frequently make use of their special properties, and in the first place use them to define other curvilinear systems.

In general, we define curvilinear coordinates x^m in terms of a system z^m of rectangular Cartesians by

$$x^m = x^m(z^1, z^2, z^3) \quad \text{with inverse} \quad z^m = z^m(x^1, x^2, x^3) \tag{I.4.1}$$

and Jacobian determinant $|\mathbf{z}/\mathbf{x}|$ (with element $\partial z^m/\partial x^n$ in the mth row and nth column).

$$|\mathbf{z}/\mathbf{x}| \equiv \left|\frac{\partial z^m}{\partial x^n}\right| \neq 0. \tag{I.4.2}$$

We assume that the three functions $z^m(\mathbf{x})$, i.e., $z^m(x^1, x^2, x^3)$, have continuous first partial derivatives with respect to x^1, x^2, x^3, and that $|\mathbf{z}/\mathbf{x}| \neq 0$. Exceptions to this assumption may occur at certain singular points or curves, but never throughout any volume.

More generally we indicate the functional transformation between any two curvilinear systems $\bar{x}^m$ and x^m by

$$\begin{gathered} \bar{x}^m = \bar{x}^m(x^1, x^2, x^3) \qquad x^m = x^m(\bar{x}^1, \bar{x}^2, \bar{x}^3) \\ |\mathbf{x}/\bar{\mathbf{x}}| = \left|\frac{\partial x^m}{\partial \bar{x}^n}\right| \qquad |\bar{\mathbf{x}}/\mathbf{x}| = \left|\frac{\partial \bar{x}^m}{\partial x^n}\right| \qquad |\mathbf{x}/\bar{\mathbf{x}}||\bar{\mathbf{x}}/\mathbf{x}| = 1, \end{gathered} \tag{I.4.3}$$

assuming again that the functions involved have continuous derivatives and express a one-to-one transformation except at possible singular points or curves. If x^1 is held constant, the three Eqs. (1b) define parametrically a surface, giving its rectangular Cartesian coordinates as a function of the two parameters x^2 and x^3. The first Eq. (1a), $x^1 = x^1(z^1, z^2, z^3)$ defines the same surface implicitly; it is an x^1-coordinate surface. For fixed x^2 another coordinate surface is obtained and a third for fixed x^3. The ***three coordinate surfaces*** intersect by pairs in ***three coordinate curves***, on each of which only one curvilinear coordinate varies. The point of intersection of all three coordinate surfaces (and of all three coordinate curves) is the point with curvilinear coordinates (x^1, x^2, x^3).

For example, in cylindrical coordinates the coordinate surfaces are the cylinders $x^1 =$ const. [or $(z^1)^2 + (z^2)^2 =$ const.], the planes $x^2 =$ const. [or $\tan^{-1}(z^2/z^1) =$ const.] and the planes $x^3 =$ const. (or $z^3 =$ const.). Figure I.2 shows for cylindrical coordinates the three surfaces through point P and the dotted coordinate curves through P.

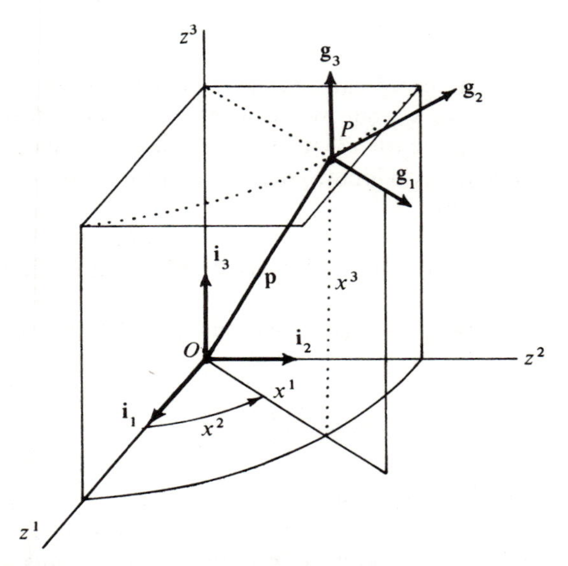

Fig. I.2 Covariant base vectors in cylindrical coordinates x^1, x^2, x^3.

Base vectors. In rectangular Cartesian coordinates, we express a vector in terms of the ***unit base vectors*** $\mathbf{i}_1$, $\mathbf{i}_2$, $\mathbf{i}_3$. For example, the position vector $\mathbf{p}$ is

$$\mathbf{p} = z^1\mathbf{i}_1 + z^2\mathbf{i}_2 + z^3\mathbf{i}_3 = z^r\mathbf{i}_r$$

and

$$d\mathbf{p} = dz^1\mathbf{i}_1 + dz^2\mathbf{i}_2 + dz^3\mathbf{i}_3 = dz^r\mathbf{i}_r. \tag{I.4.4}$$

We now introduce a set of base vectors $\mathbf{g}_s$ (not necessarily unit vectors) in the curvilinear system, called the ***natural basis*** of the curvilinear system (also called ***covariant base vectors***), such that the same position-vector increment $d\mathbf{p}$ will be similarly given in terms of the curvilinear coordinate increments dx^s by

$$d\mathbf{p} = dx^s \mathbf{g}_s. \tag{I.4.5}$$

Observe that if we move along a coordinate curve, only one of the three increments differs from zero, and $d\mathbf{p}$ is tangent to the curve. For example, if $dx^2 = dx^3 = 0$, then $d\mathbf{p} = dx^1\mathbf{g}_1$. Evidently the covariant base vector $\mathbf{g}_1$ is tangent to the coordinate curve along which x^1 varies. Similarly $\mathbf{g}_2$ and $\mathbf{g}_3$ are tangent to the coordinate curves of varying x^2 and x^3, respectively. For a general increment, Eq. (5) thus displays $d\mathbf{p}$ as the sum of three vectors, one tangent to each coordinate curve through the point. The fact that the $\mathbf{g}_s$ are not, in general, unit vectors is apparent from the case of cylindrical coordinates, where, if the position vector is resolved into vector components tangent to the coordinate curves, the vector component in the direction of increasing x^2 has magnitude $x^1\, dx^2$ and not merely dx^2; hence $\mathbf{g}_2 = x^1\hat{\mathbf{e}}_2$, if $\hat{\mathbf{e}}_2$ is the unit vector in the direction of increasing x^2. The three covariant base vectors are shown at P in Fig. I.2 for cylindrical coordinates.

From the definition implied by Eq. (5) it is apparent that

$$\mathbf{g}_s = \frac{\partial \mathbf{p}}{\partial x^s}, \tag{I.4.6}$$

for example, $\mathbf{g}_2 = d\mathbf{p}/dx^2$ when $dx^1 = dx^3 = 0$. Equation (6) may be used to express $\mathbf{g}_s$ in terms of the rectangular Cartesian unit base vectors $\mathbf{i}_r$. Thus by the chain rule of differentiation

$$\mathbf{g}_s = \frac{\partial \mathbf{p}}{\partial z^r}\frac{\partial z^r}{\partial x^s} = \frac{\partial z^r}{\partial x^s}\,\mathbf{i}_r \tag{I.4.7}$$

and, inversely,

$$\mathbf{i}_r = \frac{\partial \mathbf{p}}{\partial x^s}\frac{\partial x^s}{\partial z^r} = \frac{\partial x^s}{\partial z^r}\,\mathbf{g}_s,$$

since $\mathbf{i}_r = \partial\mathbf{p}/\partial z^r$. And indeed the relation between the covariant base vectors of any two curvilinear coordinate systems follows from Eq. (6) applied to each system:

$$\bar{\mathbf{g}}_s = \frac{\partial x^r}{\partial \bar{x}^s}\,\mathbf{g}_r \quad \text{and inversely} \quad \mathbf{g}_r = \frac{\partial \bar{x}^s}{\partial x^r}\,\bar{\mathbf{g}}_s, \tag{I.4.8}$$

i.e.,

$$a^r_s = \frac{\partial x^r}{\partial \bar{x}^s} \qquad b^s_r = \frac{\partial \bar{x}^s}{\partial x^r}$$

give the transformation coefficients a^r_s and b^s_r defined in Sec. I.2. Note that ***a superscript on a variable in the denominator is considered a subscript on the***

fraction in applying the summation convention to diagonally repeated indices. Each one of the Eqs. (8) represents three vector equations; the subscript on $\mathbf{g}_r$, for example, identifies the vector and not a component. In rectangular Cartesian coordinates, the natural base vectors are simply the unit vectors $\mathbf{i}_r$.

In rectangular Cartesian coordinates, the base vectors are also normals to the coordinate surfaces and at the same time tangent to coordinate curves. In nonorthogonal coordinates the covariant base vectors are tangent to coordinate curves, but are not, in general, normals to the coordinate surfaces. Another set of three base vectors at a point, called ***contravariant base vectors***, is defined in Eq. (9) below. Each contravariant base vector is normal to one of the three coordinate surfaces through the point. The set of contravariant base vectors is the ***dual basis*** or ***reciprocal basis*** to the natural basis, as defined in Sec. I.1. The contravariant base vectors are denoted $\mathbf{g}^r$ using a superscript, whereas the covariant vectors $\mathbf{g}_r$ of the natural basis use a subscript. For rectangular Cartesians then, $\mathbf{i}^r = \mathbf{i}_r$. The normal to the curvilinear coordinate surface $x^1 =$ const. is given by the gradient vector grad x^1, with Cartesian representation

$$\text{grad } x^1 = \frac{\partial x^1}{\partial z^1}\mathbf{i}^1 + \frac{\partial x^1}{\partial z^2}\mathbf{i}^2 + \frac{\partial x^1}{\partial z^3}\mathbf{i}^3.$$

Hence, in general, we may define $\mathbf{g}^r$ by

$$\mathbf{g}^r = \frac{\partial x^r}{\partial z^m}\mathbf{i}^m. \tag{I.4.9}$$

For any other coordinate system $\bar{x}^r$, the contravariant base vectors will be similarly defined by $\bar{\mathbf{g}}^r = (\partial \bar{x}^r/\partial z^s)\mathbf{i}^s$. By using the chain rule this can be put in the form

$$\bar{\mathbf{g}}^r = \frac{\partial \bar{x}^r}{\partial x^m}\left(\frac{\partial x^m}{\partial z^s}\mathbf{i}^s\right).$$

The quantity in parentheses is $\mathbf{g}^m$; hence we have the following transformation law. The transformation law is a set of vector equations expressing the contravariant base vectors in one system in terms of the contravariant base vectors of another system at the same point:

$$\bar{\mathbf{g}}^r = \frac{\partial \bar{x}^r}{\partial x^m}\mathbf{g}^m \quad \text{and inversely} \quad \mathbf{g}^m = \frac{\partial x^m}{\partial \bar{x}^r}\bar{\mathbf{g}}^r. \tag{I.4.10}$$

These contravariant vectors are not, in general, unit vectors; instead each $\mathbf{g}^r$ has the magnitude of the gradient of the function $x^r(z^1, z^2, z^3)$ at the point. A relationship between the magnitudes of the covariant and contravariant base vectors can be obtained by considering their dot products as follows:†

†This relationship was used to define the dual or reciprocal basis to an arbitrary basis in a vector space. See Sec. I.1.

$$\mathbf{g}^r \cdot \mathbf{g}_s = \left(\frac{\partial x^r}{\partial z^m}\frac{\partial z^n}{\partial x^s}\right)\mathbf{i}^m \cdot \mathbf{i}_n = \frac{\partial x^r}{\partial z^m}\frac{\partial z^n}{\partial x^s}\,\delta^m_n = \delta^r_s. \tag{I.4.11}$$

Thus, for example, if θ is the angle between $\mathbf{g}^2$ and $\mathbf{g}_2$ at a point, $|g^2||g_2| \cos\theta = 1$. In ***orthogonal curvilinear coordinates***, $\cos\theta = 1$; then $|\mathbf{g}^2| = 1/|\mathbf{g}_2|$, and similarly for each r:

$$\text{In orthogonal coordinates:} \quad |\mathbf{g}^r| = \frac{1}{|\mathbf{g}_r|}\,. \tag{I.4.12}$$

For example, in cylindrical coordinates, when $x^1 > 0$,

$$|\mathbf{g}_2| = x^1 \quad \text{and} \quad |\mathbf{g}^2| = \frac{1}{x^1}\,.$$

Any vector $\mathbf{a}$ may be represented in terms of either set of base vectors,

$$\mathbf{a} = a^r\mathbf{g}_r = a_r\mathbf{g}^r. \tag{I.4.13}$$

The coefficients a^r and a_r are not, in general, equal; they are equal in rectangular Cartesian coordinates, where they are each equal to the usual Cartesian components of the vector. The a^r are the ***contravariant components*** relative to the natural basis while the a_r are the ***covariant components***. The names "contravariant" and "covariant" are associated with the different ways the components transform under coordinate changes. These transformations were studied in Sec. I.2. We note that the contravariant vector components are the coefficients of the covariant base vectors, and vice versa, and emphasize again the convention that subscripts denote covariance, while superscripts denote contravariance or else identify the coordinates. Since the coordinates transform nonlinearly, in general, the coordinates are not tensor components at all, either covariant or contravariant.

Note that the curvilinear tensor components do not, in general, all have the same dimensions. For example the cylindrical-coordinate contravariant components of the position-vector differential $d\mathbf{p}$ of Eq. (5) are $dx^1 = dr$, $dx^2 = d\theta$, and $dx^3 = dz$; thus dx^1 and dx^3 have dimensions of length, while dx^2 is dimensionless. In applications, it is often preferable to work with the so-called ***physical components*** $(dr, r\,d\theta, dz)$ of $d\mathbf{p}$. The relationship of the physical components to tensor components will be given in Sec. I.5. A simpler development directly in terms of physical components is given in App. II.

Tensor component transformations in curvilinear coordinates follow immediately from Eq. (8) and the results of Sec. I.2 with

$$a^r_s = \frac{\partial x^r}{\partial \bar{x}^s} \qquad b^s_r = \frac{\partial \bar{x}^s}{\partial x^r}\,.$$

Thus

$$\left.\begin{aligned} \bar{v}_r &= \frac{\partial x^m}{\partial \bar{x}^r} v_m \qquad & v_n &= \frac{\partial \bar{x}^r}{\partial x^n} \bar{v}_r \\ \bar{v}^r &= \frac{\partial \bar{x}^r}{\partial x^m} v^m \qquad & v^n &= \frac{\partial x^n}{\partial \bar{x}^r} \bar{v}^r . \end{aligned}\right\} \tag{I.4.14}$$

$$\left.\begin{aligned} \bar{T}_{mn} &= \frac{\partial x^r}{\partial \bar{x}^m} \frac{\partial x^s}{\partial \bar{x}^n} T_{rs} \qquad & \bar{T}^{mn} &= \frac{\partial \bar{x}^m}{\partial x^r} \frac{\partial \bar{x}^n}{\partial x^s} T^{rs} \\ \bar{T}^m_{.n} &= \frac{\partial \bar{x}^m}{\partial x^r} \frac{\partial x^s}{\partial \bar{x}^n} T^r_{.s} \qquad & \bar{T}_m^{.n} &= \frac{\partial x^r}{\partial \bar{x}^m} \frac{\partial \bar{x}^n}{\partial x^s} T_r^{.s}, \end{aligned}\right\} \tag{I.4.15}$$

etc. for higher-order tensors. These transformations apply ***at a point***, where each coefficient is a number, the value of the partial derivative evaluated at that point. These formulas are especially easy to remember. Every coefficient is a partial derivative of a coordinate of one system with respect to a coordinate of the other. Attention to placing of free indices the same on both sides of equation, and summation on diagonally repeated indices then assures correctness. Remember that a superscript on a variable in the denominator counts as a subscript on the fraction. Note, for example, that since the free index m on the coefficient of the first Eq. (15) belongs to the $\bar{T}_{mn}$ on the left, it must belong to a barred coordinate as in $\bar{x}^m$ on the right.

Examples. A ***scalar point function*** has a single component in any coordinate system, which has the same numerical value in all coordinate systems when it is evaluated at the same point. Thus, for example,

$$\bar{F}(\bar{x}^1, \bar{x}^2, \bar{x}^3) = F(x^1, x^2, x^3), \tag{I.4.16}$$

where the x^i are related to the $\bar{x}^j$, by Eqs. (3), so that

$$\bar{F}(\bar{x}^1, \bar{x}^2, \bar{x}^3) \equiv F[x^1(\bar{x}^1, \bar{x}^2, \bar{x}^3), x^2(\bar{x}^1, \bar{x}^2, \bar{x}^3), x^3(\bar{x}^1, \bar{x}^2, \bar{x}^3)].$$

[See Eq. (2.5.16), in rectangular Cartesian coordinates.] Equation (16) is often written without the overbar on $\bar{F}$, but it should be remembered that the functional dependence on the $\bar{x}^k$ is different from the dependence on the x^m. For example, the scalar point function $F(z^1, z^2, z^3) \equiv 25[(z^1)^2 - (z^2)^2] + (z^3)^2$ has the form $F(x^1, x^2, x^3) \equiv 25(x^1)^2 \cos(2x^2) + (x^3)^2$ in the cylindrical coordinates defined at the beginning of this section.

Covariant components of gradient of a scalar function. By the chain rule of calculus, we have

$$\frac{\partial \bar{F}}{\partial \bar{x}^r} = \frac{\partial F}{\partial x^m} \frac{\partial x^m}{\partial \bar{x}^r} . \tag{I.4.17}$$

This expresses the three derivatives $\partial \bar{F}/\partial \bar{x}^r$ in the $\bar{x}^r$ system of curvilinear coordinates in terms of the three derivatives $\partial F/\partial x^m$ in the x^m system at the same point; the coefficients of the transformation are $\partial x^m/\partial \bar{x}^r$. Hence the partial derivatives of a scalar point function are first-order tensor covariant components. These tensor components are, in general, different from the physical components of the gradient given in textbooks on vector analysis. For example, in cylindrical coordinates r, θ, z, the covariant tensor components of grad F are

$$\left(\frac{\partial F}{\partial r}, \frac{\partial F}{\partial \theta}, \frac{\partial F}{\partial z}\right) \quad \text{while the physical components are} \quad \left(\frac{\partial F}{\partial r}, \frac{1}{r}\frac{\partial F}{\partial \theta}, \frac{\partial F}{\partial z}\right).$$

The differential dx^m. An example of a set of contravariant tensor components is furnished by the differential components dx^m of the position-vector differential increment $d\mathbf{p} = dx^m \mathbf{g}_m$, since

$$d\bar{x}^r = \frac{\partial \bar{x}^r}{\partial x^m}\, dx^m \tag{I.4.18a}$$

or, equivalently, the derivatives of x^m with respect to arc length (or any parameter s) along the same curve evaluated at the same point in two coordinate systems are related by the contravariant transformation

$$\frac{d\bar{x}^r}{ds} = \frac{\partial \bar{x}^r}{\partial x^m}\frac{dx^m}{ds}. \tag{I.4.18b}$$

Warning: Although the differentials dx^m are tensor components, the curvilinear coordinates x^m are not, since the coordinate transformations are general functional transformations and not the linear homogeneous transformations required for tensor components.

The metric tensor, raising and lowering indices, vector products, the ϵ_{mnr} and ϵ^{mnr} tensors. The element of arc length ds will now be expressed in terms of the increments dx^m of the curvilinear coordinates. By Eq. (5) the position-vector increment $d\mathbf{p}$ (whose magnitude is ds) is $d\mathbf{p} = dx^m \mathbf{g}_m$, where the $\mathbf{g}_m$ are the covariant base vectors. Hence

$$(ds)^2 = d\mathbf{p}\cdot d\mathbf{p} = (dx^m \mathbf{g}_m)\cdot(dx^n \mathbf{g}_n)$$

or

$$\left.\begin{aligned} (ds)^2 &= (\mathbf{g}_m\cdot\mathbf{g}_n)dx^m\,dx^n = g_{mn}\,dx^m\,dx^n, \\ \text{where the} \quad g_{mn} &= \mathbf{g}_m\cdot\mathbf{g}_n. \end{aligned}\right\} \tag{I.4.19}$$

The quantities g_{mn} (symmetric in the indices m and n) are the unit tensor covariant components relative to the natural basis (see Sec. I.2). The scalar-invariant quadratic form $g_{mn}\,dx^m\,dx^n$ is called the ***metric*** of the space; it is positive-definite for a Euclidean space. The unit tensor **1** whose components g_{mn} in any curvilinear coordinate system are the coefficients of the metric quadratic form is also called the ***metric tensor*** of the space.

When the curvilinear coordinates are orthogonal, the scalar products $\mathbf{g}_m \cdot \mathbf{g}_n$ vanish for $m \neq n$, and there are only three nonzero components g_{mn}, the diagonal elements of the matrix, which can be very simply written down when the expression for $(ds)^2$ is known. For example, in cylindrical coordinates r, θ, z, we have $(ds)^2 = (dr)^2 + (r\,d\theta)^2 + (dz)^2$ or in indicial notation

$$(ds)^2 = dx^1\,dx^1 + (x^1)^2\,dx^2\,dx^2 + dx^3\,dx^3.$$

Hence,

$$\text{In cylindrical coordinates:} \quad [g_{mn}] = \begin{bmatrix} 1 & 0 & 0 \\ 0 & (x^1)^2 & 0 \\ 0 & 0 & 1 \end{bmatrix}. \tag{I.4.20}$$

In nonorthogonal coordinates there will in general be nine components of the metric tensor, though only six are independent because the symmetry implies $g_{mn} = g_{nm}$.

In rectangular Cartesian coordinates,

$$(ds)^2 = (dz^1)^2 + (dz^2)^2 + (dz^3)^2 = \delta_{rs}\,dz^r\,dz^s \tag{I.4.21}$$

with coefficients $g_{rs} = \delta_{rs}$, the rectangular Cartesian components of the unit tensor. The covariant tensor transformation formula of Eq. (15) thus shows that the components g_{mn} in any curvilinear system are given by

$$\begin{aligned} g_{mn} &= \frac{\partial z^r}{\partial x^m}\frac{\partial z^s}{\partial x^n}\delta_{rs} \\ &= \frac{\partial z^1}{\partial x^m}\frac{\partial z^1}{\partial x^n} + \frac{\partial z^2}{\partial x^m}\frac{\partial z^2}{\partial x^n} + \frac{\partial z^3}{\partial x^m}\frac{\partial z^3}{\partial x^n}, \end{aligned} \tag{I.4.22}$$

as could also have been shown by substituting $dz^1 = (\partial z^1/\partial x^m)\,dx^m$, etc. in Eq. (21). In rectangular Cartesians all three Kronecker deltas δ_{rs}, δ^{rs}, and δ^r_s denote the same thing, namely

$$\delta_{rs} = \delta^{rs} = \delta^r_s = \begin{cases} 1 & \text{if} \quad r = s \\ 0 & \text{if} \quad r \neq s. \end{cases} \tag{I.4.23}$$

However, if these are taken to be components of tensors of the type indicated by the index positions, their components in other curvilinear coordinate systems will be different. Thus

$$g^{mn} = \frac{\partial x^m}{\partial z^r}\frac{\partial x^n}{\partial z^s}\delta^{rs} \tag{I.4.24}$$

$$g^m_n = \frac{\partial x^m}{\partial z^r}\frac{\partial z^s}{\partial x^n}\delta^r_s = \frac{\partial x^m}{\partial z^r}\frac{\partial z^r}{\partial x^n} = \frac{\partial x^m}{\partial x^n} = \delta^m_n. \tag{I.4.25}$$

Thus g^m_n has the same components in all coordinate systems, namely 1 if $m = n$ and 0 if $m \neq n$. All three of the metric tensor matrices $[g_{ij}]$, $[g^{ij}]$, $[g^i_j]$ are symmetric. The determinants g and $1/g$ of Eqs. (I.2.11) and (I.2.12) satisfy

$$g \equiv |g_{mn}| = |\mathbf{z}/\mathbf{x}|^2 \qquad \frac{1}{g} = |g^{mn}| = \frac{1}{|\mathbf{z}/\mathbf{x}|^2}. \tag{I.4.26}$$

That in fact the determinant g equals the square of the Jacobian determinant $|\mathbf{z}/\mathbf{x}|$ of Eq. (2) follows from Eq. (I.3.25). A similar discussion shows that $|g^{mn}| = |\mathbf{x}/\mathbf{z}|^2 = 1/|\mathbf{z}/\mathbf{x}|^2$.

When the curvilinear coordinates are orthogonal,

$$g^{11} = \frac{1}{g_{11}} \qquad g^{22} = \frac{1}{g_{22}} \qquad g^{33} = \frac{1}{g_{33}}. \tag{I.4.27}$$

The calculus of curvilinear tensor components will be developed in Sec. I.6. First, however, we consider how the physical components are related to the tensor components.

EXERCISES

1. For the cylindrical coordinates x^m defined by $z^1 = x^1 \cos x^2$, $z^2 = x^1 \sin x^2$, $z^3 = x^3$, tabulate the matrix of components $\partial z^m/\partial x^n$ and show that its determinant $|\mathbf{z}/\mathbf{x}|$ is equal to x^1.
2. Show that dx^m/dt are first-order contravariant tensor components. (*Remark:* These contravariant components v^m of velocity differ from the usual vector velocity components. For example, in cylindrical coordinates the second vector velocity component is $x^1\,dx^2/dt$ instead of simply dx^2/dt.)
3. If S_1, S_2, S_3 denote the spherical-coordinate covariant components of a vector, express them in terms of the rectangular Cartesian components C_1, C_2, C_3 of the same vector and the spherical coordinates x^1, x^2, x^3 defined by $z^1 = x^1 \sin x^2 \cos x^3$, $z^2 = x^1 \sin x^2 \sin x^3$, and $z^3 = x^1 \cos x^2$. Write out the three expressions, since one indicial expression will not do for all three components.
4. (a) If F is the scalar point function defined in rectangular Cartesians by $F = 25[(z^1)^2 - (z^2)^2] + (z^3)^2$ as in the example following Eq. (I.4.16), verify that the cylindrical-coordinate form given in the text is correct.

(b) Compute the cylindrical-coordinate covariant components of the gradient of F, and verify that they are related to the Cartesian components by the appropriate transformation.

5. Prove that if the v_k are covariant components of a vector, then $\partial v_m/\partial x^n - \partial v_n/\partial x^m$ are covariant components of a skew-symmetric second-order tensor.

6. By examining a sketch of the element, determine the coefficients of the expression for $(ds)^2$ in spherical coordinates, and hence show that the nonzero covariant components of the metric tensor are $g_{11} = 1$, $g_{22} = (x^1)^2$, and $g_{33} = (x^1 \sin x^2)^2$.

7. What are the contravariant components g^{mn} of the metric tensor in: (a) cylindrical coordinates? (b) spherical coordinates?

8. If v^1, v^2, v^3 are contravariant components of a vector in cylindrical coordinates, express the covariant components of the same vector in terms of the coordinates and the v^k.

9. Repeat Ex. 8 for spherical coordinates.

10. For cylindrical coordinates, write out explicit formulas in terms of the rectangular Cartesian unit vectors $\mathbf{i}_1$, $\mathbf{i}_2$, and $\mathbf{i}_3$ for: (a) $\mathbf{g}_1$, $\mathbf{g}_2$, and $\mathbf{g}_3$, (b) $\mathbf{g}^1$, $\mathbf{g}^2$, and $\mathbf{g}^3$.

11. Show in a sketch the three base vectors $\mathbf{g}_r$ for spherical coordinates. How are the $\mathbf{g}^r$ related to the $\mathbf{g}_r$ in direction? in magnitude?

I.5 Physical components of vectors and tensors

The contravariant or covariant tensor components of a vector do not have the same kind of physical significance in a curvilinear coordinate system as they have in a rectangular Cartesian system; in fact, they often have different dimensions. For example, the differential $d\mathbf{p}$ of the position vector has in cylindrical coordinates (r, θ, z) the contravariant tensor components $(dr, d\theta, dz)$, and $d\theta$ does not have the same dimensions as the others. The ***physical components*** in this example are $(dr, r\,d\theta, dz)$, the components of the infinitesimal vector differential $d\mathbf{p}$ in a local rectangular Cartesian system with axes tangent to the cylindrical coordinate curves at the point under consideration. We now proceed to define physical components of any tensor. When the curvilinear coordinates are orthogonal, the physical components of a vector or tensor will be just the Cartesian components in a local rectangular Cartesian system with axes tangent to the coordinate curves (see App. II). The more general definitions will be given here first, however, since these definitions apply whether or not the curvilinear coordinates are orthogonal.

In a general curvilinear coordinate system the physical components of a vector at a point are the vector components parallel to the covariant base vectors. Let $\hat{\mathbf{e}}_r$ denote dimensionless unit vectors parallel to $\mathbf{g}_r$; then, by Eqs. (I.4.5) and (I.4.19),

$$|\mathbf{g}_1| = \sqrt{g_{11}} \qquad |\mathbf{g}_2| = \sqrt{g_{22}} \qquad |\mathbf{g}_3| = \sqrt{g_{33}}$$
$$\hat{\mathbf{e}}_1 = \frac{\mathbf{g}_1}{\sqrt{g_{11}}} \qquad \hat{\mathbf{e}}_2 = \frac{\mathbf{g}_2}{\sqrt{g_{22}}} \qquad \hat{\mathbf{e}}_3 = \frac{\mathbf{g}_3}{\sqrt{g_{33}}}. \tag{I.5.1}$$

Hence, if $\mathbf{v}$ is any vector,

$$\mathbf{v} = v^k \mathbf{g}_k = v^1\sqrt{g_{11}}\,\hat{\mathbf{e}}_1 + v^2\sqrt{g_{22}}\,\hat{\mathbf{e}}_2 + v^3\sqrt{g_{33}}\,\hat{\mathbf{e}}_3. \tag{I.5.2}$$

Hence the ***physical component*** $v^{\langle k\rangle}$ of $\mathbf{v}$ parallel to $\mathbf{g}_k$ is

$$v^{\langle k\rangle} = v^k\sqrt{g_{kk}} \quad \text{and inversely} \quad v^k = \frac{v^{\langle k\rangle}}{\sqrt{g_{kk}}} \qquad \text{(no sum)}, \tag{I.5.3}$$

and in, the particular coordinate system,

$$\mathbf{v} = v^{\langle k\rangle}\hat{\mathbf{e}}_k \qquad \text{(vector sum on } k\text{)}. \tag{I.5.4}$$

Since the covariant components v_m of the same vector are related to the contravariant components v^k by $v_m = g_{mk}v^k$, the covariant components v_m may be expressed in terms of the physical components as

$$v_m = \sum_{k=1}^{3} g_{mk}\frac{v^{\langle k\rangle}}{\sqrt{g_{kk}}}. \tag{I.5.5}$$

In the case of orthogonal curvilinear coordinates, $g_{mk} = 0$ for $k \neq m$, and therefore Eq. (5) reduces to

$$\text{For orthogonal coordinates:} \quad \left\{ \begin{aligned} v_m &= \sqrt{g_{mm}}\, v^{\langle m\rangle} \\ v^{\langle m\rangle} &= \frac{v_m}{\sqrt{g_{mm}}} \end{aligned} \right\} \quad \text{(no sum)}, \tag{I.5.6}$$

as could also have been seen from Eq. (3), since $v^m = v_m/g_{mm}$ (not summed) for orthogonal coordinates.

For higher-order tensors, physical components have been defined differently by different authors,† but the various definitions reduce to the same thing in orthogonal curvilinear coordinates, where they are the Cartesian components in a local system of rectangular Cartesian coordinates with axes tangent to the coordinate curves at the point (parallel to the covariant base vectors). Physical components of higher-order tensors are most conveniently defined in terms of mixed tensor components; the relations with covariant or contravariant components can always be obtained then by raising or lowering indices on the

†See Truesdell (1953).

mixed components, using the metric tensor. A reasonable definition of physical components of a tensor is based on the way tensors appear in physics. In classical physics, only scalars and vectors have primary physical significance and tensor components are determined by means of equations relating them to measured vector components and scalars. For second-order tensors we distinguish between two cases depending on whether the tensor **T** operates as a linear vector operator of the form

$$\mathbf{u} = \mathbf{T}\cdot\mathbf{v} \tag{I.5.7a}$$

or alternatively,

$$\mathbf{u} = \mathbf{v}\cdot\mathbf{T}. \tag{I.5.7b}$$

Two different kinds of physical components arise naturally from the two cases; these components will be called right physical components and left physical components. The first case leads to the definition of ***right physical components***, where **T** operates to the right on **v**, while the second leads to the definition of left physical components.

Right physical components of the tensor **T** of Eq. (7a) are nine quantities $T^{\langle rs\rangle}$ such that inner multiplication on the right by the physical components $v^{\langle s\rangle}$ of **v** yields the physical components $u^{\langle r\rangle}$ of **u**:

$$\left.\begin{aligned} u^{\langle r\rangle} &= T^{\langle rs\rangle}v^{\langle s\rangle} \qquad \text{(summed on } s) \\ \text{or}\quad [u^{\langle r\rangle}] &= [T^{\langle rs\rangle}][v^{\langle s\rangle}] \end{aligned}\right\} \tag{I.5.8}$$

in matrix notation. (With equations between physical components, as with Cartesian components, repeated indices even on the same level imply summation. When both physical and tensor components appear, the summations will be explicitly called for as in the next equation below.) From Eq. (7), using the definitions of $u^{\langle r\rangle}$ and $v^{\langle s\rangle}$, according to Eq. (3), we have

$$\frac{u^{\langle r\rangle}}{\sqrt{g_{rr}}} = \sum_s T^r_{\cdot s}\frac{v^{\langle s\rangle}}{\sqrt{g_{ss}}}$$

whence

$$u^{\langle r\rangle} = \sum_s \left(\sqrt{\frac{g_{rr}}{g_{ss}}}\,T^r_{\cdot s}\right)v^{\langle s\rangle}. \tag{I.5.9}$$

Hence, if we define the right physical components $T^{\langle rs\rangle}$ by

$$T^{\langle rs\rangle} = \sqrt{\frac{g_{rr}}{g_{ss}}}\,T^r_{\cdot s} \qquad \text{(no sum)}, \tag{I.5.10}$$

we obtain the desired relationship of Eq. (8).

Left physical components ${}^{\langle sr\rangle}T$ are defined so that inner multiplication on the left by $v^{\langle s\rangle}$ yields $u^{\langle r\rangle}$. Thus

$$u^{\langle r\rangle} = v^{\langle s\rangle}\,{}^{\langle sr\rangle}T \quad \text{or} \quad [u^{\langle r\rangle}] = [v^{\langle s\rangle}][{}^{\langle sr\rangle}T] \tag{I.5.11}$$

in matrix notation. From the tensor component equation $u^r = v^s T_s^{.r}$ and the definitions of $u^{\langle r\rangle}$ and $v^{\langle s\rangle}$, we get in a manner similar to that for the right components the following definition for the left physical components

$$^{\langle sr\rangle}T = \sqrt{\frac{g_{rr}}{g_{ss}}}\,T_s^{.r} \qquad \text{(no sum).} \tag{I.5.12}$$

When the tensor is symmetric, $T^r_{.s} = T_s^{.r}$, and it follows that

$$T^{\langle rs\rangle} = {}^{\langle sr\rangle}T, \tag{I.5.13}$$

so that ***for a symmetric tensor the matrix of left physical components is the transpose of the matrix of right physical components.***

When the tensor is not symmetric we can derive a relationship between ${}^{\langle sr\rangle}T$ and $T^{\langle qp\rangle}$ as follows:

$$T_s^{.r} = g_{sq}T^{qr} = g_{sq}g^{rp}T^q_{.p} \qquad \text{(sum on } p, q)$$

and

$$T^q_{.p} = \sqrt{\frac{g_{pp}}{g_{qq}}}\,T^{\langle qp\rangle} \qquad \text{by Eq. (10) (no sum).}$$

Therefore Eq. (12) yields

$$^{\langle sr\rangle}T = \sum_{p,q}\sqrt{\frac{g_{rr}g_{pp}}{g_{ss}g_{qq}}}\,g_{sq}g^{rp}\,T^{\langle qp\rangle}. \tag{I.5.14}$$

For ***orthogonal coordinates,*** $g_{km} = 0$ for $k \neq m$. Hence Eq. (14) reduces to

$$^{\langle sr\rangle}T = T^{\langle sr\rangle}$$

and ***for orthogonal coordinates the two matrices of right and left components are identical.*** It is still necessary to operate with the matrix of physical components in the appropriate way, i.e., multiplying to the right or to the left, unless the tensor is symmetric. ***If the tensor is symmetric, then its matrix of physical components in orthogonal coordinates is symmetric,*** by Eq. (13). It follows from Eqs. (10) and (12) that

For orthogonal coordinates:

$$\left.\begin{aligned} {}^{\langle rs\rangle}T = T^{\langle rs\rangle} &= \sqrt{\frac{g_{rr}}{g_{ss}}}\, T^{r}_{.s} = \sqrt{\frac{g_{ss}}{g_{rr}}}\, T_{r}^{.s} \\ &= \sqrt{g_{rr}g_{ss}}\, T^{rs} = \frac{T_{rs}}{\sqrt{g_{rr}g_{ss}}} \end{aligned}\right\} \quad \text{(no sum).} \tag{I.5.15}$$

The common value of $T^{\langle rs\rangle}$ and ${}^{\langle rs\rangle}T$ is denoted $T_{\langle rs\rangle}$ in App. II, where $v_{\langle r\rangle}$ is also used for $v^{\langle r\rangle}$ in orthogonal coordinates.

As an example of a higher-order tensor consider the fourth-order tensor coefficients C^{mnrs} of the generalized Hooke's law:

$$T^{mn} = C^{mnrs} E_{rs},$$

where T^{mn} are contravariant stress components and E_{rs} are covariant components of the small-strain tensor **E**. (See Sec. 6.2 for the rectangular Cartesian components.) We desire

$$T^{\langle mn\rangle} = C^{\langle mnrs\rangle} E^{\langle rs\rangle} \qquad \text{(sum on } rs)$$

and are led to

$$C^{\langle mnrs\rangle} = \sqrt{g_{mm}g_{nn}g_{rr}g_{ss}}\, C^{mnrs} \qquad \text{(no sum).} \tag{I.5.16}$$

Principal axes and principal values. A ***proper value or eigenvalue*** α and ***left proper vector*** (***eigenvector***) components λ^r of a matrix with elements $a_r^{.s}$ satisfy

$$\lambda^r a_r^{.s} = \alpha\lambda^s, \tag{I.5.17}$$

and μ_r are ***right proper vector*** components, if

$$a_r^{.s} \mu_s = \alpha\mu_r. \tag{I.5.18}$$

In either case, the following determinant equation holds:

$$|a_r^{.s} - \alpha\delta_r^s| = 0.$$

Compare this with Sec. 3.3, where it is shown that a symmetric tensor always has at least three mutually orthogonal principal axes (proper vectors† of the matrix of rectangular Cartesian components). It can be shown‡ that the four matrices of elements $T_r^{.s}$, $T^r_{.s}$, ${}^{\langle rs\rangle}T$, $T^{\langle rs\rangle}$ have the same proper values. The left proper vectors of $[T_r^{.s}]$ coincide with the proper vectors of the matrix of the

†In this context, "vector" means a row matrix or a column matrix.

‡See Truesdell (1953), p. 352.

left physical components $[{}^{\langle rs\rangle}T]$, and the right proper vectors of $[T^{r}_{.s}]$ coincide with the proper vectors of the right physical components $[T^{\langle rs\rangle}]$. If **T** is symmetric, all four matrices of elements $T_{r}^{.s}$, $T^{r}_{.s}$, $T^{\langle rs\rangle}$, ${}^{\langle rs\rangle}T$ have common proper vectors. Be sure to observe that no statement is made here about the proper values or vectors of the matrices of covariant or contravariant components. It is only the ***mixed components*** that have been discussed.

Calculations with physical components can sometimes be made as if they were mixed tensor components. Truesdell† lists, for example, the following:

1. The matrix of right physical components of the inner product $\mathbf{P} = \mathbf{T}\cdot\mathbf{S}$ or $P^{r}_{.s} = T^{r}_{.p}S^{p}_{.s}$ is $[T^{\langle rp\rangle}S^{\langle ps\rangle}]$, i.e.,

$$P^{\langle rs\rangle} = \sqrt{\frac{g_{rr}}{g_{ss}}}\,T^{r}_{.p}S^{p}_{.s} = T^{\langle rp\rangle}\,S^{\langle ps\rangle} \qquad \text{(sum on } p \text{ only).}$$

   Similarly, with the left physical components of $T_{r}^{.p}S_{p}^{.s}$,

$${}^{\langle rs\rangle}P = \sqrt{\frac{g_{ss}}{g_{rr}}}\,T_{r}^{.p}S_{p}^{.s} = {}^{\langle rp\rangle}T\,{}^{\langle ps\rangle}S \qquad \text{(sum on } p \text{ only).}$$

2. The inverse of the matrix of right physical components $[T^{\langle rs\rangle}]$, if it exists, is the matrix of right physical components of the inverse of $T^{r}_{.s}$.
3. The matrix of right physical components of any power, positive or negative, of the matrix of mixed components $T^{r}_{.s}$ of a tensor is the same power of the matrix of right physical components.
4. Statements 2 and 3 remain true when "left" is substituted for "right," if also $T_{r}^{.s}$ is substituted for $T^{r}_{.s}$.
5. Any scalar function of the mixed components $T^{r}_{.s}$ and $T_{p}^{.q}$ of a tensor field may be regarded alternatively as the same function of the physical components $T^{\langle rs\rangle}$ and ${}^{\langle pq\rangle}T$. For example,

$$T^{r}_{.s}T_{r}^{.s} = T^{\langle rs\rangle}\,{}^{\langle rs\rangle}T \qquad \text{(sum on } r, s\text{).}$$

6. In fact, at any single point it is possible to choose a coordinate system $\bar{x}_r$ such that at that point

$$(\bar{T}^{r}_{.s})_0 = (\bar{T}^{\langle rs\rangle})_0 \quad \text{and} \quad (\bar{T}_{p}^{.q})_0 = ({}^{\langle pq\rangle}\bar{T})_0,$$

   where the subscript zero emphasizes that equality holds in the special coordinate system denoted by the bars only at the one point. Since a different special coordinate system is required at each point, this equation cannot be differentiated.

Example: Skew Cartesian coordinates. As a simple example of nonorthogonal coordinates consider skew Cartesian coordinates x^r with x^2-axis at 60° to x^1 instead of 90°. If x^1 and x^3 coincide with rectangular z^1 and z^3, the transformation is (see Fig. I.3)

†*loc. cit.*, p. 352.

$$\left.\begin{aligned} x^1 &= z^1 - \frac{1}{\sqrt{3}} z^2 \qquad & x^2 &= \frac{2}{\sqrt{3}} z^2 \qquad & x^3 &= z^3 \\ z^1 &= x^1 + \frac{1}{2} x^2 \qquad & z^2 &= \frac{\sqrt{3}}{2} x^2 \qquad & z^3 &= x^3 \end{aligned}\right\} \tag{I.5.19}$$

$$[a^m_n] = \left[\frac{\partial z^m}{\partial x^n}\right] = \begin{bmatrix} 1 & \frac{1}{2} & 0 \\ 0 & \frac{\sqrt{3}}{2} & 0 \\ 0 & 0 & 1 \end{bmatrix} \qquad [b^m_n] = \left[\frac{\partial x^m}{\partial z^n}\right] = \begin{bmatrix} 1 & -\frac{1}{\sqrt{3}} & 0 \\ 0 & \frac{2}{\sqrt{3}} & 0 \\ 0 & 0 & 1 \end{bmatrix} \tag{I.5.20}$$

$$\left.\begin{aligned} \mathbf{g}_1 &= \mathbf{i}_1 \qquad & \mathbf{g}_2 &= \frac{1}{2}\mathbf{i}_1 + \frac{\sqrt{3}}{2}\mathbf{i}_2 \qquad & \mathbf{g}_3 &= \mathbf{i}_3 \\ \mathbf{g}^1 &= \mathbf{i}_1 - \frac{1}{\sqrt{3}}\mathbf{i}_2 \qquad & \mathbf{g}^2 &= \frac{2}{\sqrt{3}}\mathbf{i}_2 \qquad & \mathbf{g}^3 &= \mathbf{i}_3 \end{aligned}\right\} \tag{I.5.21}$$

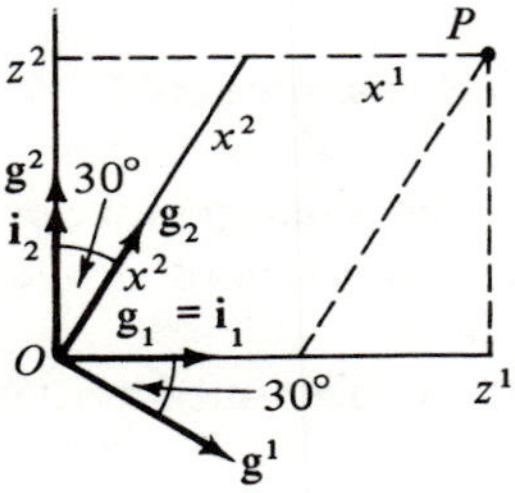

Fig. I.3 Skew Cartesian coordinates x^1, x^2.

$$[g_{rs}] = \begin{bmatrix} 1 & \frac{1}{2} & 0 \\ \frac{1}{2} & 1 & 0 \\ 0 & 0 & 1 \end{bmatrix} \qquad [g^{rs}] = \begin{bmatrix} \frac{4}{3} & -\frac{2}{3} & 0 \\ -\frac{2}{3} & \frac{4}{3} & 0 \\ 0 & 0 & 1 \end{bmatrix}. \tag{I.5.22}$$

Consider plane stress $\sigma_x, \sigma_y, \tau_{xy} = \tau$. We need consider only a 2×2 matrix. Let $\sigma^p_{.q}$ denote the Cartesian rectangular components and $T^r_{.s}$ the skew Cartesian components. Since $g_{11} = g_{22} = g_{33} = 1$, we have ${}^{\langle sr\rangle}T = T_s^{.r}$. Hence

$$[{}^{\langle sr\rangle}T] = [T_s^{.r}] = \left[\frac{\partial z^p}{\partial x^s}\right]^T [\sigma_p^{.q}] \left[\frac{\partial x^r}{\partial z^q}\right]^T \qquad \text{by Eq. (I.2.16)}$$

$$= \begin{bmatrix} 1 & 0 \\ \frac{1}{2} & \frac{\sqrt{3}}{2} \end{bmatrix} \begin{bmatrix} \sigma_x & \tau \\ \tau & \sigma_y \end{bmatrix} \begin{bmatrix} 1 & 0 \\ -\frac{1}{\sqrt{3}} & \frac{2}{\sqrt{3}} \end{bmatrix}$$

$$= \begin{bmatrix} \sigma_x - \frac{1}{\sqrt{3}}\tau & \frac{2}{\sqrt{3}}\tau \\ \frac{1}{2}(\sigma_x - \sigma_y) + \frac{1}{\sqrt{3}}\tau & \frac{1}{\sqrt{3}}\tau + \sigma_y \end{bmatrix}. \tag{I.5.23}$$

Now consider an arbitrary unit vector $\hat{\mathbf{n}} = \cos\theta\mathbf{i}_1 + \sin\theta\mathbf{i}_2$.

$$n^{\langle 1\rangle} = n^1 = \cos\theta - \frac{1}{\sqrt{3}}\sin\theta \qquad n^{\langle 2\rangle} = n^2 = \frac{2}{\sqrt{3}}\sin\theta.$$

$$[t^{\langle r\rangle}] = [n^{\langle s\rangle}][^{\langle sr\rangle}T]$$

$$\left.\begin{aligned} t^{\langle 1\rangle} &= \left(\sigma_x - \frac{\tau}{\sqrt{3}}\right)\cos\theta + \left(\tau - \frac{\sigma_y}{\sqrt{3}}\right)\sin\theta, \\ t^{\langle 2\rangle} &= \frac{2}{\sqrt{3}}\tau\cos\theta + \frac{2}{\sqrt{3}}\sigma_y\sin\theta \end{aligned}\right\} \tag{I.5.24}$$

or with $\mathbf{t} = t^{\langle 1\rangle}\mathbf{g}_1 + t^{\langle 2\rangle}\mathbf{g}_2$, since $\mathbf{g}_1$ and $\mathbf{g}_2$ are unit vectors, this is equivalent to the rectangular Cartesian representation

$$\mathbf{t} = (\sigma_x\cos\theta + \tau\sin\theta)\mathbf{i}_1 + (\tau\cos\theta + \sigma_y\sin\theta)\mathbf{i}_2. \tag{I.5.25}$$

Remarks: Some authors define the stress tensor as the transpose of the one defined in Chap. 3; then right physical components should be used. If the transposed definition is used, the second index on a rectangular Cartesian component σ_{ij} identifies the plane on which the traction vector with components $[\sigma_{1j}, \sigma_{2j}, \sigma_{3j}]$ acts, etc., so that the subscripts would be interchanged on our Fig. 3.6.

We now turn our attention to tensor calculus and covariant differentiation in Sec. I.6.

EXERCISES

1. Express the physical components of velocity in cylindrical coordinates in terms of the coordinates r, θ, and z of a particle and the time derivatives of the coordinates of the particle dr/dt, $d\theta/dt$, and dz/dt.
2. (a) If the physical components of stress in cylindrical coordinates are denoted by T_{rr}, $T_{r\theta}$, etc., express the covariant components T_{11}, T_{12}, etc., in terms of the physical components and the coordinates r, θ, z.
   (b) Repeat for contravariant components.
3. Express the physical components E_{rr}, $E_{r\theta}$, etc., of small strain in cylindrical coordinates in terms of the covariant components E_{11}, E_{12}, etc., and r, θ, z.
4. Express the covariant components u_1, u_2, u_3 of displacement in cylindrical coordinates in terms of r, θ, z and the physical components u, v, w in the r-, θ-, and z-directions, respectively.
5. Carry out the details leading to Eq. (I.5.12) for left physical components.
6. For the example of skew coordinates, verify the calculation of $\mathbf{g}_k$, $\mathbf{g}^k$, g_{rs}, and g^{rs} given in Eqs. (I.5.21) and (22).
7. Carry out the calculation to obtain Eqs. (I.5.24), and check that they are equivalent to Eq. (25).

I.6 Tensor calculus; covariant derivative and absolute derivative of a tensor field; Christoffel symbols; gradient, divergence, and curl; Laplacian

All the tensor properties discussed in previous sections were properties at a point. As a rule we deal with tensors which are themselves functions of position in some region. Such a tensor function is called a ***tensor field***, and when we study its variation from point to point we are led to the study of the coordinate derivatives of a tensor. We saw that the partial derivatives of a tensor of order zero (scalar function) are covariant components of a first-order tensor. Unfortunately this does not carry over to higher-order tensors. In general the partial derivatives of first-order tensor components with respect to curvilinear coordinates are not components of a second-order tensor.

In rectangular Cartesians, the partial derivatives of first-order tensor components are in fact Cartesian components of a second-order tensor. For example, if $\mathbf{v} = v_m \mathbf{i}_m$ in rectangular Cartesians, then

$$d\mathbf{v} = v_m \overleftarrow{\partial}_n \, dz_n \, \mathbf{i}_m = dz_n \partial_n v_m \mathbf{i}_m$$

and the quantities

$$v_m \overleftarrow{\partial}_n \equiv v_{m,n} \equiv \partial v_m / \partial z_n \quad \text{and} \quad \partial_n v_m \equiv (\partial / \partial z_n) v_m$$

are rectangular Cartesian components of the two second-order tensors which are gradients of the vector $\mathbf{v}$, namely $\mathbf{v}\overleftarrow{\nabla}$ with components $v_m \overleftarrow{\partial}_n$ and its transpose $\overrightarrow{\nabla}\mathbf{v}$ with components $\partial_n v_m$, such that

$$d\mathbf{v} = \mathbf{v}\overleftarrow{\nabla} \cdot d\mathbf{z} = d\mathbf{z} \cdot \overrightarrow{\nabla}\mathbf{v}$$

where

$$d\mathbf{z} = dz_m \mathbf{i}_m$$

[$d\mathbf{z}$ is the position-vector differential with components dz_m.] Recall that in rectangular Cartesians summation is implied by repeated letter subscripts.

But, if v^m are contravariant components relative to the natural basis $\mathbf{g}_r$ of a curvilinear system x^s, then the partial derivatives $\partial v^m / \partial x^n$ are ***not in general*** components of $\mathbf{v}\overleftarrow{\nabla}$ or $\overrightarrow{\nabla}\mathbf{v}$ relative to the $\mathbf{g}_r$. The mixed components of $\mathbf{v}\overleftarrow{\nabla}$ relative to the $\mathbf{g}_r$ are the ***covariant derivatives*** of v^m with respect to x^n, to be defined in Eq. (11) below and denoted $v^m \overleftarrow{\nabla}_n$ [or sometimes $v^m|_n$] in analogy to the partial-derivative notations $v^m \overleftarrow{\partial}_n$ [or $v^m{}_{,n}$]. The components of the other gradient $\overrightarrow{\nabla}\mathbf{v}$ are also the covariant derivatives of v^m with respect to x^n, but now denoted $\overrightarrow{\nabla}_n v^m$ in analogy to the partial-derivative notation $\partial_n v^m$.

The ***covariant derivatives*** of a tensor component will be defined in such a way that they are tensor components which will reduce to the usual partial derivatives in rectangular Cartesians. The ***absolute derivative*** is a similar generalization of the total derivative with respect to a parameter along a curve in space. The covariant derivative will appear naturally when we take the partial derivative of a vector, and in the process we will see how certain non-tensor, three-index quantities, called ***Christoffel symbols***, arise naturally when we take the partial derivative of the base vectors. The Christoffel symbols will then appear in the formulas for covariant derivatives of tensors.

To introduce the Christoffel symbols we first make use of reference rectangular Cartesian coordinates z^p in addition to the general curvilinear coordinates x^m we are interested in, but later we will define the Christoffel symbols in terms of the metric tensor components of the curvilinear system, so that they do not in fact depend upon our choice of Cartesian reference system. Since the base vectors are functions of position, they cannot be treated as constants in differentiation. For example, $\mathbf{g}_m = \partial \mathbf{p}/\partial x^m$. Hence, writing $\mathbf{p} = z^p \mathbf{i}_p$ in rectangular Cartesians, where the $\mathbf{i}_p$ can be treated as constants, we obtain

$$\frac{\partial \mathbf{g}_m}{\partial x^n} = \frac{\partial^2 z^p}{\partial x^m \, \partial x^n} \mathbf{i}_p.$$

Since

$$\mathbf{i}_p = \frac{\partial x^s}{\partial z^p} \mathbf{g}_s,$$

this becomes

$$\frac{\partial \mathbf{g}_m}{\partial x^n} = \frac{\partial^2 z^p}{\partial x^m \, \partial x^n} \frac{\partial x^s}{\partial z^p} \mathbf{g}_s,$$

or the ***derivative of the covariant base vector*** is given by the vector

$$\frac{\partial \mathbf{g}_m}{\partial x^n} = \begin{Bmatrix} s \\ m \quad n \end{Bmatrix} \mathbf{g}_s, \tag{I.6.1}$$

where the (nontensor) three-index quantities

$$\begin{Bmatrix} s \\ m \quad n \end{Bmatrix} = \frac{\partial^2 z^p}{\partial x^m \, \partial x^n} \frac{\partial x^s}{\partial z^p} \tag{I.6.2}$$

are called ***Christoffel symbols of the second kind***. They are also sometimes called the ***components of the connection***.

Christoffel symbols of the first kind are denoted $[mn, r]$ and may be defined in terms of the second kind by

$$[mn, r] = g_{rs}\begin{Bmatrix} s \\ m \quad n \end{Bmatrix} \quad \text{or inversely} \quad \begin{Bmatrix} s \\ m \quad n \end{Bmatrix} = g^{sr}[mn, r] \tag{I.6.3}$$

(summed on s in the first Eq. (3) and on r in the second). Since

$$g_{rs} = \mathbf{g}_r \cdot \mathbf{g}_s = \frac{\partial z^q}{\partial x^r}\frac{\partial z^k}{\partial x^s}\delta_{qk},$$

Eqs. (2) and (3) yield

$$[mn, r] = \frac{\partial z^q}{\partial x^r}\frac{\partial z^k}{\partial x^s}\delta_{qk}\frac{\partial^2 z^p}{\partial x^m\,\partial x^n}\frac{\partial x^s}{\partial z^p}.$$

But

$$\frac{\partial z^k}{\partial x^s}\frac{\partial x^s}{\partial z^p} = \delta_p^k \quad \text{and} \quad \delta_p^k\delta_{qk} = \delta_{pq}.$$

Hence

$$[mn, r] = \frac{\partial^2 z^p}{\partial x^m\,\partial x^n}\frac{\partial z^q}{\partial x^r}\delta_{pq}. \tag{I.6.4}$$

Since Eq. (2) defines $\begin{Bmatrix} s \\ m \quad n \end{Bmatrix}$ by using a particular rectangular Cartesian system z^p, it might be thought that the values of the Christoffel symbols depend on the choice of the z^p system. We now show that the components $[mn, r]$ can be expressed in terms of partial derivatives of the metric tensor components with respect to the curvilinear coordinates. Then the second form of Eq. (3) gives $\begin{Bmatrix} s \\ m \quad n \end{Bmatrix}$ without any reference to a Cartesian system. Since

$$g_{nr} = \mathbf{g}_n \cdot \mathbf{g}_r = \frac{\partial z^p}{\partial x^n}\frac{\partial z^q}{\partial x^r}\delta_{pq},$$

we have

$$\frac{\partial g_{nr}}{\partial x^m} = \frac{\partial^2 z^p}{\partial x^n\,\partial x^m}\frac{\partial z^q}{\partial x^r}\delta_{pq} + \frac{\partial z^p}{\partial x^n}\frac{\partial^2 z^q}{\partial x^r\,\partial x^m}\delta_{pq}.$$

Similarly, by permuting indices, we find expressions for $\partial g_{rm}/\partial x^n$ and $\partial g_{mn}/\partial x^r$, and can verify that

$$\frac{1}{2}\left(\frac{\partial g_{nr}}{\partial x^m} + \frac{\partial g_{rm}}{\partial x^n} - \frac{\partial g_{mn}}{\partial x^r}\right) = \frac{\partial^2 z^p}{\partial x^m\,\partial x^n}\frac{\partial z^q}{\partial x^r}\delta_{pq}. \tag{I.6.5}$$

Comparison of this with Eq. (4) then shows that

$$[mn, r] = \frac{1}{2}\left(\frac{\partial g_{nr}}{\partial x^m} + \frac{\partial g_{rm}}{\partial x^n} - \frac{\partial g_{mn}}{\partial x^r}\right). \tag{I.6.6}$$

This defines $[mn, r]$ without any reference to the rectangular Cartesian system, and Eq. (3) can then be used to define $\left\{ {s \atop m \quad n} \right\}$ without use of z^p. Since the metric tensor components are constant in a Cartesian reference system (rectangular or skew), Eq. (6) shows that ***all the Christoffel symbols vanish in a Cartesian system***. Both kinds of Christoffel symbols are symmetric in the paired indices mn, as is seen by Eq. (6), since g_{mn} is symmetric.

$$[mn, r] = [nm, r] \qquad \left\{ {s \atop m \quad n} \right\} = \left\{ {s \atop n \quad m} \right\}. \tag{I.6.7}$$

Another common notation for the Christoffel symbol of the second kind is

$$\Gamma^s_{mn} \equiv \left\{ {s \atop m \quad n} \right\}, \tag{I.6.8}$$

but we will not use it in this book because the Γ^s_{mn} look like mixed components of a third-order tensor, which they are not.

The ***derivative of the contravariant base vector*** can be expressed by an equation similar to Eq. (1), expect that a minus sign appears

$$\frac{\partial \mathbf{g}^m}{\partial x^n} = -\left\{ {m \atop n \quad s} \right\} \mathbf{g}^s. \tag{I.6.9}$$

Note also the appropriate positions of the free and summation indices in Eqs. (1) and (9). Equation (9) can be derived by differentiating $\mathbf{g}^m \cdot \mathbf{g}_n = \delta^m_n$ and using Eq. (1). Now that we have expressions for the partial derivatives of contravariant and covariant base vectors in terms of the Christoffel symbols of the second kind we can proceed to the derivative of any vector, and we will be led naturally to the covariant derivative of a first-order tensor.

Covariant derivatives. Let $\mathbf{v} = v^s \mathbf{g}_s$ be an arbitrary vector. Then

$$\frac{\partial \mathbf{v}}{\partial x^n} = \frac{\partial v^m}{\partial x^n} \mathbf{g}_m + v_m \frac{\partial \mathbf{g}_m}{\partial x^n} = v^m{}_{,n} \mathbf{g}_m + v^m \left\{ {s \atop m \quad n} \right\} \mathbf{g}_s.$$

The last form follows from Eq. (1) for $\partial \mathbf{g}_m / \partial x^n$. By interchanging the summation indices m and s in the last term we may write the vector $\partial \mathbf{v} / \partial x^n$ as

$$\frac{\partial \mathbf{v}}{\partial x^n} = \left[v^m{}_{,n} + \left\{ {m \atop s\quad n} \right\} v^s \right] \mathbf{g}_m = v^m \overleftarrow{\nabla}_n \mathbf{g}_m, \tag{I.6.10}$$

where the quantities in square brackets are defined to be the ***covariant derivatives*** of v^m with respect to x^n and denoted $v^m \overleftarrow{\nabla}_n$ or $v^m|_n$, which we will see below are the mixed components of the gradient $\mathbf{v}\overleftarrow{\nabla}$. The same expressions in brackets are also equal to $\overrightarrow{\nabla}_n v^m$, the indicated mixed components of the other gradient $\overrightarrow{\nabla}\mathbf{v}$, the transpose of $\mathbf{v}\overleftarrow{\nabla}$, but they should be written with s and n interchanged in the Christoffel symbol (recall that the Christoffel symbol is symmetric in sn) to give the indices for $\overrightarrow{\nabla}_n v^m$.

$$v^m \overleftarrow{\nabla}_n \equiv v^m|_n = v^m{}_{,n} + \left\{ {m \atop s\quad n} \right\} v^s \qquad \overrightarrow{\nabla}_n v^m = \partial_n v^m + \left\{ {m \atop n\quad s} \right\} v^s \tag{I.6.11}$$

In a Cartesian system

$$v^m \overleftarrow{\nabla}_n = v^m{}_{,n} \text{ and } \overrightarrow{\nabla}_n v^m = \partial_n v^m,$$

since all the

$$\left\{ {m \atop s\quad n} \right\} = 0$$

in Cartesians. That the covariant derivatives of the contravariant tensor components v^m are in fact second-order tensor mixed components of the type indicated by the index positions can be proved by forming the differential:

$$d\mathbf{v} = \frac{\partial \mathbf{v}}{\partial x^n} dx^n = [v^m \overleftarrow{\nabla}_n \, dx^n] \, \mathbf{g}_m = [dx^n \overrightarrow{\nabla}_n v^m] \mathbf{g}_m. \tag{I.6.12a}$$

The quantities in square brackets are thus the ***contravariant components*** δv^m ***of the vector*** $d\mathbf{v}$. Thus the δv^m given by

$$\delta v^m = v^m \overleftarrow{\nabla}_n \, dx^n = dx^n \overrightarrow{\nabla}_n v^m \tag{I.6.12b}$$

are first-order contravariant tensor components for arbitrary choice of the dx^n. Hence, the $v^m \overleftarrow{\nabla}_n$ and $\overrightarrow{\nabla}_n v^m$ are mixed tensor components of types indicated by the index positions, by the quotient rule of Sec. I.2.

Equation (12b) is a component form of

$$d\mathbf{v} = \mathbf{v}\overleftarrow{\nabla} \cdot d\mathbf{x} = d\mathbf{x} \cdot \overrightarrow{\nabla}\mathbf{v}, \tag{I.6.13}$$

where the mixed components of $\mathbf{v}\overleftarrow{\nabla}$ and $\overrightarrow{\nabla}\mathbf{v}$ are

$$(\mathbf{v}\overleftarrow{\nabla})^m_{\cdot n} \equiv v^m \overleftarrow{\nabla}_n \equiv v^m|_n \qquad (\overrightarrow{\nabla}\mathbf{v})_n^{\cdot m} = \overrightarrow{\nabla}_n v^m.$$

The tensors $\mathbf{v}\overleftarrow{\nabla}$ and $\overrightarrow{\nabla}\mathbf{v}$ are the gradients of the vector $\mathbf{v}$ formed by operating with $\overleftarrow{\nabla}$ from the right or with $\overrightarrow{\nabla}$ from the left, as in Eq. (2.5.32).

Note that the contravariant components δv^m of the vector $d\mathbf{v}$ are not simply equal to $d(v^m) = (\partial v^m/\partial x^n)\, dx^n$ because the base vectors are not constants. The notation δv^m is used instead of dv^m to emphasize the difference.

The absolute differential of v^m is δv^m, defined by Eqs. (12). If a curve is defined parametrically by $x^m = x^m(t)$, then the ***absolute derivative with respect to the parameter*** t is

$$\frac{\delta v^m}{\delta t} = v^m \overleftarrow{\nabla}_n \frac{dx^n}{dt} = \frac{dx^n}{dt} \overrightarrow{\nabla}_n v^m, \tag{I.6.14a}$$

where $\delta t = dt$, but the notation† $\delta v^m/\delta t$ is used to distinguish the absolute derivative from the usual total derivative $dv^m/dt = (\partial v^m/\partial x^n)dx^n/dt$. By substituting into Eq. (14a) the expression given in Eq. (11) for $v^m \overleftarrow{\nabla}_n$ we see that

$$\frac{\delta v^m}{\delta t} = \left[\frac{\partial v^m}{\partial x^n} + \left\{\begin{matrix} m \\ s \quad n \end{matrix}\right\} v^s\right] \frac{dx^n}{dt} = \frac{dv^m}{dt} + \left\{\begin{matrix} m \\ s \quad n \end{matrix}\right\} v^s \frac{dx^n}{dt}. \tag{I.6.14b}$$

The absolute derivatives of the components of a tensor are components of another tensor of the same order as the one whose absolute derivative is being taken. But the usual total derivatives of tensor components are not in general components of a tensor.

The notation $v^m|_n$ is commonly used by many writers for the covariant derivative without specifying whether it means $v^m \overleftarrow{\nabla}_n$ as in the formulation above or $\overrightarrow{\nabla}_n v^m$. Of course, for each choice of specific numbers m and n, the component $\overrightarrow{\nabla}_n v^m$ equals $v^m|_n$, but we have seen that the index placement is important in indicating how the whole tensor operates. In the following, we will frequently follow the common usage of using the vertical-bar notation for all the covariant derivatives when we are simply discussing the derivatives themselves. But when we wish to emphasize their mixed-component tensorial character we will employ the symbols $\overrightarrow{\nabla}_n$ or $\overleftarrow{\nabla}_n$. As usual, the arrows over the symbols could be omitted if the direction of the operation is clear. Usually $\nabla_n v^m$ will be understood to mean $\overrightarrow{\nabla}_n v^m$, but the often-used $v^m \overleftarrow{\nabla}_n$ might be ambiguous if the arrow were omitted when some other symbol followed.

The covariant and absolute derivatives of first-order covariant components v_m appear naturally in a derivation similar to that of Eq. (10) for the contravariant components v^m. We begin by taking the partial derivative of the vector $\mathbf{v}$, represented this time as $\mathbf{v} = v_k \mathbf{g}^k$, and are led to define the covariant derivatives $v_m \overleftarrow{\nabla}_n \equiv v_m|_n$ and $\overrightarrow{\nabla}_n v_m$ by

†Some authors denote the absolute derivative by Dv^m/Dt.

$$v_m \overleftarrow{\nabla}_n \equiv v_m|_n = v_{m,n} - \begin{Bmatrix} s \\ m \quad n \end{Bmatrix} v_s \qquad \vec{\nabla}_n v_m = \partial_n v_m - \begin{Bmatrix} s \\ n \quad m \end{Bmatrix} v_s \tag{I.6.15a}$$

in order that

$$\frac{\partial \mathbf{v}}{\partial x^n} = v_m \overleftarrow{\nabla}_n \mathbf{g}^m = (\vec{\nabla}_n v_m)\mathbf{g}^m \tag{I.6.15b}$$

and

$$\frac{\delta v_m}{\delta t} = v_m|_n \frac{dx^n}{dt} = \frac{dv_m}{dt} - \begin{Bmatrix} s \\ m \quad n \end{Bmatrix} v_s \frac{dx^n}{dt}. \tag{I.6.16}$$

Note that for the derivative of covariant vector components v_m the correction terms, added to $\partial v_m/\partial x^n$ in Eq. (15a) and to dv_m/dt in Eq. (16), are prefixed by a minus sign, while for the contravariant vector components of Eqs. (11) and (14b) the sign is plus. The Christoffel symbol in the correction terms contains one index to identify the tensor free index m, one to identify the variable x^n of differentiation, and one for summation with the factor v_s or v^s. In a Cartesian system, the covariant and absolute derivatives reduce to the usual partial and total derivatives.

Covariant derivatives of higher-order tensor components are given by similar formulas with one correction term for each index of the tensor. Each is a tensor component of the type indicated by the position of the indices. For example,

$$\left.\begin{aligned}
T_{rs}|_n &= \frac{\partial T_{rs}}{\partial x^n} - \begin{Bmatrix} m \\ r \quad n \end{Bmatrix} T_{ms} - \begin{Bmatrix} m \\ s \quad n \end{Bmatrix} T_{rm} \\
T^{rs}|_n &= \frac{\partial T^{rs}}{\partial x^n} + \begin{Bmatrix} r \\ m \quad n \end{Bmatrix} T^{ms} + \begin{Bmatrix} s \\ m \quad n \end{Bmatrix} T^{rm} \\
T^{r}_{.s}|_n &= \frac{\partial T^{r}_{.s}}{\partial x^n} + \begin{Bmatrix} r \\ m \quad n \end{Bmatrix} T^{m}_{.s} - \begin{Bmatrix} m \\ s \quad n \end{Bmatrix} T^{r}_{.m} \\
T^{r}_{.st}|_n &= \frac{\partial T^{r}_{.st}}{\partial x^n} + \begin{Bmatrix} r \\ m \quad n \end{Bmatrix} T^{m}_{.st} - \begin{Bmatrix} m \\ s \quad n \end{Bmatrix} T^{r}_{.mt} - \begin{Bmatrix} m \\ t \quad n \end{Bmatrix} T^{r}_{.sm}.
\end{aligned}\right\} \tag{I.6.17}$$

The covariant derivative formulas are very important and can be remembered rather easily as follows. The formula contains the usual partial derivative plus ***for each contravariant index a term*** containing a Christoffel symbol in which that index has been inserted on the upper level, multiplied by the tensor component with that index replaced by a dummy summation index which also appears in the Christoffel symbol. The formula also contains ***for***

each covariant index a term prefixed by a minus sign and containing a Christoffel symbol in which that index has been inserted on the lower level, multiplied by the tensor with that index replaced by a dummy which also appears in the Christoffel symbol. The remaining index in all of the Christoffel symbols is the index of the variable with respect to which the covariant derivative is taken.

If we represent a second-order tensor $\mathbf{T}$ as a dyadic using the natural base vectors of a curvilinear coordinate system, the covariant derivatives of the components appear quite naturally when we take the partial derivative of the dyadic. For example, if

$$\mathbf{T} = T^{r}_{.s}\mathbf{g}_r\mathbf{g}^s,$$

then

$$\frac{\partial \mathbf{T}}{\partial x^n} = \frac{\partial T^{r}_{.s}}{\partial x^n}\mathbf{g}_r\mathbf{g}^s + T^{r}_{.s}\frac{\partial \mathbf{g}_r}{\partial x^n}\mathbf{g}^s + T^{r}_{.s}\mathbf{g}_r\frac{\partial \mathbf{g}^s}{\partial x^n}.$$

By Eqs. (1) and (9),

$$\frac{\partial \mathbf{g}_r}{\partial x^n} = \begin{Bmatrix} p \\ r \quad n \end{Bmatrix}\mathbf{g}_p \qquad \frac{\partial \mathbf{g}^s}{\partial x^n} = -\begin{Bmatrix} s \\ m \quad q \end{Bmatrix}\mathbf{g}^q.$$

Hence

$$\frac{\partial \mathbf{T}}{\partial x^n} = \frac{\partial T^{r}_{.s}}{\partial x^n}\mathbf{g}_r\mathbf{g}^s + T^{r}_{.s}\begin{Bmatrix} p \\ r \quad n \end{Bmatrix}\mathbf{g}_p\mathbf{g}^s - T^{r}_{.s}\mathbf{g}_r\begin{Bmatrix} s \\ m \quad q \end{Bmatrix}\mathbf{g}^q.$$

We interchange the summation indices r and p in the second term and interchange s and q in the last term to obtain

$$\frac{\partial \mathbf{T}}{\partial x^n} = \left[\frac{\partial T^{r}_{.s}}{\partial x^n} + T^{p}_{.s}\begin{Bmatrix} r \\ p \quad n \end{Bmatrix} - T^{r}_{.q}\begin{Bmatrix} q \\ m \quad s \end{Bmatrix}\right]\mathbf{g}_r\mathbf{g}^s = T^{r}_{.s}|_n\mathbf{g}_r\mathbf{g}^s, \tag{I.6.18}$$

showing that the ***covariant derivative*** of $T^{r}_{.s}$ with respect to x^n is the coefficient of $\mathbf{g}_r\mathbf{g}^s$ in the dyadic representation of the ***partial derivative*** of the dyadic $\mathbf{T} = T^{r}_{.s}\mathbf{g}_r\mathbf{g}^s$.

Absolute derivatives of higher-order tensors are given by generalizations of Eqs. (13) and (16). For example,

$$\frac{\delta T^{r}_{.sq}}{\delta t} = T^{r}_{.sq}|_n\frac{dx^n}{dt}. \tag{I.6.19}$$

Because of the summation indices, each term after the first one in Eqs. (17) really represents a sum of three terms so that, for example, the last Eq. (17)

has a sum of 10 terms on the right-hand side. The calculation of the covariant derivatives in specific examples would therefore be quite lengthy, except for the fact that in the usual coordinate systems only a few of the 27 Christoffel symbols are nonzero.

Orthogonal coordinates. When the coordinate curves are orthogonal, $g_{mn} = 0$ unless $m = n$, and it follows from Eq. (6) that $[mn, r]$ vanishes unless two of the indices are equal, and that for orthogonal curvilinear coordinates we have then only

$$\left.\begin{aligned} [mm, r] &= -\frac{1}{2}\frac{\partial g_{mm}}{\partial x^r} \qquad (m \neq r, \text{ no sum}) \\ [mr, m] = [rm, m] &= +\frac{1}{2}\frac{\partial g_{mm}}{\partial x^r} \qquad (\text{no sum}). \end{aligned}\right\} \tag{I.6.20}$$

Also $\left\{ \begin{matrix} r \\ m \quad n \end{matrix} \right\}$ vanishes for orthogonal curvilinear coordinates unless two indices are equal, and then we have

$$\left.\begin{aligned} \left\{ \begin{matrix} r \\ m \quad m \end{matrix} \right\} &= -\frac{1}{2g_{rr}}\frac{\partial g_{mm}}{\partial x^r} \qquad (m \neq r, \text{ no sum}) \\ \left\{ \begin{matrix} r \\ r \quad m \end{matrix} \right\} = \left\{ \begin{matrix} r \\ m \quad r \end{matrix} \right\} &= \frac{1}{2g_{rr}}\frac{\partial g_{rr}}{\partial x^m} \qquad (\text{no sum}). \end{aligned}\right\} \tag{I.6.21}$$

We remark that the covariant derivatives of the metric tensor components g_{mn}, g^{mn}, and g^m_n are all zero, since the covariant derivatives of these metric tensor components are components of a tensor and these components vanish in Cartesian coordinates where they are simply the partial derivatives of the (constant) Kronecker deltas. Also, if ϵ^{pqr} and ϵ_{pqr} are the contravariant and covariant components of a tensor defined by

$$\epsilon_{pqr} = \sqrt{g}\, e_{pqr} \qquad \epsilon^{pqr} = \frac{1}{\sqrt{g}} e^{pqr}, \tag{I 6.22}$$

where

$$e^{pqr} = e_{pqr} = \begin{cases} 0 & \text{when any two indices are equal} \\ +1 & \text{when } pqr \text{ are an even permutation of } 1\ 2\ 3 \\ -1 & \text{when } pqr \text{ are an odd permutation of } 1\ 2\ 3, \end{cases}$$

and g is the determinant of the metric tensor matrix $[g_{ij}]$, it can be shown that all the covariant derivatives of ϵ_{pqr} and ϵ^{pqr} vanish. It follows therefore that such quantities as g_{mn} and ϵ_{mnr} may be treated as constants in taking the derivative of a linear combination of tensors. It can be shown that absolute and covariant differentiation follow the usual rules of elementary calculus for the derivative of the sum or product of two tensors.

Higher-order covariant derivatives are defined by repeated application of the first-order derivative formula. This is a straightforward process, but it can rapidly lead to computational complications. For example, we have

$$v_r|_{mn} = \frac{\partial v_r|_m}{\partial x^n} - \left\{ {s \atop r \quad n} \right\} v_s|_m - \left\{ {s \atop m \quad n} \right\} v_r|_s \tag{I.6.23}$$

or, after substitution for $v_r|_m$, $v_s|_m$, and $v_r|_s$,

$$\begin{aligned} v_r|_{mn} = {} & \frac{\partial^2 v_r}{\partial x^m \, \partial x^n} - \left\{ {s \atop r \quad m} \right\} \frac{\partial v_s}{\partial x^n} - \left\{ {s \atop r \quad n} \right\} \frac{\partial v_s}{\partial x^m} - \left\{ {s \atop m \quad n} \right\} \frac{\partial v_r}{\partial x^s} \\ & - v_s \left[\frac{\partial}{\partial x^n} \left\{ {s \atop r \quad m} \right\} - \left\{ {p \atop r \quad n} \right\} \left\{ {s \atop p \quad m} \right\} - \left\{ {q \atop m \quad n} \right\} \left\{ {s \atop r \quad q} \right\} \right]. \end{aligned} \tag{I.6.24}$$

Riemann-Christoffel curvature tensor. If we rewrite Eq. (24) with m and n interchanged and subtract the two, we obtain (making use of the symmetry of the Christoffel symbols)

$$v_r|_{mn} - v_r|_{nm} = R^s_{.rmn} v_s, \tag{I.6.25}$$

where

$$R^s_{.rmn} = \frac{\partial}{\partial x^m} \left\{ {s \atop r \quad n} \right\} - \frac{\partial}{\partial x^n} \left\{ {s \atop r \quad m} \right\} + \left\{ {p \atop r \quad n} \right\} \left\{ {s \atop p \quad m} \right\} - \left\{ {t \atop r \quad m} \right\} \left\{ {s \atop t \quad n} \right\}. \tag{I.6.26}$$

Since the v_s are covariant components of an arbitrary vector, and the left-hand sides of Eqs. (23) are covariant tensor components, it follows by the quotient rule of Sec. I.2 that $R^s_{.rmn}$ has the mixed tensor component character indicated by the index positions. The fourth-order tensor with mixed components $R^s_{.rmn}$ is called the ***Riemann-Christoffel curvature tensor***. Since it vanished identically in a Cartesian coordinate system, because all of the Cartesian $\left\{ {i \atop j \quad k} \right\}$ are zero, all of its components vanish† in any coordinate system, which expresses the fact that Euclidean space, i.e., one which admits a Cartesian system, is a ***flat space***. In more general spaces, e.g., the two-dimensional space on the surface of a sphere or the four-dimensional space-time of the general theory of relativity, the curvature tensor may not vanish. Such curved spaces will not

†The vanishing of all the $R^s_{.rmn}$ ensures, by Eq. (25), that the order of differentiation does not affect the mixed second covariant derivatives.

be considered in this book, but they are important in some branches of continuum mechanics. The intrinsic geometry of curved surfaces in three-dimensional space has applications in the theory of shell structures and in fluid flow in boundary layers near curved surfaces.† The geometry of spaces of more than three dimensions is also important in the theory of plasticity, where yield conditions are represented as hypersurfaces in a nine-dimensional stress-representation space in which the coordinates are stress components. There are, of course, many other uses of higher-order derivatives besides calculating the curvature. Any of the partial differential equations of physics has its counterpart as an equation involving covariant derivatives in general curvilinear coordinates.

Since covariant derivatives are tensors, all their indices may be raised or lowered as in Sec. I.2 by inner multiplication by the metric tensor. For example,

$$v_m|^r = g^{rn} v_m|_n \qquad v^r|^s_{\cdot\, t} = g^{ps} v^r|_{pt} \qquad U|^m_m = g^{mn} U|_{mn} = \delta^{mn} \frac{\partial^2 U}{\partial z^m \partial z^n}. \tag{I.6.27}$$

(For the scalar point function U, we have $U|_m \equiv \partial U/\partial x^m$, but $U|_{mn} \neq \partial^2 U/\partial x^m \partial x^n$.) The last expression of Eq. (27) in Cartesian coordinates is recognized as the Laplacian $\nabla^2 U$. Laplace's equation is therefore

$$\nabla^2 U \equiv U|^m_m = 0. \tag{I.6.28}$$

Some other notations for the covariant derivative, used by various authors, e.g., for $T_m^{\cdot\, n}|_r$, are

$$T_{m\ ;r}^{\cdot\, n} \qquad T_{m\ ,r}^{\cdot\, n} \qquad \delta_r T_m^{\cdot\, n} \qquad \nabla_r T_m^{\cdot\, n}$$

in addition to our

$$T_m^{\cdot\, n}|_r \qquad \vec{\nabla}_r T_m^{\cdot\, n} \quad \text{and} \quad T_m^{\cdot\, n} \overleftarrow{\nabla}_r.$$

The second notation, using the comma, is more often used to represent the usual partial derivative. The operators δ_r or ∇_r represent the covariant derivative operator with respect to x^r. We will reserve δ_r for the total covariant derivative of the two-point tensor field, to be defined in Sec. I.7, but we will often use one of the three forms employing ∇_r as the operator indicating the covariant derivative with respect to x^r. The arrow on $\vec{\nabla}_r$ or $\overleftarrow{\nabla}_r$ may be used to indicate whether the operator operates on the following or preceding quantity when there is any danger of confusion. Thus

†See, for example, Aris (1962).

$$\vec{\nabla}_n v^m = v^m|_n$$

but the matrices are transposes

$$[\vec{\nabla}_n v^m] = [v^m|_n]^T = [v^m \overleftarrow{\nabla}_n]^T. \qquad \text{(I.6.29)}$$

($\vec{\nabla}$ and $\overleftarrow{\nabla}$ are sometimes written $\underset{\rightarrow}{\nabla}$ and $\underset{\leftarrow}{\nabla}$.)

The gradient of a scalar, a vector, or a tensor is formed by an open product with the ***vector operator del***, denoted ∇, $\vec{\nabla}$, or $\overleftarrow{\nabla}$,

$$\nabla = \vec{\nabla} = \mathbf{g}^n \frac{\partial}{\partial x^n} \quad \text{or} \quad \overleftarrow{\nabla} = \frac{\overleftarrow{\partial}}{\partial x^n}\mathbf{g}^n, \qquad \text{(I.6.30)}$$

where the arrow indicating the direction of the quantity to be operated on may be omitted when there is no danger of confusion. Note that the partial derivative operators $\partial/\partial x^n$ and $\overleftarrow{\partial}/\partial x^n$ operate on all the factors of the expression they operate on. Thus

$$\text{Scalar } f\text{:} \quad \nabla f = \frac{\partial f}{\partial x^n}\mathbf{g}^n \qquad \text{(I.6.31)}$$

$$\underset{\mathbf{v}=v^m\mathbf{g}_m}{\text{Vector:†}} \quad \mathbf{v}\overleftarrow{\nabla} = \frac{\partial \mathbf{v}}{\partial x^n}\mathbf{g}^n = v^m\overleftarrow{\nabla}_n \mathbf{g}_m \mathbf{g}^n = v_m \overleftarrow{\nabla}_n \mathbf{g}^m \mathbf{g}^n \qquad \text{(I.6.32a)}$$

$$\vec{\nabla}\mathbf{v} = \mathbf{g}_n \frac{\partial \mathbf{v}}{\partial x^n} = (\vec{\nabla}_n v^m)\mathbf{g}^n \mathbf{g}_m = (\vec{\nabla}_n v_m)\mathbf{g}^n \mathbf{g}^m \qquad \text{(I.6.32b)}$$

$$\underset{\mathbf{T}=T^{rs}\mathbf{g}_r\mathbf{g}_s}{\text{Tensor:†}} \quad \mathbf{T}\overleftarrow{\nabla} = \frac{\partial \mathbf{T}}{\partial x^n}\mathbf{g}^n = T^{rs}\overleftarrow{\nabla}_n \mathbf{g}_r \mathbf{g}_s \mathbf{g}^n = T^{r}_{\cdot s}\overleftarrow{\nabla}_n \mathbf{g}_r \mathbf{g}^s \mathbf{g}^n \qquad \text{(I.6.33a)}$$

$$\vec{\nabla}\mathbf{T} = \mathbf{g}^n \frac{\partial \mathbf{T}}{\partial x^n} = (\vec{\nabla}_n T^{rs})\mathbf{g}^n \mathbf{g}_r \mathbf{g}_s = (\vec{\nabla}_n T^{r}_{\cdot s})\mathbf{g}^n \mathbf{g}_r \mathbf{g}^s. \qquad \text{(I.6.33b)}$$

Other component forms of the gradient of a tensor are available, beginning with other components of **T**. The gradient is always a polyadic of order one higher than the polyadic whose gradient is taken. (See Sec. I.3.)

Divergence of a vector or a tensor is a contraction of the gradient, of order two less than the gradient and one less than the polyadic whose divergence is taken. The contraction is most conveniently performed on an expression for the gradient where $\mathbf{g}^n$ is adjacent to a covariant base vector, since $\mathbf{g}_m \cdot \mathbf{g}^n = \mathbf{g}^n \cdot \mathbf{g}_m = \delta^n_m$.

†The notation $|_n$ should not be used in place of $\vec{\nabla}_n$ in Eqs. (32b) and (33b), because it does not have the indices appearing in the proper order for the dyadic coefficients.

$$\left.\begin{array}{ll}\text{Vector } \mathbf{v}: & \mathbf{v}\cdot\overleftarrow{\nabla} = v^m\overleftarrow{\nabla}_n\mathbf{g}_m\cdot\mathbf{g}^n = v^m\overleftarrow{\nabla}_m = v^m|_m \\ & \overrightarrow{\nabla}\cdot\mathbf{v} = \overrightarrow{\nabla}_n v^m\mathbf{g}^n\cdot\mathbf{g}_m = \overrightarrow{\nabla}_m v^m \\ & \mathbf{v}\cdot\overleftarrow{\nabla} = \overrightarrow{\nabla}\cdot\mathbf{v} \quad \text{is a scalar.}\end{array}\right\} \qquad \text{(I.6.34)}$$

$$\left.\begin{array}{ll}\text{Tensor } \mathbf{T}: & \mathbf{T}\cdot\overleftarrow{\nabla} = T^{rs}\overleftarrow{\nabla}_n\mathbf{g}_r\mathbf{g}_s\cdot\mathbf{g}^n = T^{rn}\overleftarrow{\nabla}_n\mathbf{g}_r \\ & \overrightarrow{\nabla}\cdot\mathbf{T} = (\overrightarrow{\nabla}_n T^{rs})\mathbf{g}^n\cdot\mathbf{g}_r\mathbf{g}_s = (\overrightarrow{\nabla}_n T^{ns})\mathbf{g}_s = T^{ns}|_n\mathbf{g}_s.\end{array}\right\} \qquad \text{(I.6.35)}$$

In general,

$$\overrightarrow{\nabla}\cdot\mathbf{T} \neq \mathbf{T}\cdot\overleftarrow{\nabla} \quad \text{but} \quad \overrightarrow{\nabla}\cdot\mathbf{T} = \mathbf{T}^T\cdot\overleftarrow{\nabla}.$$

For symmetric $\mathbf{S}$,

$$\overrightarrow{\nabla}\cdot\mathbf{S} = \mathbf{S}\cdot\overleftarrow{\nabla}.$$

Again, if there is danger of confusion, we place an arrow on ∇. For example, in the material derivative of Sec. 4.3,

$$\frac{df}{dt} = \frac{\partial f}{\partial t} + \mathbf{v}\cdot(\overrightarrow{\nabla}f);$$

the ∇ operates forward on f and not backward on $\mathbf{v}$, as has been doubly signified by the arrow and the parentheses. The divergence of a second-order tensor is a vector. Note that ***the form $\overrightarrow{\nabla}\cdot\mathbf{T}$ is the appropriate one to use in the equation of motion*** with our convention about the way the stress tensor operates.

Also note that all components appearing in Eqs. (30) through (35) are tensor components of the types indicated by the indices and ***not physical components***. The physical components of ∇f are the quantities $\nabla f^{\langle m\rangle}$ such that $\nabla f = \nabla f^{\langle m\rangle}\hat{\mathbf{e}}_m$, where $\hat{\mathbf{e}}_m$ is a unit vector in the direction of $\mathbf{g}_m$. In an orthogonal coordinate system $\mathbf{g}_m$ is parallel to $\mathbf{g}^m$ and $\hat{\mathbf{e}}_m = \mathbf{g}^m\sqrt{g_{mm}}$ (no sum). Thus the physical components in an orthogonal system are given by

$$\left.\begin{array}{ll}\text{Orthogonal coordinates:} & \nabla f^{\langle m\rangle} = \dfrac{\partial f}{\partial x^m}\dfrac{1}{\sqrt{g_{mm}}} \quad \text{(no sum)} \\ & \text{and} \quad \nabla f = \displaystyle\sum_m \frac{1}{\sqrt{g_{mm}}}\frac{\partial f}{\partial x^m}\hat{\mathbf{e}}_m.\end{array}\right\} \qquad \text{(I.6.36)}$$

Similar, but more complicated, substitutions give expressions for the other quantities discussed in terms of physical components in orthogonal coordinates.

The curl operators $\overrightarrow{\nabla}\times$ and $\times\overleftarrow{\nabla}$ are defined to operate by symbolically forming the cross product of the operator ∇ with the expression following or preceding. Thus

$$\vec{\nabla} \times \mathbf{v} = \mathbf{g}^m \frac{\partial}{\partial x^m} \times \mathbf{v} = \mathbf{g}^m \times \frac{\partial \mathbf{v}}{\partial x^m}$$

$$= \mathbf{g}^m \times (\vec{\nabla}_m v_n)\mathbf{g}^n \qquad \text{by Eq. (15b)}$$

$$= (\vec{\nabla}_m v_n)\mathbf{g}^m \times \mathbf{g}^n = (\vec{\nabla}_m v_n) \frac{1}{\sqrt{g}} e^{mnr} \mathbf{g}_r \qquad \text{by Eq. (I.3.27b).}$$

Hence

$$\vec{\nabla} \times \mathbf{v} = \frac{1}{\sqrt{g}} e^{mnr}(\vec{\nabla}_m v_n)\mathbf{g}_r = \frac{1}{\sqrt{g}} \begin{vmatrix} \mathbf{g}_1 & \mathbf{g}_2 & \mathbf{g}_3 \\ \nabla_1 & \nabla_2 & \nabla_3 \\ v_1 & v_2 & v_3 \end{vmatrix}, \tag{I.6.37}$$

while, by a similar derivation,

$$\mathbf{v} \times \overleftarrow{\nabla} = \frac{1}{\sqrt{g}} e^{mnr}(v_m \overleftarrow{\nabla}_n)\mathbf{g}_r = \frac{1}{\sqrt{g}} \begin{vmatrix} \mathbf{g}_1 & \mathbf{g}_2 & \mathbf{g}_3 \\ v_1 & v_2 & v_3 \\ \nabla_1 & \nabla_2 & \nabla_3 \end{vmatrix} = -\vec{\nabla} \times \mathbf{v} \tag{I.6.38}$$

in a ***right-handed coordinate system***. In the formal expansions of the determinants the differential operators ∇_r operate only on the v_r and not on the $\mathbf{g}_r$; the resulting covariant derivatives appear in coefficients multiplying the $\mathbf{g}_r$. Again when there is no danger of confusion the arrows may be omitted. If ***left-handed coordinate*** systems are used in this book, a minus sign will be prefixed to the formula. The curl components at a point then transform like true vector components. Some authors omit the minus sign; the curl is then an axial vector and the transformation formula for the curl components must be prefixed by the sign of the determinant of the transformation matrix.

The curl may also operate on a dyadic. If

$$\mathbf{T} = T_{rs}\mathbf{g}^r \mathbf{g}^s,$$

then

$$\vec{\nabla} \times \mathbf{T} = \mathbf{g}^n \times \frac{\partial \mathbf{T}}{\partial x^n} = \mathbf{g}^n \times (\vec{\nabla}_n T_{rs})\mathbf{g}^r \mathbf{g}^s$$

$$= (\vec{\nabla}_n T_{rs})(\mathbf{g}^n \times \mathbf{g}^r)\mathbf{g}^s = (\vec{\nabla}_n T_{rs})\left(\frac{1}{\sqrt{g}} e^{nrt} \mathbf{g}_t\right)\mathbf{g}^s.$$

Thus

$$\vec{\nabla} \times \mathbf{T} = \frac{1}{\sqrt{g}} e^{nrt}(\vec{\nabla}_n T_{rs})\mathbf{g}_t \mathbf{g}^s \tag{I.6.39}$$

and, similarly,

$$\mathbf{T} \times \overleftarrow{\nabla} = \frac{1}{\sqrt{g}}(T_{rs}\overleftarrow{\nabla}_n)e^{snq}\mathbf{g}^r\mathbf{g}_q, \tag{I.6.40}$$

which is not merely the negative of $\vec{\nabla} \times \mathbf{T}$ but a completely different dyadic.† We can say that the mixed tensor components are

$$\left.\begin{aligned} C^{t}_{.s} &= \frac{1}{\sqrt{g}}e^{nrt}\vec{\nabla}_n T_{rs} \quad \text{if} \quad \mathbf{C} = \vec{\nabla} \times \mathbf{T} \quad \text{or} \\ C^{.q}_{r} &= \frac{1}{\sqrt{g}}(T_{rs}\overleftarrow{\nabla}_n)e^{snq} \quad \text{if} \quad \mathbf{C} = \mathbf{T} \times \overleftarrow{\nabla}. \end{aligned}\right\} \tag{I.6.41}$$

In Sec. I.8 are summarized a number of the equations of continuum mechanics in terms of general curvilinear tensor components. Physical component forms in orthogonal coordinates are given in Sec. II.3, and some of them are written out in cylindrical and spherical coordinates in Sec. II.4. Most of these equations can be obtained by a straightforward application of the ideas of Secs. I.3 through I.6. An alternative derivation of the physical component forms for orthogonal coordinates is given in App. II. Some of the general curvilinear tensor component forms use the concept of a two-point tensor, which is explained in Sec. I.7.

EXERCISES

1. Show that in cylindrical coordinates the nonvanishing Christoffel symbols are

$$[12, 2] = [21, 2] = x^1 \qquad [22, 1] = -x^1$$

$$\begin{Bmatrix} 2 \\ 1 \quad 2 \end{Bmatrix} = \begin{Bmatrix} 2 \\ 2 \quad 1 \end{Bmatrix} = \frac{1}{x^1} \qquad \begin{Bmatrix} 1 \\ 2 \quad 2 \end{Bmatrix} = -x^1$$

2. Carry out the derivation of Eq. (I.6.9) for $\partial \mathbf{g}^m/\partial x^n$.
3. Show that $v_m\overleftarrow{\nabla}_n - \vec{\nabla}_n v_m = \partial v_m/\partial x^n - \partial v_n/\partial x^m$ in any coordinate system, if the v_k are covariant vector components.
4. Carry out the details leading to Eqs. (I.6.15) for the covariant derivatives of first-order covariant components.
5. Make a derivation for the covariant derivative of the contravariant components T^{rs} comparable to the derivation leading to Eq. (I.6.18) for the mixed components $T^{r}_{.s}$.
6. Evaluate the expressions $T^{m1}|_m$, $T^{m2}|_m$, and $T^{m3}|_m$ in cylindrical coordinates in

†In general, $\vec{\nabla} \times \mathbf{T} = -(\mathbf{T}^T \times \overleftarrow{\nabla})^T$, where the superscript T denotes the transpose. If S is symmetric, then $\vec{\nabla} \times \mathbf{S} = -(\mathbf{S} \times \overleftarrow{\nabla})^T$.

terms of the contravariant stress components T^{pq} and the coordinates. Hence write in expanded form the tensor component equations of motion in cylindrical coordinates.

7. Combine the results of Ex. 6 above with the results of Ex. 2(b) of Sec. I.5 to obtain the cylindrical-coordinate equations of motion in terms of physical components T_{rr}, $T_{r\theta}$, etc. See Eq. (II.4.C11).

8. Write out an expression for each small-strain covariant component in cylindrical coordinates in terms of the covariant displacement components u_1, u_2, u_3.

9. Combine the results of Ex. 8 above with the results of Ex. 4 of Sec. I.5 to obtain the physical components of strain in cylindrical coordinates. See Eq. (8.3.1).

10. Write a general curvilinear tensor component form of the Navier-Stokes equations for a compressible fluid, Eq. (7.1.17b).

11. Write out in expanded form the results of Ex. 10 in cylindrical coordinates.

12. Express the cylindrical-coordinate Navier-Stokes equations of Ex. 11 in terms of physical components.

Additional practice in the use of general curvilinear tensor components may be obtained by deriving some of the equations of continuum mechanics summarized in Sec. I.8 and by writing out some of them in expanded form for particular curvilinear coordinate systems. Physical-component forms are summarized in Secs. II.3 and II.4.

I.7 Deformation; two-point tensors; base vectors; metric tensors; shifters; total covariant derivative

The deformation discussed in Secs. 4.3 to 4.5 is defined by

$$x^m = x^m(X^1, X^2, X^3, t). \tag{I.7.1}$$

The X^K are the initial coordinates of a particle (called ***material coordinates*** in Sec. 4.3, where the discussion is limited to Cartesian coordinates), and the x^m are the coordinates of the same particle at time t (called ***spatial coordinates***). It is not necessary for the two sets of coordinates to be measured with reference to the same coordinate system. In fact, they may be two different frames of reference in relative motion; the two frames could be specified by giving two rectangular Cartesian reference systems moving as rigid bodies relative to each other. Figure I.4 shows a body with initial configuration B in a rectangular Cartesian system Z^K with unit base vectors $\mathbf{I}_K$, and instantaneous configuration b at time t in the rectangular Cartesian system z^k with unit base vectors $\mathbf{i}_k$. In the initial configuration, curvilinear coordinates X^K with covariant base vectors $\mathbf{G}_K$ are shown, while in the instantaneous configuration another set of curvilinear

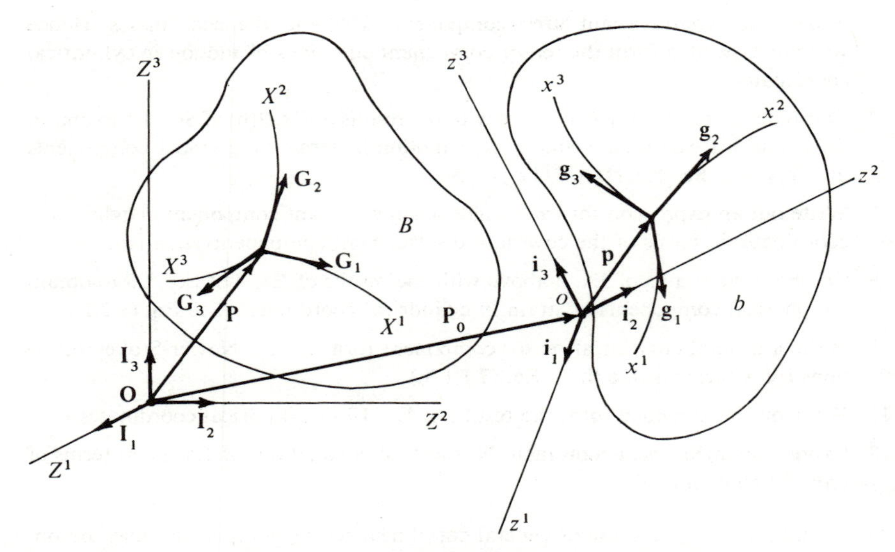

Fig. I.4 Coordinate systems in deformed body *b* and undeformed body *B*.

coordinates x^m may be chosen with covariant base vectors g_m. In each system, the whole machinery of Secs. I.1 to I.6 is applicable, leading for example to a metric tensor G_{MN} in the initial configuration (related to the initial arc length dS) and Christoffel symbols $[MN, R]$ and $\left\{ {S \atop M \quad N} \right\}$. Capital letters are used for indices in the initial system and also for the coordinates and, for example, for the position vector **P** in the initial Cartesian system.

Complete specification of the deformation requires not only the (one-to-one) functions of Eqs. (1), but also information about the definitions of the curvilinear coordinate systems. For example, each curvilinear system might be defined in terms of its Cartesian reference system, and the relation between the two Cartesian reference systems stated.

Coordinate transformations may be made in either the initial configuration or the instantaneous configuration. We may be dealing with transformations involving four coordinate systems, say $\bar{x}^k$ and x^m for present positions and $\bar{X}^K$ and X^M for initial positions. The spatial coordinate transformations are defined by functions

$$\bar{x}^k = \bar{x}^k(x^1, x^2, x^3), \tag{I.7.2}$$

while the material coordinate transformations are defined by functions

$$\bar{X}^K = \bar{X}^K(X^1, X^2, X^3). \tag{I.7.3}$$

Equations (1) describe a (time-dependent) deformation, associating two different space points but the same material particle, while the transformation of Eqs. (2) represents a change of coordinate system to describe the same space point at time t. Equations (3) are also coordinate transformations. In the new coordinate systems $\bar{x}^k$ and $\bar{X}^K$ the same deformation as that given by Eqs. (1) for time t can be represented by expressing the $\bar{x}^k$ as functions of $\bar{X}^K$ and t.

Consider how the set of nine ***deformation-gradient*** components $\partial \bar{x}^k / \partial \bar{X}^K$ of the deformation gradient $\mathbf{F}$ is related to the $\partial x^m / \partial X^M$ at time t. By the chain rule

$$\frac{\partial \bar{x}^k}{\partial \bar{X}^K} = \frac{\partial \bar{x}^k}{\partial x^m} \frac{\partial x^m}{\partial X^M} \frac{\partial X^M}{\partial \bar{X}^K}$$

or

$$\frac{\partial \bar{x}^k}{\partial \bar{X}^K} = \left(\frac{\partial \bar{x}^k}{\partial x^m} \frac{\partial X^M}{\partial \bar{X}^K} \right) \frac{\partial x^m}{\partial X^M}. \tag{I.7.4}$$

The transformation coefficient in parentheses indicates that the deformation gradient transforms contravariantly on the upper index for changes in the spatial coordinates x and covariantly on the lower index for changes in the material coordinates X. This is an example of a ***two-point tensor*** $\mathbf{F}$ with components of type $F^m_{.N}$ transforming by

$$\bar{F}^r_{.S} = \frac{\partial \bar{x}^r}{\partial x^m} \frac{\partial X^N}{\partial \bar{X}^S} F^m_{.N} \tag{I.7.5}$$

for changes in the coordinate systems describing the two points.

F ***transforms as a vector or a first-order tensor when only one of the two coordinate systems is changed***, since

$$\frac{\partial \bar{x}^r}{\partial x^m} = \delta^r_m \quad \text{if} \quad \bar{x}^r \equiv x^r,$$

while $\partial \bar{X}^N / \partial X^S = \delta^N_S$ when $\bar{X}^N \equiv X^N$. Thus

$$\bar{F}^r_{.S} = \frac{\partial X^N}{\partial \bar{X}^S} F^r_{.N} \quad \text{for} \quad \bar{x}^r \equiv x^r \tag{I.7.6a}$$

$$\bar{F}^r_{.S} = \frac{\partial \bar{x}^r}{\partial x^m} F^m_{.S} \quad \text{for} \quad \bar{X}^N \equiv X^N \tag{I.7.6b}$$

a covariant vector component transformation for changes in the material coordinates only, and a contravariant vector component transformation for changes in the spatial coordinates only. Other types of two-point tensors and higher-order tensors may also be defined. Two-point tensors are of importance in connection with finite-strain definitions. In many discussions of finite strain, however, the only coordinate transformations considered are with respect to the material coordinates (initial coordinates), so that two-point tensors are not needed; it is then quite common to use lower-case indices on the material coordinates, e.g., X^k instead of X^K.

Another example of a two-point tensor is a ***shifter***, which shifts a vector from one coordinate system to another. For example, the position vector $\mathbf{p}$ has contravariant components p^k in the x^k system such that

$$\mathbf{p} = p^k \mathbf{g}_k. \tag{I.7.7}$$

Note that

$$p^k = \mathbf{p} \cdot \mathbf{g}^k \qquad P^K = \mathbf{P} \cdot \mathbf{G}^K. \tag{I.7.8}$$

In order to compare the two vectors $\mathbf{p}$ and $\mathbf{P}$, we may wish to calculate the components of $\mathbf{p}$ in the X^K system, such that

$$\mathbf{p} = p^K \mathbf{G}_K \qquad \text{where} \quad p^K = \mathbf{p} \cdot \mathbf{G}^K. \tag{I.7.9}$$

We now substitute Eq. (7) in the second Eq. (9) to obtain

$$p^K = p^k(\mathbf{g}_k \cdot \mathbf{G}^K), \tag{I.7.10}$$

relating p^K to p^k. The quantities $g^K_{.k}$ defined by

$$g^K_{.k} = \mathbf{G}^K \cdot \mathbf{g}_k \tag{I.7.11}$$

are the components of a ***shifter***. It is a two-point tensor contravariant with respect to coordinate changes in the X^K and covariant with respect to coordinate changes in the x^k. Other types of shifters may be defined as two-point tensors of the types indicated by the indices:

$$g^k_{.K} = \mathbf{g}^k \cdot \mathbf{G}_K \tag{I.7.12}$$

$$g^{Kk} = g^{kK} = \mathbf{g}^k \cdot G^K \tag{I.7.13}$$

$$g_{Kk} = g_{kK} = \mathbf{g}_k \cdot \mathbf{G}_K. \tag{I.7.14}$$

(Note that $g_k^{\ K} = g^K_{.k}$ and $g_K^{\ k} = g^k_{.K}$.)

In all four definitions, Eqs. (11) to (14), the base vectors $\mathbf{g}_k$ and $\mathbf{g}^k$ are the base vectors in the x^k system at the point x^m associated by Eq. (1) with the point X^M, where the base vectors $\mathbf{G}_K$ and $\mathbf{G}^K$ are defined in the X^K system. (Here we are interpreting the material coordinate surfaces in the undeformed configuration B.) Raising and lowering of indices on two-point tensor components to form associated two-point tensor components is performed in the usual way, using G_{KM} and G^{KM} for indices pertaining to the X^K system, and g_{km} and g^{km} for indices of the x^k system.

When the rectangular Cartesian components themselves are used ($X^K \equiv Z^K$, and $x^k \equiv z^k$), the shifters of Eqs. (11) to (14) are $\mathbf{I}^K \cdot \mathbf{i}_k = \mathbf{i}^k \cdot \mathbf{I}_K = \mathbf{i}^k \cdot \mathbf{I}^K = \mathbf{i}_k \cdot \mathbf{I}_K$. In particular, ***when the two rectangular Cartesian systems are parallel*** the shifters are merely $\delta^K_k = \delta^k_K = \delta_{kK}$, i.e., in two parallel rectangular Cartesian systems, parallel transport does not change the components of a vector. But, as we have seen, in curvilinear coordinates, parallel transport does change the covariant or contravariant components of a vector $\mathbf{v}$.

$$v^K = g^K_{\cdot k} v^k \quad \text{and} \quad v_K = g_{\dot{K}}^{\ k} v_k \tag{I.7.15}$$

The ***partial covariant derivative*** of a two-point tensor of type $F^m_{\cdot N}$ may be taken either with respect to one of the x^k or one of the X^K and is defined as in Sec. I.6. For example,

$$F^m_{\cdot N}|_q = \frac{\partial F^m_{\cdot N}}{\partial x^q} + \begin{Bmatrix} m \\ p \quad q \end{Bmatrix} F^p_{\cdot N} \tag{I.7.16}$$

$$F^m_{\cdot N}|_R = \frac{\partial F^m_{\cdot N}}{\partial X^R} - \begin{Bmatrix} P \\ N \quad R \end{Bmatrix} F^m_{\cdot P}. \tag{I.7.17}$$

Additional notations analogous to those of Sec I.6 are

$$\vec{\nabla}_q F^m_{\cdot N} = F^m_{\cdot N} \overleftarrow{\nabla}_q = F^m_{\cdot N}|_q \quad \text{and} \quad \vec{\nabla}_R F^m_{\cdot N} = F^m_{\cdot N} \overleftarrow{\nabla}_R = F^m_{\cdot N}|_R. \tag{I.7.18}$$

As usual the arrows may be omitted if there is no danger of confusion. The ***total covariant derivative*** with respect to X^R when the x^r are related to the X^K by Eq. (1) will be denoted $\delta_R F^m_{\cdot N}$ here and defined by

$$\delta_R F^m_{\cdot N} = F^m_{\cdot N}|_R + F^m_{\cdot N}|_q \frac{\partial x^q}{\partial X^R}\cdot \tag{I.7.19}$$

Total covariant differentiation obeys the usual rules of covariant differentiation with the metric tensors and shifters behaving as constants.

Notations for these derivatives have not been standardized. Eringen (1962) uses $F^m_{\cdot N;R}$ and $F^m_{\cdot N;n}$ for the partial covariant derivatives and $F^m_{\cdot N:R}$ for the total covariant derivative. Ericksen (1958) uses a comma in place of Eringen's semicolon for the partial covariant derivative, and a semicolon in place of the colon for the total covariant derivative. It seems preferable to avoid the danger of confusing the comma, semicolon, and colon by using the notation above or possibly using $\vec{\nabla}_r$ and $\vec{\nabla}_R$ or $\overleftarrow{\nabla}_r$ and $\overleftarrow{\nabla}_R$ for covariant partial derivative operators, so that, for example,

$$\delta_R F^m_{\cdot N} = \vec{\nabla}_R F^m_{\cdot N} + \frac{\partial x^q}{\partial X^R} \vec{\nabla}_q F^m_{\cdot N}. \tag{I.7.20}$$

For further discussion of two-point tensors and for references, see Eringen (1962) and Ericksen (1960).

Except for the summary of some of the general tensor curvilinear component forms of field equations of continuum mechanics in Sec. I.8, this concludes our introduction to general curvilinear components of tensors. In App. II an independent derivation will be given for physical components in orthogonal curvilinear coordinates.

EXERCISES

1. For cylindrical coordinates X^K and x^k referred to the same reference axes, calculate the shifters $g^K_{\cdot k}$ and g^{Kk} and show that they are all functions of x^1, X^1, and the difference $x^2 - X^2$.
2. For cylindrical coordinates X^K and x^k write out in expanded form the expression for the partial covariant derivative $F^m_{\cdot N}|_R$ of the deformation gradient for the two cases: (a) $m = 1, N = 2, R = 1$, (b) $m = 1, N = 2, R = 2$.
3. For cylindrical coordinates write out in expanded form the expression for the total covariant derivative $\delta_R F^m_{\cdot N}$ of the deformation gradient for the case $m = 2$, $N = 1, R = 2$.

I.8 Summary of general-tensor curvilinear-component forms of selected field equations of continuum mechanics

Section and equation numbers are the same as in the text.

Sec. 3.2 Stress

(3.2.8) $\qquad \mathbf{t} = \hat{\mathbf{n}} \cdot \mathbf{T} \qquad t^m = n^k T_k^{\cdot m} = n^k g_{kr} T^{rm}$, etc.

Sec. 3.3

For a tensor **T** the invariants are:

(3.3.9)
$$\begin{cases} \mathrm{I} = \operatorname{tr} \mathbf{T} = T^k_{\cdot k} = T_k^{\cdot k} = g^{km} T_{km} = g_{km} T^{km} \\ \dagger\mathrm{II} = \tfrac{1}{2}[\mathbf{T} : \mathbf{T} - \mathrm{I}^2] = \tfrac{1}{2}[T^{ij} T_{ij} - (T_k^{\cdot k})^2], \quad \text{for example} \\ \mathrm{III} = \det \mathbf{T} = \det [T^k_{\cdot m}] = \det [T_m^{\cdot k}] \quad \text{for symmetric } \mathbf{T}. \end{cases}$$

†Some authors call this −II.

The ***deviators*** are:

(3.3.15)
$$T'{}_k^{\cdot m} = T_k^{\cdot m} - \tfrac{1}{3}T_r^{\cdot r}\delta_k^m, \quad \text{etc., or}$$
$$\mathbf{T} = \mathbf{T}' + (\tfrac{1}{3}\,\mathrm{tr}\,\mathbf{T})\mathbf{1}.$$

Sec. 4.2 Small Strain and Rotation

(4.2.1)
$$\begin{cases} \dfrac{d\mathbf{u}}{ds} = (\mathbf{u}\overleftarrow{\nabla})\cdot\hat{\mathbf{n}} = \hat{\mathbf{n}}\cdot(\overrightarrow{\nabla}\mathbf{u}) \\[2ex] \dfrac{du_k}{ds} = u_k\overleftarrow{\nabla}_r n^r = n^r\overrightarrow{\nabla}_r u_k \end{cases}$$

(4.2.2)
$$\begin{cases} \mathbf{E} = \tfrac{1}{2}(\mathbf{u}\overleftarrow{\nabla} + \overrightarrow{\nabla}\mathbf{u}) \qquad [\text{recall } \overrightarrow{\nabla}\mathbf{u} = (\mathbf{u}\overleftarrow{\nabla})^T] \\ E_{km} = \tfrac{1}{2}(u_k|_m + u_m|_k) = \tfrac{1}{2}(u_k\overleftarrow{\nabla}_m + \overrightarrow{\nabla}_k u_m) \end{cases}$$

(4.2.3)
$$\begin{cases} \mathbf{\Omega} = \tfrac{1}{2}(\mathbf{u}\overleftarrow{\nabla} - \overrightarrow{\nabla}\mathbf{u}) \\ \Omega_{km} = \tfrac{1}{2}(u_k|_m - u_m|_k) = \tfrac{1}{2}(u_k\overleftarrow{\nabla}_m - \overrightarrow{\nabla}_k u_m) \end{cases}$$

Sec. 4.3 Kinematics

$$\mathbf{p} = \mathbf{p}(X,t) \qquad \mathbf{v} = \frac{d\mathbf{p}}{dt} \qquad \mathbf{a} = \frac{d\mathbf{v}}{dt}$$

$$x^r = x^r(X, t) \qquad (x^r \text{ not vector components}) \quad v^m = \frac{dx^m}{dt}$$

(4.3.6a)
$$a^m = \frac{\delta v^m}{\delta t} = v^m|_n \frac{dx^n}{dt} = \frac{dv^m}{dt} + \begin{Bmatrix} m \\ s \quad n \end{Bmatrix} v^s \frac{dx^n}{dt}$$

(4.3.10a)
$$\frac{df}{dt} = \frac{\partial f}{\partial t} + \mathbf{v}\cdot(\overrightarrow{\nabla}f) \quad \text{or} \quad \frac{df}{dt} = \frac{\partial f}{\partial t} + v^k\frac{\partial f}{\partial x^k}$$

(4.3.10c)
$$\frac{d\mathbf{T}}{dt} = \frac{\partial \mathbf{T}}{\partial t} + \mathbf{v}\cdot\overrightarrow{\nabla}\mathbf{T} = \frac{\delta T_k^{\cdot m}}{\delta t}\,\mathbf{g}^k\,\mathbf{g}_m, \qquad \text{for example,}$$

where

$$\frac{\delta T_k^{\cdot m}}{\delta t} = T_k^{\cdot m}|_n \frac{dx^n}{dt}$$

Sec. 4.4 Rate of Deformation

(4.4.5)
$$\begin{cases} \mathbf{D} = \tfrac{1}{2}(\mathbf{v}\overleftarrow{\nabla} + \overrightarrow{\nabla}\mathbf{v}) & \mathbf{W} = \tfrac{1}{2}(\mathbf{v}\overleftarrow{\nabla} - \overrightarrow{\nabla}\mathbf{v}) \\ D_{km} = \tfrac{1}{2}(v_k|_m + v_m|_k) & W_{km} = \tfrac{1}{2}(v_k|_m - v_m|_k) \\ \quad = \tfrac{1}{2}(v_k\overleftarrow{\nabla}_m + \overrightarrow{\nabla}_k v_m) & \quad = \tfrac{1}{2}(v_k\overleftarrow{\nabla}_m - \overrightarrow{\nabla}_k v_m) \end{cases}$$

(4.4.9a)
$$(ds)^2 = d\mathbf{x}\cdot d\mathbf{x} = g_{mn}\,dx^m\,dx^n$$

(4.4.11) $$\frac{d}{dt}[(ds)^2] = 2\,d\mathbf{x}\cdot\mathbf{D}\cdot d\mathbf{x} = 2\,dx^m\,D_{mn}\,dx^n$$

Sec. 4.5 Finite Strain and Deformation

(4.5.2a)
$$d\mathbf{x} = \mathbf{F}\cdot d\mathbf{X} = d\mathbf{X}\cdot\mathbf{F}^T,$$

where $\mathbf{F}$ is the two-point tensor with mixed components

$$F^k_{.M} = \frac{\partial x^k}{\partial X^M} = x^k_{,M}$$

$$[dx^k] = [x^k_{,M}][dX^M] \quad \text{or} \quad [dx^k] = [dX^M][\partial_M x^k]$$

(4.5.2b)
$$\begin{cases} d\mathbf{X} = \mathbf{F}^{-1}\cdot d\mathbf{x} = d\mathbf{x}\cdot(\mathbf{F}^{-1})^T \\ [dX^M] = [X^M_{,k}][dx^k] \quad \text{or} \quad [dX^M] = [dx^k][\partial_k X^M] \end{cases}$$

(4.5.3)
$$\begin{cases} (ds)^2 - (dS)^2 = 2\,d\mathbf{X}\cdot\mathbf{E}\cdot d\mathbf{X} = 2\,dX^M\,E_{MN}\,dX^N \\ \qquad = 2\,d\mathbf{x}\cdot\mathbf{E}^*\cdot d\mathbf{x} = 2\,dx^k\,E^*_{km}\,dx^m, \end{cases}$$

where $\mathbf{E}^*$ is the "Eulerian" strain sometimes denoted $\mathbf{e}$.

(4.5.4)
$$\begin{cases} (ds)^2 = d\mathbf{X}\cdot\mathbf{C}\cdot d\mathbf{X} = dX^M\,C_{MN}\,dX^N \\ (dS)^2 = d\mathbf{x}\cdot\mathbf{B}^{-1}\cdot d\mathbf{x} = dx^k\,B^{-1}_{km}\,dx^m \end{cases}$$

(4.5.5) $$2E_{IJ} = C_{IJ} - G_{IJ} \qquad 2E^*_{ij} = g_{ij} - B^{-1}_{ij},$$

where the $G_{IJ} = \mathbf{G}_I\cdot\mathbf{G}_J$ are metric tensor components of the material coordinates at $\mathbf{X}$, while $g_{ij} = \mathbf{g}_i\cdot\mathbf{g}_j$ are metric tensor components of the spatial coordinates at $\mathbf{x}$ (not necessarily the same kind of curvilinear coordinates).†

(4.5.7)
$$\begin{cases} \mathbf{C} = \mathbf{F}^T\cdot\mathbf{F} & \mathbf{B}^{-1} = (\mathbf{F}^{-1})^T\cdot\mathbf{F}^{-1} \\ \quad = \mathbf{F}^T\cdot\mathbf{1}\cdot\mathbf{F} & \quad = (\mathbf{F}^{-1})^T\cdot\mathbf{1}\cdot\mathbf{F}^{-1} \qquad \text{whence} \\ [C_{IJ}] = [\partial_I x^k][g_{km}][x^m_{,J}] & [B^{-1}_{ij}] = [\partial_i X^K][G_{KJ}][X^J_{,j}] \end{cases}$$

It is necessary to interpose the appropriate matrix of the unit tensor $\mathbf{1}$ in order to have indices in the correct positions for the summations.

(4.5.13)
$$\begin{cases} \dfrac{d\mathbf{E}}{dt} = \mathbf{F}^T\cdot\mathbf{D}\cdot\mathbf{F} \qquad \text{where} \quad \dfrac{d\mathbf{E}}{dt} = \dfrac{\delta E_{IJ}}{\delta t}\mathbf{G}^I\mathbf{G}^J \\ \left[\dfrac{\delta E_{IJ}}{\delta t}\right] = [\partial_I x^m][D_{mn}][x^n_{,J}] \end{cases}$$

(4.5.24)–(25) $$\frac{\rho_0}{\rho} = \sqrt{\mathrm{III}_C} = \det\mathbf{F} = J = \text{Jacobian}\ \frac{\partial(x^1, x^2, x^3)}{\partial(X^1, X^2, X^3)}$$

†$\mathbf{B}^{-1}$ is sometimes denoted $\mathbf{c}$.

(4.5.29)
$$\begin{cases} n_k\,dS = \dfrac{\rho_0}{\rho}\,N_K X^K{}_{,k}\,dS_0 \quad \text{or} \\ g_{km}n^m\,dS = \dfrac{\rho_0}{\rho}\,G_{KM}N^M X^K{}_{,k}\,dS_0 \end{cases}$$

(4.5.30)
$$\mathbf{dA} = J\,\mathbf{dA}_0\boldsymbol{\cdot}\mathbf{F}^{-1}$$

Sec. 4.6 Polar Decomposition

(4.6.1a)
$$\mathbf{F} = \mathbf{R}\boldsymbol{\cdot}\mathbf{U} = \mathbf{V}\boldsymbol{\cdot}\mathbf{R}$$

(4.6.1b)
$$x^k{}_{,K} = R^k{}_{.M}U^M{}_{.K} = V^k{}_{.m}R^m{}_{.K},$$

where the components $R^p_{.Q}$ represent both translation and rotation. For example, if $g^m_{.M} = \mathbf{g}^m\boldsymbol{\cdot}\mathbf{G}_M$ is the shifter of Eq. (I.7.12), then

$$R^k{}_{.M} = R^k{}_{.m}g^m{}_{.M} \quad \text{and} \quad R^m{}_{.K} = g^m{}_{.M}R^M{}_{.K}$$

(4.6.1c)
$$R^m{}_{.K}R_k{}^{.K} = \delta^m_k \quad \text{or} \quad \mathbf{R}\boldsymbol{\cdot}\mathbf{R}^T = \mathbf{1}$$

See the equation after Eq. (10*) below. Note that the matrix of mixed components $[R_k^{.K}]$ of $\mathbf{R}^T$ is not just the transpose of the matrix of mixed components $[R^k_{.K}]$ of $\mathbf{R}$.

(4.6.10a)
$$\hat{\mathbf{n}}_\alpha = \mathbf{R}\boldsymbol{\cdot}\hat{\mathbf{N}}_\alpha \quad \text{or} \quad \hat{\mathbf{n}}^\alpha = \mathbf{R}\boldsymbol{\cdot}\hat{\mathbf{N}}^\alpha$$

(4.6.10b)
$$\begin{cases} n^k_\alpha = R^k_{.K}N^K_\alpha = R^k{}_{.m}g^m{}_{.K}N^K_\alpha \quad \text{or} \\ n^\alpha_k = R_k{}^{.K}N^\alpha_K = R_k{}^{.m}g_m{}^{.K}N^\alpha_K \end{cases}$$

Greek indices α, β, etc. identify the unit vectors, while their components are denoted by k, K, etc. For the orthonormal triads, the reciprocal or dual triads [see Eqs. (I.1.9) to (I.1.12)] coincide with the original triads, $\mathbf{N}^\alpha = \mathbf{N}_\alpha$ and $\mathbf{n}^\alpha = \mathbf{n}_\alpha$, but $N^\alpha_K \neq N^K_\alpha$ since the $\mathbf{N}^\alpha_K$ are covariant components while the N^K_α are contravariant components with respect to the natural basis $\mathbf{G}_K$ at $\mathbf{X}$ and the n^α_k are covariant components while n^k_α are contravariant components relative to the natural basis $\mathbf{g}_k$ at $\mathbf{x}$. For example, if the $\mathbf{N}_\alpha$ are a right-handed system, then by Eq. (I.3.27)

$$\mathbf{N}^{(1)} = \mathbf{N}_{(2)} \times \mathbf{N}_{(3)} = \sqrt{G}\,e_{IJK}N^I_{(2)}N^J_{(3)}\mathbf{G}^K \quad \text{or}$$
$$N^{(1)}_K = \sqrt{G}\,e_{IJK}N^I_{(2)}N^J_{(3)}$$

(numerical indices in parentheses identify vectors instead of components). It can be shown that

(4.6.10*)
$$N^\alpha_K N^M_\alpha = \delta^M_K \quad \text{and} \quad n^\alpha_k n^m_\alpha = \delta^m_k \qquad \text{(sum on } \alpha\text{).}$$

[Note that this is not the usual orthonormality condition $N^{\alpha}_{K}N^{K}_{\beta} = \delta^{\alpha}_{\beta}$, because the sum is on α in (10*). To verify (10*), calculate the N^{α}_{K} as indicated above for $N^{(1)}_{K}$. Substitution into (10*) gives an expression vanishing identically for $K \neq M$ and representing $\mathbf{N}_{(1)} \cdot \mathbf{N}_{(2)} \times \mathbf{N}_{(3)} = 1$ for $K = M$.] Substitution of expressions for n^{α}_{k} and n^{m}_{α}, according to Eqs. (10b), into the second Eq. (10*) yields

$$\delta^{m}_{k} = R_{k}^{.K} N^{\alpha}_{K} R^{m}_{.M} N^{M}_{\alpha} = R_{k}^{.K} R^{m}_{.M} \delta^{M}_{K}$$

whence Eq. (1c) follows.

By Eqs. (10b) and (10*)

$$n^{k}_{\alpha} N^{\alpha}_{M} = R^{k}_{.K} N^{K}_{\alpha} N^{\alpha}_{M} = R^{k}_{.K} \delta^{K}_{M} \quad \text{and}$$
$$n^{\alpha}_{k} N^{M}_{\alpha} = R_{k}^{.K} N^{\alpha}_{K} N^{M}_{\alpha} = R_{k}^{.K} \delta^{M}_{K}$$

whence

(4.6.11)
$$\begin{cases} R^{k}_{.M} = n^{k}_{\alpha} N^{\alpha}_{M} \quad \text{and} \quad R_{k}^{.M} = n^{\alpha}_{k} N^{M}_{\alpha} \qquad (\text{sum on } \alpha) \\ R_{k}^{.M} = G^{MQ} g_{km} R^{m}_{.Q} \end{cases}$$

(4.6.12)
$$\hat{\mathbf{N}}_{\alpha} = \mathbf{R}^{-1} \cdot \hat{\mathbf{n}}_{\alpha} \qquad N^{K}_{\alpha} = (R^{-1})^{K}_{.k} n^{k}_{\alpha} = g^{K}_{.m} (R^{-1})^{m}_{.k} n^{k}_{\alpha}$$

(4.6.13)
$$(R^{-1})^{K}_{.k} = N^{K}_{\alpha} n^{\alpha}_{k} \qquad (\text{sum on } \alpha)$$

(4.6.18)
$$dx^{k} = x^{k}_{,K} dX^{K} = g^{k}_{.P} R^{P}_{.M} U^{M}_{.K} dX^{K} = R^{k}_{.m} g^{m}_{.M} U^{M}_{.K} dX^{K}$$

(4.6.21)
$$dx^{k} = x^{k}_{,K} dX^{K} = V^{k}_{.m} R^{m}_{.p} g^{p}_{.K} dX^{K} = V^{k}_{.m} g^{m}_{.M} R^{M}_{.K} dX^{K}$$

Sec. 4.7 Compatibility for Small Strain **E**

(4.7.5b)
$$\mathbf{S} \equiv \overrightarrow{\nabla} \times \mathbf{E} \times \overleftarrow{\nabla} = 0$$

Let

$$\mathbf{P} = \mathbf{E} \times \overleftarrow{\nabla} \qquad \mathbf{S} = \nabla \times \mathbf{P} \qquad P_{rs} = g_{sq} P_{r}^{.q} \quad \text{and} \quad S^{tp} = g^{ps} S^{t}_{.s}.$$

Then, since $g^{ps} g_{sq} = \delta^{p}_{q}$, Eqs. (I.6.39) to (I.6.41) lead to

(4.7.5)
$$\begin{cases} S^{tp} = \dfrac{1}{g} e^{nrt} e^{mkg} \overrightarrow{\nabla}_{n} E_{rm} \overleftarrow{\nabla}_{k} = 0. \qquad \text{For example,} \\ S^{33} = \dfrac{1}{g} [2E_{12}|_{12} - E_{11}|_{11} - E_{22}|_{22}] = 0 \\ S^{12} = \dfrac{1}{g} [E_{33}|_{12} + E_{23}|_{33} + E_{31}|_{32} - E_{23}|_{13}] = 0. \end{cases}$$

For finite strain see Eqs. (4.7.26) to (4.7.36).

Sec. 5.2 Continuity Equations

(5.2.5) $$\frac{\partial \rho}{\partial t} + \operatorname{div} \rho \mathbf{v} = 0 \quad \text{or} \quad \frac{\partial \rho}{\partial t} + (\rho v^k)|_k = 0$$

(5.2.13) $$\rho_0 = \rho J = \rho \sqrt{\mathrm{III}_C}$$

Sec. 5.3 Equations of Motion

(5.3.4) $$\nabla \cdot \mathbf{T} + \rho \mathbf{b} = \rho \mathbf{a} \quad \text{or} \quad T^{km}|_k + \rho b^m = a^m,$$

where

$$\rho a^m = \frac{\delta v^m}{\delta t} = v^m|_n \frac{dx^n}{dt} = \frac{dv^m}{dt} + \begin{Bmatrix} m \\ s \quad n \end{Bmatrix} v^s \frac{dx^n}{dt}$$

[see Eq. (4.3.6a) above].

(5.3.12) $$\mu^{jr}|_j + \rho c^r + \sqrt{g}\; e^{rmn} T_{mn} = \rho \frac{\delta l^r}{\delta t}$$

Piola-Kirchhoff tensors are

(5.3.22) $$\begin{aligned} T_0^{Ji} &= \frac{\rho_0}{\rho} X^J{}_{,r} T^{ri} \qquad \text{where } \mathbf{T}_0 \equiv \mathbf{T}^0 \\ (T_R)^{kJ} &= \frac{\rho_0}{\rho} X^J{}_{,m} T^{mk} \qquad \mathbf{T}_R = (\mathbf{T}^0)^T \end{aligned}$$

[Truesdell and Noll, Eq. (43A.11), have T^{km} instead of T^{mk} because their $\mathbf{T}$ is the transpose of ours.]

(5.3.23) $$\tilde{T}^{JI} = \frac{\rho_0}{\rho} X^J_{,s} T^{si}(\partial_i X^I) = T_0^{Ji} \partial_i X^I$$

(5.3.24) $$T^{ji} = \frac{\rho}{\rho_0} x^j{}_{,J} T_0^{Ji} = \frac{\rho}{\rho_0} (T_R)^{iJ} \partial_J x^j$$

(5.3.25) $$T_0^{Ji} = \tilde{T}^{JI} \partial_I x^i \qquad (T_R)^{iJ} = x^i{}_{,I} \tilde{T}^{IJ}$$

(5.3.26) $$T^{ji} = \frac{\rho}{\rho_0} x^j{}_{,J} \tilde{T}^{JI} (\partial_I x^i)$$

(5.3.27) $$\delta_J T_0^{Ji} + \rho_0 b_0^i = \rho_0 a^i$$

where δ_J denotes the total covariant derivative of Eq. (I.7.19).

(5.3.28) $$\begin{cases} \delta_J(\tilde{T}^{JI} \partial_I x^i) + \rho_0 b_0^i = \rho_0 a^i \quad \text{or} \\ (\tilde{T}^{JI} \partial_I x^i)_{,J} + \left[\begin{Bmatrix} i \\ m \quad k \end{Bmatrix} x^m{}_{,I} x^k{}_{,J} + \begin{Bmatrix} M \\ M \quad J \end{Bmatrix} x^i{}_{,I} \right] \tilde{T}^{JI} + \rho_0 b_0^i = \rho_0 a^i \end{cases}$$

Sec. 5.4 Energy Equation

$$\rho \frac{du}{dt} = T^{mn} D_{mn} + \rho r - q^j|_j \qquad \text{(nonpolar case)} \tag{5.4.7}$$

$$\rho \frac{du}{dt} = T^{(ij)} D_{ij} + \mu^{ji} w_i|_j + \rho r - q^j|_j \tag{5.4.14}$$

Sec. 6.2 Linearized Elasticity—Small Strain **E**

$$T^{km} = c^{kmpq} E_{pq} \tag{6.2.2}$$

$$c^{kmpq} = c^{mkpq} = c^{kmqp} \tag{6.2.3}$$

$$\begin{cases} T^k_m = \lambda E^q_q \delta^k_m + 2\mu E^k_m \qquad \text{(isotropic)} \\ \text{(mixed components of symmetric tensors } \mathbf{T} \text{ and } \mathbf{E}) \\ T^{rk} = g^{mr} T^k_m \\ \qquad = \lambda [g^{qp} E_{pq}] g^{rk} + 2\mu g^{mr} g^{ks} E_{ms} \end{cases} \tag{6.2.4}$$

$$\begin{cases} T^k_m = c^{k.p}_{.m.q} E^q_p = \dfrac{\partial W}{\partial E^m_k} \qquad \text{where} \\ c^{k.p}_{.m.q} = \frac{1}{2}(\bar{c}^{k.p}_{.m.q} + \bar{c}^{p.k}_{.q.m}) \\ c^{k.p}_{.m.q} = c^{.kp}_{m..q} = c^{k..p}_{.mq} = c^{p.k}_{.q.m} \end{cases} \tag{6.2.18}$$

$$W(\mathbf{E}) = \tfrac{1}{2}\, \bar{c}^{k.p}_{.m.q} E^q_p E^m_k = \tfrac{1}{2}\, c^{k.p}_{.m.q} E^q_p E^m_k \tag{6.2.19}$$

$$W(\mathbf{E}) = \tfrac{1}{2}\, T^k_m E^m_k = \tfrac{1}{2}\, T^{km} E_{mk} \tag{6.2.21}$$

Sec. 6.3 Newtonian Fluids

$$\begin{cases} T^k_m = -p\delta^k_m + \lambda D^q_q \delta^k_m + 2\mu D^k_m \\ T^{rk} = -pg^{rk} + \lambda [g^{qp} D_{pq}] g^{rk} + 2\mu g^{mr} g^{ks} D_{ms} \end{cases} \tag{6.3.10}$$

$$T^k_m = -p\delta^k_m - \tfrac{2}{3}\, \mu D^q_q \delta^k_m + 2\mu D^k_m \tag{6.3.23}$$

with no bulk viscosity (Stokes condition) in Eq. (6.3.23).

This completes our discussion of general tensor curvilinear components. In App. II we make an independent presentation of the physical components of tensors in orthogonal curvilinear coordinates, not requiring the general procedures of App. I.

APPENDIX II

Orthogonal Curvilinear Coordinates, Physical Components of Tensors

II.1 Coordinate definitions; scale factors; physical components; derivatives of unit base vectors and of dyadics

In the three-dimensional Euclidean space of elementary geometry we can define a system of curvilinear coordinates by specifying three one-to-one functions of a reference system of rectangular Cartesian coordinates. For example, the equations

$$\left.\begin{array}{lll} r = \sqrt{x^2 + y^2 + z^2} & \theta = \cos^{-1}\left[\dfrac{z}{\sqrt{x^2 + y^2 + z^2}}\right] & \phi = \tan^{-1}\dfrac{y}{x} \\ \text{or, inversely,} & & \\ x = r \sin\theta \cos\phi & y = r\sin\theta\sin\phi & z = r\cos\theta \end{array}\right\} \tag{II.1.1}$$

define the spherical coordinates shown in Fig. II.1. More generally, if u, v, w are three one-to-one functions of x, y, z, then each of the three functions has a level surface passing through every point P. If the functions are such that no two of the level surfaces coincide and furthermore not all three intersect along a curve, then the values of u, v, w on three intersecting level surfaces determine a point P, and we may take the values of the functions as the curvilinear coordinates. The level surfaces are coordinate surfaces, while their curves of intersection are the coordinate curves along which only one coordinate can vary. If at each point P the coordinate curves through P are mutually orthogonal, we have a system of ***orthogonal curvilinear coordinates***. For example, the spherical coordinates of Eqs. (1) are orthogonal coordinates whose coordinate surfaces are the spheres $r =$ const., the circular cones $\theta =$ const. (around the z-axis), and planes $\phi =$ const. (containing the z-axis).

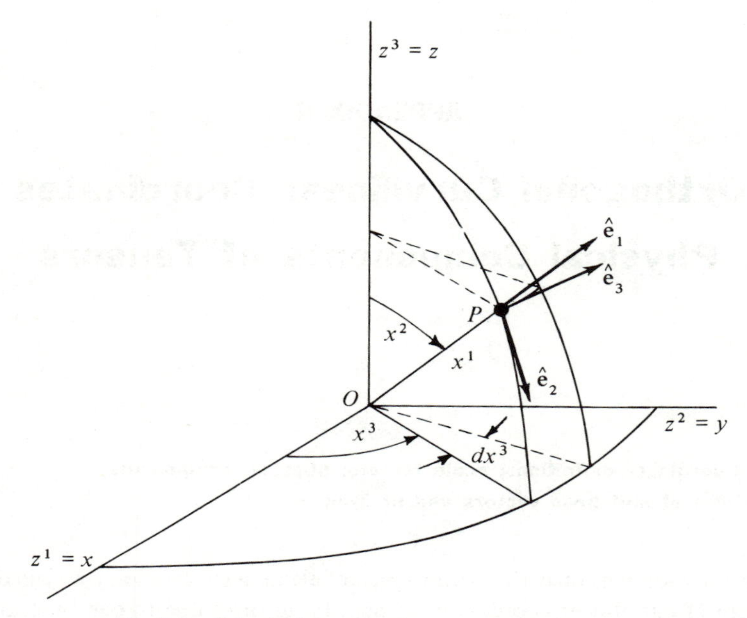

Fig. II.1 Spherical coordinates $r = x^1$, $\theta = x^2$, $\phi = x^3$.

In indicial notation, we denote the three curvilinear coordinates by x^1, x^2, x^3 and the three rectangular coordinates in a fixed system by z^1, z^2, z^3. For spherical coordinates then

$$x^1 = r \qquad x^2 = \theta \qquad x^3 = \phi \qquad \text{while} \quad z^1 = x \qquad z^2 = y \qquad z^3 = z.$$

The superscripts are not exponents; when a superscripted variable is raised to a power, it is enclosed in parentheses, e.g., $(x^3)^2$ means x^3 squared.

Unit base vectors $\hat{\mathbf{e}}_1, \hat{\mathbf{e}}_2, \hat{\mathbf{e}}_3$ tangent to the three coordinate curves through P are shown in Fig. II.1. In what follows, we shall consider only orthogonal curvilinear coordinates, and, unless otherwise mentioned, will order the three coordinates so that when each base vector $\hat{\mathbf{e}}_r$ points in the direction of increasing x^r, the three base vectors form a right-handed system. The basis $\hat{\mathbf{e}}_r$ is then a right-handed orthonormal basis.

General curvilinear coordinates x^m are defined in terms of the rectangular Cartesians by functional equations of the form

$$x^m = x^m(z^1, z^2, z^3) \quad \text{and, inversely,} \quad z^m = z^m(x^1, x^2, x^3) \qquad \text{(II.1.2)}$$

with Jacobian determinant

$$|\mathbf{z}/\mathbf{x}| \equiv \left|\frac{\partial z^m}{\partial x^n}\right| \neq 0. \qquad \text{(II.1.3)}$$

We assume that the three functions z^m of x^r have continuous first partial derivatives and that $|\mathbf{z}/\mathbf{x}| \neq 0$; exceptions may occur at singular points or curves but never throughout any volume.

Scale factors h_1, h_2, h_3 are defined as follows. They are nonnegative functions of position for a given curvilinear coordinate system, such that at any point P the infinitesimal element of arc length ds on an arbitrary curve through P is given by†

$$(ds)^2 = (h_1\,dx^1)^2 + (h_2\,dx^2)^2 + (h_3\,dx^3)^2. \tag{II.1.4}$$

In particular, we may take the coordinate curves for three successive choices of the arbitrary curve through P, to obtain

$$\left.\begin{array}{llll} ds_1 = h_1\,dx^1 & \text{for} & dx^1 \geq 0 & dx^2 = dx^3 = 0 \\ ds_2 = h_2\,dx^2 & \text{for} & dx^2 \geq 0 & dx^3 = dx^1 = 0 \\ ds_3 = h_3\,dx^3 & \text{for} & dx^3 \geq 0 & dx^1 = dx^2 = 0. \end{array}\right\} \tag{II.1.5}$$

From Eqs. (2), we can obtain expressions for the unit base vectors $\hat{\mathbf{e}}_r$ in terms of the basis $\mathbf{i}_k$ of the fixed Cartesian system. The unit vector $\hat{\mathbf{e}}_1$, for example, tangent to the curvilinear coordinate curve along which x^1 varies, has rectangular components dz^k/ds_1, the directional derivative of z^k in the direction of increasing x^1. Hence

$$\hat{\mathbf{e}}_1 = \frac{dz^k}{ds_1}\mathbf{i}_k = \frac{\partial z^k}{\partial x^1}\frac{dx^1}{ds_1}\mathbf{i}_k = \frac{1}{h_1}\frac{\partial z^k}{\partial x^1}\mathbf{i}_k = \frac{1}{h_1}\frac{\partial \mathbf{p}}{\partial x_1},$$

where $\mathbf{p} = z^k\mathbf{i}_k$ is the position vector OP. The three unit vectors are

$$\left.\begin{aligned} \hat{\mathbf{e}}_1 &= \frac{1}{h_1}\frac{\partial \mathbf{p}}{\partial x^1} = \frac{1}{h_1}\frac{\partial z^k}{\partial x^1}\mathbf{i}_k \\ \hat{\mathbf{e}}_2 &= \frac{1}{h_2}\frac{\partial \mathbf{p}}{\partial x^2} = \frac{1}{h_2}\frac{\partial z^k}{\partial x^2}\mathbf{i}_k \\ \hat{\mathbf{e}}_3 &= \frac{1}{h_3}\frac{\partial \mathbf{p}}{\partial x^3} = \frac{1}{h_3}\frac{\partial z^k}{\partial x^3}\mathbf{i}_k. \end{aligned}\right\} \tag{II.1.6}$$

Hence

$$\left.\begin{array}{l} (h_1)^2 = \dfrac{\partial z^k}{\partial x^1}\dfrac{\partial z^k}{\partial x^1} \qquad (h_2)^2 = \dfrac{\partial z^k}{\partial x^2}\dfrac{\partial z^k}{\partial x^2} \qquad (h_3)^2 = \dfrac{\partial z^k}{\partial x^3}\dfrac{\partial z^k}{\partial x^3} \quad \text{while} \\ \dfrac{\partial z^k}{\partial x^r}\dfrac{\partial z^k}{\partial x^s} = 0 \quad \text{for} \quad r \neq s \qquad (\text{sum on } k \text{ in each case}). \end{array}\right\} \tag{II.1.7}$$

†In terms of the metric tensor of Sec. I.4 each $h_r = \sqrt{g_{rr}}$ (no sum).

Equation (7) can be used to compute the quantities h_1, h_2, h_3 from the known Eqs. (2), but it is often simpler to use a geometric argument as follows.

For the spherical coordinates of Fig. II.1, since the radial coordinate curve is straight, $ds_1 = dr = dx^1$, whence $h_1 = 1$. For varying $\theta = x^2$ only, the coordinate curve is a circle of radius $r = x^1$; hence $ds_2 = x^1\,dx^2$ and $h_2 = x^1$. For varying $\phi = x^3$ only, the curve is a circle of radius $x^1 \sin x^2$; hence $ds_3 = x^1 \sin x^2\, dx^3$ and $h_3 = x^1 \sin x^2$. Thus

$$\text{For sphericals:}\quad h_1 = 1 \qquad h_2 = x^1 \qquad h_3 = x^1 \sin x^2 \tag{II.1.8}$$

while for the cylindrical coordinates ($r = x^1$, $\theta = x^2$, $z = x^3$) introduced in Sec. 2.5, we have

$$\text{For cylindricals:}\quad h_1 = 1 \qquad h_2 = x^1 \qquad h_3 = 1. \tag{II.1.9}$$

Equations (6) show that the nine components

$$e_r^k = \frac{1}{h_r}\frac{\partial z^k}{\partial x^r} \qquad \text{(no sum)} \tag{II.1.10}$$

are the transformation coefficients a_r^k for expressing the orthonormal basis $\hat{\mathbf{e}}_r$ in terms of the orthonormal basis $\mathbf{i}_k$ [see Eqs. (2.4.1)].

The ***physical components*** of a vector or second-order tensor relative to a system of orthogonal curvilinear coordinates are simply the Cartesian components in a local set of Cartesian axes, say $\bar{z}^k$, tangent to the coordinate curves through P (so that $\bar{\mathbf{i}}_r \equiv \hat{\mathbf{e}}_r$). Hence if v_k and T_{km} denote the rectangular Cartesian components of $\mathbf{v}$ and $\mathbf{T}$ in the fixed Cartesian system, while the ***physical components in the curvilinear system*** are $v_{\langle r\rangle}$ and $T_{\langle rs\rangle}$ then

$$v_{\langle r\rangle} = e_r^k\, v_k \qquad T_{\langle rs\rangle} = e_r^k\, e_s^m\, T_{km}, \tag{II.1.11}$$

according to Eqs. (2.4.6.) and (2.4.11), with the transformation coefficients $a_r^k \equiv e_r^k$ given by Eq. (10) above. When indicial notation is not used, the brackets $\langle\ \rangle$ are usually omitted. For example, in spherical coordinates

$$\mathbf{v} = \sum_k v_{\langle k\rangle}\hat{\mathbf{e}}_k \equiv v_r\hat{\mathbf{e}}_r + v_\theta\,\hat{\mathbf{e}}_\theta + v_\phi\,\hat{\mathbf{e}}_\phi,$$

and the stress and small-strain matrices are usually written as follows for ***physical components in spherical coordinates:***

$$\begin{bmatrix} \sigma_{rr} & \sigma_{r\theta} & \sigma_{r\phi} \\ \sigma_{\theta r} & \sigma_{\theta\theta} & \sigma_{\theta\phi} \\ \sigma_{\phi r} & \sigma_{\phi\theta} & \sigma_{\phi\phi} \end{bmatrix} \qquad \begin{bmatrix} \epsilon_{rr} & \epsilon_{r\theta} & \epsilon_{r\phi} \\ \epsilon_{\theta r} & \epsilon_{\theta\theta} & \epsilon_{\theta\phi} \\ \epsilon_{\phi r} & \epsilon_{\phi\theta} & \epsilon_{\phi\phi} \end{bmatrix}.$$

Any tensor equation involving only tensor values at one point and no derivatives with respect to the coordinates has the same form in terms of the physical components in orthogonal curvilinear coordinates as it has in terms of rectangular Cartesians. For example, the Hooke's law of Eq. (6.2.2) becomes

$$T_{\langle ij \rangle} = c_{\langle ijrs \rangle} E_{\langle rs \rangle},$$

where the $c_{\langle ijrs \rangle}$ are related to the elastic coefficients in the fixed Cartesian system by a fourth-order version of Eq. (11). Repeated letter indices indicate summation unless otherwise noted. When an index appears three or more times in one term the summations will be explicitly indicated by $\sum$, otherwise no summation is implied. If the material is elastically isotropic, the Hooke's law becomes

$$T_{\langle ij \rangle} = \lambda E_{\langle kk \rangle} \delta_{ij} + 2\mu E_{\langle ij \rangle}$$

and the elastic strain-energy function is

$$W = \frac{1}{2} \lambda (E_{\langle kk \rangle})^2 + \mu E_{\langle ij \rangle} E_{\langle ij \rangle} \qquad \text{(sum on } i, j, k\text{)},$$

as in Eqs. (6.2.4) and (6.2.40). But any equations involving derivatives with respect to the coordinates will have to be modified, because the base vectors are no longer constants as they were in the Cartesian system.

Derivatives of unit base vectors. Since the unit vectors have constant magnitudes, their only change is in direction. Hence, the rate of change of unit vector $\hat{\mathbf{e}}_k$ has no component in the direction of $\hat{\mathbf{e}}_k$. We may think of $\hat{\mathbf{e}}_k$ as a part of a rigid body which rotates as it moves; then

$$d\hat{\mathbf{e}}_k = \mathbf{d\Omega} \times \hat{\mathbf{e}}_k, \qquad \text{(II.1.12)}$$

where $\mathbf{d\Omega}$ is the infinitesimal vector rotation of the rigid body. [Compare Eq. (12) with Eq. (2.5.9)]. In cylindrical and spherical coordinates it is easy to represent $\mathbf{d\Omega}$ in terms of infinitesimal coordinate changes for motion along a coordinate curve. For example, in the spherical coordinates of Fig. II.1

$$\begin{aligned} \mathbf{d\Omega} &= 0 \qquad \text{for} \quad d\phi = d\theta = 0 \\ \mathbf{d\Omega} &= d\theta \hat{\mathbf{e}}_\phi \quad \text{for} \quad d\phi = 0 \\ \mathbf{d\Omega} &= d\phi \hat{\mathbf{e}}_z = d\phi(\cos\theta \hat{\mathbf{e}}_r - \sin\theta \hat{\mathbf{e}}_\theta) \quad \text{for} \quad d\theta = 0. \end{aligned}$$

Thus, for any k, by Eq. (12)

$$\frac{\partial \hat{\mathbf{e}}_k}{\partial r} = 0$$

$$\frac{\partial \hat{\mathbf{e}}_k}{\partial \theta} \equiv \left(\frac{d\hat{\mathbf{e}}_k}{d\theta}\right)_{d\phi=0} = \hat{\mathbf{e}}_\phi \times \hat{\mathbf{e}}_k$$

$$\frac{\partial \hat{\mathbf{e}}_k}{\partial \phi} \equiv \left(\frac{d\hat{\mathbf{e}}_k}{d\phi}\right)_{d\theta=0} = (\cos\theta\hat{\mathbf{e}}_r - \sin\theta\hat{\mathbf{e}}_\theta) \times \hat{\mathbf{e}}_k$$

or

Spherical coordinates r, θ, ϕ:

$$\left.\begin{cases} \dfrac{\partial \hat{\mathbf{e}}_r}{\partial r} = 0 & \dfrac{\partial \hat{\mathbf{e}}_r}{\partial \theta} = \hat{\mathbf{e}}_\theta & \dfrac{\partial \hat{\mathbf{e}}_r}{\partial \phi} = +\sin\theta\hat{\mathbf{e}}_\phi \\ \dfrac{\partial \hat{\mathbf{e}}_\theta}{\partial r} = 0 & \dfrac{\partial \hat{\mathbf{e}}_\theta}{\partial \theta} = -\hat{\mathbf{e}}_r & \dfrac{\partial \hat{\mathbf{e}}_\theta}{\partial \phi} = \cos\theta\hat{\mathbf{e}}_\phi \\ \dfrac{\partial \hat{\mathbf{e}}_\phi}{\partial r} = 0 & \dfrac{\partial \hat{\mathbf{e}}_\phi}{\partial \theta} = 0 & \dfrac{\partial \hat{\mathbf{e}}_\phi}{\partial \phi} = -\cos\theta\hat{\mathbf{e}}_\theta - \sin\theta\hat{\mathbf{e}}_r. \end{cases}\right\} \quad \text{(II.1.13)}$$

The cylindrical coordinate results may be derived similarly. Indeed for cylindrical coordinates, $\hat{\mathbf{e}}_z$ is constant, while $\hat{\mathbf{e}}_r$ and $\hat{\mathbf{e}}_\theta$ vary only with θ, so that in

$$\text{Cylindrical coordinates:} \quad \frac{\partial \hat{\mathbf{e}}_r}{\partial \theta} = \hat{\mathbf{e}}_\theta \qquad \frac{\partial \hat{\mathbf{e}}_\theta}{\partial \theta} = -\hat{\mathbf{e}}_r. \qquad \text{(II.1.14)}$$

For arbitrary, orthogonal curvilinear coordinates the representation of $\mathbf{d\Omega}$ for use in Eq. (12) may be difficult, and it may be easier to use Eqs. (20) below, which will be derived by using the rectangular Cartesian coordinates as follows.

From Eq. (6),

$$h_1 \frac{\partial \hat{\mathbf{e}}_1}{\partial x^1} + \frac{\partial h_1}{\partial x^1}\hat{\mathbf{e}}_1 = \frac{\partial}{\partial x^1}(h_1\hat{\mathbf{e}}_1) = \frac{\partial^2 z^k}{\partial x^1 \partial x^1}\mathbf{i}_k.$$

We know that $\partial\hat{\mathbf{e}}_1/\partial x_1$ has no component parallel to $\hat{\mathbf{e}}_1$. To obtain its components in the $\hat{\mathbf{e}}_2$ and $\hat{\mathbf{e}}_3$ directions we can dot the last equation into $\hat{\mathbf{e}}_2$ and $\hat{\mathbf{e}}_3$, respectively. Since the basis is orthogonal, this yields

$$h_1 \frac{\partial \hat{\mathbf{e}}_1}{\partial x^1}\cdot\hat{\mathbf{e}}_2 = \frac{\partial^2 z^k}{\partial x^1 \partial x^1} e_2^k = \frac{\partial^2 z^k}{\partial x^1 \partial x^1}\frac{\partial z^k}{\partial x^2}\frac{1}{h_2}$$

$$h_1 \frac{\partial \hat{\mathbf{e}}_1}{\partial x^1}\cdot\hat{\mathbf{e}}_3 = \frac{\partial^2 z^k}{\partial x^1 \partial x^1} e_3^k = \frac{\partial^2 z^k}{\partial x^1 \partial x^1}\frac{\partial z^k}{\partial x^3}\frac{1}{h_3}$$

whence

$$h_1 \frac{\partial \hat{\mathbf{e}}_1}{\partial x^1} = \frac{\partial^2 z^k}{\partial x^1 \partial x^1}\left[\frac{\partial z^k}{\partial x^2}\frac{1}{h_2}\hat{\mathbf{e}}_2 + \frac{\partial z^k}{\partial x^3}\frac{1}{h_3}\hat{\mathbf{e}}_3\right]. \qquad \text{(II.1.15a)}$$

A similar derivation yields

$$h_1 \frac{\partial \hat{\mathbf{e}}_1}{\partial x_2} = \frac{\partial^2 z^k}{\partial x^1 \partial x^2} \left[\frac{\partial z^k}{\partial x^2} \frac{1}{h_2} \hat{\mathbf{e}}_2 + \frac{\partial z^k}{\partial x^3} \frac{1}{h_3} \hat{\mathbf{e}}_3 \right] \tag{II.1.15b}$$

and expressions for $\partial \hat{\mathbf{e}}_1 / \partial x^3$ and for the derivatives of $\hat{\mathbf{e}}_2$ and $\hat{\mathbf{e}}_3$ can be written down in similar fashion. These expressions each contain two terms, in each of which appears an expression of the form

$$\sum_k \frac{\partial^2 z^k}{\partial x^m \partial x^n} \frac{\partial z^k}{\partial x^r}.$$

This is the Christoffel symbol $[mn, r]$ of Sec. I.6, but without using the results of Sec. I.6 we can evaluate the expressions as follows. Let

$$g_{pq} = \sum_k \frac{\partial z^k}{\partial x^p} \frac{\partial z^k}{\partial x^q}. \tag{II.1.16}$$

Then

$$\frac{\partial g_{nr}}{\partial x^m} = \sum_k \left[\frac{\partial^2 z^k}{\partial x^n \partial x^m} \frac{\partial z^k}{\partial x^r} + \frac{\partial z^k}{\partial x^n} \frac{\partial^2 z^k}{\partial x^r \partial x^m} \right],$$

and by permuting the indices in this last equation we can obtain similar expressions for $\partial g_{rm} / \partial x^n$ and $\partial g_{mn} / \partial x^r$. The three expressions can be combined to yield

$$\sum_k \frac{\partial^2 z^k}{\partial x^m \partial x^n} \frac{\partial z^k}{\partial x^r} = \frac{1}{2} \left[\frac{\partial g_{nr}}{\partial x^m} + \frac{\partial g_{rm}}{\partial x^n} - \frac{\partial g_{mn}}{\partial x^r} \right]. \tag{II.1.17}$$

Since by Eqs. (7), in orthogonal coordinates

$$g_{pq} = h_p h_q \delta_{pq} \qquad \text{(no sum)}, \tag{II.1.18}$$

Eq. (17) takes the form

$$\sum_k \frac{\partial^2 z^k}{\partial x^m \partial x^n} \frac{\partial z^k}{\partial x^r} = \begin{cases} 0 & \text{if } m, n, r \text{ are all different} \\ -h_m \dfrac{\partial h_m}{\partial x^r} & \text{(no sum) if } m = n \neq r \\ +h_m \dfrac{\partial h_m}{\partial x^m} & \text{(no sum) if } m = n = r \\ +h_r \dfrac{\partial h_r}{\partial x^m} & \text{(no sum) if } n = r \neq m \\ +h_r \dfrac{\partial h_r}{\partial x^n} & \text{(no sum) if } m = r \neq n. \end{cases} \tag{II.1.19}$$

For example, Eq. (17) yields

$$\sum_k \frac{\partial^2 z^k}{\partial x^1 \partial x^2} \frac{\partial z^k}{\partial x^1} = \frac{1}{2}\left[0 + \frac{\partial (h_1)^2}{\partial x^2} - 0\right] = + \, h_1 \frac{\partial h_1}{\partial x^2}.$$

Equation (19) can now be used to transform Eqs. (15) into a form no longer containing explicit reference to the Cartesian coordinates. Thus

$$h_1 \frac{\partial \hat{\mathbf{e}}_1}{\partial x^1} = - \frac{h_1}{h_2} \frac{\partial h_1}{\partial x^2} \hat{\mathbf{e}}_2 - \frac{h_1}{h_3} \frac{\partial h_1}{\partial x_3} \hat{\mathbf{e}}_3$$

and

$$h_1 \frac{\partial \hat{\mathbf{e}}_1}{\partial x^2} = \frac{h_2}{h_2} \frac{\partial h_2}{\partial x^1} \hat{\mathbf{e}}_2 = \frac{\partial h_2}{\partial x^1} \hat{\mathbf{e}}_2 , \quad \text{etc.}$$

Seven additional equations can be written by appropriate permutations of the indices. But there are just two types of equations, depending on whether it is $\partial \hat{\mathbf{e}}_m / \partial x^m$ (no sum), or $\partial \hat{\mathbf{e}}_m / \partial x^n$ $(m \neq n)$.

Derivatives of Unit Base Vectors
in Orthogonal Coordinates

No Summations m, n, r All Different:

$$\frac{\partial \hat{\mathbf{e}}_m}{\partial x^m} = - \frac{1}{h_n} \frac{\partial h_m}{\partial x^n} \hat{\mathbf{e}}_n - \frac{1}{h_r} \frac{\partial h_m}{\partial x^r} \hat{\mathbf{e}}_r \qquad \text{(II.1.20a)}$$

$$m \neq n: \quad \frac{\partial \hat{\mathbf{e}}_m}{\partial x^n} = + \frac{1}{h_m} \frac{\partial h_n}{\partial x^m} \hat{\mathbf{e}}_n. \qquad \text{(II.1.20b)}$$

The reader can verify that these equations give the same results as Eqs. (13) in spherical coordinates, when the h_r are given by Eqs. (8)

With these formulas available for differentiating the unit base vectors we can write the partial derivatives of any vector or tensor as follows.

$$\text{For } \mathbf{v} = \sum_m v_{\langle m \rangle} \hat{\mathbf{e}}_m \qquad \frac{\partial \mathbf{v}}{\partial x^n} = \sum_m \frac{\partial v_{\langle m \rangle}}{\partial x^n} \hat{\mathbf{e}}_m + \sum_m v_{\langle m \rangle} \frac{\partial \hat{\mathbf{e}}_m}{\partial x^n}, \qquad \text{(II.1.21a)}$$

which, by use of Eqs. (20), becomes

$$\frac{\partial \mathbf{v}}{\partial x^n} = \sum_m \frac{\partial v_{\langle m \rangle}}{\partial x^n} \hat{\mathbf{e}}_m + \sum_{\substack{m \\ m \neq n}} v_{\langle m \rangle} \frac{1}{h_m} \frac{\partial h_n}{\partial x^m} \hat{\mathbf{e}}_n$$

$$- \left[\frac{v_{\langle n \rangle}}{h_s} \frac{\partial h_n}{\partial x^s} \hat{\mathbf{e}}_s + \frac{v_{\langle n \rangle}}{h_r} \frac{\partial h_n}{\partial x^r} \hat{\mathbf{e}}_r \right]_{\text{no sum}} \qquad \text{(II.1.21b)}$$

(n, s, r all different in last bracket) .

This is considerably more complicated than the corresponding expression in Sec. I.6 in terms of covariant derivatives of general tensors. But this is the price we have to pay for avoiding the study of general tensors. The actual computations involved in either formulation for explicit evaluation come to about the same thing, but the general-tensor formula is more compact. For theoretical study, however, it is usually best to avoid components altogether and work with the symbolic forms of gradient, divergence, and curl.

For a second-order tensor, a comparable formula to Eq. (21b) could be developed by beginning with the dyadic representation. Thus for

$$\mathbf{T} = \sum_{m,n} T_{\langle mn\rangle}\hat{\mathbf{e}}_m\hat{\mathbf{e}}_n \qquad \frac{\partial \mathbf{T}}{\partial x^r} = \sum_{m,n}\left[\frac{\partial T_{\langle mn\rangle}}{\partial x^r}\hat{\mathbf{e}}_m\hat{\mathbf{e}}_n + T_{\langle mn\rangle}\frac{\partial \hat{\mathbf{e}}_m}{\partial x^r}\hat{\mathbf{e}}_n + T_{\langle mn\rangle}\hat{\mathbf{e}}_m\frac{\partial \hat{\mathbf{e}}_n}{\partial x^r}\right]. \tag{II.1.22}$$

We could now substitute for the derivatives of the unit base vectors the expressions given by Eqs. (20), separating out for special treatment by (20a) the cases $m = r$ and $n = r$. For computational purposes, it is probably better to compute a table of the derivatives first, as in Eqs. (13) for spherical coordinates, and substitute them as needed, thus avoiding the complicated formula. Higher-order tensors can be treated in the same way by writing them as polyadics. Higher derivatives can also be evaluated in specific cases, although the computation is long.

The results of this section on derivatives of unit vectors, arbitrary vectors, and tensors will be used in Sec. II.2 to formulate expressions for the gradient, divergence, and curl in orthogonal curvilinear coordinates.

Certain identities involving the scale factors may be proved. (They are related to the vanishing of the Riemann-Christoffel curvature tensor of Sec. I.6. See Aris (1962), p. 174) The identities are a sufficient condition that

$$\frac{\partial^2 \mathbf{v}}{\partial x^m\,\partial x^n} = \frac{\partial^2 \mathbf{v}}{\partial x^n\,\partial x^m}$$

for arbitrary sufficiently differentiable $\mathbf{v}$, i.e., they ensure the independence of the order of differentiation for the mixed derivatives of $\mathbf{v}$. The identities are equivalent to the vanishing of a certain symmetric matrix

$$[S^{ij}] = 0,$$

where

For m, n, r All Different *No Summations :*

$$\left.\begin{aligned} S^{rr} &= -\frac{1}{h_r^2 h_m h_n}\left[\frac{\partial}{\partial x^m}\left(\frac{1}{h_m}\frac{\partial h_n}{\partial x^m}\right) + \frac{\partial}{\partial x^n}\left(\frac{1}{h_n}\frac{\partial h_m}{\partial x^n}\right) + \frac{1}{h_r^2}\frac{\partial h_m}{\partial x^r}\frac{\partial h_n}{\partial x^r}\right] \\ S^{mn} &= \frac{1}{h_m^2 h_n^2 h_r}\left[\frac{\partial^2 h_r}{\partial x^m\,\partial x^n} - \frac{1}{h_m}\frac{\partial h_r}{\partial x^m}\frac{\partial h_m}{\partial x^n} - \frac{1}{h_n}\frac{\partial h_r}{\partial x^n}\frac{\partial h_n}{\partial x^m}\right]. \end{aligned}\right\} \tag{II.1.23}$$

EXERCISES

1. Compute the scale factors of Eq. (II.1.9) for cylindrical coordinates by using Eqs. (7), and verify that the same results are obtained as by using a geometric argument similar to that leading to Eqs. (8).
2. Compute the scale factors of Eqs. (II.1.8) for spherial coordinates by using Eqs. (1) and (7).
3. Verify the calculations of Eqs. (II.1.13) for the partial derivatives of the spherical-coordinate unit vectors.
4. Carry out the derivation of Eq. (II.1.15b) for $h_1 \partial \hat{\mathbf{e}}_1 / \partial x^2$.
5. Verify Eqs. (II.1.18), $g_{pq} = h_p h_q \delta_{pq}$ (no sum), for orthogonal coordinates.
6. Verify Eq. (II.1.19) for the case $m = n = 2$, $r = 3$.
7. Verify that Eqs. (II.1.20) give the same results as Eqs. (13) for the case of spherical coordinates.
8. Verify that the S^{ij} of Eq. (II.1.23) all vanish for the case of: (a) cylindrical coordinates, (b) spherical coordinates.

II.2 Gradient, divergence, and curl in orthogonal curvilinear coordinates

The gradient of a scalar f is defined to be the vector, denoted grad f or ∇f, such that for any unit vector $\hat{\mathbf{u}}$, the directional derivative df/ds in the direction of $\hat{\mathbf{u}}$ is given by

$$\frac{df}{ds} = \nabla f \cdot \hat{\mathbf{u}}, \tag{II.2.1}$$

where

$$\hat{\mathbf{u}} = \frac{d\mathbf{p}}{ds}.$$

Note that this definition makes no reference to any coordinate system. The gradient is thus a vector invariant independent of the coordinate system. To find its components in any orthogonal coordinate system, we note that

$$\hat{\mathbf{u}} = \frac{d\mathbf{p}}{ds} = \sum_r h_r \frac{dx^r}{ds} \hat{\mathbf{e}}_r . \tag{II.2.2a}$$

Now by the chain rule of calculus,

$$\frac{df}{ds} = \sum \frac{\partial f}{\partial x^r}\frac{dx^r}{ds}\,. \qquad \text{(II.2.2b)}$$

If $(\nabla f)_{\langle r\rangle}$ denotes the physical components of ∇f, then

$$\nabla f\cdot\hat{\mathbf{u}} = \sum_r (\nabla f)_{\langle r\rangle} h_r \frac{dx^r}{ds} = \frac{\partial f}{\partial x^r}\frac{dx^r}{ds}\,,$$

by Eqs. (2), or

$$\sum_r \left[h_r(\nabla f)_{\langle r\rangle} - \frac{\partial f}{\partial x^r}\right]\frac{dx^r}{ds} = 0$$

for arbitrary choice of the dx^r/ds (arbitrary direction $\hat{\mathbf{u}}$). Hence

$$(\nabla f)_{\langle r\rangle} = \frac{1}{h_r}\frac{\partial f}{\partial x^r} \quad \text{or} \quad \nabla f = \sum_r \frac{1}{h_r}\frac{\partial f}{\partial x^r}\hat{\mathbf{e}}_r\,. \qquad \text{(II.2.3)}$$

Thus

$$\nabla = \sum_r \hat{\mathbf{e}}_r \frac{1}{h_r}\frac{\partial}{\partial x^r}\,. \qquad \text{(II.2.4)}$$

The gradients $\vec{\nabla}\mathbf{v}$ ***and*** $\mathbf{v}\overleftarrow{\nabla} = (\vec{\nabla}\mathbf{v})^T$ ***of a vector*** $\mathbf{v}$ are similarly defined as second-order tensors such that

$$\frac{d\mathbf{v}}{ds} = (\mathbf{v}\overleftarrow{\nabla})\cdot\hat{\mathbf{u}} = \hat{\mathbf{u}}\cdot(\vec{\nabla}\mathbf{v})\,. \qquad \text{(II.2.5)}$$

Now, in any orthogonal coordinate system,

$$\begin{aligned}\frac{d\mathbf{v}}{ds} &= \sum_n \frac{\partial \mathbf{v}}{\partial x^n}\frac{dx^n}{ds}\\ &= \sum_n \left(h_n\frac{dx^n}{ds}\right)\left(\frac{1}{h_n}\frac{\partial \mathbf{v}}{\partial x^n}\right)\\ &= \left(\sum_m h_m\frac{dx^m}{ds}\hat{\mathbf{e}}_m\right)\cdot\left(\sum_n \hat{\mathbf{e}}_n\frac{1}{h_n}\frac{\partial \mathbf{v}}{\partial x^n}\right)\end{aligned}$$

since $\hat{\mathbf{e}}_m\cdot\hat{\mathbf{e}}_n = \delta_{mn}$. Hence, by Equation (2a),

$$\begin{aligned}\frac{d\mathbf{v}}{ds} &= \hat{\mathbf{u}}\cdot\left[\left(\sum_n \hat{\mathbf{e}}_n\frac{1}{h_n}\frac{\partial}{\partial x^n}\right)\mathbf{v}\right]\\ &= \hat{\mathbf{u}}\cdot(\vec{\nabla}\mathbf{v})\,,\end{aligned}$$

where

$$\vec{\nabla}\mathbf{v} = \left(\sum_n \hat{\mathbf{e}}_n \frac{1}{h_n}\frac{\partial}{\partial x^n}\right)\mathbf{v}. \tag{II.2.6}$$

Alternatively,

$$\begin{aligned}\frac{d\mathbf{v}}{ds} &= \sum_n \left(\frac{1}{h_n}\frac{\partial \mathbf{v}}{\partial x^n}\right)\left(h_n \frac{dx^n}{ds}\right) \\ &= \left(\sum_n \frac{1}{h_n}\frac{\partial \mathbf{v}}{\partial x^n}\hat{\mathbf{e}}_n\right)\cdot\left(\sum_m h_m \frac{dx^m}{ds}\hat{\mathbf{e}}_m\right) \\ &= (\mathbf{v}\overleftarrow{\nabla})\cdot\hat{\mathbf{u}}\,,\end{aligned}$$

where

$$\mathbf{v}\overleftarrow{\nabla} = \sum_n \frac{1}{h_n}\frac{\partial \mathbf{v}}{\partial x^n}\hat{\mathbf{e}}_n = \mathbf{v}\left(\sum_n \frac{\overleftarrow{\partial}}{\partial x^n}\frac{1}{h_n}\hat{\mathbf{e}}_n\right) \tag{II.2.7}$$

and the notation $\overleftarrow{\partial}/\partial x^n$ implies the derivative of the preceding quantity instead of the following one. [$\partial/\partial x^n$ and $\overleftarrow{\partial}/\partial x^n$ are sometimes denoted ∂_n and $\overleftarrow{\partial}_n$, respectively.] This somewhat awkward notation is needed only if it is desired to make a symbolic representation of the del operators $\vec{\nabla}$ and $\overleftarrow{\nabla}$ as

$$\vec{\nabla} = \left(\sum_n \hat{\mathbf{e}}_n \frac{1}{h_n}\frac{\partial}{\partial x^n}\right) \qquad \overleftarrow{\nabla} = \left(\sum_n \frac{\overleftarrow{\partial}}{\partial x^n}\frac{1}{h_n}\hat{\mathbf{e}}_n\right). \tag{II.2.8}$$

Be sure to write the factors $\hat{\mathbf{e}}_n$ and $1/h_n$ in the operator in a position where it is clear that the $\partial/\partial x^n$ does not operate on them. Of course, in order to evaluate Eqs. (6) or (7) we must make use of Eq. (II.1.21) to calculate the derivatives $\partial \mathbf{v}/\partial x^n$.

The gradients of a second-order tensor **T** are similarly defined as the quantities $\vec{\nabla}\mathbf{T}$ and $\mathbf{T}\overleftarrow{\nabla}$ such that

$$\frac{d\mathbf{T}}{ds} = (\mathbf{T}\overleftarrow{\nabla})\cdot\hat{\mathbf{u}} = \hat{\mathbf{u}}\cdot(\vec{\nabla}\mathbf{T})\,. \tag{II.2.9}$$

Since

$$\begin{aligned}\frac{d\mathbf{T}}{ds} &= \sum_n \frac{\partial \mathbf{T}}{\partial x^n}\frac{dx^n}{ds} = \sum_n \left(\frac{1}{h_n}\frac{\partial \mathbf{T}}{\partial x^n}\right)\left(h_n\frac{dx^n}{ds}\right) \\ &= \left(\sum_n \frac{1}{h_n}\frac{\partial \mathbf{T}}{dx^n}\hat{\mathbf{e}}_n\right)\cdot\left(\sum_m h_m\frac{dx^m}{ds}\hat{\mathbf{e}}_m\right) = \left(\sum_m h_m \frac{dx^m}{ds}\hat{\mathbf{e}}_m\right)\cdot\left(\sum_n \hat{\mathbf{e}}_n\frac{1}{h_n}\frac{\partial \mathbf{T}}{\partial x^n}\right),\end{aligned}$$

we obtain

$$\mathbf{T}\overleftarrow{\nabla} = \mathbf{T}\left(\frac{\overleftarrow{\partial}}{\partial x^n}\frac{1}{h_n}\hat{\mathbf{e}}_n\right) \quad \text{and} \quad \vec{\nabla}\mathbf{T} = \left(\hat{\mathbf{e}}_n\frac{1}{h_n}\frac{\partial}{\partial x^n}\right)\mathbf{T} \tag{II.2.10}$$

in agreement with formal use of the del operators of Eqs. (8). The required derivatives $\partial \mathbf{T}/\partial x^n$ can be evaluated by use of Eq. (II.1.22) and a table of the derivatives of the unit base vectors for the particular coordinate system.

Invariance of the form of the del operator under changes of coordinate systems was assured in the formation of the gradient of a scalar, vector, or tensor because these gradients were first given a definition independent of the coordinate system and then the form of the components found for an arbitrary orthogonal coordinate system. That the del operator is also invariant in form when operating, for example, as $(\nabla \cdot)$ or $(\nabla \times)$ to form the divergence or curl of a vector or a tensor can be demonstrated by giving definitions of divergence and curl independent of the coordinate system. For example, we could use the divergence theorem of Sec. 2.5 to define div $\mathbf{v}$ as that scalar point function whose value at any point P is the limit of $(1/V_k)\int_{S_k} \mathbf{v}\cdot\mathbf{n}\, dS$ for a sequence of arbitrary volumes V_k with bounding surfaces S_k shrinking to the point P. A similar definition of curl $\mathbf{v}$ could be based on Stokes theorem. For brevity we omit the proofs of form-invariance, which are implicit in the general-tensor treatment of Sec. I.6, and merely show how the del operator can be used (assuming that it is form-invariant) to get representations of the divergence and curl. For the divergence formulas it will be useful to have the following formulas, each obtained as the scalar product of a unit vector by one of the Eqs. (II.1.20).

Components of Unit Base Vector Derivatives

No Summations:

$$\frac{\partial \hat{\mathbf{e}}_m}{\partial x^m}\cdot\hat{\mathbf{e}}_m = 0 \qquad \frac{\partial \hat{\mathbf{e}}_m}{\partial x^m}\cdot\hat{\mathbf{e}}_r = -\frac{1}{h_r}\frac{\partial h_m}{\partial x^r} \qquad (r \neq m)$$

For $m \neq n$:

$$\frac{\partial \hat{\mathbf{e}}_m}{\partial x^n}\cdot\hat{\mathbf{e}}_n = \frac{1}{h_m}\frac{\partial h_n}{\partial x^m} \qquad \frac{\partial \hat{\mathbf{e}}_m}{\partial x^n}\cdot\hat{\mathbf{e}}_r = 0 \qquad (r \neq n)\,. \tag{II.2.11}$$

Divergence of a vector div $\mathbf{v} = \vec{\nabla}\cdot\mathbf{v} = \mathbf{v}\cdot\overleftarrow{\nabla}$ is a scalar point function, given in any orthogonal curvilinear coordinate system by

$$\begin{aligned}\vec{\nabla}\cdot\mathbf{v} &= \Big(\sum_n \hat{\mathbf{e}}_n \frac{1}{h_n}\frac{\partial}{\partial x^n}\Big)\cdot\mathbf{v} = \sum_n \hat{\mathbf{e}}_n \frac{1}{h_n}\frac{\partial}{\partial x^n}\big(\sum_m v_{\langle m\rangle}\,\hat{\mathbf{e}}_m\big)\\ &= \sum_n\sum_m \Big(\frac{1}{h_n}\frac{\partial v_{\langle m\rangle}}{\partial x^n}\hat{\mathbf{e}}_n\cdot\hat{\mathbf{e}}_m + \frac{1}{h_n}v_{\langle m\rangle}\hat{\mathbf{e}}_n\cdot\frac{\partial \hat{\mathbf{e}}_m}{\partial x^n}\Big).\end{aligned}$$

Since $\hat{\mathbf{e}}_n\cdot\hat{\mathbf{e}}_m = \delta_{mn}$ the first term reduces to a single sum on n, while use of Eqs. (11) simplifies the second term. Thus we obtain

$$\vec{\nabla}\cdot\mathbf{v} = \sum_n \frac{1}{h_n}\frac{\partial v_{\langle n\rangle}}{\partial x^n} + \sum_n \left(\sum_{\substack{m \\ m\neq n}} \frac{v_{\langle m\rangle}}{h_m h_n}\frac{\partial h_n}{\partial x^m}\right). \qquad \text{(II.2.12)}$$

When the summations are written out, and $1/h_1 h_2 h_3$ factored out, this takes the form

$$\operatorname{div}\mathbf{v} = \frac{1}{h_1 h_2 h_3}\left[\frac{\partial}{\partial x^1}(h_2 h_3 v_{\langle 1\rangle}) + \frac{\partial}{\partial x^2}(h_3 h_1 v_{\langle 2\rangle}) + \frac{\partial}{\partial x^3}(h_1 h_2 v_{\langle 3\rangle})\right]. \qquad \text{(II.2.13a)}$$

(The same form also results if we begin with $\mathbf{v}\cdot\vec{\nabla}$.) Since, by Eq. (II.1.18)

$$h_1 h_2 h_3 = \sqrt{g}, \qquad \text{where} \quad g = \det|g_{ij}|,$$

Eq. (13a) can be written as

$$\operatorname{div}\mathbf{v} = \frac{1}{\sqrt{g}}\sum_n \frac{\partial}{\partial x^n}\left(\frac{\sqrt{g}}{h_n}v_{\langle n\rangle}\right). \qquad \text{(II.2.13b)}$$

The Laplacian operator $\nabla^2 \equiv \vec{\nabla}\cdot\vec{\nabla}$ is obtained by the indicated repeated application of the operator $\vec{\nabla}$. When the Laplacian operates on a twice-differentiable scalar point function the result is a scalar point function. Thus

$$\nabla^2 f \equiv \vec{\nabla}\cdot\vec{\nabla}f \equiv \operatorname{div}(\operatorname{grad} f) = \frac{1}{\sqrt{g}}\sum_n \frac{\partial}{\partial x^n}\left[\frac{\sqrt{g}}{(h_n)^2}\frac{\partial f}{\partial x^n}\right]. \qquad \text{(II.2.14)}$$

The Laplacian may also operate on a vector. For example, $\vec{\nabla}\cdot\vec{\nabla}\mathbf{v}$ is obtained by evaluating $\vec{\nabla}\cdot\mathbf{T}$ according to Eq. (II.1.16) below with the tensor $\mathbf{T}$ chosen as $\mathbf{T} = \vec{\nabla}\mathbf{v}$. The result obtained by applying the Laplacian to a vector is another vector.

Divergence of a second-order tensor $\vec{\nabla}\cdot\mathbf{T} \neq \mathbf{T}\cdot\vec{\nabla}$. The divergence $\vec{\nabla}\cdot\mathbf{T}$ of a second-order tensor is a vector, given in orthogonal curvilinear coordinates as follows.

$$\begin{aligned}\vec{\nabla}\cdot\mathbf{T} &= \sum_r \left(\hat{\mathbf{e}}_r \frac{1}{h_r}\frac{\partial}{\partial x^r}\right)\cdot\sum_m\sum_n T_{\langle mn\rangle}\,\hat{\mathbf{e}}_m\hat{\mathbf{e}}_n \\ &= \sum_r\sum_n \frac{1}{h_r}\frac{\partial T_{\langle rn\rangle}}{\partial x^r}\hat{\mathbf{e}}_n + \sum_r\sum_m\sum_n \frac{1}{h_r}T_{\langle mn\rangle}\left(\hat{\mathbf{e}}_r\cdot\frac{\partial\hat{\mathbf{e}}_m}{\partial x^r}\right)\hat{\mathbf{e}}_n + \sum_r\sum_n \frac{1}{h_r}T_{\langle rn\rangle}\frac{\partial\hat{\mathbf{e}}_n}{\partial x^r},\end{aligned} \qquad \text{(II.2.15)}$$

since $\hat{\mathbf{e}}_r\cdot\hat{\mathbf{e}}_m = \delta_{rm}$. We now use Eqs. (11) to evaluate the second term as

$$\sum_n\sum_r \frac{1}{h_r}\left(\sum_{m\neq r} T_{\langle mn\rangle}\frac{1}{h_m}\frac{\partial h_r}{\partial x^m}\right)\hat{\mathbf{e}}_n$$

while Eqs. (II.1.20) give the last term as

$$\sum_r \left\{\frac{1}{h_r}\left(\sum_{n\neq r} T_{\langle rn\rangle}\frac{1}{h_n}\frac{\partial h_r}{\partial x^n}\right)\hat{\mathbf{e}}_r + \frac{1}{h_r}T_{\langle rr\rangle}\left(-\frac{1}{h_p}\frac{\partial h_r}{\partial x^p}\hat{\mathbf{e}}_p - \frac{1}{h_q}\frac{\partial h_r}{\partial x^q}\hat{\mathbf{e}}_q\right)\right\}$$

(p, q, r all different).

Hence Eq. (15) becomes

$$\begin{aligned}\vec{\nabla}\cdot\mathbf{T} = &\sum_n \left[\sum_r \left(\frac{1}{h_r}\frac{\partial T_{\langle rn\rangle}}{\partial x^r} + \sum_{m\neq r}\frac{T_{\langle mn\rangle}}{h_r h_m}\frac{\partial h_r}{\partial x_m}\right)\right]\hat{\mathbf{e}}_n \\ &+ \sum_r \left[\sum_{n\neq r}\frac{T_{\langle rn\rangle}}{h_r h_n}\frac{\partial h_r}{\partial x^n}\right]\hat{\mathbf{e}}_r - \frac{1}{h_1}T_{\langle 11\rangle}\left(\frac{1}{h_2}\frac{\partial h_1}{\partial x^2}\hat{\mathbf{e}}_2 + \frac{1}{h_3}\frac{\partial h_1}{\partial x^3}\hat{\mathbf{e}}_3\right) \\ &- \frac{1}{h_2}T_{\langle 22\rangle}\left(\frac{1}{h_3}\frac{\partial h_2}{\partial x^3}\hat{\mathbf{e}}_3 + \frac{1}{h_1}\frac{\partial h_2}{\partial x^1}\hat{\mathbf{e}}_1\right) - \frac{1}{h_3}T_{\langle 33\rangle}\left(\frac{1}{h_1}\frac{\partial h_3}{\partial x^1}\hat{\mathbf{e}}_1 + \frac{1}{h_2}\frac{\partial h_3}{\partial x^2}\hat{\mathbf{e}}_2\right).\end{aligned} \tag{II.2.16}$$

The coefficient of $\hat{\mathbf{e}}_1$, for example, is

$$(\vec{\nabla}\cdot\mathbf{T})_{\langle 1\rangle} = \sum_r \left(\frac{1}{h_r}\frac{\partial T_{\langle r1\rangle}}{\partial x^r} + \sum_{m\neq r}\frac{T_{\langle m1\rangle}}{h_r h_m}\frac{\partial h_r}{\partial x^m}\right) + \sum_{n\neq 1}\frac{T_{\langle 1n\rangle}}{h_1 h_n}\frac{\partial h_1}{\partial x^n} - \frac{T_{\langle 22\rangle}}{h_2 h_1}\frac{\partial h_2}{\partial x^1} - \frac{T_{\langle 33\rangle}}{h_3 h_1}\frac{\partial h_3}{\partial x^1}.$$

When the summations are written out this can be brought into the form

$$\boxed{\begin{aligned}(\vec{\nabla}\cdot\mathbf{T})_{\langle 1\rangle} = &\frac{1}{h_1 h_2 h_3}\left[\frac{\partial}{\partial x^1}(h_2 h_3 T_{\langle 11\rangle}) + \frac{\partial}{\partial x^2}(h_3 h_1 T_{\langle 21\rangle}) + \frac{\partial}{\partial x^3}(h_1 h_2 T_{\langle 31\rangle})\right] \\ &+ \frac{T_{\langle 12\rangle}}{h_1 h_2}\frac{\partial h_1}{\partial x^2} + \frac{T_{\langle 13\rangle}}{h_1 h_3}\frac{\partial h_1}{\partial x^3} - \frac{T_{\langle 22\rangle}}{h_1 h_2}\frac{\partial h_2}{\partial x^1} - \frac{T_{\langle 33\rangle}}{h_1 h_3}\frac{\partial h_3}{\partial x^1}.\end{aligned}} \tag{II.2.17}$$

The other two coefficients $(\vec{\nabla}\cdot\mathbf{T})_{\langle 2\rangle}$ and $(\vec{\nabla}\cdot\mathbf{T})_{\langle 3\rangle}$ can be written down by cyclic permutation of the indices in Eq. (17). Note that up to this point the matrix $[T_{\langle ij\rangle}]$ has not been assumed symmetric, so that is necessary to distinguish, for example, between $T_{\langle 21\rangle}$ and $T_{\langle 12\rangle}$ in writing the formula. When $\mathbf{T}$ is the stress tensor defined in Chap. 3, the equation of motion is $\vec{\nabla}\cdot\mathbf{T} + \rho\mathbf{b} = \rho\, d\mathbf{v}/dt$. (Some authors define the stress tensor $\mathbf{T}$ as the transpose of the one we have used, so that $T_{\langle ij\rangle}$ is the traction component in the $\hat{\mathbf{e}}_i$ direction on a surface perpendicular to $\hat{\mathbf{e}}_j$; with this alternative convention the first term of the equation of motion would be $\vec{\nabla}\cdot\mathbf{T}^T$, which coincides with our form in the nonpolar case where $\mathbf{T}$ is symmetric.)

A similar derivation can be made for $\mathbf{T}\cdot\overleftarrow{\nabla}$, and it turns out that

$$\mathbf{T}\cdot\overleftarrow{\nabla} = \vec{\nabla}\cdot\mathbf{T}^T, \tag{II.2.18}$$

where the superscript T denotes the transpose, so that the two operations $\mathbf{T}\cdot\overleftarrow{\nabla}$ and $\vec{\nabla}\cdot\mathbf{T}$ give the same vector when $\mathbf{T} = \mathbf{T}^T$, that is, ***when* T *is symmetric*.**

$$\mathbf{T}\cdot\overleftarrow{\nabla} = \vec{\nabla}\cdot\mathbf{T} \quad \text{for} \quad \mathbf{T} = \mathbf{T}^T. \tag{II.2.19}$$

Curl of a vector $\nabla \times \mathbf{v}$ is a vector, given in orthogonal curvilinear coordinates by

$$\begin{aligned}\nabla \times \mathbf{v} &= \left(\sum_n \hat{\mathbf{e}}_n \frac{1}{h_n}\frac{\partial}{\partial x^n}\right) \times \left(\sum_m v_{\langle m\rangle}\hat{\mathbf{e}}_m\right)\\
&= \sum_n \sum_m \frac{1}{h_n}\frac{\partial v_{\langle m\rangle}}{\partial x^n}\hat{\mathbf{e}}_n \times \hat{\mathbf{e}}_m + \frac{v_{\langle m\rangle}}{h_n}\hat{\mathbf{e}}_n \times \frac{\partial \hat{\mathbf{e}}_m}{\partial x^n}\\
&= \sum_n \sum_m \frac{1}{h_n}\frac{\partial v_{\langle m\rangle}}{\partial x^n}\hat{\mathbf{e}}_n \times \hat{\mathbf{e}}_m + \sum_n \left[\sum_{m\neq n} \frac{v_{\langle m\rangle}}{h_n}\hat{\mathbf{e}}_n \times \left(\frac{1}{h_m}\frac{\partial h_n}{\partial x^m}\hat{\mathbf{e}}_n\right)\right]\\
&\quad - \frac{v_{\langle 1\rangle}}{h_1}\hat{\mathbf{e}}_1 \times \left(\frac{1}{h_2}\frac{\partial h_1}{\partial x^2}\hat{\mathbf{e}}_2 + \frac{1}{h_3}\frac{\partial h_1}{\partial x^3}\hat{\mathbf{e}}_3\right)\\
&\quad - \frac{v_{\langle 2\rangle}}{h_2}\hat{\mathbf{e}}_2 \times \left(\frac{1}{h_3}\frac{\partial h_2}{\partial x^3}\hat{\mathbf{e}}_3 + \frac{1}{h_1}\frac{\partial h_2}{\partial x^1}\hat{\mathbf{e}}_1\right)\\
&\quad - \frac{v_{\langle 3\rangle}}{h_3}\hat{\mathbf{e}}_3 \times \left(\frac{1}{h_1}\frac{\partial h_3}{\partial x^1}\hat{\mathbf{e}}_1 + \frac{1}{h_2}\frac{\partial h_3}{\partial x^2}\hat{\mathbf{e}}_2\right). \qquad \text{(II.2.20)}\end{aligned}$$

For a right-handed system

$$\hat{\mathbf{e}}_n \times \hat{\mathbf{e}}_m = e_{nmp}\hat{\mathbf{e}}_p, \qquad \text{(II.2.21)}$$

where the e_{nmp} are the three-index permutation symbols introduced in Eq. (2.3.22). Thus, for example, the coefficient of $\hat{\mathbf{e}}_1$ in Eq. (20) is

$$(\nabla \times \mathbf{v})_{\langle 1\rangle} = \frac{1}{h_2}\frac{\partial v_{\langle 3\rangle}}{\partial x^2} - \frac{1}{h_3}\frac{\partial v_{\langle 2\rangle}}{\partial x^3} - \frac{v_{\langle 2\rangle}}{h_2 h_3}\frac{\partial h_2}{\partial x^3} + \frac{v_{\langle 3\rangle}}{h_3 h_2}\frac{\partial h_3}{\partial x^2}$$

or

$$(\nabla \times \mathbf{v})_{\langle 1\rangle} = \frac{1}{h_2 h_3}\left[\frac{\partial(v_{\langle 3\rangle}h_3)}{\partial x^2} - \frac{\partial(v_{\langle 2\rangle}h_2)}{\partial x^3}\right]. \qquad \text{(II.2.22)}$$

The other coefficients can be written down by cyclic permutation of the indices, and the whole formula can be written as

$$\nabla \times \mathbf{v} = \sum_r \left[\sum_m \sum_n e_{mnr}\frac{1}{h_m h_n}\frac{\partial(v_{\langle m\rangle}h_m)}{\partial x^n}\right]\hat{\mathbf{e}}_r. \qquad \text{(II.2.23)}$$

It can be shown that $\mathbf{v} \times \overleftarrow{\nabla} = -\overrightarrow{\nabla} \times \mathbf{v}$.

The curl operator can also operate on a second-order tensor, as follows:

$$\begin{aligned}\overrightarrow{\nabla} \times \mathbf{T} &= \left(\sum_r \hat{\mathbf{e}}_r \frac{1}{h_r}\frac{\partial}{\partial x^r}\right) \times \left(\sum_m \sum_n T_{\langle mn\rangle}\hat{\mathbf{e}}_m\hat{\mathbf{e}}_n\right)\\
&= \sum_n \sum_r \sum_m \frac{1}{h_r}\left\{\frac{\partial T_{\langle mn\rangle}}{\partial x^n}(\hat{\mathbf{e}}_r \times \hat{\mathbf{e}}_m)\,\hat{\mathbf{e}}_n + T_{\langle mn\rangle}\left(\hat{\mathbf{e}}_r \times \frac{\partial \hat{\mathbf{e}}_m}{\partial x^r}\right)\hat{\mathbf{e}}_n\right.\\
&\quad \left. + T_{\langle mn\rangle}(\hat{\mathbf{e}}_r \times \hat{\mathbf{e}}_m)\frac{\partial \hat{\mathbf{e}}_n}{\partial x^r}\right\}. \qquad \text{(II.2.24)}\end{aligned}$$

A rather long computation shows that the resulting second-order tensor has physical components of three types, for example,

$$(\vec{\nabla} \times \mathbf{T})_{\langle 11\rangle} = \frac{1}{h_2 h_3}\left[\frac{\partial(h_3 T_{\langle 31\rangle})}{\partial x^2} - \frac{\partial(h_2 T_{\langle 21\rangle})}{\partial x^3}\right] + \frac{T_{\langle 23\rangle}}{h_3 h_1}\frac{\partial h_3}{\partial x^1} - \frac{T_{\langle 32\rangle}}{h_2 h_1}\frac{\partial h_2}{\partial x^1}$$

$$(\vec{\nabla} \times \mathbf{T})_{\langle 12\rangle} = \frac{1}{h_2 h_3}\left[\frac{\partial(h_3 T_{\langle 32\rangle})}{\partial x^2} - \frac{\partial(h_2 T_{\langle 22\rangle})}{\partial x^3}\right] + \frac{T_{\langle 31\rangle}}{h_2 h_1}\frac{\partial h_2}{\partial x^1} + \frac{T_{\langle 23\rangle}}{h_3 h_2}\frac{\partial h_3}{\partial x^2} + \frac{T_{\langle 33\rangle}}{h_2 h_3}\frac{\partial h_2}{\partial x^3}$$

$$(\vec{\nabla} \times \mathbf{T})_{\langle 21\rangle} = \frac{1}{h_3 h_1}\left[\frac{\partial(h_1 T_{\langle 11\rangle})}{\partial x^3} - \frac{\partial(h_3 T_{\langle 31\rangle})}{\partial x^1}\right] - \frac{T_{\langle 32\rangle}}{h_1 h_2}\frac{\partial h_1}{\partial x^2} - \frac{T_{\langle 13\rangle}}{h_3 h_1}\frac{\partial h_3}{\partial x^1} - \frac{T_{\langle 33\rangle}}{h_1 h_3}\frac{\partial h_1}{\partial x^3}. \quad \text{(II.2.25)}$$

The other diagonal $\langle 22\rangle$ and $\langle 33\rangle$ components may be obtained by cyclic permutation of the indices in the $\langle 11\rangle$ component. Also cyclic permutation of the indices in the $\langle 12\rangle$ component yields the $\langle 23\rangle$ and $\langle 31\rangle$ components, while acyclic permutation of the indices of $\langle 21\rangle$ yields $\langle 13\rangle$ and $\langle 32\rangle$. (For acyclic permutation replace 1 by 3, 3 by 2, and 2 by 1.)

The complete formula is summarized below in dyadic notation, but Eq. (25) may be more convenient for explicit evaluation. The e_{mnr} and e_{rst} are the three-index symbols introduced in Sec. 2.3.

$$\begin{aligned}\vec{\nabla} \times \mathbf{T} = {} & \sum_r \sum_s \left[\sum_m \sum_n e_{mnr} \frac{1}{h_m h_n}\frac{\partial(h_n T_{\langle ns\rangle})}{\partial x^m}\right]\hat{\mathbf{e}}_r \hat{\mathbf{e}}_s \\ & + \sum_r \sum_s e_{rst}\left[\frac{T_{\langle tr\rangle}}{h_s h_r}\frac{\partial h_s}{\partial x^r} + \frac{T_{\langle st\rangle}}{h_t h_s}\frac{\partial h_t}{\partial x^s} + \frac{T_{\langle tt\rangle}}{h_s h_t}\frac{\partial h_s}{\partial x^t}\right]\hat{\mathbf{e}}_r \hat{\mathbf{e}}_s \\ & + \sum_r \sum_s e_{rst}\frac{T_{\langle st\rangle}}{h_t h_r}\frac{\partial h_t}{\partial x^r}\hat{\mathbf{e}}_r \hat{\mathbf{e}}_r. \end{aligned} \quad \text{(II.2.26)}$$

The first line of Eq. (26) has for each fixed s the form of the curl of a vector with components $v_{\langle m\rangle} = T_{\langle ms\rangle}$. The second line gives the extra terms for the off-diagonal components, while the last line gives the extra terms for the diagonal components.

The curl operator can also operate from the right on **T**, as

$$\mathbf{T} \times \overleftarrow{\nabla} = \left(\sum_m \sum_n T_{\langle mn\rangle}\hat{\mathbf{e}}_m \hat{\mathbf{e}}_n\right) \times \left(\sum_r \frac{\overleftarrow{\partial}}{\partial x^r}\frac{1}{h_r}\hat{\mathbf{e}}_r\right) \quad \text{(II.2.27)}$$

and it turns out that

$$\mathbf{T} \times \overleftarrow{\nabla} = -[\vec{\nabla} \times \mathbf{T}^T]^T. \quad \text{(II.2.28)}$$

Thus, interchanging r and s in the coefficients of $\hat{\mathbf{e}}_r\hat{\mathbf{e}}_s$ in Eq. (26) and also interchanging each pair of indices on $T_{\langle ij\rangle}$, we obtain

$$\begin{aligned}\mathbf{T}\times\overleftarrow{\nabla} = &-\sum_r\sum_s\left[\sum_m\sum_n e_{mns}\frac{1}{h_m h_n}\frac{\partial(h_n T_{\langle rn\rangle})}{\partial x^m}\right]\hat{\mathbf{e}}_r\hat{\mathbf{e}}_s\\ &-\sum_r\sum_s e_{srt}\left[\frac{T_{\langle st\rangle}}{h_r h_s}\frac{\partial h_r}{\partial x^s}+\frac{T_{\langle tr\rangle}}{h_t h_r}\frac{\partial h_t}{\partial x^r}+\frac{T_{\langle tt\rangle}}{h_r h_t}\frac{\partial h_r}{\partial x^t}\right]\hat{\mathbf{e}}_r\hat{\mathbf{e}}_s\\ &-\sum_s\left(\sum_r e_{srt}\frac{T_{\langle tr\rangle}}{h_t h_s}\frac{\partial h_t}{\partial x^s}\right)\hat{\mathbf{e}}_s\hat{\mathbf{e}}_s.\end{aligned} \tag{II.2.29}$$

For example,

$$\left.\begin{aligned}(\mathbf{T}\times\overleftarrow{\nabla})_{\langle 11\rangle} &= \frac{1}{h_2 h_3}\left[\frac{\partial(h_2 T_{\langle 12\rangle})}{\partial x^3}-\frac{\partial(h_3 T_{\langle 13\rangle})}{\partial x^2}\right]\\ &\quad+\frac{T_{\langle 23\rangle}}{h_2 h_1}\frac{\partial h_2}{\partial x^1}-\frac{T_{\langle 32\rangle}}{h_3 h_1}\frac{\partial h_3}{\partial x^1}\\ (\mathbf{T}\times\overleftarrow{\nabla})_{\langle 12\rangle} &= \frac{1}{h_3 h_1}\left[\frac{\partial(h_3 T_{\langle 13\rangle})}{\partial x^1}-\frac{\partial(h_1 T_{\langle 11\rangle})}{\partial x^3}\right]\\ &\quad+\frac{T_{\langle 23\rangle}}{h_1 h_2}\frac{\partial h_1}{\partial x^2}+\frac{T_{\langle 31\rangle}}{h_3 h_1}\frac{\partial h_3}{\partial x^1}+\frac{T_{\langle 33\rangle}}{h_1 h_3}\frac{\partial h_1}{\partial x^3}\\ (\mathbf{T}\times\overleftarrow{\nabla})_{\langle 21\rangle} &= \frac{1}{h_2 h_3}\left[\frac{\partial(h_2 T_{\langle 22\rangle})}{\partial x^3}-\frac{\partial(h_3 T_{\langle 23\rangle})}{\partial x^2}\right]\\ &\quad-\frac{T_{\langle 13\rangle}}{h_2 h_1}\frac{\partial h_2}{\partial x^1}-\frac{T_{\langle 32\rangle}}{h_3 h_2}\frac{\partial h_3}{\partial x^2}-\frac{T_{\langle 33\rangle}}{h_2 h_3}\frac{\partial h_2}{\partial x^3}.\end{aligned}\right\} \tag{II.2.30}$$

Again the other diagonal $\langle 22\rangle$ and $\langle 33\rangle$ components may be obtained by cyclic permutation of the indices in the $\langle 11\rangle$ component, and cyclic permutation of $\langle 12\rangle$ yields $\langle 23\rangle$ and $\langle 31\rangle$, while acyclic permutation of $\langle 21\rangle$ yields $\langle 13\rangle$ and $\langle 32\rangle$.

We now consider some examples of the application of the results of this appendix to the equations of continuum mechanics. Section II.3 gives examples in arbitrary orthogonal curvilinear coordinates, while Sec. II.4 involves cylindrical and spherical coordinates.

EXERCISES

1. Obtain the forms of ∇f according to Eq. (II.2.3) for: (a) cylindrical coordinates, (b) spherical coordinates. See Sec. II.4.
2. Obtain the matrices of physical components of $\mathbf{v}\overleftarrow{\nabla}$ according to Eq. (II.2.7) for: (a) cylindrical coordinates, (b) spherical coordinates. See Sec. II.4.

3. Show that Eq. (II.2.14) gives the forms for $\nabla^2 f$ listed in Sec. II.4 for: (a) cylindrical coordinates, (b) spherical coordinates.
4. Show that Eq. (II.2.17) and the two additional equations obtained by permuting the indices give the results of Eq. (II.4.C6) for cylindrical coordinates.
5. Verify that Eqs. (II.2.20) and (21) lead to (22) for the first component of $\nabla \times \mathbf{v}$.
6. Verify that Eqs. (II.2.22) or (23) lead to the formulas for $\nabla \times \mathbf{v}$ given in Sec. II.4 for: (a) cylindrical coordinates, (b) spherical coordinates.
7. Verify that Eq. (II.2.24) leads to the expressions for the physical components of $\nabla \times \mathbf{T}$ given in Eq. (25) for the following index choices: (a) 11, (b) 12, (c) 21.

II.3 Examples of field equations of continuum mechanics, using physical components in orthogonal curvilinear coordinates

Section numbers and equation numbers below correspond to the numbering in Chaps. 4 and 5 of the text.

Sec. 4.2 Displacement Gradients:

$$\mathbf{u}\overleftarrow{\nabla} = \sum_k \sum_m \left(\frac{1}{h_m} \frac{\partial u_{\langle k \rangle}}{\partial X^m} \hat{\mathbf{e}}_k \hat{\mathbf{e}}_m + \frac{u_{\langle k \rangle}}{h_m} \frac{\partial \hat{\mathbf{e}}_k}{\partial X^m} \hat{\mathbf{e}}_m \right)$$

$$\overrightarrow{\nabla}\mathbf{u} = \sum_k \sum_m \left(\frac{1}{h_k} \frac{\partial u_{\langle m \rangle}}{\partial X^k} \hat{\mathbf{e}}_k \hat{\mathbf{e}}_m + \frac{u_{\langle m \rangle}}{h_k} \hat{\mathbf{e}}_k \frac{\partial \hat{\mathbf{e}}_m}{\partial X^k} \right)$$

Small Strain and Rotation:

$$\mathbf{E} = \frac{1}{2}(\mathbf{u}\overleftarrow{\nabla} + \overrightarrow{\nabla}\mathbf{u}) = \sum_k \sum_m \left\{ \frac{1}{2} \left[\frac{1}{h_m} \frac{\partial u_{\langle k \rangle}}{\partial X^m} + \frac{1}{h_k} \frac{\partial u_{\langle m \rangle}}{\partial X^k} \right] \hat{\mathbf{e}}_k \hat{\mathbf{e}}_m + \frac{1}{2} \left[\frac{u_{\langle k \rangle}}{h_m} \frac{\partial \hat{\mathbf{e}}_k}{\partial X^m} \hat{\mathbf{e}}_m + \frac{u_{\langle m \rangle}}{h_k} \hat{\mathbf{e}}_k \frac{\partial \hat{\mathbf{e}}_m}{\partial X^k} \right] \right. \tag{4.2.2}$$

$\boldsymbol{\Omega} = \frac{1}{2}(\mathbf{u}\overleftarrow{\nabla} - \overrightarrow{\nabla}\mathbf{u})$ [Replace + by − in each bracket of (8).] For example

$$E_{\langle 11 \rangle} = \frac{1}{h_1} \frac{\partial u_{\langle 1 \rangle}}{\partial X^1} + \frac{u_{\langle 2 \rangle}}{h_2 h_1} \frac{\partial h_1}{\partial X^2} + \frac{u_{\langle 3 \rangle}}{h_3 h_1} \frac{\partial h_1}{\partial X^3}$$

$$E_{\langle 12 \rangle} = \frac{1}{2} \left[\frac{1}{h_2} \frac{\partial u_{\langle 1 \rangle}}{\partial X^2} + \frac{1}{h_1} \frac{\partial u_{\langle 2 \rangle}}{\partial X^1} - \frac{1}{h_1 h_2} \left(u_{\langle 1 \rangle} \frac{\partial h_1}{\partial X^2} + u_{\langle 2 \rangle} \frac{\partial h_2}{\partial X^1} \right) \right]$$

$$\Omega_{\langle 12 \rangle} = \frac{1}{2} \left[\frac{1}{h_2} \frac{\partial u_{\langle 1 \rangle}}{\partial X^2} - \frac{1}{h_1} \frac{\partial u_{\langle 2 \rangle}}{\partial X^1} + \frac{1}{h_1 h_2} \left(u_{\langle 1 \rangle} \frac{\partial h_1}{\partial X^2} - u_{\langle 2 \rangle} \frac{\partial h_2}{\partial X^1} \right) \right].$$

Other components may be written by cyclic permutation of the indices [$\Omega_{\langle 11 \rangle} = \Omega_{\langle 22 \rangle} = \Omega_{\langle 33 \rangle} = 0$]. The capital letters X^1, X^2, X^3 denote material coordinates, but for small displacements the distinction between X^1 and x^1, etc. is usually ignored.

Sec. 4.3

$$\text{(4.3.6)} \qquad \mathbf{a} = \frac{d\mathbf{v}}{dt} = \frac{\partial \mathbf{v}}{dt} + \mathbf{v}\cdot\overrightarrow{\nabla}\mathbf{v} = \frac{\partial \mathbf{v}}{\partial t} + \sum_k \sum_m \left[\frac{v_{\langle k\rangle}}{h_k}\frac{\partial v_{\langle m\rangle}}{\partial x^k}\hat{\mathbf{e}}_m + \frac{v_{\langle k\rangle}v_{\langle m\rangle}}{h_k}\frac{\partial \hat{\mathbf{e}}_m}{\partial x^k}\right]$$

For example,

$$a_{\langle 1\rangle} = \frac{\partial v_{\langle 1\rangle}}{\partial t} + \frac{v_{\langle 1\rangle}}{h_1}\frac{\partial v_{\langle 1\rangle}}{\partial x^1} + \frac{v_{\langle 2\rangle}}{h_2}\frac{\partial v_{\langle 1\rangle}}{\partial x^2} + \frac{v_{\langle 3\rangle}}{h_3}\frac{\partial v_{\langle 1\rangle}}{\partial x^3}$$
$$+ \frac{v_{\langle 1\rangle}}{h_1}\left[\frac{v_{\langle 2\rangle}}{h_2}\frac{\partial h_1}{\partial x^2} + \frac{v_{\langle 3\rangle}}{h_3}\frac{\partial h_1}{\partial x^3}\right] - \left[\frac{v_{\langle 2\rangle}^2}{h_1 h_2}\frac{\partial h_2}{\partial x^1} + \frac{v_{\langle 3\rangle}^2}{h_1 h_3}\frac{\partial h_3}{\partial x^1}\right]$$

Scalar f:

$$\text{(4.3.10)} \qquad \frac{df}{dt} = \frac{\partial f}{\partial t} + \mathbf{v}\cdot\nabla f = \frac{\partial f}{\partial t} + \sum_k \frac{v_{\langle k\rangle}}{h_k}\frac{\partial f}{\partial x^k}$$

Sec. 4.4

Formulas for $D_{\langle ij\rangle}$ and $W_{\langle ij\rangle}$ may be written by replacing $\mathbf{u}$ by $\mathbf{v}$ in the formulas of Sec. 4.2 and replacing X by x. Then

$$\text{(4.4.5)} \qquad \mathbf{D} = \tfrac{1}{2}(\mathbf{v}\overleftarrow{\nabla} + \overrightarrow{\nabla}\mathbf{v}) \quad \text{and} \quad \mathbf{W} = \tfrac{1}{2}(\mathbf{v}\overleftarrow{\nabla} + \overrightarrow{\nabla}\mathbf{v})$$

yield, for example,

$$D_{\langle 11\rangle} = \frac{1}{h_1}\frac{\partial v_{\langle 1\rangle}}{\partial x^1} + \frac{v_{\langle 2\rangle}}{h_2 h_1}\frac{\partial h_1}{\partial x^2} + \frac{v_{\langle 3\rangle}}{h_3 h_1}\frac{\partial h_1}{\partial x^3}$$
$$D_{\langle 12\rangle} = \frac{1}{2}\left[\frac{1}{h_2}\frac{\partial v_{\langle 1\rangle}}{\partial x^2} + \frac{1}{h_1}\frac{\partial v_{\langle 2\rangle}}{\partial x^1} - \frac{1}{h_1 h_2}\left(v_{\langle 1\rangle}\frac{\partial h_1}{\partial x^2} + v_{\langle 2\rangle}\frac{\partial h_2}{\partial x^1}\right)\right]$$
$$W_{\langle 12\rangle} = \frac{1}{2}\left[\frac{1}{h_2}\frac{\partial v_{\langle 1\rangle}}{\partial x^2} - \frac{1}{h_1}\frac{\partial v_{\langle 2\rangle}}{\partial x^1} + \frac{1}{h_1 h_2}\left(v_{\langle 1\rangle}\frac{\partial h_1}{\partial x^2} - v_{\langle 2\rangle}\frac{\partial h_2}{\partial x^1}\right)\right]$$

$$\text{(4.4.11)} \qquad \frac{d}{dt}[(ds)^2] = 2d\mathbf{p}\cdot\mathbf{D}\cdot d\mathbf{p} = 2[dp_{\langle i\rangle}]\,[D_{\langle ij\rangle}]\,[dp_{\langle j\rangle}] \quad \text{where}$$

$$dp_{\langle k\rangle} = h_k\, dx^k \qquad \text{(no sum)}$$

(4.4.18) Natural strain increments $d\epsilon_{\langle ij\rangle}$ are formed similarly to $D_{\langle ij\rangle}$, with $\overline{d\mathbf{u}}$ replacing $\mathbf{v}$.

Sec. 4.5

In this section, both spatial point **x** and material particle **X** appear. When it is necessary to distinguish these in the gradient, use $\nabla_{\mathbf{X}} f$ for the material gradient and $\nabla_{\mathbf{x}} f$ for the spatial gradient. We will use $\hat{\mathbf{e}}_1^0$ and h_1^0, etc. for the unit vectors and scale factors of orthogonal, curvilinear material coordinates if both sets of unit vectors and scale factors appear in the same discussion, but will omit the superscript zero when a capital letter subscript is used or when the context makes it clear that material coordinates are used.

The notation $d\mathbf{x}$ is used for the relative position vector $d\mathbf{p}$ of two points separated by distance ds, while $d\mathbf{X}$ denotes $d\mathbf{P}$ of magnitude dS.

(4.5.2a)
$$d\mathbf{x} = (\mathbf{x}\overleftarrow{\nabla})\cdot d\mathbf{X} = d\mathbf{X}\cdot\overrightarrow{\nabla}\mathbf{x}$$
$$d\mathbf{x} = \sum_k h_k \, dx^k \, \hat{\mathbf{e}}_k \qquad d\mathbf{X} = \sum_m h_M \, dX^M \, \hat{\mathbf{e}}_M$$

Deformation Gradients:

$$\mathbf{F} \equiv \mathbf{x}\overleftarrow{\nabla}_{\mathbf{X}} \quad \text{and} \quad \mathbf{F}^T \equiv \overrightarrow{\nabla}_{\mathbf{X}}\mathbf{x}$$

are defined as the two-point tensors such that Eq. (2a) gives

$$d\mathbf{x} = \mathbf{F}\cdot d\mathbf{X} \quad \text{or} \quad d\mathbf{x} = d\mathbf{X}\cdot\mathbf{F}^T.$$

Note that since the x^r are not vector components, we do not obtain formulas comparable to those for $\mathbf{u}\overleftarrow{\nabla}$ and $\overrightarrow{\nabla}\mathbf{u}$ given above for Sec. 4.2, but in fact simpler formulas. Since

$$dx_{\langle k\rangle} = h_k \, dx^k = \sum_M h_k \frac{\partial x^k}{\partial X^M} dX^M = \sum_M \frac{h_k}{h_M} \frac{\partial x^k}{\partial X^M} dX_{\langle M\rangle} \qquad \text{(no sum on } k\text{)}$$

we require

$$F_{\langle kM\rangle} = \frac{h_k}{h_M}\frac{\partial x^k}{\partial X^M} = (F^T)_{\langle Mk\rangle}$$

or

(4.5.2a)
$$(\mathbf{x}\overleftarrow{\nabla}_{\mathbf{X}})_{\langle kM\rangle} = \frac{h_k}{h_M}\frac{\partial x^k}{\partial X^M} = (\overrightarrow{\nabla}_{\mathbf{X}}\mathbf{x})_{\langle Mk\rangle} \qquad \text{(no sums)}$$

to obtain the matrix equations

$$[dx_{\langle k\rangle}] = [F_{\langle kM\rangle}]\,[dX_{\langle M\rangle}] = [dX_M]\,[F_{\langle kM\rangle}]^T.$$

The spatial gradients $\mathbf{F}^{-1} = \mathbf{X}\overleftarrow{\nabla}_{\mathbf{x}}$ and $(\mathbf{F}^{-1})^T = \overrightarrow{\nabla}_{\mathbf{x}}\mathbf{X}$ of the inverse deformation equation

$$\mathbf{X} = \mathbf{X}(\mathbf{x}, t)$$

are similarly defined, yielding

$$(4.5.2b)\quad \begin{cases} [F^{-1}]_{\langle Mk\rangle} \equiv (\mathbf{X}\overleftarrow{\nabla}_{\mathbf{x}})_{\langle Mk\rangle} = \dfrac{h_M}{h_k}\dfrac{\partial X^M}{\partial x^k} = (\overrightarrow{\nabla}_{\mathbf{x}}\mathbf{X})_{\langle kM\rangle} \equiv [(F^{-1})^T]_{\langle kM\rangle} \\ [dX_{\langle M\rangle}] = [(\mathbf{X}\overleftarrow{\nabla}_{\mathbf{x}})_{\langle Mk\rangle}]\,[dx_{\langle k\rangle}] = [dx_{\langle k\rangle}]\,[(\overrightarrow{\nabla}_{\mathbf{x}}\mathbf{X})_{\langle kM\rangle}] \quad \text{for} \\ d\mathbf{X} = \mathbf{X}\overleftarrow{\nabla}_{\mathbf{x}}\cdot d\mathbf{x} = d\mathbf{x}\cdot\overrightarrow{\nabla}_{\mathbf{x}}\mathbf{X}. \end{cases}$$

Deformation Tensors $\mathbf{C}$ and $\mathbf{B}^{-1}$:†

$$(4.5.4)\qquad (ds)^2 = d\mathbf{X}\cdot\mathbf{C}\cdot d\mathbf{X} \qquad (dS)^2 = d\mathbf{x}\cdot\mathbf{B}^{-1}\cdot d\mathbf{x}$$

$$(4.5.7)\quad \begin{cases} \mathbf{C} = \mathbf{F}^T\cdot\mathbf{F} = (\overrightarrow{\nabla}_{\mathbf{X}}\mathbf{x})\cdot(\mathbf{x}\overleftarrow{\nabla}_{\mathbf{X}}) & \mathbf{B}^{-1} = (\overrightarrow{\nabla}_{\mathbf{x}}\mathbf{X})\cdot(\mathbf{X}\overleftarrow{\nabla}_{\mathbf{x}}) \\ C_{\langle IJ\rangle} = \sum_k \dfrac{h_k}{h_I}\dfrac{\partial x^k}{\partial X^I}\dfrac{h_k}{h_J}\dfrac{\partial x^k}{\partial X^J} & B^{-1}_{\langle ij\rangle} = \sum_K \dfrac{h_K}{h_i}\dfrac{\partial X^K}{\partial x^i}\dfrac{h_K}{h_j}\dfrac{\partial X^K}{\partial x^j} \end{cases}$$

(no sums on I, J, or on i, j)

Finite-Strain Tensors $\mathbf{E}$ and $\mathbf{E}^*$:‡

$$(4.5.8)\qquad E_{\langle IJ\rangle} = \tfrac{1}{2}[C_{\langle IJ\rangle} - \delta_{IJ}] \qquad E^*_{\langle ij\rangle} = \tfrac{1}{2}[\delta_{ij} - B^{-1}_{\langle ij\rangle}]$$

$$(4.5.3)\qquad (ds)^2 - (dS)^2 = 2d\mathbf{X}\cdot\mathbf{E}\cdot d\mathbf{X} = 2d\mathbf{x}\cdot\mathbf{E}^*\cdot d\mathbf{x}$$

Sec. 4.7

The compatibility equations for small strain $\mathbf{E}$, expressed in terms of physical components have the form

$$(4.7.5)\qquad S_{\langle mn\rangle} = 0 \qquad \text{where} \quad \mathbf{S} = \overrightarrow{\nabla}\times\mathbf{E}\times\overleftarrow{\nabla}.$$

Only six distinct equations will be obtained because $S_{\langle nm\rangle} = S_{\langle mn\rangle}$; although they will not appear identical, use of Eqs. (II.1.23) will show that $S_{\langle mn\rangle} - S_{\langle nm\rangle} \equiv 0$ for any sufficiently differentiable functions $E_{\langle rs\rangle}$.

Let

$$\mathbf{P} = \mathbf{E}\times\overleftarrow{\nabla}.$$

Then

$$\mathbf{S} = \overrightarrow{\nabla}\times\mathbf{P}$$

†$\mathbf{B}^{-1}$ was denoted by $\mathbf{c}$ in Truesdell and Toupin (1960). The tensor $\mathbf{B} = \mathbf{F}\cdot\mathbf{F}^T$ is used in Truesdell and Noll (1965).

‡$\mathbf{E}^*$ is sometimes denoted by $\mathbf{e}$.

and the two cross products can be evaluated by using Eqs. (II.2.25) through (II.2.30). For example,

$$S_{\langle 11\rangle} = \frac{1}{h_2 h_3}\left[\frac{\partial(h_3 P_{\langle 31\rangle})}{\partial X^2} - \frac{\partial(h_2 P_{\langle 21\rangle})}{\partial X^3}\right] + \frac{P_{\langle 23\rangle}}{h_3 h_1}\frac{\partial h_3}{\partial X^1} - \frac{P_{\langle 32\rangle}}{h_2 h_1}\frac{\partial h_2}{\partial X^1}$$

and

$$S_{\langle 12\rangle} = \frac{1}{h_2 h_3}\left[\frac{\partial(h_3 P_{\langle 32\rangle})}{\partial X^2} - \frac{\partial(h_2 P_{\langle 22\rangle})}{\partial X^3}\right] + \frac{P_{\langle 31\rangle}}{h_2 h_1}\frac{\partial h_2}{\partial X^1} + \frac{P_{\langle 23\rangle}}{h_3 h_2}\frac{\partial h_3}{\partial X^2} + \frac{P_{\langle 33\rangle}}{h_2 h_3}\frac{\partial h_2}{\partial X^3}$$

whence

$$\begin{aligned}
S_{\langle 11\rangle} = {} & \frac{1}{h_2 h_3}\frac{\partial}{\partial X^2}\left\{\frac{1}{h_2}\left[\frac{\partial(h_2 E_{\langle 32\rangle})}{\partial X^3} - \frac{\partial(h_3 E_{\langle 33\rangle})}{\partial X^2}\right]\right. \\
& \left. + \frac{E_{\langle 12\rangle}}{h_1}\frac{\partial h_3}{\partial X^1} + \frac{E_{\langle 23\rangle}}{h_2}\frac{\partial h_2}{\partial X^3} + \frac{E_{\langle 22\rangle}}{h_2}\frac{\partial h_3}{\partial X^2}\right\} \\
& - \frac{1}{h_2 h_3}\frac{\partial}{\partial X^3}\left\{\frac{1}{h_3}\left[\frac{\partial(h_2 E_{\langle 22\rangle})}{\partial X^3} - \frac{\partial(h_3 E_{\langle 23\rangle})}{\partial X^2}\right]\right. \\
& \left. - \frac{E_{\langle 32\rangle}}{h_3}\frac{\partial h_3}{\partial X^2} - \frac{E_{\langle 13\rangle}}{h_1}\frac{\partial h_2}{\partial X^1} - \frac{E_{\langle 33\rangle}}{h_3}\frac{\partial h_2}{\partial X^3}\right\} \\
& + \frac{1}{h_1^2 h_2 h_3}\frac{\partial h_3}{\partial X^1}\left[\frac{\partial(h_1 E_{\langle 21\rangle})}{\partial X^2} - \frac{\partial(h_2 E_{\langle 22\rangle})}{\partial X^1}\right] \\
& + \frac{1}{h_3 h_1}\frac{\partial h_3}{\partial X^1}\left[\frac{E_{\langle 31\rangle}}{h_2 h_3}\frac{\partial h_2}{\partial X^3} + \frac{E_{\langle 12\rangle}}{h_1 h_2}\frac{\partial h_1}{\partial X^2} + \frac{E_{\langle 11\rangle}}{h_1 h_2}\frac{\partial h_2}{\partial X^1}\right] \\
& - \frac{1}{h_1^2 h_2 h_3}\frac{\partial h_2}{\partial X^1}\left[\frac{\partial(h_3 E_{\langle 33\rangle})}{\partial X^1} - \frac{\partial(h_1 E_{\langle 31\rangle})}{\partial X^3}\right] \\
& + \frac{1}{h_2 h_1}\frac{\partial h_2}{\partial X^1}\left[\frac{E_{\langle 21\rangle}}{h_2 h_3}\frac{\partial h_3}{\partial X^2} + \frac{E_{\langle 13\rangle}}{h_3 h_1}\frac{\partial h_1}{\partial X^3} + \frac{E_{\langle 11\rangle}}{h_3 h_1}\frac{\partial h_3}{\partial X^1}\right]
\end{aligned}$$

and

$$\begin{aligned}
S_{\langle 12\rangle} = {} & \frac{1}{h_2 h_3}\frac{\partial}{\partial X^2}\left\{\frac{1}{h_1}\left[\frac{\partial(h_3 E_{\langle 33\rangle})}{\partial X^1} - \frac{\partial(h_1 E_{\langle 31\rangle})}{\partial X^3}\right]\right. \\
& \left. - \frac{E_{\langle 21\rangle}}{h_2}\frac{\partial h_3}{\partial X^2} - \frac{E_{\langle 13\rangle}}{h_1}\frac{\partial h_1}{\partial X^3} - \frac{E_{\langle 11\rangle}}{h_1}\frac{\partial h_3}{\partial X^1}\right\} \\
& - \frac{1}{h_2 h_3}\frac{\partial}{\partial X^3}\left\{\frac{h_2}{h_1 h_3}\left[\frac{\partial(h_3 E_{\langle 23\rangle})}{\partial X^1} - \frac{\partial(h_1 E_{\langle 21\rangle})}{\partial X^3}\right]\right. \\
& \left. + \frac{E_{\langle 31\rangle}}{h_3}\frac{\partial h_3}{\partial X^2} - \frac{E_{\langle 13\rangle}}{h_1}\frac{\partial h_1}{\partial X^2}\right\}
\end{aligned}$$

(Equation continued on next page.)

$$+\frac{1}{h_1h_2^2h_3}\frac{\partial h_2}{\partial X^1}\left[\frac{\partial(h_2E_{\langle 32\rangle})}{\partial X^3}-\frac{\partial(h_3E_{\langle 33\rangle})}{\partial X^2}\right]$$

$$+\frac{1}{h_2h_1}\frac{\partial h_2}{\partial X^1}\left[\frac{E_{\langle 12\rangle}}{h_3h_1}\frac{\partial h_3}{\partial X^1}+\frac{E_{\langle 23\rangle}}{h_2h_3}\frac{\partial h_2}{\partial X^3}+\frac{E_{\langle 22\rangle}}{h_3h_2}\frac{\partial h_3}{\partial X^2}\right]$$

$$+\frac{1}{h_1h_2^2h_3}\frac{\partial h_3}{\partial X^2}\left[\frac{\partial(h_1E_{\langle 21\rangle})}{\partial X^2}-\frac{\partial(h_2E_{\langle 22\rangle})}{\partial X^1}\right]$$

$$+\frac{1}{h_3h_2}\frac{\partial h_3}{\partial X^2}\left[\frac{E_{\langle 31\rangle}}{h_2h_3}\frac{\partial h_2}{\partial X^3}+\frac{E_{\langle 12\rangle}}{h_1h_2}\frac{\partial h_1}{\partial X^2}+\frac{E_{\langle 11\rangle}}{h_2h_1}\frac{\partial h_2}{\partial X^1}\right]$$

$$+\frac{1}{h_1h_2^2h_3}\frac{\partial h_2}{\partial X^3}\left[\frac{\partial(h_1E_{\langle 31\rangle})}{\partial X^2}-\frac{\partial(h_2E_{\langle 32\rangle})}{\partial X^1}\right]$$

$$+\frac{1}{h_2h_3}\frac{\partial h_2}{\partial X^3}\left[\frac{E_{\langle 12\rangle}}{h_1h_3}\frac{\partial h_1}{\partial X^3}-\frac{E_{\langle 21\rangle}}{h_2h_3}\frac{\partial h_2}{\partial X^3}\right].$$

$S_{\langle 22\rangle}$ and $S_{\langle 33\rangle}$ may be written down by cyclic permutation of the indices in $S_{\langle 11\rangle}$, while cyclic permutation of the indices in $S_{\langle 12\rangle}$ yields $S_{\langle 23\rangle}$ and $S_{\langle 32\rangle}$. Or, for example, in cylindrical coordinates write Eqs. (5a) and (5b) three times, successively choosing:

1. $X^1 = r \qquad X^2 = \theta \qquad X^3 = z \qquad h_2 = r \qquad h_1 = h_3 = 1,$
2. $X^1 = \theta \qquad X^2 = z \qquad X^3 = r \qquad h_1 = r \qquad h_2 = h_3 = 1,$
3. $X^1 = z \qquad X^2 = r \qquad X^3 = \theta \qquad h_3 = r \qquad h_1 = h_2 = 1.$

[See Eqs. (II.4.C15).]

It should be recalled that the six compatibility equations so obtained represent only three independent conditions, as was pointed out in Sec. 4.7.

Sec. 5.1

To evaluate the integral theorems in integral form we need in addition to the expressions for the integrands the following representations of dV and $\mathbf{n}\,dS$.

$$dV = h_1h_2h_3\,dx^1\,dx^2\,dx^3.$$

If a surface is represented parametrically by equations

$$x^1 = x^1(\alpha,\beta) \qquad x^2 = x^2(\alpha,\beta) \qquad x^3 = x^3(\alpha,\beta)$$

in such a way that the outer unit normal $\mathbf{n}$ is collinear with $(\partial\mathbf{p}/\partial\alpha)\times(\partial\mathbf{p}/\partial\beta)$, where $\mathbf{p}$ is the position vector and $\partial\mathbf{p}/\partial\alpha$ and $\partial\mathbf{p}/\partial\beta$ are vectors in the direction of increasing α and β, respectively, on the surface, then

$$\mathbf{n}\,dS = \Sigma\left\{h_2 h_3 \frac{\partial(x^2, x^3)}{\partial(\alpha, \beta)}\,\mathbf{e}_1\,d\alpha\,d\beta\right\},$$

where the notation $\Sigma\{\ \ \}$ indicates the sum of three expressions obtained by cyclic permutation of the indices inside the braces $\{\ \ \}$.† Actual evaluations of such integrals are much simplified if the surface is composed of coordinate surfaces. For example, on a coordinate surface $x^3 =$ const., with $\mathbf{n}$ in the direction of $\mathbf{e}_3$,

$$\mathbf{n}\,dS = h_1 h_2\,dx^1\,dx^2\,\mathbf{e}_3.$$

Sec. 5.2 Continuity Equations

In spatial coordinates:

$$(5.2.7)\quad \begin{cases} \dfrac{d\rho}{dt} + \rho\,\mathrm{div}\,\mathbf{v} = 0, \quad \text{or} \\ \dfrac{\partial\rho}{\partial t} + \sum_k \dfrac{v_{\langle k\rangle}}{h_k}\dfrac{\partial\rho}{\partial x^k} + \dfrac{\rho}{h_1 h_2 h_3}\,\Sigma\left\{\dfrac{\partial}{\partial x^1}(h_2 h_3 v_{\langle 1\rangle})\right\} = 0, \end{cases}$$

where again $\Sigma\{\ \ \}$ indicates the sum of three terms obtained by cyclic permutation of the indices.

In material coordinates where

$$(5.2.11)\qquad |\mathbf{x}/\mathbf{X}| = \det|F_{\langle kM\rangle}| = \sqrt{\mathrm{III}_C} \quad \text{and} \quad F_{\langle kM\rangle} = \frac{h_k}{h_M}\frac{\partial x^k}{\partial X^M} \quad (\text{no sum}),$$

$$(5.2.13)\qquad \rho|\mathbf{x}/\mathbf{X}| = \text{const.}$$

Sec. 5.3 Equations of Motion

$$(5.3.4)\quad \begin{cases} \nabla\cdot\mathbf{T} + \rho\mathbf{b} = \rho\dfrac{d\mathbf{v}}{dt} \\ (\nabla\cdot\mathbf{T})_{\langle k\rangle} + \rho b_{\langle k\rangle} = \rho a_{\langle k\rangle}, \end{cases}$$

where $(\nabla\cdot\mathbf{T})_{\langle k\rangle}$ is given by Eq. (II.2.17) and $a_{\langle k\rangle}$ is given by the equation above for Sec. 4.3.

$$(5.3.7)\qquad \text{Nonpolar case:}\quad T_{\langle mn\rangle} = T_{\langle nm\rangle}$$

Piola-Kirchhoff Stresses (no sums on K or M):

$$(5.3.22)\qquad \mathbf{T}^0 = \frac{\rho_0}{\rho}(\mathbf{X}\overleftarrow{\nabla}_{\mathbf{x}})\cdot\mathbf{T} \quad \text{or} \quad T^0_{\langle Km\rangle} = \frac{\rho_0}{\rho}\sum_r \frac{h_K}{h_r}\frac{\partial X_K}{\partial x_r}T_{\langle rm\rangle}$$

†$\partial(x^2, x^3)/\partial(\alpha, \beta)$ is the Jacobian determinant.

(5.3.23)
$$\begin{cases} \tilde{\mathbf{T}} = \mathbf{T}^0 \cdot (\vec{\nabla}_x \mathbf{X}) \quad \text{or} \quad \tilde{T}_{\langle KM \rangle} = \sum_r T^0_{\langle Kr \rangle} \dfrac{h_M}{h_r} \dfrac{\partial X^M}{\partial x^r} \\ \tilde{\mathbf{T}} = \dfrac{\rho_0}{\rho} (\mathbf{X} \overleftarrow{\nabla}_x) \cdot \mathbf{T} \cdot (\vec{\nabla}_x \mathbf{X}) \quad \text{or} \\ \tilde{T}_{\langle KM \rangle} = \dfrac{\rho_0}{\rho} \sum_r \sum_s \dfrac{h_K}{h_r} \dfrac{\partial X^K}{\partial x^r} T_{\langle rs \rangle} \dfrac{h_M}{h_s} \dfrac{\partial X^M}{\partial x^s} \end{cases}$$

Sec. 5.4 Energy Equation (nonpolar case)

$$\mathbf{T} : \mathbf{D} + \rho h = \nabla \cdot \mathbf{q} + \rho \frac{du}{dt} \quad \text{or}$$

(5.4.7)

$$\sum_i \sum_j T_{\langle ij \rangle} D_{\langle ij \rangle} + \rho h = \frac{1}{h_1 h_2 h_3} \sum \left\{ \frac{\partial}{\partial x^1} (h_2 h_3) q_{\langle 1 \rangle} \right\} + \rho \left[\frac{\partial u}{\partial t} + \sum_k \frac{v_{\langle k \rangle}}{h_k} \frac{\partial u}{\partial x^k} \right],$$

where again $\sum \{ \quad \}$ denotes the sum of three terms obtained by cyclic permutation of the indices.

Sec. II.4 summarizes some of the results for cylindrical and spherical coordinates.

EXERCISES

1. Verify that Eq. (4.3.6) of Sec. II.3 leads to the formulas for a_r given in Sec. II.4 for: (a) cylindrical coordinates, (b) spherical coordinates.
2. Verify that Eq. (4.4.5) of Sec. II.3 for $D_{\langle 12 \rangle}$ gives the formula for $D_{r\theta}$ in cylindrical coordinates listed in Eq. (II.4.C9).
3. Verify that Eq. (4.4.5) of Sec. II.3 for $D_{\langle 12 \rangle}$ with indices permuted to give $D_{\langle 31 \rangle}$ leads to the formula for $D_{r\phi}$ in spherical coordinates listed in Eq. (II.4.S9).
4. Show that equations following Eqs. (4.7.5) of Sec. II.3 give the results listed for S_{rr}, $S_{r\theta}$, and S_{zr} in Eq. (II.4.C15).
5. Write out the continuity equation of Eq. (5.2.7), Sec. II.3, in: (a) cylindrical coordinates, (b) spherical coordinates.
6. Show that Eqs. (5.3.4) of Sec. II.3 lead to the equations of motion in cylindrical coordinates given in Eq. (II.4.C11).
7. Show that Eqs. (5.3.4) of Sec. II.3 lead to the equations of motion in spherical coordinates given in Eq. (II.4.S11).
8. Write out the energy equation, Eq. (5.4.8) of Sec. II.3, in: (a) cylindrical coordinates, (b) spherical coordinates.

II.4 Summary of differential formulas in cylindrical and spherical coordinates

v_r, v_θ, $T_{r\theta}$, etc., denote physical components.

Cylindrical coordinates (r, θ, z).

$$\left.\begin{array}{lll} h_r = 1 & h_\theta = r & h_z = 1 \\ \dfrac{\partial \hat{\mathbf{e}}_r}{\partial \theta} = +\hat{\mathbf{e}}_\theta & \dfrac{\partial \hat{\mathbf{e}}_\theta}{\partial \theta} = -\hat{\mathbf{e}}_r & \text{other } \dfrac{\partial \hat{\mathbf{e}}_m}{\partial x^n} = 0 \end{array}\right\} \tag{II.4.C1}$$

$$\nabla = \hat{\mathbf{e}}_r \frac{\partial}{\partial r} + \hat{\mathbf{e}}_\theta \frac{1}{r}\frac{\partial}{\partial \theta} + \hat{\mathbf{e}}_z \frac{\partial}{\partial z} \tag{II.4.C2}$$

$$\nabla f = \frac{\partial f}{\partial r}\hat{\mathbf{e}}_r + \frac{1}{r}\frac{\partial f}{\partial \theta}\hat{\mathbf{e}}_\theta + \frac{\partial f}{\partial z}\hat{\mathbf{e}}_z \tag{II.4.C3}$$

$$\nabla^2 f = \frac{1}{r}\frac{\partial}{\partial r}\left(r\frac{\partial f}{\partial r}\right) + \frac{1}{r^2}\frac{\partial^2 f}{\partial \theta^2} + \frac{\partial^2 f}{\partial z^2} \tag{II.4.C4}$$

$$\overrightarrow{\nabla}\cdot\mathbf{v} = \mathbf{v}\cdot\overleftarrow{\nabla} = \frac{1}{r}\frac{\partial}{\partial r}(rv_r) + \frac{1}{r}\frac{\partial v_\theta}{\partial \theta} + \frac{\partial v_z}{\partial z} \tag{II.4.C5}$$

$$\begin{aligned} \nabla\cdot\mathbf{T} = {} & \left[\frac{1}{r}\frac{\partial(rT_{rr})}{\partial r} + \frac{1}{r}\frac{\partial T_{\theta r}}{\partial \theta} + \frac{\partial T_{zr}}{\partial z} - \frac{1}{r}T_{\theta\theta}\right]\hat{\mathbf{e}}_r \\ & + \left[\frac{1}{r}\frac{\partial(T_{\theta\theta})}{\partial \theta} + \frac{\partial T_{z\theta}}{\partial z} + \frac{1}{r}\frac{\partial}{\partial r}(rT_{r\theta}) + \frac{1}{r}T_{\theta r}\right]\hat{\mathbf{e}}_\theta \\ & + \left[\frac{\partial T_{zz}}{\partial z} + \frac{1}{r}\frac{\partial}{\partial r}(rT_{rz}) + \frac{1}{r}\frac{\partial T_{\theta z}}{\partial \theta}\right]\hat{\mathbf{e}}_z \end{aligned} \tag{II.4.C6}$$

$$\nabla \times \mathbf{v} = \left(\frac{1}{r}\frac{\partial v_z}{\partial \theta} - \frac{\partial v_\theta}{\partial z}\right)\hat{\mathbf{e}}_r + \left(\frac{\partial v_r}{\partial z} - \frac{\partial v_z}{\partial r}\right)\hat{\mathbf{e}}_\theta + \left(\frac{1}{r}\frac{\partial}{\partial r}(rv_\theta) - \frac{1}{r}\frac{\partial v_r}{\partial \theta}\right)\hat{\mathbf{e}}_z \tag{II.4.C7}$$

Matrices of Physical Components:

$$[\mathbf{v}\overleftarrow{\nabla}] = \begin{vmatrix} \dfrac{\partial v_r}{\partial r} & \dfrac{1}{r}\dfrac{\partial v_r}{\partial \theta} - \dfrac{v_\theta}{r} & \dfrac{\partial v_r}{\partial z} \\ \dfrac{\partial v_\theta}{\partial r} & \dfrac{1}{r}\dfrac{\partial v_\theta}{\partial \theta} + \dfrac{v_r}{r} & \dfrac{\partial v_\theta}{\partial z} \\ \dfrac{\partial v_z}{\partial r} & \dfrac{1}{r}\dfrac{\partial v_z}{\partial \theta} & \dfrac{\partial v_z}{\partial z} \end{vmatrix} \tag{II.4.C8}$$

$$[\overrightarrow{\nabla}\mathbf{v}] = [\mathbf{v}\overleftarrow{\nabla}]^T$$

Rate of Deformation **D** and Spin **W**:

$$\left.\begin{aligned}
&[\mathbf{D}] = \tfrac{1}{2}\{[\mathbf{v}\nabla] + [\mathbf{v}\nabla]^T\} \qquad [\mathbf{W}] = \tfrac{1}{2}\{[\mathbf{v}\nabla] - [\mathbf{v}\nabla]^T\}\\
&D_{rr} = \frac{\partial v_r}{\partial r} \qquad D_{\theta\theta} = \frac{1}{r}\frac{\partial v_\theta}{\partial \theta} + \frac{v_r}{r} \qquad D_{zz} = \frac{\partial v_z}{\partial z}\\
&D_{r\theta} = \frac{1}{2}\left[\frac{1}{r}\frac{\partial v_r}{\partial \theta} + \frac{\partial v_\theta}{\partial r} - \frac{v_\theta}{r}\right] \qquad D_{rz} = \frac{1}{2}\left(\frac{\partial v_r}{\partial z} + \frac{\partial v_z}{\partial r}\right)\\
&D_{\theta z} = \frac{1}{2}\left(\frac{\partial v_\theta}{\partial z} + \frac{1}{r}\frac{\partial v_z}{\partial \theta}\right)
\end{aligned}\right\} \qquad \text{(II.4.C9)}$$

Accelerations in Spatial Coordinates $\mathbf{a} = \dfrac{\partial \mathbf{v}}{\partial t} + \mathbf{v}\cdot\vec{\nabla}\mathbf{v}$:

$$\left.\begin{aligned}
a_r &= \frac{\partial v_r}{\partial t} + v_r\frac{\partial v_r}{\partial r} + \frac{v_\theta}{r}\frac{\partial v_r}{\partial \theta} + v_z\frac{\partial v_r}{\partial z} - \frac{1}{r}v_\theta^2\\
a_\theta &= \frac{\partial v_\theta}{\partial t} + v_r\frac{\partial v_\theta}{\partial r} + \frac{v_\theta}{r}\frac{\partial v_\theta}{\partial \theta} + v_z\frac{\partial v_\theta}{\partial z} + \frac{1}{r}v_r v_\theta\\
a_z &= \frac{\partial v_z}{\partial t} + v_r\frac{\partial v_z}{\partial r} + \frac{v_\theta}{r}\frac{\partial v_z}{\partial \theta} + v_z\frac{\partial v_z}{\partial z}
\end{aligned}\right\} \qquad \text{(II.4.C10)}$$

Equations of Motion (acceleration component formulas are given above):

$$\left.\begin{aligned}
&\frac{\partial T_{rr}}{\partial r} + \frac{1}{r}\frac{\partial T_{\theta r}}{\partial \theta} + \frac{\partial T_{zr}}{\partial z} + \frac{1}{r}(T_{rr} - T_{\theta\theta}) + \rho b_r = \rho a_r\\
&\frac{\partial T_{r\theta}}{\partial r} + \frac{1}{r}\frac{\partial T_{\theta\theta}}{\partial \theta} + \frac{\partial T_{z\theta}}{\partial z} + \frac{1}{r}(T_{r\theta} + T_{\theta r}) + \rho b_\theta = \rho a_\theta\\
&\frac{\partial T_{rz}}{\partial r} + \frac{1}{r}\frac{\partial T_{\theta z}}{\partial \theta} + \frac{\partial T_{zz}}{\partial z} + \frac{1}{r}T_{rz} + \rho b_z = \rho a_z
\end{aligned}\right\} \qquad \text{(II.4.C11)}$$

Displacement Equations of Small Motion in Isotropic Elasticity:
(In material coordinates $a_r = \partial^2 u_r/\partial t^2$, $a_\theta = \partial^2 u_\theta/\partial t^2$, $a_z = \partial^2 u_z/\partial t^2$ and for small motions the difference between evaluations of the left-hand sides in material and spatial coordinates is negligible.) Thus Eq. (8.1.10) gives

$$\left.\begin{aligned}
&(\lambda + 2G)\frac{\partial e}{\partial r} - \frac{2G}{r}\frac{\partial \omega_z}{\partial \theta} + 2G\frac{\partial \omega_\theta}{\partial z} + \rho b_r = \rho\frac{\partial^2 u_r}{\partial t^2}\\
&(\lambda + 2G)\frac{1}{r}\frac{\partial e}{\partial \theta} - 2G\frac{\partial \omega_r}{\partial z} + 2G\frac{\partial \omega_z}{\partial r} + \rho b_\theta = \rho\frac{\partial^2 u_\theta}{\partial t^2}\\
&(\lambda + 2G)\frac{\partial e}{\partial z} - \frac{2G}{r}\frac{\partial}{\partial r}(r\omega_\theta) + \frac{2G}{r}\frac{\partial \omega_r}{\partial \theta} + \rho b_z = \rho\frac{\partial^2 u_z}{\partial t^2},
\end{aligned}\right\} \qquad \text{(II.4.C12)}$$

where

$$e = \frac{1}{r}\frac{\partial}{\partial r}(ru_r) + \frac{1}{r}\frac{\partial u_\theta}{\partial \theta} + \frac{\partial u_z}{\partial z} \qquad \text{(II.4.C13)}$$

and

$$\left.\begin{aligned}\omega_r &= -\Omega_{\theta z} = \frac{1}{2}\left(\frac{1}{r}\frac{\partial u_z}{\partial \theta} - \frac{\partial u_\theta}{\partial z}\right)\\ \omega_\theta &= -\Omega_{zr} = \frac{1}{2}\left(\frac{\partial u_r}{\partial z} - \frac{\partial u_z}{\partial r}\right)\\ \omega_z &= -\Omega_{r\theta} = \frac{1}{2}\left(\frac{1}{r}\frac{\partial(ru_\theta)}{\partial r} - \frac{1}{r}\frac{\partial u_r}{\partial \theta}\right).\end{aligned}\right\} \qquad \text{(II.4.C14)}$$

Compatibility Equations in Cylindricals [See Eqs. (4.7.5), p.662.]: ($\epsilon_{r\theta}$, etc., denote physical components of small strain E.)

$$\left.\begin{aligned}S_{rr} &\equiv \frac{2}{r}\frac{\partial^2\epsilon_{z\theta}}{\partial z\partial\theta} - \frac{1}{r^2}\frac{\partial^2\epsilon_{zz}}{\partial\theta^2} - \frac{\partial^2\epsilon_{\theta\theta}}{\partial z^2} + \frac{2}{r}\frac{\partial\epsilon_{rz}}{\partial z} - \frac{1}{r}\frac{\partial\epsilon_{zz}}{\partial r} = 0\\ S_{\theta\theta} &\equiv 2\frac{\partial^2\epsilon_{rz}}{\partial r\partial z} - \frac{\partial^2\epsilon_{rr}}{\partial z^2} - \frac{\partial^2\epsilon_{zz}}{\partial r^2} = 0\\ S_{zz} &\equiv \frac{2}{r}\frac{\partial^2\epsilon_{\theta r}}{\partial\theta\partial r} - \frac{\partial^2\epsilon_{\theta\theta}}{\partial r^2} - \frac{1}{r^2}\frac{\partial^2\epsilon_{rr}}{\partial\theta^2} + \frac{1}{r}\frac{\partial\epsilon_{rr}}{\partial r} + \frac{2}{r^2}\frac{\partial\epsilon_{r\theta}}{\partial\theta}\\ &\quad - \frac{2}{r}\frac{\partial\epsilon_{\theta\theta}}{\partial r} = 0\\ S_{r\theta} &\equiv \frac{1}{r}\frac{\partial^2\epsilon_{zz}}{\partial r\partial\theta} - \frac{\partial}{\partial z}\left[\frac{1}{r}\frac{\partial\epsilon_{zr}}{\partial\theta} + \frac{\partial\epsilon_{\theta z}}{\partial r} - \frac{\partial\epsilon_{\theta r}}{\partial z}\right]\\ &\quad + \frac{1}{r}\frac{\partial\epsilon_{z\theta}}{\partial z} - \frac{1}{r^2}\frac{\partial\epsilon_{zz}}{\partial\theta} = 0\\ S_{\theta z} &\equiv \frac{1}{r}\frac{\partial^2\epsilon_{rr}}{\partial\theta\partial z} - \frac{\partial}{\partial r}\left[\frac{\partial\epsilon_{\theta r}}{\partial z} + \frac{1}{r}\frac{\partial\epsilon_{zr}}{\partial\theta} - \frac{\partial\epsilon_{z\theta}}{\partial r}\right]\\ &\quad - \frac{2}{r}\frac{\partial\epsilon_{r\theta}}{\partial z} + \frac{1}{r}\frac{\partial\epsilon_{z\theta}}{\partial r} - \frac{1}{r^2}\epsilon_{z\theta} = 0\\ S_{zr} &\equiv \frac{\partial^2\epsilon_{\theta\theta}}{\partial z\partial r} - \frac{1}{r}\frac{\partial}{\partial\theta}\left[\frac{\partial\epsilon_{\theta z}}{\partial r} + \frac{\partial\epsilon_{r\theta}}{\partial z} - \frac{1}{r}\frac{\partial\epsilon_{rz}}{\partial\theta}\right]\\ &\quad + \frac{2}{r}\frac{\partial\epsilon_{zr}}{\partial r} - \frac{1}{r^2}\frac{\partial\epsilon_{\theta z}}{\partial\theta} - \frac{1}{r}\frac{\partial}{\partial z}(\epsilon_{rr} - \epsilon_{zz}) = 0\end{aligned}\right\} \qquad \text{(II.4.C15)}$$

Spherical coordinates (r, θ, ϕ)

$$h_r = 1 \qquad h_\theta = r \qquad h_\phi = r\sin\theta$$

$$\left.\begin{aligned}&\frac{\partial\hat{\mathbf{e}}_r}{\partial r} = \frac{\partial\hat{\mathbf{e}}_\theta}{\partial r} = \frac{\partial\hat{\mathbf{e}}_\phi}{\partial r} = 0\\ &\frac{\partial\hat{\mathbf{e}}_r}{\partial\theta} = \hat{\mathbf{e}}_\theta \qquad \frac{\partial\hat{\mathbf{e}}_\theta}{\partial\theta} = -\hat{\mathbf{e}}_r \qquad \frac{\partial\hat{\mathbf{e}}_\phi}{\partial\theta} = 0\\ &\frac{\partial\hat{\mathbf{e}}_r}{\partial\phi} = +\sin\theta\hat{\mathbf{e}}_\phi \qquad \frac{\partial\hat{\mathbf{e}}_\theta}{\partial\phi} = \cos\theta\hat{\mathbf{e}}_\phi \qquad \frac{\partial\hat{\mathbf{e}}_\phi}{\partial\phi} = -\sin\theta\hat{\mathbf{e}}_r - \cos\theta\hat{\mathbf{e}}_\theta\end{aligned}\right\} \qquad \text{(II.4.S1)}$$

$$\nabla = \hat{\mathbf{e}}_r \frac{\partial}{\partial r} + \hat{\mathbf{e}}_\theta \frac{1}{r}\frac{\partial}{\partial \theta} + \hat{\mathbf{e}}_\phi \frac{1}{r \sin\theta}\frac{\partial}{\partial \phi} \tag{II.4.S2}$$

$$\nabla f = \frac{\partial f}{\partial r}\hat{\mathbf{e}}_r + \frac{1}{r}\frac{\partial f}{\partial \theta}\hat{\mathbf{e}}_\theta + \frac{1}{r\sin\theta}\frac{\partial f}{\partial \phi}\hat{\mathbf{e}}_\phi \tag{II.4.S3}$$

$$\nabla^2 f = \frac{1}{r^2}\frac{\partial}{\partial r}\left(r^2 \frac{\partial f}{\partial r}\right) + \frac{1}{r^2 \sin\theta}\frac{\partial}{\partial \theta}\left(\sin\theta \frac{\partial f}{\partial \theta}\right) + \frac{1}{r^2 \sin^2\theta}\frac{\partial^2 f}{\partial \phi^2} \tag{II.4.S4}$$

$$\nabla \cdot \mathbf{v} = \frac{1}{r^2}\frac{\partial}{\partial r}(r^2 v_r) + \frac{1}{r\sin\theta}\frac{\partial}{\partial \theta}(v_\theta \sin\theta) + \frac{1}{r\sin\theta}\frac{\partial v_\phi}{\partial \phi} \tag{II.4.S5}$$

$$\begin{aligned}
\nabla \cdot \mathbf{T} = &\left[\frac{1}{r^2}\frac{\partial}{\partial r}(r^2 T_{rr}) + \frac{1}{r\sin\theta}\frac{\partial}{\partial \theta}(T_{\theta r}\sin\theta) + \frac{1}{r\sin\theta}\frac{\partial T_{\phi r}}{\partial \phi}\right. \\
&\left. - \frac{1}{r}(T_{\theta\theta} + T_{\phi\phi})\right]\hat{\mathbf{e}}_r \\
&+ \left[\frac{1}{r^2}\frac{\partial}{\partial r}(r^2 T_{r\theta}) + \frac{1}{r\sin\theta}\frac{\partial}{\partial \theta}(T_{\theta\theta}\sin\theta) + \frac{1}{r\sin\theta}\frac{\partial T_{\phi\theta}}{\partial \phi}\right. \\
&\left. + \frac{1}{r}T_{\theta r} - \frac{\cot\theta}{r}T_{\phi\phi}\right]\hat{\mathbf{e}}_\theta \\
&+ \left[\frac{1}{r^2}\frac{\partial}{\partial r}(r^2 T_{r\phi}) + \frac{1}{r}\frac{\partial}{\partial \theta}(T_{\theta\phi}\sin\theta) + \frac{1}{r\sin\theta}\frac{\partial T_{\phi\phi}}{\partial \phi}\right. \\
&\left. + \frac{1}{r}(T_{\phi r} + T_{\phi\theta})\right]\hat{\mathbf{e}}_\phi
\end{aligned} \tag{II.4.S6}$$

$$\begin{aligned}
\nabla \times \mathbf{v} = &\left[\frac{1}{r\sin\theta}\frac{\partial}{\partial \theta}(v_\phi \sin\theta) - \frac{1}{r\sin\theta}\frac{\partial v_\theta}{\partial \phi}\right]\hat{\mathbf{e}}_r \\
&+ \left[\frac{1}{r\sin\theta}\frac{\partial v_r}{\partial \phi} - \frac{1}{r}\frac{\partial}{\partial r}(r v_\phi)\right]\hat{\mathbf{e}}_\theta \\
&+ \left[\frac{1}{r}\frac{\partial}{\partial r}(r v_\theta) - \frac{1}{r}\frac{\partial v_r}{\partial \theta}\right]\hat{\mathbf{e}}_\phi
\end{aligned} \tag{II.4.S7}$$

Matrices of Physical Components:

$$[\mathbf{v}\overleftarrow{\nabla}] = \begin{bmatrix}
\dfrac{\partial v_r}{\partial r} & \dfrac{1}{r}\dfrac{\partial v_r}{\partial \theta} - \dfrac{v_\theta}{r} & \dfrac{1}{r\sin\theta}\dfrac{\partial v_r}{\partial \phi} - \dfrac{v_\phi}{r} \\
\dfrac{\partial v_\theta}{\partial r} & \dfrac{1}{r}\dfrac{\partial v_\theta}{\partial \theta} + \dfrac{v_r}{r} & \dfrac{1}{r\sin\theta}\dfrac{\partial v_\theta}{\partial \phi} - \dfrac{v_\phi}{r}\cot\phi \\
\dfrac{\partial v_\phi}{\partial r} & \dfrac{1}{r}\dfrac{\partial v_\phi}{\partial \theta} & \dfrac{1}{r\sin\theta}\dfrac{\partial v_\phi}{\partial \phi} + \dfrac{v_\theta}{r}\cot\theta + \dfrac{v_r}{r}
\end{bmatrix} \tag{II.4.S8}$$

$[\overrightarrow{\nabla}\mathbf{v}] = [\mathbf{v}\overleftarrow{\nabla}]^T$

$$
\left.
\begin{aligned}
&[\mathbf{D}] = \tfrac{1}{2}\{[\mathbf{v}\overleftarrow{\nabla}] + [\mathbf{v}\overleftarrow{\nabla}]^T\} \qquad [\mathbf{W}] = \tfrac{1}{2}\{[\mathbf{v}\overleftarrow{\nabla}] - [\mathbf{v}\overleftarrow{\nabla}]^T\} \\
&D_{rr} = \frac{\partial v_r}{\partial r} \qquad D_{\theta\theta} = \frac{1}{r}\frac{\partial v_\theta}{\partial \theta} + \frac{v_r}{r} \\
&D_{\phi\phi} = \frac{1}{r\sin\theta}\frac{\partial v_\phi}{\partial \phi} + \frac{v_\theta}{r}\cot\theta + \frac{v_r}{r} \\
&D_{r\theta} = \frac{1}{2}\left[\frac{1}{r}\frac{\partial v_r}{\partial \theta} + \frac{\partial v_\theta}{\partial r} - \frac{v_\theta}{r}\right] \\
&D_{r\phi} = \frac{1}{2}\left[\frac{1}{r\sin\theta}\frac{\partial v_r}{\partial \phi} + \frac{\partial v_\phi}{\partial r} - \frac{v_\phi}{r}\right] \\
&D_{\theta\phi} = \frac{1}{2}\left[\frac{1}{r\sin\theta}\frac{\partial v_\theta}{\partial \phi} + \frac{1}{r}\frac{\partial v_\phi}{\partial \theta} - \frac{v_\phi}{r}\cot\phi\right]
\end{aligned}
\right\} \qquad \text{(II.4.S9)}
$$

Accelerations in Spatial Coordinates $\mathbf{a} = \partial\mathbf{v}/\partial t + \mathbf{v}\cdot\overrightarrow{\nabla}\mathbf{v}$:

$$
\left.
\begin{aligned}
a_r &= \frac{\partial v_r}{\partial t} + v_r\frac{\partial v_r}{\partial r} + \frac{v_\theta}{r}\frac{\partial v_r}{\partial \theta} + \frac{v_\phi}{r\sin\theta}\frac{\partial v_r}{\partial \phi} - \frac{v_\phi^2 + v_\theta^2}{r} \\
a_\theta &= \frac{\partial v_\theta}{\partial t} + v_r\frac{\partial v_\theta}{\partial r} + \frac{v_\theta}{r}\frac{\partial v_\theta}{\partial \theta} + \frac{v_\phi}{r\sin\phi}\frac{\partial v_\theta}{\partial \phi} + \frac{v_r v_\theta}{r} - \frac{v_\phi^2}{r}\cot\theta \\
a_\phi &= \frac{\partial v_\phi}{\partial t} + v_r\frac{\partial v_\phi}{\partial r} + \frac{v_\theta}{r}\frac{\partial v_\phi}{\partial \theta} + \frac{v_\phi}{r\sin\phi}\frac{\partial v_\phi}{\partial \phi} + \frac{v_r v_\phi}{r} + \frac{v_\theta v_\phi\cot\theta}{r}
\end{aligned}
\right\} \qquad \text{(II.4.S10)}
$$

Equations of Motion:
(Acceleration component formulas are given above.)

$$
\left.
\begin{aligned}
&\frac{\partial T_{rr}}{\partial r} + \frac{1}{r}\frac{\partial T_{\theta r}}{\partial \theta} + \frac{1}{r\sin\theta}\frac{\partial T_{\phi r}}{\partial \phi} + \frac{1}{r}[2T_{rr} - T_{\theta\theta} - T_{\phi\phi} \\
&\qquad\qquad + T_{\theta r}\cot\phi] + \rho b_r = \rho a_r \\
&\frac{\partial T_{r\theta}}{\partial r} + \frac{1}{r}\frac{\partial T_{\theta\theta}}{\partial \theta} + \frac{1}{r\sin\theta}\frac{\partial T_{\phi\theta}}{\partial \phi} + \frac{1}{r}[2T_{r\theta} + T_{\theta r} \\
&\qquad\qquad + (T_{\theta\theta} - T_{\phi\phi})\cot\theta] + \rho b_\theta = \rho a_\theta \\
&\frac{\partial T_{r\phi}}{\partial r} + \frac{1}{r}\frac{\partial T_{\theta\phi}}{\partial \theta} + \frac{1}{r\sin\theta}\frac{\partial T_{\phi\phi}}{\partial \phi} + \frac{1}{r}[2T_{r\phi} + T_{\phi r} \\
&\qquad\qquad + (T_{\phi\theta} + T_{\theta\phi})\cot\theta] + \rho b_\phi = \rho a_\phi
\end{aligned}
\right\} \qquad \text{(II.4.S11)}
$$

For the usual nonpolar case, in the last two equations $2T_{r\theta} + T_{\theta r} = 3T_{r\theta}$, $2T_{r\phi} + T_{\phi r} = 3T_{r\phi}$ and $T_{\phi\theta} + T_{\theta\phi} = 2T_{\theta\phi}$.

Displacement Equations of Small Motion in Isotropic Elasticity:
(In material coordinates $a_r = \partial^2 u_r/\partial t^2$, $a_\theta = \partial^2 u_\theta/\partial t^2$, $a_\phi = \partial^2 u_\phi/\partial t^2$; and for small motions the difference between evaluations of the left-hand sides in material and spatial coordinates is negligible.) Thus Eq. (8.1.10) gives

$$\left.\begin{aligned}
&(\lambda + 2G)\frac{\partial e}{\partial r} - \frac{2G}{r\sin\theta}\frac{\partial}{\partial\theta}(\omega_\phi \sin\theta) + \frac{2G}{r\sin\theta}\frac{\partial\omega_\theta}{\partial\phi} + \rho b_r = \rho\frac{\partial^2 u_r}{\partial t^2}\\
&\frac{(\lambda + 2G)}{r}\frac{\partial e}{\partial\theta} - \frac{2G}{r\sin\theta}\frac{\partial\omega_r}{\partial\phi} + \frac{2G}{r\sin\theta}\frac{\partial}{\partial r}(r\omega_\phi \sin\theta) + \rho b_\theta = \rho\frac{\partial^2 u_\theta}{\partial t^2}\\
&\frac{(\lambda + 2G)}{r\sin\theta}\frac{\partial e}{\partial\phi} - \frac{2G}{r}\frac{\partial}{\partial r}(r\omega_\theta) + \frac{2G}{r}\frac{\partial\omega_r}{\partial\theta} + \rho b_\phi = \rho\frac{\partial^2 u_\phi}{\partial t^2},
\end{aligned}\right\} \qquad \text{(II.4.S12)}$$

where

$$e = \frac{1}{r^2\sin\theta}\left[\frac{\partial}{\partial r}(r^2 u_r \sin\theta) + \frac{\partial}{\partial\theta}(r u_\theta \sin\theta) + \frac{\partial}{\partial\phi}(r u_\phi)\right] \qquad \text{(II.4.S13)}$$

and

$$\left.\begin{aligned}
\omega_r &= -\Omega_{\theta\phi} = \frac{1}{2}\left[\frac{1}{r\sin\theta}\frac{\partial}{\partial\theta}(u_\phi \sin\theta) - \frac{1}{r\sin\theta}\frac{\partial u_\theta}{\partial\phi}\right]\\
\omega_\theta &= -\Omega_{\phi r} = \frac{1}{2}\left[\frac{1}{r\sin\theta}\frac{\partial u_r}{\partial\phi} - \frac{1}{r}\frac{\partial(r u_\phi)}{\partial r}\right]\\
\omega_\phi &= -\Omega_{r\theta} = \frac{1}{2}\left[\frac{1}{r}\frac{\partial}{\partial r}(r u_\theta) - \frac{1}{r}\frac{\partial u_r}{\partial\theta}\right]
\end{aligned}\right\} \qquad \text{(II.4.S14)}$$

Compatibility equations for small strain may be written by using the general formulas following Eqs. (4.7.5) in Sec. II.3.

BIBLIOGRAPHY

Twentieth-Century Authors Cited in the Text

See also the suggested references for collateral reading at the ends of Secs. 1.1, 2.5, 6.4, 6.5, 7.5, 8.4, I.1 and I.3. For bibliography on writings before the twentieth century see, for example, Truesdell and Toupin (1960) and treatises on elasticity or fluid dynamics.

Adkins, J. E., "Symmetry relations for orthotropic and transversely isotropic materials," *Arch. Rational Mech. Anal.*, **4**, 193-213, 1960a.

Adkins, J. E., "Further symmetry relations for transversely isotropic materials," *Arch. Rational Mech. Anal.*, **5**, 263-274, 1960b.

Aero, E. L., and E. V. Kuvshinskii, "Fundamental equations of the theory of elastic media with rotationally interacting particles," *Soviet Physics Solid State*, **2,** 1272-1281, 1961.

Alfrey, T., *Mechanical Behavior of High Polymers.* New York: Interscience Publishers, Inc., 1948.

Alfrey, T., and E. F. Gurnee, "Dynamics of Viscoelastic Behavior," in *Rheology*, ed. F. R. Eirich, Vol. 1, 387-429. New York: Academic Press, 1956.

Alfrey, T., and E. F. Gurnee, *Organic Polymers.* Englewood Cliffs, N. J.: Prentice-Hall, Inc., 1967.

Almansi, E., "Sulle deformazione finite dei solidi elastici isotropi, I," *Rend. Accad. Naz. Lincei*, (Ser. 5), **20**, 705-714, 1911.

Aris, R., *Vectors, Tensors, and the Basic Equations of Fluid Mechanics.* Englewood Cliffs, N. J.: Prentice-Hall, Inc., 1962.

Batdorf, S. B., and B. Budiansky, *A Mathematical Theory of Plasticity Based on the Concept of Slip*, N. A. C. A. TN 1871, April 1949.

Batdorf, S. B., and B. Budiansky, "Polyaxial stress-strain relations of a strain-hardening metal," *J. Appl. Mech.*, **21**, 323-326, 1954.

Berker, R., "Integration des équations du mouvement d'un fluide visqueux incompressible," in *Encyclopedia of Physics*, ed. S. Flügge, Vol. 8/2, 1-384. Berlin: Springer-Verlag, 1963.

Bertsch, P. K., and W. N. Findley, "An experimental study of yield surfaces—corners, normality, Bauschinger and allied effects," *Proc. 4th U. S. Natl. Congr. Appl. Mech.*, 893-907. New York: A.S.M.E., 1962.

Biezeno, C. B., and R. Grammel, *Engineering Dynamics*, Vol. 1: *Theory of Elasticity* (translated from the German by M. L. Meyer). London: Blackie and Son, Ltd., 1955.

Bingham, E. C., *Fluidity and Plasticity*. New York: McGraw-Hill Book Company, 1922.

Biot, M. A., *Mechanics of Incremental Deformations*. New York: John Wiley & Sons, Inc., 1965.

Bland, D. R., *The Theory of Linear Viscoelasticity*. Oxford: Pergamon Press Ltd., 1960.

Blasius, H., "Grenzschichten in Flüssigkeiten mit kleiner Reibung," *Zeits. Math. u. Phys.*, **56**, 1-37, 1908 (see also translation by J. Vanier, N.A.C.A. TM 1256, 1950).

Boley, B. A., and M. B. Friedman, "On the viscous flow around the leading edge of a flat plate," *J. Aero/Space. Sci.*, **26**, 453-454, 1959.

Boley, B. A., and J. H. Weiner, *Theory of Thermal Stresses*. New York: John Wiley & Sons, Inc., 1960.

Boresi, A. P., *Elasticity in Engineering Mechanics*. Englewood Cliffs, N. J.: Prentice-Hall, Inc., 1965.

Brand, L., "The pi theorem of dimensional analysis," *Arch. Rational Mech. Anal.*, **1**, 34-45, 1957.

Bridgman, P. W., *Dimensional Analysis*. New Haven: Yale University Press, 1931.

Bridgman, P. W., *Studies in Large Plastic Flow and Fracture*. New York: McGraw-Hill Book Company, 1952.

Buckingham, E., "On physically similar systems: illustrations of the use of dimensional analysis," *Phys. Rev.*, **4**, 345-376, 1914.

Budiansky, B., *Fundamental Theorems and Consequences of the Slip Theory of Plasticity*, Ph. D. Thesis, Brown University, Providence, R. I., 1950.

Budiansky, B., "A reassessment of deformation theories of plasticity," *J. Appl. Mech.*, **26**, 259-264, 1959.

Carrier, G. F., and C. C. Lin, "On the nature of the boundary layer near the leading edge of a flat plate," *Quart. Appl. Math.*, **6**, 63-68, 1948.

Cesaro, E., "Sulle formole del Volterra, fondamentali nella teoria delle distorsioni elastiche," *Rend. Accad. Napoli*, **12**, 311-321, 1906.

Cicala, P., "Sobre la teoria de Batdorf y Budiansky de la deformación plástica," *Revista de la Universidad de Córdoba* (Argentina), Ano XII, No. 2, 1950.

Coburn, N., *Vector and Tensor Analysis*. New York: The Macmillan Company, 1955.

Coleman, B. D., "Thermodynamics of materials with memory," *Arch. Rational Mech. Anal.*, **17**, 1-46, 1964.

Coleman, B. D., "Simple liquid crystals," *Arch. Rational Mech. Anal.*, **20**, 41-58, 1965.

Coleman, B. D., and V. J. Mizel, "Existence of caloric equations of state in thermodynamics," *J. Chem. Phys.*, **40**, 1116-1125, 1964.

Coleman, B. D., and W. Noll, "Foundations of linear viscoelasticity," *Rev. Mod. Phys.*, **33**, 239-249, 1961 (reprinted in *Foundations of Elasticity Theory*, ed. C. Truesdell. New York: Gordon & Breach, 1965).

Combebiac, G., "Sur les équations générales de l'élasticité," *Bull. Soc. Math. France*, **30**, 108-110, 242-247, 1902.

Courant, R., and D. Hilbert, *Methods of Mathematical Physics*, Vol. 2. New York: Interscience Publishers, 1962.

Daily, J. W., and D. R. F. Harleman, *Fluid Dynamics*. Reading, Mass.: Addison-Wesley Publishing Co., 1966.

De Groot, S. R. (see Groot, S. R. de).

Drew, T. B., *Handbook of Vector and Polyadic Analysis*. New York: Reinhold Publishing Corp., 1961.

Drucker, D. C., "Relation of experiments to mathematical theories of plasticity," *J. Appl. Mech.*, **16**, 349-357, 1949.

Drucker, D. C., "A more fundamental approach to plastic stress-strain relations," *Proc. 1st U.S. Natl. Congr. Appl. Mech.*, 487-491. New York: A.S.M.E., 1951.

Drucker, D. C., "Stress-strain relations in the plastic range of metals—experiments and basic concepts," in *Rheology*, ed. F. R. Eirich, Vol. 1, 97-119. New York: Academic Press, 1956.

Drucker, D. C., "A definition of stable inelastic material," *J. Appl. Mech.*, **26**, 101-106, 1959.

Duhem, P., "Recherches sur l'hydrodynamique," *Ann. Toulouse* (2), **3**, 315-377, 379-431, 1901.

Eirich, F. R., ed., *Rheology, Theory and Applications*. New York: Academic Press, Vol. 1, 1956; Vol. 2, 1958; Vol. 3, 1960.

Ericksen, J. L., "Tensor fields," in *Encyclopedia of Physics*, ed. S. Flügge, Vol. 3/1, 794-858, 1960.

Eringen, A. C., *Nonlinear Theory of Continuous Media*. New York: McGraw-Hill Book Company, 1962.

Eringen, A. C., *Mechanics of Continua*. New York: John Wiley & Sons, Inc., 1967.

Eubanks, R. A., and E. Sternberg, "On the completeness of the Boussinesq-Papkovich stress functions," *J. Rational Mech. Anal.*, **5,** 735-746, 1956.

Eucken, A., "Über das Wärmeleitvermögen, die spezifische Wärme und die innere Reibung der Gase," *Phys. Zeits.*, **14,** 324-332, 1913.

Ferry, J. D., *Viscoelastic Properties of Polymers*. New York: John Wiley & Sons, Inc., 1961.

Fitts, D. D., *Nonequilibrium Thermodynamics*. New York: McGraw-Hill Book Company, 1962.

Flügge, W., *Viscoelasticity*. Waltham, Mass.: Blaisdell Publishing Co., 1967.

Frederick, D., and T. S. Chang, *Continuum Mechanics*. Boston: Allyn and Bacon, Inc., 1965.

Fredrickson, A. G., *Principles and Applications of Rheology*. Englewood Cliffs, N. J.: Prentice-Hall, Inc., 1964.

Freudenthal, A. M., and H. Geiringer, "The mathematical theories of the inelastic continuum," in *Encyclopedia of Physics*, ed. S. Flügge, Vol. 6, 229-433. Berlin: Springer-Verlag, 1958.

Fung, Y. C., *Foundations of Solid Mechanics*. Englewood Cliffs, N. J.: Prentice-Hall, Inc., 1965.

Galerkin, B., "Contribution à la solution générale du problème de la théorie de l'élasticité dans le cas de trois dimensions," *Comptes Rendus*, **190,** 1047-1048, 1930; *Comptes rendus (Doklady) acad. sci. U.R.S.S.*, Ser. A, **14,** 353, 1930.

Goodier, J. N., and P. G. Hodge, Jr., *Elasticity and Plasticity*. New York: John Wiley & Sons, Inc., 1958.

Green, A. E., and J. E. Adkins, *Large Elastic Deformations and Non-linear Continuum Mechanics*. Oxford: Clarendon Press, 1960.

Green, A. E., and P. M. Naghdi, "A general theory of an elastic-plastic continuum," *Arch. Rational Mech. Anal.*, **18,** 251-281, 1965.

Green, A. E., and R. S. Rivlin, "The mechanics of non-linear materials with memory, Part I," *Arch. Rational Mech. Anal.*, **1,** 1-21, 470, 1957 (reprinted in *Rational Mechanics of Materials*, ed. C. Truesdell. New York: Gordon & Breach, 1965).

Green, A. E., R. S. Rivlin, and A. J. M. Spencer, "The mechanics of non-linear materials with memory, Part II," *Arch. Rational Mech. Anal.*, **3,** 82-90, 1959.

Green, A. E., and W. Zerna, *Theoretical Elasticity*. Oxford: Clarendon Press, 1954.

Greenberg, H. J., W. S. Dorn, and A. H. Wetherell, "A comparison of flow and deformation theories in plastic torsion," in *Plasticity*, eds. E. H. Lee and P. S. Symonds, 279-296. Oxford: Pergamon Press Ltd., 1960.

Groot, S. R. de, *Thermodynamics of Irreversible Processes*. Amsterdam: North-Holland Publishing Co.; New York: Interscience Publishers, Inc., 1951.

Groot, S. R. de, and P. Mazur, *Non-Equilibrium Thermodynamics*. Amsterdam: North-Holland Publishing Co., 1962.

Gross, B., *Mathematical Structures of the Theories of Viscoelasticity*. Paris: Hermann, 1953.

Gurtin, M., and E. Sternberg, "On the linear theory of viscoelasticity," *Arch. Rational Mech. Anal.*, **11**, 291-356, 1962.

Hamel, G., *Elementare Mechanik*, Leipzig and Berlin, 1912.

Hencky, H., "Zur Theorie plastischer Deformationen und der hierdurch hervorgerufenen Nachspannungen," *Zeits. angew. Math. u. Mech.*, **4**, 323-334, 1924. See also *Proc. 1st Internat. Congr. Appl. Mech.*, Delft (1924), pp. 312-317.

Hill, R., "A theory of the yielding and plastic flow of anisotropic metals," *Proc. Roy. Soc.* (Lond.), **A193**, 281-297, 1948.

Hill, R., *The Mathematical Theory of Plasticity*. Oxford: Clarendon Press, 1950.

Hodge, P. G., Jr., "Piecewise linear plasticity," *Proc. 9th Internat. Congr. Appl. Mech.* (Brussels 1956), Vol. 8, 65-72.

Hodge, P. G., Jr., Discussion of Prager (1956), *J. Appl. Mech.*, **24**, 482-483, 1957.

Hohenemser, K., and W. Prager, "Über die Ansätze der Mechanik isotroper Kontinua," *Zeits. angew. Math. u. Mech.*, **12**, 216-226, 1932.

Howarth, L., "On the solution of the laminar boundary layer equations," *Proc. Roy. Soc.* (Lond.), **A164**, 547-579, 1938.

Hu, L. W., "Determination of the Plastic Stress Strain Relations in Tension for Nittany No. 2 Brass under Hydrostatic Pressure," *Proc. 3rd. U. S. Natl. Congr. Appl. Mech.*, 557-562, New York: A.S.M.E., 1958.

Hu, L. W., "Plastic Stress-Strain Relations and Hydrostatic Stress," in *Plasticity*, eds. E. H. Lee and P. S. Symonds, 194-201. Oxford: Pergamon Press Ltd., 1960.

Hu, L. W., J. Markovitz, and T. A. Bartush, "A triaxial stress experiment on yield conditions in plasticity," *Experimental Mechanics*, **6**, 58-64, 1966.

Ivey, H. J., "Plastic stress-strain relations and yield surfaces for aluminum alloys," *J. Mechanical Engr. Sci.*, **3**, 15-31, 1961.

Iyengar, K. T. S. R., and C. V. Yogananda, "The end problem of hollow cylinders," *J. Appl. Mech., Trans. A.S.M.E.*, **33**, 685-686, 1966.

Jaunzemis, W., *Continuum Mechanics*. New York: The Macmillan Company, 1967.

Kármán, Th. von, "Über laminare und turbulente Reibung," *Zeits. angew. Math. u. Mech.*, **1**, 244-247, 1921.

Kline, S. J., *Similitude and Approximation Theory*. New York: McGraw-Hill Book Company, 1965.

Kliushnikov, V. D., "On a possible manner of establishing the plasticity relations," *TPMM* (Translation of *Prikl. Mat. i. Mekh.*), **23**, 405-418, 1959.

Koiter, W. T., "Stress-strain relations, uniqueness and variational theorems for elastic, plastic materials with a singular yield surface," *Q. Appl. Math.*, **11**, 350-354, 1953.

Kubaschewski, O., and E. L. Evans, *Metallurgical Thermochemistry*, 2nd ed. New York: John Wiley & Sons, Inc., 1956. (See also following entry.)

Kubaschewski, O., E. L. Evans, and C. B. Alcock, *Metallurgical Thermochemistry*, 4th ed. Oxford: Pergamon Press Ltd., 1967.

Lamb, H., *Hydrodynamics* (reprint of 6th ed., Cambridge University Press, 1932). New York: Dover Publications, Inc., 1945.

Landau, L. D., and E. M. Lifshitz, *Fluid Mechanics* (translated from Russian by J. B. Sykes and W. H. Reid). Oxford: Pergamon Press Ltd.; Reading, Mass.: Addison-Wesley Publishing Co., 1959.

Leaderman, H., "Viscoelasticity phenomena in amorphous high polymeric systems," in *Rheology*, ed. F. R. Eirich, Vol. 2, 1-61, 1958.

Lee, E. H., "Viscoelastic stress analysis," in *Structural Mechanics*, eds. J. N. Goodier and N. J. Hoff, 456-482. Oxford: Pergamon Press Ltd., 1960.

Lee, E. H., and T. C. Rogers, "Solution of viscoelastic stress analysis problems using measured creep or relaxation properties," *J. Appl. Mech.*, **30**, 127-133, 1963.

Lee, E. H., and P. Symonds, eds., *Plasticity*. Oxford: Pergamon Press Ltd., 1960.

Lee, E. H., "Elastic-Plastic Waves of One-Dimensional Strain," *Proc. 5th U. S. Natl. Congr. Appl. Mech.*, 405-420. New York: A.S.M.E., 1966.

Lee, E. H., *Elastic-Plastic Deformation at Finite Strains*, Technical Report No. 183, Contract No. DA-04-200-AMC-659(X) to Ballistic Research Laboratory. Department of Engineering Mechanics, Stanford University, June 1968.

Leigh, D. C., *Nonlinear Continuum Mechanics*, New York: McGraw-Hill Book Company, 1968.

Lenoe, E. M., and C. J., Martin, "A technique for the formulation of meaningful viscoelastic constitutive equations," *Proc. 5th U. S. Natl. Congr. Appl. Mech.*, 483-510. New York: A.S.M.E., 1966.

Lin, T. H., "On stress-strain relations based on slips," *Proc. 3rd U. S. Natl. Congr. Appl. Mech.*, 581-587 New York: A.S.M.E., 1958.

Lindsay, R. B., *Mechanical Radiation*. New York: McGraw-Hill Book Company, 1960.

Loitsyanskii, L. G., *Mechanics of Liquids and Gases*. Oxford: Pergamon Press Ltd., 1966.

Love, A. E. H., *The Mathematical Theory of Elasticity*, 2nd ed. London: Cambridge University Press, 1906; 4th ed., Cambridge University Press, 1927; 4th ed. reprinted, New York: Dover Publications, Inc., 1944.

Lubahn, J. D., and R. P. Felgar, *Plasticity and Creep of Metals*. New York: John Wiley & Sons, Inc., 1961.

Ludwik, P., *Elemente der technologischen Mechanik*. Berlin: Springer-Verlag, 1909.

Lur'e, A. I., *Three-Dimensional Problems of the Theory of Elasticity* (translated from Russian by D. B. McVean). New York: Interscience Publishers, Inc., 1964.

Marin, J., *Mechanical Behavior of Engineering Materials.* Englewood Cliffs, N. J.: Prentice-Hall, Inc., 1962.

Meares, P., *Polymers, Structure and Bulk Properties.* Princeton, N. J.: D. Van Nostrand Co., Inc., 1965.

Meredith, R., "The rheology of fibers," in *Rheology*, ed. F. R. Eirich, Vol. 2, 261-312, 1958.

Miller, P. M., *Numerical Analysis of Combined Bending and Torsion of a Prismatic Bar Using Rigid-Plastic and Work-Hardening Plasticity Theories*, Ph. D. Thesis, Michigan State University, East Lansing, Mich., 1966.

Miller, P. M., and L. E. Malvern, "Numerical analysis of combined bending and torsion of a work-hardening square bar," *J. Appl. Mech., Trans. A.S.M.E.*, **34,** 1005-1010, 1967.

Milne-Thomson, L. M., *Theoretical Hydrodynamics*, 4th ed. New York: The Macmillan Company, 1960.

Milne-Thomson, L. M., *Plane Elastic Systems.* Berlin: Springer-Verlag, 1960.

Milne-Thomson, L. M., *Antiplane Elastic Systems.* Berlin: Springer-Verlag, 1962.

Mindlin, R. D., "Note on the Galerkin and Papkovich stress functions," *Bull. Am. Math. Soc.*, **42,** 373-376, 1936.

Mindlin, R. D., and H. F. Tiersten, "Effects of couple-stresses in linear elasticity," *Arch. Rational Mech. Anal.*, **11,** 415-448, 1962.

Moriguti, S., "Fundamental theory of dislocation of elastic bodies," *Oyo Sugaku Rikigaku* (Applied Mathematics and Mechanics), **1,** 87-90, 1947.

Muskhelishvili, N. I., *Some Basic Problems of the Mathematical Theory of Elasticity* (translated from Russian by J. R. M. Radok). Groningen, The Netherlands: P. Noordhoff, Ltd., 1963.

Nadai, A., "Der Beginn des Fliessvorganges in einem tortierten Stab," *Zeits. angew. Math. u. Mech.*, **3,** 442-454, 1923.

Naghdi, P. M., "Stress-strain relations in plasticity and thermoplasticity," in *Plasticity*, eds. E. H. Lee and P. Symonds, 121-169. Oxford: Pergamon Press Ltd., 1960.

Naghdi, P. M., F. Essenburg, and W. Koff, "An experimental study of initial and subsequent yield surfaces in plasticity," *J. Appl. Mech.*, **25,** 201-209, 1958.

Neou, C. Y., "A direct method for determining Airy polynomial stress functions," *J. Appl. Mech.*, **24,** 387-390, 1957.

Neuber, H., "Ein neuer Ansatz zur Lösung räumlicher Probleme der Elastizitätstheorie," *Zeits. angew. Math. u. Mech.*, **14,** 203-212, 1934.

Nielsen, L. E., *Mechanical Properties of Polymers.* New York: Reinhold Publishing Corp., 1962.

Noll, W., "On the continuity of the solid and fluid states," *J. Rational Mech. Anal.*, **4,** 3-81, 1955 (reprinted in *Rational Mechanics of Materials*, ed. C. Truesdell. New York: Gordon & Breach, 1965).

Noll, W., "A mathematical theory of the mechanical behavior of continuous media," *Arch. Rational Mech. Anal.*, **2,** 1958 (reprinted in *Rational Mechanics of Materials*, ed. C. Truesdell. New York: Gordon & Breach, 1965).

Oldroyd, J. G., "On the formulation of rheological equations of state," *Proc. Roy. Soc.* (Lond.), **A200,** 523-541, 1950 (reprinted in *Rational Mechanics of Materials*, ed. C. Truesdell. New York: Gordon & Breach, 1965).

Onsager, L., "Reciprocal relations in irreversible processes," I, *Phys. Rev.*, **37,** 405-426, 1931; II, *Phys. Rev.*, **38,** 2265-2279, 1931.

Pai, S.-I., *Viscous Flow Theory*, Vol. 1, *Laminar Flow*, 1956; Vol. 2, *Turbulent Flow*, 1957. Princeton, N. J.: D. Van Nostrand Co., Inc.

Pai, S.-I., *Introduction to the Theory of Compressible Flow*. Princeton, N. J.: D. Van Nostrand Co., Inc., 1959.

Papkovich, P. F., "Solution générale des équations différentielles fondamentales d'élasticité, exprimée par trois fonctions harmoniques," *Comptes Rendus*, **195,** 513-515, 1932; "Expressions générales des composantes des tensions, ne renfermant comme fonctions arbitraires que des fonctions harmoniques," *Comptes Rendus*, **195,** 754-756, 1932.

Patterson, G. N., *Molecular Flow of Gases*. New York: John Wiley & Sons, Inc., 1956.

Perzyna, P., "The constitutive equations for rate-sensitive plastic materials," *Q. Appl. Math.*, **20,** 321-332, 1963.

Phillips, A., *Introduction to Plasticity*. New York: The Ronald Press Company, 1956.

Phillips, A., and G. A. Gray, "Experimental investigation of corners in the yield surface," *J. Basic Engineering, Trans. A.S.M.E.*, **83,** Ser. D, 275-289, 1961.

Pickett, G., "Application of the Fourier method to the solution of certain boundary problems in the theory of elasticity," *J. Appl. Mech., Trans. A.S.M.E.*, **66,** A-176 to A-182, 1944.

Pipkin, A. C., and R. S. Rivlin, "The formulation of constitutive equations in continuum physics, I," *Arch. Rational Mech. Anal.*, **4,** 129-144, 1959.

Pipkin, A. C., and A. S. Wineman, "Material symmetry restrictions on non-polynomial constitutive equations," *Arch. Rational. Mech. Anal.*, **12,** 420-426, 1963.

Pohlhausen, K., "Zur näherungsweisen Integration der Differentialgleichung der laminaren Grenzschicht," *Zeits. angew. Math. u. Mech.*, **1,** 252-268, 1921.

Polakowski, N. H., and E. J. Ripling, *Strength and Structure of Engineering Materials*. Englewood Cliffs, N. J.: Prentice-Hall, Inc., 1966.

Prager, W., "Strain hardening under combined stress," *J. Appl. Phys.*, **16,** 837-840, 1945.

Prager, W., "The theory of plasticity: a survey of recent achievements" (James Clayton Lecture), *Proc. Inst. Mech. Engrs.*, **169,** 41-57, 1955.

Prager, W., "A new method of analyzing stresses and strains in work-hardening plastic solids," *J. Appl. Mech.*, **23,** 493-496, 1956.

Prager, W., *An Introduction to Plasticity*. Reading, Mass.: Addison-Wesley Publishing Co., 1959.

Prager, W., *Introduction to Mechanics of Continua.* New York: Ginn and Co., 1961.

Prager, W., and P. G. Hodge, Jr., *Theory of Perfectly Plastic Solids.* New York: John Wiley & Sons, Inc., 1951.

Prandtl, L., "Über Flüssigkeitsbewegung bei sehr kleiner Reibung," *Proc. 3rd Internat. Congr. Math.* (Heidelberg 1904), 484-491, Leipzig, 1905 (Kraus Reprint Ltd., Nendeln/Liechtenstein, 1967).

Prandtl, L., "Spannungsverteilung in plastischen Koerpern," *Proc. 1st Internat. Congr. Appl. Mech.* (Delft 1924), 43-54.

Quinney, H., and G. I. Taylor, "The emission of the latent energy due to previous cold working when a metal is heated," *Proc. Roy. Soc.* (Lond.), **A163,** 157-181, 1937.

Ramberg, W., and W. R. Osgood, *Description of Stress-Strain Curves by Three Parameters*, N.A.C.A. TN 902, 1943.

Reuss, E., "Beruecksichtigung der elastischen Formaenderungen in der Plasizitaetstheorie," *Zeits. angew. Math. u. Mech.*, **10,** 266-274, 1930.

Reynolds, O., *The Sub-Mechanics of the Universe.* Cambridge, England: Roy. Soc. (Lond.), 1903.

Rivlin, R. S., "Further remarks on the stress-deformation relations for isotropic materials," *J. Rational Mech. Anal.*, **4,** 681-702, 1955.

Rivlin, R. S., "The formulation of constitutive equations in continuum physics, II," *Arch. Rational Mech. Anal.*, **4,** 262-272, 1960.

Rivlin, R. S., "Nonlinear viscoelastic solids," *SIAM Review*, **7,** 323-340, 1964.

Rivlin, R. S., "The fundamental equations of nonlinear continuum mechanics," in *Dynamics of Fluids and Plasmas* (Burgers Anniversary Volume), ed. S.-I. Pai. New York: Academic Press, 1966.

Rivlin, R. S., and J. L. Ericksen, "Stress-deformation relations for isotropic materials," *J. Rational Mech. Anal.*, **4,** 323-425, 1955 (reprinted in *Rational Mechanics of Materials*, ed. C. Truesdell. New York: Gordon & Breach, 1965).

Rivlin, R. S., and G. F. Smith, "Orthogonal integrity bases for *N* symmetric tensors," in *Contributions to Mechanics* (Reiner Anniversary Volume), ed. S. Abir. Oxford: Pergamon Press Ltd. (in press, 1969).

Rohsenow, W. M., and H. Choi, *Heat, Mass and Momentum Transfer*, Englewood Cliffs, N. J.: Prentice-Hall, Inc., 1961.

Sanders, J. L., Jr., "Plastic stress-strain relations based on linear loading functions," *Proc. 2nd U. S. Natl. Congr. Appl. Mech.*, 455-460. New York: A. S. M. E., 1954.

Savin, G. N., *Stress Concentrations Around Holes* (translated from Russian by E. Gros). New York: Pergamon Press Ltd., 1961.

Schapery, R. A., "Application of thermodynamics to thermomechanical, fracture, and birefringent phenomena in viscoelastic media," *J. Appl. Phys.*, **35**, 1451-1465, 1964.

Schapery, R. A., "A theory of nonlinear thermoviscoelasticity based on irreversible thermodynamics," *Proc. 5th U. S. Nat. Congr. Appl. Mech.*, 511-530. New York: A.S.M.E., 1966.

Schlichting, H., *Boundary Layer Theory* (translated from German by J. Kestin). New York: McGraw-Hill Book Company, 4th ed., 1960; 6th ed., 1968.

Sedov, L. I., *Similarity and Dimensional Methods in Mechanics* (translated by M. Friedman from 4th Russian edition). New York: Academic Press Inc., 1959.

Shapiro, A. H., *The Dynamics and Thermodynamics of Compressible Fluid Flow*, 2v. New York: The Ronald Press Company, 1953, 1954.

Shield, R. T., and H. Ziegler, "On Prager's hardening rule," *Zeits. angew. Math. u. Phys.*, **9**, 260-276, 1958.

Signorini, A., "Sulle deformazione finite dei sistemi a trasformazioni reversibili," *Rend. Accad. Naz. Lincei*, Ser. 6, **18**, 388-394, 1933.

Smith, G. F., "Further results on the strain-energy function for anisotropic elastic materials, "*Arch. Rational Mech. Anal.*, **10**, 108-110, 1962.

Smith, G. F., and R. S. Rivlin, "The strain-energy function for anisotropic elastic materials," *Trans. Amer. Math. Soc.*, **88**, 175-193, 1958.

Smith, G. F., and R. S. Rivlin, "Integrity bases for vectors. The crystal classes," *Arch. Rational Mech. Anal.*, **15**, 169-221, 1964.

Smith, G. F., M. M. Smith, and R. S. Rivlin, "Integrity bases for a symmetric tensor and a vector—The crystal classes," *Arch. Rational Mech. Anal.*, **12**, 93-133, 1963.

Sneddon, I. N. and D. S. Berry, "The Classical Theory of Elasticity," pp. 1-126 in *Encyclopedia of Physics*, v. 6, ed. S. Flügge, Berlin: Springer Verlag, 1958.

Sokolnikoff, I. S., *Mathematical Theory of Elasticity*, 2nd ed. New York: McGraw-Hill Book Company, 1956.

Sommerfeld, A., *Thermodynamics and Statistical Mechanics* (translated from German by J. Kestin). New York: Academic Press Inc., 1964.

Southwell, R. V., "Castigliano's principle of minimum strain-energy and the conditions of compatibility for strain," in *S. Timoshenko 60th Anniversary Volume*, 211-217. New York: The Macmillan Company, 1938.

Spencer, A. J. M., "The invariants of six symmetric 3×3 matrices," *Arch. Rational Mech. Anal.*, **11**, 357-367, 1961.

Spencer, A. J. M., and R. S. Rivlin, "The theory of matrix polynomials and its application to the mechanics of isotropic continua," *Arch. Rational Mech. Anal.*, **2**, 309–336, 1959a.

Spencer, A. J. M., and R. S. Rivlin, "Finite integrity bases for five or fewer symmetric 3 × 3 matrices," *Arch. Rational Mech. Anal.*, **2**, 435–446, 1959b.

Spencer, A. J. M., and R. S. Rivlin, "Further results on the theory of matrix polynomials," *Arch. Rational Mech. Anal.*, **4**, 214–230, 1960.

Spielrein, J., *Lehrbuch der Vektorrechnung nach den Bedürfnissen in der Technischen Mechanik und Elektrizitätslehre*, Stuttgart, 1916.

Sternberg, E., "On Saint-Venant's principle," *Q. Appl. Math.*, **11**, 393–402, 1954.

Taylor, G. I., and H. Quinney, "The latent energy remaining in a metal after coldworking," *Proc. Roy. Soc.* (Lond.), **A143**, 307–326, 1934. (See also Quinney and Taylor, 1937.)

Templin, R. L., and R. G. Sturm, "Some stress-strain studies of metals," *J. Aero. Sci.*, **7**, 189–198, 1940.

Thomsen, E. G., C. T. Yang, and S. Kobayashi, *Mechanics of Plastic Deformation in Metal Processing*. New York: The Macmillan Company, 1965.

Timoshenko, S., and J. N. Goodier, *Theory of Elasticity*, 2nd ed. New York: McGraw-Hill Book Company, 1951.

Timoshenko, S., and S. Woinowsky-Krieger, *Theory of Plates and Shells*, 2nd ed. New York: McGraw-Hill Book Company, 1959.

Timpe, A., "Probleme der Spannungsverteilung in ebenen Systemen, einfach gelost mit Hilfe der Airyschen Funktion," *Zeits. Math. u. Phys.*, **52**, 348–383, 1905.

Tobolsky, A. V., "Stress relaxation studies of the viscoelastic properties of polymers," in *Rheology*, ed. F. R. Eirich, Vol. 2, 63–81. New York: Academic Press Inc., 1958.

Tobolsky, A. V., *Properties and Structures of Polymers*. New York: John Wiley & Sons, Inc., 1960.

Toupin, R. A., "The elastic dialectric," *J. Rational Mech. Anal.*, **5**, 849–915, 1956.

Toupin, R. A., "Elastic materials with couple-stresses," *Arch. Rational Mech. Anal.*, **11**, 385–414, 1962.

Townsend, A. A., *The Structure of Turbulent Shear Flow*. London: Cambridge University Press, 1956.

Truesdell, C., "A new definition of a fluid, II: The Maxwellian fluid," *J. Math. pures appl.*, **30**, 111–158, 1951.

Truesdell, C., "The mechanical foundations of elasticity and fluid dynamics," *J. Rational Mech. Anal.*, **1**, 125–300, 1952 (corrected reprint in *The Mechanical Foundations of Elasticity and Fluid Dynamics*. New York: Gordon & Breach, 1966).

Truesdell, C., "Corrections and additions to 'The mechanical foundations of elasticity and fluid dynamics'" (see Truesdell, 1952), *J. Rational Mech. Anal.*, **2**, 505-616, 1953.

Truesdell, C., "The physical components of vectors and tensors," *Zeits. angew. Math. u. Mech.*, **33**, 345-356, 1953. (See also Truesdell's remarks on this paper, same journal, **34**, 69-70, 1954.)

Truesdell, C., *The Elements of Continuum Mechanics*. New York: Springer-Verlag New York, Inc., 1965.

Truesdell, C., ed., *Continuum Mechanics* (reprints). New York: Gordon & Breach, Vol. 1, 1966; Vols. 2, 3, 4, 1965.

Truesdell, C., *Six Lectures on Modern Natural Philosophy*. New York: Springer-Verlag New York, Inc., 1966.

Truesdell, C., and W. Noll, "The non-linear field theories of mechanics," in *Encyclopedia of Physics*, ed. S. Flügge, Vol. 3/3. Berlin: Springer-Verlag, 1965.

Truesdell, C., and R. A. Toupin, "The classical field theories," in *Encyclopedia of Physics*, ed. S. Flügge, Vol. 3/1, 226-793. Berlin: Springer-Verlag, 1960.

Veblen, O., *Invariants of Quadratic Differential Forms*. London: Cambridge University Press, 1927.

Vennard, J. K., *Elementary Fluid Mechanics*, 3rd ed. New York: John Wiley & Sons, Inc., 1954.

Voigt, W., *Lehrbuch der Kristallphysik*. Leipzig and Berlin: B. G. Teubner, 1910 (reprinted 1928).

Wang, C. T., *Applied Elasticity*. New York: McGraw-Hill Book Company, 1953.

Washizu, K., "A note on the conditions of compatibility," *J. Math. Phys.*, **36**, 306-312, 1958.

Westergaard, H. M., *Theory of Elasticity and Plasticity*. Cambridge, Mass.: Harvard University Press, 1952; New York: Dover Publications, Inc., 1952.

Wineman, A. S., and A. C. Pipkin, "Material symmetry restrictions on constitutive equations," *Arch. Rational Mech. Anal.*, **17**, 184-214, 1964.

Yuan, S. W., *Foundations of Fluid Mechanics*. Englewood Cliffs, N. J.: Prentice-Hall, Inc., 1967.

Ziegler, H., "A modification of Prager's hardening rule," *Q. Appl. Math.*, **17**, 55-65, 1959.

Ziegler, H., "Some extremum principles in irreversible thermodynamics, with application to continuum mechanics," in *Progress in Solid Mechanics*, eds. I. N. Sneddon and R. Hill, Vol. 4, 91-193. Amsterdam: North-Holland Publishing Co. and New York: Interscience Publishers, Inc., 1963.

Author Index†

†See also Bibliography, p. 673 ff.

Y

Z

Subject Index

C

D

F

U

V

W